Springer Theses

Recognizing Outstanding Ph.D. Research

Aims and Scope

The series "Springer Theses" brings together a selection of the very best Ph.D. theses from around the world and across the physical sciences. Nominated and endorsed by two recognized specialists, each published volume has been selected for its scientific excellence and the high impact of its contents for the pertinent field of research. For greater accessibility to non-specialists, the published versions include an extended introduction, as well as a foreword by the student's supervisor explaining the special relevance of the work for the field. As a whole, the series will provide a valuable resource both for newcomers to the research fields described, and for other scientists seeking detailed background information on special questions. Finally, it provides an accredited documentation of the valuable contributions made by today's younger generation of scientists.

Theses may be nominated for publication in this series by heads of department at internationally leading universities or institutes and should fulfill all of the following criteria

- They must be written in good English.
- The topic should fall within the confines of Chemistry, Physics, Earth Sciences, Engineering and related interdisciplinary fields such as Materials, Nanoscience, Chemical Engineering, Complex Systems and Biophysics.
- The work reported in the thesis must represent a significant scientific advance.
- If the thesis includes previously published material, permission to reproduce this must be gained from the respective copyright holder (a maximum 30% of the thesis should be a verbatim reproduction from the author's previous publications).
- They must have been examined and passed during the 12 months prior to nomination.
- Each thesis should include a foreword by the supervisor outlining the significance of its content.
- The theses should have a clearly defined structure including an introduction accessible to new PhD students and scientists not expert in the relevant field.

Indexed by zbMATH.

More information about this series at https://link.springer.com/bookseries/8790

Nathalie Nagl

A New Generation of Ultrafast Oscillators for Mid-Infrared Applications

Doctoral Thesis accepted by
Ludwig-Maximilians-Universität München, Munich,
Germany

Springer

Author
Dr. Nathalie Nagl
Ludwig-Maximilians-Universität München
Max Planck Institute of Quantum Optics
Garching, Germany

Supervisor
Prof. Dr. Ferenc Krausz
Ludwig-Maximilians-Universität München
Max Planck Institute of Quantum Optics
Garching, Germany

ISSN 2190-5053 ISSN 2190-5061 (electronic)
Springer Theses
ISBN 978-3-030-89753-6 ISBN 978-3-030-89754-3 (eBook)
https://doi.org/10.1007/978-3-030-89754-3

This Springer imprint is published by the registered company Springer Nature Switzerland AG
The registered company address is: Gewerbestrasse 11, 6330 Cham, Switzerland

Für meine Eltern und Großeltern,

für meine Geschwister,

und für Markus.

Supervisor's Foreword

The ability to detect and analyze the composition of materials via spectroscopy is arguably one of the most essential tools in chemistry and biology. The performance of any spectroscopic technique, in turn, is critically dependent on the characteristics of the light source applied.

Recent advances in lasers capable of emitting ultrashort pulses lasting a mere few femtoseconds (one millionth of one billionth second) has spearheaded the development of novel and powerful spectroscopy techniques. One such technique is field-resolved spectroscopy, which can concurrently characterize the chemical composition over a vast range of different concentrations and with unprecedented sensitivity. If such techniques can be implemented with light that simultaneously covers the entire mid-infrared (MIR) spectral range—where the unique absorption signatures of biologically and medically relevant molecules reside—it will enable the parallel detection and evaluation of numerous chemical specimens and allow the simultaneous tracking of their changes at different time scales. Applied to medical samples—blood droplets from humans for example—it will open a new path for the full and timely monitoring of the health state of individuals, promising the detection of even the most minuscule trace of early illnesses that is impossible to spot with existing technologies. This possibility of identifying abnormalities in seemingly healthy individuals—as considered by current standard of care—will allow treatment to be administered well before any symptoms and suffering would arise, paving the way toward a new personalized and preventive medicine.

A roadblock hindering the advancement of such technology has been the lack of suitable light sources. The ideal source not only needs to fulfill technical criteria including bandwidth, power, pulse duration, stability and efficiency, it also needs to be reliable, user-friendly, and above all: cost-effective to facilitate its wide proliferation and penetration into daily clinical use. Thus far, all potential candidates have been suffering from one or more shortcomings.

Dr. Nagl's demonstration of the first directly diode-pumped Cr^{2+}-doped II–VI few-cycle laser oscillator as a driving source for MIR generation represents the desired breakthrough. It satisfies all of the above-mentioned requirements and provides the much sought-after solution to the impasse. The first demonstration of the new all-solid-state infrared source technology already outperforms its predecessors by a factor of sixty in terms of peak power. Subsequent improvements and power scaling have enabled the generation, via nonlinear frequency conversion, of MIR light that spans the entire "molecular fingerprint" region, all the way from a few hundred to several thousand wavenumbers, offering an unprecedented combination of spectral coverage and peak/average power. By investigating and solving the numerous intricate problems in generating ultrashort light pulses, especially in the uncommon wavelength range and using diode lasers as pump sources, Dr. Nagl built a compact, cost-effective broadband MIR light source that matches and even exceeds the performance of state-of-the-art systems based on more costly and complex pump sources.

The development of this technology is comprehensively and lucidly documented in this thesis. Together with a thorough review of literature and applications, and an extensive analysis of the theoretical foundations behind ultrafast laser oscillators, the thesis will be particularly useful for those working on the development and advancement of next-generation femtosecond mid-infrared light sources.

Garching, Germany
June 2021

Prof. Dr. Ferenc Krausz

Abstract

The mid-infrared (MIR) spectral range with wavelengths between 2 μm and 20 μm holds tremendous potential for the study of complex biological systems, given the abundance of intense and unique molecular absorption lines that can be detected. Consequently, spectroscopic applications of mid-infrared radiation have garnered enormous attention in recent years. A particularly striking example is the combination of multi-MHz-repetition-rate, few-cycle MIR light sources with electric-field-resolved techniques, enabling the recording of amplitude- and phase-resolved molecular signals with unparalleled specificity and shot-noise-limited sensitivity. Despite the ever-growing research demand, their widespread use is severely hampered by the lack of low-noise, compact, and ultrafast laser systems.

In this dissertation, a new generation of table-top mid-infrared laser sources is presented, bringing cutting-edge laser diode technology and few-cycle Cr^{2+}-doped II-VI solid-state oscillators together for the first time. Not only have these laser systems proven to reliably provide coherent radiation in the 2–3 μm region, the simultaneous reduction in size and complexity, accompanied by an improved overall efficiency and—most importantly—noise performance, renders this approach as pioneering for future MIR applications.

In total, three laser systems are developed, each of them pushing the frontiers of directly diode-pumped laser technology. The first one is driven by a single-emitter indium phosphide laser diode and delivers more than 500 mW of output power combined with pulse durations as short as 45 fs. With this first ever directly diode-pumped Kerr-lens mode-locked (KLM) Cr^{2+}-doped II-VI oscillator, experimental results confirm a highly stable operation. In addition, amplitude noise measurements reveal an excellent low-noise performance of the mode-locked laser output.

Driven by the desire to match and exceed the performance of more mature fiber-laser-pumped counterparts and also to boost the efficiency of the envisaged downstream applications, the output of two single-emitter pump laser diodes is carefully combined and implemented into a second-generation design. The achieved peak powers are almost three-times higher compared to before, while the low-noise performance of the KLM output is maintained.

Typically, the design architecture of laser systems used for generating mid-infrared radiation up to several tens of microns includes a sophisticated chain of amplification, pulse compression and parametric conversion stages. Using a powerful few-cycle mid-infrared oscillator as driving laser source instead not only significantly improves the effectiveness of these nonlinear schemes, but could even supersede the need for initial amplification. The presented third-generation system brings the directly diode-pumped Cr^{2+}-doped II-VI solid-state laser technology to a new level; with output peak powers reaching 1 MW and pulse durations as short as 28 fs, direct generation of CEP-stable mid-infrared pulses in a nonlinear optical crystal (ZGP) becomes feasible and results in a multi-octave-spanning spectrum between 4.5 μm and 14 μm more than 20 mW of average power.

The successful development and realization of the three novel and powerful directly diode-pumped mid-infrared laser systems serves as a foundation for a new generation of few-cycle MIR light sources, capable of performing spectroscopic measurements at unprecedented efficiency and shot-noise-limited sensitivity, while paving the way towards a more accessible alternative to synchrotron-like infrared radiation.

Acknowledgements

At the beginning of my thesis, I would like to take the opportunity to express my heartfelt gratitude to all those people who, one way or another, have supported me during my time as a Ph.D. student:

- First of all, I would like to thank my Ph.D. advisor, Ferenc Krausz, for his continuous encouragement and support on the project. I really enjoyed to be part of the big *Attoworld* research group and it was very enriching to work together with colleagues from so many different fields of laser physics. Moreover, I would like to thank him for the inspiration that he gave me based on the concept of a Just Cause.
- A very special thanks goes out to my group leader Kafai Mak, who has teached me the secrets behind Cr:ZnS/ZnSe laser systems based on his own fiber-laser-pumped oscillator. He always had an open ear for my ideas and concerns, and I cannot say whether the progress of my experimental research would have been that fast without his mental and also scientific support—ranging from nonlinear optics to python programming and presentation skills. In addition, I also want to thank Oleg Pronin, who was my former group leader during the first year of my Ph.D., for all his support. I am especially grateful that he and Ferenc Krausz recommended me to the Cusanuswerk.
- Thanks to my (former) team members of the "AG HFS" group and office-mates around Sebastian Gröbmeyer, Markus Pötzlberger, Kilian Fritsch, Jinwei Zhang, Qing Wang, Philipp Steinleitner, Aleksandar Sebesta, Shizhen Qu, Maciej Kowalczyk, and Naeem Ahmad. I really enjoyed working together with all of you! Moreover, it was a great pleasure to join our daily walk to our canteen next door and the yearly Christmas activities, ranging from lasertag and go-carting to skiing and city walks. Though I was surrounded by male colleagues only, it was so much fun with you, guys!
- Also, I want to thank all my colleagues around Ioachim Pupeza's and Mihaela Zigman's group for the regular exchange of knowledge during our team meetings, and also thanks for lending experimental equipment to us. In particular, I am

thankful to Alexander Weigel and Marinus Huber for the useful discussions on amplitude noise measurements.

- The generation of few-cycle pulses from Cr:ZnS/ZnSe oscillators would not have been possible without the dedicated multi-layer optics provided by Vladimir Pervak and his team. It was really helpful to have you sitting just a couple of rooms away for discussing all the various customized mirror designs that emerged during my time as a Ph.D.
- Particular thanks also goes to all the organizers of the *International Max Planck Research School of Advanced Photon Science* (IMPRS-APS). I really enjoyed being part of the graduate school, especially because of the great training opportunities through seminars, workshops and the block lectures. Also, the annual meetings at Ringberg Castle were legendary!
- In addition, I would like to acknowledge the financial and immaterial support that I received as part of my doctoral scholarship by the *Bischöfliche Studienförderung Cusanuswerk*. It was a great honor for me to meet so many inspiring people from all academic disciplines and I will always remember the discussions that we had and the time that we spent during our meetings.
- A special thanks goes out to all the administrative and technical staff at our institute, in particular Klaus Franke, Katharina Adler, Martin Groß, Jana Gerding, Kerstin Schmidt-Buchmann, Hans-Friedrich Wirth, Florian Saran, Thorsten Naeser, and the mechanical workshop.
- I would also like to thank Prof. Dr. Ursula Keller for offering to review my thesis.
- I also want to thank all the people who contributed to this thesis in one way or another but whom I forgot to mention.
- Ein ganz herzlicher Dank gilt schließlich auch meiner Familie, insbesondere meinen Eltern und Großeltern sowie meinen Geschwistern, dass sie mich zu jeder Zeit unterstützt haben. Ohne euch wäre dieser Weg so bestimmt nicht möglich gewesen! Auch möchte ich mich bei Markus bedanken, der mir als Partner in jeder Situation zur Seite stand und immer ein offenes Ohr für mich hatte.

Contents

1 Motivation 1
1.1 *MIRanda*—A New Ultrafast Workhorse for Mid-Infrared Applications 1
1.2 The Power of Directly Diode-Pumped, Few-Cycle Mid-Infrared Lasers 5
1.3 Thesis Outline 8
References 9

2 Fundamentals of Ultrashort Pulse Generation 13
2.1 Formation of Laser Pulses 13
2.1.1 Locking the Longitudinal Modes of a Laser Cavity 15
2.1.2 Experimental Realization Methods 17
2.2 Soliton Pulse Propagation 22
2.2.1 Dispersion 22
2.2.2 Self-Phase Modulation 24
2.2.3 The Fundamental Soliton 26
2.3 Soliton Mode-Locking 27
2.3.1 Gain, Loss, and the Spectral Filtering 27
2.3.2 Self-Amplitude Modulation 28
2.3.3 The Master Equation of Mode-Locking 29
2.3.4 The Effect of Different Dispersion Regimes 30
2.4 Summary 31
References 32

3 The First Directly Diode-Pumped Few-Cycle Cr-Doped II-VI Laser 35
3.1 Chromium-Doped II-VI—*The Titanium Sapphire of the Mid-Infrared* 36
3.2 Challenges of Diode-Pumping 38
3.2.1 The Right Choice of Laser Diode Type 39
3.2.2 Properties of Single-Emitter Diodes 40

3.2.3 Effects on the Laser System and the Few-Cycle Pulse Generation ... 43
3.3 Few-Cycle Pulse Generation via Kerr-Lens Mode-Locking ... 46
3.3.1 Pump Setup and Resonator Design ... 47
3.3.2 The Importance of Intra-cavity Dispersion Management ... 51
3.3.3 Characterization of the Mode-Locked Output ... 55
3.3.4 The Success of Powerful Diode-Pumped Operation ... 62
3.4 Summary ... 65
References ... 66

4 Reaching the Performance of State-of-the-Art Fiber-Pumped Systems ... 71
4.1 Challenges of Reaching Higher Average and Peak Powers ... 72
4.1.1 Pump Setup Considerations ... 73
4.1.2 Effects on the Mode-Locking Performance ... 74
4.2 Comparison and Improvements over *MIRanda1* ... 80
4.2.1 Pump and Cavity Design ... 80
4.2.2 Mode-Locked Performance ... 83
4.3 Summary ... 90
References ... 91

5 Pushing the Peak Power for Direct Mid-Infrared Generation ... 95
5.1 Challenges of Reaching Higher Peak Powers ... 99
5.2 Experimental Achievements ... 102
5.2.1 Pump and Oscillator Setup ... 102
5.2.2 Mode-Locked Performance and the Effect of Air Dispersion ... 107
5.2.3 Towards Even Higher Peak Power Levels ... 115
5.3 Frequency Down-Conversion into the Mid-Infrared ... 118
5.3.1 Intra-Pulse Difference Frequency Generation in a ZGP Crystal ... 118
5.3.2 Characterization of the Generated Radiation ... 120
5.4 Summary ... 124
References ... 125

6 Conclusion and Outlook ... 129
6.1 Summary ... 129
6.2 A New Ultrafast Workhorse for Spectroscopic Applications? ... 131
6.3 Closing Remarks ... 133
References ... 134

Appendix ... 137

List of Acronyms

AM	Amplitude modulation
AOI	Angle of incidence
AR	Anti-reflective
BaF_2	Barium fluoride
BBO	Beta barium borate, β-BaB_2O_4
CEO	Carrier-envelope offset
CEP	Carrier-envelope phase
CM	Chirped mirror
Cr:ZnS	Chromium-doped zinc sulfide, Cr^{2+}:ZnS
Cr:ZnSe	Chromium-doped zinc selenide, Cr^{2+}:ZnSe
CSP	Chirped solitary pulse
CW	Continuous-wave
DC	Direct current
DFG	Difference frequency generation
DM	Dispersive mirror
EOS	Electro-optical sampling
Er	Erbium
FM	Frequency modulation
FOM	Figure of merit
FROG	Frequency-resolved optical gating
FRS	Field-resolved spectroscopy
FS	Fused silica
FTIR	Fourier-transform infrared
FTL	Fourier-transform limit
FWHM	Full width at half maximum
GaSe	Gallium selenide
GD	Group delay
GDD	Group delay dispersion
GVD	Group velocity dispersion
GVM	Group velocity mismatch
HITRAN	High-resolution transmission molecular absorption database

HR	Highly-reflective
IDFG	Intra-pulse difference frequency generation
InGaAs	Indium gallium arsenide
InP	Indium phosphide
IR	Infrared
ISO	International organization for standardization
KLM	Kerr-lens mode-locking/mode-locked
LGS	Lanthanum gallium silicate
$LiIO_3$	Lithium iodate
MgF_2	Magnesium fluoride
MIR	Mid-infrared
ML	Mode-locking
MOPA	Master oscillator power amplifier
ND	Neutral density
NIR	Near-infrared
OC	Output coupler/coupling
OPA	Optical parametric amplifier/amplification
OPO	Optical parametric oscillator/oscillation
PBC	Polarizing beam combiner
PbS	Lead sulfide
PD	Photodetector
pf-SNOM	Peak force scattering-type scanning near-field optical microscopy
PM	Phase-matching
PPLN	Periodically poled lithium niobate
QCL	Quantum cascade semiconductor laser
RBW	Resolution bandwidth
RF	Radio-frequency
RH	Relative humidity
RIN	Relative intensity noise
RMS	Root mean square
RO	Relaxation oscillation
ROC	Radius of curvature
RQPM	Random quasi-phase-matching
SAM	Self-amplitude modulation
SESAM	Semiconductor saturable absorber mirror
SH	Second harmonic
SHG	Second-harmonic generation
SNOM	Scanning near-field optical microscopy
SNR	Signal-to-noise ratio
SPM	Self-phase modulation
s-SNOM	Scattering-type scanning near-field optical microscopy
TE	Transverse electromagnetic
TEC	Thermoelectric cooling
TEM	Transverse electromagnetic mode
Ti:Sa	Titanium-doped sapphire, $Ti:Al_2O_3$

TM	Transition metal
Tm	Thulium
TOD	Third-order dispersion
U:CaF	Uranium-doped calcium fluoride, $U^{3+}:CaF_2$
UV	Ultraviolet
VA	Voltage amplifier
XFROG	Cross-correlation frequency-resolved optical gating
YAG	Yttrium aluminum garnet
ZBLAN	ZrF_4-BaF_2-LaF_3-AlF_3-NaF
ZGP	Zinc Germanium Phosphide, $ZnGeP_2$
ZnS	Zinc sulfide
ZnSe	Zinc selenide

Chapter 1
Motivation

1.1 *MIRanda*—A New Ultrafast Workhorse for Mid-Infrared Applications

The Advent of Mid-Infrared Laser Sources and Spectroscopy

The second half of the last century was marked by profound technological developments and most of them are now taken for granted in daily life. The foundations for one of the most outstanding inventions have been layed in 1917, when Albert Einstein predicted the phenomenon of "stimulated emission" [1]—a fundamental concept that became the backbone of laser physics. However, it took more than 40 years before the first working prototype of a laser was finally demonstrated by Theodore Maiman in 1960 [2], putting lasers onto a successful path that persists until today.

Inspired by the first discovery of laser action in a ruby crystal, an intense search for other suitable material was triggered, and Peter Sorokin and Mirek Stevenson were the first to demonstrate a 4-level solid-state laser based on uranium-doped calcium fluoride (U^{3+}:CaF_2) [3]. Notably, this world's second laser was emitting infrared light at a wavelength of 2.49 μm, unlike the clearly visible red light generated by a ruby laser (0.694 μm). Even though it never proved practical, it opened the door to an enormously promising region for spectroscopic applications—the mid-infrared (MIR).

Often referred to as the **molecular "finger-print" region** [4, 5], the MIR range between 2.5 μm and 20 μm (500-4000 cm^{-1}) is home to a grand variety of intense molecular absorption lines of organic and inorganic substances [6] (see Fig. 1.1). Owing to the energy associated with infrared radiation, which is too low to excite electrons to higher-lying energy levels, electromagnetic radiation can resonate with rotational and/or vibrational modes of chemical bonds or functional groups within the molecule, such that information on the chemical arrangement as well as the intramolecular interactions can be retrieved [7]. Probing these fundamental modes *directly* in the mid-infrared is a great advantage over near-infrared (NIR) spec-

N. Nagl, *A New Generation of Ultrafast Oscillators for Mid-Infrared Applications*, Springer Theses, https://doi.org/10.1007/978-3-030-89754-3_1

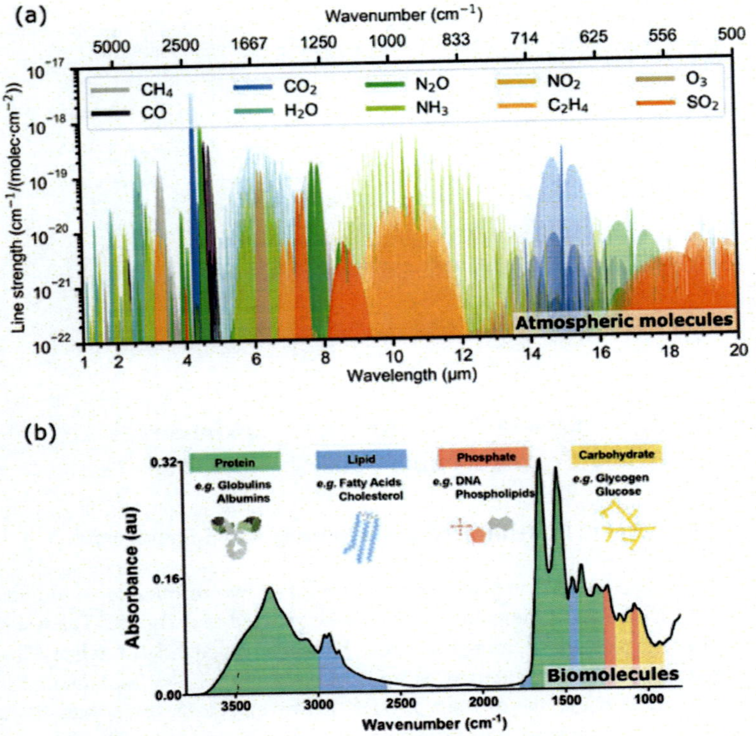

Fig. 1.1 Absorption spectra of atmospheric (gaseous) molecules and human blood serum in the MIR region. **a** Narrow-band absorption lines of typical atmospheric molecules in the range between 1 μm and 20 μm. The data were obtained from HITRAN simulations. **b** Absorption spectrum of human blood serum and identifiable biomolecules analyzed with FTIR spectroscopy [9]. Compared to measurements in the gas-phase (see **a**), where the molecules are isolated and do not interact with each other, the individual absorption features in the liquid phase start to broaden and smear out, eventually resulting in a continuous appearance of the absorbance

troscopy, where molecular absorption is due to anharmonic overtone and combination bands, making NIR spectra very complex and more difficult to interpret [8].

Spectroscopic measurements aiming to address as many different molecules as possible benefit greatly from **broadband MIR radiation** such as those emitted from synchrotrons [10–12] or globars. The latter is a thermal source, which emits blackbody radiation from an electrically heated material. Even though this method is readily available and straightforward to use, with several watts of average power being generated, thermal radiation lacks both spatial and temporal coherence. A high degree of coherence, however, is an important prerequisite for a broad range of mid-infrared applications, including the emerging fields of nano-imaging [13–15] as well as time-domain [14, 16] and frequency comb spectroscopy [17–19]. Therefore, synchrotrons are often preferred over globars because their higher brilliance and thus highly spatially confined photon flux allows for an improved signal-to-noise ratio [12]—especially important for measurements requiring good spatial resolution. Synchrotron radiation is generated by electrons that are accelerated and stored in large ring structures. These electrons travel in bunches of about 1 mm to 10 mm length, and produce short pulses of light when deflected by magnetic fields. With emitted wavelengths continuously extending from the very far-infrared to the high-energy X-ray regime, the light is confined to temporal widths of tens of picoseconds to a few nanoseconds (depending on the ring and its respective mode of operation) [12]. Consequently, phenomena can also be investigated in a time-resolved manner [20, 21], but with restricted access to nonlinear interactions happening on shorter time scales [22]. Other serious drawbacks of synchrotrons arise from the very limited beam time, the extremely high costs associated with their elaborate infrastructure as well as their MIR radiation power being limited to a few milliwatts only [23]. In this respect, broadband coherent MIR systems based on laser technology are a highly sought after alternative.

The search for **suitable mid-infrared laser sources** is, unfortunately, not straightforward. To address various molecular species simultaneously, MIR generation systems need to provide direct lasing in the 2.5 μm to 20 μm range, while covering a broad portion of the mid-IR at once or in a broadly tunable manner. Among others, quantum cascade semiconductor lasers (QCLs) are used for highly accurate and quantitative measurements, given their very narrow emission linewidths [24, 25]. Though they can be made quite compact in design, the limited wavelength tunability impedes parallel sensing of multiple molecular species, and several QCLs would need to be used together. Another promising approach is to exploit **nonlinear frequency down-conversion** into the MIR using mature near-infrared laser sources. Thus, optical parametric oscillators (OPOs) and amplifiers (OPAs) [26–30] have become very popular in recent years. Pulse energies reaching up to at least the μJ-level combined with an extremely large tunability allow to cover spectral domains, where laser options are typically scarce or require cryogenic cooling of the gain material. However, their sensitivity to misalignment [31] and crystal orientation due to phase-matching and energy conservation requirements increases the overall complexity. By contrast, a relatively simple method is based on collinear difference frequency generation (DFG), and is known as **intra-pulse DFG (IDFG)** [32–37]. Difference frequencies are generated when different frequency components of a single ultrabroadband laser pulse are mixed together, resulting in an intrinsically carrier-envelope phase (CEP)-stabilized signal regardless of the CEP-stability of the incident pulse. In recent years, the technique of IDFG was shown to result in the generation of several tens of milliwatts of MIR power with super-octave spectral coverage [32–37], providing ideal prerequisites for spectroscopic measurements. Moreover, due to the intrinsic CEP-stability of the mid-infrared radiation [38], new spectroscopic techniques can be exploited [39, 40].

About Ultrafast Lasers and Their Impact on MIR Spectroscopy

Coherent light sources have enabled methods such as Fourier-transform infrared (FTIR) spectroscopy [6] to extensively advance the progress in wide-ranging fields from atmospheric and trace-gas analysis [41] to biology and medicine [42–44]. Compared to measurements performed with conventional dispersive spectrometers, FTIR-based techniques are superior due to the higher spectral accuracy, better signal-to-noise ratio (SNR) and sensitivity, as well as optical throughput [45, 46]. Moreover, switching from continuous-wave (CW) to pulsed laser operation comes along with further advantages: Firstly, pulsed laser light is much more intense to exploit frequency down-conversion schemes in nonlinear optical crystals [47], such that a much wider portion in the mid-infrared can be covered—often in a single and compact optical arrangement [36]. Secondly, the temporal structure of the pulse—especially when its CEP and repetition rate are well controlled in the form of a frequency comb [48]—opens up new detection schemes like **electro-optical sampling (EOS)** [39, 49] as well as time-resolved spectroscopic techniques [16, 17, 40, 50]. Among them, **field-resolved spectroscopy (FRS)** in the molecular fingerprint region relies on EOS to gain direct access to the spectral amplitude and phase of the recorded waveforms. Moreover, it allows temporally separating the vibrational response of a sample from the high-power ultrashort excitation event, making measurements with an unparalleled specificity and sensitivity over the entire molecular fingerprint region feasible.

The potential of lasers for ultrafast spectroscopic applications, however, is strongly correlated with the pulse duration they can deliver [14]. Indeed, ultrashort light pulses with durations of less than a hundred femtoseconds (1 fs $= 10^{-15}$ s) enable scientists to study and control motions on ever shorter timescales and with increasing temporal resolution [51–53], with the shortest pulses reaching down to the attosecond scale [54]. The generation of ultrashort pulses has its origin in 1964, when the principle of mode-locking (ML) was discovered [55, 56]. Since then, new techniques—in particular semiconductor saturable absorber ML [57] and **Kerr-lens ML** [58]—greatly advanced the frontiers of femtosecond solid-state laser technology [59, 60], and pulse durations down to few optical cycles can now be readily generated [61–64].

MIRanda: Advance a Just Cause

Nowadays, laser systems often involve a chain of complicated setups and many of them are not likely to leave the research laboratories due to their complexity and difficult maintenance. Consequently, laser-based applications will struggle to be widely adopted, thus severely hampering the rate of new scientific discoveries in the emerging field of mid-infrared spectroscopy and microscopy. In this PhD thesis, *MIRanda*, a new era of MIR laser systems is presented, generating highly-intense optical few-cycle pulses[1] at multi-MHz repetition rates directly in the spectral region between

[1] At a given wavelength, a single-cycle pulse is the shortest possible waveform that occurs when the electric field performs only one oscillation period within the pulse envelope. For a laser with output pulse duration τ_0 and central frequency ν_0, the number of optical cycles is calculated as: $N = \tau_0 \cdot \nu_0$ [65].

2 μm and 3 μm. Thanks to the innovative combination of direct diode-pumping with ultrafast Kerr-lens ML Cr:ZnS/ZnSe laser technology, high photon-flux can be generated at rather exotic wavelengths and with intrinsic low amplitude noise performance, vastly enhancing the efficiency, sensitivity, and statistics in spectroscopic experiments. Combined with a highly compact design, it will help to advance next-generation molecular diagnostics performed in-house, aiming at health monitoring and disease detection via laser-based human blood analysis.[2] Moreover, underpinning all these developments can be formulated as a *Just Cause*[3]: Developing means to monitor and preserve the health of every human being in the best possible way and to the benefit of everyone.

1.2 The Power of Directly Diode-Pumped, Few-Cycle Mid-Infrared Lasers

The study of fundamental molecular vibrations in the range between 2.5 μm and 20 μm requires laser systems that can provide broadband MIR laser light. Direct generation of coherent emission in the molecular fingerprint region is, however, still rather challenging, even though much work has been undertaken to push the emission of ultrafast solid-state oscillators toward ever longer wavelengths [26, 47, 67, 68].

In the 1.9 μm to 3.0 μm region, pulsed laser oscillators and amplifier systems based on thulium- [69, 70], holmium- [71, 72] and chromium- [33, 64, 73] doped gain media have proven to provide peak power levels sufficient for further nonlinear frequency down-conversion schemes into the MIR. Besides fiber and bulk architectures, thin-disk technology has attracted growing interest in these materials, mainly due to its power and energy scalability [74]. Owing to the efficient cooling geometry of the gain medium and the reduced thermal lensing effects, record output powers of more than several watts in mode-locked operation can be achieved [71, 75]. However, the high intra-cavity powers make these systems prone to detrimentally high heat load and optical component damage, while suffering from long initial pulse durations that are attributed to the limited emission bandwidth of the disk material. Attempts have been made to incorporate broadband gain media such as Cr:ZnS/ZnSe into a thin-disk geometry, but it is impeded by the huge thermal load associated to their intrinsically large quantum defect [76]. Though thermal lensing could be reduced in a highly elaborate spinning ring gain element approach [77], its applicability to a mode-locked oscillator is still uncertain.

To access the longer wavelengths, most frequency down-conversion schemes thus rely on more mature and powerful near-infrared driving lasers, albeit with the need for pulse compression or even parametric conversion stages to increase pulse energies

---

[2] Lasers4Life project: www.lasers4life.de, www.mukkozpont.hu.

[3] "A *Just Cause* is a specific vision of a future state that does not yet exist. [...] It describes the world we hope to live in and will commit to help build [...] and the goal is not to win, but to keep playing for advancing something bigger than ourselves." [66].

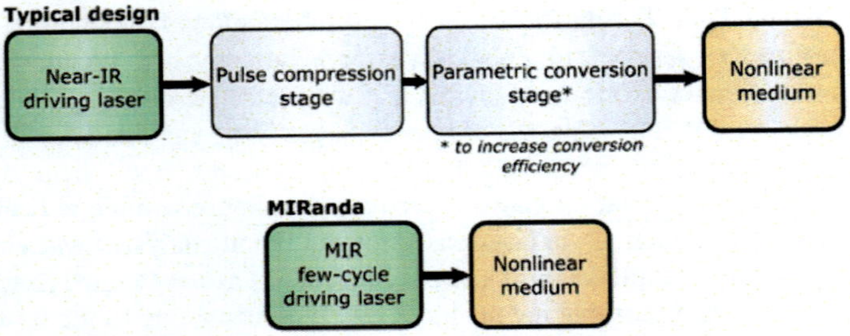

Fig. 1.2 Comparison of the typical scheme used for mid-infrared generation and its vast simplification offered by *MIRanda*

and improve conversion efficiencies (see Fig. 1.2). In the following, the advantages of directly diode-pumped as well as few-cycle laser systems with direct emission in the MIR will be presented which—when all characteristics are united—can dramatically simplify the overall scheme.

Direct Lasing in the Mid-Infrared Region

MIR radiation can be generated in a relatively straightforward way by means of nonlinear optical crystals in an IDFG arrangement, and non-oxide materials are preferred over oxide-based ones, whose performance is greatly affected by the onset of multi-phonon and multi-photon absorption [26]. However, pumping non-oxide crystals with 1-µm near-infrared driving systems is extremely difficult and limitations arise from residual absorption loss, unfavourable transparency range or phase-matching (PM), while crystal quality and availability can cause further problems [78]. By contrast, moving to longer pump wavelengths not only results in a smaller quantum defect and a crystal transparency penetrating deeper into the mid-IR, but can also improve the PM bandwidths as well as the overall conversion efficiency [26, 32, 67]. Consequently, direct lasing in the MIR region beyond a wavelength of 2 µm is of great interest for applications and few-cycle driving systems like those based on Cr^{+2}:ZnS/ZnSe gain media are highly attractive in this respect. As an example, the phase-matching (PM) curve for frequency down-conversion in a LGS crystal, pumped by a typical 1-µm Yb:YAG laser [79, 80], is compared with that in a GaSe crystal, which can be driven by Cr:ZnS/ZnSe-laser technology [33] (Fig. 1.3). The simulations clearly illustrate the broader span of mid-infrared wavelengths that can be covered via 2 µm-based driving systems.

Direct Generation of Few-Cycle Laser Pulses

Since frequency down-conversion techniques like IDFG arise from quadratic [$\chi^{(2)}$-type] nonlinear optical effects, the intensity of the driving pump pulse and its associated pulse duration have a strong impact on the conversion efficiency of the process

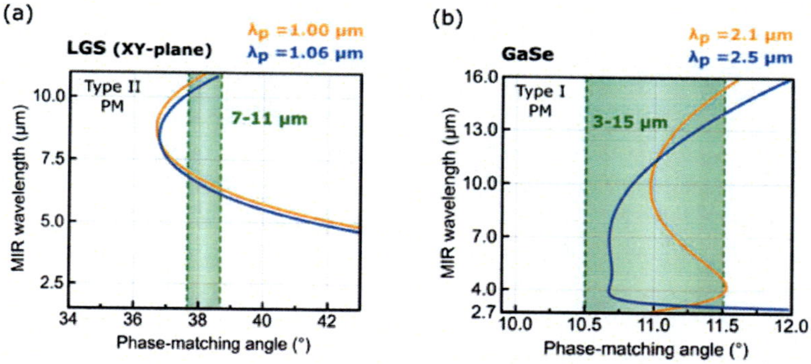

Fig. 1.3 Simulated phase-matching (PM) curves for the generated signal (idler) wavelengths in **a** LGS and **b** GaSe via frequency down-conversion. To resemble an IDFG geometry, where frequency components of a single pump beam mix together and generate mid-IR radiation, two different pump wavelengths (λ_p) were assumed. They correspond to the shortest (orange) and longest (blue) wavelength components of an incident pump beam as they are readily emitted from ultrafast Yb:YAG or Cr:ZnS/ZnSe laser systems. For a slight variation in angle of ±0.5°, the achievable MIR wavelengths are shaded in green. They indicate an "ideal" angle of about 38° for LGS and 11° for GaSe, which is in good agreement with the PM angles reported for LGS [79, 80] and GaSe [33]. Also, the phase-matching bandwidth offered by a 2 μm-pumped GaSe crystal is larger than obtained with a 1 μm-pumped LGS crystal, which allows a broader portion of the MIR to be covered via frequency down-conversion. Note that, compared to GaSe, LGS is biaxial and the propagation is simulated in one of the crystal's principal planes (here: XY). The calculations were performed with the open-source software *SNLO*

[81]. Using the few-cycle pulses directly available from Cr:ZnS/ZnSe oscillators, as demonstrated in this thesis, will thus provide significant advantage over existing femtosecond pump sources like Er/Tm-fiber lasers [82, 83] and optical parametric generators/amplifiers [84] due to their intrinsically much shorter pulse durations. The need for intricate and alignment-sensitive post-compression chains based on multi-pass (Herriott) cells [85] or nonlinear fibers [35] is eliminated, greatly simplifying the overall design. Moreover, the broader the generated output spectrum of the laser oscillator, the shorter the mid-IR wavelengths that can ultimately be reached.[4] This allows to address a larger set of molecules, with biomolecules like lipids and proteins facing strong absorption even below 4 μm (> 2500 cm^{-1}, see Fig. 1.1). Combining the few-cycle emission with high average powers and MHz-repetition rates can thus eliminate the need for post-amplification, as will be demonstrated in Chap. 5.

Direct Diode-Pumping

Applications of coherent mid-infrared radiation such as FTIR-microscopy, FRS, or scattering-type scanning near-field optical microscopy (s-SNOM) [15] strongly benefit from high-efficiency and compact high-brightness light sources. In addition, a

[4] The shortest wavelengths that can be generated via DFG are determined by the shortest (λ_1) and longest (λ_2) wavelength components of the incident laser pulse: $\lambda_3^{-1} = \lambda_1^{-1} - \lambda_2^{-1}$.

low-noise performance is desired since intensity fluctuations of the laser output will set an ultimate limit to the sensitivity of the measurements. Fortunately, all these characteristics are inherent to systems in which the gain medium can be pumped directly by semiconductor laser diodes. Even though there are several major challenges arising from the combination of direct diode-pumping and the technique of (soft-aperture) Kerr-lens mode-locking (KLM), a sophisticated pump setup and cavity design were realized and three laser systems could be successfully set up in the framework of this thesis.

1.3 Thesis Outline

This dissertation sheds new light on the question of how state-of-the-art solid-state laser systems can be improved to greatly advance downstream MIR applications. The focus lies on the first demonstration and further development of powerful diode-pumped femtosecond Cr^{2+}:ZnS/ZnSe laser oscillators based on KLM.

The thesis is organized as follows:

Chapter 2 lays the theoretical foundations of ultrashort pulse generation and introduces the basic concepts of soliton propagation and soliton mode-locking. First, the origin of laser pulse formation will be explained in Sect. 2.1 together with the experimental technique of Kerr-lens mode-locking, which allows to generate few-cycle pulses directly from a laser cavity. In Sects. 2.2 and 2.3, the principal factors influencing soliton dynamics are examined and summarized in the master equation of mode-locking. Finally, the effect of positive and negative net dispersion on the pulse formation process is addressed with respect to two specific mode-locking regimes.

Chapter 3 starts with introducing the advantages of Cr^{2+}-doped II-VI gain media (Sect. 3.1) and continues with examining the major challenges associated with direct diode-pumping (Sect. 3.2). In Sect. 3.3, the experimental results on the first directly diode-pumped KLM Cr^{2+}-doped II-VI oscillator are presented, highlighting the major breakthrough achieved upon soft-aperture KLM combined with a careful intra-cavity dispersion management. Moreover, measurements on the amplitude noise of the mode-locked output reveal a highly promising low-noise performance.

The subject of **Chap.** 4 is to demonstrate the power-scalability of the results from Chap. 3, driven by the overall goal to reach and surpass the performance of state-of-the-art fiber-laser-pumped systems. In Sect. 4.1, a focus is placed on the challenges when going towards higher intra-cavity peak powers with respect to mode-locking performance and pump setup design. The achieved output results will then be presented in Sect. 4.2, revealing the performance improvements over the first generation.

Chapter 5 focuses on pushing the ultimate limits of ultrafast diode-pumped Cr^{2+}-doped II-VI lasers towards ever higher peak powers. The resulting mode-matching issues between pump and fundamental cavity beam are addressed in Sect. 5.1, and

Sect. 5.2 reports a series of experimental findings on the cavity design and mode-locking performance as well as the achieved output results. The power of the third-generation system is finally illustrated through direct generation of mid-infrared radiation in a ZGP crystal (Sect. 5.3), with a summary of the results achieved concluding this last experimental chapter (Sect. 5.4).

In **Chap.** 6, a summary is presented together with an outlook, highlighting and discussing future developments that are initiated by the achievements presented in the framework of this thesis.

References

1. Einstein A (1917) Zur Quantentheorie der Strahlung. Physikalische Zeitschrift 18:121–128
2. Maiman TH (1960) Stimulated optical radiation in ruby. Nature 187:493–494
3. Sorokin PP, Stevenson MJ (1960) Stimulated infrared emission from trivalent uranium. Phys Rev Lett 5(12):557–559
4. Cossel KC, Waxman EM, Finneran IA, Blake GA, Ye J, Newbury NR (2017) Gas-phase broadband spectroscopy using active sources: progress, status, and applications. J Opt Soc Am B 34(1):104–129
5. Li J, Chen W, Fischer H (2013) Quantum cascade laser spectrometry techniques: a new trend in atmospheric chemistry. Appl Spectrosc Rev 48(7):523–559
6. Griffiths PR, de Haseth JA (2006) Fourier transform infrared spectrometry, 2nd edn. Wiley, Inc
7. Movasaghi Z, Rehman S, Rehman I (2008) Fourier transform infrared (FTIR) spectroscopy of biological tissues. Appl Spectrosc Rev 43(2):134–179
8. Eldin AB (2011) Near infra red spectroscopy. In: Akyar I (ed) Wide spectra of quality control. IntechOpen, pp 237–248
9. Gray E et al (2018) Health economic evaluation of a serum-based blood test for brain tumour diagnosis: exploration of two clinical scenarios. BMJ Open 8(5)
10. Petibois C, Deleris G, Piccinini M, Cestelli-Guidi M, Marcelli A (2009) A bright future for synchrotron imaging. Nat Photonics 3:179
11. Martin MC, Schade U, Lerch P, Dumas P (2010) Recent applications and current trends in analytical chemistry using synchrotron-based Fouriertransform infrared microspectroscopy. Trends Anal Chem 29(6):453–463
12. Marcelli A, Cinque G (2011) Infrared synchrotron radiation beamlines: high brilliance tools for IR spectromicroscopy. RSC analytical spectroscopy series, pp 67–104
13. Amenabar I, Poly S, Goikoetxea M, Nuansing W, Lasch P, Hillenbrand R (2017) Hyperspectral infrared nanoimaging of organic samples based on Fourier transform infrared nanospectroscopy. Nat Commun 8(14402)
14. Eisele M et al (2014) Ultrafast multi-terahertz nano-spectroscopy with sub-cycle temporal resolution. Nat Photonics 8:841–845
15. Wang L, Xu XG (2015) Scattering-type scanning near-field optical microscopy with reconstruction of vertical interaction. Nat Commun 6:8973
16. Lanin AA, Voronin AA, Fedotov AB, Zheltikov AM (2014) Timedomain spectroscopy in the mid-infrared. Sci Rep 4:6670
17. Piqué N, Hänsch TW (2019) Frequency comb spectroscopy. Nat Photonics 13:146–157
18. Schliesser A, Picqué N, Hänsch TW (2012) Mid-infrared frequency combs. Nat Photonics 6:440–449
19. Coddington I, Newbury N, Swann W (2016) Dual-comb spectroscopy. Optica 3(4):414–426
20. Lobo RPSM et al (2005) Photoinduced time-resolved electrodynamics of superconducting metals and alloys. Phys Rev B 72(024510)

21. Lobo RPSM, LaVeigne JD, Reitze DH, Tanner DB, Carr GL (2002) Subnanosecond, time-resolved, broadband infrared spectroscopy using synchrotron radiation. Rev Sci Instrum 73(1)
22. Roeser CAD, Mazur E (2005) Light-matter interactions on the femtosecond time scale. In: Di Bartolo B, Forte O (eds) Frontiers of optical spectroscopy. Springer, pp 29–54
23. Dumas P, Polack F, Lagarde B, Chubar O, Giorgetta JL, Lefrançois S (2006) Synchrotron infrared microscopy at the French synchrotron facility SOLEIL. Infrared Phys Technol 49:152–160
24. Vitiello MS, Scalari G, Williams B, De Natale P (2015) Quantum cascade lasers: 20 years of challenges. Opt Express 23(4):5167–5182
25. Lee BG et al (2007) Widely tunable single-mode quantum cascade laser source for mid-infrared spectroscopy. Appl Phys Lett 91(23):231101
26. Petrov V (2012) Parametric down-conversion devices: the coverage of the midinfrared spectral range by soild-state laser sources. Opt Mater 34(3):536–554
27. Peng J (2014) Developments of mid-infrared optical parameteric oscillators for spectroscopic sensing: a review. Opt Eng 53(6):061613
28. Keilmann F, Amarie S (2013) Mid-infrared frequency comb spanning an octave based on an Er fiber laser and difference-frequency generation. In: 2013 conference on lasers and electro-optics - international quantum electronics conference. Optical Society of America, CC.P.2
29. Steinle T, Mörz F, Steinmann A, Giessen H (2016) Ultra-stable high average power femtosecond laser system tunable from 1.33 to 20 μm. Opt Lett 41(21):4863–4866
30. Smolski VO, Vasilyev S, Schunemann PG, Mirov SB, Vodopyanov KL (2015) Cr:ZnS laser-pumped subharmonic GaAs optical parameteric oscillator with the spectrum spanning 3.6-5.6 μm. Opt Lett 40(12):2906–2908
31. Fix A et al (2011) Optical parametric oscillators and amplifiers for airborne and spaceborne active remote sensing of CO2 and CH4. In: Singh UN, Pappalardo G (eds) Lidar technologies, techniques, and measurements for atmospheric remote sensing VII, vol 8182. SPIE, pp 28–37
32. Gaida C et al (2018) Watt-scale super-octave mid-infrared intrapulse difference frequency generation. Light Sci Appl 7(1):94
33. Wang Q et al (2019) Broadband mid-infrared coverage (2–17 μm) with few-cycle pulses via cascaded parametric processes. Opt Lett 44(10):2566–2569
34. Wang Q et al (2019) Highly efficient broadband mid-infrared generation (2.8-12.5 μm) based on a compact Cr:ZnS laser. In: 2019 conference on lasers and electro-optics Europe and European quantum electronics conference. CF-7.5. IEEE
35. Zhang J et al (2018) Multi-mW, few-cycle mid-infrared continuum spanning from 500 to 2250 cm^{-1}. Light Sci Appl 7:17180
36. Zhang J, Fritsch K, Wang Q, Krausz F, Mak K, Pronin O (2019) Intra-pulse difference-frequency generation of mid-infrared (2.7-20 μm) by random quasi-phase-matching. Opt Lett 44(12):2986–2989
37. Vasilyev S, Moskalev I, Mirov M, Mirov S, Gapontsev V (2016) Multi- Watt mid-Ir femtosecond polycrystalline Cr^{2+}:ZnS and Cr^{2+}:ZnSe laser amplifiers with the spectrum spanning 2.0-2.6 μm. Opt Soc Am 24(2):1616–1623
38. Baltuška A, Fuji T, Kobayashi T (2002) Controlling the carrier-envelope phase of ultrashort light pulses with optical parametric amplifiers. Phys Rev Lett 88(13):133901
39. Pupeza I et al (2015) High-power sub-two-cycle mid-infrared pulses at 100 MHz repetition rate. Nat Photonics 9:721–724
40. Pupeza I et al (2020) Field-resolved infrared spectroscopy of biological systems. Nature 577(7788):52–59
41. Bacsik Z, Mink J, Keresztury G (2004) FTIR spectroscopy of the atmosphere. I. Principles and methods. Appl Spectrosc Rev 39(3):295–363
42. Barth A (2007) Infrared spectroscopy of proteins. Biochimica et Biophysica Acta (BBA) - Bioenerg 1767(9):1073–1101
43. Petibois C, Deleris G (2006) Chemical mapping of tumor progression by FTIR imaging: towards molecular histopathology. Trends Biotechnol 24(10):455–462

44. Sala A et al (2020) Biofluid diagnostics by FTIR spectroscopy: a platform technology for cancer detection. Cancer Lett 477:122–130
45. Jacquinot P (1960) New developments in interference spectroscopy. Rep Prog Phys 23(1):267–312
46. Connes P (1970) Astronomical Fourier spectroscopy. Annu Rev Astron Astrophys 8:209–230
47. Tittel FK, Richter D, Fried A (2003) Solid-state mid-infrared laser sources, chap 11. Springer, pp 445–497
48. Hänsch TW (2006) Nobel lecture: passion for precision. Rev Mod Phys 78(4):1297–1309
49. Keiber S et al (2016) Electro-optic sampling of near-infrared waveforms. Nat Photonics 10:159–162
50. Mörz F et al (2017) Nearly diffraction limited FTIR mapping using an ultrastable broadband femtosecond laser tunable from 1.33 to 8 μm. Opt Express 25(26):32355–32363
51. Diddams SA, Hollberg L, Mbele V (2007) Molecular fingerprinting with the resolved modes of a femtosecond laser frequency comb. Nature 445:627–630
52. Sibbett W, Lagatsky AA, Brown CTA (2012) The development and application of femtosecond laser systems. Opt Express 20(7):6989–7001
53. Sorokin E, Sorokina IT, Mandon J, Guelachvili G, Picqué N (2007) Sensitive multiplex spectroscopy in the molecular fingerprint 2.4 μm region with a Cr^{2+}:ZnSe femtosecond laser. Opt Express 15(25):16540–16545
54. Krausz F, Ivanov M (2009) Attosecond physics. Rev Mod Phys 81(1):163–234
55. DiDomenico M (1964) Small-signal analysis of internal (coupling-type) modulation of lasers. J Appl Phys 35(10):2870–2876
56. Hargrove LE, Fork RL, Pollack MA (1964) Locking of He-Ne laser modes induced by synchronous intracavity modulation. Appl Phys Lett 5(1):4–5
57. Keller U, Miller DAB, Boyd GD, Chiu TH, Ferguson JF, Asom MT (1992) Solid-state low-loss intracavity saturable absorber for Nd:YLF lasers: an antiresonant semiconductor Fabry-Perot saturable absorber. Opt Lett 17(7):505–507
58. Spence DE, Kean PN, Sibbett W (1991) 60-fsec pulse generation from a self-mode-locked Ti:sapphire laser. Opt Lett 16(1):42–44
59. Krausz F et al (1992) Femtosecond solid-state lasers. IEEE J Quantum Electron 28(10):2097–2122
60. Keller U (2003) Recent developments in compact ultrafast lasers. Nature 424:831–838
61. Ell R et al (2001) Generation of 5-fs pulses and octave-spanning spectra directly from a Ti:sapphire laser. Opt Lett 26(6):373–375
62. Wagenblast PC et al (2002) Generation of sub-10-fs pulses from a Kerr-lens mode-locked Cr^{3+}:LiCAF laser oscillator by use of third-order dispersioncompensating double-chirped mirrors. Opt Lett 27(19):1726–1728
63. Chudoba C et al (2001) All-solid-state Cr:forsterite laser generating 14-fs pulses at 1.3 μm. Opt Lett 26(5):292–294
64. Mirov S et al (2018) Frontiers of mid-IR lasers based on transition metal doped chalcogenides. IEEE J Sel Top Quantum Electron 24(5):1–29
65. Kärtner FX (ed) (2004) Few-cycle laser pulse generation and its applications, vol 95. Springer
66. Sinek S (2019) The infinite game. Portfolio/Penguin
67. Sorokina IT, Vodopyanov KL (2003) Solid-state mid-infrared laser sources, vol 89. Topics in applied physics. Springer
68. Godard A (2007) Infrared (2–12 μm) solid-state laser sources: a review. Comptes Rendus Physique 8(10):1100–1128
69. Zhang J et al (2018) High-power, high-efficiency Tm:YAG and Ho:YAG thin-disk lasers. Laser Photonics Rev 12(3):1700273
70. Gaida C, Gebhardt M, Heuermann T, Stutzki F, Jauregui C, Limpert J (2018) Ultrafast thulium fiber laser system emitting more than 1 kW of average power. Opt Lett 43(23):5853–5856
71. Zhang J, Mak K, Pronin O (2018) Kerr-lens mode-locked 2 μm thin-disk lasers. IEEE J Sel Top Quantum Electron 24(5)

72. Li P, Ruehl A, Grosse-Wortmann U, Hartl I (2014) Sub-100 fs passively mode-locked holmium-doped fiber oscillator operating at 2.06 μm. Opt Lett 39(24):6859–6862
73. Nagl N, Mak K, Wang Q, Pervak V, Krausz F, Pronin O (2019) Efficient femtosecond mid-infrared generation based on a Cr:ZnS oscillator and stepindex fluoride fibers. Opt Lett 44(10):2390–2393
74. Giesen A, Hügel H, Voss A, Wittig K, Brauch U, Opower H (1994) Scalable concept for diode-pumped high-power solid-state lasers. Appl Phys B 58:365–372
75. Saraceno CJ, Sutter D, Metzger T, Ahmed MA (2019) The amazing progress of high-power ultrafast thin-disk lasers. J Eur Opt Soc-Rapid Publ 15(15)
76. Renz G, Speiser J, Giesen A, Sorokina IT, Sorokin E (2013) Cr:ZnSe thin disk cw laser. In: Solid state lasers XXII: technology and devices, vol 8599. SPIE, pp 310–316
77. Moskalev I et al (2016) 140 W Cr:ZnSe laser system. Opt Express 24(18):21090–21104
78. Hildenbrand A et al (2014) Laser damage of the nonlinear crystals CdSiP2 and ZnGeP2 studied with nanosecond pulses at 1064 and 2090 nm. Opt Eng 53(12):122511
79. Seidel M et al (2018) Multi-watt, multi-octave, mid-infrared femtosecond source. Sci Adv 4(4)
80. Chen B-H, Nagy T, Baum P (2018) Efficient middle-infrared generation in $LiGaS_2$ by simultaneous spectral broadening and difference-frequency generation. Opt Lett 43(8):1742–1745
81. Boyd R (2008) Nonlinear optics, 3rd edn. Academic Press
82. Vazquez-Zuniga LA, Jeong Y (2013) Wavelength-tunable, passive mode-locked erbium doped fiber master-oscillator incorporating a semiconductor saturable absorber mirror. J Opt Soc Korea 17(2):117–129
83. Sorokina IT, Dvoyrin VV, Tolstik N, Sorokin E (2014) Mid-IR ultrashort pulsed fiber-based lasers. IEEE J Sel Top Quantum Electron 20(5)
84. Petrov V (2015) Frequency down-conversion of solid-state laser sources to the mid-infrared spectral range using non-oxide nonlinear crystals. Prog Quantum Electron 42:1–106
85. Fritsch K, Pötzlberger M, Pervak V, Brons J, Pronin O (2018) All-solidstate multipass spectral broadening to sub-20 fs. Opt Lett 43(19):4643–4646

Chapter 2
Fundamentals of Ultrashort Pulse Generation

The generation of few-cycle laser pulses —as emitted directly from a diode-pumped solid-state laser —is the result of a fascinating journey of mode-locked laser development, that has been started more than 50 years ago. In order to understand the series of linear and nonlinear mechanisms involved in the process, this chapter serves as a comprehensive introduction to the basic ideas behind ultrashort pulse generation (Sect. 2.1), with a focus devoted to Kerr-lens mode-locking as well as solitary pulse propagation (Sect. 2.2) and formation (Sect. 2.3). The latter has become one of the most important pulse-shaping processes in nowaday's ultrafast laser systems, and is of great importance for the three laser systems presented in this thesis. In the following, the respective physical concepts will be explained and combined with mathematical derivations, based on the content of a series of excellent graduate-level textbooks [1–5].

2.1 Formation of Laser Pulses

Compared to other sources of radiation, lasers are devices that produce light of remarkable properties, ranging from high directionality and coherence to spectral purity and intensity. Broadly speaking, only three components are required to make a laser functional, namely (i) a gain medium for population inversion, (ii) a pump source to provide the energy needed for the inversion process, and (iii) a resonator structure that serves as an optical feedback mechanism for the amplification of light. More precisely, laser action is a result of stimulated emission in the gain medium, in which the power of incoming photons gets amplified during the pumping process. However, this requires more atoms of the medium to be excited into some higher quantum-mechanical energy level than are located in lower-lying states. As shown in Fig. 2.1, a 4-level system is characterized by radiative and fast non-radiative energy

N. Nagl, *A New Generation of Ultrafast Oscillators for Mid-Infrared Applications*, Springer Theses, https://doi.org/10.1007/978-3-030-89754-3_2

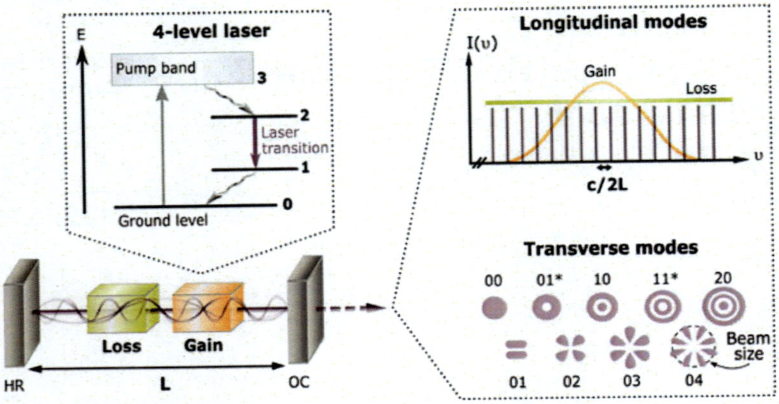

Fig. 2.1 Illustration of a two-mirror cavity of length L, including a highly-reflective (HR) mirror and a partially transmissive output coupler (OC). The energy diagram of the gain medium is represented by a simplified 4-level structure, which involves radiative transitions and non-radiative decays (denoted as wavy arrows). Also depicted are cylindrical transverse and longitudinal mode distributions supported by the resonator. In the frequency domain, the standing waves of the cavity are separated by $\Delta\nu = c/2L$, and only a certain number of modes gets amplified when the gain of the lasing medium exceeds the resonator losses. By contrast, the transverse modes refer to the electromagnetic field patterns in the spatial domain

transitions, resulting in a sufficiently long lifetime of the upper laser level 2 and a short lifetime of the lower level 1. Hence, these kind of gain media are ideal to achieve a reasonably large population inversion and thus efficient laser action.

Typically, a classification is made based on the type of gain medium used and the temporal and spatial behaviour of the light emission. For the three lasers systems presented in this thesis, laser action is obtained by optically pumping a solid-state crystal with the emission of laser diodes. A part of the radiation of this **solid-state laser** is then coupled out through one of the end-mirrors (OC), eventually providing the coherent output beam for downstream applications. In addition, a laser is also characterized according to its longitudinal and transverse mode properties (see Fig. 2.1). As will be discussed in the next section, a multitude of longitudinal modes, which are electromagnetic plane waves that propagate in opposite directions and with different frequencies along the cavity axis, can oscillate within a laser resonator. Also, modes with different sizes and divergences can emerge from the cavity, being associated with the transverse electromagnetic mode (TEM_{mn}) intensity distributions [6]. However, for the single-transversal-mode lasers presented in this thesis, only the fundamental TEM_{00} mode is supported, and—as we shall see later—a good spatial overlap of the pump beam and the fundamental Gaussian mode inside the laser crystal is essential to maintain this mode of operation.

Up to now, only continuous-wave (CW) laser emission was considered, with the output power being nominally constant over time. In the next section, the concept of mode-locking together with its experimental techniques will be introduced, which is essential to realize a pulsed laser output.

2.1.1 Locking the Longitudinal Modes of a Laser Cavity

In the very general case of a standing-wave cavity made from two parallel reflecting surfaces, only a specific set of resonant frequencies ν_n is supported, since the standing waves are forced to form nodes at the end-mirrors:

$$\nu_n = \frac{c}{\lambda} = \frac{c}{2L}n \quad (n \in \mathbb{N}). \tag{2.1}$$

Here, c is the speed of light, L indicates the length of the cavity, and n is the mode number. As depicted in Fig. 2.1, the frequency difference between two consecutive modes is $\Delta\nu = c/2L = 1/t_{rep}$ and is given by the inverse of the cavity round-trip time $1/t_{rep}$. In principle, an infinite number of standing waves could oscillate within the resonator, but, depending on the gain bandwidth of the laser medium, only a certain fraction experiences sufficient gain to account for losses. For a free-running laser, there is no fixed mode-to-mode phase and amplitude relationship and the total electric field $E(t)$ can therefore be described as [2]:

$$E(t) = \sum_{n=-m}^{n=m} E_n e^{i(w_0+n2\pi/t_{rep})t+i\varphi_n}, \tag{2.2}$$

where φ_n and E_n correspond to the electric field and phase of the nth mode, with w_0 being the mode frequency at the center of the gain bandwidth. Consequently, the output laser intensity fluctuates in time and, at first sight, it looks like a completely random spiking behaviour with the mode phases varying between $-\pi$ and $+\pi$. However, since the spikes originate from the sum of $N = 2m + 1$ evenly spaced frequency components, the resulting waveform reveals some characteristics of a Fourier series. As illustrated in Fig. 2.2a, it has some kind of periodicity with a period of $1/\Delta\nu$, with each spike having a duration of roughly $1/N\Delta\nu$ [1].

However, in order to generate temporally well-behaved output pulses, the modes have to be "coupled" together. In a process referred to as *mode-locking*, the spectral amplitudes and phases get correlated with each other, and Eq. 2.2 evolves into [2]:

$$E(t) = E_0 \sum_{n=-m}^{n=m} e^{i(w_0+n2\pi/t_{rep})t} = E_0 \left[\frac{1 - e^{iN(2\pi/t_{rep})t}}{1 - e^{i(2\pi/t_{rep})t}}\right] e^{iw_0 t}. \tag{2.3}$$

For this, all modes were assumed to have equal field strengths $E_n = E_0$ and, by a shift of our choice of time, we have also set $\varphi_n = \varphi_0 = 0$. The corresponding

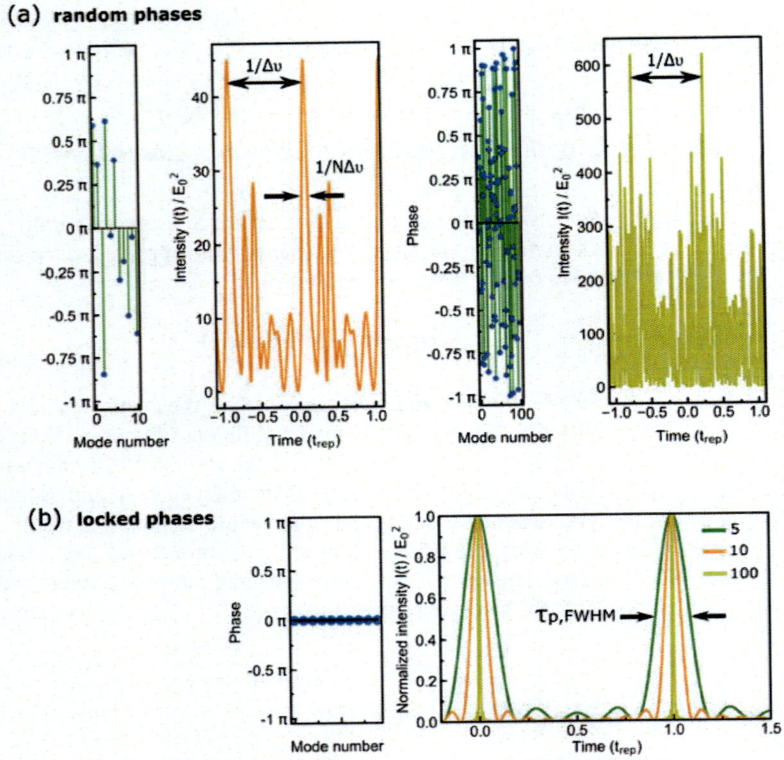

Fig. 2.2 Simulated temporal behaviour of the squared amplitude of the total electric field for the case of 10 and 100 oscillating modes (**a**), all having the same amplitude $E_0 = 1$ but random phases. The spike with the highest intensity will experience the highest gain during a cavity roundtrip and will eventually lead to the initiation of the mode-locking process. For simplicity, the amplitudes were assumed to be constant (i.e. flat gain), which is not the case in reality. Note that the response time of conventional photodetectors is typically too slow to resolve this complex behaviour in the temporal domain. Instead, its average value is measured. **b** Same parameters as before, but now for the case of 5, 10, and 100 modes having constant phases. The more modes are oscillating, the narrower the FWHM, the higher the amplitude, and the longer the dead time in between. Note that the intensities were normalized to an interval of [0, 1]

output intensity is then simply the product of the electric field $E(t)$ with its complex conjugate $E^*(t)$:

$$I(t) = |E(t)|^2 = E(t)E^*(t) = E_0^2 \frac{\sin^2(N\pi t/t_{rep})}{\sin^2(\pi t/t_{rep})}. \tag{2.4}$$

The result is a train of short but intense pulses separated by the round-trip time $t = t_{rep}$, and with a FWHM pulse duration given approximately by $t_p \approx t_{rep}/N \approx 1/\Delta\nu_g$, where $\Delta\nu_g$ is the gain bandwidth of the laser [2]. As depicted in Fig. 2.2b, the main peaks are accompanied by $N - 2$ much weaker subsidiary peaks, and the larger the number of oscillating modes, the shorter will be the generated pulse duration. In Chap. 3, we will see that the first diode-pumped laser system presented in this thesis, *MIRandal*, is able to emit pulses as short as 45 fs at a repetition rate of 65 MHz. With an output spectral bandwidth (3 dB) of about 185 nm (10.04 THz) at a central wavelength of 2.35 μm, an impressively large number of $N = 10.04$ THz/65 MHz $\approx$ 154461 modes are thus locked in phase.

2.1.2 Experimental Realization Methods

In general, different techniques exist to force the laser into pulsed operation. They have in common that a fixed-phase relationship among the longitudinal modes is established by introducing some kind of mode modulation, with the modulation frequency being linked to the repetition rate of the laser cavity. Here, locked modes in the form of a more intense radiation peak suffer lower propagation losses inside the resonator than any other combination of (unlocked) modes, and are therefore favoured.

Typically, a distinction is made between active and passive methods. **Active mode-locking** was first theoretically described by Kuizenga and Siegman [7] and requires the resonator losses to be modulated by an externally driven amplitude (AM) or phase (FM) modulator. However, the commonly used electro-optic or acousto-optic component adds to the cost and complexity of the setup since the modulation frequency must be matched precisely to the spacing of the cavity modes. Also, the external modulation is limited to electronic speed, therefore constraining the minimum attainable pulse duration (~ps-range).

These drawbacks can be overcome by **passive mode-locking** techniques, which do not require an external source but only rely on the nonlinear loss modulation triggered by the laser pulse itself. This modulation is automatically synchronized with the travelling pulses, and is accomplished by incorporating a saturable absorber into the laser cavity, whose absorption decreases with increasing intensity. A very prominent example are the semiconductor saturable absorber mirrors (SESAMs) [8] that have found widespread use in research and industry. They can be properly taylored for optimal performance with respect to the magnitude of modulation, the time needed for the system to recover, and the radiation fluence required to bring about the absorption saturation effect.

Rather than exploiting real saturable absorption, the few-cycle laser pulses generated in the framework of this thesis exploit so-called artificial saturable absorber action based on the electronic Kerr-effect. First demonstrated and explained in 1991 [9, 10], the Kerr-lens mode-locking (KLM) technique has enabled the generation of one of the shortest pulses directly from a laser cavity, with pulse durations of less than two optical cycles [11]. Similarly short pulse durations could also be obtained upon SESAM-assisted Kerr-lens mode-locking [12], however, pure KLM operation allows for a fast saturable-absorber-like action with almost instantaneous relaxation, response and no real absorption.

Kerr-Lens Mode-Locking

The general principle behind KLM relies on a beam self-focusing effect arising from a third-order $\chi^{(3)}$-nonlinearity inside a nonlinear medium. When the electric field of a laser $E(t)$ is applied to some medium, the existing dipoles are aligned along the direction of the incoming field or new electric dipoles are induced. The resulting effect is termed polarization $P(t)$ [13, 14] and can be expressed by a linear $P^{(1)}(t)$ and nonlinear term $P^{NL}(t)$ in a Taylor-series expansion:

$$\begin{aligned} P(t) &= \epsilon_0(\chi^{(1)}E(t) + \chi^{(2)}E(t)E(t) + \chi^{(3)}E(t)E(t)E(t) + [...]) \\ &= \epsilon_0\chi^{(1)}E(t) + \epsilon_0\sum_{n=2}^{\infty}\chi^n E(t)^n \\ &\equiv P^{(1)}(t) + P^{NL}(t), \end{aligned} \tag{2.5}$$

where χ^n is referred to as the nonlinear optical susceptibility[1] of order n and ϵ_0 is the vacuum permittivity. For the power levels considered in this thesis, we only consider $\chi^{(2)}$- and $\chi^{(3)}$-related processes, and an overview of some related optical phenomena is given in Table 2.1 [13–16].

The Kerr effect is a third-order ($\chi^{(3)}$) effect and originates from the intensity dependence of a material's refractive index. It can be written as [13]:

$$n(I) = \sqrt{1+\chi^{(1)}} + \frac{3}{4}\frac{\chi^{(3)}}{\epsilon_0 c n_0^2} \cdot I \equiv n_0 + n_2 \cdot I, \tag{2.6}$$

where n_0 and n_2 are the linear and nonlinear refractive indices, respectively. Since the intensity is highest in the center of a laser pulse, this part will experience a higher refractive index than the wings in transverse direction, leading to a focusing effect in media with $n_2 > 0$. For pulses with peak power P_0 and $(1/e^2)$-beam radius w, this Kerr-lens has an approximate focal length of [17][2]:

$$f_{Kerr} = \frac{\pi w^4}{4 n_2 P_0 L}, \tag{2.7}$$

with L denoting the length of the medium. When adding a Kerr lens into a laser cavity, the laser mode diameter for pulsed operation changes but does not auto-

[1] Note that the susceptibility has the form of a tensor, and $\chi^{(2)}$ and $\chi^{(3)}$ are tensors of the third and fourth rank, respectively.

[2] Equation 2.7 is only valid for a circular beam with $w_x = w_y = w$. If $w_x \neq w_y$, both directions will experience a different strength of focusing.

Table 2.1 Important second- and third-order nonlinear optical susceptibilities and corresponding nonlinear processes. The first argument in the susceptibility represents the frequency of the generated wave, while the terms after the semicolon are related to the frequencies of the incoming waves, with $\omega_3 = \omega_1 + \omega_2$ and $\omega_4 = \omega_1 + \omega_2 + \omega_3$

Nonlinear process	Acronym	Susceptibility
Second-harmonic generation	SHG	$\chi^{(2)}(2\omega; \omega, \omega)$
Sum frequency generation	SFG	$\chi^{(2)}(\omega_3; \omega_1, \omega_2)$
Difference frequency generation	DFG	$\chi^{(2)}(\omega_2; \omega_3, -\omega_1)$
Optical rectification	OR	$\chi^{(2)}(0; \omega, -\omega)$
Kerr effect (nonlinear refr. index)		$\chi^{(3)}(\omega; \omega, -\omega, \omega)$
Four-wave mixing	FWM	$\chi^{(3)}(\omega_4; \omega_1, \omega_2, \omega_3)$
Third-harmonic generation	THG	$\chi^{(3)}(3\omega; \omega, \omega, \omega)$
Two-photon absorption	TPA	$\mathrm{Im}[\chi^{(3)}(\omega; \omega, \omega, -\omega)]$

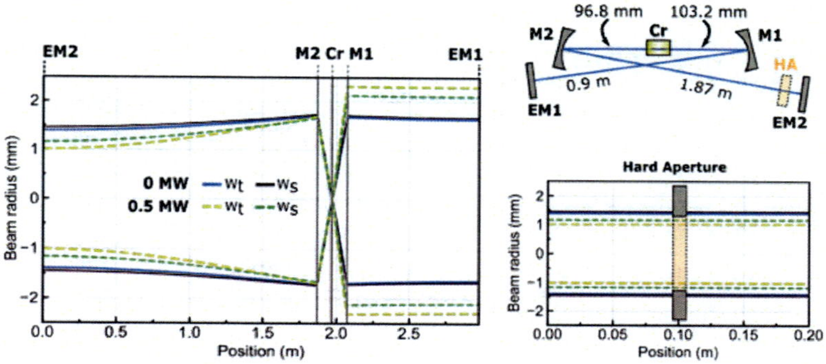

Fig. 2.3 Using the nonlinear ABCD-model described in Sect. 4.1.2, the intra-cavity beam radius of an asymmetrical x-folded cavity, similar to *MIRanda2*, was calculated for two different peak powers. The self-focusing Kerr effect takes place inside the 5 mm long Cr:ZnS crystal (Cr), which is surrounded by the spherical mirrors M1 and M2. The angle of incidence on the latter was assumed to be 2° and the arm lengths correspond to a repetition rate of 50 MHz. To initiate hard-aperture KLM, a suitable position was identified close to the end-mirror EM2 in the long arm, where the diameter of the high-power beam is decreased

matically impose a preference for mode-locked operation. Only the combination of self-focusing together with an intra-cavity aperture provides a fast saturable absorber-like action resulting in a so-called *self-amplitude modulation* (SAM, see Sect. 2.3.2) [18].

As shown in Fig. 2.3, one could place a physical hard aperture into the beam path to induce higher losses for continuous-wave operation. On the other hand, the Kerr effect inside the gain crystal results in an intensity-dependent gain distribution and can lead to a better overlap between the pump and fundamental laser mode [19] (Fig. 2.4).

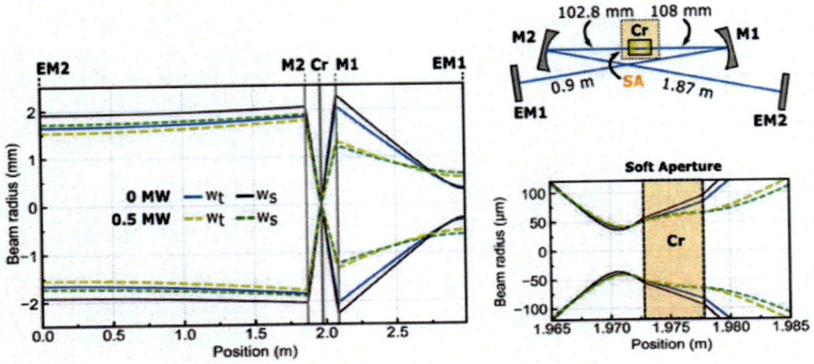

Fig. 2.4 Same parameters as in Fig. 2.3 but with different distances between the curved mirrors to illustrate soft-aperture KLM. In this case, the beam diameter of the high-power beam is slightly decreased inside the gain medium. Depending on the pump beam geometry, this can lead to a better overlap between the pump and fundamental cavity beam, therefore favoring pulsed operation

As will be demonstrated in the experimental chapters, this **soft-aperture KLM** technique is a very elegant way to start and stabilize the mode-locking mechanism, in which the crystal plays both the role of a gain and Kerr medium.

However, this additional nonlinear lensing is only drastically enhanced when the cavity is operated close to one of the stability zone boundaries. In this respect, a useful parameter is the so-called **Kerr strength**, which is a measure of the effect of the Kerr lens on the laser cavity. It can be expressed as the change in cavity mode size due to a small increase in the intra-cavity peak power P_0 [20, 21]:

$$\gamma_{Kerr} = \left(\frac{1}{\omega_{cw}} \frac{d\omega}{dp} \right)_{p=0}, \tag{2.8}$$

where ω_{cw} refers to the continuous-wave mode size ($P_0 = 0$) at a given plane inside the oscillator and $p = P_0/P_{cr}$ is the normalized power with P_{cr} being the critical power for self-focusing. The analytic form of the latter has been shown to be of the form $P_{cr} = \alpha(\lambda^2/4\pi n_0 n_2)$, where α is a material-independent constant[3] [22].

When going to higher and higher intra-cavity peak powers, suitable locations for a soft (or hard) aperture need to be identified to still ensure sufficient SAM-action. Therefore, the variation of γ_{Kerr} will be investigated in more detail for the second-generation laser system, *MIRanda2*, which was meant to be optimized for higher power levels. More information on the performed **nonlinear ABCD-matrix simulations** that were also used to plot Figs. 2.3 and 2.4 can be found in Sect. 4.1.2. To provide a first impression of the Kerr strength variations for different gain medium positions and curved mirror separations, a typical Ti:sapphire laser resonator was simulated based on the parameters presented in Ref. [20]. The result is plotted in

[3] In [22], a theoretical lower bound for self-focusing in bulk media was found to be $\alpha \approx 1.86225$.

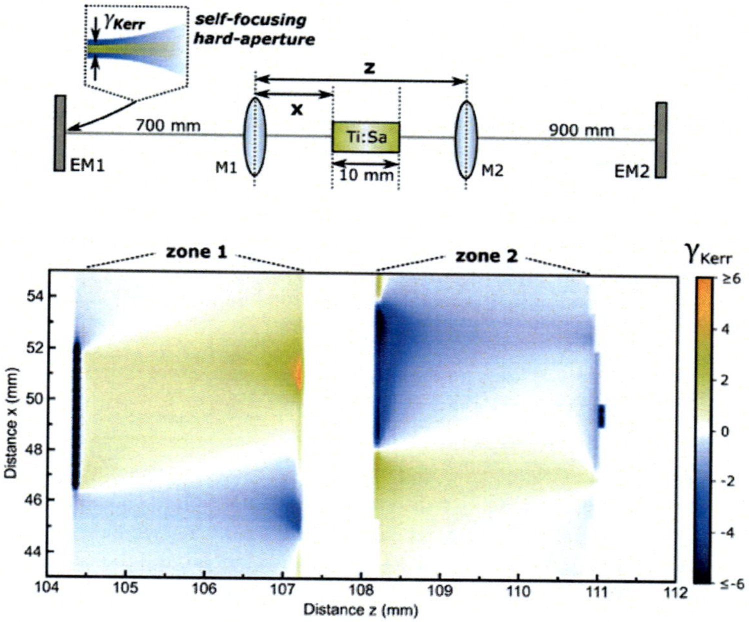

Fig. 2.5 Two-dimensional simulation of the Kerr strength parameter for varying crystal positions x and distances z between the focusing lenses $M1$ and $M2$ (f = 50 mm). The γ_{Kerr}-values are calculated for spot sizes on mirror $EM1$, revealing the two stability zones of a typical laser resonator

Fig. 2.5 and is in close agreement with the simulation presented by Magni et al. Small deviations are most likely due to uncertainties in the simulation parameters as well as the slightly different approach used for modeling the Kerr-lensing effect in the gain medium (see Sect. 4.1.2). Nonetheless, this example serves to illustrate the typical operation characteristics of a Kerr-lens mode-locked laser: (i) To obtain a significant nonlinear loss modulation for KLM, the laser cavity needs to be operated close to the stability zone boundaries, at which the Kerr strength and thus the nonlinear effects are enhanced. (ii) Moreover, zero values not only indicate positions of constant beam sizes but also denote unstable laser configurations, revealing the two characteristic zones of stability. (iii) Also, the crystal position needs to be adjusted, which can result in either decreasing ($\gamma_{Kerr} < 0$) or increasing ($\gamma_{Kerr} > 0$) beam diameters upon pulsed operation. Hence, these two-dimensional simulations provide a useful guideline for oscillator design and optimization.

Before we conclude this part, it should be noted that the intensity of a laser pulse is not only a function of space but also time. As will be seen in the following Sect. 2.2.2, another important effect termed self-phase modulation (SPM) plays a crucial role.

2.2 Soliton Pulse Propagation

The concept of solitons already came up more than a century ago [23], and it would be difficult to find a field of theoretical physics where it is not used for explaining important phenomena, including the fields of hydrodynamics, particle physics and astronomy. In laser physics, the soliton regime allows to generate and sustain the shortest achievable pulses, which are referred to as *temporal optical solitons* or *fundamental solitons*. Upon propagation along nonlinear dispersive media, their temporal and spectral shape remains unchanged through a proper balance of group velocity dispersion (GVD) and self-phase modulation (SPM) and negligible losses.

2.2.1 *Dispersion*

In mode-locked lasers, an abundance of frequency components is travelling around, each of them experiencing a different change in absolute phase $\phi(w)$ when passing through material or being reflected by specific mirrors (see Sect. 3.3.2). The phase-term $\phi(w)$ is typically expanded in a Taylor series around the central frequency w_0 of the pulse [24, 25]:

$$\phi(w) = \sum_{m=0}^{\infty} \frac{(w - w_0)^m}{m!} \underbrace{\left(\frac{\partial^m}{\partial w^m} \phi(w) \right)}_{D_m} \quad (m \in \mathbb{N}). \tag{2.9}$$

Together with the relationship between the phase and the wave vector $k(w)$, $\phi(w) = k(w)z = wn(w)z/c$, the coefficients D_m provide a means to describe the effect of dispersion on a pulse:

- $\mathbf{D_0}$: an overall phase accumulation.
- $\mathbf{D_1}$**(GD)**: the *group delay* (GD) denotes a temporal shift of the pulse with respect to a pulse propagating in free space, and contains the inverse of the group velocity $1/v_g = \partial k/\partial w$.
- $\mathbf{D_2}$**(GDD)**: the *group delay dispersion* (GDD) term is the most famous term. It refers to the variation of the GD with respect to w and is usually specified in fs^2 for a given length of a medium. The so-called *group velocity dispersion* (GVD, $\partial^2 k/\partial w^2$) is the GDD per unit length (fs^2/mm). In Table 2.2, the GVD-values of selected materials relevant for this thesis are listed, with GVD > 0 indicating positive (normal) dispersion, while substrates with GVD < 0 are said to exhibit negative (anomalous) dispersion.
- $\mathbf{D_3}$**(TOD)**, $\mathbf{D_4}$, $\mathbf{D_5}$, ...: third- and other higher-order dispersion terms. For narrow-band laser pulses, they can often be neglected, however, third-order dispersion (TOD) becomes significant when going to even shorter pulse durations. There-

Table 2.2 Linear and nonlinear ((cm^2/W) $\cdot 10^{-16}$) refractive index of selected materials and their group velocity dispersion (fs^2/mm). Unless otherwise stated, the values correspond to a wavelength of 2.3 μm. *Estimated value based on ZnSe

Material	n_0	n_2	GVD	Ref.
ZnS	2.26	<100* (@1.95 μm)	+128	[26]
ZnSe	2.44	~100 (@1.95 μm)	+218	[27, 28]
YAG ($Y_3Al_5O_{12}$)	1.80	8±2 (@1.3 μm)	−124	[29, 30]
Sapphire (Al_2O_3)	1.73	3.3 (@1.55 μm)	−210	[31, 32]
Fused silica (SiO_2)	1.43	2.8±0.1 (@1.55 μm)	−180	[33, 34]
MgF_2	1.37	1.1±0.2 (@1.06 μm)	−61	[35, 36]
BaF_2	1.46	2.08-3.22 (@0.532 μm)	−14	[37, 38]
ZGP ($ZnGeP_2$)	3.14	~550 (@2.2 μm)	+511	[39, 40]

fore, specifically taylored dispersive mirrors are needed to account for TOD in few-cycle laser systems (see Sect. 3.3.2).

Controlling chromatic dispersion in a laser cavity is a major challenge in ultrashort pulse generation since every optical element—and even air—is dispersive. To include the effect of dispersion on a laser pulse, its (1 + 1)-dimensional field envelope $A(z, \tau)$ can be described in the local pulse-frame $\tau = t - z/v_{g,w_0}$, traveling with the central group velocity v_{g,w_0} and along the propagation distance $z \in [0, N \cdot L_{\text{cav}}]$ [41]. Here, L_{cav} is the length corresponding to one complete round-trip through the cavity, and N is the number of round-trips. To model the propagation and formation of pulses, a slowly-varying envelope approximation is typically employed [42]. It implies that the change of the pulse amplitude during an optical cycle is negligible when compared to the magnitude of the amplitude itself. Moreover, it can be applied down to the single-cycle regime, when some additional corrections are included [43]. The effect of dispersion on the laser field envelope is then described as [41]:

$$\frac{\partial A(z,\tau)}{\partial z} = i\frac{D_2}{2}\frac{\partial^2 A(z,\tau)}{\partial \tau^2} + [...], \tag{2.10}$$

where the square brackets refer to the higher-order dispersion terms.

2.2.2 Self-Phase Modulation

As we have seen in Sect. 2.1.2, the material's refractive index does not only depend on frequency but also intensity. Besides the explained effect of spatial beam self-focusing, the time-dependent intensity $I(t)$ of a laser pulse can thus result in modulations of the laser pulses' temporal phase $\Phi(t)$ [5, 13]:

$$\Phi(t) = w_0 t - \frac{2\pi}{\lambda_0} n(I) L. \tag{2.11}$$

Consequently, the instantaneous frequency $w(t)$ of a laser pulse changes as:

$$w(t) = \frac{d\Phi}{dt} = w_0 - \frac{2\pi}{\lambda_0}\frac{dn(I)}{dt} L = w_0 \left(1 - \frac{n_2}{c}\frac{dI(t)}{dt} L\right). \tag{2.12}$$

To visualize the effect on the temporal and spectral intensity distribution of a laser pulse, spectrograms were calculated for different pulse peak intensities (see Fig. 2.6). A spectrogram is the time convolution of the pulse electric field $E(t)$ with a variable-delay gate function $g(t-\tau)$: $S(w,\tau) = |\int_{-\infty}^{+\infty} E(t)g(t-\tau)\exp(-iwt)dt|^2$ [44, 45], and resembles the outcome of a cross-correlation frequency-resolved optical gating (XFROG) setup [46]. Here, two pulses—a Fourier-transform limited gate pulse and a probe pulse—perform sum-frequency generation (SFG) inside a nonlinear medium, with one pulse being variably delayed with respect to the other. In the performed simulations, both pulses had a central wavelength of 2300 nm (130 THz), corresponding to a sum-frequency of 260 THz. The gate pulse duration was set to be Fourier-transform limited (40 fs), however, the phase of the second (probe) pulse was changed to investigate the effects of either pure SPM (Fig. 2.6a, b) or negative GDD (Fig. 2.6c) on the shape of the spectrogram. For the sake of illustration, the additional chirp acquired by material dispersion was neglected in (a) and (b). In addition, the marginals are plotted on the sides, and are obtained through integration over the delay and frequency axis, respectively. When comparing the two spectrograms related to SPM with the one of a negatively chirped pulse, one can see that the effect of negative GDD can account for the linear (positive) chirp in the center of a SPM-broadened pulse.

Interestingly, strong self-phase modulation can lead to pronounced oscillations in the optical spectrum, as shown in the frequency marginal of (b). According to Eq. 2.12, SPM changes the instantaneous frequency of the pulse as $\Delta w \propto dI(t)/dt$. Therefore, there are two distinct points in time that take the same frequency value. Depending on their relative phase difference, constructive or destructive interference takes place and results in the observed multipeak structure [24].

Finally, to also include the effect of self-phase modulation on the laser field envelope $A(z,\tau)$, the following expression is used [41]:

$$\frac{\partial A(z,\tau)}{\partial z} \approx -i\gamma |A(z,\tau)|^2 A(z,\tau). \tag{2.13}$$

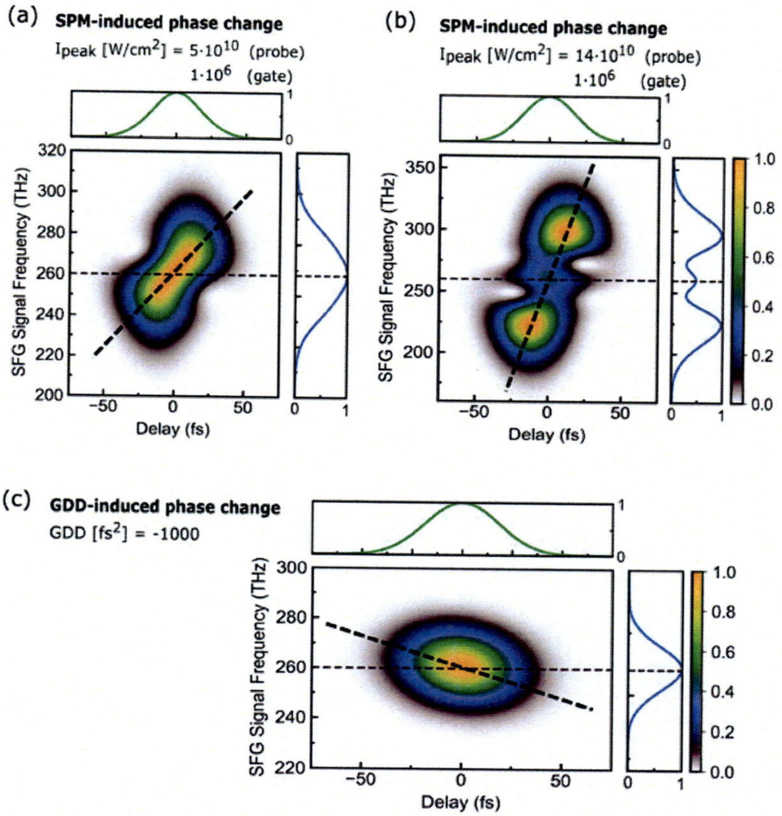

Fig. 2.6 Simulated spectrograms to illustrate the effect of SPM and negative GDD on the temporal and spectral pulse shape. **a**, **b** SPM-induced nonlinear phase change arising from different peak intensities of the probe and gate pulse when passing through a 2 mm thick ZnS-plate. The higher the peak intensity, the stronger the frequency modulation in the spectrogram. The temporal pulse envelope is, however, not directly affected by SPM. Also, the tilted black dashed lines indicate that the chirp in the center of the SPM-broadened pulse is (mostly) linear. **c** To only model the effect of second-order dispersion, the temporal phase of the probe pulse was changed according to [47], and results in a linear chirp of opposite sign. Therefore, negative GDD can account for the linear (positive) chirp in the center of a SPM-broadened pulse

Here, γ is the so-called SPM coefficient and can be estimated as $\gamma = 8\pi^2 n_0 n_2/\lambda^2$ for the case of strong beam focusing inside a nonlinear medium.

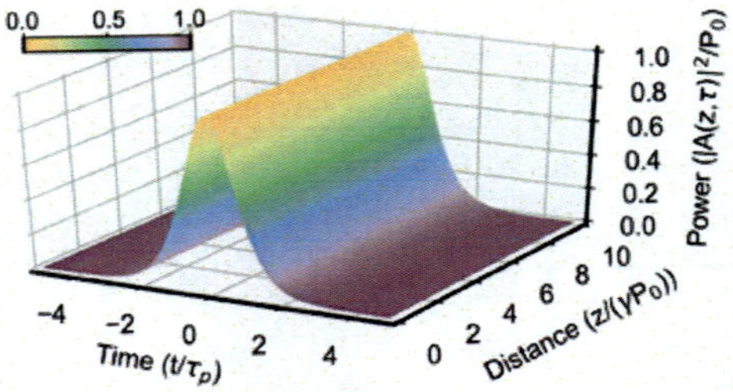

Fig. 2.7 Propagation of a fundamental soliton

2.2.3 *The Fundamental Soliton*

When considering only the effects of GDD and SPM, the propagation of laser pulses can be described by an equation that is known as the *nonlinear Schrödinger equation* (NLSE)[4] [48–50]:

$$\frac{\partial A(z,\tau)}{\partial z} = \underbrace{\boxed{i\frac{D_2}{2}\frac{\partial^2 A(z,\tau)}{\partial \tau^2}}}_{\text{GDD}} - \underbrace{\boxed{i\gamma |A(z,\tau)|^2 A(z,\tau)}}_{\text{SPM}}. \tag{2.14}$$

For the case of negative dispersion with $D_2 < 0$, the NLSE supports a stable solution —the unchirped *fundamental soliton* [24, 48, 51]:

$$A(z,\tau) = \sqrt{P_0} \cdot \operatorname{sech}\left(\frac{\tau}{\tau_0}\right) e^{i\gamma P_0 z/2}. \tag{2.15}$$

Its amplitude exhibits a characteristic hyperbolic secant profile with peak power $P_0 = |A_0(z,\tau)|^2$, while the soliton's temporal (and spectral) shape remains unchanged upon propagation. The full width at half maximum (FWHM) of the soliton's pulse width can then be calculated as $\tau_{FWHM} \approx 1.76 \cdot \tau_0$ [24]. In order to obtain a stable solution, the soliton peak power and pulse duration are related to the dispersion and nonlinearity via the area theorem[5] [24, 48]:

$$\tau_0 = \sqrt{\frac{|D_2|}{\gamma P_0}}. \tag{2.16}$$

Hence, when the effects of both SPM and negative dispersion are properly balanced, the optical bandwidth of an initially unchirped pulse is left unaffected and a stable soliton pulse is formed as shown in Fig. 2.7.

[4] It is very similar to the Schrödinger equation in quantum mechanics that describes how a wave function spreads out in space as time goes by.

[5] For a particular pulse solution to exist, certain conditions must be satisfied that are expressed in the form of so-called area theorems [52].

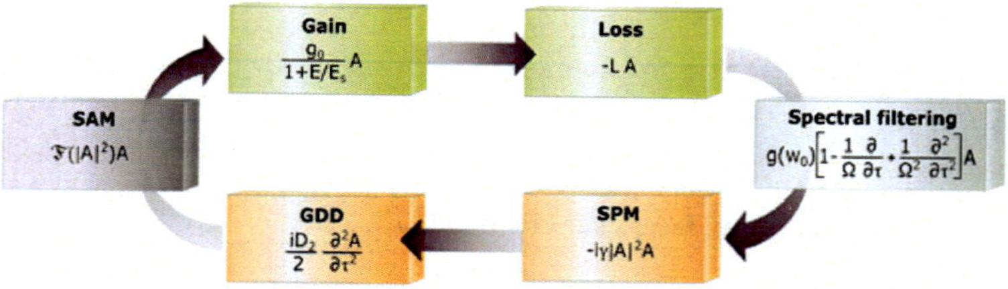

Fig. 2.8 Principal factors influencing the generation and evolution of ultrashort pulses

2.3 Soliton Mode-Locking

Laser oscillators are complex nonlinear systems that can support the formation of stable optical pulse trains. However, besides dispersion and SPM, also other effects need to be considered that contribute to the soliton-like pulse shaping (see Fig. 2.8). Upon a cavity roundtrip, the soliton is periodically perturbed by the action of gain, loss, bandwidth limiting and SAM, and stable pulse generation relies on a careful laser design to achieve the desired balance between the various effects. Their combined action on the laser field envelope is eventually summarized in the *master equation of mode-locking* in Sect. 2.3.3.

2.3.1 Gain, Loss, and the Spectral Filtering

The effect of loss is represented by a linear factor l and also the saturable gain g is assumed to have no effect on the pulse dynamics or shape [41, 53]:

$$\frac{\partial A(z,\tau)}{\partial z} = \left(\underbrace{\frac{g_0}{1+E(z)/E_s}}_{g} - l \right) A(z,\tau). \tag{2.17}$$

Here, g_0 is the unsaturated (small signal) gain, $E(z) = \int |A(z,\tau)|^2 dt$ is the laser pulse energy, and $E_s = \hbar w_0 S/\sigma_{em}$ is the gain saturation energy with the emission cross-section σ_{em} and the laser mode area S.

For solid-state lasers such as those based on Cr:ZnS/ZnSe, the constant gain is a valid approximation since their upper-state lifetimes τ_f (~μs-ms) are typically much longer than the pulse repetition rates (~ns), with $\sigma_{em} \propto 1/\tau_f$ for gaussian-shaped gain spectra [2]. In an intuitive picture, the long lifetime of the excited ions is associated with a high energy storage capacity of the gain medium, and only a sufficiently large number of photons that are passing by results in a reasonable chance to de-excite the ions back to the ground state. Hence, substantial gain saturation only occurs, when the time-integrated optical intensity, the so-called pulse fluence, approaches the saturation fluence $F_s = E_s/S$. In other words, only if $E/E_s \rightarrow 1$, dynamic gain saturation sets in. Here, the gain gets shaped by a pulse-induced and fast saturation, which recovers between consecutive pulses [54]. This effect can be included by a time-dependent gain function $g(\tau)$, taking into account the characteristics of a four-level laser gain medium [55]:

$$\frac{dg}{d\tau} = \sigma_{abs}(g_0 - g)\frac{I_p}{h\nu_p} - \frac{\sigma_{em}}{h\nu}g|A|^2 - \frac{g}{\tau_g}. \tag{2.18}$$

Here, σ_{abs} is the absorption cross-section of the gain medium, I_p and ν_p are the pump intensity and frequency, respectively, and τ_g is the characteristic time related to gain relaxation. In general, to obtain stable soliton mode-locking, the net saturated gain $g - l$ must be negative, otherwise the continuous-wave background radiation will also experience amplification.

Besides the effect of gain and loss, the finite amplification bandwidth of the laser medium results in a spectral filtering action that curtails the pulse spectral bandwidth. Usually, a filter profile with a Lorentzian line shape and filter bandwidth Ω is assumed [56], which reads in the time domain as follows [41]:

$$\frac{\partial A(z,\tau)}{\partial z} \approx g(w_0)\left[1 - \frac{1}{\Omega}\frac{\partial}{\partial\tau} + \frac{1}{\Omega^2}\frac{\partial^2}{\partial\tau^2}\right]A(z,\tau). \tag{2.19}$$

2.3.2 *Self-Amplitude Modulation*

As discussed in Sect. 2.1.2, KLM arises from an interplay between an intra-cavity aperture and beam self-focusing, in which intense pulsed operation is favored over CW lasing. This power discrimination effect is termed self-amplitude modulation and can be modelled as a saturable absorber-like action with a response function $\mathfrak{F}$ [41]:

$$\frac{\partial A(z,\tau)}{\partial z} = \mathfrak{F}(|A(z,\tau)|^2)\cdot A(z,\tau), \tag{2.20}$$

$$\begin{aligned} \textbf{HA}: \quad & \mathfrak{F} \approx -\frac{l'}{1+\kappa|A|^2}, \\ \textbf{SA}: \quad & \mathfrak{F} \approx -l' + \kappa(1-\zeta|A|^2)|A|^2, \end{aligned} \tag{2.21}$$

where the κ- and ζ- parameters are the coefficients of self-amplitude modulation and alteration, respectively, and l' is the low-power saturable loss. The (loss) modulation functions of the saturable absorber depend on the intensity $I(z,\tau) = |A(z,\tau)|^2$, and hence they act instantaneously on the pulse.

For the case of hard-aperture (HA) KLM, this function can be described by a perfectly saturable absorber action, in which an increasing intra-cavity power reduces

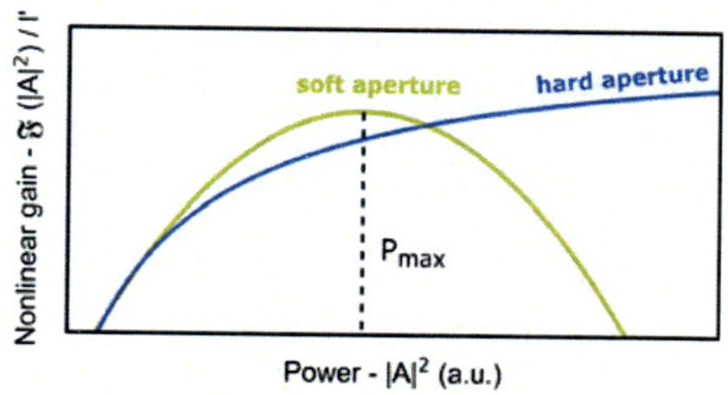

Fig. 2.9 Qualitative behaviour of the SAM functions $\mathfrak{F}(|A(z,\tau)|^2)$ for hard- and soft-aperture KLM

the beam size on a hard aperture. This self-focusing effect will eventually cause the diffraction loss l' to monotonically decrease, as shown in Fig. 2.9. However, if a soft-aperture (SA) is used, the behaviour changes: even though the loss is initially reduced by a growth in power, there is a turning point at which the loss gets enhanced again. The point of alteration is associated with a maximized overlapping between the pump and fundamental cavity beam in the gain medium. If the power is increased further, the self-focusing action will worsen the overlap and result in reduced gain.

2.3.3 *The Master Equation of Mode-Locking*

The effects governing the pulse evolution in a laser cavity can be summarized by the *master equation of mode-locking*, that joins Eqs. 2.10, 2.13, 2.17, 2.19 and 2.20 together [41]:

$$\frac{\partial A}{\partial z} = [\underbrace{\boxed{g-l}}_{\text{gain/loss}} + \underbrace{\boxed{\frac{g(w_0)}{\Omega^2}\frac{\partial^2}{\partial \tau^2}}}_{\text{spectr. filter}} + \underbrace{\boxed{\mathfrak{F}(|A|^2)}}_{\text{SAM}}]A + i[\underbrace{\boxed{\frac{D_2}{2}\frac{\partial^2}{\partial \tau^2}}}_{\text{GDD}} - \underbrace{\boxed{\gamma|A|^2}}_{\text{SPM}}]A, \tag{2.22}$$

where it was assumed that only second-order dispersion (D_2) is present. The equation was first introduced by Haus et al. [57] and provides a general description of the ultrashort pulses generated in solid-state [57] and fiber lasers [58]. It has secant-hyperbolic solutions of type [50, 57]:

$$A(\tau) = A_0 \cdot \text{sech}\left(\frac{\tau}{\tau_0}\right) e^{(i\beta\cdot\log(\text{sech}(\tau/\tau_0)))} = A_0 \cdot \text{sech}^{1+i\beta}\left(\frac{\tau}{\tau_0}\right), \tag{2.23}$$

with pulse duration τ_0, chirp parameter β and pulse amplitude A_0. The pulse duration and chirp depend on D_2, Ω, γ, κ and the pulse energy E in a rather complicated fashion and are described in more detail in [50, 53, 57].

In Fig. 2.10, the qualitative behaviour of the chirp, the pulse width and the stability ($g_{\text{pulse}} - g_{\text{cw}}$) is plotted as a function of group velocity dispersion for different values

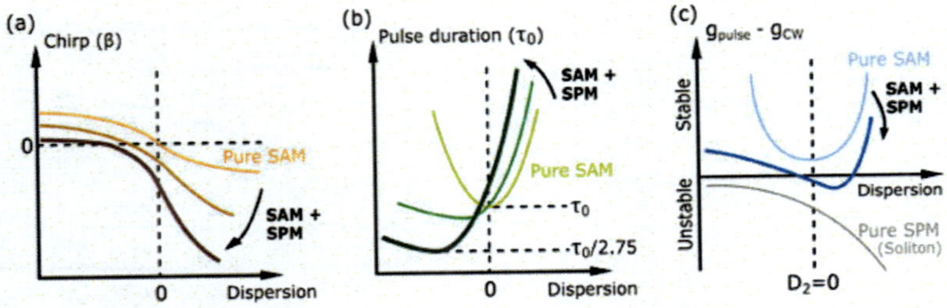

Fig. 2.10 Effect of dispersion (D_2) and SPM on **a** the chirp, **b** the pulse width and **c** the stability for a fixed pulse energy and gain bandwidth but different values of SPM. As indicated in **c**, only sufficient negative or positive dispersion can drive the system into stable operation. The figure was adapted from [53]

of SPM (γ). One can see that for an increasing amount of SPM, the point of minimum pulse duration is moved towards the negative dispersion regime. Moreover, when the effects of GVD and SPM are properly balanced, the pulse duration is reduced by a factor of 2.75. Also, the solutions predicted by the master equation are not always stable and the competing CW oscillations can experience a greater gain g_{cw} than the laser pulses g_{pulse}. In this respect, a pure soliton laser without the effect of self-amplitude modulation is not stable against the CW background.

Interestingly, the start-up of mode-locked lasers can be captured in real-time via the use of time-stretch dispersive Fourier transform. In [59], a 20 fs KLM Ti:sapphire laser with a repetition rate of 90 MHz was used to obtain a spectrally resolved view of the onset of femtosecond pulsing, evolving continuously from the initial picosecond fluctuations after several thousand cavity round-trips.

2.3.4 The Effect of Different Dispersion Regimes

Pulse formation, evolution and shaping are significantly affected by the intra-cavity dispersion and its interplay with nonlinear effects [60, 61]. In this section, two commonly applied mode-locking regimes are introduced that play an important role in various ultrafast solid-state and fiber laser systems.

Dispersion-Managed Soliton Formation [51]:

So far, we have assumed the dispersion and self-phase modulation to be continuously distributed along pulse propagation. Strictly speaking, this is not true for typical femtosecond lasers. In the case of a solid-state bulk laser, for example, the discrete action of positive SPM and GDD in the laser crystal as well as the discrete action of negative dispersion introduced by chirped mirrors or material plates (see Sect. 3.3.2) cannot be neglected anymore. In 1994, Spielmann et al. found that the precise arrangement of the cavity elements has a major impact on the pulse formation process, such that even the positioning of the output coupler plays a role in extracting the spectrally broadest pulses [62]. This knowledge has then enabled the generation of pulses with durations of less than two optical cycles (5 fs) in Ti:sapphire lasers [63].

As a general guideline for a solid-state laser, an equal distribution of negative dispersion into both arms of the cavity is beneficial and allows to exploit the KLM action twice per round-trip [11]. Moreover, when reaching shorter and shorter pulse durations, the dispersion swings between the negative and positive dispersion sections start to affect the pulse shape. While having a sech-shape at longer pulse durations, the pulse spectra start to get Gaussian-shaped and increasingly more flattopped for shorter pulse widths.

Chirped-Pulse Formation [41, 64]:

Many applications benefit from both ultrashort and powerful laser systems. Scaling the intra-cavity peak power, however, unavoidably results in ever stronger nonlinearities in the gain medium, manifesting themselves in the fission of single coherent pulses to form chaotic multi-pulse emission. The saturable absorber action is then said to be "overdriven". In this respect, the regime of chirped-pulse formation can help to mitigate the limitations of classical soliton lasers operating in the net negative dispersion regime ($D_2 < 0$). As shown in Fig. 2.10c, highly chirped-pulse solutions accompanied by strongly increased pulse durations are found for $D_2 > 0$ while the system is still running stably.

Whereas solitonic pulse-shaping due to a strict balance between SPM and negative GDD occurs for oscillators operating in the negative dispersion regime, the description of mode-locking in the positive dispersion regime is more challenging for the theoretical analysis. In [65], a chirped solitary pulse (CSP) solution was found analytically, revealing that the CSP has a Lorentzian spectral profile with truncations at some defined frequency Δ away from the carrier frequency. Even though the peak power can be kept below the instability threshold, the heavily chirped output pulses (~ps-range) of such laser systems need to be compressed externally to the few-cycle femtosecond regime again [66]. Also, because of the reduced pulse intensities, it might get difficult to initiate the pulse buildup process due to an insufficient saturation of the SAM.

2.4 Summary

In this chapter, the theoretical foundations behind ultrashort pulse generation have been laid, with a focus on soliton pulse propagation and soliton mode-locking. Starting with the concept of mode-locking in a simple Fabry-Pérot laser resonator, the advantageous technique of Kerr-lens mode-locking (KLM) was described and analyzed in detail, enabling the generation of few-cycle femtosecond pulses directly from a laser cavity. Conditions under which fundamental solitons are formed were also discussed on the basis of the nonlinear Schrödinger equation. Finally presented was the master equation of mode-locking which characterizes the complex interplay between the many effects governing the pulse evolution in a laser cavity, including self-amplitude and self-phase modulation.

References

1. Svelto O (2009) Principles of lasers, 5th edn. Springer
2. Koechner W (2006) Solid-state laser engineering. Springer
3. Anthony E (1986) Siegman. University Science Books, Lasers
4. Eichler HJ, Eichler J, Lux O (2018) Lasers - basics, advances and applications, 1st edn. Springer, Berlin
5. Weiner AM (2009) Ultrafast optics, 1st edn. Wiley
6. Koechner W, Bass M (2003) Optical resonator. Solid-state lasers. Advanced texts in physics. Springer, New York, pp 149–186
7. Kuizenga DJ, Siegman AE (1970) FM and AM mode locking of the homogeneous laser - part I: theory. IEEE J Quantum Electron 6(11):694–708
8. Keller U et al (1996) Semiconductor saturable absorber mirrors (SESAM's) for femtosecond to nanosecond pulse generation in solid-state lasers. IEEE J Sel Top Quantum Electron 2(3):435–453
9. Spence DE, Kean PN, Sibbett W (1991) 60-fsec pulse generation from a self-mode-locked Ti:sapphire laser. Opt Lett 16(1):42–44
10. Keller U, 'tHooft GW, Knox WH, Cunningham JE (1991) Femtosecond pulses from a continuously self-starting passively mode-locked Ti:sapphire laser. Opt Lett 16(13):1022–1024
11. Morgner U et al (1999) Sub-two-cycle pulses from a Kerr-lens mode-locked Ti:sapphire laser. Opt Lett 24(6):411–413
12. Sutter DH et al (1999) Semiconductor saturable-absorber mirror-assisted Kerrlens mode-locked Ti:sapphrie laser producing pulses in the two-cycle regime. Opt Lett 24(9):631–633
13. Boyd R (2008) Nonlinear optics, 3rd edn. Academic Press
14. Dussauze M, Cardinal T (2019) Nonlinear optical properties of glass. In: Musgraves JD, Hu J, Calvez L (eds) Springer handbook of glass. Springer, pp 193–225
15. Prussel L (2017) Ab-initio description of optical nonlinear properties of semiconductors under an electrostatic field. PhD thesis. Université Paris-Saclay
16. Li C (2017) Nonlinear optics - principles and applications, chap 2, 1st edn. Springer, pp 23–50
17. Degiorgio V, Cristiani I (2014) Photonics - a short course. Undergraduate lecture notes in physics. Springer
18. Herrmann J (1994) Theory of Kerr-lens mode locking: role of self-focusing and radially varying gain. J Opt Soc Am B 11(3):498–512
19. Piché M, Salin F (1993) Self-mode locking of solid-state lasers without apertures. Opt Lett 18(13):1041–1043
20. Magni V, Cerullo G, De Silvestri S (1993) Closed form gaussian beam analysis of resonators containing a Kerr medium for femtosecond lasers. Opt Commun 101(5–6):365–370
21. Yefet S, Pe'er A (2013) A review of cavity design for Kerr Lens mode-locked solid-state lasers. Appl Sci 3(4):694–724
22. Fibich G, Gaeta AL (2000) Critical power for self-focusing in bulk media and in hollow waveguides. Opt Lett 25(5):335–337
23. Remoissenet M (1996) Waves called solitons, 2nd edn. Springer
24. Agrawal GP (2009) Nonlinear fiber optics, 4th edn. Elsevier (Singapore) Pte Ltd
25. Buck JA (2004) Fundamentals of optical fibers, 2nd edn. Wiley-Interscience

26. Debenham M (1984) Refractive indices of zinc sulfide in the 0.405-13 μm wavelength range. Appl Opt 23(14):2238–2239
27. Major A, Aitchison JS, Smith PWE, Sorokin E, Sorokina IT (2005) Z-scan characterization of the nonlinear refractive index of single crystal ZnSe in the 1.20-1.95 μm wavelength range. In: Proceedings of SPIE 5971.59710H
28. Marple DTF (1964) Refractive index of ZnSe, ZnTe, and CdTe. J Appl Phys 35(3), 539–542
29. Zelmon DE, Small DL, Page R (1998) Refractive-index measurements of undoped yttrium aluminum garnet from 0.4 to 5.0 μm. Appl Opt 37(21):4933–4935
30. Samuel P, Ensley TR, Hu H, Hagan DJ, Van Stryland EW, Gaume R (2015) Nonlinear refractive index measurement on pure and Nd doped YAG ceramic by dual arm Z-scan technique. AIP Conf Proc 1665, 060010
31. Malitson IH (1962) Refraction and dispersion of synthetic sapphire. J Opt Soc Am 52(12):1377–1379
32. Major A, Yoshino F, Nikolakakos I, Aitchison JS, Smith PWE (2004) Dispersion of the nonlinear refractive index in sapphire. Opt Lett 29(6):602–604
33. Malitson IH (1965) Interspecimen comparison of the refractive index of fused silica. J Acoust Soc Am 55(10):1205–1209
34. Milam D (1998) Review and assessment of measured values of the nonlinear refractive-index coefficient of fused silica. Appl Opt 37(3):546–550
35. Dodge MJ (1984) Refractive properties of magnesium fluoride. Appl Opt 23(12):1980–1985
36. DeSalvo R, Said AA, Hagan DJ, Van Stryland EW, Sheik-Bahae M (1996) Infrared to ultraviolet measurements of two-photon absorption and n_2 in wide bandgap solids. IEEE J Quantum Electron 32(8):1324–1333
37. Li HH (1980) Refractive index of alkaline earth halides and its wavelength and temperature derivatives. J Phys Chem Ref Data 9:161–289
38. DeSalvo R, Sheik-Bahae M, Said AA, Hagan DJ, Van Stryland EW (1993) Z-scan measurements of the anisotropy of nonlinear refraction and absorption in crystals. Opt Lett 18(3):194–196
39. Das S, Bhar GC, Gangopadhyay S, Ghosh C (2003) Linear and nonlinear optical properties of ZnGeP2 crystal for infrared laser device applications: revisited. Appl Opt 42(21):4335–4340
40. Li W, Li Y, Xu Y (2017) Measurements of nonlinear refraction in the mid-infrared materials $ZnGeP_2$ and $AgGaS_2$. Appl Phys B 123(82)
41. Kalashnikov VL (2012) Chirped-pulse oscillators: route to the energy-scalable femtosecond pulses. In: Al-Khursan A (ed) Solid state laser. In- Tech, pp 145–184
42. Butcher PN, Cotter D (1990) The elements of nonlinear optics. Cambridge University Press, Cambridge
43. Brabec T, Krausz F (1997) Nonlinear optical pulse propagation in the single-cycle regime. Phys Rev Lett 78(17):3282–3285
44. Trebino R, DeLong KW, Fittinghoff DN, Sweetser JN, Krumbügel MA, Richman BA (1998) Measuring ultrashort laser pulses in the time-frequency domain using frequency-resolved optical gating. Rev Sci Instrum 68(9)
45. Monmayrant A, Weber S, Chatel B (2010) A newcomer's guide to ultrashort pulse shaping and characterization. J Phys B: At Mol Opt Phys 43(10):103001
46. Trebino R (2000) Frequency-resolved optical gating: the measurement of ultrashort laser pulses. Springer
47. Borzsonyi A, Kovacs AP, Osvay K (2013) What we can learn about ultrashort pulses by linear optical methods. Appl Sci 3(2):515–544
48. Mitschke F, Mahnke C, Hause A (2017) Soliton content of fiber-optic light pulses. Appl Sci 7(6):635
49. Hasegawa A, Matsumoto M (2003) Optical solitons in fibers. Springer series in photonics. Springer
50. Haus HA (2000) Mode-locking of lasers. IEEE J Sel Top Quantum Electron 6(6):1173–1185
51. Kärtner FX (ed) (2004) Few-cycle laser pulse generation and its applications, vol 95. Springer

52. Renninger WH, Chong A, Wise FW (2010) Area theorem and energy quantization for dissipative optical solitons. J Opt Soc Am B 27(10):1978–1982
53. Ippen EP (2002) Femtosecond pulse generation: principles and fiber applications. In: Di Bartolo B, Gambarota G (eds) Ultrafast dynamics of quantum systems: physical processes and spectroscopic techniques, vol 372. Springer, pp 213–231
54. Keller U (2007) Ultrafast solid-state lasers. In: Herziger G, Weber H, Poprawe R (eds) Landolt-Börnstein - group VIII/1B1. Springer, pp 33–167
55. Jasapara J, Rudolph W, Kalashnikov VL, Krimer DO, Poloyko IG, Lenzner M (2000) Automodulations in Kerr-lens mode-locked solid-state lasers. J Opt Soc Am B 17(2):319–326
56. Haus HA (1975) Theory of mode locking with a fast saturable absorber. J Appl Phys 46(7):3049–3058
57. Haus HA, Fujimoto JG, Ippen EP (1991) Structures for additive pulse mode locking. J Opt Soc Am B 8(10):2068–2076
58. Ding E, Kutz JN (2009) Operating regimes, split-step modeling, and the Haus master mode-locking model. J Opt Soc Am B 26(12):2290–2300
59. Herink G, Jalali B, Ropers C, Solli DR (2016) Resolving the build-up of femtosecond mode-locking with single-shot spectroscopy at 90 MHz frame rate. Nat Photonics 10:321–326
60. Woodward RI (2018) Dispersion engineering of mode-locked fibre lasers. J Opt 20:033002
61. Baumgartl M, Ortaç B, Limpert J, Tünnermann A (2012) Impact of dispersion on pulse dynamics in chirped-pulse fiber lasers. Appl Phys B 107:263–274
62. Spielmann C, Curley PF, Brabec T, Krausz F (1994) Ultrabroadband femtosecond lasers. IEEE J Quantum Electron 30(4):1100–1114
63. Ell R et al (2001) Generation of 5-fs pulses and octave-spanning spectra directly from a Ti:sapphire laser. Opt Lett 26(6):373–375
64. Fernandez A, Fuji T, Poppe A, Fürbach A, Krausz F, Apolonski A (2004) Chirped-pulse oscillators: a route to high-power femtosecond pulses without external amplification. Opt Lett 29(12):1366–1368
65. Kalashnikov VL, Podivilov E, Chernykh A, Apolonski A (2006) Chirped-pulse oscillators: theory and experiment. Appl Phys B 83:503–510
66. Cho SH et al (2001) Generation of 90-nJ pulses with a 4-MHz repetition-rate Kerr-lens mode-locked Ti:Al_2O_3 laser operating with net positive and negative intracavity dispersion. Opt Lett 26(8):560–562

Chapter 3
The First Directly Diode-Pumped Few-Cycle Cr-Doped II-VI Laser

In recent years, applications of mid-infrared radiation have proliferated tremendously, resulting in an ever-growing search for suitable laser sources that can directly provide the few-cycle pulses needed for efficient downstream MIR generation. To achieve the high sensitivity and specificity required for the spectroscopic investigation of (bio-)molecular samples in the MIR fingerprint region, sophisticated detection schemes such as electro-optic sampling (EOS) have to be combined with ultrafast laser technology to detect subtle changes encoded in the mid-infrared electromagnetic waveform. However, the probe pulse duration in EOS-based measurements is a critical parameter since it limits the temporal resolution and thus the spectral bandwidth that can be measured at high frequencies. Therefore, ultrashort probe pulses are highly sought after to increase the detection bandwidth. Moreover, if few-cycle laser pulses can be directly generated in the short-wave mid-infrared region at around 2–3 μm, the conversion efficiency for frequency downconversion into the long-wavelength MIR beyond 3 μm is significantly improved, while a broader selection of suitable nonlinear crystals is available also.

As outlined in Chap. 1, crystalline mid-infrared lasers based on Cr^{+2}:ZnS/ZnSe gain media are highly attractive candidates for directly providing coherent few-cycle radiation in the 2–3 μm range and—especially when combined with diode-pumping—offer an intrinsic low-noise performance and a much more compact alternative over commonly used systems. In this chapter, the first directly diode-pumped, Kerr-lens mode-locked Cr^{+2}-doped II-VI oscillator, *MIRandal*, will be presented, and a focus will be placed on the favourable material and spectroscopic properties of Cr^{+2}:ZnS/ZnSe (Sect. 3.1) as well as on the major challenges associated with direct diode-pumping (Sect. 3.2). Finally, the main findings will be summarized in Sect. 3.4, highlighting the important breakthrough achieved with this laser system.

N. Nagl, *A New Generation of Ultrafast Oscillators for Mid-Infrared Applications*, Springer Theses, https://doi.org/10.1007/978-3-030-89754-3_3

3.1 Chromium-Doped II-VI—*The Titanium Sapphire of the Mid-Infrared*

Today's most popular femtosecond bulk lasers are certainly represented by the large number of reported and even commercially available Ti:sapphire ($Ti^{3+}Al_2O_3$) laser systems. These near-infrared lasers commonly deliver few-optical-cycle pulse durations and the recent advents of turn-key and diode-pumped versions have resulted in a diversification of application areas [1]. With transition-metal (Cr^{2+})-doped II-VI crystals (ZnS, ZnSe), a new class of laser materials was found in the 1990s to provide analogous properties in the 2–3 μm mid-IR spectral range [2–4], combining high emission cross-sections, fairly good mechanical and chemical stability as well as a similar thermal conductivity and ultrabroad bandwidth (see Table 3.1).

Octahedrally coordinated Ti^{3+}-doped Al_2O_3 and tetrahedrally coordinated Cr^{2+}-doped ZnS/ZnSe are both crystals doped with transition metal (TM) ions. These kind of laser gain media exhibit a particularly strong interaction between the electronic states and lattice vibrations—the phonons [7, 8]. The vibrational-electronic ("vibronic") interaction is associated with a pronounced homogeneous line broadening of the laser transition and results in very broad absorption and emission bands. It gives rise to a rather complicated energy band level scheme which can hardly be described by the simple three- or four-level picture.[1] In fact, vibronic line broadening also takes place in prominent rare-earth ions such as trivalent lanthanides (Tm^{3+}, Ho^{3+}, Er^{3+}) [9], which also provide radiation in the MIR spectral range. The interaction with the lattice crystalline field is, however, much weaker than for TM-ions.

In the case of Cr:ZnS/ZnSe, a broad absorption band centered at around 1.6–1.8 μm corresponds to the transition between the $^5T_2$ ground state and the ^5E excited state, and is the only spin-allowed transition for this system [5, 10]. In addition, both involved levels are split due to the Jahn-Teller effect, which is stronger in the ground state (see Fig. 3.1). Therefore, the absorption spectrum is single-peaked, whereas the fluorescence gets double-peaked, with Cr^{2+}-doped materials exhibiting one of the largest relative bandwidths $\Delta\lambda/\lambda_0$ of more than 0.35. Also, given the fundamental scaling rule for the required pump intensity at threshold, $I_{th} \propto (\Delta\lambda/\lambda_0)\lambda_0^{-4}$ [9], the longer infrared wavelengths compared to Ti:sapphire are of additional benefit.

Typically, mid-IR vibronic laser materials are characterized by high nonradiative decay rates at room temperature [9]. The onset of strong multiphonon relaxation processes leads to a pronounced decline of their upper-state lifetimes and therefore decreases ("quenches") the upper-level population. As a consequence, the gain efficiency will be drastically reduced, requiring cryogenic cooling. Cr^{2+}-doped ZnS/ZnSe, however, can be conveniently operated at room temperature [15]. The fluorescence lifetime measurements of Cr:ZnSe reveal a constant lifetime well beyond 300 K, while the one of Cr:ZnS shortens by 20% at 300 K—similar to Ti:sapphire.

[1] The first ruby laser, for example, belongs to the class of three-level lasers and the lasing transition happens between two specific electronic states. These sharp electronic transitions result in a very narrow-linewidth laser emission.

Table 3.1 Material and spectroscopic characteristics of the broadband laser materials Cr^{2+}:ZnS/ZnSe and Ti^{3+}:Sapphire. $\Delta\lambda/\lambda_0$: relative emission bandwidth; λ_0 (nm): maximum gain wavelength; $\Delta\lambda$ (nm): gain bandwidth (FWHM); τ (μs): fluorescence lifetime; σ_{abs} (10^{-20} cm^2): peak absorption cross-section; σ_{em} (10^{-20} cm^2): peak emission cross-section; κ (W/(m · K)): thermal conductivity; dn/dT (10^{-6} 1/°C): thermo-optic coefficient; bandgap energy (eV). All parameters refer to operation at room temperature (~300 K)

	Cr^{2+}:ZnS	Cr^{2+}:ZnSe	Ti^{3+}:Al_2O_3
$\Delta\lambda/\lambda_0$	0.38	0.37	0.3
λ_0	2350	2450	790
$\Delta\lambda$	890	900	230
τ	4.3	5.5	3.2
σ_{abs}	100	110	6.5
σ_{em}	140	130	39
κ	27	18	27
dn/dT	46	70	12
Hardness (Knoop)	160	120	2000
Bandgap	3.8	2.8	6
Refs.	[10–12]	[9, 10, 12]	[12–14]

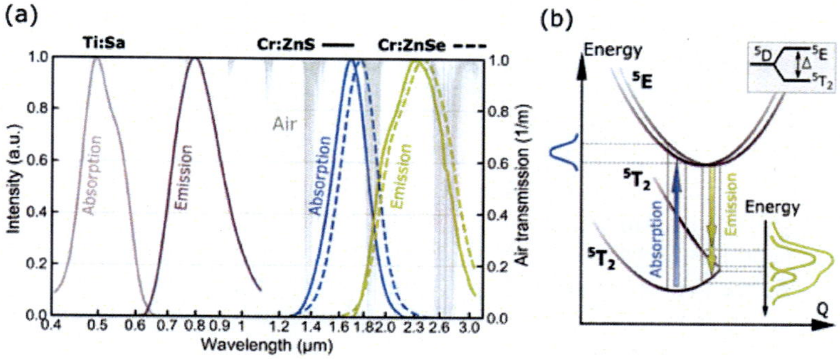

Fig. 3.1 **a** Absorption and emission spectra of Ti:sapphire and Cr:ZnS/ZnSe crystals together with the transmission of air obtained from HITRAN simulations. The wavelength axis is plotted on a logarithmic scale to preserve the relative bandwidth $\Delta\lambda/\lambda_0$. **b** Coordinational diagram of the laser transition in Cr:ZnS/ZnSe, including the Jahn-Teller splitting. (Data taken from [5, 6])

Since zinc-blende II-VI semiconductor compounds like Cr:ZnS/ZnSe do not possess inversion symmetry and have a high nonlinear index of refraction (see Table 2.2)—even 30-times higher than that of sapphire—strong second- and third-order nonlinearity are expected [10, 16]. In this respect, **polycrystalline Cr:ZnS/ZnSe crystals** play a special role as they were also used in the three diode-pumped systems presented in this thesis. Hampered by sublimation during the growth process and suffering from the resulting poor uniformity of single-crystal Cr:ZnS/ZnSe, the simpler production of polycrystalline material has relaxed the requirements on crystal quality while attracting growing interest in recent years [17]. The latter is certainly driven by the appearance of random quasi-phase-matching (RQPM) [18]: because the average grain size within the polycrystalline medium is on the order of the coherence length of typical nonlinear optical processes, efficient generation of higher-order harmonics such as the second harmonic (SH) were already demonstrated [16, 19, 20].

In 2009, the first Kerr-lens mode-locked operation of a Cr^{2+}:ZnS/ZnSe laser was realized [21, 22], and since then ever shorter output pulse durations were achieved—finally rendering their broadband and ultrafast performance as ideal starting point to exploit further nonlinear frequency conversion [23–26]. As will be demonstrated in the last experimental Sect. 5.3, the 2–3 μm output radiation can be efficiently converted through intra-pulse difference frequency generation (IDFG) into the deeper MIR range (here: 4.5–14 μm) in a relatively straightforward and simple arrangement.

3.2 Challenges of Diode-Pumping

With their ultrabroad absorption bands, Cr:ZnS/ZnSe lasers offer great flexibility in terms of pump wavelength selection. Ideally, the crystals are pumped close to their respective absorption peaks near 1.695 μm (Cr:ZnS) and 1.770 μm (Cr:ZnSe), but only very few CW laser sources exist that can provide sufficient pump power in this spectral range. Over the past years, a large variety of different pump laser sources was successfully implemented, including rare-earth doped fiber and solid-state lasers (Er^{3+} @1.6 μm [11, 19, 27], Tm^{3+} @1.9 μm [28–30], Ho^{3+} @2.1 μm [31]), Raman-shifted Nd^{3+} lasers (@1.6 μm [32, 33]), NaCl:OH color center lasers (@1.58 μm [34]), Raman-fiber lasers (@1.6–1.8 μm [35]) as well as TM doped solid-state lasers (Co:MgF_2 @1.86 μm [3]). To date, virtually all ultrafast Cr:ZnS/ZnSe lasers are pumped by mature and high-power Er or Tm fiber lasers, mainly thanks to their near diffraction-limited output beam characteristics and ease of use. Consequently, up to 2 W of average power could already be demonstrated with Kerr-lens mode-locked oscillators [36].

On the other hand, the low pump threshold of Cr:ZnS/ZnSe gain crystals allows for direct diode-pumping and represents an excellent alternative that combines the low-noise performance, compactness, high-efficiency, and affordability needed for laser-based applications. Therefore, bringing cutting-edge laser diode technology and few-cycle 2–3 μm solid-state lasers together will help meet the growing demand for accessible and reliable table-top mid-IR sources [37]. The first diode-pumped operation of Cr:ZnS/ZnSe oscillators was demonstrated in 1997 in a side-pumped configuration[2] [38], while a longitudinally pumped continuous-wave cavity was realized only a few years later [29]. When using single-emitter laser diodes, output powers of more than 400 mW could be achieved from the Cr:ZnS/ZnSe lasers so far [39],

[2] The output of this laser cannot be considered as truly continuous-wave, given that the laser diode array was operated in a pulsed mode (~50 μs pulse width).

and a record output power of even 3.7 W was demonstrated when the gain crystal was pumped by a 30 W diode bar [40]. Finally, the first femtosecond operation of a diode-pumped Cr^{2+}-doped II-VI oscillator was realized in 2012 [41], resulting in average output powers of 50 mW. By using a 3.5 W single-emitter laser diode at 1550 nm as well as a SESAM device to allow for self-started mode-locking (see Sect. 2.1.2), a pulse duration of 180 fs was achieved, which was significantly longer than what is supported by the gain medium's bandwidth.

On the contrary, the technique of Kerr-lens mode-locking (see Sect. 2.1.2) is known to support much shorter pulse durations [42], which can even exceed the emission bandwidth-limit of the gain medium [43, 44]. Since it relies on the action of a Kerr lens inside the laser crystal, a good mode-matching between pump and fundamental cavity mode is crucial to obtain sufficient gain and a good power discrimination between CW and pulsed operation. Selecting the most suitable laser diode type is therefore of utmost importance.

3.2.1 The Right Choice of Laser Diode Type

Laser diodes are available in a myriad of different designs and output beam characteristics. To choose the right type, one can start with a set of parameters defined by the corresponding application. For this thesis, the desired characteristics are summarized in Table 3.2, indicating the need for a laser diode beam with the highest possible beam quality to ensure good focusability. Even though single-mode laser diodes with $M^2 \approx 1$ exist, their output power is limited to a few 100 mW only. This is a main drawback of laser diodes because a higher beam quality usually comes along with a very much reduced output power. Therefore, multi-mode laser diodes need to be used, and a good compromise between output power and beam quality can be provided by **single-emitter laser diodes** (see Sect. 3.2.2). However, if several single-emitter diodes are stacked up vertically to increase the total output power in the form of a laser diode bar, the brightness will deteriorate dramatically. The resulting poor beam quality is reflected by the high M^2 values of >100 or even >1000 spatial ("in both transverse directions"). For comparison, a diffraction-limited Gaussian beam is characterized by an M^2 factor of 1. Hence, using diode bars for pumping ultrafast KLM oscillators is not an option.

In terms of output wavelength, semiconductor laser diodes based on **indium phosphide** (InP) are the most promising candidates to provide lasing beyond 1.6 μm [45]. Single-frequency operation is not specifically demanded since the absorption band of Cr:ZnS/ZnSe crystals is broadband enough to sustain variations of about $\Delta\lambda < 10$ nm without a significant drop in gain. Finally, multi-mode fiber-coupled packaging should be avoided to maintain the high unidirectional beam quality provided by free-space single-emitter diodes. Putting all these desired characteristics together, **C-Mount packaged** single-emitter laser diodes provide an ideal solution and were used for the three laser systems presented in the remaining part of this thesis.

Table 3.2 Desired output characteristics of a laser diode to be used for the few-cycle oscillators presented in this thesis

Parameter	Desired value
Output power	Several watts
Beam quality	Ideally single-mode, $M^2 = 1$
Wavelength	$\lambda = 1.6$–1.8 μm
Wavelength stability	$\Delta\lambda < 10$ nm
Beam delivery	Free space, not fiber-coupled

3.2.2 *Properties of Single-Emitter Diodes*

Semiconductor laser diodes are typically operated in a forward-biased double-hetero p-i-n structure, in which an undoped junction acts as the laser-active layer [46, 47]. Because of its higher refractive index and reflective ends, light gets tightly confined and amplified inside this region when an external voltage is applied.

Output Beam Properties:

The active layer of the utilized single-emitter diodes has a rectangular shaped cross section with a horizontal ("slow axis") and vertical ("fast axis") dimension of 95 μm and 1 μm, respectively. Compared to true single-mode diodes with an emitter width of only a few microns, the increased volume allows to generate and extract more laser power. However, it comes at the expense of a reduced brightness in the slow axis direction, where multiple transverse modes are emitted in a closely spaced manner [48] (see Fig. 3.2b). Typical beam quality factors are $M^2 \simeq 1$ (fast) and $M^2 \gg 1$ (slow). The strong diffraction at the output facet results in a **highly elliptic** and **divergent** emission, with the TE modes gradually merging together in the horizontal direction to form a light line. Depending on the number and intensity distribution of the mixed modes, the diode emission can have a variety of different shapes, and a pronounced dependance on the applied pump current was found experimentally (see Fig. 3.3).

Even though wide stripe laser diodes can now deliver up to a few watts of average power, temperature-related problems have significantly diminished the maximum output of diodes emitting around 1.5–1.7 μm [49–51]. But, since the output of laser diodes is **linearly polarized** along the slow axis direction, polarization multiplexing schemes can be used to increase the total effective power (see Sect. 4.1.1).

Noise Properties:

The amplitude noise of most solid-state and semiconductor lasers is dominated by a characteristic relaxation oscillation (RO), which only appears when the upper-

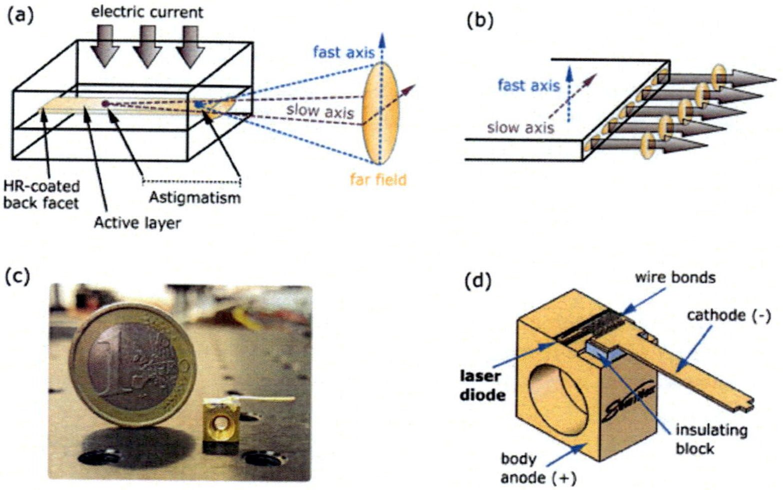

Fig. 3.2 **a** The anisotropic output emission of a laser diode. For the sake of clarity, the degree of astigmatism is much exaggerated. **b** The output emission of a wide stripe laser diode containing multiple TE modes in the slow axis (horizontal) direction. **c** Picture of a C-Mount diode next to a 1€-coin to illustrate the small dimensions. **d** Components of a C-Mount diode (figure provided by *SemiNex Corp.*)

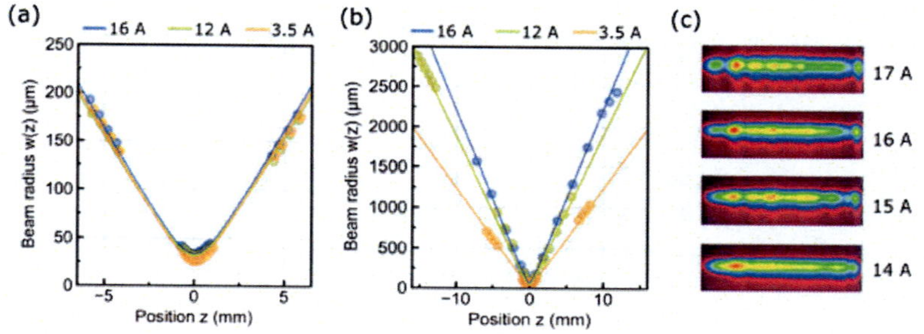

Fig. 3.3 Beam quality and beam profile measurements of the laser diode output for different driving currents. **a** Results for the fast axis, with calculated M^2-values of 1.5 (3.5 A), 1.8 (12 A) and 2.1 (16 A). **b** Results for the slow axis, with calculated M^2-values of 9.1 (3.5 A), 17 (12 A) and 28 (16 A). In **c**, the output beam was collimated and measured with a pyroelectric camera (*Spiricon Ophir Photonics, PY-III-C-B*) and reveals spatial changes in the beam profile for rather small steps in pump current

state lifetime τ_2 is substantially longer[3] than the laser cavity decay time τ_c [46, 52]. Initially, when these lasers are suddenly switched on, the laser output power may exhibit several spikes due to fluctuations in the population inversion and laser gain that will settle down fairly quickly into a quasi-sinusoidal decaying oscillation about the steady-state value. After a given amount of time and in the case of a very quiet and stable environment, the output intensity can eventually be treated as constant. However, when the laser is perturbed by any sudden change in gain, pump power, cavity alignment or cavity loss, another transient relaxation oscillation will be triggered and will again evolve in a exponentially decaying manner.

Cr:ZnS/ZnSe as well as rare-earth doped fiber lasers are both solid-state laser sources that are prone to relaxation oscillations. In addition, the noise performance of a Cr:ZnS/ZnSe laser system is greatly affected by the noise of its pump source since pump noise can be transferred to oscillator noise below the oscillator's relaxation oscillation frequency ν_{ro} [53]. The latter can be written as [46, 52]:

$$\nu_{ro} = \frac{1}{2\pi}\sqrt{(x-1)/(\tau_2\tau_c)}, \tag{3.1}$$

where $\tau_c = -\nu_{rep}^{-1}\log[R_1R_2(1-T_i)^2]$ and $x = P_{pump}/P_{th}$ with the pump power P_{pump} and the threshold power P_{th}.

Let us assume that we have set up a Cr:ZnS laser with a repetition rate of $\nu_{rep} = 50$ MHz, an end-mirror reflectivity of $R_1 = 1$, an output coupler reflectivity of $R_2 = 0.75$, and an internal loss per pass of $T_i = 0.05$. Then, $\tau_c \approx 7.8$ ns and together with $\tau_2 = 4.3$ µs (see Table 3.1), the relaxation oscillation frequency will amount to $\nu_{ro,Cr:ZnS} \approx$ **870 kHz** for a pump rate of $x = 2$. Rare-earth doped fiber lasers, which are the most common pump source of Cr:ZnS/ZnSe lasers, have RO frequencies that lie in the few 10 kHz to a few 100 kHz range [54–57]. Therefore, their noise characteristics will be transferred and added to the Cr:ZnS laser noise if they are not suppressed by sophisticated feedback schemes [58].

By contrast, semiconductor laser diodes also exhibit relaxation oscillations but in a very different frequency domain [46, 52]. A GaAs laser diode, for example, with a cavity length of L = 300 µm, a pump rate of $x = 1.5$, a cavity round-trip time of $\tau_{rep} \approx 6.7$ ps and lifetime $\tau_c \approx 1.1$ ps, and an upper-state lifetime in the p-n junction of $\tau_2 \approx 3$ ns, has a RO frequency[4] of about $\nu_{ro,GaAs} \approx$ **2 GHz**. This means that the relaxation oscillations of laser diodes are located in the multiple GHz-range and thus experience stronger damping by the Cr^{2+}-doped ZnS/ZnSe gain medium because $\nu_{ro,GaAs} \gg \nu_{ro,Cr:ZnS}$.

In Fig. 3.4, the amplitude noise measurement of a laser diode is compared to the results obtained with three different rare-earth doped fiber lasers, which have been available at our research laboratory. The noise signals were recorded in a 1 MHz-span in the baseband region and clearly reveal the intrinsic low-noise characteristics

[3] This condition does not hold for most gas lasers, so that spiking and relaxation oscillations are in general not observed.

[4] Note that the condition $\tau_2 \gg \tau_c$ is met in all considered cases.

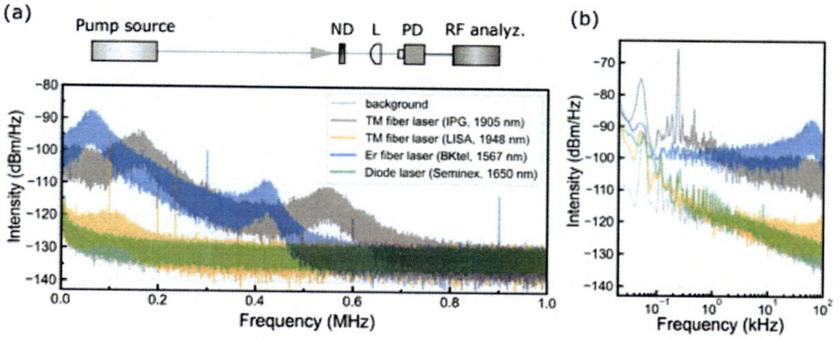

Fig. 3.4 Qualitative comparison of the amplitude noise of available pump sources for Cr:ZnS/ZnSe oscillators. The measurements reveal the strong relaxation oscillations in the kHz-range, which are subject to standard rare-earth doped TM (*IPG, TLR-120-WC-Y12*) and Er fiber lasers (*Bktel*, HPFL-370-1567-FCAPC). But, fiber lasers can also be designed for lower noise operation (for example: Tm laser, *LISA laser products*, 101 030 010). By contrast, semiconductor laser diodes (here: *SemiNex Corp.*, C-132) reveal an intrinsic low noise performance, which is very close to the performed background noise measurement taken without any incoming signal. For the measurements, the pump light was focused by a lens (L) onto a biased InGaAs photodetector (PD, *Thorlabs DET10D2*), which is sensitive from 0.9–2.6 μm. The DC voltage generated by the PD was kept constant for all measurements with the aid of a neutral density (ND) filter wheel, and the amplitude noise signal was measured by a high dynamic-range radio-frequency (RF) spectrum analyzer (*Agilent*, E4447A) in a 1 MHz-range (see **a**). **b** The low-frequency noise is plotted on a logarithmic x-axis for frequencies up to 100 kHz

of a laser diode. This is of great interest for mid-infrared applications such as near-field scanning optical microscopy (s-SNOM, PF-SNOM) [59, 60], which typically operate at tapping frequencies in the kHz-domain and are highly sensitive to noise in this range.

3.2.3 Effects on the Laser System and the Few-Cycle Pulse Generation

The lower brightness of laser diodes requires a series of precautions when designing and setting up a few-cycle laser system, especially when the technique of soft-aperture Kerr-lens mode-locking is used. Compared to fiber laser-pumped systems, some major challenges arise that will be discussed in the following.

3.2.3.1 Mode-Matching Considerations

Adapting the pumped volume of the crystal to the volume occupied by the fundamental laser resonator mode is known as mode-matching and plays a crucial role when selecting lenses for pump beam collimation. The optimization of the overlap

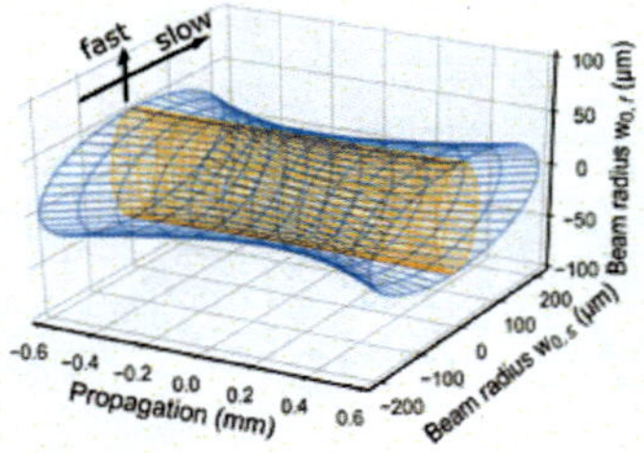

Fig. 3.5 Simulation of the overlapped volume between the laser diode beam (blue) and the fundamental cavity beam (orange) along the pump absorption length

is not only important for extracting the highest possible gain and thus output power of the cavity, but also to exploit efficient self-amplitude modulation for KLM (see Sect. 2.3.2). Dealing with the highly anisotropic and non-Gaussian like pump beam of a single-emitter laser diode is a demanding task since the brightness of the fast and slow axis emission differs significantly and greatly affects the pumped crystal volume. For optimized mode-matching, a set of aspheric and/or cylindrical lenses is therefore needed to reshape and collimate each axis in a separate manner.

In literature, several guidelines have been reported on improving a laser system's performance by an appropriate choice of pump and laser beam diameter [61–67]. However, most of them cannot directly be applied to the cases considered in this thesis. For example, when assuming a perfectly uniform pump distribution of radius w_p and a Gaussian-shaped resonator mode of 1/e-radius w, both are ideally chosen to be equal in size inside the gain crystal ($w_p/w \simeq 1$) [61]. While this can be readily fulfilled with single-mode pump sources, it does not hold for the multi-transverse-mode emission of a single-emitter laser diode. Also, most theoretical investigations already start with a given pump beam size/ellipticity and aim to find the most suitable mode size for the laser cavity to be built [62, 65, 66]. Consequently, the laser cavity mode is also allowed to be of elliptic shape, which can be realized through cylindrical cavity optics [63, 64]. Building a laser cavity around a given pump setup is, yet, not practical for the laser systems considered in this thesis, in which the laser mode sizes are already defined by the envisaged resonator geometry.

Nevertheless, a good approximation can be obtained when simulating the overlap volume between a perfect Gaussian cavity beam ($M^2 = 1$) and an elliptic pump beam (Fig. 3.5). Even though it is a simplified model and does not take thermal and spatial effects like self-focusing or gain saturation into account, it provides a general idea of the optimal diameter ratio between the fast and slow axis of the pump laser diode. A useful guidance is provided in [63], in which the pump absorption length L is taken as a measure to optimize the pump beam and cavity mode diameters. After the propagation length L, namely, the intensity of an incident light beam has already decayed to 1/e (≈37%) of its initial value and the intensity after a certain distance can be written as [68]:

$$I(z) = I_0 \cdot e^{-z/L} = I_0 \cdot e^{-N_{dop}\sigma_{abs}z}, \tag{3.2}$$

Table 3.3 Best simulation results for given M^2-values of the pump beam and focus size $w_{0,f}$ for the fast axis. The overlap is given with respect to the pumped volume ($\pm 0.5L$)

***$M^2_{f,s}$= 1.2/7.6**	**$w_{0,s}/w_{0,f}$**	**Overlap**
$w_{0,f} = 30$ μm	**3.2**	22%
$w_{0,f} = 50$ μm	**1.9**	36%
$w_{0,f} = 55$ μm	**1.8**	39%
****$M^2_{f,s} = 1.5/9$**		
$w_{0,f} = 30$ μm	**3.9**	17%
$w_{0,f} = 50$ μm	**2.3**	30%
$w_{0,f} = 55$ μm	**2.1**	33%

where N_{dop} and σ_{abs} are the crystal's doping concentration and absorption cross section, respectively. In this chapter, a 5 mm-long Cr:ZnSe crystal with $N_{dop} = 7.5 \cdot 10^{18}\text{cm}^{-3}$ and $\sigma_{abs} = 110 \cdot 10^{-20}\text{cm}^2$ (see Table 3.1) was used, resulting in a pump absorption length of $L \approx 1.2$ mm. Since the fast axis emission of the diode is of near diffraction-limited beam quality ($M^2 \approx 1$) , its waist radius $w_{0,f}$ in the focus was set equal to the mode waist radius w_0 of the fundamental cavity beam. This, we believe, is essential for efficient soft-aperture KLM (see also Ref. [69]). In Table 3.3, the simulation results are depicted for different waist sizes $w_{0,f}$, which are chosen to resemble the three diode-pumped laser systems of this thesis.

The simulations predict that an axis ratio of about 2–2.5 in the focus might be ideal to achieve the highest gain, which also agrees well with the ratios reported for diode-pumped KLM Ti:sapphire systems [69–71]. In general, keeping the slow axis diameter as confined as possible also minimizes the risk to excite other than the fundamental cavity mode.

3.2.3.2 Technical Challenges

Apart from a proper selection of aspherical and cylindrical lenses to shape the diode output beam, some technical—but not insignificant—challenges arise that need to be considered in the laser design:

- **Pump beam optics:** A spot size ratio of about 2:1 requires the collimated slow axis diameter to be much larger in size than the fast axis diameter before being focused into the laser crystal. In fact, it was found experimentally that the slow axis diameter had to be close to 1 in. (25.4 mm) in size, which is the standard size for commonly available optics. Therefore, customized rectangular-shaped mirrors were used and highly-reflective (HR) coated for the pump wavelength.
- **Pump diode handling:** The laser diodes had been delivered without any heatsink or mount attached to it. To ensure an adequate heat removal for the waste heat produced and to avoid damage by electrostatic discharge, a special mount had to be designed (see Fig. 3.6a). Since a typical InP single-emitter diode delivers up to

4 W of output power at 15 A/1.8 V, only 15% of electricity gets converted into light. The waste heat of 23 W is generated within a small area of only 7x6 mm defined by the C-mount body. In the first design, water-cooled components as well as a Peltier element were implemented to stabilize the diode's output power via thermoelectric cooling.

- **Plane input coupler mirror:** A standard four-mirror cavity consists of two flat end-mirrors and two curved mirrors with the gain crystal placed in between (see Fig. 3.6b). The pump beam is then focused by a suitable lens through one of the curved mirrors. This "classic" bow-tie design is, however, hard to realize for diode-pumped systems, since it limits the selection of suitable pump lenses needed to achieve the required tight focus in the gain crystal. For practical reasons, a "folded" bow-tie cavity design had to be implemented in the first two laser systems.[5] In this case, the pump beam enters the cavity through a plane input coupler mirror, and pump focusing optics with a focal length equal or shorter than the focal length of the cavitie's curved mirrors can be used.
- **Crystal dimensions:** The low brightness of the slow axis causes the pump beam to diverge rapidly when focused to a tiny spot. To avoid clipping at the crystal end facets, a gain medium of suitable dimension is required. The Cr:ZnS/ZnSe gain media available in our lab are typically 5 mm and 9 mm long. Therefore, taking the measured M^2-value of about 7.6 for the slow axis and the corresponding pump spot diameter of about 90 μm, which is true for the first laser system built, a slow axis radius of $w(\pm 2.5\ \text{mm}) = 0.7$ mm and $w(\pm 4.5\ \text{mm}) = 1.25$ mm is calculated at the crystal end facets.[6] Therefore, crystals with clear apertures of at least 2–3 mm are needed (see Fig. 3.6c).
- **Cavity alignment**: To build and align the laser oscillator, the pump beam typically serves as guidance. Using the large collimated pump beam of a wide stripe laser diode is, however, not practical, especially when it diverges rapidly after passing through a focal plane. In this thesis, a solution was found by either relying on the single-mode emission of an Erbium fiber laser ($\lambda = 1567$ nm), which was temporarily installed as an alternative pump source, or by using a red alignment laser diode. The latter has the advantage to provide radiation visible to the human eye, therefore facilitating the setup of the cavity.

3.3 Few-Cycle Pulse Generation via Kerr-Lens Mode-Locking

In the following, the results of the first directly diode-pumped KLM Cr^{2+}-doped II-VI laser are presented along with a focus on the intra-cavity dispersion management, the success of the soft-aperture Kerr-lens mode-locking technique, and the measured

---

[5] For the third laser system, a workaround could be found based on a novel pump focusing scheme (see Sect. 5.2.1).

[6] This assumes that the waist is directly positioned at the center of the crystal.

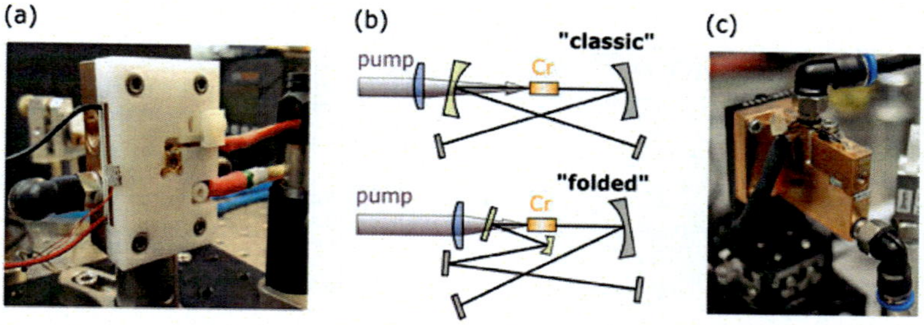

Fig. 3.6 **a** Picture of the laser diode mount, which was equipped with a peltier element. **b** Schematic sketch of a "classic" and "folded" bow-tie cavity. **c** The 5 mm long gain crystal (clear aperture: 2 mm by 2 mm) was placed inside a water-cooled copper mount

amplitude noise performance. Even though this first demonstration has already been published in Ref. [72], the following subsections serve to provide a deeper insight into the experimental observations and the results achieved.

3.3.1 Pump Setup and Resonator Design

A single indium phosphide (InP)-based C-Mount laser diode (*SemiNex Corp.*, C-132) was used, providing up to 3.8 W at 1650 nm from a 95 μm wide stripe. To stabilize the diode's output power and wavelength and mitigate the thermal power roll-off, which is characteristic for these laser diodes [50], simultaneous thermoelectric and water cooling was implemented. The measured output power is shown in Fig. 3.7a, clearly indicating the unavoidable onset of the roll-off at higher pump current levels. To evaluate the output beam quality in the horizontal and vertical direction, the $D_{4\sigma}$ beam radii were measured with the help of a scanning slit profiler (*Ophir*, NanoScan 2s) at different positions along the optical axis z. Following the ISO standard[7] 11146 [73], the measured beam radii $w(z)$ were related to the M^2-factor, the wavelength λ, and the measured radius at the beam waist, $w_0 = w(z = z_0)$, according to [74]:

$$w(z)^2 = w_0^2 \left[1 + (z - z_0)^2 \left(\frac{M^2 \lambda}{\pi w_0^2} \right)^2 \right]. \tag{3.3}$$

A fit through the data points revealed M^2-values of 1.2 and 7.6 for the fast and slow axis of the diode, respectively (see Fig. 3.7b). Based on the mode-matching considerations and the technical challenges listed in the previous section, the anisotropic

[7] It demands that half of the measured data points must be located at a distance greater than two Rayleigh lengths away from the beam focus, whereas the other half of the points is measured around the focused beam waist.

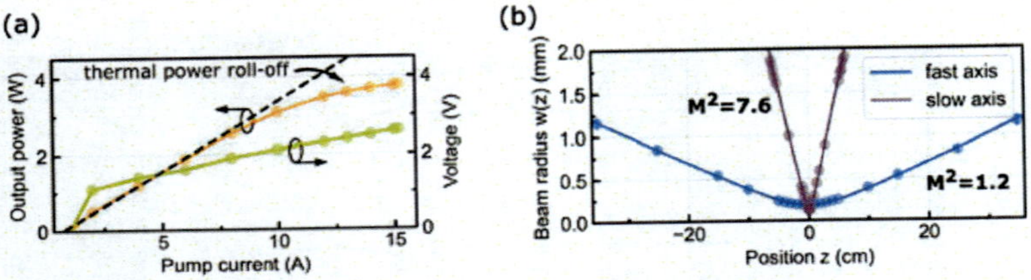

Fig. 3.7 **a** Laser diode output power and voltage versus pump current. **b** M^2-measurement of the fast (vertical) and slow (horizontal) axis emission. Note that this data was taken at a rather low pump current (2–3 A) since the change in beam quality for increasing pump currents was not yet known at that point

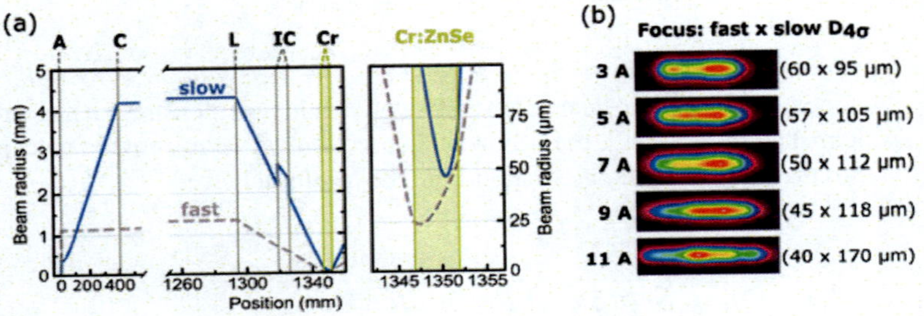

Fig. 3.8 **a** Simulated pump beam propagation for the fast and slow axis using the open source software *reZonator*. To shape the output emission, an aspheric (A, f = 4.51 mm) and cylindrical lens (C, f = 400 mm) were used, with the beam being focused by an aspheric pump lens (L, f = 50 mm) trough a 6 mm thick plane-plane input coupler mirror (IC) placed at 45°. The green shaded area marks the focal area inside the gain crystal. **b** Change in pump spot diameters for increasing current levels

and highly-divergent output of the diode was then reshaped with a set of AR-coated aspheric and cylindrical lenses.

To identify the most suitable lens combination for a desired pump spot size ratio of ~1:2 (fast:slow axis) and a theoretical cavity mode diameter of ~60 µm in the focus, ABCD-matrix calculations had been performed, with the final result being depicted in Fig. 3.8a. Taking into account the measured beam divergences and radii, it predicts a theoretical pump spot diameter of 48 µm (fast) × 94 µm (slow) and is in good agreement with the measured spot diameter of approximately 60 µm × 95 µm. The latter, however, is only true for a very modest pump current of 3 A (see Fig. 3.8b). In fact, later measurements revealed that this deviation becomes larger for increasing pump currents. The particularly strong change in pump beam characteristics beyond 11 A manifested itself in a drop in continuous-wave oscillator output power, when pump currents larger than 11 A were applied. At that point, corresponding thermal problems related to the ever increasing heat load inside the crystal could be excluded by monitoring the crystal's surface temperature via a NTC sensor. The latter has been attached to the copper mount of the gain medium.

(a)

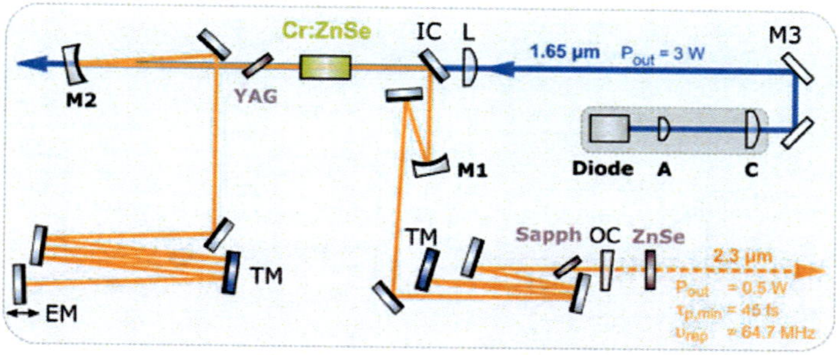

(b)

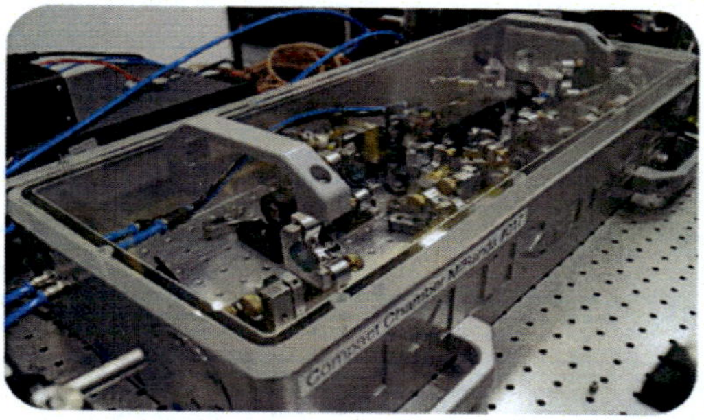

Fig. 3.9 **a** Schematic of the experimental setup of *MIRandal* and **b** a picture of the real laser system in the lab. M1 and M2: curved mirrors ($ROC_{M1} = -100$ mm, $ROC_{M2} = -200$ mm), M3: highly-reflective rectangular-shaped mirrors for the pump laser (1.4 µm–1.7 µm), A: aspheric lens (f = 4.51 mm), C: cylindrical lens (f = 400 mm), L: aspheric pump lens (f = 50 mm), Cr:ZnSe: gain medium (5 mm), Brewster-oriented material (3 mm YAG—for astigmatism compensation, 3 mm Sapphire—for dispersion compensation), OC: output coupler (34%), TM: third-order dispersive mirror (blue-shaded), ZnSe: 5 mm AR-coated substrate to temporally re-compress the output pulses. Non-labelled mirrors, including the input coupler (IC) and the end-mirror (EM), are HR-coated (2.0–2.6 µm) and transmit the pump light

Figure 3.9 shows a schematic of the experimental setup along with a picture of the real system in the laboratory environment. The oscillator contained a 5 mm long anti-reflection (AR)-coated Cr:ZnSe polycrystalline gain element (*IPG Photonics*) with a clear aperture of 2 mm by 2 mm and a doping concentration of $7.5 \cdot 10^{18}$ cm^{-3}. It was mounted inside a water-cooled copper heat sink (T = 12 °C) and the single-pass absorption at 1650 nm was measured to be ∼90%. Compared to a gain medium placed at Brewster's angle, in which a strong beam expansion in the tangential (horizontal) plane arises from a refraction at the material's surfaces [75, 76], an AR-coated crystal allows for a more homogeneous spot size while reducing asymmetric thermal

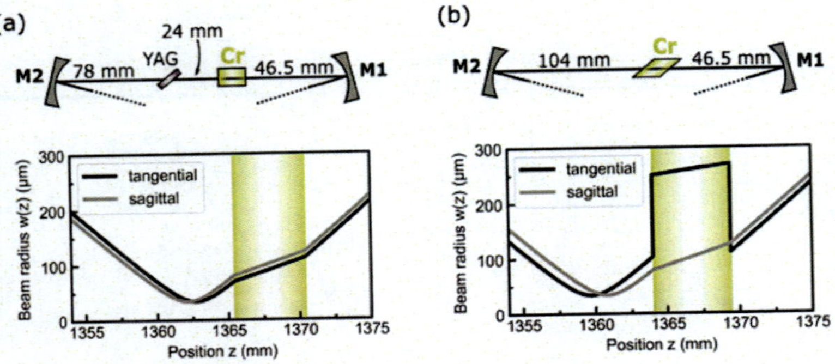

Fig. 3.10 Illustration of the difference in beam radii for **a** an AR-coated crystal and **b** a crystal placed at Brewster's angle. For the ABCD-matrix calculations, similar parameters were used, however, an additional 3 mm YAG-plate was placed inside the cavity of figure **a** to compensate for astigmatic distortions caused by an oblique angle of incidence on the curved mirrors

lensing effects [36] (see Fig. 3.10).[8] Moreover, when using an AR-coated crystal, the pump beam does not need to be p-polarized for minimal reflection losses and can theoretically take arbitrary polarization directions. This results from the fact that Cr:ZnS/ZnSe crystals do not exhibit any considerable polarization dependence with regard to their absorption cross-sections [77].

As illustrated in Fig. 3.9, the Cr:ZnSe crystal was placed between two curved mirrors M1 and M2 inside an asymmetric X-fold cavity with an arm length ratio of $\overline{\text{M1-OC}}/\overline{\text{M2-EM}} = 0.7$. This is in accordance with the KLM design procedure suggested in Ref. [78], in which a reasonable trade-off between stable laser operation and high modulation efficiency can be achieved for arm ratios between 2/3 and 3/4. Since the selection of suitable optics was limited at the time of the oscillator construction, the radii of curvature (ROC) for M1 and M2 were different, but chosen to increase the focal spot diameter in the Cr:ZnSe crystal. With this, the nonlinear effects in the gain crystal are reduced and the onset of multi-pulsing for increasing pump powers can be prevented [79]. In addition, astigmatic distortions are introduced due to the oblique angle of incidence on the curved mirrors, and a 3 mm-thick Brewster plate (YAG) is placed beside the gain medium to avoid a corresponding degradation of the laser performance. Using multiple highly-reflective 45-degree mirrors inside the laser oscillator allowed to assemble the whole cavity—excluding the pump setup—inside a monolithic housing with a footprint of only 30 cm by 70 cm.

[8] Note that the pronounced jump in the tangential beam size—as it is depicted in Fig. 3.10b—is not an exact representation of the real beam distortions at the crystal's surfaces.

3.3.2 The Importance of Intra-cavity Dispersion Management

Even though the ultrabroad emission cross-section of Cr^{2+}-doped II-VI gain crystals provides an ideal condition for the direct generation of few-cycle laser light, the pulse duration of KLM laser systems will ultimately be limited by the degree of higher-order dispersion control inside the oscillator. As explained in the theoretical Sect. 2.2, the self-phase modulation arising from the Kerr effect in the gain crystal needs to be balanced by an appropriate amount of negative GDD, such that solitonic pulse formation can appear in ultrafast solid-state laser systems. In the first diode-pumped Cr:ZnSe laser, material plates made of sapphire were incorporated into the cavity. As listed in Table 2.2 of Sect. 2.2.1, this material is able to provide suitable negative GVD, which can also account for the large positive second-order dispersion of the gain crystal. For generating the shortest pulses, however, a small but constant total GDD per round-trip is required over the entire spectral range [80], with the pulse duration of the soliton scaling as $\tau_0 \propto \sqrt{|D_2|}$ for a given value $D_2 < 0$ of the intra-cavity GDD (see theory Sect. 2.2.3). Therefore, the additional third-order dispersion accumulated by the gain crystal and added material plates needs to be addressed as well when operation close to zero dispersion is desired. In this respect, the design and usage of tailored **dispersive mirrors** (DM) has become quite popular.

Introduced in 1994 [81], laser mirrors based on chirped multilayer coatings turned out to be indispensible for the few-cycle pulse generation and octave-spanning spectra demonstrated with Ti:sapphire laser systems [82, 83]. When designed carefully, these mirrors uniquely combine ultra-high reflectivity (typically >99.5%) with well controlled dispersion characteristics over a broad spectral operation range [84]. The frequency-dependent chirp caused upon mirror reflection arises from the varying thickness of the multiple layers, such that different wavelengths penetrate different path lengths into the stack (see inset of Fig. 3.11b). Beside this chirped mirror (CM) technology, there is also the principle of resonant storage, in which nanoscale Fabry-Pérot-like cavities are embedded into the multilayer structure to resonantly trap selected wavelengths [85, 86]. The combination of wavelength-dependent penetration and resonance effects is realized in modern dispersive mirrors to provide high reflectivity and GDD over a broad wavelength range without increasing the thickness of the coating. Recently, this technology has also become available for ultrafast 2–3 μm Cr:ZnS/ZnSe-laser systems [87] and was directly applied to the laser systems presented in this thesis.

In Fig. 3.11, the reflectivity and second-order dispersion of various intra-cavity elements, including ZnSe, YAG, sapphire and a dispersive mirror, are depicted. The latter was designed to only compensate for the materials' third-order dispersion (TOD) without adding additional GDD at the central wavelength ($\lambda \sim 2.3$ μm). Also plotted is the calculated theoretical round-trip dispersion for the combination of mirrors and material plates that experimentally resulted in the shortest pulse duration. However, it is important to point out that the exact dispersion values were

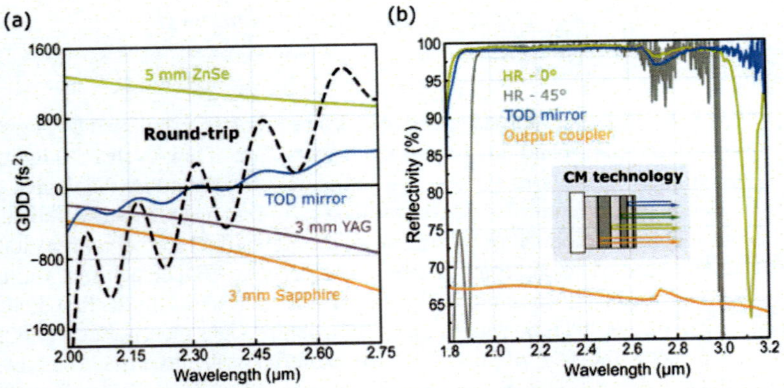

Fig. 3.11 **a** Theoretical dispersion values of intra-cavity elements including the single-pass GDD of incorporated material plates as well as the dispersion introduced by a single TOD mirror. The calculated theoretical round-trip dispersion for the case of eight total bounces on the TOD mirror is shown as a dashed line. **b** Measured reflectivity curves (p-polarized) of the oscillator mirrors that were designed to operate in a broadband spectral range (2.0–2.6 μm)

not experimentally measured, and each individual dispersive mirror may exhibit a slightly different oscillation pattern.

For efficient Kerr-lens mode-locking, laser operation close to the limits of one of the oscillator's stability ranges is required [78, 88, 89]. The stability zones for *MIRandal* are shown in Fig. 3.12a, and were calculated by evaluating the stability parameter of the ABCD round-trip cavity matrix [52, 90] for a varying distance L between the curved mirror $M2$ and the YAG plate. Following the guidelines on hard-aperture KLM presented in [78], the laser was adjusted to operate closer to the 2nd edge of stability zone I ($L1 = 78$ mm) with the Cr:ZnSe crystal being slightly shifted towards mirror $M1$ of the shorter resonator arm (see Fig. 3.12a). Stable soft-aperture KLM operation could then be initiated by slightly mis-aligning and rapidly moving[9] the end-mirror (EM) that was mounted on a mechanical translation stage. Because of the polycrystalline nature of the gain medium and the effect of random quasi-phase-matching (see Sect. 3.1), the onset of second-harmonic light was a clear indicator for mode-locked operation, when the gain crystal's surface was monitored by an infrared-sensitive camera. In addition, the theoretical beam radii for CW and pulsed operation have been calculated and are depicted in Fig. 3.12b.

In *MIRandal*, a first optimization of the intra-cavity dispersion was performed when inserting sapphire plates into the shorter resonator arm. Varying the substrate thickness from 2 mm to 6 mm in small steps (0.5 mm), the corresponding mode-locked output spectra were analyzed with a grating-based PbS array spectrometer (*B&W Tek*, $\lambda = 1500$–2750 nm). As illustrated in Fig. 3.13a, the slight changes in

[9] This is a commonly used technique to introduce an intensity spike and initiate mode-locked operation (see simulated spiking behaviour in Fig. 2.2).

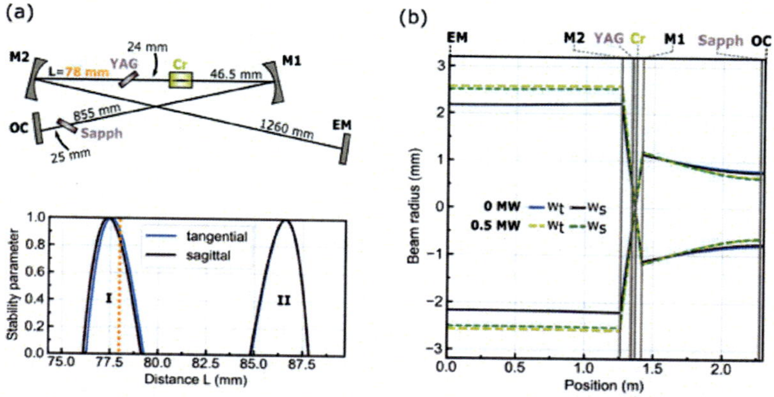

Fig. 3.12 **a** Schematic of the laser cavity and the experimental length scales. To determine the location of the oscillator's stability zones, the distance L between mirror $M2$ and the YAG plate, and also the distance between $M2$ and EM was varied, with the final position for KLM marked in orange. **b** Theoretically calculated beam radii for CW ($P_0 = 0$ MW) and mode-locked ($P_0 = 0.5$ MW) laser performance based on nonlinear ABCD-matrix analysis

round-trip dispersion manifest themselves in varying output spectral shapes and coverage, with their full width at half maxima (FWHM) being inversely proportional to the generated pulse duration ($\tau_0 \propto 1/\Delta\lambda$). Therefore, the measured FWHM[10] is plotted for different sapphire thicknesses, which reveals that the broadest spectral bandwidth is generated with a 3.5 mm thick plate.

To achieve an even broader spectral coverage, however, further dispersion tuning was needed. This became apparent when analyzing the spectral shapes for 2 mm and 3 mm thick sapphire plates more closely: here, the round-trip GDD is decreased by so much that it eventually crosses the zero dispersion line at certain wavelengths, resulting in the appearance of a resonant spectral sideband—a so-called "**dispersive wave**". Owing to the presence of relatively strong third-order dispersion, the soliton undergoes dispersive and periodic perturbation, and energy can be transferred from the soliton to a narrowband resonance in the normal dispersion regime [91]. The resulting emergence of a dispersive wave has already been reported for a number of ultrafast laser systems, including KLM bulk solid-state lasers [15, 92–94] and mode-locked fiber lasers [91, 95]. In *MIRandal*, the generation of dispersive waves at around 2.1–2.2 µm is attributed to the strong uncompensated TOD inside the cavity, which causes the round-trip dispersion to become positive at the shorter wavelengths.

To overcome this limitation, specific dispersive mirrors manufactured in-house were used. These multi-layer optics had been designed to compensate for the material dispersion of 2 mm sapphire and 2 mm ZnSe, which theoretically results in net zero dispersion at the central wavelength ($GDD_{2.3\mu m} \approx 0$ fs^2). Thus, they were referred

[10] The optical resolution of the fiber-coupled spectrometer was rather limited and specified as 2–10 nm. Hence, the measured FWHM spectra were only used for a qualitative comparison.

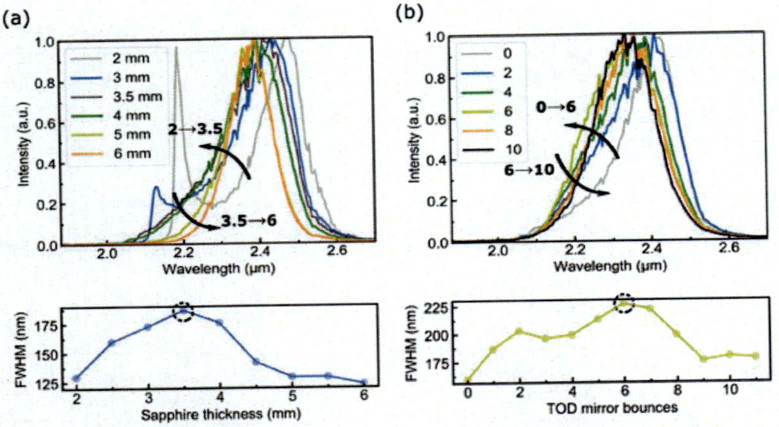

Fig. 3.13 Illustration of the intra-cavity dispersion tuning and its effect on the mode-locked spectral shape and FWHM. **a** Selected mode-locked output spectra for different thicknesses of sapphire placed into the oscillator. The pronounced sidebands for 2 mm and 3 mm are a clear indication of dispersive wave formation and a domination of strong TOD in the cavity. Also plotted are the spectral FWHM for each measurement to identify the most suitable configuration. **b** Variation of the number of TOD mirror bounces per cavity round-trip for a fixed sapphire plate thickness of 3.5 mm. An odd number of bounces could be achieved when one of the TOD mirrors was used as an end-mirror. Similar to **a**, the FWHM of the generated spectra is displayed as well

to as "**TOD mirrors**" since they mainly addressed the higher-order dispersion of the laser resonator. To identify the amount of mirrors needed for the given cavity configuration, the mode-locked spectral FWHM was measured for a varying number of TOD mirror bounces per round-trip. Finally, even further optimization was feasible when the thickness of the sapphire plates and the number of TOD mirrors was successively adjusted until convergence was reached. Also, additional fluoride-based substrates (CaF_2, BaF_2) were tried for finer steps in dispersion, but no further improvement was observed.

In this section, all presented spectra were taken at the same output coupling ratio of 14% with measured average output powers in the range between 200 and 250 mW. Therefore, slight fine-tuning was needed when going towards ever higher output coupling rates, most likely because of the additional dispersion introduced by the different OC mirror coatings. In the end, the laser was operated at an OC ratio as high as 34%, and delivered the shortest output pulse durations when 3 mm sapphire and 8 bounces on the TOD mirrors were used.

3.3.3 *Characterization of the Mode-Locked Output*

Pulse Duration and Average Power

At a repetition rate of 64.7 MHz, the diode-pumped system delivered average output powers of over 700 mW (CW) and 500 mW (ML) in a fundamental single-mode operation.[11] At the latter, mode-locked root mean square (RMS) power fluctuations of less than 0.12% were recorded over 3 h when measured with a thermal power meter (Fig. 3.16a). To specify the generated FWHM pulse duration, the dispersion of the 6.35 mm-thick fused-silica output coupler substrate was compensated by a 5 mm thick ZnSe plate placed into the optical beam path. With a self-built second-harmonic frequency-resolved optical gating device (SHG-FROG, Fig. 3.14) and a 10 μm thick BBO crystal used for frequency doubling, the pulses were determined to be as short as 45 fs with a Fourier-transform limited (FTL) duration of 41 fs. The respective FROG error amounted to $0.7 \cdot 10^{-3}$ when retrieving on a grid size of $512 \cdot 512$ points. Correspondingly, this first KLM diode-pumped system is able to deliver pulse energies of up to 8 nJ and peak powers as high as 151 kW at pulse durations shorter than six optical cycles. Compared to the previous demonstration of a SESAM mode-locked diode-pumped Cr:ZnSe laser [41], it represents a ten-fold increase in average power as well as a four-fold reduction in pulse duration. Moreover, it constitutes an over 2.5-times increase in peak power over more mature fiber-laser-pumped Cr:ZnSe oscillators [19, 54].

Slope Efficiency and Output Beam Quality

Similar to Sect. 3.3.1, the mode-locked output beam quality (M^2) was evaluated, and the horizontal and vertical M^2-factors were characterized to be nearly diffraction limited with $M^2 < 1.1$ (Fig. 3.16b). By means of a wideband IR polarizer (*Innovation Photonics*, PGC-5), which consists of two precision polished Germanium plates at Brewster's angle and ensures a high extinction ratio of $2 \cdot 10^5$, the beam was measured to be polarized in the horizontal direction with an extinction ratio of at least 300. Since the oscillator's polarization cleanness is enhanced by the two intra-cavity Brewster plates, this ratio was most likely higher but hampered by the limited sensitivity at low powers for the power meter used.

When plotting the output power P_{out} of a laser for different pump powers P_p, the resulting curve is typically close to a straight line above the lasing threshold P_{th}. The slope efficiency η can then be extracted as the slope of this line: $P_{out} = \eta(P_p - P_{th})$. Varying the pump power through simply adjusting the laser diode current, however, turned out to be not practical in this case. As shown in Fig. 3.15, the CW slope efficiency curve reveals a pronounced break at around 2.5 W of pump power. At this occasion, the diode output beam characteristics were examined in more detail and a

[11] The maximum CW output power corresponds to laser operation at the center of stability zone I, whereas mode-locking was only obtained when moving towards the stability zone edge at decreased CW power levels of ~450 mW.

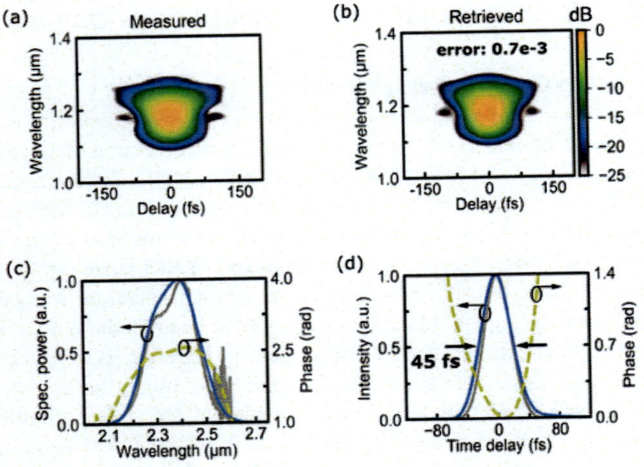

Fig. 3.14 SHG-FROG measurement of the mode-locked laser output. **a–b** Measured and retrieved FROG traces plotted on a normalized logarithmic intensity scale. **c** Retrieved FROG results in the spectral domain, being in good agreement with the measured output spectrum shown in grey. The latter was recorded with a high-resolution Fourier-transform-based laser spectrum analyzer (*Bristol Instruments*, 771B-IR) and the SHG spectra were recorded with a fiber-coupled grating-based spectrometer (*Ocean Optics*, NIR-512). **d** Temporal profile of the retrieved output pulse, indicating a measured FWHM pulse duration of 45 fs (blue) and a FTL of 41 fs (grey). Adapted with permission from [72] ©The Optical Society

Fig. 3.15 Recorded CW slope efficiency when different pump currents were applied to increase the pump power. The break in the measured slope at around 2.5 W is attributed to the changing output beam characteristics of the laser diode

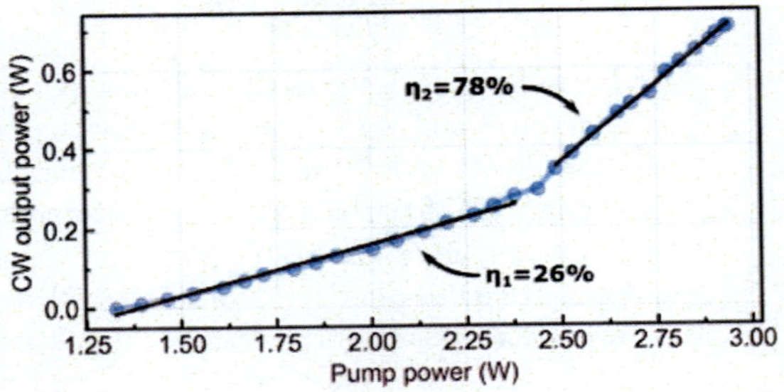

changing behaviour was found for different driving currents (see Sect. 3.3.1). Therefore, the observed break is most likely attributed to the changing mode-matching efficiency between the pump and cavity mode. To verify this assumption and to ensure a constant pump beam quality, a ND filter wheel was placed in front of the pump focusing lens. The CW slope efficiency was then measured at a fixed diode current of 10.2 A. With the incident pump power being varied through filter wheel rotation, a clean slope efficiency of $\eta = 50\%$ was measured together with a slight change in the oscillator output wavelength (see Fig. 3.16c).

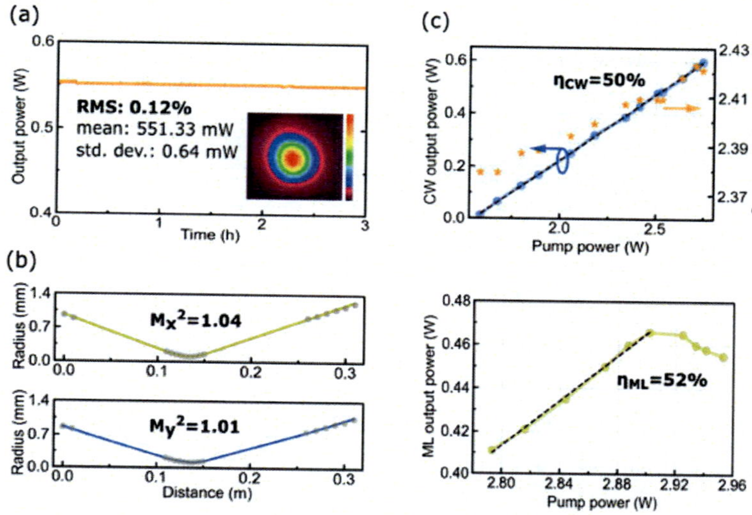

Fig. 3.16 Measured output characteristics and slope efficiencies of *MIRanda1*. **a** Mode-locked output power stability recorded with a thermal power meter (*Coherent*, PS19) and corresponding beam profile measured with a pyroelectric camera (*Ophir Optronics*, Pyrocam III). **b** M^2-measurement performed for the horizontal and vertical axis of the mode-locked output. **c** CW and ML slope efficiency curves. For the CW slope efficiency, a ND filter wheel had to be used in the beam path to ensure constant pump beam quality (see main text)

Finally, to record a slope in mode-locked operation, the filter wheel had to be removed since it caused undesirably high losses. Only for incident pump powers >2.8 W, mode-locking could be initiated, and a slope efficiency of $\eta = 52\%$ was found. Beyond 2.9 W of pump power, however, the ML output power started to drop. Besides a degradation in mode-matching efficiency, it was attributed to the excessive heating of the gain medium, which was cooled mainly from one side only. Moreover, Cr:ZnSe shares similar spectroscopic properties with Cr:ZnS but has an about 1.5-times higher thermo-optic coefficient (see Table 3.1). Therefore, an improved cooling architecture as well as a change to Cr:ZnS as gain medium was implemented in the subsequent two laser systems *MIRanda2* and *MIRanda3*.

Amplitude Noise Measurements

As highlighted in the introductory Sect. 1.2, various applications of femtosecond MIR radiation greatly benefit from a low-noise laser performance. The output of real laser systems exhibits power (and thus intensity) fluctuations, which may be categorized as short-term or long-term [96, 97]. A degrading pump laser source, contaminated optical surfaces, or mechanical misalignment due to thermal effects are potential sources of long-term changes. In this section, however, the focus is placed on short-term fluctuations that happen within a period of less than a second.

At the most fundamental level, the discrete quantum nature of the photons results in a contribution from **shot noise** [98], and arises from the fact that photons are spontaneously emitted from excited atoms or ions. Consequently, these photons arrive at irregular intervals at a photodetector, and some of which are absorbed to generate photocarriers. The resulting photocurrent can then be seen as the sum of all carriers, being characterized by fluctuations in the average detected current level. Other sources of short-term fluctuations include variations in the pump laser intensity, cavity mode beating and hopping, thermal instabilities in the gain medium, mechanical vibrations of optical components, and **relaxation oscillations** (RO). The RO noise is characteristic for most solid-state and semiconductor lasers since small disturbances can result in large output power oscillations when the perturbing frequency components are located at the vicinity of the RO's resonance frequency ν_{ro}. Therefore, excess RO noise can dominate other sources of noise, and signals of interest in spectroscopic applications get easily overwhelmed. Below ν_{ro}, pump laser noise can also severely add to the noise of the oscillator. As found in Sect. 3.2.2, rare-earth doped fiber lasers or solid-state lasers, such as those based on Cr:ZnS/ZnSe, have RO frequencies in the few 10 kHz to a few 100 kHz range. By contrast, semiconductor laser diodes exhibit relaxation oscillations in the GHz-domain. Using laser diodes rather than a fiber laser to pump the Cr:ZnS/ZnSe gain medium is therefore highly advantageous.

For quantifying the short-term noise of a laser system, let us consider its optical output $P(t)$ as a function of time: $P(t) = P_{av} + \Delta P(t)$, with $P_{av} = \langle P \rangle$ being the average output power and $\Delta P(t)$ denoting the output power fluctuations. According to the ISO Standard 11554 [99], a so-called **relative intensity noise (RIN)** can then be defined in the frequency domain:

$$\mathrm{RIN}(\nu) = \frac{\langle \Delta P(\nu)^2 \rangle}{\langle P(\nu)^2 \rangle} \cdot \frac{1}{\Delta \nu}, \tag{3.4}$$

where the mean square power fluctuations are normalized to the mean square power as well as a noise bandwidth $\Delta\nu$. The laser RIN is fundamentally limited by the **shot noise RIN**, which can be formulated as [100, 101]:

$$\mathrm{RIN_{sn}} = \frac{2h\nu_0}{\eta P_{av}} = \frac{2q}{I}, \tag{3.5}$$

where $h\nu_0$ is the photon energy, P_{av} denotes the average optical power, and η is the quantum efficiency of the detector. In terms of electrical units, the generated photocurrent is $I = P_{av} R(\nu_0) = P_{av}\eta q / h\nu_0$ including the detector's responsivity $R(\nu_0)$. Since shot noise is constant with frequency, it may be considered as a white noise source.

Typically, amplitude noise (also referred to as intensity noise) is plotted in relative terms to allow measurements taken under different operation conditions to be compared in a straightforward manner. The noise measurements presented in this thesis are therefore plotted in **logarithmic units,**[12] and normalized to the carrier (DC level) [102, 103]:

$$\mathrm{RIN}(\nu)\ [\mathrm{dBc/Hz}] = \mathrm{RIN}(\nu)\ [\mathrm{dBm}] - 10 \cdot \log(\Delta\nu) - 10 \cdot \log\left(\frac{U_{DC}^2}{R_{load}} \frac{1}{1\ \mathrm{mW}}\right). \tag{3.6}$$

In the experiments, RIN(ν)[dBm] is the noise being measured by a radio-frequency spectrum analyzer at a certain resolution bandwidth (RBW) $\Delta\nu$. Since the power fluctuations are detected with a fast photodetector, and optical power is transferred into electrical power ($P_{el} \propto P_{opt}^2$), the measured RMS electrical power fluctuations were normalized to the DC voltage U_{DC}. The latter is proportional to the average optical power incident on the photodetector [102], and was determined with the aid of a DC-coupled oscilloscope. Integrating the relative intensity noise over a stated frequency interval, a **root mean square (RMS) value** can be obtained [102, 103]

$$\mathrm{RIN}\ [\%\ \mathrm{RMS}] = \sqrt{\int_{\nu_1}^{\nu_2} \mathrm{RIN}(\nu)\ [\mathrm{dBc/Hz}]\ d\nu}. \tag{3.7}$$

For the measurements, the amplitude noise characteristics of the mode-locked pulse trains were carefully analyzed with a photodetector attached to a high dynamic-range radio-frequency (RF) spectrum analyzer (see Fig. 3.17a). Even though the selection of appropriate photodetectors was rather limited to record the laser output centered at 2.3–2.4 μm, a biased InGaAs detector was found, being sensitive at the 0.9–2.6 μm wavelengths. Also, a high signal-to-noise ratio of ~90 dB could be achieved for all measurements using a 50 Ω termination for impedance matching. To avoid cross-coupling of beam-pointing fluctuations into the relative intensity noise (RIN) measurements [104], the beam was focused onto the detector area (⌀ 1 mm). Mixing artefacts generated by the RF spectrum analyzer were reduced by filtering the electronic DC signal at the analyzer input. Also, a neutral density (ND) filter wheel was placed into the beam path to keep the optical signal just below the detector's saturation level.[13] By setting the measurement step size to 0.5 · RBW, a good overlap of the data points was ensured, and a better estimate of the mean-squared quantities was obtained when averaging over 20 traces taken within a certain RBW.

The laser amplitude noise was then analyzed in the low-frequency range between 20 Hz to 1 MHz, also referred to as the baseband [103]. Here, an average optical power of less than 1 mW and a corresponding DC voltage of 47 mV were sufficient to obtain high signal strengths without saturation. As defined in formula (3.6), the amplitude noise of the direct oscillator output was measured first and plotted in relative units of

---

[12] Conversion formula: RIN [dB] $= 10 \cdot \log_{10}(\mathrm{RIN})$.

[13] When the photodetector signal was monitored with a DC-coupled oscilloscope (*Teledyne LeCroy*, HDO4034), a saturation of the detector manifested itself in a deformation of the resolved pulse train shape.

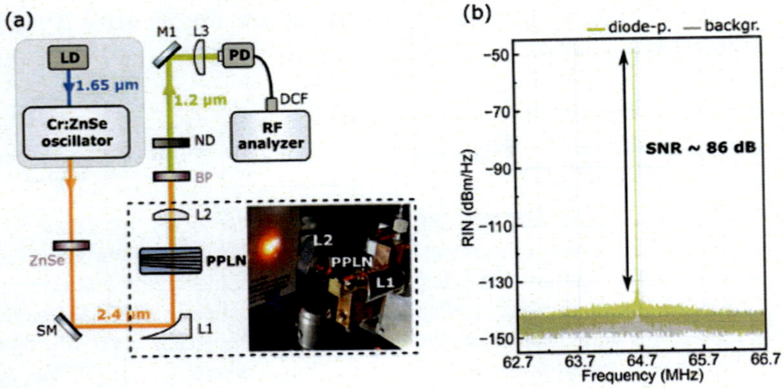

Fig. 3.17 **a** Setup for measuring the mode-locked amplitude noise of the frequency-doubled oscillator output. LD: laser diode, ZnSe: 5 mm AR-coated substrate for temporal pulse re-compression, SM: silver mirror, L1: parabolic silver mirror (f = 7 mm), PPLN: periodically-poled lithium niobate crystal for second-harmonic generation (*HC Photonics*, SC1705100107, fan-out grating $\Lambda = 21$–35 μm), L2: N-BK7 lens (f = 15 mm), BP: band-pass filter at 1.2 μm (*Thorlabs*, FB1200-10), ND: neutral density filter wheel, M1: dielectric mirror (*Thorlabs*, BB1-E03), L3: AR-coated UVFS lens (f = 50 mm), PD: biased InGaAs photodetector (*Thorlabs*, DET10D2), DCF: DC block filter (*Thorlabs*, EF500), RF analyzer: radio-frequency spectrum analyzer (*Agilent*, E4447A). The SHG generation in the PPLN crystal was captured with a camera. **b** RF spectrum of the carrier region, where the grey curve indicates the background noise floor of the measurement. High signal-to-noise ratios close to 90 dB were obtained for the frequency-doubled output, comparable to the measurements at the fundamental wavelengths

dBc/Hz. Similarly, the background noise of the measurement devices was obtained when blocking the optical signal into the detector. Even though the photodetector was already operating at the limit of the linear regime, the laser amplitude noise was surprisingly close to the background noise floor. Therefore, a periodically poled lithium niobate (PPLN) crystal was used to take additional data on the frequency-doubled output after second-harmonic generation (see Fig. 3.17). The quadratic dependence of the SHG frequency conversion process on the optical input power should enhance the RIN by a factor of two [102, 105], corresponding to an increase of 6 dB in electrical power. Moreover, to only measure the noise centered at 1.2 μm and block contributions from the fundamental and other wavelengths, a suitable bandpass filter was placed right behind the crystal. Similar to the measurements at the fundamental wavelength, it resulted in a DC-voltage of about 47 mV, which corresponds to an optical power of $P_{av} = U_{DC}/(R(\lambda)R_{load}) = 47mV/(0.39\text{A/W} \cdot 50\Omega) \approx 2.4$ mW at 1.2 μm. In Fig. 3.17b, the RF spectrum of the carrier region is depicted, revealing a correspondingly high signal-to-noise ratio of close to 90 dB similar to the measurements at the fundamental output.

The measurement results are shown in Fig. 3.18 (green curves) and were obtained for two different resolution bandwidths and frequency spans. At the very low frequencies up to about 200 Hz, the noise spectrum exhibits a f^{-1}-behaviour, as it has been observed as resistance fluctuations of metals or semiconductors [106–109]. Beyond 1 kHz, a more "plateau"-like noise structure without clear noise peak is seen instead, which falls to the background noise near 2 MHz. This can be attributed to the short fluorescence lifetime of Cr:ZnSe, which is on the order of a few microseconds (see Table 3.1). As calculated in Sect. 3.2.2, the relaxation oscillation frequency of Cr:ZnS/ZnSe is located at hundreds of kHz, and any noise at higher frequencies will be damped because of the filtering effect of the gain medium. Pump laser noise at lower frequencies, however, can be transferred to oscillator noise, and it is worthwhile to note that the usual noise peak originating from relaxation oscillations in the rare-earth doped (erbium/thulium) pump fiber lasers at hundreds of kHz [54] is absent in this directly diode-pumped system. In fact, a similar plateau-like structure has also been observed for diode-pumped Ti:sapphire oscillators, and was attributed to the relaxation oscillations of the Ti:sapphire laser itself [102, 103].

To further investigate the transfer of pump noise, the pump source of *MIRanda1* was changed to a commercial Erbium fiber laser (*Bktel*, HPFL-370-1567-FCAPC) that has been connected to the same power supply as the InP pump diode. The direct noise output of this fiber laser has already been shown in Fig. 3.4 when comparing the amplitude noise of different available pump laser sources. It should be noted, however, that this laser is not meant to represent state-of-the-art low-noise fiber lasers, and it was also not optimized for low-optical-noise performance. Nevertheless, it serves to illustrate the effect of a noisy pump source on the output noise characteristics of a Cr:ZnS/ZnSe laser. The respective measurements (blue curves, Fig. 3.18) were now taken at the fundamental wavelength without further noise-enhancing frequency doubling. The resulting amplitude noise of the fiber-laser-pumped Cr:ZnSe laser is now notably higher, with spectral features such as the noise peak at 40 kHz clearly originating from fiber laser noise being transferred to oscillator noise. A similar observation is also presented in Ref. [54].

Finally, for a quantitative comparison between the diode- and fiber-pumped Cr:ZnSe laser, the integrated RMS noise was calculated for the measured frequency span from 20 Hz to 1 MHz according to formula (3.7). The lower integration limit was set to 20 Hz since the RF spectrum analyzer itself exhibited undesirably high noise below 20 Hz, even when the signal input was simply terminated with a 50 Ω resistor. The integrated RMS noise level was then calculated to be 0.095% for the fiber-pumped system, with 0.043% being attributed to the background noise floor. By contrast, the integrated RIN of the directly diode-pumped laser system was measured to be 0.068% after frequency doubling, and should be even lower for the fundamental wavelengths. This clearly illustrates the highly promising low-noise performance and potential of directly diode-pumped Cr:ZnS/ZnSe lasers. Since the calculated shot noise level was at -154 dBc/Hz and the background noise floor was measured to be -130 dBc/Hz, further optimization in the noise measurement setup was required for future noise studies. In the noise measurements for *MIRanda2* and *MIRanda3*, an improved measurement setup will be presented, eventually revealing the mode-locked noise of the diode-pumped system directly at the 2–3 μm wavelengths.

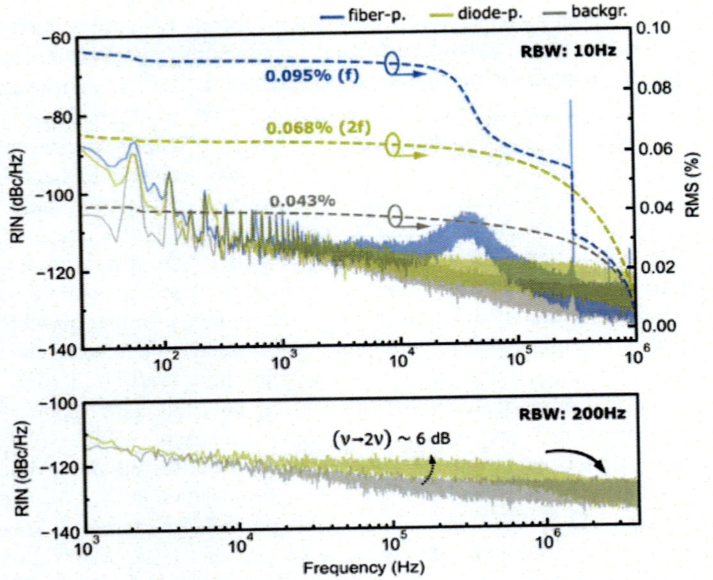

Fig. 3.18 Amplitude noise measurements and integrated RMS noise of the frequency-doubled (2f) output of *MIRanda1* (green, diode-pumped). For comparison, the data of the fundamental (f) output is added when the oscillator was pumped by a commercial fiber laser instead (blue). Both the fiber laser and laser diode were driven by the same low-noise current source (*Delta Elektronika*, SM 70-AR-24). The RIN was also measured in a broader span up to 4 MHz, and illustrates the drop in noise signal to the background noise floor at frequencies around 2 MHz—attributed to the filtering effect of the gain medium

3.3.4 The Success of Powerful Diode-Pumped Operation

At first sight, it sounds remarkable that high continuous-wave average powers and soft-aperture KLM—both achieved in a clean spatial TEM_{00} output mode—can be supported by a solid-state laser cavity which is pumped by the non-diffraction-limited output of a single wide stripe laser diode. According to the conventional understanding [110], the self-focusing inside the gain crystal allows the cavity mode to be smaller in size when the oscillator is generating ultrashort pulses. This results in an improved overlap between the pump and cavity mode, typically yielding higher output powers in mode-locked operation. In the case of *MIRanda1*,[14] however, the pump beam diameter inside the Cr:ZnSe crystal roughly equals the laser mode in the fast axis direction, but is much larger in the slow axis direction. Therefore, it cannot per se provide a higher gain for a shrinking cavity mode. Experimentally, soft-aperture mode-locking was successfully implemented together with high average powers in

[14] This is also true for the other two laser systems that will be presented in this thesis.

CW and ML laser operation, and some potential reasons are listed below to help explain the observed efficiency and stability (see also Refs. [69, 70]).

Mode Diameters in the Crystal:

In Sect. 3.3.1, an ABCD-matrix simulation of the experimental setup was presented for continuous-wave (0 MW) and pulsed (0.5 MW) operation, and an enlarged section of the focal region is now shown in Fig. 3.19a. Interestingly, the pulsed laser mode diameter inside the gain crystal is about 30–32 μm (15–18%) larger than the CW mode in the sagittal and tangential direction for the given set of parameters. By a proper adjustment of the pump focus position, the pump mode is therefore able to act as a soft aperture: because the cavity mode in the Cr:ZnSe crystal increases upon mode-locked operation, it allows for a slightly improved overlap with the pump beam and thus increased laser gain. This explanation is supported by the experimental observation that the oscillator output power increased by a factor of 10–15% upon mode-locking.

Pump Focusing Characteristics:

Another effect is depicted in Fig. 3.19b: after the cavity and pump setup parameters had been optimized for highest CW output power and best KLM performance—with the corresponding output results shown in the previous section—the actual focusing characteristics of the pump beam were analyzed in more detail. For practical reasons, the pump beam was deflected and sent into a separate setup, mimicking the beam path up to the focusing region. With the scanning slit profiler in place, the $D_{4\sigma}$ beam diameters were measured around the focus for both beam directions. Even though a slight difference was observed when operating the diode at two different pump currents, which is attributed to the changing M^2-value and beam divergence, it reveals a clear discrepancy of $\Delta > 1$ mm in the actual focus position of the fast and slow axis. Indeed, this might be more favorable with respect to mode-matching and mode-locking since it helps to increase the effective Rayleigh length of the pump beam. Consequently, the resulting improvement in the overlap volume between the fundamental cavity beam and the pump beam allows for higher average output powers.[15]

Output Power Versus Cavity/Pump Mode Ratio:

Under certain conditions, it might actually be advantageous to have a pumped area that is larger than the one spanned by the fundamental laser mode in the gain medium. In fact, Chap. 7 of Ref. [46] presents a relationship between the continuous-wave output power P_{out} and the pump power P_p for a given Gaussian pump beam radius w_p and cavity mode size w_0. For the sake of simplicity, it is assumed that both beam radii remain nearly constant over the crystal length. Also, when assuming a Gaussian pump

[15] Note that the position of the cylindrical pump lenses was optimized to reach the highest CW output power in single-mode laser operation, and was also found to be ideal for mode-locked operation.

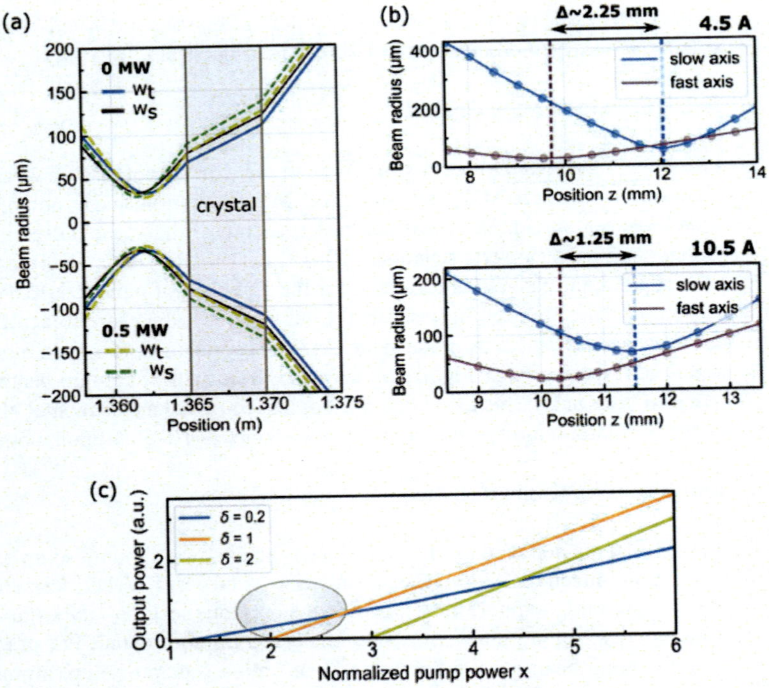

Fig. 3.19 **a** Simulated tangential and sagittal beam radii of the CW and ML cavity mode in the focusing region around the crystal position. The calculations are based on (non)linear ABCD-matrix simulations for two different intra-cavity peak powers. **b** Measured pump beam radii around the focus when the diode is pumped at low current (4.5 A) as well as at the current that was finally used for operation of *MIRanda1* (10.5 A). **c** Theoretical laser output power versus normalized pump power for different relative cavity and pump mode sizes ($\delta = (w_0/w_p)^2$)

distribution, a spatial dependence of the gain on radial and longitudinal coordinates is included in the mathematical calculations. For the analysis, a normalized output power $y = P_{out}/P_S$ and a normalized pump power $x = P_p/P_{mth}$ are defined, where P_S and P_{mth} are the saturation power and the minimum threshold power according to:

$$P_S = \frac{\gamma_2 \pi w_0^2}{4} I_S, \qquad P_{mth} = \left(\frac{\gamma}{\eta_p}\right)\left(\frac{h\nu_p}{\tau_2}\right)\left(\frac{\pi w_p^2}{2\sigma_{em}}\right) \quad \text{for } w_0 \ll w_p. \tag{3.8}$$

These quantities depend on the single pass loss γ of the cavity, which includes the logarithmic loss γ_2 due to the transmission of the output coupling mirror, as well as the pump efficiency η_p, the saturation intensity I_S, the upper-state lifetime τ_2, and the emission cross-section σ_{em}. After some mathematical calculations (see Appendix E in Ref. [46]), the resulting relation between x and y can be written as:

$$\frac{1}{x} = \int_0^1 \frac{t^\delta}{1+yt} dt, \tag{3.9}$$

with $\delta = (w_0/w_p)^2$. This equation was solved analytically for three different cases, and the results are plotted in Fig. 3.19c: (i) an equal pump and laser mode size ($\delta = 1$), (ii) a slightly larger laser mode than pump mode ($\delta = 2$), and (iii) a pump mode being almost twice as large as the cavity mode ($\delta = 0.2$). Even though lasers are typically operated at values $x \gg 1$, this does not necessarily hold for the rather low pump powers considered in this thesis, and $x \approx 2$ is a good approximation then. As indicated by the grey circle in Fig. 3.19c, the lasing threshold could not only be reduced for $\delta < 1$, but also the output power is possibly larger when operating near the minimum threshold (around $x \approx 2$)—even though the CW and ML cavity mode are smaller than the pump mode.

3.4 Summary

With the non-diffraction limited output of a single-emitter laser diode, a directly diode-pumped few-cycle Cr^{2+}-doped II-VI oscillator was demonstrated for the first time. Delivering a mode-locked output power of 500 mW and pulse durations of 45 fs, it constitutes a sixty-fold increase in peak power over the previous record based on SESAM mode-locking [41] (see Table 3.4), and a more than 2.5-times improvement when compared to more mature fiber-laser-pumped versions [19, 54]. Moreover, noise measurements confirmed an excellent amplitude noise performance, with a measured integrated RIN of less than 0.068% RMS (20 Hz–1 MHz) for the frequency-doubled output.

In Sect. 3.1, Cr-doped II-VI crystals were presented to be highly attractive candidates for the direct generation of ultrashort femtosecond pulses in the 2–3 μm spectral range. With favorable material and spectroscopic characterstics similar to Ti:sapphire—one of the most popular near-infrared laser materials—they are commonly referred to as the "Ti:sapphire of the mid-infrared" [4]. As presented in Sect. 3.3, pulse durations shorter than six optical cycles at 2.4 μm were directly generated via soft-aperture Kerr-lens mode-locking. The latter, however, relies on the action of a Kerr lens inside the gain crystal, which renders the approach of direct diode-pumping highly challenging for ultrafast and stable mode-locking. When careful considerations regarding laser diode selection, mode-matching as well as the overall mechanical design (Sect. 3.2) are made, the highly divergent and elliptic output beam of a low-brightness single-emitter laser diode can be successfully implemented in an ultrafast KLM-based Cr:ZnS/ZnSe laser. *MIRanda1* thus represents a

Table 3.4 Comparison of the output characteristics of the previous record versus the first demonstration of diode-pumped Cr^{2+}-doped II-VI femtosecond oscillators. τ_0: measured output pulse duration, E_0: pulse energy, P_0: peak power. For calculating the peak power of solitonic pulses ($sech^2$-shape), the following formula was used: $P_0 = 0.88 \cdot P_{ML}/(\tau_0 \cdot \nu_{rep})$ [111]

	Previous record [41]	This work [72]
ML mechanism	**SESAM ML**	**Kerr-lens ML**
Gain medium	Cr:ZnSe (4 mm)	Cr:ZnSe (5 mm)
P_{pump}	3.5 W	3 W
λ_{pump}	1.55 μm	1.65 μm
P_{CW}	340 mW	700 mW
P_{ML}	50 mW	500 mW
τ_0	180 fs	45 fs
ν_{rep}	100 MHz	64.7 MHz
OC ratio	5%	34%
E_0	**0.50 nJ**	**7.7 nJ**
P_0	**2.44 kW**	**151 kW**

new class of few-cycle, low-noise, and table-top laser systems, which can serve as a highly attractive and accessible route to generate coherent radiation in the 2–3 μm range, and also in the deeper MIR (see Chap. 5).

References

1. Haight R, Carr AV (2018) Industrial applications of ultrafast lasers, vol 11. Materials and energy. WSPC
2. Page RH et al (1997) Cr^{2+}-doped zinc chalcogenides as efficient widely tunable mid-infrared lasers. IEEE J Quantum Electron 33(4):609–619
3. DeLoach LD, Page RH, Wilke D, Payne SA, Krupke WF (1996) Transition metal-doped zinc chalcogenides: spectroscopy and laser demonstration of a new class of gain media. IEEE J Quantum Electron 32(6):885–895
4. Payne SA, Krupke WF (1996) A glimpse into the laser-crystal ball. Opt Photonics News 7(8):31–35
5. Kärtner FX (ed) (2004) Few-cycle laser pulse generation and its applications, vol 95. Springer
6. Peppers J, Konak T, Martyshkin DV, Fedorov V, Mirov SB (2014) Spectroscopy and mid-IR lasing of Cr^{2+} ions in ZnSe/ZnS crystals under visible excitation. In: Proceedings of SPIE - the international society for optical engineering, 8959
7. Schwendimann P, Sigmund E, Zeile K (1988) Model for laser action in vibronic systems. Phys Rev A 37(8):3018–3027
8. Kück S (2001) Laser-related spectroscopy of ion-doped crystals for tunable solidstate lasers. Appl Phys B 72:515–562
9. Sorokina IT, Vodopyanov KL (2003) Solid-state mid-infrared laser sources, vol 89. Topics in applied physics. Springer
10. Sorokina IT (2004) Cr^{2+}-doped II-VI materials for lasers and nonlinear optics. Opt Mater 26(4):395–412

11. Sorokina IT et al (2002) Continuous-wave tunable Cr^{2+}:ZnS laser. Appl Phys B: Lasers Opt 74(6):607–611
12. Landolt-Börnstein (1999) Numerical data and functional relationships in science and technology - new series, vol 41. Semiconductors, subvolume B. Springer
13. Moulton PF (1986) Spectroscopic and laser characteristics of Ti:Al_2O_3. J Opt Soc Am B 3(1):125–133
14. Kusuma HH, Ibrahim Z, Saidin MK (2011) Optical energy gap of Ti:Al_2O_3 single crystals. J Appl Sci 11(5):888–891
15. Sorokina IT, Sorokin E (2015) Femtosecond Cr^{2+}-based lasers. IEEE J Sel Top Quantum Electron 21(1):1601519
16. Mirov S et al (2018) Frontiers of mid-IR lasers based on transition metal doped chalcogenides. IEEE J Sel Top Quantum Electron 24(5):1–29
17. Mirov SB, Fedorov VV, Martyshkin DV, Moskalev LS, Mirov MS, Gapontsev VP (2011) Progress in mid-IR Cr^{2+} and Fe^{2+} doped II-VI materials and lasers [Invited]. Opt Mater Express 1(5)
18. Baudrier-Raybaut M, Haïdar R, Kupecek P, Lemasson P, Rosencher E (2004) Random quasi-phase-matching in bulk polycrystalline isotropic nonlinear materials. Nature 432:374–376
19. Vasilyev S, Mirov M, Gapontsev V (2014) High power Kerr-Lens mode- locked femtosecond mid-IR laser with efficient second harmonic generation in polycrystalline Cr^{2+}:ZnS and Cr^{2+}:ZnSe. In: Advanced solid state lasers (ASSL) AM3A.3
20. Vasilyev S, Moskalev I, Mirov M, Smolski V, Mirov S, Gapontsev V (2016) Mid-IR Kerr-lens mode-locked polycrystalline Cr:ZnS and Cr:ZnSe lasers with intracavity frequency conversion via random quasi-phase-matching. In: Proceedings of the SPIE 9731, 97310B
21. Cizmeciyan MN, Cankaya H, Kurt A, Sennaroglu A (2009) Kerr-lens mode-locked femtosecond Cr^{2+}:ZnSe laser at 2420 nm. Opt Lett 34(20):3056–3058
22. Sorokin E, Sorokina IT (2009) Ultrashort-pulsed Kerr-lens modelocked Cr:ZnSe laser. In: CLEO Europe - EQEC CF1.3
23. Nagl N, Mak K, Wang Q, Pervak V, Krausz F, Pronin O (2019) Efficient femtosecond mid-infrared generation based on a Cr:ZnS oscillator and step-index fluoride fibers. Opt Lett 44(10):2390–2393
24. Wang Q et al (2019) Broadband mid-infrared coverage (2-17 μm) with few-cycle pulses via cascaded parametric processes. Opt Lett 44(10):2566–2569
25. Vasilyev S et al (2019) Super-octave longwave mid-infrared coherent transients produced by optical rectification of few-cycle 2.5-μm pulses. Optica 6(1):111–114
26. Smolski VO, Vasilyev S, Schunemann PG, Mirov SB, Vodopyanov KL (2015) Cr:ZnS laser-pumped subharmonic GaAs optical parameteric oscillator with the spectrum spanning 3.6-5.6 μm. Opt Lett 40(12):2906–2908
27. Mirov SB, Fedorov VV, Graham K, Moskalev IS, Badikov VV, Panyutin V (2002) Erbium fiber laser-pumped continuous-wave microchip Cr^{2+}:ZnS and Cr^{2+}:ZnSe lasers. Opt Lett 27(11):909–911
28. Wagner GJ et al (1999) Continuous-wave broadly tunable Cr^{2+}:ZnSe laser. Opt Lett 24(1): 19–21
29. Podlipensky AV et al (2001) Efficient laser operation and continuous-wave diode pumping of Cr^{2+}:ZnSe single crystals. Appl Phys B 72(2):253–255
30. Moskalev I, Fedorov V, Mirov S, Berry PA, Schepler K (2009) 12-Watt CW polycrystalline Cr^{2+}:ZnSe laser pumped by Tm-fiber laser. In: Advanced solid-state photonics. OSA
31. Wu Y, Zhou F, Larsen EW, Zhuang F, Yin Y, Chang Z (2020) Generation of few-cycle multi-millijoule 2.5 μm pulses from a single-stage Cr^{2+}:ZnSe amplifier. Sci Rep 10(7775)
32. Podlipensky AV et al (1999) Pulsed laser operation of diffusion-doped Cr^{2+}:ZnSe. Opt Commun 167(1–6):129–132
33. Graham K et al (2004) Pulsed mid-IR Cr^{2+}:ZnS and Cr^{2+}:ZnSe lasers pumped by Raman-shifted Q-switched neodymium lasers. Quantum Electron 34(1):8–14
34. Sennaroglu A, Konca AO, Pollock CR (2000) Continuous-wave power performance of a 2.47-μm Cr^{2+}:ZnSe laser: experiment and modeling. IEEE J Quantum Electron 36(10): 1199–1205

35. Champert PA, Popov SV, Taylor JR (2002) Efficient lasing of Cr^{2+}:ZnSe at 2.2 μm pumped by all-fibre-format seeded Raman source. Electron Lett 38(10):448–449
36. Mirov SB, Fedorov VV, Martyshkin D, Moskalev IS, Mirov M, Vasilyev S (2015) Progress in mid-IR lasers based on Cr and Fe-doped II-VI chalcogenides. IEEE J Sel Top Quantum Electron 21(1):292–310
37. Ebrahim-Zadeh M, Sorokina IT (eds) (2008) Mid-infrared coherent sources and applications. Springer
38. Page RH, Skidmore JA, Schaffers KI, Beach RJ, Payne SA, Krupke WF (1997) Demonstrations of diode-pumped and grating-tuned ZnSe : Cr^{2+} lasers. OSA Trends Opt Photonics Ser Adv Solid State Lasers 10:208–210
39. Moskalev IS, Fedorov VV, Mirov SB (2010) InP diode-pumped Cr:ZnS and Cr:ZnSe highly-efficient, widely-tunable, mid-IR lasers. In: Proceedings of SPIE 7578
40. Wagner GJ, Schober AM, Bennett GT, Bruns DL, Marquardt JH, Carrig TJ (2012) Multi-watt broadly-tunable diode-pumped Cr:ZnSe laser. In: Conference on lasers and electro-optics 2012. CTu2D.2. Optical society of America. OSA technical digest
41. Slobodtchikov E, Moulton PF (2012) Progress in ultrafast Cr:ZnSe lasers. In: Lasers, sources, and related photonic devices. AW5A.4. Optical society of America. OSA technical digest
42. Sorokin E, Tolstik N, Schaffers KI, Sorokina IT (2012) Femtosecond SESAM-modelocked Cr:ZnS laser. Opt Express 20(27), 28947–28952
43. Zhang J et al (2015) 49-fs Yb:YAG thin-disk oscillator with distributed Kerr-lens mode-locking. In: 2015 European conference on lasers and electro-optics - European quantum electronics conference (CLEO/Europe-EQEC) PD.A.1
44. Tokurakawa M et al (2008) Diode-pumped 65 fs Kerr-lens mode-locked Yb^{3+}:Lu_2O_3 and nondoped Y_2O_3 combined ceramic laser. Opt Lett 33(12):1380–1382
45. Phelan R et al Optical fiber applications. In: Huerta-Cuellar G, Imani R (eds) Mid-infrared InP-based discrete mode laser diodes. IntechOpen
46. Svelto O (2009) Principles of lasers, 5th edn. Springer, Berlin
47. Diehl R (ed) (2000) High-power diode lasers - fundamentals, technology, applications, vol 78. Springer, Berlin
48. Sun H (2015) A practical guide to handling laser diode beams, 2nd edn. Springer briefs in physics. Springer, Berlin
49. Kot M, Žd'ánský K (1992) Measurement of radiative and nonradiative recombination rate in InGaAsP-InP LED's. IEEE J Quantum Electron 28(8):1746–1750
50. Piprek J, White JK, SpringThorpe AJ (2002) What limits the maximum output power of long-wavelength AlGaInAs/InP laser diodes. IEEE J Quantum Electron 38(9):1253–1259
51. Qing K et al (2015) Fabrication and optimization of 1.55-μm InGaAsP/InP high-power semiconductor diode laser. J Semicond 36(9)
52. Anthony E (1986) Siegman. University Science Books, Lasers
53. Harb CC, Ralph TC, Huntington EH, McClelland DE, Bachor H-A, Freitag I (1997) Intensity-noise dependence of Nd:YAG lasers on their diodelaser pump source. J Opt Soc Am B 14(11):2936–2945
54. Wang Y, Fernandez TT, Coluccelli N, Gambetta A, Laporta P, Galzerano G (2017) 47-fs Kerr-lens mode-locked Cr:ZnSe laser with high spectral purity. Opt Express 25(21):25193–25200
55. Yang S, Ponomarev EA, Bao X (2005) Experimental study on relaxation oscillation in a detuned FM harmonic mode-locked Er-doped fiber laser. Opt Commun 245(1–6):371–376
56. Cheng Y, Kringlebotn JT, Loh WH, Laming RI, Payne DN (1995) Stable single-frequency traveling-wave fiber loop laser with integral saturable-absorber-based tracking narrow-band filter. Opt Lett 20(8):875–877
57. Zhang W-N et al (2012) A compact low noise single frequency linearly polarized dbr fiber laser at 1550 nm. Chin Phys Lett 29(8):084205
58. Zhu G, Wang Q, Dong H, Dutta NK (2004) Fiber laser relaxation oscillation noise suppression through the use of self-biased intracavity loss modulator. J Appl Phys 96(4):1790–1793
59. Wang L, Xu XG (2015) Scattering-type scanning near-field optical microscopy with reconstruction of vertical interaction. Nat Commun 6:8973

60. Wang H, Wang L, Jakob DS, Xu XG (2018) Tomographic and multimodal scattering-type scanning near-field optical microscopy with peak force tapping mode. Nat Commun 9(2005)
61. Hall DG (1981) Optimum mode size criterion for low-gain lasers. Appl Opt 20(9):1579–1583
62. Kopf D, Aus der Au J, Keller U, Bona GL, Roentgen P (1995) 400-mW continuous-wave diode-pumped Cr:LiSAF laser based on a power-scalable concept. Opt Lett 20(17):1782–1784
63. Kopf D et al (1997) High-average-power diode-pumped femtosecond Cr:LiSAF lasers. Appl Phys B 65:235–243
64. Krausz F, Zehetner J, Brabec T, Wintner E (1991) Elliptic-mode cavity for diode-pumped lasers. Opt Lett 16(19):1496–1498
65. Laporta P, Brussard M (1991) Design criteria for mode size optimization in diode-pumped solid-state lasers. IEEE J Quantum Electron 27(10):2319–2326
66. Salin F, Squier J (1991) Geometrical optimization of longitudinally pumped solid-state lasers. Opt Commun 86:397–400
67. Sanchez F, Brunel M, Aït-Ameur K (1998) Pump-saturation effects in end-pumped solid-state lasers. J Opt Soc Am B 15(9):2390–2394
68. Boyd R (2008) Nonlinear optics, 3rd edn. Academic Press
69. Backus S, Kirchner M, Durfee C, Murnane M, Kapteyn H (2017) Direct diode-pumped Kerr Lens 13 fs Ti:sapphire ultrafast oscillator using a single blue laser diode. Opt Express 25(11):12469–12477
70. Durfee CG et al (2012) Direct diode-pumped Kerr-lens mode-locked Ti:sapphire laser. Opt Express 20(13):13677–13683
71. Gürel K et al (2015) Green diode-pumped femtosecond Ti:sapphire laser with up to 450 mW average power. Opt Express 20(23):30043–30048
72. Nagl N, Gröbmeyer S, Pervak V, Krausz F, Pronin O, Mak K (2019) Directly diode-pumped, Kerr-lens mode-locked, few-cycle Cr:ZnSe oscillator. Opt Express 27(17):24445–24454
73. Lasers and laser-related equipment - Test methods for laser beam widths, divergence angles and beam propagation ratios. ISO Standard 11146 (2005)
74. Siegman AE (1993) Defining, measuring, and optimizing laser beam quality. In: Proceedings of SPIE 1868
75. Wang X, Hu G, Li Y, Yao J (1994) Numerical analysis of beam parameters and stability regions in a folded or ring cavity. J Opt Soc Am A 11(8):2265–2270
76. Tushar SNB, Nath SD, Akhtar J, Reja MI (2018) Modeling and analysis of Z folded solid state laser cavity with two curved mirrors. Int J Microw Opt Technol 13(3):244–253
77. Mirov SB et al (2003) Diode and fibre pumped Cr^{2+}:ZnS mid-infrared external cavity and microchip lasers. IEE Proc Optoelectron 150(4):340–345
78. Brabec T, Curley F, Spielmann C, Wintner E, Schmidt JA (1993) Hard-aperture Kerr-lens mode locking. J Opt Soc Am B 10(6):1029–1034
79. Tolstik N, Sorokoin E, Sorokina IT (2013) Kerr-lens mode-locked Cr:ZnS laser. Opt Lett 38(3):299–301
80. Haus HA, Fujimoto JG, Ippen EP (1992) Analytic theory of additive pulse and Kerr Lens mode locking. IEEE J Quantum Electron 28(10):2086–2096
81. Szipöcs R, Ferencz K, Spielmann C, Krausz F (1994) Chirped multilayer coatings for broadband dispersion control in femtosecond lasers. Opt Lett 19(3):201–203
82. Ell R et al (2001) Generation of 5-fs pulses and octave-spanning spectra directly from a Ti:sapphire laser. Opt Lett 26(6):373–375
83. Morgner U et al (2000) Ultrabroadband double-chirped mirror pairs covering one octave of bandwidth. In: Conference on lasers and electro-optics, CThE5
84. Kärtner FX et al (1997) Design and fabrication of double-chirped mirrors. Opt Lett 22(11):831–833
85. Szipöcs R et al (2000) Negative dispersion mirrors for dispersion control in femtosecond lasers: chirped dielectric mirrors and multi-cavity Gires-Tournois interferometers. Appl Phys B 70:S51–S57
86. Golubovic B et al (2000) Double Gires-Tournois interferometer negative-dispersion mirrors for use in tunable mode-locked lasers. Opt Lett 25(4):275–277

87. Pervak V, Amotchkina T, Wang Q, Pronin O, Mak K, Trubetskov M (2019) 2/3 octave Si/SiO_2 infrared dispersive mirrors open new horizons in ultrafast multilayer optics. Opt Express 27(1):55–62
88. Herrmann J (1994) Theory of Kerr-lens mode locking: role of self-focusing and radially varying gain. J Opt Soc Am B 11(3):498–512
89. Brabec T, Spielmann Ch, Curley PF, Krausz F (1992) Kerr lens mode locking. Opt Lett 17(18):1292–1294
90. Kogelnik H, Li T (1966) Laser beams and resonators. Appl Opt 5(10):1550–1567
91. Agrawal GP (2009) Nonlinear fiber optics, 4th edn. Elsevier (Singapore) Pte Ltd
92. Spielmann C, Curley PF, Brabec T, Krausz F (1994) Ultrabroadband femtosecond lasers. IEEE J Quantum Electron 30(4):1100–1114
93. Cundiff ST, Soto-Crespo JM, Akhmediev N (2002) Experimental evidence for soliton explosions. Phys Rev Lett 88(7):073903
94. Müller A et al (2011) Frequency-doubled DBR-tapered diode laser for direct pumping of Ti:sapphire lasers generating sub-20 fs pulses. Opt Express 19(13):12156–12163
95. Dennis ML, Duling IN III (1994) Experimental study of sideband generation in femtosecond fiber lasers. IEEE J Quantum Electron 30(6):1469–1477
96. Koechner W (2006) Solid-state laser engineering. Springer
97. Agrawal GP, Dutta NK (1993) Semiconductor lasers. In: Springer US. Chapter Noise characteristics, pp. 261–269
98. Henry CH, Kazarinov RF (1996) Quantum noise in photonics. Rev Mod Phys 68(3):801–853
99. Optics and photonics - Lasers and laser-related equipment - Test methods for laser beam power, energy and temporal characteristics. ISO Standard 11554 (2017)
100. Paschotta R et al (2005) Relative timing jitter measurements with an indirect phase comparison method. Appl Phys B 80:185–192
101. Obarski GE, Splett JD (2001) Transfer standard for the spectral density of relative intensity noise of optical fiber sources near 1550 nm. J Opt Soc Am B 18(6):750–761
102. Tawfieq M, Kragh Hansen A, Bjarlin Jensen O, Marti D, Sumpf B, Andersen PE (2018) Intensity noise transfer through a diode-pumped titanium sapphire laser system. IEEE J Quantum Electron 54(1):1700209
103. Scott RP, Langrock C, Kolner BH (2001) High-dynamic-range laser amplitude and phase noise measurement techniques. IEEE J Sel Top Quantum Electron 7(4):641–655
104. Seifert F, Kwee P, Heurs M, Willke B, Danzmann K (2006) Laser power stabilization for second-generation gravitational wave detectors. Opt Lett 31(13):2000–2002
105. Wang Y, Fonseca-Campos J, Liang W-G, Xu C-Q, Vargas-Baca I (2008) Noise analysis of second-harmonic generation in undoped and MgO-Doped periodically poled lithium niobate. Adv OptoElectron 2008(428971)
106. Hooge FN, Kleinpenning TGM, Vandamme LKJ (1981) Experimental studies on 1/f noise. Rep Prog Phys 44:479–531
107. Dutta P, Horn PM (1981) Low-frequency fluctuations in solids: 1/f noise. Rev Mod Phys 53(3):497–516
108. Keshner MS (1982) 1/f noise. Proc IEEE 70(3):212–218
109. Llopis O, Azaizia S, Saleh K, Slimane AA, Fernandez A (2013) Photodiode 1/f noise and other types of less known baseband noises in optical telecommunications devices. In: 22nd international conference on noise and fluctuations (ICNF) 13710909
110. Piché M, Salin F (1993) Self-mode locking of solid-state lasers without apertures. Opt Lett 18(13):1041–1043
111. Mollenauer LF, Stolen RH (1984) The soliton laser. Opt Lett 9(1):13–15

Chapter 4
Reaching the Performance of State-of-the-Art Fiber-Pumped Systems

The advancement of femtosecond solid-state laser technology along with its downstream applications is greatly affected by the average and peak power scalability of existing laser systems. With powerful few-cycle oscillators, not only the effectiveness of laser-driven nonlinear optical processes can be dramatically enhanced but also their respective threshold gets notably reduced—being of fundamental interest for the envisaged mid-IR generation. In Table 2.1 of the theoretical Sect. 2.1.2, an overview of relevant nonlinear processes based on $\chi^{(2)}$- and $\chi^{(3)}$-related effects was provided, which are triggered by a material's nonlinear response under intense optical excitation.

Considering state-of-the-art fiber-laser-pumped Cr:ZnS/ZnSe laser systems, their average output and peak power levels are already sufficient to drive a series of nonlinear $\chi^{(2)}$ optical processes, including recently demonstrated intra-pulse DFG [1, 2]. First reported in Ref. [3], this robust single-beam technique provides one of the most promising routes for generating broadband coherent mid-infrared radiation, which is passively CEP-stabilized. Moreover, a powerful few-cycle Cr:ZnS/ZnSe driving laser not only improves the MIR bandwidth and power levels that can be obtained upon IDFG, but also improves the effectiveness of nonlinear spectral broadening in fibers [4]. For example, the extent of frequency shifts for Raman-induced soliton self-frequency shifting is inversely proportional to the 4th power of the driving pulse duration, while a simultaneous reduction in soliton order (see Fig. 4.1) results in compressed MIR pulses with less pedestal as well as supercontinua featuring higher spectral coherence [5]. With demonstrated conversion efficiencies around 50% [4], fiber spectral broadening is particularly interesting for covering the shorter-wave mid-infrared around (3–5 μm), being host to the molecular fingerprints of many hydrocarbons, or greenhouse and atmospheric gases [6].

Combining the salient advantages of powerful and ultrafast driving laser systems with the benefits of directly diode-pumped Kerr-lens mode-locked Cr^{2+}-doped II–VI

N. Nagl, *A New Generation of Ultrafast Oscillators for Mid-Infrared Applications*, Springer Theses, https://doi.org/10.1007/978-3-030-89754-3_4

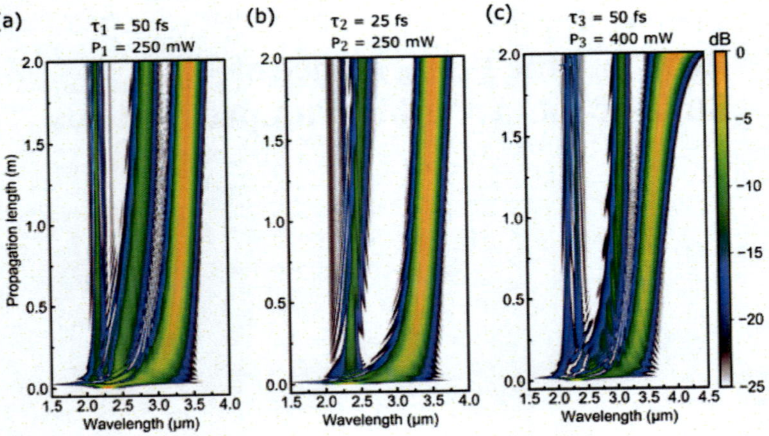

Fig. 4.1 Simulated pulse propagation in a ZBLAN fiber (6.8 µm core diameter, 2 m), illustrating the effect of input power and pulse duration on Raman-induced soliton self-frequency shifting. The simulations were performed with the software *fiberdesk*, and the parameters were taken from [4]. In **a**, a solitonic shift from 2.3 to 3.5 µm is shown, along with a second but substantially weaker soliton as well as some dispersive wave generation around 2.1 µm. In **b**, an input pulse with half the pulse duration is used, resulting in a slightly more red-shifted (3.6 µm) soliton. Most importantly, less energy is shed into dispersive waves or secondary solitons, given the reduction in soliton order. In **c**, a simulation was performed for a higher input power level, resulting in an even stronger shift up to 4.0 µm. This clearly illustrates the advantage of using higher average powers combined with shorter pulse durations for nonlinear optical processes such as the nonlinear spectral broadening in fibers

oscillators has motivated the set up of a second diode-pumped system—*MIRanda2*, which will be presented in this chapter. Driven by the results achieved in Chap. 3 as well as by the goal to reach and surpass the performance of state-of-the-art and more mature fiber-laser-pumped systems, the following sections serve to illustrate the power-scalability of direct diode-pumping. Also, major challenges will be presented along with design and performance improvements achieved over the first generation.

4.1 Challenges of Reaching Higher Average and Peak Powers

The emission of a single-emitter laser diode was sufficient to demonstrate the successful combination of soft-aperture Kerr-lens mode-locking with diode-pumping, resulting in 500 mW of stable and low-noise output pulse trains. To approach the Watt-level, however, several additional challenges arise and need to be addressed. In this section, a focus will be placed on the increase in continuous-wave average power via pump beam polarization-multiplexing, along with the effects of higher intra-cavity peak powers on the KLM performance.

4.1.1 Pump Setup Considerations

Scaling the continuous-wave output power of fiber-laser-pumped Cr:ZnS/ZnSe systems is typically accomplished through an increase in the pump laser's output power. Thanks to their applicability to telecommunications and their widespread commercial availability, Erbium-doped fiber lasers ($\lambda = 1.5\,\mu$m) have become one of the most attractive pump sources for Cr:ZnS/ZnSe lasers. With up to 10 W of available pump power, watt-level CW and mode-locked systems have already been demonstrated [7, 8]. By contrast, the power scalability of directly diode-pumped Cr:ZnS/ZnSe oscillators is strongly hampered by the relatively low output power of single-emitter laser diodes. Taking into account the decreased beam quality at higher driving currents (see Sect. 3.3), the effective output power of one single InP diode ($\lambda \approx 1.55$–$1.65\,\mu$m) is limited to 2.5–3 W. Therefore, the only reasonable way to increase the overall pump power is the incorporation of **two single-emitter laser diodes** into the pump setup.

Compared to Ti:sapphire crystals, which are known to exhibit a polarization-dependent absorption cross-section [9, 10], the absorption of Cr:ZnS/ZnSe gain media can be regarded as optically isotropic. Since no significant change is observed for varying pump polarization [11], the linearly polarized emission of two single-emitter laser diodes can be combined into a single beam geometry to pump the Cr:ZnS/ZnSe crystal. Moreover, **polarization-multiplexing** is a great advantage over diode-pumped Ti:sapphire systems, which typically rely on spatial multi-plexing [12] or counter-propagating pump schemes [12–14], with the latter potentially causing detrimental feedback from one diode into the other.

To experimentally realize a polarization-multiplexed setup, the horizontally and vertically (p-/s-) polarized beams of two single-emitter diodes can be united via a polarizing beam combiner (PBC). Because single-emitter diodes feature intrinsic polarization along the slow axis (horizontal) direction, the polarization of the second diode needs to be rotated via a $\lambda/2$-wave plate. In principle, the same effect could also be achieved when rotating the diode base by 90° [15]. But, as indicated in Fig. 4.2a, the combined emission of two elliptical pump beams would then form a cross rather than a line. Even though this configuration was not experimentally tested, it might be less ideal for soft-aperture Kerr-lens mode locking, which requires a tight focus in at least one direction for a better mode-matching in pulsed operation [16]. Beside the effect of polarization, also wavelength-multiplexing [14] of commercially available InP diodes lasing at $\lambda = 1650$ nm and $\lambda = 1565$ nm would be possible (see Fig. 4.2b). However, this approach was not considered for the second-generation system, since the absorption cross section and thus the total absorbed pump power inside the gain medium is higher when two laser diodes at $1.65\,\mu$m are combined. Nevertheless, it might be of particular interest for future diode-pumped laser amplifiers, in which simultaneous polarization- and wavelength-multiplexing can boost the average power

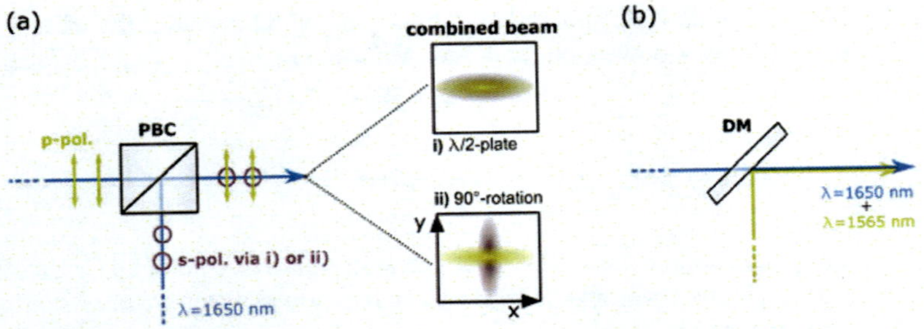

Fig. 4.2 Two ways to increase the effective pump power in a Cr:ZnS/ZnSe system pumped by the combined emission of single-emitter laser diodes: **a** polarization-multiplexing with the aid of a polarization beam combiner (PBC) or **b** wavelength-multiplexing with a dichroic mirror (DM). To rotate the polarization of one diode with respect to the other, either a λ/2-wave plate can be used or one diode mount is turned by 90°. These two methods, however, result in completely different combined pump beam shapes, as illustrated in (**a**)

levels of diode-pumped Cr:ZnS/ZnSe systems to ever higher levels, while retaining the overall low-noise performance.

Once set up, the beam divergences of both diodes and both axis directions need to be matched to focus to the same plane inside the gain crystal. This can be achieved by slight variations of the cylindrical and aspheric lens positions of the two laser diodes. Also, one pump beam path needs to be equipped with two additional steering mirrors to overlap the focal spot positions. Otherwise, the effective pump area is increased and the excitation of unwanted higher-order modes inside the cavity is highly likely.

4.1.2 Effects on the Mode-Locking Performance

From a very basic point of view and without thermal issues, one might expect that the intra-cavity pulse energy and peak power in a single-pulse and fundamental TEM_{00} laser operation depend linearly on the incident pump power. While this is certainly true for moderate power levels or a large net-negative intra-cavity dispersion with $\tau_0 \propto \sqrt{|D_2|}$ (see Sect. 2.2.3), excessive Kerr nonlinearities and decreased KLM action at higher power and shorter pulse durations can drive soliton mode-locking unstable. Once this limit is reached, any further increase in peak power might either result in a breakthrough of a continuous-wave component and its further enhancement or the appearance of multiple pulses—setting a limit to the shortest achievable pulse duration.

In the theory Sect. 2.3, the action of gain and loss as well as the spectral filtering (gain dispersion) were presented as perturbations to the nonlinear Schrödinger Eq. 2.14. In fact, they cause the soliton pulse to loose and shed energy into a so-called "**continuum**" [17, 18], which manifests itself in a small contribution to the solution of the master Eq. 2.22 and is not directly coupled to the soliton [19]. As indicated in

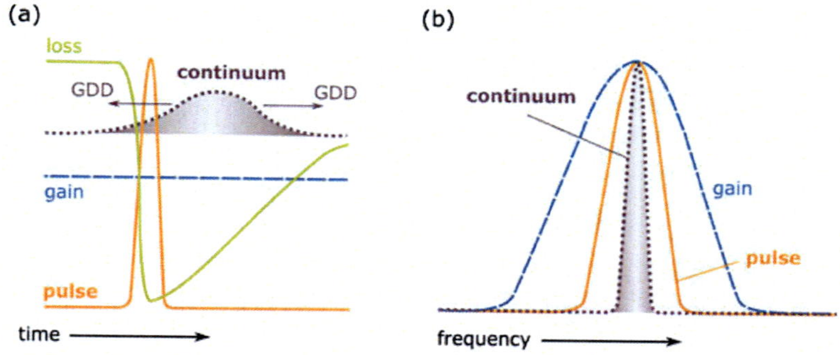

Fig. 4.3 Soliton mode-locking in **a** the temporal and **b** the spectral domain accompanied by the appearance of a background radiation. As illustrated in **a**, a mode-locked laser with strong soliton-like pulse shaping is characterized by a slow absorber action that leaves a long net gain window open behind the main pulse. Even though it might allow a background radiation to grow, the intra-cavity GDD causes the initially weak continuum to spread in time, thus experiencing increasing losses. **b** On the other hand, the spectrum of the continuum is significantly narrower than that of the ultrashort soliton pulse. Hence, the spectral filtering action of the gain medium results in a gain advantage for the narrowband continuum. The figure is adapted from [20]

Fig. 4.3a, this lost energy is initially confined to a background pulse of low intensity, which is too weak to experience significant nonlinear effects like SPM. While the soliton pulse itself is shaped by a balance between negative GDD and SPM (see Sect. 2.2), the dispersion is not simultaneously balanced for the continuum. Therefore, it is temporally spread into regions where the loss already slowly recovers. The resulting "cleanup" of the continuum radiation due to the group delay mismatch will eventually stabilize the soliton as long as its losses are smaller than the continuum loss.

When evaluating the continuum formation in the frequency domain, however, it reveals a gain advantage for the narrower background radiation when compared to a broadband soliton pulse (see Fig. 4.3b). This is related to the limited gain bandwidth and the spectral filtering action of the gain medium. The same is true for pulses with a longer pulse duration and smaller spectrum. Therefore, when a laser cavity is operated at very high pulse energies or decreased amount of negative dispersion, a single pulse can split its energy into two or even more pulses, having individually much longer pulse durations than the original one ($E_p \propto 1/\tau_0$, see Sect. 2.2.3) [21, 22]. To prevent the continuum from reaching threshold and avoid the onset of **multi-pulsing**, as was recorded in Fig. 4.4b, a large modulation depth of the saturable absorber can be helpful to maintain stable soliton mode-locking. Also, it can add additional loss for the growing continuum [17]. In this respect, large values of the Kerr strength are seeked after for KLM-based oscillators.

The second diode-pumped oscillator presented in this chapter is meant to operate at higher power levels and shorter pulse durations than before, rendering the cavity more susceptible to the aforementioned instabilities. Hence, a closer investigation

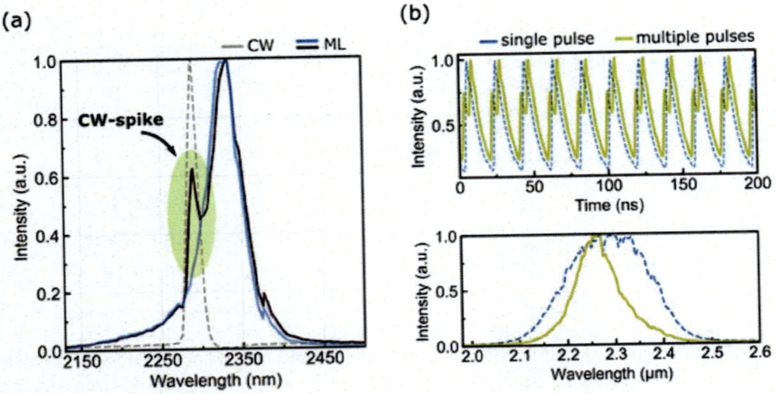

Fig. 4.4 Recorded mode-locking instabilities in the form of **a** a continuous-wave background and **b** multi-pulsing. The data was taken during the buildup and subsequent optimization phase of *MIRanda2*. As indicated in **a**, a CW-breakthrough (dark blue) was observed in mode-locked operation, and could be removed by a decrease in pump power or a shift in gain medium position (light blue). Moreover, the CW-spike position coincides with the spectrum in continuous-wave operation (dashed grey line), which is attributed to the wavelength at the maximum absorption cross section of the Cr:ZnS crystal. In figure **b**, multi-pulsing was captured with a fast photodetector (*Thorlabs*, DET10D2), while a simultaneous reduction in pulse bandwidth was observed in the spectral domain—attributed to the longer pulse durations. It should be noted that the spectra were recorded with the same grating-based and fiber-coupled spectrometer as in Fig. 3.13, being only of limited spectral resolution. Hence, the fringe-like interference pattern in the mode-locked spectrum, which is typically observed for multi-pulsing, could not be resolved here

of the γ_{Kerr}-parameter was performed via **nonlinear ABCD-matrix simulations** to identify potentially favorable operation conditions and guidelines for the experimental fine adjustment of the new cavity.

As defined in Eq. 2.8, the **Kerr strength** is a measure of the small signal relative spot size variation and denotes the change in cavity mode size due to a small increase in the intra-cavity peak power. For example, if the Kerr strength is positive/negative, the mode diameter increases/decreases in mode-locked laser operation. Moreover, the magnitude of γ_{Kerr} can be evaluated as a function of the distance z between the two focusing mirrors as well as for varying crystal positions. Because Gaussian laser beams of non-vanishing peak power experience a lensing effect inside nonlinear media such as the gain crystal, one needs to derive nonlinear ABCD-matrices that account for the peak-power-dependent self-focusing. In this respect, a useful approach is presented in [23, 24], in which the crystal is divided into many thin Kerr-lens slices, whose thickness l_{cr} is chosen small enough to ensure a negligible change in beam size per plate. The crystal can then be described as a stack of thin-lens matrices, where each matrix M_{slice} can be represented as a superposition of a thin-lens effect with nonlinear focal length f_{Kerr} (Eq. 2.7) and a propagation through a single slice of thickness l_{cr} with refractive index n_{cr} [23]:

$$M_{slice} = \begin{pmatrix} 1 & 0 \\ -\frac{1}{f_{Kerr}(\omega)} & 1 \end{pmatrix} \begin{pmatrix} 1 & l_{cr}/n_{cr} \\ 0 & 1 \end{pmatrix}. \tag{4.1}$$

This definition is similar to the one presented by Magni et al. [25, 26], however, with the difference that they split the crystal into two blocks of equal length. The Kerr nonlinearity is then grouped into one single matrix, which is surrounded by the linear propagation segments inside the crystal blocks.

To numerically evaluate the Kerr strength over the cavities' two stability zones, the following **procedure** is applied: first, after setting up the system's ABCD round-trip matrix, the beam size w at each Kerr-lens plate is calculated for zero intra-cavity peak power ($P_0 = 0$) according to [27, 28]:

$$q = \frac{Aq + B}{Cq + D}. \tag{4.2}$$

Here, A, B, C and D denote the ABCD matrix elements for a complete cavity round-trip and q refers to the beam parameter defined as $q^{-1} = R^{-1} - i\lambda/(\pi w^2)$ with the (1/e^2) Gaussian beam radius w and the wavefront curvature R. From this initial set of beam sizes, the ABCD-matrices for the Kerr slices can be determined at a certain intra-cavity power level ($P_0 > 0$), resulting in a new round-trip matrix for the system. With Eq. 4.2, the new set of mode sizes is then derived at each Kerr-lens plate again. Next, the ABCD-matrices for the Kerr slices are calculated again for $P_0 > 0$, with the new mode diameters arising from the corresponding system matrix. This step is repeated in an iterative manner for the same power $P_0 > 0$ until a steady state is reached; at this point, the beam sizes at the Kerr-lens plates effectively do not change anymore, and Eq. 2.8 can be written as a difference equation to extract the Kerr strengths for a certain normalized power $p = P_0/P_{cr}$ [29]:

$$\gamma_{Kerr} = \frac{1}{p}\frac{\omega(p) - \omega(0)}{\omega(0)}. \tag{4.3}$$

For soft-aperture KLM, the γ_{Kerr}-parameter was averaged over the crystal to include contributions from the whole crystal region [24]. By contrast, the Kerr strength analysis in hard-aperture KLM would rather focus on a certain position inside the cavity, which is typically chosen to be the position of the mechanical slit close to one of the end-mirrors [30, 31]. The same was also true for the simulation presented in the theoretical Sect. 2.1.2.

For the simulations performed in this chapter, beam astigmatism was neglected and the curved cavity mirrors were modeled as thin lenses (AOI = 0°), resulting in equal Kerr strengths for the tangential and sagittal plane. A schematic representation of the simulation parameters as well as an overview of the respective values can be found in Fig. 4.5 and Table 4.1. Following the procedure described above, a 2D-plot of the Kerr strength behaviour over the two stability zones was simulated for $p = 0.0001$ and plotted in Fig. 4.6. Here, negative/positive values indicate a decrease/increase in beam size for $P_0 > 0$, while zero values refer to either unstable cavity configurations

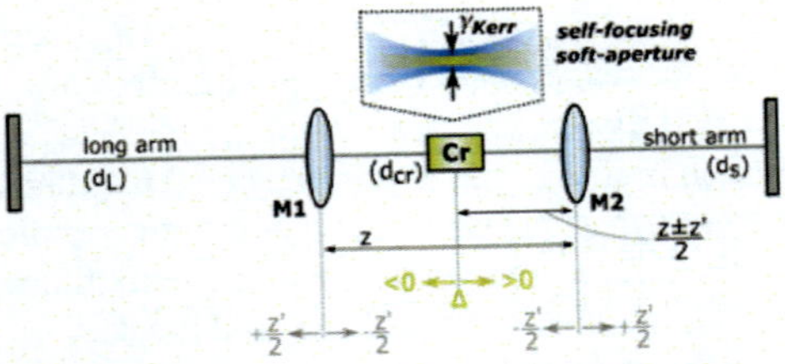

Fig. 4.5 Representative model for the nonlinear ABCD-matrix calculations, with z: fixed distance between the focusing optics, z': variation of z, d_{cr}: crystal length, $d_{l/s}$: cavity arm lengths, Δ: deviation of the crystal location from the center position

Table 4.1 Simulation parameters, with $f_{\mathrm{M1,\,M2}}$: focal length of lenses M1 and M2

	MIRanda2
Cr	Cr:ZnS
d_{cr}	5 mm
n	2.26
n_2	$100{\cdot}10^{-16}$ cm²/W
Δ	$\pm$10 mm
$f_{\mathrm{M1,\,M2}}$	100 mm
ν_{rep}	50 MHz
d_s/d_l	0.5
z	212.5 mm
z'	$\pm$10 mm

or positions of insignificant change in mode diameter. Also, two additional simulation runs were performed to further investigate the effect of different arm length ratios d_s/d_l on the Kerr strength. The simulations were conducted for $d_s/d_l = 0.3$ and 0.8 and are compared with the results for $d_s/d_l = 0.5$ in Fig. 4.7, where all other parameters, including the repetition rate and step size, were kept constant for better comparability. Moreover, the spot size in the crystal was confirmed to stay almost unaffected at $\varnothing \approx 85–90\,\mu$m for $P_0 = 0$ and different arm ratios.

Having a closer look at the 2D-maps, some general conclusions can be drawn:

- To achieve large variations in spot size and obtain strong nonlinear loss modulations, the cavity needs to be operated close to one of the stability zone boundaries (indicated with dashed circles). However, a large magnitude of the Kerr strength needs to be traded off against reduced cavity stability and laser gain. This observation is similar to the findings presented in [23, 24, 26, 32].
- For soft-aperture KLM, the simulation predicts operation at the first edge of stability zone 1 and zone 2 to be most favorable—assuming that the beam size for $P_0 > 0$ needs to decrease for stable mode-locking (see blue-shaded areas with $\gamma_{Kerr} < 0$). The corresponding enhancement of γ_{Kerr}, however, seems to be more

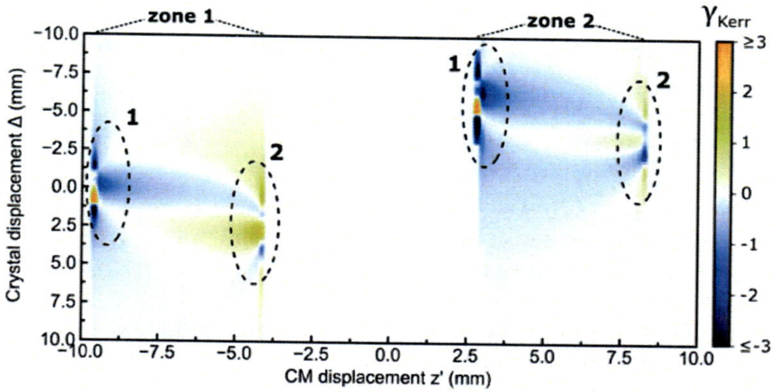

Fig. 4.6 Two-dimensional representation of the Kerr strength for varying crystal and curved mirror (CM) displacements. In the plot, two stable regions can be identified, which represent the two stability zones of the oscillator. Also, a noticeable enhancement of γ_{Kerr} is apparent at the zone boundaries, with the respective edge positions being marked with dashed lines. When the crystal is exactly centered between the two focusing lenses, the displacement is $\Delta = 0$

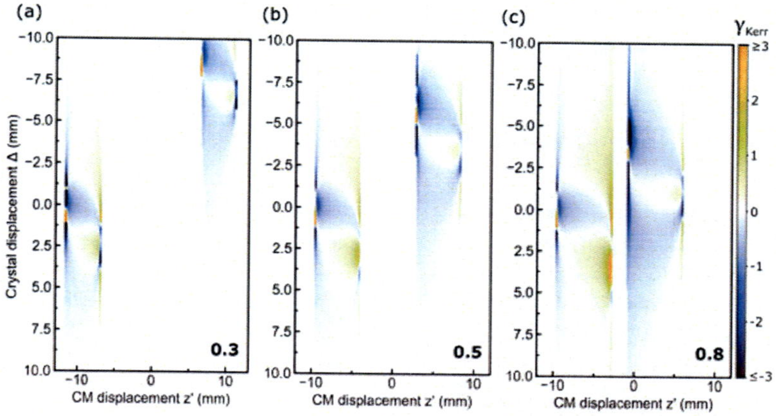

Fig. 4.7 Kerr strength simulations for varying arm length ratios, with $d_s/d_l = 0.3$ (**a**), 0.5 (**b**), and 0.8 (**c**). The step size was kept constant for all calculations, indicating a slight increase in nonlinear loss modulation for converging arm lengths

confined to the very edge in the first zone, while the conditions in the second zone appear to be slightly more relaxed. This renders the second stability zone more promising for efficient KLM, as also found in [24, 32].

- Investigating the Kerr strength with respect to crystal displacement reveals a strong positional dependence. Also, the direction of crystal movement seems to matter since $\Delta < 0$ implies the crystal to be moved towards the longer arm, while $\Delta > 0$ indicates a position more closely located to the shorter arm of the cavity.

- When varying the arm length ratio d_s/d_l at a fixed repetition rate, the simulations predict a slightly increased nonlinear loss modulation for the higher ratios. Similar conclusions were also drawn in [33] for nearly symmetric resonators, which might even allow for self-starting operation [34].

The small signal simulations performed here serve to illustrate some general laser characteristics in pulsed operation when operating within different stability zone regions. In reality, however, additional factors will influence the lasing behaviour of the cavity, making it significantly harder to give accurate predictions and run simulations. As shown in [35], for example, the stable regions will get continuously deformed when increasing the intra-cavity peak power, while thermal distortions arising from the comparably large quantum defect in Cr:ZnS/ZnSe gain media are likely to appear at the higher pump powers. Therefore, strong nonlinear effects in the gain crystal as well as thermal lensing effects might require the crystal to be shifted further away from the focal region than predicted by simulations. Moreover, the saturable absorber action for soft-aperture KLM might get less efficient beyond a certain power level, as was shown in Fig. 2.9, and is greatly affected by the mode-matching between the pump and cavity mode. If a crystal with a non-uniform transverse gain profile is used as a gain medium, also gain guiding effects would need to be taken into account [36].

4.2 Comparison and Improvements over *MIRanda1*

In this section, the experimental realization of a more powerful diode-pumped Cr:ZnS system will be presented, starting with an overview of the new pump setup and oscillator design. For KLM operation and its subsequent optimization, a clear strategy will be reported in Sect. 4.2.2, based on the results achieved with *MIRanda1* and the conclusions drawn from Kerr strength simulations.

4.2.1 Pump and Cavity Design

As introduced in Sect. 4.1.1, polarization-multiplexing of two single-emitter laser diodes was applied. While this scheme appears straightforward to be implemented, the combination of two separate pump units adds complexity, while increasing the demand on mechanical stability and compactness. Therefore, a new pump assembly was realized, uniting all collimation lenses and the pump diode in a single optical cage system (see picture Fig. 4.8a). Also, a new cooling architecture based on direct water-cooling was successfully adopted, rendering thermoelectric cooling of the diode base obsolete. To ensure a constant cooling temperature of 12 °C, the setup was equipped with an external circulating bath unit (*Thermo Fisher Scientific*, NESLAB RTE-7), which is specified for temperature stabilities of ±0.01 °C at a maximum cooling

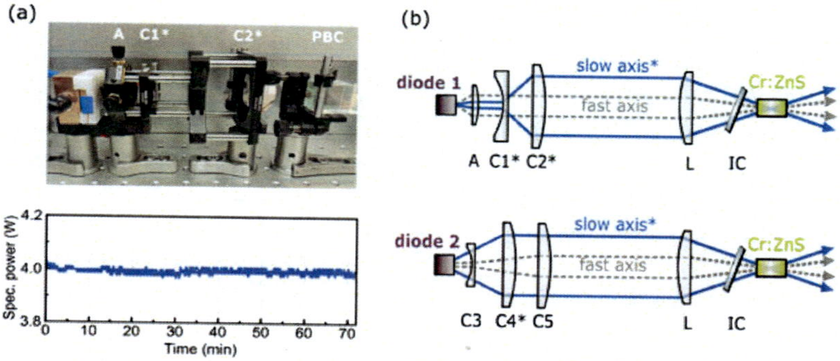

Fig. 4.8 Performance of the new diode mount and schematic of the two pump collimation assemblies. **a** Picture of the pump unit, with the diode mount being attached to a lens cage system. Also, direct water-cooling of the diode base was implemented, resulting in stable output power operation as shown in the figure below. PBC: polarizing beam combiner. **b** Schematic of the lens combinations used to address the slow and fast axis emission of diode 1 and 2, respectively. A: aspheric lens (f = 3.10 mm), C: cylindrical lenses ($f_{C1*} = -15$ mm, $f_{C2*} = 100$ mm, $f_{C3} = -25$ mm, $f_{C4*} = 70$ mm, $f_{C5} = 150$ mm), L: achromatic pump lens (f = 100 mm), IC: input coupler. Asterisks(*) denote lenses that were used to address the slow axis only

capacity of 500 W at 20 °C. As a result, a long term measurement of the direct laser diode output revealed a stable performance with an emitted power of 4 W (Fig. 4.8a).

Similar to *MIRanda1*, the slow and fast axis emission of the diodes was manipulated with a set of aspheric and cylindrical lenses to achieve a focal spot diameter of around 90–95 µm for the fast axis. This number was obtained through linear ABCD-matrix analysis, when simulating a laser cavity that was envisaged to operate at a repetition rate of 50 MHz and be equipped with curved mirrors of 200 mm focal length. For the first diode assembly, the same C-Mount type was used as before (*SemiNex Corp.*, C-132), however, with the 400 mm cylindrical lens being replaced by a pair of lenses (C1*, C2*). This allowed to expand and re-collimate the diode's slow axis emission in a more compact fashion (see Fig. 4.8b). For the second pump unit, a lensed version of the diode (*SemiNex Corp.*, C-132-118) was tested, whose highly-divergent fast axis was already collimated to 10 mrad with a f = 274 µm lens. Hence, a single cylindrical lens (C4*) was needed to collimate the slow axis, while a cylindrical lens telescope (C3, C5) expanded the collimated beam diameter of the fast axis. Both diodes were operated at supply currents around 12 A, resulting in $2 \cdot 2.5$ W of average power when measured in front of the pump lens (L). Despite all optics being AR-coated for the pump wavelength, about 10% of total power were lost in the lens collimation systems, while an additional 6% of the s-polarized diode beam were reflected at the 45°-input coupler mirror. Finally, a diameter of 85–90 µm and 170–190 µm was determined for the fast and slow axis in the focus, being in good agreement with the envisaged spot sizes and the spot size ratio of 1:2 (fast:slow axis).

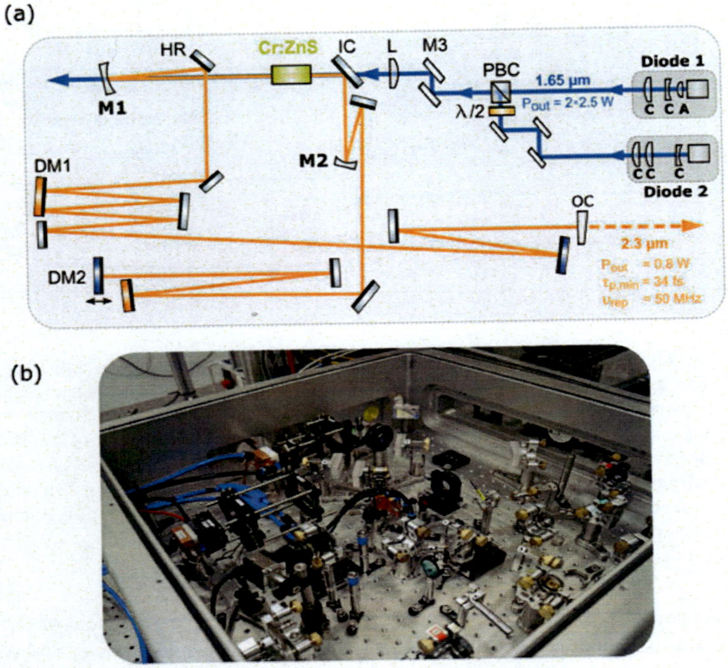

Fig. 4.9 **a** Experimental setup of *MIRanda2* with **b** a picture of the laser system. Diode 1: Type C-132, Diode 2: Type C-132-118, C: cylindrical lens, A: aspheric lens, PBC: polarizing beam combiner, $\lambda/2$: half-wave plate, M3: highly-reflective rectangular-shaped mirrors (1.4–1.7 µm), L: achromatic pump lens (f = 100 mm), IC: input coupler, M1 and M2: ROC = −200 mm, DM1 and DM2 (orange/blue): dispersive mirrors, HR (grey): highly-reflective mirrors optimized for 2.0–2.6 µm, Cr:ZnS: gain medium (5 mm), OC: 25% output coupler. Mode-locked operation was started by a fast translation of the end-mirror

For the laser oscillator, a 5 mm long AR-coated polycrystalline gain medium (*IPG Photonics*) with a clear aperture of 1.9 mm · 4 mm was used and placed inside a X-fold cavity at normal-incidence angle. To avoid thermal instabilities arising from the increased heat load at higher pump powers, Cr:ZnS was preferred over Cr:ZnSe, which has an about 1.5 times higher thermo-optic coefficient (see Table 3.1). At a doping concentration of $3.5 \cdot 10^{18}$ cm^{-3}, the single-pass absorption was measured to be ∼81%, with the water-cooled copper heat sink being connected to the same circulating chiller as the diodes. To further improve the system's performance and minimize the pump beam astigmatism arising from the 45°-input coupler mirror, the thickness of the latter was reduced from 6 to 3 mm, while the incident angle on the curved mirrors was kept low at about 4–5 degrees. In contrast to *MIRanda1*, also the pump setup could be incorporated into the monolithic chamber, resulting in a higher stability and a footprint of only 60 cm × 60 cm for the entire system (Fig. 4.9).

4.2.2 Mode-Locked Performance

To initiate and further optimize Kerr-lens mode-locked operation, a clear strategy was pursued: First, the cavity parameters were kept similar to the ones of *MIRanda1*, given its successful realization of few-cycle pulse generation. Therefore, only one laser diode (C-132) was operated initially, and an arm length ratio of 0.7 was used for the cavity design. Also, to start soft-aperture KLM, the resonator was adjusted to operate at the 2nd edge of stability zone 1, with the crystal being shifted towards the shorter arm. This agrees well with the simulation results shown in Figs. 4.6 and 4.7, indicating high relative magnitudes of the Kerr strength for crystal displacements $\Delta > 0$, very similar to the design procedure described in [37].

Once pulsed operation was initiated by a slight detuning and rapid movement of the end miror, the crystal and curved mirror positions were fine-adjusted to enable reliable and stable mode-locked operation. The latter could be further improved through an optimization of the pump focus location, with the pump lens being positioned on a mechanical translation stage. As a next step, the pump and intra-cavity power was increased with the aid of the second laser diode (C-132-118). However, this resulted in mode-locked spectra being accompanied by a **pronounced CW-background** or being affected by the onset of **multi-pulsing** (see Fig. 4.4). Ways to cope with these kind of instabilities are to shift the gain medium out of focus or to reduce the pump power. While a stronger misalignment of the cavities' end-mirror can also help to maintain clean single-pulse operation, they inevitably reduce, at fixed output coupling (OC) ratio, the intra-cavity peak power and the average output power of the oscillator.

In the experiment, the OC ratio was increased step-wise starting at 11%, with the achieved mode-locked output power and corresponding intra-cavity power being depicted in Fig. 4.10. Even though the output power was improved by a factor of three, the second pump diode was at no time operated at its full capacity due to the appearance of a CW-spike at higher powers. This was rather surprising, given the very high output coupling ratios of up to 64%, which is—to the best of the authors knowledge—the largest reported for Cr:ZnS/ZnSe lasers so far [8]. Also, one would expect that the output coupling rate attains an optimum value, beyond which the output power decreases for a given pump rate [38, 39]. This optimum can be explained by the fact that the output power tends to increase for higher OC transmission, but also tends to decrease beyond a certain value due to the reduced number of cavity photons at increasing cavity losses. Experimentally, however, it was not possible to clearly identify the point of alteration at which the highest mode-locked output power can be achieved. Instead, when using OC ratios larger than 64%, the oscillator was not mode-lockable anymore.

To investigate whether the strong tendency for CW-spiking and multi-pulsing is caused by the low brightness of the pump source, an Erbium fiber laser (*Bktel*, HPFL-370-1567-FCAPC) was installed to provide a diffraction-limited pump beam quality ($M^2 \approx 1$). Interestingly, no change in behaviour was observed upon mode-locking. Hence, further strengthening of the self-amplitude modulation action in the gain medium was expected to be necessary for maintaining single-pulse operation

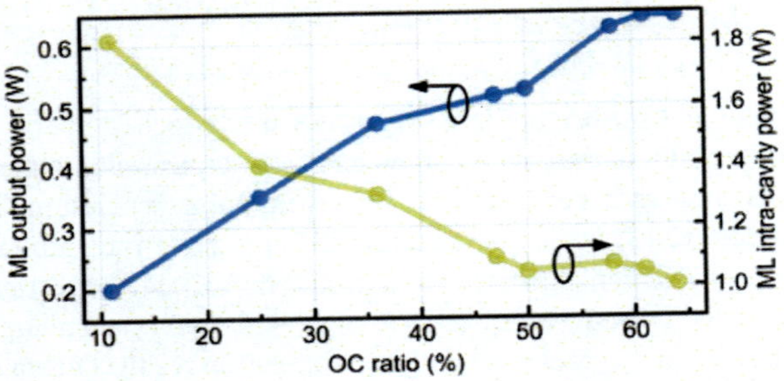

Fig. 4.10 Increase of the output coupling ratio in mode-locked operation. By combining several output couplers (3%, 11%, 14%, 25%), very high rates of up to 64% could be achieved. The intra-cavity power levels were deduced from the measured output powers in mode-locked operation. It should be noted that the intra-cavity dispersion and thus the generated pulse durations and peak power levels slightly changed when different output couplers were inserted. Also, the power values were recorded for clean single-pulse operation without CW-spike

at the higher pump levels. Since simulations predict soft-aperture KLM to be more favorable at the inner boundary of **stability zone 2** (see Fig. 4.6), the cavitie's stability behaviour had to be characterized first. When measuring the CW output power at varying curved mirror distances, as plotted in Fig. 4.11a, a suitable position for KLM was identified in the second zone. Also, it was experimentally confirmed to result in highly reliable mode-locked operation—even without the need for a shift in crystal position or misalignment of the cavity. The latter might be due to the increased SAM action inside the Cr:ZnS crystal, while the former result was already predicted by the simulations (see Fig. 4.6): when increasing the distance between the curved mirrors by about 7–8 mm to move the cavity operation away from the second edge in stability zone 1 to the first edge in stability zone 2, the crystal displacement Δ amounts to a very similar value of around 7 mm. Therefore, no strong change in crystal position is expected, which was also confirmed experimentally.

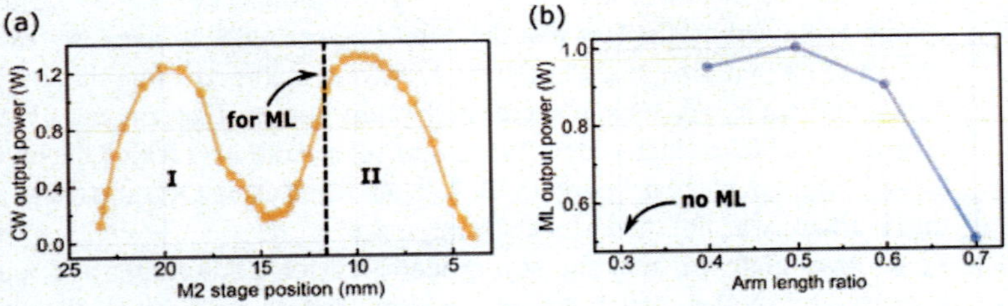

Fig. 4.11 **a** Stability zone scan and **b** KLM performance at varying arm ratios d_s/d_l for an output coupling rate of 25% and a repetition rate of 50 MHz. In **a** decreasing values on the x-axis correspond to a shift of the curved mirror M2 away from the crystal

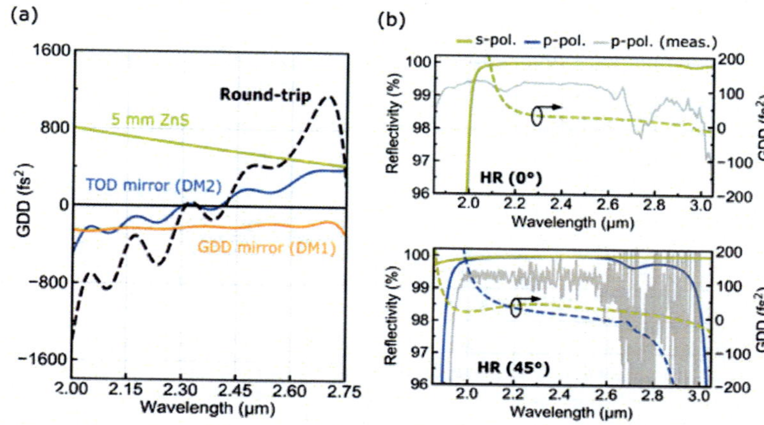

Fig. 4.12 **a** Theoretical dispersion values of the 5 mm Cr:ZnS gain medium as well as the dispersion arising from a single GDD (DM1) and TOD (DM2) mirror, which was also used for *MIRanda1*. Also, the total round-trip dispersion is shown for the combination of mirrors that experimentally resulted in the shortes pulse duration. **b** Theoretical reflectivity and dispersion curves for the two sets of highly-reflective (HR) mirrors, being designed for AOI = 0° or AOI = 45°. For the AOI = 0° case, the s-polarized and p-polarized curves are equivalent. Also, the measured reflectivity for p-polarized light is added

Though the CW-spiking turned out to be much weaker in the 2nd stability zone, further improvement was needed. Because the arm length ratio was shown to have some impact on the Kerr strength (see Fig. 4.7), different d_s/d_l-values were tested. The results are plotted in Fig. 4.11b and reveal an ideal ratio of 0.5 for extracting the highest power in single-pulse operation. Mode-locked output powers of up to 1 W were finally measured at the full pump power and without the onset of pulse instabilities.

After having optimized the mode-locked output power, the intra-cavity dispersion was fine-tuned to further improve the output pulse duration. For the second-generation system, a new set of dispersive mirrors (DM1) was tested, being designed to compensate for the GDD and TOD of the 5 mm long Cr:ZnS crystal. In theory, 4 bounces on this mirror fully balance the gain medium's dispersion while adding slight negative dispersion of GDD ≈ -250 fs^2 for soliton mode-locking. Similar to the optimization steps performed for *MIRanda1*, different combinations of mirrors were tested in a systematic manner, revealing the optimal configuration to be 6 bounces on DM1 and 3 bounces on the TOD mirror DM2 per round-trip. This resulted in the shortest output pulse durations for an output coupling ratio of 25%, with the corresponding round-trip dispersion being calculated and plotted in Fig. 4.12. Also, it seems that other cavity optics like the output coupler or HR mirrors have caused extra dispersion, explaining the need for additional TOD mirrors (DM2) and the discrepancy with the theoretical design.

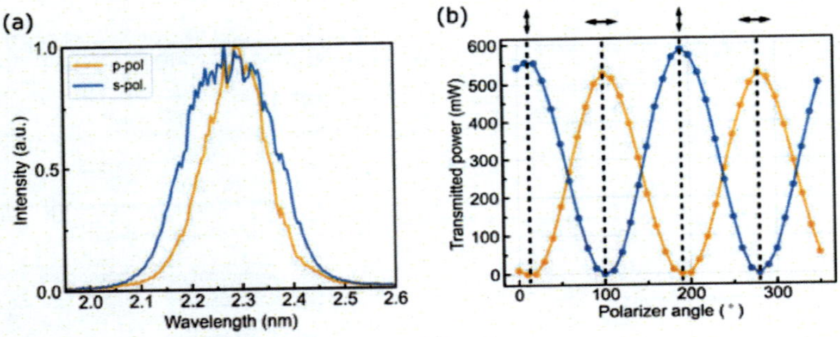

Fig. 4.13 Observed jump in the ML output polarization when measured at constant cavity parameters. Corresponding measurements of the output spectrum and polarization are shown in (**a**) and (**b**), respectively. The polarization was determined via measuring the transmitted power through a wideband IR polarizer (*Innovation Photonics*, PGC-5)

Since no intra-cavity element was placed at Brewster's angle in order to avoid further addition of material dispersion, the oscillator's polarization was not fixed. Therefore, for a given combination of cavity mirrors, the output pulses were measured to be either purely p- or s-polarized upon ML operation. Respective measurements are depicted in Fig. 4.13, which reveal a slight change in output spectral bandwidth for different polarization states. This behaviour is attributed to the highly-reflective coating of the 45°-mirrors, revealing very different dispersion characteristics and an improved reflectivity for s-polarized light (Fig. 4.12b). As will be seen in Chap. 5, these optics were therefore not considered for the third-generation system.

In its final state, the oscillator was providing stable s-polarized output pulses with pulse durations as short as 34 fs (Fourier-transform limit: 32 fs), corresponding to **nearly four optical cycles**. Combined with mode-locked output powers of over 800 mW at a 25% OC ratio and at a repetition rate of 49.9 MHz, this laser system delivered almost three-times higher peak powers compared to the first KLM diode-pumped Cr^{2+}-doped II–VI laser presented in Chap. 3. In the continuous-wave regime, about 1.2 W of average power were achieved when operating at the center of the stability zones. Interestingly, a 5–10% drop in power was observed when the laser switched to pulsed operation. According to numerical simulations presented in Ref. [32], this drop might occur when the cavity is operated at higher Kerr nonlinearities. In this case, CW oscillations experience increased residual gain.

A respective SHG-FROG measurement is depicted in Fig. 4.14a, and was taken without further post-compression of the output pulses. Since the 25% output coupling mirror was made of 3 mm BaF_2, it caused only very little dispersion when compared to other commonly used materials (see Table 2.2). For frequency doubling, a 20 µm thick GaSe crystal was used, since the transmission of BBO starts to drop in an unsmooth fashion beyond 2 µm, which will likely cause artifacts in few-cycle pulse measurements. To also illustrate the effect of higher intra-cavity power on the pulse spectral bandwidth, another FROG measurement was performed at a very low out-

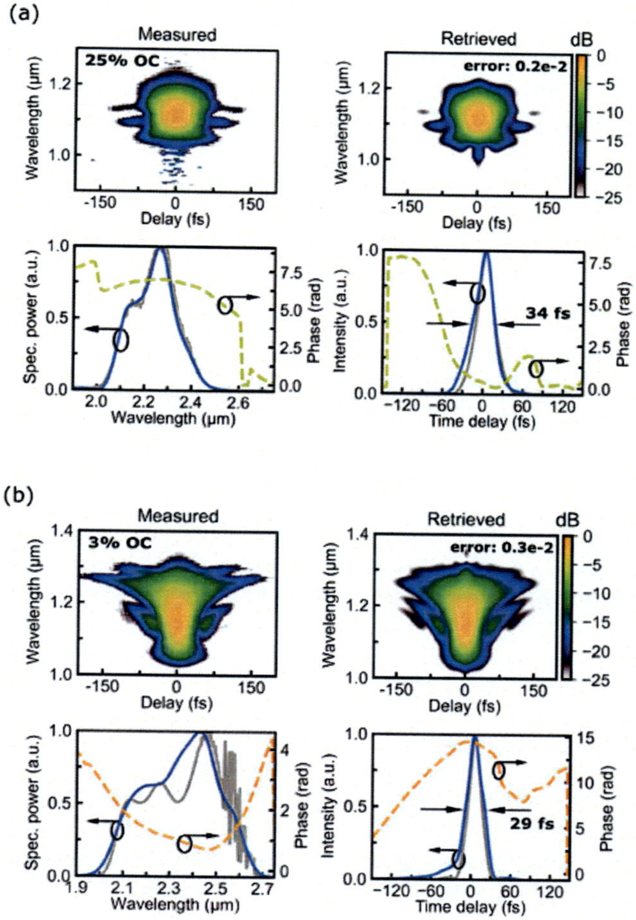

Fig. 4.14 **a** SHG-FROG measurement of the output pulses at a 25% output coupling ratio. **b** For comparison, another FROG trace was taken at a very low OC rate of 3% to illustrate the effect of higher intra-cavity power on the generated pulse duration. Both FROG traces were retrieved on a grid size of 512·512 points, with the measured spectra and calculated FTL intensity profiles indicated as grey lines

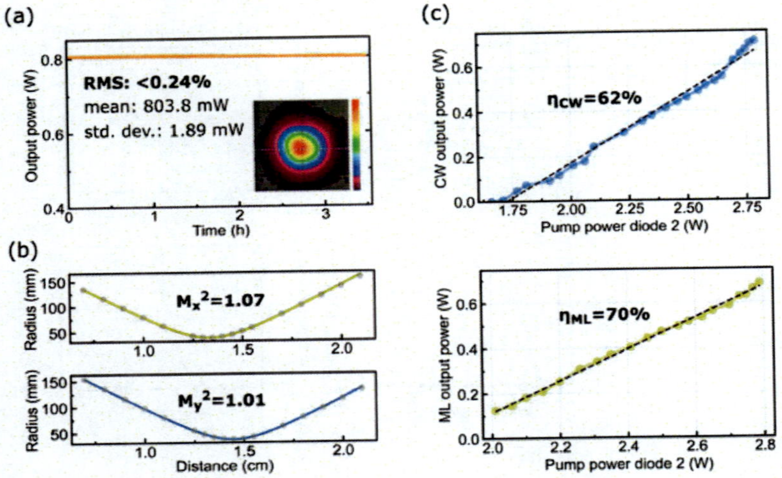

Fig. 4.15 Output performance of *MIRanda2*. **a** Mode-locked output power stability recorded with the thermal power meter (*Coherent*, PS19) and corresponding single-mode beam profile. **b** M^2-measurement of the horizontal and vertical beam axis of the mode-locked output. **c** Slope efficiency curves measured for a decrease in pump current of the second diode

put coupling rate of only 3% (see Fig. 4.14b). After accounting for the additional dispersion of the OC mirror (fused silica, 6 mm), a very broad spectrum could be measured at an average power of about 150 mW. It resulted in a retrieved pulse duration of 29 fs with a Fourier-transform limit of 27 fs. The shape of the FROG trace, however, displays a change in phase in the longer wavelength range. This is potentially due to the cavity being operated at very low negative dispersion. In this regime, spectral oscillations in the round-trip dispersion, which are caused by mirror dispersion or air absorption lines around 2.5–2.7 μm, can result in slight crossings of the zero-dispersion line, hence disturbing clean soliton mode-locking in the negative GDD regime.

To fully characterize the performance of the laser system, similar to *MIRanda1*, the mode-locked output power stability was measured and depicted for a 3.5 h snapshot in Fig. 4.15a. With corresponding RMS power fluctuations of less than 0.24%, the oscillator was operating reliably for the whole working day and with a clean single-mode output beam profile. In addition, the horizontal and vertical beam quality factor M^2 of the linearly polarized mode-locked output pulse trains were proven to be nearly diffraction limited with $M^2 < 1.1$ (Fig. 4.15b). Since the oscillator was pumped by two laser diodes simultaneously, the slope efficiency for CW and ML operation was measured for decreasing pump powers of the second diode (C-132-118), revealing high efficiencies of 62% and 70%, respectively (Fig. 4.15c).

Amplitude Noise Measurements

In the measurements presented in the previous chapter, the directly diode-pumped oscillator exhibited a surprisingly low noise performance, which was hardly detectable when measured at the fundamental output wavelengths. Instead, a workaround via second-harmonic generation in a PPLN crystal was needed to enhance the oscillator's output noise. To determine the noise characteristics of the second-generation Cr:ZnS system, however, the measurement setup was equipped with a new wideband voltage amplifier (VA), rendering analysis of the direct laser output feasible.

The setup is sketched in Fig. 4.16a, and includes the same RF spectrum analyzer and photodetector (PD) as used in Sect. 3.3.3. To enlarge the signal-to-noise ratio (SNR) between the oscillator's low-frequency noise and the background noise of the measurement devices, the detector signal was sent through a low-noise and AC-coupled variable gain VA, which was operated at the highest gain level of 60 dB. Though its respective cut-off frequency was set to 10 MHz, the strong carrier frequency around 50 MHz was filtered by an additional low-pass filter (LPF) placed in front. To plot the measured noise levels in logarithmic units of dBc/Hz and evaluate the integrated RMS noise, the same formulae were applied as in the previous noise analysis, with the exception that the theoretically generated DC-voltage level[1] of $U_{DC,60dB} = 44$ mV $\cdot 10^{60/20} = 44$ V was now taken as reference level in Eq. 3.6.

The results are plotted in Fig. 4.16b along with the corresponding background noise of the photodetector as well as the voltage amplifier and the spectrum analyzer. The latter two were obtained for signal inputs that had been terminated with a 50 Ω resistor, while the noise floor of the PD was recorded when the optical signal to the detector was blocked. Obviously, the noise floor of the photodetector was slightly lower than the noise floor of the VA, resulting in very similar integrated RMS noise values of 0.0049% and 0.0048%. The shot noise level was also plotted and calculated according to Eq. 3.5, resulting in a value of −154 dBc/Hz.

Compared to the output characteristics of *MIRanda1*, a noise peak at frequencies around 400–600 kHz is apparent, which is most likely related to slight relaxation oscillations of the Cr:ZnS gain crystal. The f^{-1}-behaviour at the very low frequencies together with the drop in noise signal around 1 MHz, however, is in accordance with the previous findings. Moreover, by using the new voltage amplifier, the SNR of the overall measurement system could be noticeably improved. With the background noise floor of the RF analyzer being located at −190 dBc/Hz and the noise level of the VA and PD approaching -148 dBc/Hz at the higher frequencies, the influence of instrument noise on the oscillator signal was minimized. For the frequency range between 20 Hz and 1 MHz, it resulted in a very low integrated RMS noise of 0.014%

[1] To calculate the DC-voltage U_2 after amplification with a certain voltage gain factor G_V [dB], the following formula was used: $U_2 = U_1 \cdot 10^{G_V/20}$.

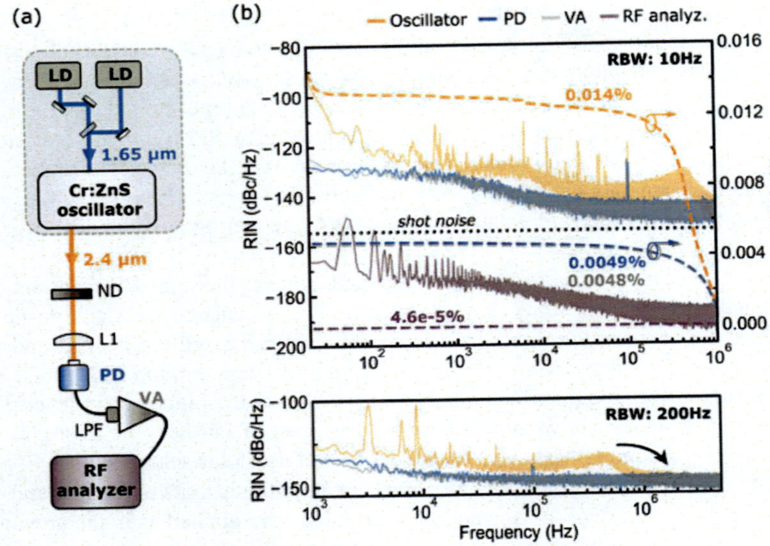

Fig. 4.16 Amplitude noise measurements and integrated RMS noise of the mode-locked output of *MIRanda2*. **a** Setup. LD: InP laser diode, ND: neutral density filter wheel, L1: N-BK7 lens (f = 25 mm), PD: biased InGaAs photodetector (*Thorlabs*, DET10D2), LPF: low-pass filter DC-11 MHz (*Mini-Circuits*, BLP-10.7+, 50Ω), VA: wideband voltage amplifier (*FEMTO*, DHPVA-101), RF analyzer: radio-frequency spectrum analyzer (*Agilent*, E4447A). **b** Measurements and integrated RMS noise of the direct mode-locked output (orange) along with data on instrument noise. Below, the RIN was measured in a broader span up to 5 MHz, and illustrates the drop in noise signal to the background noise floor at frequencies around 1 MHz

for the mode-locked output—comparable to reported [40] and even commercially available[2] ultrafast Ti:sapphire laser systems.

4.3 Summary

With *MIRanda2*, a powerful KLM Cr:ZnS oscillator pumped by two single-emitter laser diodes was demonstrated, delivering mode-locked average powers of over 800 mW combined with pulse durations down to nearly four optical cycles. Despite the highly elliptic collimated pump beam shapes and the higher intra-cavity peak powers, stable soft-aperture mode-locking performance was achieved in accordance with nonlinear ABCD-matrix simulations. Moreover, the low amplitude noise of

[2] The Ti:sapphire oscillator "Mai Tai" from *Spectra-Physics* has a specified noise of <0.15% RMS for a 10 Hz–10 MHz bandwidth. For more information on the product, the reader is referred to the website: https://www.spectra-physics.com/products/ultrafast-lasers/mai-tai (25.10.2020).

Table 4.2 Comparison of the laser performance and the respective improvements achieved over the first directly diode-pumped Cr^{2+}-doped II–VI laser system

	MIRanda1 [41]	MIRanda2 [42]
ML mechanism	Kerr-lens ML	Kerr-lens ML
Gain medium	Cr:ZnSe (5 mm)	Cr:ZnS (5 mm)
P_{pump}	3 W	2 × 2.5 W
λ_{pump}	1.65 µm	1.65 µm
P_{CW}	700 mW	1200 mW
P_{ML}	500 mW	800 mW
τ_0	45 fs	34 fs
ν_{rep}	64.7 MHz	49.9 MHz
OC ratio	34%	25%
E_0	**7.7 nJ**	**16 nJ**
P_0	**151 kW**	**415 kW**

the directly diode-pumped oscillator could be retained, revealing a measured integrated RIN [20 Hz–1 MHz] of less than 0.014% RMS for the fundamental wavelengths. Compared to the first-generation system, an almost three-fold increase in peak power was achieved via polarization-multiplexing of two pump laser diodes (see Table 4.2). In addition, the pulse duration was shortened further with the aid of newly designed dispersive cavity mirrors. This places the power performance of directly diode-pumped Cr^{2+}:ZnS/ZnSe oscillators on the same level as of state-of-the-art fiber-laser-pumped versions [8], with the diode-pumped systems representing one of the most attractive driving sources for efficient mid-infrared generation via nonlinear $\chi^{(2)}$ optical processes.

References

1. Wang Q et al (2019) Broadband mid-infrared coverage (2–17 µm) with few-cycle pulses via cascaded parametric processes. Optics Lett 44(10):2566–2569
2. Vasilyev S, et al (2019) Super-octave longwave mid-infrared coherent transients produced by optical rectification of few-cycle 2.5-µm pulses. Optica 6(1):111–114
3. Wang Q, et al (2018) Complete mid-infrared coverage (2-20 µm) via cascaded parametric processes. Laser Congress 2018 (ASSL) ATu6A.4
4. Nagl N, Mak K, Wang Q, Pervak V, Krausz F, Pronin O (2019) Efficient femtosecond mid-infrared generation based on a Cr:ZnS oscillator and step-index fluoride fibers. Optics Lett 44(10):2390–2393
5. Dudley JM (2006) Supercontinuum generation in photonic crystal fiber. Rev Modern Phys 78(4):1135–1184
6. Grassani D et al (2019) Mid infrared gas spectroscopy using efficient fiber laser driven photonic chip-based supercontinuum. Nature Commun 10:1553

7. Vasilyev S, Moskalev I, Mirov M, Smolski V, Mirov S, Gapontsev V (2017) Ultrafast middle-IR lasers and amplifiers based on polycrystalline Cr:ZnS and Cr:ZnSe. Optics Mater Express 7(7):2636–2650
8. Mirov S et al (2018) Frontiers of mid-IR lasers based on transition metal doped chalcogenides. IEEE J Selected Topics Quantum Electron 24(5):1–29
9. Moulton PF (1986) Spectroscopic and laser characteristics of Ti:Al_2O_3. J Optical Soc Amer B 3(1):125–133
10. Backus S et al (2017) Direct diode pumped Ti:sapphire ultrafast regenerative amplifier system. Optics Express 25(4):3666–3674
11. Mirov SB et al (2003) Diode and fibre pumped Cr^{2+}:ZnS mid-infrared external cavity and microchip lasers. IEE Proc Optoelectron 150(4):340–345
12. Durfee CG et al (2012) Direct diode-pumped Kerr-lens mode-locked Ti:sapphire laser. Optics Express 20(13):13677–13683
13. Gürel K et al (2015) Green diode-pumped femtosecond Ti:sapphire laser with up to 450 mW average power. Optics Express 20(23):30043–30048
14. Sugiyama N, Tanaka H, Kannari F (2018) Mode-locked Ti:sapphire laser oscillators pumped by wavelength-multiplexed laser diodes. Jpn J Appl Phys 57(5):052701
15. Byrd M et al (2015) Wavelength selection and polarization multiplexing of blue laser diodes. IEEE Photon Technol Lett 27(20):2166–2169
16. Backus S, Kirchner M, Durfee C, Murnane M, Kapteyn H (2017) Direct diode-pumped Kerr Lens 13 fs Ti:sapphire ultrafast oscillator using a single blue laser diode. Optics Express 25(11):12469–12477
17. Keller U (1994) Ultrafast all-solid-state laser technology. Appl Phys B 58:347–363
18. Kärtner FX, Aus der Au J, Keller U (1998) Mode-locking with slow and fast saturable absorbers - What's the difference? IEEE J Selected Topcis Quantum Electron 4(2):159–168
19. Kärtner FX, Jung ID, Keller U (1996) Soliton mode-locking with saturable absorbers. IEEE J Select Topics Quantum Electron 2(3):540–556
20. Keller U (2007) Ultrafast solid-state lasers. In: Herziger G, Weber H, Poprawe R (eds) Landolt-Börnstein - Group VIII/1B1. Springer, Berlin, pp 33–167
21. Aus der Au J, Kopf D, Morier-Genoud F, Moser M, Keller U (1997) 60-fs pulses from a diode-pumped Nd:glass laser. Optics Lett 22(5):307–309
22. Kalashnikov VL, Sorokin E, Sorokina IT (2003) Multipulse operation and limits of the Kerr-Lens mode-locking stability. IEEE J Quantum Electron 39(2):323–336
23. Yefet S, Pe'er A (2013) A review of cavity design for Kerr Lens mode-locked solid-state lasers. Appl Sci 3(4):694–724
24. Lee YW, Yi JH, Cha YH, Rhee YJ, Lee BC, Yoo BD (2005) Numerical analysis of soft-aperture Kerr-lens mode locking in Ti:Sapphire laser cavities by using nonlinear ABCD matrices. J Korean Phys Soc 46(5):1131–1136
25. Magni V, Cerullo G, De Silvestri S (1993) ABCD matrix analysis of propagation of gaussian beams through Kerr media. Optics Commun 96(4–6):348–355
26. Magni V, Cerullo G, De Silvestri S (1993) Closed form gaussian beam analysis of resonators containing a Kerr medium for femtosecond lasers. Optics Commun 101(5–6):365–370
27. Anthony E (1986) Siegman. University Science Books, Lasers
28. Kogelnik H, Li T (1996) Laser beams and resonators. Appl Optics 5(10):1550–1567
29. Grace EJ, Ritsataki A, French PMW, New GHC (2000) New optimization criteria for slit-apertured and gain-apertured KLM all-solidstate lasers. Optics Commun 183:249–264
30. Dymott MJP, Ferguson AI (1994) Self-mode-locked diode-pumped Cr:LiSAF laser. Optics Lett 19(23):1988–1990
31. Sorokina IT, Sorokin E, Wintner E, Cassanho A, Jenssen HP, Szipöcs R (1997) Sub-20 fs pulse generation from the mirror dispersion controlled Cr:LiSGaF and Cr:LiSAF lasers. Appl Phys B 65:245–253
32. Piché M, Salin F (1993) Self-mode locking of solid-state lasers without apertures. Opt Lett 18(13):1041–1043
33. Xu L et al (1997) High-power sub-10-fs Ti:sapphire oscillators. Appl Phys B 65(2):151–159

34. Cerullo G, De Silvestri S, Magni V (1994) Self-starting Kerr-lens mode locking of a Ti:sapphire laser. Optics Lett 19(14):1040–1042
35. Xia P, Kuwata-Gonokami M, Yoshioka K (2020) Geometrical analysis of Kerr-lens mode-locking for high-peak-power ultrafast oscillators. Jpn J Appl Phys 59:062002
36. Salin F, Squier J (1992) Gain guiding in solid-state lasers. Optics Lett 17(19):1352–1354
37. Brabec T, Curley F, Spielmann C, Wintner E, Schmidt JA (1993) Hard-aperture Kerr-lens mode locking. J Optical Soc Amer B 10(6):1029–1034
38. Yariv A (1963) Energy and power considerations in injection and optically pumped lasers. Proceed IEEE 51(12):1723–1731
39. Hodgson N, Weber H (2005) Output power of laser resonators. In: Laser resonators and beam propagation. Springer series in Optical sciences, vol 108, pp 393–412. Springer, New York
40. Tawfieq M, Kragh Hansen A, Bjarlin Jensen O, Marti D, Sumpf B, Andersen PE (2018) Intensity noise transfer through a diode-pumped titanium sapphire laser system. IEEE J Quantum Electron 54(1):1700209
41. Nagl N, Gröbmeyer S, Pervak V, Krausz F, Pronin O, Mak K (2019) Directly diode-pumped, Kerr-lens mode-locked, few-cycle Cr:ZnSe oscillator. Optics Express 27(17):24445–24454
42. Nagl N, Gröbmeyer S, Pötzlberger M, Pervak V, Krausz F, Mak KF (2020) Directly diode-pumped few-optical-cycle Cr:ZnS laser at 800 mW of average power. In: Conference on lasers and electro-optics. Optical Society of America, SF3H.5

Chapter 5
Pushing the Peak Power for Direct Mid-Infrared Generation

With *MIRanda2*, directly diode-pumped few-cycle Cr^{2+}-doped II–VI lasers have now reached the power level of state-of-the-art fiber-laser-pumped versions, hence opening the door to nonlinear frequency conversion into the mid-IR driven by 2–3 μm lasers. Among the various techniques that can be exploited, two main categories are typically employed: nonlinear spectral broadening in waveguide structures [1–4] and parametric frequency conversion processes in nonlinear crystals [5–8]. In recent years, both types have been successfully applied to Cr:ZnS/ZnSe pump systems [9–14], resulting in 1.7-octave-spanning supercontinua (1.6–5.1 μm) at 350 mW average power via step-index fluoride fibers [9]. To access even longer wavelengths, however, the technique of intra-pulse difference frequency generation (IDFG) holds great potential, and spectral components up to 18 μm could already be demonstrated with the chromium-based driving lasers [10, 11, 13].

DFG is a nonlinear $\chi^{(2)}$ optical process and allows for a **self-stabilization of the carrier-envelope phase (CEP)** when two pulses sharing the same shot-to-shot CEP fluctuations mix together in a nonlinear medium [15, 16]. The carrier-envelope phase is an important property of ultrashort pulses[1] and has a strong effect on the electric field waveform of the pulse, especially when going down to few-optical-cycle pulse durations. Moreover, emerging mid-infrared applications such as field-resolved infrared spectroscopy [17, 18] require the CEP of the laser pulse to be stable in time. Otherwise, when exciting molecular vibrations with pulsed laser light, the consecutive waveforms emitted by the sample can hardly be interpreted via electro-optic sampling (EOS), whose dynamic range increases with averaging and thus longer measurement time ($\sim$10s of seconds [18]). For example, for a driving laser with a specified repetition rate of 28 MHz, 28 million electric-field waveforms are generated

[1] The frequency components of a laser pulse can be written as $\nu_m = \nu_{CEO} + m \cdot \nu_{rep}$, with m denoting the number of the mode. A pulse is said to be CEP-stable when the carrier-envelope offset (CEO) frequency ν_{CEO} is well controlled.

N. Nagl, *A New Generation of Ultrafast Oscillators for Mid-Infrared Applications*, Springer Theses, https://doi.org/10.1007/978-3-030-89754-3_5

per second, and a multiple of them would need to be detected. If the waveform is not identical for every pulse, it is impossible to draw reliable conclusions from the measured EOS traces after averaging.

Without sophisticated feedback mechanisms acting on the pulses' phase, mode-locked lasers generally fail to emit phase-stable pulses. In fact, laser intensity variations, mechanical vibrations, temperature or air turbulences cause the CEP to fluctuate randomly from pulse to pulse [16]. For lasers generating soliton-like pulses, this time-dependent phase change $\delta\psi(t)$ can be expressed as [19]:

$$\delta\psi(t) = \phi_{Kerr}\left(\frac{dP_0}{dW}\right)_{W_0} \cdot \delta W(t) - \omega_0 D \delta\omega_0(t), \tag{5.1}$$

with peak power P_0, pulse energy W ($W_0 = W_{t=0}$), center frequency ω_0, net intracavity GDD D, and soliton phase shift ϕ_{Kerr} due to the optical Kerr effect in the gain medium. Hence, changes in the center frequency $\delta\omega_0(t)$ and pulse energy $\delta W(t)$ together with external noise sources perturbing the peak power P_0 govern the overall phase fluctuations. Consequently, the DFG scheme is highly sought after for delivering the phase-stabilized pulses needed in applications: when mixing two pulses with the same shot-to-shot CEP fluctuations in a nonlinear crystal, these fluctuations are automatically cancelled for the generated difference frequency components [20, 21].

Compared to standard inter-pulse DFG, in which the interacting pump and seed beam originate from separate laser beams, **intra-pulse DFG** has become quite popular in this respect. Also referred to as optical rectification, its robust and compact single-beam geometry involves the mixing between different spectral components of the same ultrabroadband pulse (see Fig. 5.1). Without doubt, IDFG is experimentally much easier to set up and does not require precise spatial and temporal overlapping

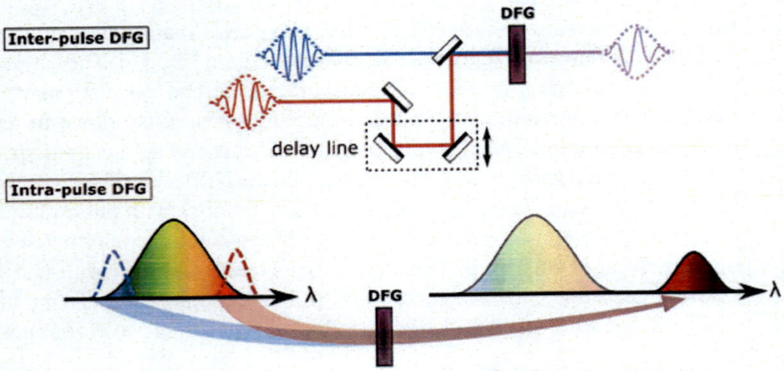

Fig. 5.1 Comparison of inter-pulse and intra-pulse difference frequency generation schemes. The figure is inspired by [21]

of two beams. Moreover, it is not affected by a timing jitter between the pump and seed pulse, which introduces additional fluctuations on the CEP.

Regarding the nonlinear conversion efficiencies of DFG-based processes, however, they are only in the range of 10^{-3}–10^{-2} [22]. Even though efficiencies as high as 50% have already been demonstrated with spectral broadening in waveguides [9], the bulk materials used for DFG are far less sensitive to beam alignment and can often handle much higher power levels than waveguide structures. Thus, to also reach large **signal strengths** via DFG, a closer inspection of the scaling law of the generated MIR intensities I_3 is helpful [22, 23]:

$$I_3(L) = \frac{1}{2}\epsilon_0 c n_3 |A_3(L)|^2 = \frac{8d_{eff}^2 \omega_3^2 I_1 I_2}{\epsilon_0^3 c^3 n_1 n_2 n_3} L^2 \cdot \mathrm{sinc}^2\left(\frac{\Delta k L}{2}\right). \tag{5.2}$$

Here, $\Delta k = k_1 - k_2 - k_3$ is the momentum mismatch, which is a scalar in the case of IDFG, L is the propagation distance in the crystal, $d_{eff} = \frac{1}{2}\chi^{(2)}$ refers to the so-called effective nonlinear coefficient, A denotes the amplitude of the optical field, and n is the frequency-dependent refractive index. The subscripts 1, 2 and 3 relate to the pump, signal and generated DFG components with frequencies w_1, w_2 and w_3, respectively. When assuming perfect phase-matching conditions ($\Delta k = 0$) and solitonic pulses (sech^2-shape), the mid-IR intensity thus scales with the input pulse parameters as follows:

$$I_3 \propto I_1 \cdot I_2 \propto I_{0,1} \cdot I_{0,2} \quad \text{with} \quad I_0 = 0.88 \frac{P_{av}}{\nu_{rep}\tau_p} \frac{1}{A_{foc}} = P_0 \frac{1}{A_{foc}}. \tag{5.3}$$

Hence, to achieve large signal strengths, high peak intensities P_0 of the driving laser pulses are desired along with small focal spot sizes A_{foc} in the nonlinear crystal.[2] Recent demonstrations of Cr:ZnS/ZnSe-based IDFG involved further scaling of the average oscillator output power P_{av} to enhance P_0. By means of additional amplification stages, these master oscillator power amplifier (MOPA) systems can boost the average power level of the few-cycle laser pulses to several watts, albeit increasing the complexity of the overall setup [10, 11, 13].

In this chapter, a **new approach** is presented, in which IDFG is directly driven with the mode-locked output of a diode-pumped Cr:ZnS oscillator for the first time. As illustrated in Fig. 5.2, the simultaneous reduction in pulse duration τ_p and repetition rate ν_{rep} also results in a peak power enhancement—potentially without the need for further amplification stages. Moreover, if the oscillator can be directly pumped with laser diodes, the low-noise characteristics of the mode-locked laser output are preserved and not affected by subsequent amplification.

In addition, the shorter pump pulse duration also corresponds to a broader spectral bandwidth. Since the shortest mid-infrared wavelengths that can be generated upon IDFG are dictated by the frequency difference between the shortest (λ_1) and longest

---

[2] It should be noted that the minimum focal spot size in the nonlinear optical crystal is limited by the damage threshold and the onset of multi-photon absorption at high peak intensities (see Sect. 5.3.2).

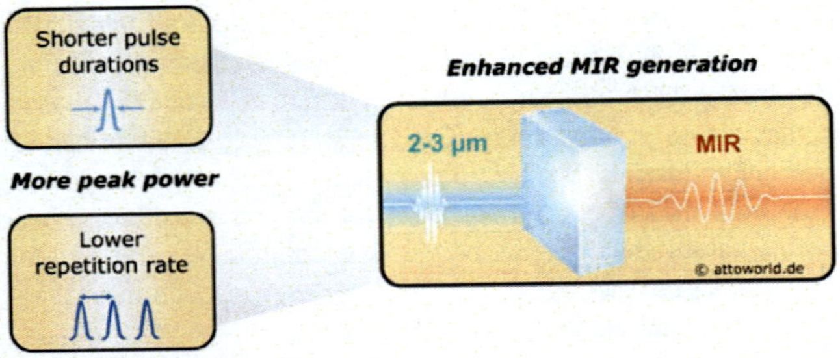

Fig. 5.2 Illustration of enhanced mid-infrared generation via shorter driving pulse durations and a lower repetition rate, denoting the operation characteristics of the third-generation laser system *MIRanda3*

Table 5.1 Dependance of the shortest mid-IR wavelengths reachable for different driving pulse durations, with the first two τ_p-values corresponding to *MIRanda1* (45 fs) and *MIRanda2* (34 fs). To obtain the shortest and longest wavelength components, the spectral limits at the 20 dB were considered for bandwidth-limited pulses

Pulse duration	20 dB-limits	$\lambda_{3,min}$ (µm)
$\tau_p = 45$ fs	$\lambda_1 = 2100$ nm	12.6
	$\lambda_2 = 2520$ nm	
$\tau_p = 34$ fs	$\lambda_1 = 2045$ nm	9.6
	$\lambda_2 = 2600$ nm	
$\tau_p = 25$ fs	$\lambda_1 = 1960$ nm	7.0
	$\lambda_2 = 2720$ nm	
$\tau_p = 15$ fs	$\lambda_1 = 1775$ nm	4.2
	$\lambda_2 = 3050$ nm	

(λ_2) wavelength components of the driving laser beam ($\lambda_{3,min}^{-1} = \lambda_1^{-1} - \lambda_2^{-1}$), oscillators delivering broader output spectra and thus shorter pulse durations are highly desired. In Table 5.1, four different pulse durations were considered, with the minimum achievable wavelength λ_3 being calculated for each case. It shows that shorter pulse durations allow to reach shorter and shorter DFG wavelengths, clearly illustrating the advantage of few-cycle laser oscillators to span ever increasing portions of the mid-infrared region.

In the following, a new generation of diode-pumped few-cycle Cr^{2+}-doped II–VI lasers is presented, operating at an unprecedented low repetition rate of $\nu_{rep} = 23$ MHz combined with pulse durations measured down to three optical cycles. This unique combination significantly enhanced the output peak power of the oscillator, enabling direct generation of mid-infrared radiation via IDFG for the first time (Sect. 5.3). However, further improvements in oscillator and pump design (Sect. 5.2.1) were needed to tackle challenges arising from higher intra-cavity peak power levels. Also, the effect of air dispersion on the mode-locked oscillator performance was studied by purging the chamber with dry nitrogen (see Sect. 5.2.2).

5.1 Challenges of Reaching Higher Peak Powers

To operate the oscillator at higher peak powers P_0, the three laser parameters P_{av}, ν_{rep} and τ_p can be addressed (see Eq. 5.3). In principle, an increase in average power would be a viable approach. However, since the output power of a single-emitter InP pump laser diode is limited to 2.5–3 W, an improvement over the second-generation system (Chap. 4) would have only been achievable by incorporating additional pump diodes into the setup. Hence, to push the peak power from 400 kW to the 1 MW level, the laser system would need to be operated at 2.5-times the average mode-locked power of *MIRanda2*, requiring at least two more pump diodes to be integrated via simultaneous polarization- and wavelength-multiplexing. Since this increases the complexity of the system, a better approach to peak power scaling is to lower the oscillator's repetition frequency. Though this approach is highly attractive, new challenges arise over the previous diode-pumped systems (Chaps. 3 and 4), which are briefly addressed in the following:

- **Enhanced nonlinearities:**
  To scale the peak power of the passively mode-locked laser, one of the major challenges is attributed to the ever stronger nonlinear effects inside the gain medium. If they become excessive, multi-pulse instabilities will develop and affect the soft-aperture Kerr-lens mode-locked operation. While this was also observed for *MIRanda2* (see Sect. 4.1.2), the effect is expected to be even stronger in the third-generation system. Also, to reduce the nonlinearities, operation of the cavity in stability regimes with lower self-amplitude modulation and thus lower Kerr strength might be inevitable. However, this would come at the expense of a mode-locked operation being much more difficult to start.

A useful measure to assess the strength of nonlinear effects is given by the so-called "B-integral" [24]:

$$B = \frac{2\pi}{\lambda} \int_0^L n_2 I_0(z) dz, \tag{5.4}$$

with n_2 and $I_0(z)$ denoting the nonlinear refractive index and the optical peak intensity along the propagation direction z. To avoid serious distortion effects due to self-phase modulation or self-focusing in a nonlinear medium, the generally accepted criterion is to keep the cumulative B-integral below the value B $\leq$ 3 to 5.
Considering the second-generation laser system presented in the previous chapter, mode-locked output parameters of $P_{av} = 800$ mW and $\tau_p = 34$ fs could be achieved at a repetition rate of 50 MHz and an output coupling ratio of 25%. With a focal spot diameter of ~90 µm and a central wavelength of $\lambda = 2.3$ µm, the B-integral for a single-pass through the $L = 5$ mm long Cr:ZnS gain crystal was calculated and plotted in Fig. 5.3a. The same parameters were also used for a calculation at half the laser repetition rate (25 MHz). Assuming a simple lengthening of the cavity with the same cavity optics, and maintaining the arm length ratio of 0.5, ABCD-matrix

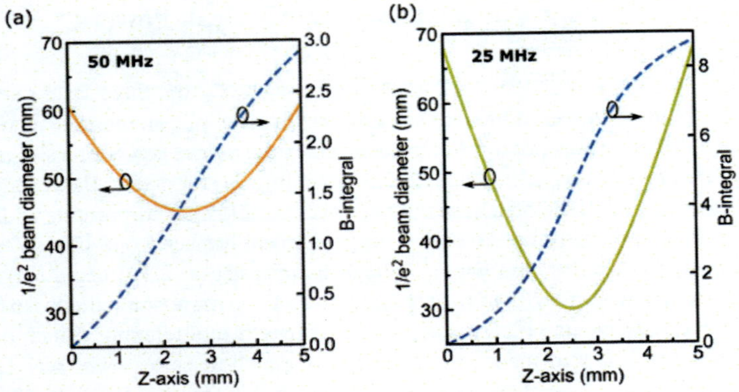

Fig. 5.3 Calculation of the single-pass B-integral inside the Cr:ZnS crystal ($n_2 = 1 \cdot 10^{-18}$ m²/W). The smaller the focal spot diameter, the higher the accumulated nonlinear phase shift (B-integral)

analysis indicates the respective focal spot diameter in the crystal to be ~60 µm. While the single-pass B-integral in *MIRanda2* was still below B = 3, the decrease in repetition rate would result in B > 8 given the smaller focus size (see Fig. 5.3b). Therefore, to avoid destabilization of the low repetition rate KLM oscillator, other approaches are required.

- **Low repetition rate design:**
  The high cumulative B-integral of >8 for the 25 MHz repetition rate case calls for larger beam sizes in the crystal. A relatively simple way to achieve this is by selecting spherical mirrors with longer focal lengths. At the same time, however, limitations arise from mode-matching issues. As will be illustrated in Sect. 5.2.1, the third-generation cavity design is meant to be based on a "classic" X-fold cavity rather than a "folded" resonator design (see Fig. 3.6 as a reminder). The classic design is much more elegant but requires the pump beam to be focused through one of the curved mirrors. Since the pump focusing lens needs to be placed in front of this spherical input coupler mirror, restrictions are imposed on the selection of the pump lens' focal length. Hence, to ensure a good mode-matching between the low-brightness pump beam and the fundamental cavity mode, the focal lengths of the spherical mirrors cannot be chosen arbitrarily long. Nevertheless, a suitable workaround could be found and will be presented in Sect. 5.2.1.
- **Air dispersion:**
  To enhance the peak power and improve the short-wavelength coverage upon IDFG, shorter durations of the driving laser pulses are of additional benefit. However, when going towards ever shorter pulse durations, the spectral width of the laser will inevitably start to overlap with the water absorption region at around 2.6–2.8 µm (see Fig. 3.1). Since the laser was operated at ambient air conditions in a laboratory environment, the relative humidity (RH) in the laser chamber amounted to about 25% in winter and 45% during summer for a room temperature of T

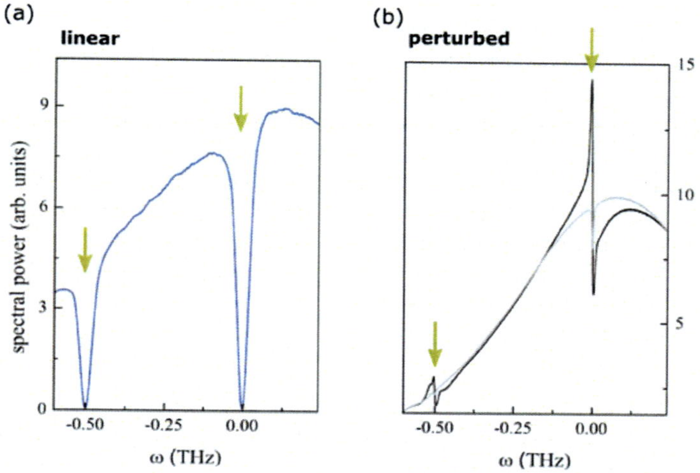

Fig. 5.4 Part of the laser pulse spectrum after linear propagation (**a**) and perturbed soliton propagation (**b**). The location of the two absorption lines is indicated with arrows. The data was taken from Ref. [25]

$\approx 20\,^{\circ}$C. If the moisture content in the air cannot be reduced, soliton perturbation theory predicts strong spectral distortions of the travelling soliton due to the presence of water absorption lines [25]. This causes the refractive index and thus the dispersion of air to be modulated by sharp Fano-shaped features of varying strength and separation [26]. This additional dispersion severely affects the spectral shape of the laser pulse, and a different effect is observed for linear (a) and perturbed (b) soliton propagation (see Fig. 5.4). The former takes place for linear propagation, for example outside the resonator structure, resulting in spectral absorption dips as expected from the classic Lambert–Beer law.[3] When the soliton is already peturbed inside the laser resonator, however, the pulse spectrum reveals sharp spectral bends with strongly suppressed absorption dips. Since the presence of absorption lines not only degrades the spatial but also the temporal pulse quality [26], the effect of water dispersion needs to be mitigated inside the laser chamber and will be analyzed in Sect. 5.2.2.

[3] It relates the intensity of the incident light I_0 to the intensity transmitted through an absorptive solution via $I = I_0 \cdot 10^{-\epsilon c l}$, with the molar absorption coefficient ϵ, the concentration c, and the optical path length l [27].

5.2 Experimental Achievements

To address the series of challenges listed above, the design of the laser system had to be modified. Moreover, when compared to the cavity design of the first- and second-generation system, no 45°-mirrors inside the oscillator were used to avoid limitations in their reflective and dispersive properties (see Sect. 4.2.2). The resulting resonator layout was based on a classic bow-tie design, with all setup components assembled inside a single monolithic housing. The latter had a footprint of only 90 cm × 60 cm and enabled proper sealing of the laser assembly. Hence, studies on the effect of air dispersion on the mode-locked laser performance could be carried out through purging experiments (Sect. 5.2.2).

5.2.1 *Pump and Oscillator Setup*

To realize the classic bow-tie design but still allow for a good mode-matching between the pump and fundamental cavity beam, a workaround was needed. Certainly, one could choose the collimated pump beam diameter arbitrarily large such that any pump spot diameter can be achieved for a certain pump lens. For practical reasons, however, this is not ideal since large pump beam diameters require large pump optics, rendering the whole setup bulky and mechanically less stable. Instead, a solution was found as shown in Fig. 5.5: By combining two lenses with negative (L1) and positive (L2) focal lengths, the distance d to the focus could be successfully increased. The reason behind is that the incident pump beam on the lens L2 was now divergent rather than collimated. To confirm that the focal spot diameters are still kept at a reasonably low level for this two-lens combination and do not get significantly distorted via spherical aberrations, measurements were performed and summarized in Table 5.2. The focal lengths of the optics were chosen with regard to the new oscillator design, which should be based on a repetition rate of 25 MHz. With arm lengths of 1900 mm and 3800 mm, spherical cavity mirrors with focal lengths of f = −200 mm and f = −300 mm were selected from the set of available cavity optics. According to ABCD-matrix analysis, this combination results in a focal spot diameter of 85–90 μm for the fundamental cavity mode—comparable to *MIRanda2*.

The measurement setup consisted of a meniscus lens with negative focal length as well as an achromat and a spherical mirror (see Fig. 5.5). This combination was exper-

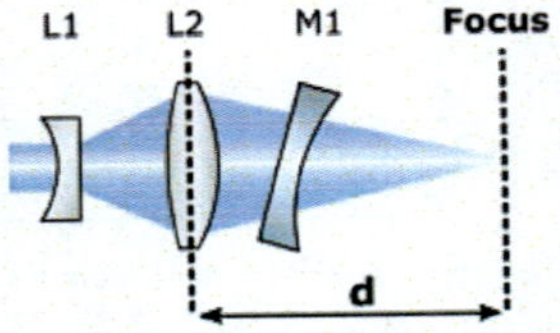

Fig. 5.5 Schematic of the two-lens pump focusing unit. L1: f = −150 mm, L2: f = 75 mm, M1: ROC = −200 mm

Table 5.2 Measured focal spot diameters and distance *d* for a given set of optics

Optics	Focus ∅	d (mm)
L2	80 μm (x)	**75**
	50 μm (y)	
L2, M1	95 μm (x)	**85**
	65 μm (y)	
L2, M1, L1	140 μm (x)	**135**
	95 μm (y)	

imentally found to provide the best performance regarding the minimum achievable focal spots. For the pump beam, the collimated output of a InP laser diode (*SemiNex Corp.*, C-132-118) was used at a diode driving current of 10 A. The results are depicted in Table 5.2 and reveal that the successive integration of the input coupler mirror (M1) and the defocusing lens (L1) into the measurement setup resulted in a 1.8-times increase in focal distance (75 mm → 135 mm). At the same time, however, the focal spot diameters increased in the horizontal (x) and vertical (y) direction, which is attributed to the spherical aberrations imposed on the pump beam. Nevertheless, a tight spot diameter of 95 μm was achieved for the fast axis direction, which is only marginally larger than the theoretical value for the cavity mode. Hence, a good mode-matching is still expected. Besides the tight focus, it is also the distance of d = 135 mm that enables a successful realization of this new design. Since the spherical mirror M1 has a focal length of 100 mm, the two-lens combination (L1, L2) can be safely placed in front of it because $d > f_{M1}$. Finally, it should be mentioned that the M^2-values of the laser diode beam were confirmed to be not seriously affected by this optics combination.

A schematic of the experimental setup is depicted in Fig. 5.6, including two $4f$-telescopes to extend the cavity arm lengths. However, note that this figure represents the oscillator in its final state, and some adjustments to the cavity design were required to achieve optimal laser performance (see Sects. 5.2.1.1 and 5.2.1.2). In addition, a new type of C-Mount laser diode was tested (*SemiNex Corp.*, C-132-134), for which the fast axis divergence was matched to the one of the slow axis. Unlike the previous laser diodes, this type required only one cylindrical lens for collimating the fast (f) and slow (s) axis, respectively, making the collimation procedure much more convenient. For diode 1 (C-132-134), cylindrical lenses with f = 20 mm (f) and f = 75 mm (s) were used, whereas a cylindrical lens combination of f = −25 mm (f), f = 100 mm (s) and f = 150 mm (f) was selected to address the output of diode 2 (C-132-118). Moreover, a rail-based assembly was chosen for the pump collimation unit, while the pump focusing lenses were mounted inside a cage system (see pictures Fig. 5.7). The gain medium consisted of a 5.1 mm long AR-coated polycrystalline Cr:ZnS crystal (*IPG Photonics*) with a clear aperture of 1.8 mm × 4.1 mm. Similar to before, a water-cooled copper mount was used as a crystal heat sink and cooled to a temperature of 13°C, resulting in a measured single-pass absorption of

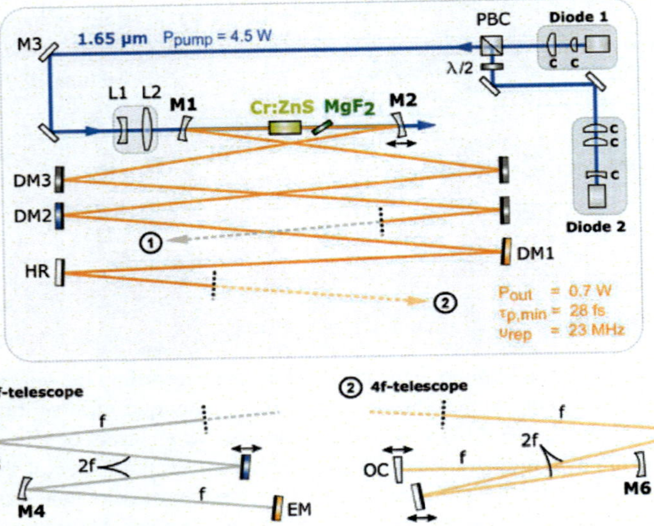

Fig. 5.6 Experimental layout of *MIRanda3*, including the additional $4f$-telescope extensions. Diode 1: Type C-132-134, Diode 2: Type C-132-118, C: cylindrical lens, PBC: polarizing beam combiner, $\lambda/2$: half-wave plate, M3: highly-reflective rectangular-shaped mirrors, L1: meniscus lens (f = −200 mm), L2: achromatic lens (f = 80 mm), M1 and M2: ROC = −200 mm, M3-M6: ROC = −800 mm, DM (orange, blue, dark grey): dispersive mirrors, HR (light grey): highly-reflective mirrors, Cr:ZnS: gain medium (5 mm), Brewster-plate: MgF_2 (3 mm), OC: 25% output coupler

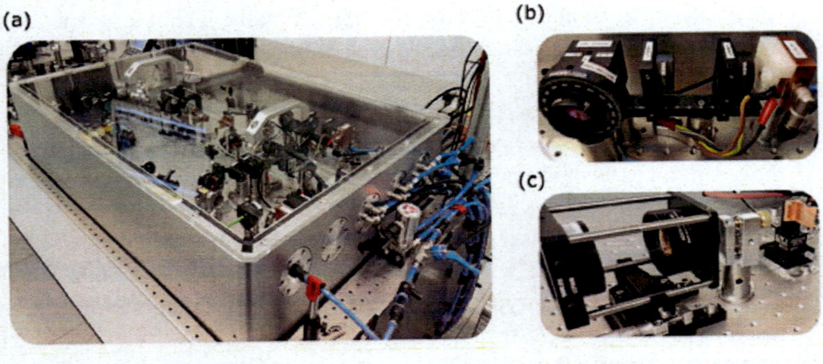

Fig. 5.7 Pictures of the actual laser assembly. **a** Laser housing. **b** Rail-based pump diode setup. **c** Modified pump focusing unit. The latter was based on a cage system and placed on a translation stage for fine-adjusting the pump focus position

about 79%. The cooling temperature was identical for the pump laser diodes, which were connected to the same external circulating bath unit (*termotek*, P10040-19192). Compared to the first- and second-generation system, *MIRanda3*'s beam path configuration is much simpler given the classic bow-tie style of the cavity.

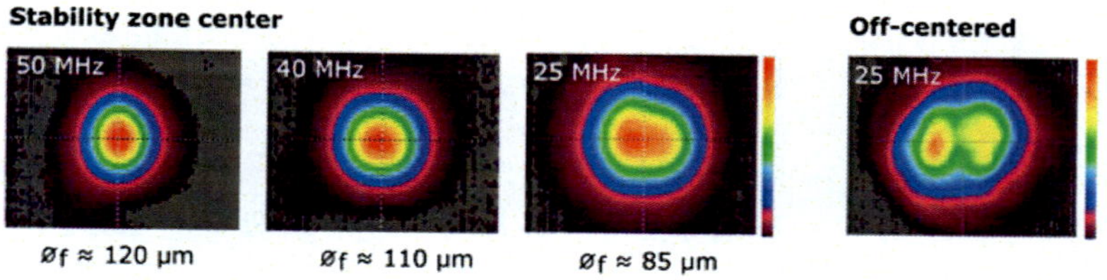

Fig. 5.8 Continuous-wave output beam corresponding to different cavity arm lengths. The theoretical repetition rate is depicted along with the theoretical cavity mode diameter $\varnothing_f$ in the focus. The pump spot was as large as $\varnothing_p$ ~95 µm, which explains the degradation in beam quality for the 25 MHz cavity ($\varnothing_p > \varnothing_f$)

5.2.1.1 Intra-Cavity 4f-Telescopes

To realize an oscillator design with a low repetition rate of about 25 MHz, the idea was to build a cavity similar to *MIRanda2*, but with elongated arm lengths of 1900 mm and 3800 mm, respectively. Also, a combination of f = 200 mm and f = 300 mm spherical mirrors was chosen such that a similar focal spot size of 85–90 µm would be obtained in the gain crystal. Experimentally, however, it turned out that the continuous-wave output beam started to significantly degrade for the longer cavity lengths. As shown in Fig. 5.8, the CW output was measured with the pyroelectric camera and looked single-mode for cavity lengths corresponding to 50 and 40 MHz repetition rates. However, it started to become larger and slightly less circular for the 25 MHz resonator. In particular, when the output beam was monitored for slightly off-centered positions (±1 mm) within the cavitie's stability zones, a multi-mode behaviour could be detected. This was attributed to a degradation in mode-matching efficiency: since the pump focal spot diameter (~95 µm) and the theoretical cavity mode diameter (~85 µm) were already slightly differing at the zone center, an operation closer to the zone boundaries might have enlarged this discrepancy. Thus, the probability of exciting higher-order modes increased.

Because the pump focus could not be reduced further, a solution was found by extending the cavity arms of a 45 MHz cavity via ***4f*-telescopes** (see Fig. 5.6). Here, the abbreviation "4f" refers to the four separate distances of length f that separate the input plane of a travelling laser beam from its output plane, and a 4f-assembly can be thought as an imaging system with unit magnification [28]. Hence, when the ends of both resonator arms are lengthened by 4f-telescopes, the beam sizes will effectively not change at the end-mirror positions. Moreover, the beam diameters in the original 45 MHz cavity will stay unaffected—including the focal spot sizes in the gain medium. In the experiment, the resonator was lengthened to a nominal repetition rate of 23 MHz by using spherical mirrors (M3-M6, f = −400 mm), resulting in a spatially Gaussian single-mode CW operation around the stability zone centers. Next, soft-aperture Kerr-lens mode-locking was tried via a rapid movement and slight misalignment of the output coupler. Since no pulsed operation was achieved, the same was done for the original cavity without the 4f-extensions. Surprisingly,

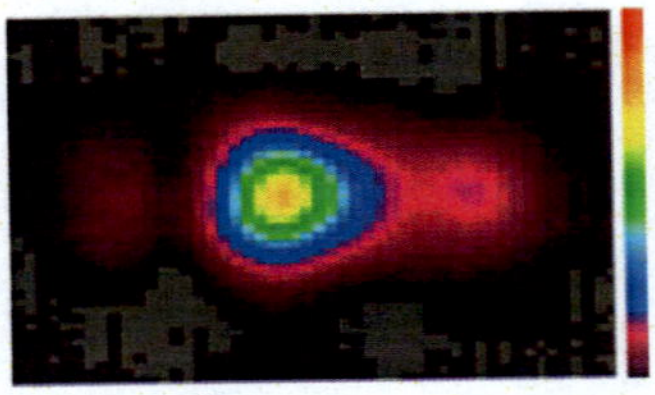

Fig. 5.9 Mode-locked output beam profile for the 45 MHz cavity

the oscillator could only be mode-locked for a strong detuning of the OC mirror, resulting in a highly non-single-mode output beam profile as depicted in Fig. 5.9. The output beam revealed three lobes in the horizontal direction, with the central lobe being the most intense one.

This observation along with the fact that the cavity mode in the crystal is larger than for the first- and second-generation system (~110 µm vs. ~60 µm for *MIRanda1* and ~90 µm for *MIRanda2*) lead to the assumption that the oscillator might favor tighter focal spot sizes for soft-aperture KLM. Moreover, it is worthwhile to mention that the low brightness of the pump laser diodes cannot explain the observed behaviour because the same mode-locked output beam shape was obtained when the oscillator was pumped by a single-mode fiber laser with diffraction-limited beam quality instead (*Bktel, HPFL-370-1567-FCAPC*). Therefore, the impact of the focal spot size on the oscillator's mode-locking performance was analyzed further.

5.2.1.2 Selection of the Curved Mirror Pair

To illustrate the effect of different focal spot diameters on the Kerr strength parameter, two-dimensional simulations were performed on the basis of the nonlinear ABCD-matrix analysis presented in Sect. 4.1.2. For this, the cavity arm lengths were kept constant and the radii of curvature (ROC) of the two spherical mirrors surrounding the 5 mm long Cr:ZnS gain crystal were chosen to be equal. When varying the ROC from −100 to −300 mm, the simulated results are shown in Fig. 5.10 and reveal a clear tendency: because the intensity scale for the Kerr strength parameter was kept the same for all figures, the more intense colouring at the shorter ROC indicates that KLM might be more favorable when tighter spot sizes are present in the gain medium.

As a consequence, the ROC = −300 mm mirror inside *MIRanda3* was replaced by a ROC = −200 mm mirror. While this decreases the focal spot diameter from ~110 to ~90 µm, it can still ensure a good mode-matching with the pump mode (~95 µm). Therefore, no further reduction in focus sizes was considered to avoid the onset of multi-mode lasing. Finally, the combination of tighter focusing along with the incorporation of $4f$-telescopes was successfully implemented in the third-generation system, resulting in a stable single-mode and Kerr-lens mode-locked operation at a low repetition rate of 23 MHz.

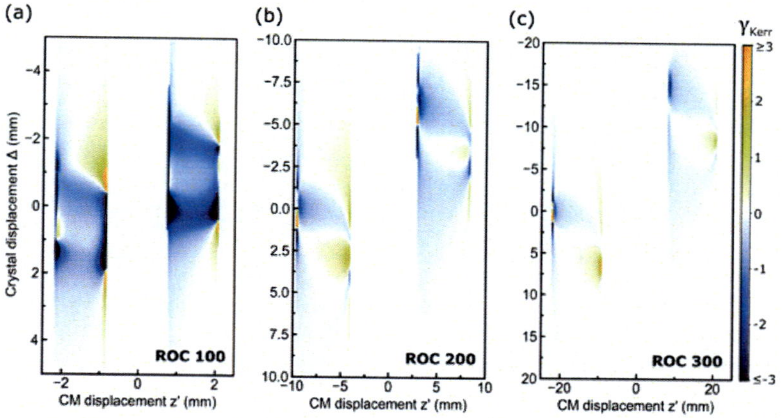

Fig. 5.10 Two-dimensional simulations of the Kerr strength parameter for different radii of curvature (ROC) of the spherical cavity mirrors. The arm lengths were chosen to be 900 mm and 1870 mm, resulting in a theoretical repetition rate of about 50 MHz. For a 5 mm long Cr:ZnS gain crystal, the results are depicted for **a** ROC $= -100$ mm, **b** ROC $= -200$ mm, and **c** ROC $= -300$ mm

5.2.2 Mode-Locked Performance and the Effect of Air Dispersion

Stability Zones

To locate the oscillator's stability zones and identify parameter ranges that are potentially favorable for soft-aperture Kerr-lens mode-locking, the CW output power of the oscillator was recorded for different mechanical stage positions of mirror M2. Since no Brewster-plate was used in the cavity, the beam astigmatism arising from a reflection on the curved mirrors was uncompensated. Hence, the measured width of the two zones was rather narrow and is plotted in Fig. 5.11a. Therefore, a 3 mm thick MgF_2-plate was placed at Brewster's angle in close proximity to the gain crystal (see Fig. 5.6). This resulted in enlarged stability zone widths that are in good agreement with the theoretically calculated zones. After this modification, the oscillator layout was not changed anymore, and a linear ABCD-matrix analysis was performed to illustrate the beam radii along the 23 MHz cavity axis (see Fig. 5.11b). The latter clearly reveals the additional foci created by mirrors M3 to M6, which arise from the extended-cavity structure.

Intra-cavity Dispersion Management

For the third-generation system, three different dispersive mirror designs (DM1-DM3) were used to fine-tune the cavity dispersion. The design of mirror DM1 was the same as for *MIRanda2*, while the TOD mirror design DM2 has also been used in both *MIRanda1* and *MIRanda2*. To generate even shorter pulse durations, how-

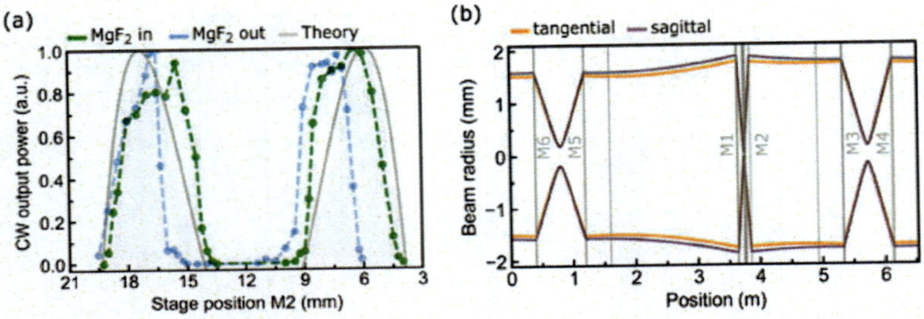

Fig. 5.11 **a** Stability zone scan performed for an output coupling ratio of 1% and normalized to the maximum CW output power. The theoretical stability zone locations were calculated via linear ABCD-matrix analysis together with **b** the beam radii along the more than 6.5 m long cavity axis

ever, the reflection bandwidth of both mirror coatings was slightly improved, which is indicated by the abbreviation "-v5" in Fig. 5.12. In addition, another dispersive mirror design (DM3) was incorporated into the oscillator, providing a moderate but constant amount of second-order dispersion (GDD ≈ -75 fs^2) per mirror bounce without adding TOD. After systematically changing the dispersive mirror combinations and recording the corresponding mode-locked output spectra (see illustration in Fig. 5.13a), it was found that a round-trip combination of 3 bounces on DM1 together with 4 bounces on DM2 and 6 bounces on DM3 was resulting in the shortest output pulse durations for the given set of cavity optics. The corresponding round-trip dispersion is plotted in Fig. 5.12a and was calculated from the theoretical dispersion values provided by our mirror coating department. Measurements of the mirrors' and output coupler's reflectivity are shown in Fig. 5.12b and reveal a promising highly-

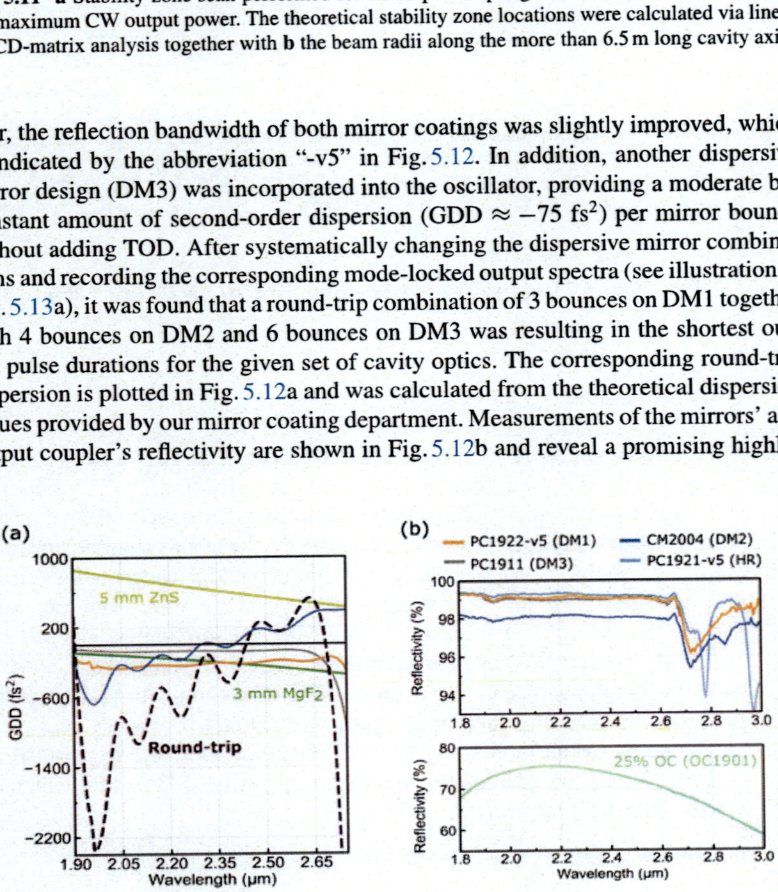

Fig. 5.12 **a** Theoretical dispersion values of intra-cavity elements including the Brewster-plate (MgF_2) and the gain crystal (Cr:ZnS) as well as the dispersive mirrors. **b** Measured reflectivity curves (p-polarized) of the highly-reflective (HR) and dispersive (DM1-DM3) mirrors along with the output coupling mirror (OC)

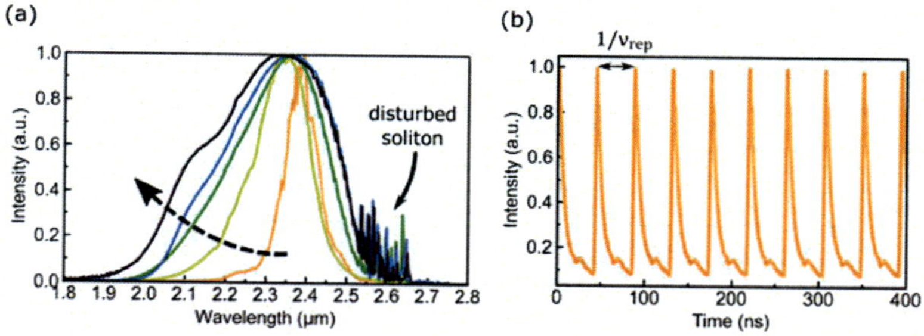

Fig. 5.13 **a** Illustration of the intra-cavity dispersion tuning and the effect on the mode-locked spectral width along with **b** the $\nu_{rep} = 23$ MHz output pulse train. The latter was recorded with a fast photodetector (*Thorlabs*, DET10D2) and oscilloscope (*Teledyne LeCroy*, HDO4034), while all measured spectra depicted in this chapter were obtained with a grating-based spectrometer (*APE* waveScan 1.5–6.3 μm, customized)

reflective behaviour in a broadband spectral range from 1.8 to 2.9 μm. The dips at around 2.7 μm are attributed to water absorption and might appear stronger than expected—most likely due to the additional effect of air humidity in the free-space reflectivity measurement setup.

Output Power and Pulse Duration

At an output coupling ratio of 25% and a repetition rate as low as 22.9 MHz (see Fig. 5.13b), the oscillator was delivering average output powers of over 1100 mW in CW and 720 mW in single-mode KLM operation when pumped by the combined emission of the two InP laser diodes (4.5 W). A further increase in mode-locked power, however, was not possible and resulted in the appearance of pulse instabilities. Note that, unlike the KLM power, the maximum CW value was obtained for an operation at the stability zone center. A SHG-FROG measurement of the output pulses is shown in Fig. 5.14 with a retrieved FWHM pulse duration of $\tau_p = 28$ fs (FTL: 28 fs), corresponding to **less than four optical cycles**. This time, a 100 μm thick uncoated lithium iodate crystal (*Eksma*, $LiIO_3$, $\theta = 20.5°$) was used for frequency doubling in the FROG device, since it has intrinsically much more favorable phase-matching properties than the BBO and GaSe crystals used so far. According to simulations performed with the open-source software *SNLO*, the SHG acceptance bandwidth of $LiIO_3$ amounts to 112.93 nm · cm at 2.3 μm, whereas it is much lower for BBO (19.05 nm · cm) and GaSe (5.6 nm · cm). Hence, for a certain crystal thickness, a much broader portion of the pulse spectrum can be phase-matched with a $LiIO_3$ crystal, which offers a high and rather smooth transmission profile for the 2–3 μm polarized light [29].

The Effect of Air Dispersion

As explained in the previous Sect. 5.1, the soliton circulating in the laser cavity acquires strong spectral modulations when encountered by some kind of absorbing

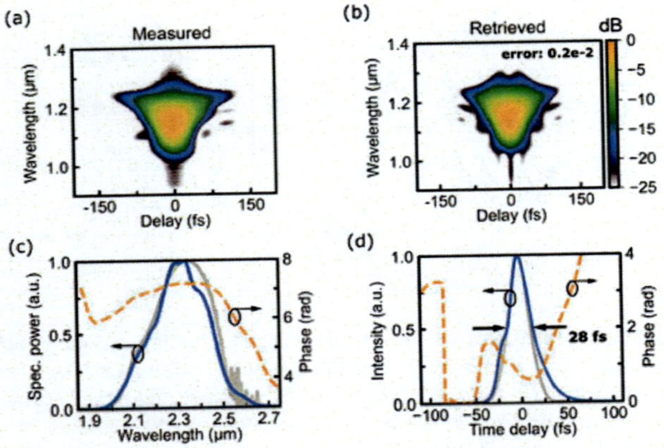

Fig. 5.14 Characterization of the generated laser pulse durations of *MIRanda3* via SHG-FROG measurements. **a** + **b** Measured and retrieved FROG trace. **c** Results in the spectral domain, with the measured spectrum depicted in grey. **d** Results in the temporal domain including the Fourier-transform limited pulse shape (grey)

substance. In the case of 2–3 μm Cr:ZnS/ZnSe oscillators, these sharp features arise from water vapor absorption in the 2.6–2.8 μm spectral region and are detectable in the mode-locked output spectra (see Fig. 5.13 and Refs. [25, 30, 31]). So far, the performance of all three diode-pumped oscillators presented in this thesis has been optimized under ambient air conditions. To illustrate the effect of air dispersion on the mode-locked laser performance, the laser chamber of this third-generation system was closed and purged with a dry gas. In principle, an evacuation of the system would have also been a viable approach to remove water vapor absorption. However, this has the disadvantage of exerting undesirably high pressure on the 30 mm thick chamber walls and the 40 mm thick base, which might cause mechanical deformations at low pressure levels. Purging with a moderate gas flow was thus considered to be more attractive. Typically, dry air or dry nitrogen are used for purging when less humid environments need to be set up. For the results presented in the following, dry nitrogen was preferred over dry air[4] to reach the lowest possible relative humidity (RH) levels. These RH-values were recorded according to a digital humidity sensor (*SENSIRION*, SHT30), which is specified for an operation range between 0 and 100% RH. Additional efforts were undertaken to seal and purge the remaining path of the laser beam outside the chamber including the pulse characterization diagnostics. However, no significant change was observed in the magnitude of the recorded spectral distortions, and thus the following results were obtained for a purged laser chamber only.

[4] "Dry air"—as available in our laboratories—might still contain some rest (though minor) humidity of a few percent.

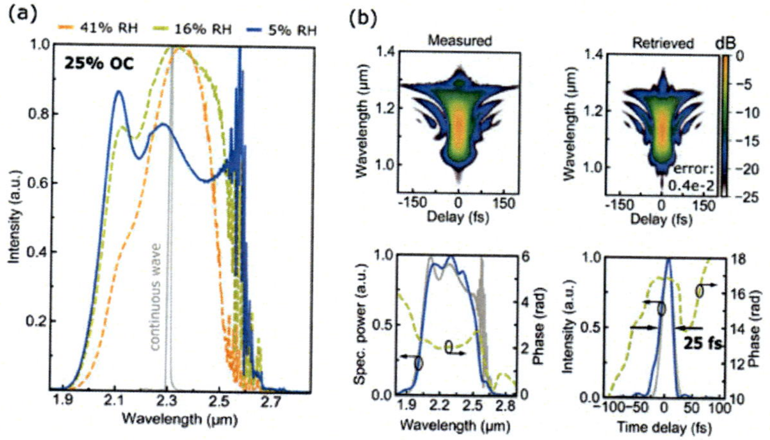

Fig. 5.15 **a** Change of the spectral shape and **b** pulse duration upon purging with dry nitrogen. The FROG trace was recorded for the lowest humidity level of 5% RH

In Fig. 5.15, the mode-locked output spectra are depicted for three different humidity levels: 41% RH (starting point), 16% RH (intermediate level), and 5% RH (final state). These measurements repeatedly revealed a significant change in the shape of the laser pulse spectrum, which could suddenly be reversed when the gas flow was stopped and the chamber lid was removed—while the laser oscillator was continuously generating ultrashort pulses. The changing behaviour of the spectral shape can be related to a variation in the total dispersion per cavity round-trip, which represents an optical path length as long as 13 m. Recalling that the diode-pumped laser systems are operated in the regime of dispersion-managed soliton formation (see Sect. 2.3.4), large swings of $\pm D$ between the negative and positive dispersion sections inside the cavity can have a pronounced effect on the solitonic pulse shape. Since the spectra measured at the very low RH-levels down to 5% became increasingly more flat-toped, a higher pulse stretching ratio might have been present. This ratio is defined as $S = LD/\tau_{FWHM}^2$ [32], with the crystal length L and dispersion swing D. Hence, to confirm that the pulse duration τ_{FWHM} of the soliton has decreased and resulted in a larger stretching ratio, a FROG trace was recorded at the lowest relative humidity level of 5% (see Fig. 5.15). It resulted in a retrieved pulse duration of 25 fs with a Fourier-transform limit of 23 fs.

However, the Kerr-lens mode-locking performance became notably unstable for a decreasing moisture content, and a reduction in intra-cavity peak power via a decrease in pump power or a shift in gain medium position was necessary to maintain stable pulsed operation. Also, KLM could not be maintained below 5% RH. This can also be related to the round-trip dispersion converging towards the GDD = 0 dispersion line, at which soliton mode-locking becomes unstable (recall Fig. 2.10 from the theory section). Interestingly, strong and sharply pronounced spectral features at the longer wavelengths (2.5–2.7 μm) were still present at the very low humidity levels. Hence,

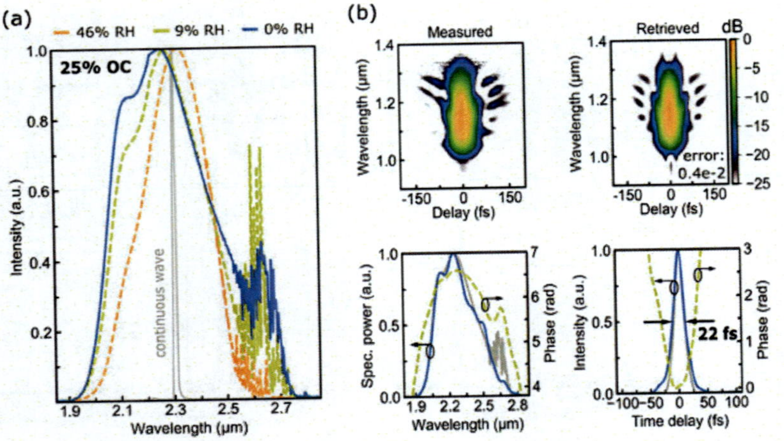

Fig. 5.16 **a** Change of the spectral shape and **b** pulse duration upon purging at slightly increased net negative intra-cavity dispersion. The FROG trace was recorded for the lowest humidity level of 0% RH. Note that the spectra and the FROG trace appear to be cleaner than those presented in Fig. 5.15

even a minor amount of water content in air can significantly perturb the circulating soliton in the resonator.

Since the measured spectral shapes started to resemble the typical output of lasers operating in the regime of chirped-pulse formation (see Sect. 2.3.4) [33–35], a **second purging experiment** was performed. This time, one of the highly-reflective (HR) cavity mirrors was replaced by a dispersive mirror (DM3) to operate the oscillator with slightly added negative dispersion and thus further away from the GDD = 0 dispersion boundary. As a result, the oscillator performed better than before when purged, revealing less flattopped spectral shapes and a cleaner SHG-FROG trace (see Fig. 5.16). Moreover, a low humidity level of 0% RH was obtained according to the sensor,[5] with the sharp spectral modulations around 2.6–2.8 μm having almost disappeared. Though a reduction in intra-cavity peak power was again necessary to maintain pulsed operation at low RH-values, a FROG measurement could be performed at the 0%-level. With a measured temporal FWHM width as low as 22 fs (FTL: 22 fs), pulses with durations of **less than three optical cycles** were generated upon purging.

Other Output Characteristics and Amplitude Noise Measurements

To fully characterize the performance of *MIRanda3* when operated at the highest (unpurged) peak power level of nearly 1 MW, further measurements were performed and depicted in Fig. 5.17. With RMS fluctuations of less than 0.10%, the oscillator

[5] Note that the measurement accuracy of the sensor was specified to be ±4% RH at relative humidity levels approaching 0% RH.

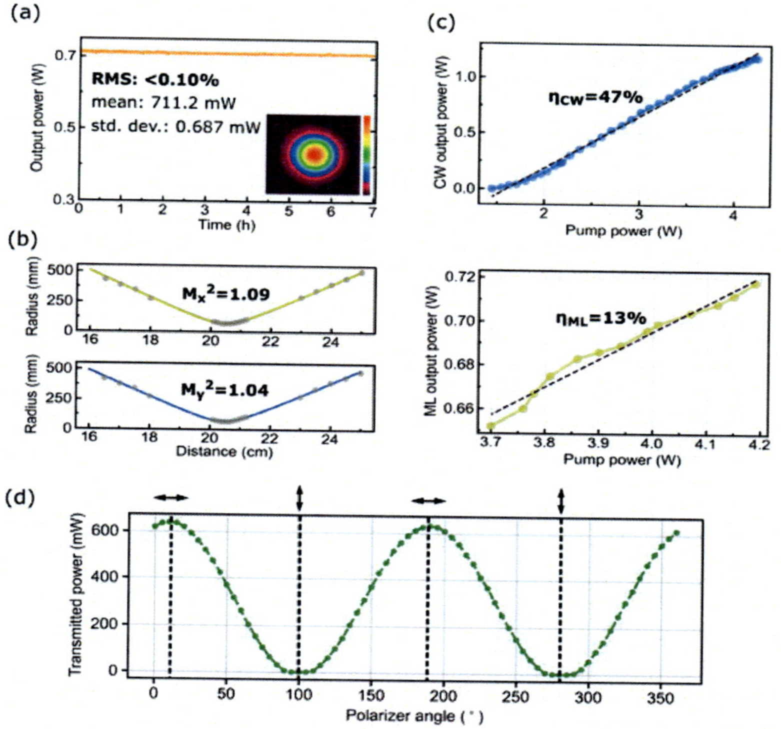

Fig. 5.17 Output characteristics of *MIRanda3*. **a** Mode-locked output power stability recorded with the thermal power meter (*Coherent*, PS19) and corresponding single-mode beam profile. **b** M^2-measurement of the horizontal and vertical beam axis of the mode-locked output. **c** Slope efficiency measured for a decrease in pump current of diode 2 and diode 1. **d** Characterization of the mode-locked output polarization via measuring the transmitted power through a wideband IR polarizer (*Innovation Photonics*, PGC-5)

revealed a stable mode-locked output and single-mode performance over the entire working day (>7 h). Moreover, the beam quality factors M^2 for the horizontal (x) and vertical (y) direction were determined to be as low as 1.09 and 1.04, respectively.

For recording the slope efficiency curves in CW and KLM operation, the output power of the cavity was measured for different pump power levels similar to the results shown for the second-generation system. With the actual power values of the combined pump beam being measured directly in front of the gain crystal, a continuous-wave slope efficiency of 47% was obtained, with a value of about 13% upon mode-locking. The latter was calculated for the limited range of pump powers where the oscillator was mode-locked. A saturation of the laser output at this power level is likely to have caused a reduction in slope efficiency compared to the previous two laser systems.

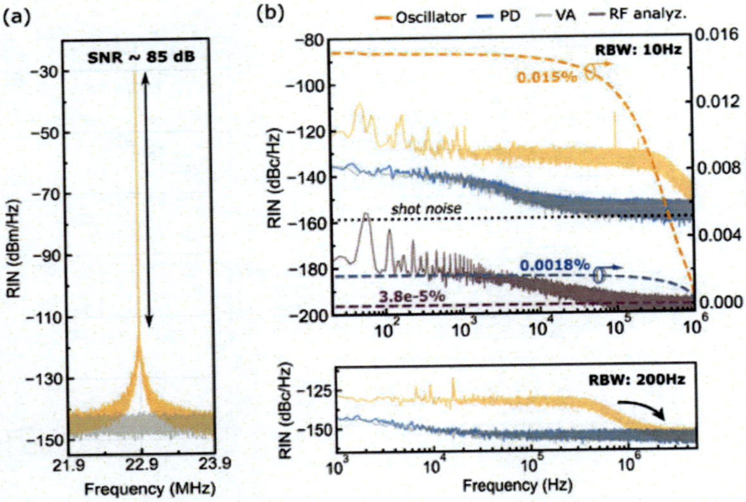

Fig. 5.18 Amplitude noise measurements and integrated RMS noise of the mode-locked output of *MIRanda3*. **a** RF Spectrum of the carrier region taken without the voltage amplifier. The grey line indicates the noise floor of the detection setup. **b** Amplitude noise measurements in the low-frequency region between 20 Hz to 1 MHz and 1 kHz to 5 MHz, revealing a 'plateau'-like noise curve without clear noise peaks. For the measurements, the same setup was used as in Fig. 4.16

Finally, the polarization of the laser output pulses turned out to be perfectly aligned along the horizontal (p-) direction, with the high polarization extinction ratio being attributed to the additional Brewster plate (MgF_2) incorporated into the cavity. To demonstrate the low-noise performance of the diode-pumped oscillator—even when running at almost 2.5-times higher intra-cavity peak power levels than before—the same amplitude noise measurement setup was used as presented in the previous chapter (see Sect. 4.2.2). The voltage amplifier (VA) was again operated at the highest gain level of 60 dB, with the theoretically generated DC-voltage[6] of $U_{DC,60\,dB} = 117.7\text{ mV} \cdot 10^{60/20} = 117.7$ V taken as reference for the dBc/Hz-conversion. As depicted in Fig. 5.18, it resulted in integrated RMS noise values of only 0.015% for the mode-locked output between 20 Hz and 1 MHz, which is almost identical to the performance of *MIRanda2* (0.014%). Moreover, the measurement curves of the oscillator noise together with the noise floor of the photodetector (PD) and the voltage amplifier (VA) were finally close to the calculated shot noise level of −159 dBc/Hz, because higher DC voltage levels could be obtained without notably saturating the photodiode. Moreover, as expected from the previous measurements, a drop in the 'plateau'-like noise signal to the background noise floor was revealed at frequencies around 1–2 MHz, which arises from the filtering action of the Cr:ZnS gain medium.

[6] To calculate the DC-voltage U_2 after amplification with a certain voltage gain factor G_V [dB], the following formula was used: $U_2 = U_1 \cdot 10^{G_V/20}$.

5.2.3 *Towards Even Higher Peak Power Levels*

Though peak powers of up to 1 MW were finally realized with directly diode-pumped Cr:ZnS/ZnSe oscillators, the question arises whether further improvements in pulse duration or intra-cavity pulse energies can boost it to even higher levels. Some of the potential methods are briefly addressed in the following:

- **Advanced dispersive mirror technology:**
  The main problem of using dispersive mirrors for few-cycle laser operation is related to the mirrors' GDD behaviour, which is characterized by unavoidable and irregular oscillations [36]. In particular, when considering the net dispersion per cavity round-trip, theses oscillations might add up dramatically for several bounces on these mirrors (see the theoretical calculations in Figs. 3.11, 4.12 and 5.12). For a mode-locked laser resonator operating at the vicinity of the GDD = 0 dispersion line, a significant effect on the pulse spectral shape can arise, as it was seen from the flattopped and wavy spectra in Fig. 5.15. To suppress these undesirable oscillations, several approaches exist regarding the design of dispersive mirrors [36]. For the few-cycle pulse durations of Cr:ZnS/ZnSe oscillators and their ultra-broad emission spectrum, the use of **dispersive mirror pairs** is considered to be a highly promising approach. Introduced about 20 years ago [37, 74], these pairs have identical GDD oscillation amplitudes but the oscillations are shifted by a half period with respect to each other, resulting in very small average GDD oscillations per mirror pair. Such a design has already been successfully realized for the wavelength range between 400 and 1200 nm [38], enabling the generation of single-cycle pulse durations from Ti:sapphire laser systems [39, 40]. Just recently, our in-house coating department also managed to fabricate dispersive mirror pairs for the 1.8–3.2 μm operation range of the Cr:ZnS/ZnSe oscillators (see Fig. 5.19). They are designed to have three mirror pairs being used for compensating the dispersion of the gain crystal (5 mm ZnS) and accounting for some additional TOD-slope. The latter could be identified from theoretical round-trip dispersion calculations as presented in this thesis. Though the slope's origin could not be completely clarified and might partially be attributed to air dispersion, its magnitude was consistently of the same strength. Currently, first tests are undertaken to assess the quality and performance of the new mirrors as well as to identify potential downsides that need to be improved in future iterations. Once these octave-spanning[7] mirror pairs proved practical, Cr:ZnS/ZnSe oscillators with pulse durations approaching the sub-two optical cycle regime ($\tau_p < 15$ fs) might be feasible.
- **Tighter focal spot diameters:**
  As shown from simulations (see Fig. 5.10), a tighter spot size in the crystal can significantly increase the Kerr strength (and thus the self-amplitude modulation) and might render the oscillator more robust against the onset of pulse instabilities—

[7] For a central wavelength of 2.3 μm, an optical octave would span the range from about 1.63 to 3.26 μm. With a designed operational range of 1.8–3.2 μm, these mirrors can be considered as "octave-spanning".

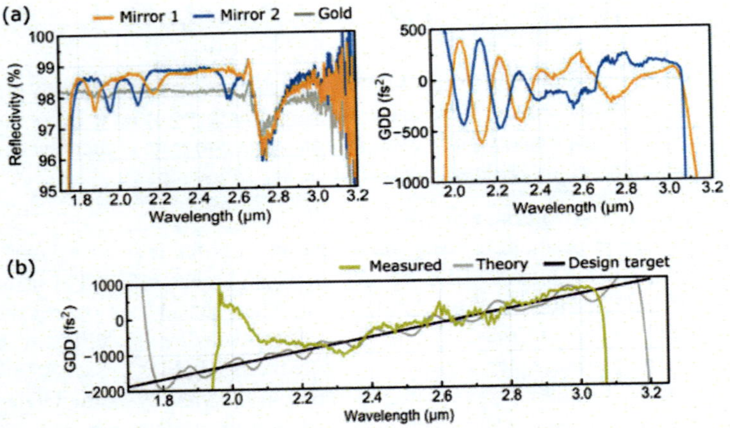

Fig. 5.19 Performance of the newly designed dispersive mirror pair for the 1.8–3.2 µm spectral range. **a** Measured reflectivity and dispersion curves of a pair consisting of mirror 1 and mirror 2. For comparison, the reflectivity curve of a gold-coated sample is shown in grey. **b** Comparison of the total dispersion provided by three dispersive mirror pairs. The black line indicates the original design target. Note that the dispersion measurements started at wavelengths around 2 µm, and measured GDD-values below 2 µm are thus not trustworthy

especially when heading towards higher peak power levels. Experimental observations obtained with the three diode-pumped laser systems can confirm this behaviour, since no CW-spiking or multi-pulsing could be observed for *MIRanda1*, given its relatively small focus sizes of only ~60 µm. Moreover, reports about powerful fiber-laser-pumped KLM Cr:ZnS/ZnSe oscillators lead to the assumption that their notably higher repetition rates ($\nu_{rep} \geq 75$ MHz) combined with spherical focusing mirrors of smaller radii of curvature (ROC≤150 mm) resulted in tighter focal spot diameters of about ≤70 µm in the gain medium than implemented in *MIRanda2* and *MIRanda3* [41–43]. For future diode-pumped cavity designs, it might therefore be advantageous to decrease the focal spot diameters further while keeping the low repetition rate. However, special care must be taken to maintain stable few-cycle KLM operation since the tighter foci automatically call for higher peak intensities and, hence, stronger interplay between nonlinear effects like SAM and SPM in the crystal [44].

- **Separate gain and Kerr medium:**
  Another potential option could be to completely decouple the gain and the Kerr medium. In traditional KLM-based oscillators, the Kerr-lensing effect and light amplification are both provided by the same crystal, imposing certain restrictions on the laser design. As has been demonstrated for KLM thin-disk lasers [45, 75] and diode-pumped KLM bulk lasers [46–48], the use of an **additional nonlinear medium** inside the cavity can strongly enhance the Kerr-lensing effect, while a complete decoupling between the gain and Kerr medium is also feasible. As a

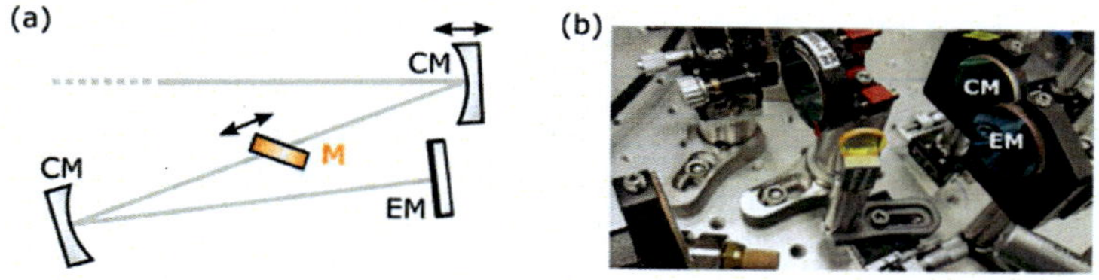

Fig. 5.20 **a** Sketch and **b** picture of the additional telescope added to the laser cavity of *MIRanda3* for test purposes. M: 2 mm ZnSe, CM: curved mirror (ROC = −100 mm), EM: highly-reflective end-mirror

consequence, the gain crystal position can be chosen for maximum laser gain, since the Kerr lensing is optimized independently by choosing suitable positions, thicknesses and materials for the additional Kerr plate. Also, the respective focal spot diameters can be freely adjusted in both the gain and Kerr medium, enabling the use of highly multi-mode fiber-coupled laser diodes ($M^2_{x,y} \gg 10$) combined with large pump spot diameters in the laser crystal [48]. However, note that the additional telescope increases the range of parameters that need to be adjusted for mode-locked operation, and changing the distances within one telescope critically affects the beam diameters in the other telescope [49].
A first step into this direction was also performed with the setup of *MIRanda3*: by replacing the end-mirror (EM) in Fig. 5.6 with a spherical mirror (CM) (see Fig. 5.20), the short arm of the cavity was extended by an additional telescope. With short radii of curvature of only ROC = −100 mm and 2 mm ZnSe chosen as nonlinear medium (M), it was possible to mode-lock the oscillator by varying the position of the ZnSe plate and one of the curved mirrors. Even though ZnSe can provide a slightly larger nonlinear coefficient n_2 than the ZnS-based gain crystal (see Table 2.2), and tighter spot sizes of about 60–70 μm were present in the additional plate, it was not possible to further improve the laser system's performance. Moreover, an even stronger tendency of the cavity towards multi-pulsing and CW-spiking was observed, most likely due to an increase in the accumulated nonlinear phase shift per resonator round-trip. Hence, instead of decoupling the laser gain and the Kerr effect, the plate might have introduced additional SPM similar to the double-Z cavities demonstrated with octave-spanning Ti:sapphire lasers [50]. Therefore, a reliable test in future generation systems would require a greatly enlarged spot size in the gain crystal to avoid contributions from Kerr lensing.

- **Further considerations:**
With the $4f$-extensions, the repetition rate of the third diode-pumped KLM system was successfully decreased to 23 MHz. A further reduction to even lower values might be achieved via cascading of several $4f$-telescopes or with the aid of a Herriott-style **multi-pass cavity** [33, 34, 51, 52]. Similar to the $4f$-imaging method, these multi-pass telescopes are designed to leave the cavity mode invari-

ant when inserted into the laser system, and repetition rates down to 4 MHz[8] were already demonstrated for KLM Ti:sapphire lasers [33, 53].
Furthermore, the use of **gain-matched output couplers**, whose transmission profile matches the gain spectrum of the laser crystal, was shown to enhance the self-amplitude modulation depth. Moreover, it reduces the need for a strong SAM effect to sustain stable pulsed operation [54, 55]. Also, to scale the peak power of diode-pumped Cr:ZnS/ZnSe lasers to ever higher values, a proper management of the thermal self-focusing in the gain medium will become increasingly important. Hence, **new crystal geometries** similar to the thin-disk [56] or rod design could be tried for a better heat extraction. Finally, a switch from dispersion-managed soliton formation to **chirped-pulse formation** in the positive dispersion regime (see Sect. 2.3.4) might need to be considered for reaching the highest possible peak power levels with Cr:ZnS/ZnSe lasers [57, 58].

5.3 Frequency Down-Conversion into the Mid-Infrared

To demonstrate the power of the third-generation diode-pumped laser system, its mode-locked output was directly sent into a nonlinear crystal (ZGP) for generating broadband mid-infrared via intra-pulse difference frequency generation. Given the high peak power levels of up to 1 MW without external amplification, initial highly promising results were obtained.

5.3.1 Intra-Pulse Difference Frequency Generation in a ZGP Crystal

Zinc Germanium Phosphide ($ZnGeP_2$, commonly referred to as ZGP) is a positive uniaxial crystal featuring a high laser damage threshold and good thermal conductivity [59]. With its high figure of merit (FOM)[9] of close to 200 pm^2/V^2, which is larger than for other well-known nonlinear crystals such as GaSe or LGS (see Fig. 1.3) [59], it is one of the most promising materials for the nonlinear frequency conversion aimed for in this thesis. Because its transparency window only spans from approximately 2 to 10 μm (see Fig. 5.21a), ZGP needs to be pumped at wavelengths longer than 2 microns. Hence, a series of 2 μm-pumped ZGP optical parametric oscillators and amplifiers has already been successfully realized over the past two decades [59, 61–

[8] A repetition frequency of $\nu_{rep} = 4$ MHz corresponds to an impressively large optical path length of 75 meters per cavity round-trip. With the Herriott-cells, this can be achieved in a rather compact manner.

[9] The FOM can be calculated as d_{eff}^2/n^3, with the crystal's refractive index n and an effective nonlinear coefficient d_{eff}. It is typically used to assess the crystal's potential with regard to nonlinear frequency conversion processes and can be taken as a measure for the expected signal conversion efficiencies [59, 60].

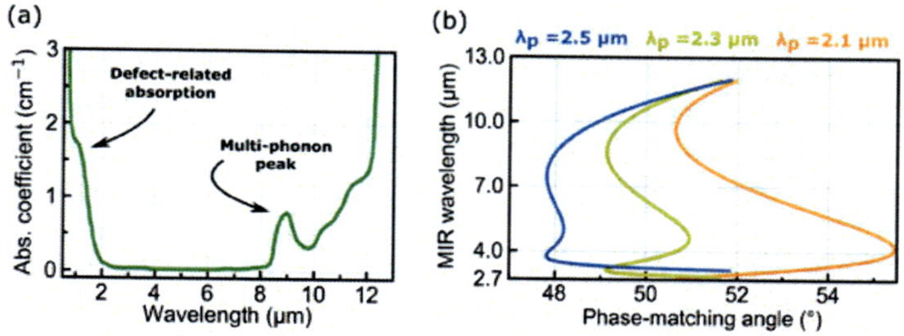

Fig. 5.21 **a** Optical absorption of ZGP [65] and **b** simulated type I phase-matching (PM) curves for the generated signal (idler) wavelengths via frequency down-conversion. To mimic the case of intra-pulse DFG for a broadband input spectrum, different pump wavelengths were assumed

64]. For example, a ZGP OPO was reported to deliver continuously tunable output wavelengths from 3.8 to 12.4 μm when pumped at 2.9 μm [61], while a combination of ZGP OPO and subsequent ZGP OPA resulted in watt-level average powers driven by a Q-switched Ho:YAG laser (2.097 μm) [64].

Just recently, ZGP was identified as a highly attractive nonlinear medium in the context of Cr:ZnS/ZnSe laser-driven intra-pulse DFG. With optical-to-optical frequency conversion efficiencies of up to a few percent [11, 13, 66], high average powers of over 128 mW were recorded for the generated MIR wavelengths between 4.5 and 12.5 μm [11]. This power level is much higher than compared to similar experiments performed with GaSe [10, 13, 76]. For the IDFG-based setup presented in this section, which was directly pumped by the oscillator output, the large conversion efficiencies were considered to be crucial. In the experiment, a ZGP crystal (*BAE Systems*) with dimensions of 6 × 7 × 2 mm (height × width × length) was used, which has been cut for type I phase-matching ($\theta = 55°$). This provides a larger nonlinear coefficient of $d_{eff} = 80.8$ pm/V than type II-based difference frequency generation with $d_{eff} = 74.0$ pm/V, allowing for slightly higher nonlinear conversion efficiencies.[10]

To select a reasonable crystal thickness, one can consider the interaction length scale over which DFG takes place. This length is mainly limited by the group velocity mismatch (GVM) between the pump and MIR pulse [13], which causes the two pulse envelopes to separate from each other after a certain distance L_{eff}. Assuming a central pump wavelength of 2.4 μm and mid-infrared wavelength of 10 μm, it can be calculated with the group velocities obtained from *SNLO*:

$$L_{eff} = \frac{\tau_p}{GVM} = \frac{\tau_p}{(v_{g,\,2.4\,\mu\mathrm{m}})^{-1} - (v_{g,\,10\,\mu\mathrm{m}})^{-1}} = \frac{28\ \mathrm{fs}}{3.22928/c - 3.21953/c} \approx 861\ \mu\mathrm{m}, \tag{5.5}$$

[10] The d_{eff}-values were obtained from simulations with the software *SNLO*, assuming a pump and signal frequency of 2.0 μm and 2.6 μm, respectively.

with c denoting the speed of light. Though only a 2 mm thick ZGP crystal was available, even thinner substrates might be considered for future experiments. Also, the crystal was provided with an AR-coating for the wavelengths between 2.04–2.2 μm and 2.8–6 μm, which was optimized for another experiment and not for the IDFG with the broadband output of the Cr:ZnS oscillator. Nonetheless, first measurements could be successfully performed and can serve as a useful reference for future experiments.

5.3.2 Characterization of the Generated Radiation

The ZGP crystal was placed into the focus of a parabolic mirror (see Fig. 5.22a) with a corresponding focal spot diameter of about 35 microns measured with a rotating-slit beam profiler (*Ophir Optronics Inc.*, NanoScan 2s). The transmitted beam was re-collimated by another parabolic mirror of the same focal length. In addition, the crystal was mounted on top of a mechanical translation stage and placed inside a rotational mount to move it along the pump focusing direction and to satisfy the phase matching condition.[11] Thus, by adjusting the orientation of the crystal, the polarization vector of the incident pump E-field was aligned to have approximately equal e and o components with respect to the crystal axis.

As a first step, the mid-infrared power levels (>4.5 μm) as well as the total transmitted power behind the ZGP crystal were analyzed at different crystal positions along the propagation axis (Z-axis). The result is depicted in Fig. 5.22b together with the corresponding **beam profiles** that were recorded with a pyroelectric camera (*Ophir Optronics*, Pyrocam III). Here, three different cases are distinguished: (i) the continuous-wave (CW) and (ii) mode-locked (ML) beam profiles measured at the fundamental wavelengths, and (iii) the mode-locked beam profiles that were analyzed after a 4.5 μm long-pass filter. In the first case, no degradation in beam quality was observed when the crystal was moved in and out of focus, and a single-mode beam profile was detected for various different Z-axis positions. In the second case, however, a spatially Gaussian beam intensity distribution could only be recorded when the crystal was moved at least ± 1 mm away from the focus ($Z \approx 5.3$ mm). As soon as the nonlinear medium entered the focal area at Z-values between 4.3 and 6.2 mm, the beam profile significantly degraded, but it recovered when the ZGP crystal was placed outside the focusing region again. Also, a similarly strong distortion was observed for the generated mid-IR radiation at wavelengths greater than 4.5 μm.

In addition to the beam profile measurements, the total **transmitted power** was analyzed directly behind the crystal and resulted in a pronounced drop around Z-positions between 4.5 and 6.0 mm. Surprisingly, no crystal damage was observed due to the recovery of the absorption and the measured transmitted powers when moving the crystal in and out of focus. Given this strong absorption in the fundamental power,

[11] For the DFG process, the type I phase-matching in a positive birefringent crystal can be written as: λ_p (o), λ_s (e) $\rightarrow$ λ_i (e), with e and o denoting the extraordinary and ordinary polarization of the pump (p), signal (s) and idler (i) beams, respectively. $\lambda_i \geq \lambda_s > \lambda_p$ [67].

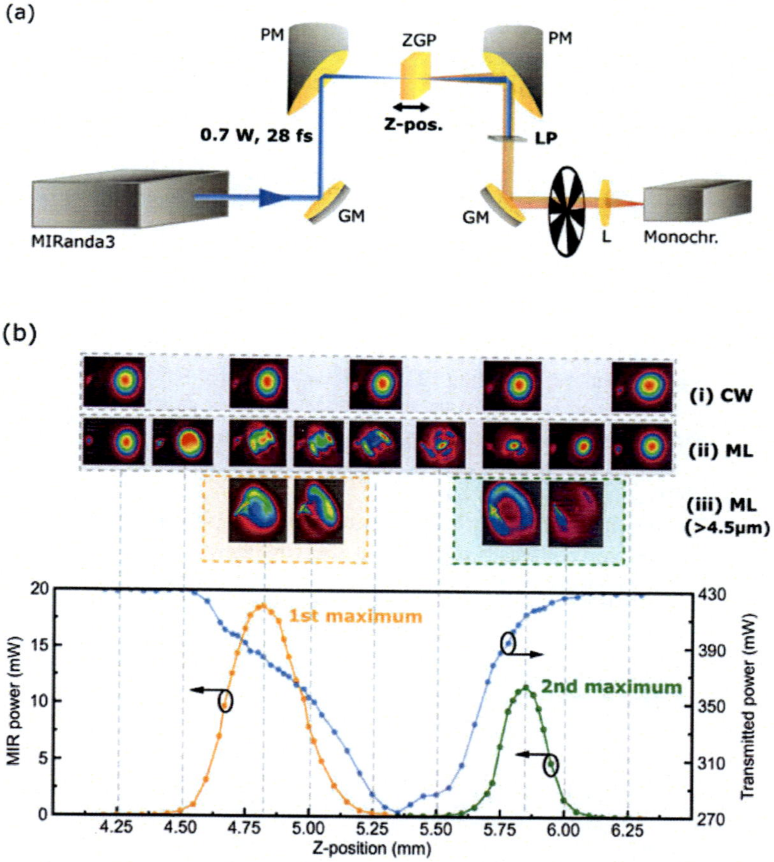

Fig. 5.22 Mid-infrared generation setup along with the measured power levels and beam profiles behind the ZGP crystal. **a** Setup, including two gold-coated 90° off-axis parabolic mirrors (PM, f = 25.4 mm), gold-coated mirrors (GM), a long-pass filter (LP), a monochromator, a ZGP crystal (d = 2 mm), and a ZnSe lens (L, f = 75 mm). **b** Generated mid-infrared power (>4.5 µm) and transmitted total power behind the 2 mm thick ZGP crystal, revealing two maxima (orange/green). Increasing Z-values denote a movement of the crystal towards the second parabolic mirror. Also, the corresponding beam profiles were recorded at the fundamental wavelengths for (i) continuous-wave (CW) and (ii) mode-locked (ML) operation as well as for (iii) the MIR radiation behind a 4.5 µm long-pass filter. Interestingly, the shape of the distorted beam profiles reveals some kind of symmetry, similar to the cylindrical transverse mode patterns (TEM) presented in the theoretical section (Fig. 2.1)

however, the appearance of mid-infared radiation was shifted to regions of reduced peak intensities, and a maximum in the generated MIR power was measured on both sides of the dip (see Fig. 5.22b).

For the fundamental wavelengths, the pronounced degradation of the spatial beam profiles and the reduction in transmitted power is most likely due to the interplay between strong multi-photon absorption and self-focusing [68]. According to Table 2.2, ZGP features a high refractive index of $n_0 = 3.14$ and a comparably large nonlinear coefficient of $n_2 \approx 550 \cdot 10^{-16}$ cm^2/W, and it has a bandgap energy of 2 eV [69]. Hence, its critical power for self-focusing amounts to only $P_{cr} = 45.4$ kW (see definition in Sect. 2.1.2), which is much lower than the mode-locked peak power $P_0 \approx 1$ MW delivered by *MIRanda3*. Indeed, it was observed that the collimating parabolic mirror had to be re-adjusted for different crystal position for proper collimation. When the crystal is placed at the focus, where the spot diameter is 35 μm, the beam should have theoretically collapsed to an arbitrarily small size after propagation through $z_f \approx 573$ μm of the 2 mm thick ZGP crystal. This propagation distance z_f was calculated according to $z_f = k \cdot a_0^2 \cdot (P_0/P_{cr} - 1)^{-0.5}$ [70], including the spot radius a_0 and the wave vector $k = 2\pi \cdot n_0/\lambda_0$ at a central wavelength of $\lambda_0 =$ 2300 nm. However, the increasing multi-photon absorption at high intensities might have arrested the collapse of the laser beam, and therefore no permanent damage of the ZGP crystal was observed.

The interaction between self-focusing and multi-photon absorption is highly sensitive to the initial intensity distribution of the incident beam. As multi-photon absorption scales with the laser peak intensity, the resulting beam profile will not be radially symmetric if the initial beam is slightly elliptical, as in our case. Furthermore, compared to the edges, the central portion of the original fundamental beam (TEM_{00}) will experience higher absorption rate and thus lower intensities after propagation through the crystal. These effects could also explain the ring- or donut-shaped beam profiles of the generated MIR light, which correlate with the intensity distribution of the fundamental beam within the crystal (see Fig. 5.22b, case (iii)). Note that laser beam filamentation associated with plasma formation is unlikely to play a role here, given the nanojoule-level of pulse energies used in this experiment [71, 72]. However, it is remarkable that a degradation in beam quality has not been reported in previous experiments [11, 13, 66]. Therefore, further measurements are planned in the future to better understand the physical origin behind the observed effects.

To analyze the generated **MIR spectra**, a Czerny–Turner monochromator (*Newport*, Cornerstone 260) was combined with a lock-in amplifier (*Stanford Research Systems*, SR810) and a mechanical chopper to modulate the intensity of the laser beam. For detection, a lithium tantalate pyroelectric detector (*Oriel Instruments*, DET-L-PYK5-R-P) was attached to the monochromator and connected with the lock-in amplifier, allowing for highly sensitive detection of the weak mid-infrared radiation. This measurement system allowed to detect wavelengths between 0.725 and 20 μm since the monochromator was equipped with three gratings to cover different spectral portions: 0.725–3.25 μm (grating 1), 2.5–11 μm (grating 2), and 4.5–20 μm (grating 3). In addition, suitable long-pass filters with cut-on wavelengths at 2.8 μm, 4.5 μm, 7.3 μm, and 11 μm were used to suppress second-order diffractions arising from the monochromator's gratings and to analyze the signals between 2.5–5.6 μm, 4.0–9.0 μm, 6.5–14 μm, and 9.5–20 μm, respectively. The measurement ranges were

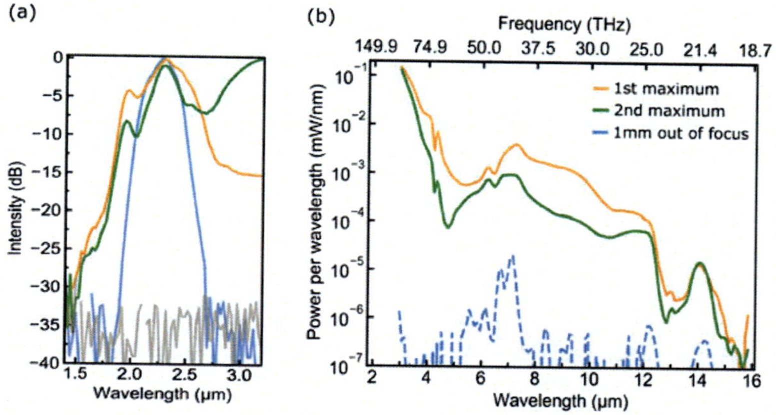

Fig. 5.23 Generated (mid-)infrared spectra with a ZGP crystal for the two observed maxima. **a** Data recorded for the fundamental wavelength region of the laser oscillator, indicating the presence of SPM-induced spectral broadening. The blue curve resembles the mode-locked spectrum of the oscillator, and the grey line denotes the detection noise floor. **b** Generated mid-infrared wavelengths via cascaded parametric processes. The measured spectra were calibrated and stitched together to be displayed at normalized units of mW/nm

chosen slightly larger to provide sufficient data points for the subsequent stitching process.

The intensities of the measured MIR spectra beyond 2.8 µm were calibrated with the aid of a silicon nitride infrared emitter (*Bentham Instruments Ltd.*, CL2-IR) to account for the wavelength-dependent efficiency variation of the monochromator gratings and the filter losses. Also, the transmission curve of the ZnSe lens placed in front of the detection system was taken into account. When stitching the intensity-corrected measurements together and normalizing the intensity axis to the respective power levels measured behind the long-pass filters, a final spectrum was obtained (Fig. 5.23b). The average powers were recorded with a thermal power meter (*Coherent*, PS19) and are summarized in Table 5.3. Note that the real values are most likely even higher, given that the transmission curve of the filters varied strongly between 70 and 90% for the measured MIR wavelengths.

The **generated MIR spectrum** covers about two optical octaves in the wavelength range from 2.8 to 14 µm (714–3571 cm^{-1}), and thus a great variety of molecular absorption lines in the gas and liquid phase could already be addressed (see Fig. 1.1). Interestingly, the spectral components below 7.3 µm cannot be explained by simple intra-pulse DFG alone, given that the mode-locked oscillator spectrum has spectral limits of $\lambda_1 = 1.95\,\mu m$ and $\lambda_2 = 2.66\,\mu m$ at the −20 dB level.[12] In fact, the recorded spectrum reveals a peak near 7.3 µm, with a substantial amount of signal being located at even shorter mid-infrared wavelengths. As explained and simulated

[12] Recall that the wavelengths generated via DFG are calculated through $\lambda_3^{-1} = \lambda_1^{-1} - \lambda_2^{-1}$, which amounts to $\lambda_3 = 7.31\,\mu m$ for the considered minimum and maximum wavelengths.

Table 5.3 Average power levels measured directly behind different long-pass filters for the 1st and 2nd maximum

Filter	1st maximum (mW)	2nd maximum (mW)
2.8 μm (2x)	75	42
4.5 μm	20.7	11.9
7.3 μm	15.8	7.5
11 μm	1.3	1.0

in Refs. [10, 73], a **cascading effect** takes place here: the generated fundamental DFG components are further mixed with the initial driving laser frequencies, and several cascading processes set in to generate an even broader range of spectral components. The corresponding background level in Fig. 5.23b was obtained by shifting the nonlinear crystal 1 mm away from the focus to suppress any nonlinear effects. Besides the nonlinear mixing, also spectral broadening of the oscillator input spectrum down to 1.5 μm was observed (see Fig. 5.23a), being attributed to self-phase modulation at the fundamental wavelengths. When taking into account the average transmission of about 85% for the 4.5 μm long-pass filter, the optical-to-optical conversion efficiencies amounted to 5.7%. This demonstrates the great potential of using powerful diode-pumped Cr:ZnS/ZnSe oscillators for direct MIR generation.

With regard to future experiments, the setup can be further improved to exploit the full potential of the IDFG-based system. For example, when analyzing the corresponding B-integral in the ZGP crystal for the measured focal spot diameter of 35 μm, a very large value of close to $B = 15$ is calculated (see Fig. 5.24a), which may also contribute to the strong degradation of the spatial beam profile. By contrast, a larger focus size of 120 μm would significantly reduce the B-integral to below 2.5 (see Fig. 5.24b). In addition, it improves the beam overlap between the fundamental and the generated mid-infrared beam given the decreased beam divergences. A thinner crystal with a slightly optimized phase-matching angle of $\theta \approx 50°$ might also be selected (see simulations presented in Fig. 5.21b) along with a more broadband AR-coating to push the conversion efficiencies to ever higher values.

5.4 Summary

With this third-generation system, the performance of directly diode-pumped Cr^{2+}-doped II–VI lasers has been brought to a new level. With the aid of a novel pump focusing approach and a classic bow-tie design of the cavity, a powerful oscillator with peak powers of up to 1 MW has been realized. This constitutes an almost 2.5-fold increase in peak power over the previous diode-pumped generation, which was pumped in a similar fashion by the combined emission of two single-emitter laser diodes. Moreover, the new intra-cavity $4f$-telescope extensions greatly reduced the

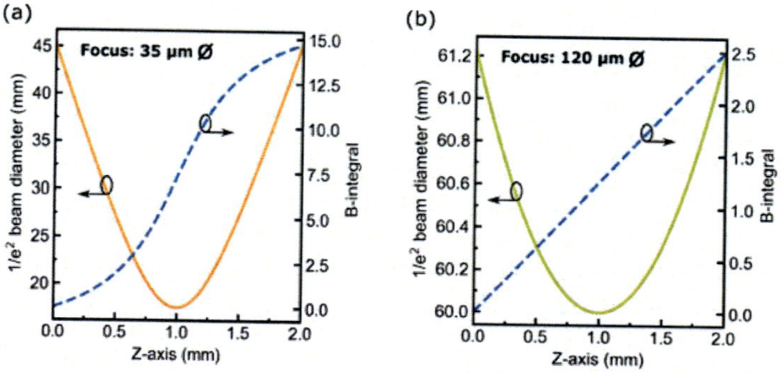

Fig. 5.24 Calculation of the single-pass B-integral inside the ZGP crystal. The smaller the focal spot diameter, the higher the accumulated nonlinear phase shift (B-integral)

repetition rate of the oscillator down to 23 MHz, while single-pulse operation in the fundamental mode was maintained. When reducing the relative humidity level in the laser chamber, a strong effect on the mode-locked spectrum was observed, with the change in intra-cavity dispersion resulting in a measured Fourier-transform limited pulse duration of less than three optical cycles ($\tau_p = 22\,\mathrm{fs}$). To the best of our knowledge, the pulse durations presented in this chapter are among the shortest reported for Cr:ZnS/ZnSe oscillators [43], and are the shortest that have been fully characterized in both amplitude and phase.

Also, given the high peak power levels of up to 1 MW, generation of mid-infared radiation was demonstrated for the first time with the direct output of a Cr:ZnS/ZnSe oscillator. Notably, no additional amplification stage was involved in the setup—transferring the low-noise characteristics of the diode-pumped laser system directly into the mid-infrared region. With the technique of intra-pulse difference frequency generation and a ZGP crystal used as nonlinear medium, a comparably high optical-to-optical conversion efficiency of 5.7% was obtained for the spectral range between 4.5 and 14 µm. The corresponding average power levels amounted to more than 20 mW when measured directly behind the 4.5 µm long-pass filter, highlighting the enormous potential of direct oscillator-driven MIR generation.

References

1. Salem R, et al (2015) Mid-infrared supercontinuum generation spanning 1.8 octaves using step-index indium fluoride fiber pumped by a femtosecond fiber laser near 2 µm. Optics Express 23(24):30592–30602
2. Tang Y, Wright LG, Charan K, Wang T, Xu C, Wise FW (2016) Generation of intense 100 fs solitons tunable from 2 to 4.3 µm in fluoride fiber. Optica 3(9):948–951
3. Ren H, et al (2019) Low-loss silicon core fibre platform for mid-infrared nonlinear photonics. Light: Sci & Appl 8:105

4. Saini TS, Tuan TH, Suzuki T, Ohishi Y (2020) Coherent Mid-IR supercontinuum generation using tapered chalcogenide step-index optical fiber: experiment and modelling. Sci Rep 10:2236
5. Zhang J, et al (2018) Multi-mW, few-cycle mid-infrared continuum spanning from 500 to 2250 cm^{-1}. Light: Sci Appl 7:17180
6. Zhang J, Fritsch K, Wang Q, Krausz F, Mak K, Pronin O (2019) Intra-pulse difference-frequency generation of mid-infrared (2.7 -20 μm) by random quasi-phase-matching. Optics Lett 44(12):2986–2989
7. Gaida C, et al (2018) Watt-scale super-octave mid-infrared intrapulse difference frequency generation. Light: Sci & Appl 7(1):94
8. Butler TP et al (2019) Watt-scale 50-MHz source of single-cycle waveform-stable pulses in the molecular fingerprint region. Optics Lett 44(7):1730–1733
9. Nagl N, Mak K, Wang Q, Pervak V, Krausz F, Pronin O (2019) Efficient femtosecond mid-infrared generation based on a Cr:ZnS oscillator and step-index fluoride fibers. Optics Lett 44(10):2390–2393
10. Wang Q et al (2019) Broadband mid-infrared coverage (2–17 μm) with few-cycle pulses via cascaded parametric processes. Optics Lett 44(10):2566–2569
11. Wang Q, et al (2019) Highly efficient broadband mid-infrared generation (2.8-12.5 μm) based on a compact Cr:ZnS laser. In: 2019 Conference on lasers and electro-optics Europe and European quantum electronics conference. CF-7.5. IEEE
12. Smolski VO, Vasilyev S, Schunemann PG, Mirov SB, Vodopyanov KL (2015) Cr:ZnS laser-pumped subharmonic GaAs optical parameteric oscillator with the spectrum spanning 3.6-5.6 μm. Optics Lett 40(12):2906–2908
13. Vasilyev S, et al (2019) Super-octave longwave mid-infrared coherent transients produced by optical rectification of few-cycle 2.5-μm pulses. Optica 6(1):111–114
14. Smolski V et al (2018) Half-watt average power femtosecond source spanning 3–8 μm based on subharmonic generation in GaAs. Appl Phys B 124:101
15. Homann C, Bradler M, Förster M, Hommelhoff P, Riedle E (2012) Carrier-envelope phase stable sub-two-cycle pulses tunable around 1.8 μm at 100 kHz. Optics Lett 37(10):1673–1675
16. Telle HR, Steinmeyer G, Dunlop AE, Stenger J, Sutter DH, Keller U (1999) Carrier-envelope offset phase control: a novel concept for absolute optical frequency measurement and ultrashort pulse generation. Appl Phys B 69:327–332
17. Pupeza I et al (2015) High-power sub-two-cycle mid-infrared pulses at 100 MHz repetition rate. Nature Photonics 9:721–724
18. Pupeza I et al (2020) Field-resolved infrared spectroscopy of biological systems. Nature 577(7788):52–59
19. Xu L, Spielmann Ch, Poppe A, Brabec T, Krausz F, Hänsch TW (1996) Route to phase control of ultrashort light pulses. Optics Lett 21(24):2008–2010
20. Baltuška A, Fuji T, Kobayashi T, (2002) Controlling the carrier-envelope phase of ultrashort light pulses with optical parametric amplifiers. Phys Rev Lett 88(13):133901
21. Cerullo G, Baltuska A, Mücke OD, Vozzi C (2011) Few-optical-cycle light pulses with passive carrier-envelope phase stabilization. Laser & Photonics Rev 5(3):323–351
22. Lee Y-S (2009) Principles of terahertz science and technology. Springer, US
23. Petrov V (2015) Frequency down-conversion of solid-state laser sources to the mid-infrared spectral range using non-oxide nonlinear crystals. Progress Quantum Electron 42:1–106
24. Anthony E (1986) Siegman. University Science Books, Lasers
25. Kalashnikov VL, Sorokin E (2010) Soliton absorption spectroscopy. Phys Rev A 81:033840
26. Gebhardt M et al (2015) Impact of atmospheric molecular absorption on the temporal and spatial evolution of ultra-short optical pulses. Optics Express 23(11):13776–13787
27. Hisaaki K (ed) (2018) Radiation Applications, vol 7. An Advanced Course in Nuclear Engineering, Springer, Singapore
28. Goodman JW (2005) Analog optical information processing. Introduction to fourier optics, 3rd edn, pp 219-290. Roberts and Company Publishers, Greenwood Village
29. Yelisseyev AP, Isaenko LI, Starikova MK (2012) Optical study of defects in lithium iodate α-LiIO3. J Optical Soc Amer B 29(6):1430–1435

30. Sorokina IT, Sorokin E (2015) Femtosecond Cr^{2+}-based lasers. IEEE J Selected Topics Quantum Electron 21(1):1601519
31. Tolstik N, Sorokin E, Sorokina IT (2014) Graphene mode-locked Cr:ZnS laser with 41 fs pulse duration. Optics Express 22(5):5564–5571
32. Kärtner FX (ed) (2004) Few-cycle laser pulse generation and its applications, vol 95. Springer, Berlin
33. Cho SH et al (2001) Generation of 90-nJ pulses with a 4-MHz repetition-rate Kerr-lens mode-locked Ti:Al_2O_3 laser operating with net positive and negative intracavity dispersion. Optics Lett 26(8):560–562
34. Fernandez A, Fuji T, Poppe A, Fürbach A, Krausz F, Apolonski A (2004) Chirped-pulse oscillators: a route to high-power femtosecond pulses without external amplification. Optics Lett 29(12):1366–1368
35. Kalashnikov VL, Podivilov E, Chernykh A, Apolonski A (2006) Chirpedpulse oscillators: theory and experiment. Appl Phys B 83:503–510
36. Pervak V, Razskazovskaya O, Angelov IB, Vodopyanov KL, Trubetskov M (2014) Dispersive mirror technology for ultrafast lasers in the range 220–4500 nm. Adv Optical Technol 3(1):55–63
37. Laude V, Tournois P (1999) Chirped-mirror-pairs for ultra-broadband dispersion control. In: Conference on Lasers and Electro-Optics. Optical Society of America, CTuR4
38. Pervak V, Tikhonravov AV, Trubetskov MK, Naumov S, Krausz F, Apolonski A (2007) 1.5-octave chirped mirror for pulse compression down to sub-3 fs. Appl Phys B 87:5–12
39. Rausch S et al (2008) Controlled waveforms on the single-cycle scale from a femtosecond oscillator. Optics Express 16(13):9739–9745
40. Sander MY, Birge J, Benedick A, Crespo HM, Kärtner FX (2009) Dynamics of dispersion managed octave-spanning titanium:sapphire lasers. J Optical Soc Amer B 26(4):743–749
41. Vasilyev S, Moskalev I, Mirov M, Mirov S, Gapontsev V (2015) Three optical cycle mid-IR Kerr-lens mode-locked polycrystalline Cr^{2+}:ZnS laser. Optics Lett 40(21):5054–5057
42. Vasilyev S, Moskalev I, Mirov M, Smolski V, Mirov S, Gapontsev V (2017) Ultrafast middle-IR lasers and amplifiers based on polycrystalline Cr:ZnS and Cr:ZnSe. Optics Mater Express 7(7):2636–2650
43. Mirov S et al (2018) Frontiers of mid-IR lasers based on transition metal doped chalcogenides. IEEE J Selected Topics Quantum Electron 24(5):1–29
44. Spielmann C, Curley PF, Brabec T, Krausz F (1994) Ultrabroadband femtosecond lasers. IEEE J Quantum Electron 30(4):1100–1114
45. Zhang J, et al (2015) 49-fs Yb:YAG thin-disk oscillator with distributed Kerr-lens mode-locking. In: 2015 European Conference on Lasers and Electro-Optics - European Quantum Electronics Conference (CLEO/Europe-EQEC) PD.A.1
46. Tian W et al (2018) Diode-pumped power scalable Kerr-lens mode-locked Yb:CYA laser. Photonics Res 6(2):127–131
47. Tian W, et al (2019) Diode-pumped high-power sub-100 fs Kerr-lens mode-locked Yb:$CaYAlO_4$ laser with 1.85 MW peak power. Optics Express 27(15):21448–21454
48. Akbari R, Major A (2018) Kerr-lens mode locking of a diode-pumped Yb:KGW laser using an additional intracavity Kerr medium. Laser Phys Lett 15(8):085001
49. Han X, Zeng H (2008) Kerr-lens mode-locked Ti:sapphire laser with an additional intracavity nonlinear medium. Optics Express 16(23):18875–18880
50. Ell R et al (2001) Generation of 5-fs pulses and octave-spanning spectra directly from a Ti:sapphire laser. Optics Lett 26(6):373–375
51. Cho SH, Bouma BE, Ippen EP, Fujimoto JG (1999) Low-repetition-rate high-peak-power Kerr-lens mode-locked Ti:Al_2O_3 laser with a multiple-pass cavity. Optics Lett 24(6):417–419
52. Kowalevicz AM, Sennaroglu A, Zare AT, Fujimoto JG (2006) Design principles of q-preserving multipass-cavity femtosecond lasers. J Optical Soc Amer B 23(4):760–770
53. Song DH, Cho WB, Lee HW, Shin DH, Ko D-K, Jung MY (2013) A compact Kerr-lens mode-locked Ti:sapphire oscillator with 330 nJ soliton-like pulses. Laser Phys Lett 10:065003

54. Chen L-J, Sander MY, Kärtner FX (2010) Kerr-lens mode locking with minimum nonlinearity using gain-matched output couplers. Optics Lett 35(17):2916–2918
55. Cihan C et al (2014) Gain-matched output couplers for efficient Kerr-Lens mode- locking of low-cost and high-peak power Cr:LiSAF Lasers. IEEE J Selected Topics Quantum Electron 21(1):94–105
56. Renz G, Speiser J, Giesen A, Sorokina IT, Sorokin E (2013) Cr:ZnSe thin disk cw laser. In: Solid state lasers XXII: technology and devices, vol 8599, pp 310–316. SPIE
57. Tolstik NA, Pospischil A, Sorokin E, Sorokina T (2014) Graphene mode-locked Cr:ZnS chirped-pulse oscillator. Optics Express 22(6):7284–7289
58. Tolstik N, Lee CSJ, Sorokin E, Sorokina IT (2018) 8.6 MHz extended cavity Cr:ZnS Chirped-pulse Oscillator. In: Conference on Lasers and Electro-Optics, SF1N.2
59. Petrov V (2012) Parametric down-conversion devices: the coverage of the mid-infrared spectral range by soild-state laser sources. Optical Mater 34(3):536–554
60. Astill AG (1991) Material figures of merit for non-linear optics. Thin Solid Films 204(1):1–17
61. Vodopyanov KL, Ganikhanov F, Maffetone JP, Zwieback I, Ruderman W (2000) $ZnGeP_2$ optical parametric oscillator with 3.8-12.4-μm tunability. Optics Lett 25(11):841–843
62. Hemming A, et al (2013) 99 W mid-IR operation of a ZGP OPO at 25% duty cycle. Optics Express 21(8):10062–10069
63. Sanchez D et al (2016) 7 μm, ultrafast, sub-millijoule-level mid-infrared optical parametric chirped pulse amplifier pumpet at 2 μm. Optica 3(2):147–150
64. Qian C, et al (2018) 11.4 W long-wave infrared source based on $ZnGeP_2$ optical parametric amplifier. Optics Express 26(23):30195–30201
65. Schunemann PG, Zawilski KT, Pomeranz LA, Creeden DJ, Budni PA (2016) Advances in nonlinear optical crystals for mid-infrared coherent sources. J Opt Soc Amer B 33(11):D36–D43
66. Vasilyev S et al (2020) Multi-octave infrared femtosecond continuum generation in Cr:ZnS-GaSe and Cr:ZnS-ZGP tandems. In: Schunemann PG, Schepler KL (eds) Nonlinear frequency generation and conversion: materials and devices XIX, vol 11264. International society for optics and photonics, SPIE, pp 13–24
67. Sorokina IT, Vodopyanov KL (2003) Solid-state mid-infrared laser sources, vol 89. Topics in applied physics, Springer, Berlin
68. Polyakov S, Yoshino F, Stegeman G (2001) Interplay between self-focusing and high-order multiphoton absorption. J Optical Soc Amer B 18(12):1891–1895
69. Vodopyanov KL (2020) Laser-based Mid-infrared sources and applications. Wiley, New York
70. Wagner WG, Haus HA, Marburger JH (1968) Large-scale self-trapping of optical beams in the paraxial ray approximation. Phys Rev 175(1):256–266
71. Couairon A, Mysyrowicz A (2006) Femtosecond filamentation in air. Progress in ultrafast intense laser science. Springer series in Chemical physics, vol 84, pp 235–258. Springer, Berlin
72. Boyd R (2008) Nonlinear optics, 3rd edn. Academic, Cambridge
73. Wang Q, et al (2018) Complete mid-infrared coverage (2-20 μm) via cascaded parametric processes. In: Laser congress 2018 (ASSL) ATu6A.4
74. Kärtner FX et al (2001) Ultrabroadband double-chirped mirror pairs for generation of octave spectra. J Optical Soc Amer B 18(6):882–885
75. Brons J et al (2014) Energy scaling of Kerr-lens mode-locked thin-disk oscillators. Optics Lett 39(22):6442–6445
76. Vasilyev S et al (2019) Multi-octave visible to long-wave IR femtosecond continuum generated in Cr:ZnS-GaSe tandem. Optics Express 27(11):16405–16413

Chapter 6
Conclusion and Outlook

In this thesis, diode-pumped few-cycle Cr^{2+}-doped II–VI oscillators have been realized for the first time, with the oscillators' performance matching and even exceeding that of state-of-the-art systems based on mature fiber-laser-pumped technologies. The innovative combination of direct diode-pumping with ultrafast Kerr-lens mode-locked Cr:ZnS/ZnSe technology led to highly stable, low-noise, and femtosecond output at wavelengths between 2 and 3 μm, facilitating the direct generation of broadband mid-infrared spectra at the milliwatt-level. This last chapter serves to highlight the main experimental findings of this thesis and strives to showcase their tremendous potential for the rapidly growing field of laser-based molecular fingerprinting.

6.1 Summary

Generating broadband and coherent radiation in the mid-infrared range between 2.5 and 20 μm is of utmost importance for a plethora of applications, ranging from environmental monitoring [1, 2] to medical diagnostics [3, 4]. Recent advancements have markedly benefited from the continuous development of ultrafast laser sources, where great interest is devoted to materials that can provide direct lasing in the MIR region. Solid-state bulk oscillators based on Cr^{2+}-doped II–VI materials have proven to be reliable sources for generating few-cycle pulses in the 2–3 μm spectral range, and rare-earth-doped fiber lasers have been the pump sources of choice for building high-power systems. This thesis has showcased the enormous potential of bringing cutting-edge laser diode technology and powerful Kerr-lens mode-locked Cr:ZnS/ZnSe lasers together for the first time to help meet the growing demand for broadband, low-noise, and accessible table-top mid-IR sources. The main findings and results are briefly summarized in the following, highlighting the major

N. Nagl, *A New Generation of Ultrafast Oscillators for Mid-Infrared Applications*, Springer Theses, https://doi.org/10.1007/978-3-030-89754-3_6

Table 6.1 Comparison of the laser performance of the three directly diode-pumped Cr^{2+}-doped II–VI laser systems presented in this thesis. The integrated RMS relative intensity noise (RIN) was calculated between 20 Hz and 1 MHz and measured in mode-locked laser operation. (*)Value corresponds to the frequency-doubled laser output

	MIRanda1 [5]	MIRanda2 [6]	MIRanda3
ML mechanism	**Kerr-lens ML**	**Kerr-lens ML**	**Kerr-lens ML**
Gain medium	Cr:ZnSe (5 mm)	Cr:ZnS (5 mm)	Cr:ZnS (5 mm)
Pump diodes	InP (1x)	InP (2x)	InP (2x)
P_{pump}	3 W	5 W	4.5 W
λ_{pump}	1.65 μm	1.65 μm	1.65 μm
P_{CW}	700 mW	1200 mW	1100 mW
P_{ML}	500 mW	800 mW	720 mW
τ_0	45 fs	34 fs	28 fs
ν_{rep}	64.7 MHz	49.9 MHz	22.9 MHz
OC ratio	34%	25%	25%
RIN	0.068%(*)	0.014%	0.015%
E_0	**7.7 nJ**	**16 nJ**	**31 nJ**
P_0	**151 kW**	**415 kW**	**988 kW**

breakthroughs and continuous refinements achieved during the development of three directly diode-pumped KLM systems (see Table 6.1).

In Chap. 2, the theoretical foundations behind ultrashort pulse generation have been laid, with a focus on soliton pulse propagation and soliton mode-locking. Starting with the concept of mode-locking in a simple Fabry–Pérot laser resonator, the advantageous technique of Kerr-lens mode-locking (KLM) was described and analyzed in detail. Conditions under which fundamental solitons are formed were also discussed. Finally, the master equation of mode-locking was presented, which characterizes the complex interplay between the many effects governing the pulse evolution in a laser cavity, including self-amplitude modulation and the spectral filtering action of the gain medium.

In Chap. 3, the favourable material and spectroscopic properties of chromium-doped zinc sulfide and zinc selenide (Cr^{2+}:ZnS/ZnSe) were highlighted, drawing parallels to the well-known titanium sapphire crystals used in the near-infrared region (<1 μm). Also, the challenges associated with direct diode-pumping and their effects on the generation of few-cycle pulses were addressed, particularly with regard to soft-aperture KLM. A sophisticated pump setup and resonator design finally led to the first realization of a directly diode-pumped Kerr-lens mode-locked Cr^{2+}-doped II–VI oscillator (*MIRanda1*), pumped by the emission of a single-emitter laser diode. With a mode-locked output power of 500 mW at a repetition rate of 65 MHz and pulse durations as short as 45 fs, a sixty-fold increase in peak power could be demonstrated over the previous record based on SESAM mode-locking. Moreover, noise measurements revealed an excellent amplitude noise performance of the mode-locked output, resulting in an integrated RMS relative intensity noise of less than 0.068% RMS (20 Hz–1 MHz) for the frequency-doubled output.

In Chap. 4, the polarization-multiplexed combination of two single-emitter pump laser diodes resulted in a powerful second-generation system (*MIRanda2*), placing the performance of directly diode-pumped Cr^{2+}:ZnS/ZnSe laser systems on the same level as state-of-the-art fiber-laser-pumped counterparts. By evaluating the Kerr strength of the oscillator via nonlinear ABCD-matrix simulations and overcoming the challenges of higher average and intra-cavity peak power levels, mode-locked output powers of over 800 mW were recorded. Combined with the improved fine-tuning of the cavities' round-trip dispersion via dedicated dispersive optics, a further reduction in pulse duration down to four optical cycles ($\tau_p = 34$ fs) was achieved. At a repetition frequency of 50 MHz, this resulted in an almost three-fold increase in peak power over *MIRanda1*, while the low amplitude noise performance of the Kerr-lens mode-locked oscillator was retained (0.014% RMS for 20 Hz–1 MHz).

In Chap. 5, the performance of directly diode-pumped Cr^{2+}-doped II–VI laser oscillators was brought to a new level. A classic bow-tie cavity was realized for direct diode-pumping by implementing a novel pump focusing approach. Moreover, the incorporation of 4f-telescope extensions into the resonator arms and a change of the focal spot diameter in the crystal allowed to operate the soft-aperture KLM laser system at very low repetition rates of only 23 MHz. Being pumped by the emission of two single-emitter laser diodes at up to 4.5 Watts of combined power, a peak power as high as 1 MW was reached at pulse durations of 28 fs and an equally low amplitude noise level as previously achieved (0.015% RMS for 20 Hz to 1 MHz). To demonstrate the power of this third-generation system (*MIRanda3*), the generation of mid-infrared radiation using the oscillator's direct output was presented for the first time. With the output focused into a nonlinear optical crystal (ZGP), a multi-octave-spanning spectrum between 4.5 and 14 μm was detected. At optical-to-optical efficiencies of up to 5.7%, the utilized technique of intra-pulse difference frequency generation can directly provide few-cycle pulses at the milliwatt-level, featuring intrinsic carrier-envelope phase stabilization.

6.2 A New Ultrafast Workhorse for Spectroscopic Applications?

To answer this question, it is worthwhile to examine the results achieved from an application-oriented perspective. Starting with the output characteristics provided by the three directly diode-pumped Cr:ZnS/ZnSe oscillators, a highly-promising low-noise performance of the few-cycle laser output has been shown. For applications such as Fourier-transform-based spectroscopy [7], it is the laser's relative intensity noise (RIN) that constitutes the main limiting factor on the overall detection sensitivity. Besides the dynamic range of the detection setup as well as the detector noise and shot noise, laser RIN sets the ultimate limit to the signal-to-noise ratio at source powers approaching the mW-range [7]. With the technique of direct diode-pumping and the resulting low-noise laser operation, we are confident that the specificity and sensitivity achieved with electric-field resolved measurements [8] can be pushed to a new level. Moreover, when increasing the peak power of diode-pumped Cr:ZnS/ZnSe lasers to the 1 MW-level, multi-octave-spanning mid-infrared generation between 4.5 and 14 μm is feasible via oscillator-driven intra-pulse DFG in a nonlinear crystal. Not only are the low-noise properties of the driving laser directly transferred into the MIR region, this alignment-insensitive single-beam geometry also allows to generate few-cycle MIR pulses with intrinsic carrier-envelope phase (CEP) stability [9].

The presented results already lay the foundation for advancing the field of infrared spectroscopy, but there are aspects that can be further improved in the future. In particular, the technique of electro-optic sampling greatly benefits from shorter sampling pulses. EOS probes the birefringence of the crystal that has been modulated by the electric field of the mid-infrared pulse, with the measurements being performed as a function of the delay between both pulses. However, the recorded EOS signal on the short-wavelength side is limited by the duration of the probe pulse [10], and broader frequency spans are only detectable for decreasing sampling pulse durations. Typically, it is said that the probe pulse has to be shorter than a half-cycle of the oscillating field to be observed [11], and a common way to achieve this is to split off part of the oscillator output for further nonlinear spectral broadening and compression [3, 10]. In nonlinear media, the propagation of few-cycle pulses is influenced by the $\chi^{(3)}$-effects of self-focusing and self-phase modulation, and a proper control of the material parameters can result in a significant reduction in pulse duration. Over the past few years, fiber-based [12] approaches as well as bulk compression schemes have already been successfully demonstrated in our group [13–15] and more experimental research is currently undertaken with Cr:ZnS/ZnSe-based systems.

In addition to shorter pulses, the carrier-envelope phase of the oscillator can be actively stabilized, such that new MIR frequencies—converted using methods other than non-cascaded IDFG [16]—will also share this important property. With the aforementioned shortening of pulse durations and the corresponding spectral broadening, traditional carrier-envelope-offset detection schemes based on interference beatings between the fundamental spectrum and its second-harmonic signal can be employed [17]. Alternatively, a recently reported scheme for Cr:ZnS/ZnSe-based master oscillator power amplifier (MOPA) systems [18] utilizing the harmonics-enhanced supercontinuum generation observed in polycrystalline ZnS and ZnSe [19] could also be implemented. By detecting these beatings between adjacent harmonics and incorporating phase-locked loop controllers, ultrabroadband and CEP-stabilized mid-infrared waveforms can finally be realized.

Therefore, one can confidently say that diode-pumped few-cycle Cr^{2+}-doped II–VI technology has enormous potential to bring us closer towards the ultimate limits of mid-IR spectroscopy. The dramatically simplified design architecture for mid-infrared generation, combined with high efficiencies, intrinsic low-noise, and direct few-cycle pulse generation, is already competitive with currently used systems, and will potentially render the complex 1 μm-based laser frontends obsolete.

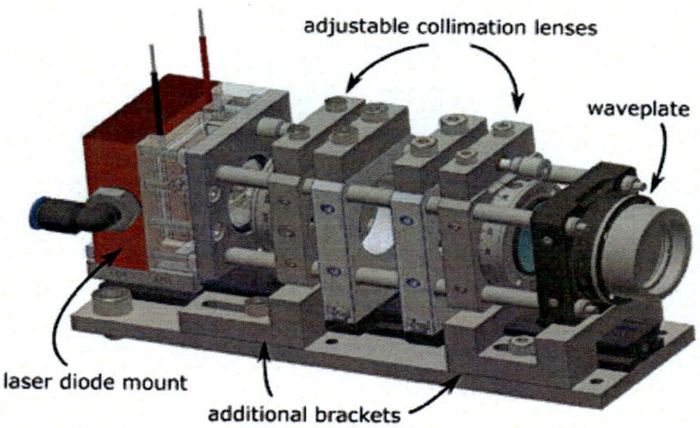

Fig. 6.1 3D mechanical design of the new laser diode collimation unit. The picture was kindly provided by our laser engineer Aleksandar Sebesta

What's Next?—Prototype Development for Mid-IR Applications

Based on the experimental findings and achievements presented in this thesis, a laser system highly optimized for stability, compactness and low-maintenance operation (see 3D designs in Figs. 6.1 and 6.2) is currently being developed at our research laboratory for upcoming spectroscopic measurements [8]. It will incorporate two pump laser diodes and a 50 MHz-oscillator structure—very similar to MIRanda2—and is meant to deliver more than 500 mW of average power at pulse durations shorter than five optical cycles ($\tau_p < 40$ fs). Moreover, the airtight and vacuum-compatible chamber environment eliminates effects such as air turbulences, dust particles and water absorption to affect the laser performance, and allows for its flexible deployment in different laboratories. Features such as live monitoring of laser parameters further help to identify potential problems before they aggravate. A first prototype is currently set up by our laser engineer, with a picture of the real laser system shown in Fig. 6.2. After all those months and years of work, it is gratifying to see how the basic experimental research performed in this thesis starts to enter a new phase, heading to explore the weakest molecular signatures in the entire fingerprint region via field-resolved spectroscopy.

6.3 Closing Remarks

The presented few-cycle Cr^{2+}-doped II–VI laser oscillators—combined with the intra-pulse difference frequency generation scheme—herald a new era of ultrafast mid-infrared coherent light sources and render directly diode-pumped Cr:ZnS/ZnSe oscillators as one of the most promising driving sources for the emerging fields of nano-imaging [20–22] as well as time-domain [21, 23] and frequency comb spectroscopy [24–26]. Moreover, with their unique capability to provide low-noise and ultrabroadband mid-infrared radiation already at the milliwatt level, the directly diode-pumped Cr:ZnS/ZnSe systems open up new avenues into the study of com-

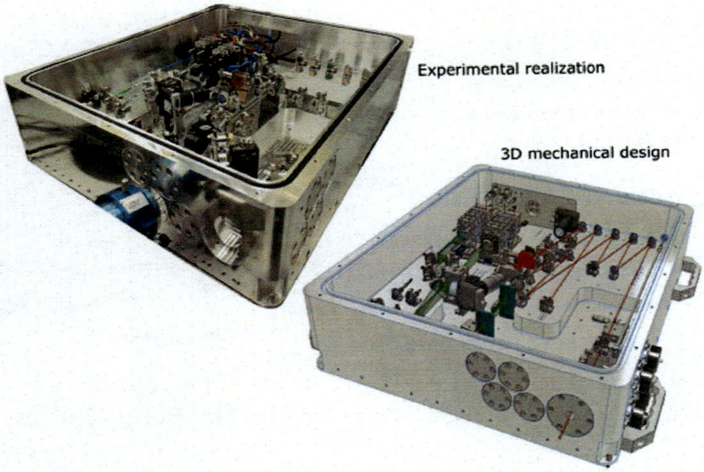

Fig. 6.2 3D mechanical design of the laser prototype and its experimental realization. The laser design is based on the second diode-pumped system *MIRanda2*. The pictures were kindly provided by our laser engineer Aleksandar Sebesta

plex biological systems in the entire molecular fingerprint region [8]. The exploration phase has certainly just begun, and there are many more specimens that are waiting to be examined by ultrafast laser technology.

Pushing the frontiers of state-of-the-art solid-state laser systems for advancements in MIR applications was the key focus and main driver of this PhD thesis. From the knowledge gained and the breakthroughs achieved, the realization of a compact and low-noise coherent mid-infrared source is now one step closer. Nonetheless, the philosophy behind this work will remain unchanged:

Advancing a Just Cause.

References

1. Grassani D et al (2019) Mid infrared gas spectroscopy using efficient fiber laser driven photonic chip-based supercontinuum. Nature Commun 10:1553
2. Willer U, Saraji M, Khorsandi A, Geiser P, Schade W (2006) Near- and mid-infrared laser monitoring of industrial processes, environment and security applications. Optics Lasers Eng 44(7):699–710
3. Pupeza I et al (2017) Field-resolved spectroscopy in the molecular fingerprint region. In: Conference on lasers and electro-optics Europe & European quantum electronics conference (CLEO/Europe-EQEC) CH-2.4 (2017)
4. Lasch P, Kneipp J (2008) Biomedical vibrational spectroscopy. Wiley, New York

5. Nagl N, Gröbmeyer S, Pervak V, Krausz F, Pronin O, Mak K (2019) Directly diode-pumped, Kerr-lens mode-locked, few-cycle Cr:ZnSe oscillator. Optics Express 27(17):24445–24454
6. Nagl N, Gröbmeyer S, Potzlberger M, Pervak V, Krausz F, Mak KF (2020) Directly diode-pumped few-optical-cycle Cr:ZnS laser at 800 mW of average power. In: Conference on lasers and electro-optics. Optical Society of America, SF3H.5
7. Newbury NR, Coddington I, Swann W (2010) Sensitivity of coherent dual-comb spectroscopy. Optics Express 18(8):7929–7945
8. Pupeza I et al (2020) Field-resolved infrared spectroscopy of biological systems. Nature 577(7788):52–59
9. Pupeza I et al (2015) High-power sub-two-cycle mid-infrared pulses at 100 MHz repetition rate. Nature Photonics 9:721–724
10. Chen B-H, Wittmann E, Morimoto Y, Baum P, Riedle E (2019) Octave-spanning single-cycle middle-infrared generation through optical parametric amplification in $LiGaS_2$. Optics Express 27(15):21306–21318
11. Keiber S et al (2016) Electro-optic sampling of near-infrared waveforms. Nature Photonics 10:159–162
12. Nagl N, Mak K, Wang Q, Pervak V, Krausz F, Pronin O (2019) Efficient femtosecond mid-infrared generation based on a Cr:ZnS oscillator and step-index fluoride fibers. Optics Lett 44(10):2390–2393
13. Fritsch K, Pötzlberger M, Pervak V, Brons J, Pronin O (2018) All-solid-state multipass spectral broadening to sub-20 fs. Optics Lett 43(19):4643–4646
14. Gröbmeyer S et al (2020) Self-compression at 1 μm wavelength in all-bulk multipass geometry. Appl Phys B 126:159
15. Seidel M, Arisholm G, Brons J, Pervak V, Pronin O (2016) All solid-state spectral broadening: an average and peak power scalable method for compression of ultrashort pulses. Optics Express 24(9):9412–9428
16. Wang Q et al (2019) Broadband mid-infrared coverage (2–17 μm) with few-cycle pulses via cascaded parametric processes. Optics Lett 44(10):2566–2569
17. Telle HR, Steinmeyer G, Dunlop AE, Stenger J, Sutter DH, Keller U (1999) Carrier-envelope offset phase control: a novel concept for absolute optical frequency measurement and ultrashort pulse generation. Appl Phys B 69:327–332
18. Vasilyev S, Moskalev I, Mirov M, Mirov S, Gapontsev V (2016) Multi-Watt Mid-IR femtosecond polycrystalline Cr^{2+}:ZnS and Cr^{2+}:ZnSe laser amplifiers with the spectrum spanning 2.0-2.6 μm. Opt Soc Amer 24(2):1616-1623
19. Šuminas R, Marcinkevičiute A, Tamošauskas G, Dubietis A (2019) Even and odd harmonics-enhanced supercontinuum generation in zinc-blende semiconductors. J Opt Soc Amer B 36(2):A22–A27
20. Amenabar I, Poly S, Goikoetxea M, Nuansing W, Lasch P, Hillenbrand R (2017) Hyperspectral infrared nanoimaging of organic samples based on Fourier transform infrared nanospectroscopy. Nature Commun 8:14402
21. Eisele M et al (2014) Ultrafast multi-terahertz nano-spectroscopy with sub-cycle temporal resolution. Nature Photonics 8:841–845
22. Wang L, Xu XG (2015) Scattering-type scanning near-field optical microscopy with reconstruction of vertical interaction. Nature Commun 6:8973
23. Lanin AA, Voronin AA, Fedotov AB, Zheltikov AM (2014) Time-domain spectroscopy in the mid-infrared. Sci Rep 4:6670
24. Piqué N, Hänsch TW (2019) Frequency comb spectroscopy. Nature Photonics 13:146–157
25. Schliesser A, Picqué N, Hänsch TW (2012) Mid-infrared frequency combs. Nature Photonics 6:440–449
26. Coddington I, Newbury N, Swann W (2016) Dual-comb spectroscopy. Optica 3(4):414–426

Appendix

A.1 Data Archiving

The experimental raw data, evaluation files, and original figures can be found on the Data Archive Server of the Laboratory for Attosecond Physics at the Max Planck Institute of Quantum Optics:

//AFS/ipp-garching.mpg.de/mpq/lap/publication_archive.

All files are organized in a folder structure that is in accordance with the numbering and the structure of the thesis. Each folder contains figures in .png format, raw data and Python scripts if needed. Also, a *ReadMe.txt* file is added to every folder, providing further explanations and information on the presented data.

N. Nagl, *A New Generation of Ultrafast Oscillators for Mid-Infrared Applications*, Springer Theses, https://doi.org/10.1007/978-3-030-89754-3

A.2 Curriculum Vitae

Personal information

Name: Nathalie Nagl
Date of birth: 4 February 1994
Place of birth: Landshut
Nationality: German
Email: nathalie.nagl@mpq.mpg.de

Current affiliation:
Max Planck Institute of Quantum Optics,
Laboratory for Attosecond Physics

Education

- **Ph.D. in Physics (Dr.rer.nat.)**, 02/2018–01/2021, *with greatest distinction*
  Ludwig-Maximilians-Universität München

  Thesis title: A New Generation of Ultrafast Oscillators for Mid-Infrared Applications
  Thesis advisor: Prof. Dr. Ferenc Krausz

- **Master of Science in Physics (M.Sc.)**, 10/2015–10/2017
  Ludwig-Maximilians-Universität München

  Thesis title: Nonlinear spectral broadening and characterization of pulses from a Cr:ZnS femtosecond oscillator
  Thesis advisor: Prof. Dr. Ferenc Krausz

- **Bachelor of Science in Physics (B.Sc.)**, 10/2012–08/2015
  Ludwig-Maximilians-Universität München

  Thesis title: Resonante Anregung von NV-Zentren und Bestimmung von Glasfaserfluoreszenzen
  Thesis advisor: Prof. Dr. Harald Weinfurter

Honors and Awards

- **2021, Springer Theses**
  Ph.D. thesis award

- **2021, Deutsche Physikalische Gesellschaft (DPG)**
  Female Physicist of the Week

- **2021, Soapbox Science Munich**
  Selected speaker at the Soapbox Science event
- **2020, Highlighted conference contribution**
  Upgrade to an invited talk at one of the world's largest international conferences in the field of photonics (CLEO US 2020)
- **2018–2021, Graduate school IMPRS-APS**
  Scholarship of the International Max Planck Research School on Advanced Photon Science
- **2018–2020, Bischöfliche Studienförderung Cusanuswerk**
  Doctoral scholarship
- **2017, MTU Aero Engines Studien-Stiftung**
  Seminar for gifted young women in scientific and technical courses of study
- **2012, Hans-Carossa-Gymnasium Landshut**
  Graduation award
- **2011, Bavarian state ministry of education and culture**
  Seminar for highly gifted, versatile and especially capable high school students

A.3 List of Publications

Peer-Reviewed Journal Articles (First Author):

- **Nathalie Nagl**, Ka Fai Mak, Qing Wang, Vladimir Pervak, Ferenc Krausz, and Oleg Pronin, "Efficient femtosecond mid-infrared generation based on a Cr:ZnS oscillator and step-index fluoride fibers," Opt. Lett. 44, 2390–2393 (2019)
  The author designed and set up the fiber-based nonlinear spectral broadening unit, prepared the fibers, performed the measurements, analyzed the experimental data, and prepared the manuscript together with K. F. Mak.
- **Nathalie Nagl**, Sebastian Gröbmeyer, Vladimir Pervak, Ferenc Krausz, Oleg Pronin, and Ka Fai Mak, "Directly diode-pumped, Kerr-lens mode-locked, few-cycle Cr:ZnSe oscillator," Opt. Express 27, 24445–24454 (2019)
  The author designed and built the laser system, performed the measurements, analyzed the experimental data, and prepared the manuscript together with K. F. Mak.

Peer-Reviewed Journal Articles (Co-Author):

- Jinwei Zhang, Ka Fai Mak, **Nathalie Nagl**, Marcus Seidel, Dominik Bauer, Dirk Sutter, Vladimir Pervak, Ferenc Krausz, and Oleg Pronin, "Multi-mW, few-cycle mid-infrared continuum spanning from 500 to 2250 cm^{-1}," Light Sci. Appl. 7, 17180 (2017)

The author built the pulse characterization unit (SHG-FROG), assisted in the pulse duration measurements, revised and discussed the manuscript.

- Qing Wang, Jinwei Zhang, Alexander Kessel, **Nathalie Nagl**, Vladimir Pervak, Oleg Pronin, and Ka Fai Mak, "Broadband mid-infrared coverage (2–17 μm) with few-cycle pulses via cascaded parametric processes," Opt. Lett. 44, 2566–2569 (2019)
  The author built the pulse characterization unit (SHG-FROG), assisted in the pulse duration measurements, revised and discussed the manuscript.
- Thomas P. Butler, Nikolai Lilienfein, Jia Xu, **Nathalie Nagl**, Christina Hofer, Daniel Gerz, Ka Fai Mak, Christian Gaida, Tobias Heuermann, Martin Gebhardt, Jens Limpert, Ferenc Krausz and Ioachim Pupeza, "Multi-octave spanning, Watt-level ultrafast mid-infrared source," JPhys Photonics 1, 044006 (2019)
  The author performed preliminary proof of concept experiments with the ZBLAN fibers, assisted in the fiber parameter selection as well as the fiber preparation, revised and discussed the manuscript.

Conference Talks:

- **Nathalie Nagl**, Ka Fai Mak, Vladimir Pervak, Ferenc Krausz, and Oleg Pronin, "200 mW-level, few-opticle-cycle tunable femtosecond mid-infrared source based on a Cr:ZnS oscillator and step-index fluoride fibers," in 2018 8th EPS-QEOD Europhoton Conference, paper FrM1.3 (2018)
- **Nathalie Nagl**, Sebastian Gröbmeyer, Vladimir Pervak, Oleg Pronin, Ka Fai Mak, and Ferenc Krausz,"Directly Diode-Pumped, Kerr-Lens Mode-Locked Cr:ZnSe Oscillator," in 2019 Conference on Lasers and Electro-Optics Europe and European Quantum Electronics Conference, OSA Technical Digest (Optical Society of America, 2019), paper CF-2.2 (2019)
- **Nathalie Nagl**, Sebastian Gröbmeyer, Markus Pötzlberger, Vladimir Pervak, Ferenc Krausz, and Ka Fai Mak, "Directly Diode-Pumped Few-Optical-Cycle Cr:ZnS Laser at 800 mW of Average Power," in 2020 Conference on Lasers and Electro-Optics, OSA Technical Digest (Optical Society of America, 2020), paper SF3H.5 (2020)
- **Nathalie Nagl**, Vladimir Pervak, Ferenc Krausz, and Ka Fai Mak, "Milliwatt-Level Multi-Octave Mid-Infrared Generation by a Diode-Pumped Cr:ZnS Oscillator," in 2021 Conference on Lasers and Electro-Optics Europe and European Quantum Electronics Conference, OSA Technical Digest (Optical Society of America, 2021), paper CF-6.1 (2021)

UHPCC Under Impact and Blast

Qin Fang · Hao Wu · Xiangzhen Kong

UHPCC Under Impact and Blast

Science Press
Beijing

Springer

Qin Fang
Army Engineering University of PLA
Nanjing, Jiangsu, China

Xiangzhen Kong
College of National Defense
Army Engineering University of PLA
Nanjing, Jiangsu, China

Hao Wu
College of Civil Engineering
Tongji University
Shanghai, China

ISBN 978-981-33-6841-5 ISBN 978-981-33-6842-2 (eBook)
https://doi.org/10.1007/978-981-33-6842-2

Jointly published with Science Press, Beijing, China
The print edition is not for sale in China Mainland. Customers from China Mainland please order the print book from: Science Press.

This Springer imprint is published by the registered company Springer Nature Singapore Pte Ltd.
The registered company address is: 152 Beach Road, #21-01/04 Gateway East, Singapore 189721, Singapore

Preface

Compared with the normal strength concrete (NSC), ultra-high performance cementitious composites (UHPCC) is a relativity new type of cementitious materials which consists of very low water-to-binder ratio, high amount of high-range water reducer, fine aggregates and high-strength steel or organic fibers. With the prominent mechanical properties, e.g., high compressive and tensile strength, high ductility, high fracture energy and very low permeability, UHPCC has been becoming the most prospective construction cement-based material for both civil and military engineering structures, such as fortifications, nuclear waste storage containments, highway bridges, high-rise buildings, to resist high-speed projectile penetration, low-velocity impact and blast, as well as earthquake loadings. Therefore, investigations on the static and dynamic mechanical properties as well as the impact and blast resistance of UHPCC are essential and important for the design and safety assessment of protective structures. In this book, the related work conducted by authors are presented and the main contents are as follows:

Chapter 1: A series of cubic/axial compressive, direct tensile, four-point and three-point flexural tests on UHPCC specimens are presented, in which the effects of six volume fractions (0 ~ 2.5%) and two types (micro-straight and hooked) of steel fibers on the static mechanical properties of UHPCC are examined.

Chapter 2: The SHPB test on UHPCC specimens is presented, in which the effects of strain rate ranges from 17.6 to 328.4 s^{-1} and two typical steel fibers (micro-straight and hooked) with three volume fractions of 0%, 1.0% and 2.0% on the dynamic compressive mechanical properties of UHPCC are analyzed.

Chapter 3: By using a 50 mm-diameter conic variable cross-sectional SHPB, the dynamic spalling test on cylindrical UHPCC specimens is given, and the influences of the tensile strain rate range from 14.3 to 214.8 s^{-1}, and two typical steel fibers (micro-straight and hooked) with three volume fractions of 0%, 1.0% and 2.0% on the dynamic tensile mechanical properties of UHPCC are studied.

Chapter 4: The triaxial compressive behavior of UHPCC under high confining pressure (up to 100 MPa) is experimentally studied, and the failure criteria and toughness of UHPCC under triaxial compression are discussed.

Chapter 5: The impact resistance of coarse aggregated UHPCC target against the medium caliber projectile is studied. The high-speed projectile penetration tests on UHPCC targets with the striking velocities at 510 to 1320 m/s as well as the comparable projectile penetration test on UHP-CASFRC are presented. The influence of coarse aggregates strength on the impact resistance of UHPCC targets is discussed based on the 3D mesoscopic concrete model.

Chapter 6: Aiming to protect person and valuable equipment from the perforated small caliber arms and scabbing fragments, the 7.62 mm API bullet impacting test on bare and rear fabric (CFRP or UHMWPE) strengthened UHP-BASFRC panels is given and the excellent impact resistance of bare and composite UHP-BASFRC panels is validated.

Chapter 7: The medium caliber projectile impact resistance of armor steel/ceramic/UHPCC-layered composite targets against 30CrMnSiNi2A steel projectiles is experimentally studied. The numerical simulations are further performed by calibrating the Johnson and Cook (JC) constitutive model parameters of the 10CrNi3MoV21A, NP450 and NP500 armor steel.

Chapter 8: The impact behavior of UHPCC-FST under transverse impact load is investigated experimentally and numerically. Three UHPCC-FST specimens subjected to low-velocity impact by using a drop-hammer impact device are examined experimentally. The model parameters of K&C model for UHPCC are calibrated by using the existing static and dynamic experimental data. Then, a FE analysis model is established to predict the dynamic responses of UHPCC-FSTs under transverse impact load.

Chapter 9: A total of eleven steel bar reinforced UHPCC and two reinforced NSC control specimens subjected to drop-hammer impact are studied. The outstanding impact resistance of UHPC members is validated and assessed quantitatively. The model parameters and the parameters generation method of continuous surface cap (CSC) model for UHPCC are calibrated and fully validated.

Chapter 10: The test on five UHPCC-FST specimens under contact explosion of TNT charges is given, and the original axial capacity of the intact columns and the residual axial capacity of the blast-damaged columns are further evaluated though the axial compression tests. Besides, the damage and failure modes of UHPCC-FST are numerically reproduced, and the related parametric influences are discussed.

Chapter 11: The field test of four circular UHPCC-FST specimens under the close-range TNT charge explosion with the scaled standoff distance of 0.12 ~ 0.14 m/kg$^{1/3}$ is presented. The dynamic responses of those specimens are analyzed by three different methods, i.e., ALE method, velocity loading method and SDOF method, and the velocity method is recommended by considering both the accuracy and efficiency of computation.

Chapter 12: The residual seismic resistance (RSR) of UHPCC-FST specimens after contact explosion is studied experimentally. The specimens are firstly subjected to blast loadings, in which the TNT charge weights are 1 ~ 3 kg and the height of bursts are set to 250 mm. Furthermore, the RSR of UHPCC-FST specimens are examined through the low-frequency horizontal cyclic loading test in two perpendicular directions. A composite damage index is proposed to evaluate the RSR of the post-blast columns.

Chapter 13: The field tests and numerical simulations of the blast-resistant behavior of one-way simply supported reinforced UHPCC panels and NSC control panels are performed. The explosion tests are in different scaled distances (0.5 ~ 1.0 m/kg$^{1/3}$) with end-detonated cylindrical charges. The blast loadings induced by the medium-range explosions and the constitutive model parameters of UHPC are mainly concerned. The proposed FE model, numerical algorithm and the calibrated model parameters are fully verified by comparing to the test data.

Chapter 14: A new constitutive model of UHPCC material under impact and blast loadings is developed, which involves proposing a new tensile damage model for UHPCC which is then incorporated into the Kong-Fang material model recently developed, and calibrating parameters of this modified material model based on existing test data. Single element tests are firstly conducted to demonstrate the performances of the proposed material model. Then, three selected experiments are numerically simulated and compared with corresponding experimental data and good agreements are observed.

Professor Qi Hu Qian is our guide though the study of protective structures. The late Prof. Zhao Yuan Chen from Tsinghua University and Prof. Wei Sun from Southeast University always encouraged us to develop new blast/impact-resistant materials and structures. We are grateful to them for introducing and helping us to continue to walk along this path with ever-increasing interest.

The present book is the result of a cooperative effort of our team. The authors would like to thank the our colleagues and postgraduates, who are Jianzhong Liu, Genmao Ren, Yong Peng, Ziguo Wang, Yangxiu Zhai, Yuehua Cheng. The authors would also like to take this opportunity to thank the long-tern support provided by National Natural Science Foundation of China. We hope this book will serve as a useful source of information for scientists, engineers and students active in protective structure research.

Nanjing, China | Qin Fang
Shanghai, China | Hao Wu
Nanjing, China | Xiangzhen Kong

Contents

1 Static Mechanical Properties of UHPCC 1
1.1 Introduction 1
1.2 Test Program 2
1.2.1 Raw Materials and Mixture Proportions 2
1.2.2 Specimen Preparation and Curing 4
1.3 Instrumentation and Loading Scheme 5
1.3.1 Cubic Compressive Test 5
1.3.2 Axial Compressive Test 6
1.3.3 Direct Tensile Test 7
1.3.4 Four-Point Flexural Test 8
1.3.5 Three-Point Flexural Test 9
1.4 Test Results and Discussion 10
1.4.1 Compression Test 10
1.4.2 Direct Tension Test 14
1.4.3 Four-Point Flexure Test 17
1.4.4 Three-Point Flexure Test 21
1.5 Summary 27
References 28

2 Dynamic Compressive Mechanical Properties of UHPCC 31
2.1 Introduction 31
2.2 Specimen Preparation 32
2.3 SHPB Test 33
2.3.1 Test Device 33
2.3.2 Test Technique 34
2.4 Test Results and Discussions 37
2.4.1 Stress Equilibrium 37
2.4.2 Strain Rate Determination 38
2.4.3 Dynamic Failure Pattern 39
2.4.4 Dynamic Stress–Strain Curve 40
2.4.5 Dynamic Increase Factor 44
2.4.6 Energy Absorption Capacity 48

2.5 Visco-Elastic Damage Model ... 49
2.5.1 Nonlinear Visco-Elastic Model ... 49
2.5.2 Model Calibration and Validation ... 51
2.6 Summary ... 51
References ... 53

3 Dynamic Tensile Mechanical Properties of UHPCC ... 55
3.1 Introduction ... 55
3.2 Specimen Preparation ... 56
3.3 Dynamic Spalling Test ... 57
3.3.1 Test Device ... 57
3.3.2 Test Technique ... 58
3.4 Test Results and Discussions ... 61
3.4.1 Dynamic Failure Patterns ... 61
3.4.2 Dynamic Spalling Strength ... 63
3.4.3 Relations Between the Critical Time to Fracture and Dynamic Spalling Strength ... 66
3.4.4 Dynamic Increase Factor ... 67
3.5 Summary ... 70
References ... 71

4 Triaxial Compressive Behavior of UHPCC and Application in the Numerical Analyses of Projectile Impact ... 73
4.1 Introduction ... 73
4.2 A Review of the Existing Works on Triaxial Behavior of Concrete ... 74
4.3 Mixing Optimization of UHPCC and Triaxial Compression Test ... 76
4.3.1 Compositions ... 76
4.3.2 Mixing Procedure ... 78
4.3.3 Triaxial Compression Test ... 79
4.4 Results and Analysis ... 80
4.4.1 Failure Pattern ... 80
4.4.2 Stress–Strain Curve ... 81
4.4.3 Failure Criteria ... 85
4.4.4 Toughness ... 90
4.5 Applications in the Numerical Analyses ... 92
4.5.1 Brief Introduction of HJC Constitutive Model ... 92
4.5.2 HJC Model Parameters for HSC ... 94
4.5.3 Validations ... 95
4.6 Summary ... 98
References ... 101

5 Projectile Penetrations into Coarse Aggregated UHPCC Targets 105
5.1 Introduction 105
5.2 Basalt Aggregated UHPCC Target 108
5.2.1 Target 108
5.2.2 Projectile 109
5.2.3 Test Setup 113
5.2.4 Test Results 114
5.2.5 Discussions 116
5.3 Corundum Aggregated UHPCC Target 121
5.3.1 Target and Projectile 121
5.3.2 Test Results 122
5.3.3 Discussions 127
5.4 Numerical Simulations Based on 3D Mesoscopic Concrete Model 138
5.4.1 3D Mesoscopic Concrete Model 138
5.4.2 Validations 148
5.4.3 Impact Resistance of Different Aggregated UHPC 153
5.5 Summary 158
References 161

6 Impact Resistance of Basalt Aggregated UHP-SFRC/Fabric Composite Panels Against Small Caliber Arm 163
6.1 Introduction 163
6.2 Bullet Perforation Test 165
6.2.1 Bullet 165
6.2.2 UHP-BASFRC Panels 165
6.2.3 Fabric Strengthening 167
6.2.4 Test Setup 169
6.3 Test Results 170
6.3.1 Damage of Target 170
6.3.2 Dimension of Crater 174
6.3.3 Perforation Limit 175
6.3.4 Recovered Bullet 176
6.3.5 Damage of Aluminum Plate 177
6.4 Discussions 177
6.4.1 Crater Dimensions 177
6.4.2 Terminal Ballistic Parameter 179
6.4.3 Fabric Effect 183
6.5 Summary 184
References 184

7 Impact Resistance of Armsector Steel/Ceramic/UHPCC Layered Composite Targets Against 30CrMnSiNi2A Steel Projectiles 187
7.1 Introduction 187
7.2 Impact Test on 10CrNi3MoV21A Armor Steel/SiC Ceramic/UHPCC Composite Targets and Numerical Simulations 190
7.2.1 Impact Test 190
7.2.2 Numerical Simulations 201
7.3 Impact Test on NP450 Armor Steel/UHPCC and NP500 Armor Steel/UHPCC Composite Targets and Numerical Simulations 217
7.3.1 Impact Test 217
7.3.2 Numerical Simulations 222
7.4 Summary 232
References 233

8 Response of UHPCC-FST Subjected to Low-Velocity Impact 237
8.1 Introduction 237
8.2 Test Program 239
8.2.1 UHPCC-FST Specimens 239
8.2.2 Axial Compression Test 240
8.2.3 Drop-Hammer Impact Test 243
8.3 Test Results and Discussions 244
8.3.1 Axial Compression 244
8.3.2 Lateral Impact Resistance 245
8.3.3 Impact Force–Time History 246
8.3.4 Deflection-Time History 247
8.4 Calibration of K&C Model Parameters for UHPCC 248
8.4.1 Brief Introduction of K&C Model 249
8.4.2 Calibration 251
8.5 Numerical Simulation 260
8.5.1 Present Test 260
8.5.2 Yoo et al. (2015) Test 264
8.6 Summary 266
References 267

9 Dynamic Responses of Reinforced UHPCC Members Under Low-Velocity Lateral Impact 271
9.1 Introduction 271
9.2 Test Program 274
9.2.1 Specimen Fabrication 274
9.2.2 Drop-Hammer Impact Test 275

9.3 Test Results and Discussions ... 277
9.3.1 Failure Mode ... 277
9.3.2 Impact Force–Time History ... 280
9.3.3 Deflection-Time History ... 284
9.3.4 Energy Dissipation ... 287
9.4 Numerical Simulation ... 288
9.4.1 FE Model ... 288
9.4.2 Calibration ... 289
9.4.3 Comparisons of Numerical Results with Test Data ... 300
9.5 Further Validations ... 304
9.5.1 Reinforced UHPC Members ... 305
9.5.2 UHPC-FST Members ... 310
9.6 Summary ... 313
References ... 315

10 Residual Axial Capacity of UHPCC-FST Column Under Contact Explosion ... 319
10.1 Introduction ... 319
10.2 Review of the Existing Works ... 321
10.3 UHPCC-FST Columns ... 322
10.3.1 Fabrications ... 323
10.3.2 Steel Tube ... 324
10.3.3 UHPCC ... 325
10.4 Field Contact Explosion Test ... 325
10.4.1 Test Setup ... 325
10.4.2 Test Results ... 327
10.5 Axial Compression Test ... 329
10.5.1 Test Setup ... 329
10.5.2 Test Results ... 330
10.6 Numerical Simulation ... 336
10.6.1 FE Model ... 336
10.6.2 Material Model ... 338
10.6.3 Loading Scheme ... 345
10.7 Comparisons with Test Data ... 346
10.7.1 Damage and Failure Modes of Columns ... 346
10.7.2 Residual Axial Capacity and Failure Mode of Columns ... 350
10.8 Parametric Study ... 355
10.8.1 Steel Tube Thickness and Strength ... 355
10.8.2 Core Concrete Strength and Cross-Sectional Diameter ... 357
10.8.3 Influence of Varied Parameters on Damage Index ... 362
10.9 Summary ... 363
References ... 364

11 Experimental and Numerical Study of UHPCC-FST Columns Subjected to Close-Range Explosion 369
11.1 Introduction 369
11.2 Explosion Test on UHPCC-FST Column 371
11.2.1 Specimens 371
11.2.2 Test Setup 373
11.2.3 Test Results and Analyses 373
11.3 Analytical Methods for Predicting the Dynamic Responses of UHPCC-FST Columns 377
11.3.1 ALE Method 377
11.3.2 Velocity Method 381
11.3.3 SDOF Method 384
11.3.4 Comparisons of Predictions by Different Methods 388
11.4 Further Numerical Analyses and Discussion 389
11.5 Summary 392
References 393

12 Experimental Study on the Residual Seismic Resistance of UHPCC Filled Steel Tube (UHPCC-FST) After Contact Explosion 397
12.1 Introduction 397
12.2 UHPCC-FST Specimens 399
12.2.1 Steel Tube 399
12.2.2 UHPCC 400
12.2.3 Fabrications 401
12.3 Contact Explosion Test 403
12.3.1 Test Setup 403
12.3.2 Test Results and Discussions 404
12.4 Low-Frequency Cyclic Loading Test 408
12.4.1 Test Setup 408
12.4.2 Test Results and Discussions 409
12.5 Assessment of Residual Seismic Resistance of the Post-blast Column 423
12.6 Summary 426
References 427

13 Experimental and Numerical Studies on Dynamic Behavior of Reinforced UHPCC Panel Under Medium-Range Explosions 431
13.1 Introduction 431
13.2 Review of the Existing Work 432
13.3 Field Blast Test 435
13.3.1 Specimen 435
13.3.2 Test Setup 436
13.3.3 Test Results and Discussions 439

13.4 Numerical Simulation 453
13.4.1 FE Model 453
13.4.2 Material Model of UHPCC 454
13.4.3 Material Model of NSC 463
13.4.4 Material Models for Rebar and Support 463
13.5 Comparisons of Numerical Results with Test Data 464
13.5.1 Overpressures-Time History 464
13.5.2 Deflection-Time History 466
13.5.3 Post-blast Damage 466
13.6 Summary 468
References 470

14 Constitutive Modelling of UHPCC Material Under Impact and Blast Loadings 475
14.1 Introduction 475
14.2 UHPCC Material Model 478
14.2.1 Brief Introduction of the Original Kong-Fang Concrete Model 478
14.2.2 New Tensile Damage Model 480
14.2.3 Parameter Calibration 484
14.3 Single Element Tests 489
14.3.1 Unconfined Uniaxial Tests 489
14.3.2 Triaxial Compression Test 490
14.4 Experimental Validation 491
14.4.1 UHPCC-FST Column Subjected to Low Speed Impact 492
14.4.2 UHPCC-FST Column Subjected to Near Explosion 495
14.4.3 Reinforced UHPCC Slab Subjected to Blast Loading 498
14.5 Summary 500
References 501

Abbreviations

AASHTO	American Association of State Highway and Transportation Officials
AEPW	Advanced earth penetration weapon
ALE	Arbitrary Lagrange–Euler
API	Armor-piercing incendiary
CFRP	Carbon fiber-reinforced polymer
CFST	Concrete-filled steel tube
CMOD	Crack mouth opening displacement
CRH	Caliber radius head
CS	Cowper–Symonds
DIF	Dynamic increase factor
DOP	Depth of the projectile penetrate
ECC	Engineered cementitious composites
EPW	Earth penetration weapon
FE	Finite element
FHWA	Federal Highway Administration
FRC	Fiber-reinforced concrete
FRCCs	Fiber-reinforced cementitious composites
FRCFST	Circular steel fiber-reinforced concrete-steel tube
FSI	Fluid-Structure Interaction
HJC	Holmquist–Johnson–Cook
HPFRC	High-performance fiber-reinforced concrete
HRWR	High-range water reducer
HSC	High-strength concrete
JC	Johnson–Cook
LOP	Limit of proportionality
LVDT	Linear variable differential transformer
LVDT	Linear voltage differential transducers
MOR	Modulus of rupture
MSCC	Multi-scale cement composite
NDRC	National Defense Research Committee
NSC	Normal strength concrete

PTCFST	Circular posttensioned concrete-filled steel tube
RC	Reinforced concrete
RPC	Reactive powder concrete
RPF	Rigid polyurethane foam
SFHSC	Steel fiber-reinforced HSC
SFRC	Steel fiber-reinforced concrete
SHPB	Split Hopkinson pressure bar
SIFCON	Slurry infiltrated fiber concrete
TI	Toughness index
UHMWPE	Ultra-high molecular weight polyethylene
UHP-BASFRC	Basalt aggregated UHP-SFRC
UHPC	Ultra-high-performance concrete
UHPCC	Ultra-high-performance cementitious composites
UHPCC-FST	UHPCC-filled steel tube
UHPFRC	Ultra-high-performance fiber-reinforced concrete
UHP-SFRC	Ultra-high-performance steel fiber-reinforced concrete
UHTCC	Ultra-high toughness cementitious composites
W/B	Water-to-binder
W/CM	Water-to-cementitious materials

Nomenclature

d_s	Length of specimen (m)
E	Compressive elastic modulus (GPa)
E_m	Elastic modulus of matrix (GPa)
V_f	Volumetric ratio of steel fibers
E_s	Elastic modulus of steel fibers (GPa)
P_{LOP}	Loads at *LOP* in four-point load-deflection curves (kN)
P_{MOR}	Loads at *MOR* in four-point load-deflection curves (kN)
δ_{LOP}	Deflections at *LOP* in four-point load-deflection curves (mm)
δ_{MOR}	Deflections at *MOR* in four-point load-deflection curves (mm)
$Toughness_{MOR}$	Toughness at *MOR* in four-point load-deflection curves (N m)
f_{MOR}	Flexural strength at *MOR* in four-point load-deflection curves (MPa)
L	Span length of specimen (m)
b	Width of specimen (m)
h	Height of specimen (m)
δ	Deflections in four-point load-deflection curves (mm)
f	Flexural strength in four-point load-deflection curves (MPa)
$Toughness$	Toughness at *MOR* in four-point load-deflection curves (N m)
K_{IC}	Fracture toughness (MPa $m^{1/2}$)
F_{un}	Maximum load in three-point load-deflection curves (MPa)
m_1	Mass of specimen between the supports (kg)
g	Gravity (m/s^2)
α	Crack-to-depth ratio
β	Span-to-depth ratio
a_c	Critical effective crack length at F_{un} (mm)
h_0	Thickness of the steel plate pasted on the specimen (mm)
V_c	Critical value of CMOD at F_{un}
$p_4(\alpha)$	Cubic polynomial
$p_\infty(\alpha)$	Cubic polynomial
G_F	Fracture energy (N/m)
W_0	Area under the load-deflection curve (J/m^3)
m_0	Mass of specimen including the mass of measuring device (kg)

δ_0	Deflection at finale fracture (mm)
a_0	Notch depth of specimen (mm)
V_0	Speeds of striker bar
t	Time (μs)
$\varepsilon_i(t)$	Incident strain
$\varepsilon_r(t)$	Reflected strain
$\varepsilon_t(t)$	Transmitted strain
$\sigma_s(t)$	Dynamic stress of the specimen (MPa)
$\varepsilon_s(t)$	Dynamic strain of the specimen
$\dot{\varepsilon}_s(t)$	Strain rate of the specimen (s^{-1})
E_b	Elastic modulus of the bars (GPa)
A_0	Cross-sectional area of the bars (mm^2)
C_0	Elastic wave velocity of the bars (m/s)
σ_t	Transmitted stress on the specimen (MPa)
σ_i	Incident stress on the specimen (MPa)
σ_r	Reflected stress on the specimen (MPa)
$\dot{\varepsilon}$	Strain rate (s^{-1})
$\dot{\varepsilon}_{ts}$	Static strain rate (s^{-1})
A_f	Coefficient
B_f	Coefficient
C_f	Coefficient
$\dot{\varepsilon}_{TR}$	Transition strain rate (s^{-1})
W	Toughness of a material (N m)
σ	Stress (MPa)
ε	Strain
W_p	Peak dynamic toughness (N m)
W_u	Ultimate dynamic toughness (N m)
E_0	Elastic constant
λ	Elastic constant
η	Time variable
E_1	Elastic modulus of the low-frequency Maxwell element (GPa)
θ_1	Relaxation time of the low-frequency Maxwell element (s)
E_2	Elastic modulus of the high-frequency Maxwell element (GPa)
θ_2	Relaxation time of the high-frequency Maxwell element (s)
σ_M	Modified stress (MPa)
k	Bearing capacity of the steel fibers in the specimen damage zone
D	Damage factor
m_d	Damage evolution coefficient
a_d	Damage evolution coefficient
ε_{th}	Ultimate elastic strain
ε_u	Peak strain
u	Particle displacement (m)
x	Longitudinal coordinate (m)
c	Elastic wave velocity of the specimen (m/s)
σ_d	Dynamic spalling strength (MPa)

τ_t	Time required from the tensile stress occurs on the specimen to the rupture (s)
E_S	Approximate tensile elastic modulus (GPa)
l	Distance between the first fractured section and the free end of specimens (mm)
$\dot{\varepsilon}_0$	Reference strain rate (s^{-1})
σ_2	Confining pressure (MPa)
σ_3	Confining pressure (MPa)
F_0	Axial deviatoric load (N)
σ_1	Actual axial stress (MPa)
A_s	Cross-sectional area of the cylindrical specimen (mm^2)
σ_3^{th}	Targeted value of confining pressure (MPa)
$\Delta\sigma$	Axial deviatoric stress (MPa)
ε_v	Volumetric strain
f_c'	Unconfined compressive strength of the specimen (MPa)
σ_1^u	Peak axial stress (MPa)
ε_1^u	Peak axial strain
ε_3^u	Peak lateral strain
ε_c^u	Peak axial strain of UHPCC under uniaxial compression
ν	Poisson's ratio
φ	Internal-friction angle of concrete
σ_m	Main normal stress (MPa)
τ_m	Main shear stress (MPa)
a_w	Willam–Warnke model coefficient
a_{w1}	Willam–Warnke model coefficient
a_{w2}	Willam–Warnke model coefficient
a_p	Coefficient
b_p	Coefficient
R	Toughness of concrete (N m)
ε_1	Axial strain
ε_3	Lateral strain
$\varepsilon_c r$	First-crack strain
σ^*	Normalized equivalent stress
P^*	Normalized hydrostatic pressure
T^*	Normalized maximum tensile hydrostatic pressure
P	Actual pressure (MPa)
T	Maximum tensile hydrostatic pressure (MPa)
$\dot{\varepsilon}^*$	Dimensionless strain rate
A	Material strength parameter
B	Material strength parameter
N	Material strength parameter
$S_{\max}$	Material strength parameter
C	Strain rate coefficient
μ	Volumetric strain
ρ	Current density (kg/m^3)

ρ_0	Initial density (kg/m^3)
K	Elastic bulk modulus (GPa)
P_{crush}	Pressure when the material begins to undergo plastic deformation (MPa)
μ_{crush}	Volumetric strain when the material begins to undergo plastic deformation (MPa)
P_{lock}	Compacting hydrostatic pressure (MPa)
μ_{plock}	Volumetric strain at the compacting hydrostatic pressure
K_1	Constant
K_2	Constant
K_3	Constant
$\Delta\varepsilon_p$	Equivalent plastic strain during a cycle of integration
$\Delta\mu_p$	Plastic volumetric strain during a cycle of integration
D_1	Damage constant
D_2	Damage constant
$EFMIN$	Material constant used to suppress fracture from weak tensile waves
D_t	Diameter of targets (cm)
H_t	Thickness of targets (cm)
V_r	Residual velocity of the projectile (m/s)
d	Diameter of projectiles (cm)
V_b	Volume fractions of basalt aggregate
h_c	Depth of the impact crater (mm)
d_c	Equivalent diameter of the impact crater (mm)
M_0	Initial mass of projectile (g)
V_d	Density of projectile (kg/m^3)
P_p	Bullets perforated both the target and the rear aluminum plate
$h_{c,f}$	Height of crater from front crater (mm)
$h_{c,r}$	Height of crater from rear crater (mm)
$d_{c,f}$	Equivalent diameter of crater from front crater (mm)
$d_{c,r}$	Equivalent diameter of crater from rear crater (mm)
α_s	Slope angle
d_s	Diameter of the hard steel core (mm)
M_{p0}	Initial masses of the panels (kg)
M_{pr}	Residual masses of the panels (kg)
T_t	Thickness of the panels (mm)
N^*	Projectile nose geometry factor
h_{per}	Perforation limit
a	Half of the aggregate size (mm)
h_c	Length of the hard core nose (mm)
ρ_t	Density of the target (kg/m^3)
V_{BL}	Ballistic limit
H_{plug}	Thickness of the rear crater (mm)
H_t	Thickness of the given target plate (mm)
$d_{a,max}$	Maximum aggregate size (mm)
ξ	Constraining factor

A_s	Cross-sectional areas of the steel tube (m^2)
A_c	Cross-sectional areas of the core concrete (m^2)
α_r	Steel ratio of CFST member
f_y	Yield strength of steel tube (MPa)
E_K	Impact energy (kJ)
m	Mass of drop hammer (kg)
H	Release height of drop hammer (m)
$\Delta\sigma_y$	Initial yield strength surface
$\Delta\sigma_m$	Maximum strength surface
$\Delta\sigma_r$	Residual strength surface
a_i	Strength constant, $i = 0, 1, 2$
a_{iy}	Strength constant, $i = 0, 1, 2$
a_{if}	Strength constant, $i = 0, 1, 2$
f_c	Unconfined uniaxial compressive strength (MPa)
f_t	Unconfined uniaxial tensile strength (MPa)
p	Pressure (kN)
ψ	Tensile-to-compressive meridian ratio
γ	Current failure surface
J_2	Second deviatoric stress invariant
r'	Ratio of the current meridian to the compressive meridian
η	Yield scale factor referring to the damage function
λ	Damage function
b_1	Damage constant
b_2	Damage constant
$d\bar{\varepsilon}_p$	Effective plastic strain increment
$d\varepsilon_{ij}^p$	Plastic strain increment tensor
r_f	Strength dynamic increase factor
E_i	Internal energy per initial volume
γ_i	Ratio of specific heat
$C(\mu)$	Tabulated pressure as functions of volumetric strain
$T(\mu)$	Tabulated temperature as functions of volumetric strain
$d\tilde{\mu}$	Proportionality constant
$d\varepsilon_{11}^p$	Axial plastic strain increment
G	Strain energy (MP m)
h_e	Element size (mm)
σ_d	Dynamic flow stress (MPa)
σ_s	Associated static flow stress (MPa)
C_c	Strain rate parameter
P_c	Strain rate parameter
E_t	Tangent modulus (GPa)
ρ_r	Reinforcement ratio

Chapter 1
Static Mechanical Properties of UHPCC

1.1 Introduction

Ultra-high performance cementitious composites (UHPCC) is a relativity new type of cementitious materials which has very low water-to-binder (W/B) ratio, high amount of high-range water reducer (HRWR), fine aggregates and high-strength steel fibers (Graybeal 2006). With the prominent mechanical properties, i.e., the high compressive and tensile strengths, high ductility as well as the high fracture energy, UHPCC becomes the most prospective construction materials for both civil and military structures, such as fortification, nuclear waste storage containment, highway bridge, high-rise building, etc. (Schmidt and Fehling 2005). Investigations of the mechanical behaviors of UHPCC are important and essential to provide valuable information for the structural design and calibration/validation of the constitutive model. In this Chapter, static mechanical properties of UHPCC are mainly concerned.

For the existing experimental studies on the influence of steel fiber content on the static behavior of UHPCC, Hassan et al. (2012) developed the simplified test methods to measure the elastic modulus, stress–strain curves and post-cracking behaviors of ultra-high performance fiber reinforced concrete (UHPFRC) specimens in both compression and tension. It was concluded that, addition of steel fibers (0% and 2.0% by volume) significantly increased the tensile strength and ductility of UHPFRC, but had relatively slight effect on the compressive strength and elastic modulus. Wille et al. (2014) conducted a series of direct tensile tests (dog-bone specimen with the length of 147 mm) to study the effect of steel fiber content (1.5 ~ 3.0% by volume) on the tensile properties of UHPFRC. The results showed that the tensile strength and the corresponding strain as well as the energy absorption capacity had a strong dependency on the steel fiber content. Liu et al. (2016) studied the effect of steel fiber content (0 ~ 2.5%) on the tensile properties of ultra-high performance concrete (UHPC) with coarse aggregates. It was concluded that, both the first crack tensile strength and the fiber-bridging stress increased with the fiber volumetric ratio rising. Abbas et al. (2015) (1.0%, 3.0% and 6.0% by volume), Kazemi and Lubell (2012) (0 ~ 5.0% by volume), Wu et al. (2016) (0 ~ 3.0% by volume), Yoo et al. (2013,

Q. Fang et al., *UHPCC Under Impact and Blast*,
https://doi.org/10.1007/978-981-33-6842-2_1

2014) (1.0 ~ 4.0% by volume) have studied the influence of steel fiber content on the compressive and flexural behaviors of UHPFRC, respectively. The results indicated that, with the increase of steel fiber content, the compressive strength gradually increased, while the flexural strength, deflection and CMOD (crack mouth opening displacement) at peak load pseudo-linearly increased correspondingly, and the steel fiber content had no noticeable influence on the first crack flexural strength and the corresponding deflection/CMOD.

As for the existing works on the influence of steel fiber type on the static behavior of UHPCC, Wille et al. (2014) also experimentally studied the effect of steel fiber type (smooth, hooked and twisted) on the tensile properties of UHPFRC, which showed that the tensile strength and the corresponding strain as well as the energy absorption capacity were less affected by the steel fiber type. Liu et al. (2016) also studied the effect of steel fiber type (smooth, spiral and hooked micro-fibers as well as hooked macro-fiber) on the tensile properties of UHPC with coarse aggregates. The results showed that the micro-hooked steel fiber had the most significant effect on the first crack tensile strength and the fiber-bridging stress. Wu et al. (2016) also investigated the effects of three typical steel fibers (straight, corrugated and hooked-end) on the compressive and flexural properties of UHPC. It was concluded that the steel fiber type had little influence on the first crack flexural strength and the corresponding deflection, and the UHPC with hooked-end steel fiber had the highest compressive and flexural strengths.

Generally, considering the time and economic cost, limited works were performed to systematically assess the effects of steel fiber content and type on the static strengths as well as the full stress–strain curves and flexural load–deflection/CMOD curves of UHPCC. In the present chapter, a series of cubic/axial compressive, direct tensile, four-point and three-point flexural tests on UHPCC specimens are conducted, in which two typical steel fibers (micro-straight and hooked) with six volume fractions of 0 ~ 2.5% are considered. The systematical full compressive and tensile stress–strain curves as well as the flexural load–deflection and flexural load-CMOD curves are obtained. The effects of steel fiber content and type on the cubic and axial compressive strengths, compressive elastic modulus and Poisson's ratio, direct tensile strength, flexural strength, load carrying capacity, energy absorption capacity, fracture toughness and fracture energy are comprehensively discussed. The present work is fundamental and the derived conclusions could provide helpful references for the analysis and design of the UHPCC structures.

1.2 Test Program

1.2.1 Raw Materials and Mixture Proportions

UHPCC was prepared in the State Key Laboratory of High Performance Civil Engineering Materials, Jiangsu Research Institute of Building Science in China, and the mixture proportions are given in Table 1.1. Chinese standard Graded 52.5 P.II type Portland cement (20 ~ 30 μm in particle diameter) from Xiaoyetian corporation, silica fume (particle size of 0.10 ~ 0.26 μm, density of 2.1 g/cm^3, specific surface area of 20500m^2/kg) from Elkem company and ultra-fine mineral admixture (density of 2.45 g/cm^3, specific surface area of 8500m^2/kg, 28d activity index of 115%) from SBT®-HDC (V) are used as the cementitious materials. Table 1.2 lists the chemical composition of cement and silica fume. The water-to-cementitious materials (W/CM) ratio and HRWR-to-cementitious materials ratio are 0.16 and 0.024, respectively. Six volume fractions (0, 0.5, 1.0, 1.5, 2.0 and 2.5%) of two typical steel fibers, including micro-straight (diameter of 0.2 mm, length of 13 mm, tensile strength of 2800 MPa) and hooked (diameter of 0.5 mm, length of 25 mm, tensile strength of 1200 MPa) are incorporated, as shown in Fig. 1.1. The water-reducing ratio of the PCA®-I polycarboxylic type HRWR is no less than 35%, which is developed by Jiangsu SOBUTE new material Co., Ltd.

Compared with the traditional high performance concrete, the ground fine quartz sand (the size is less than 0.6 mm) is substituted by easily obtained natural river sand (maximal diameter of 2.5 mm). Also the partial silica fume is replaced by the ultra-fine mineral admixture which includes fly ash and ultra-fine slag (the specific proportions is undisclosed for commercial confidentiality). Since the production of cement and fine quartz sand consumes much more energy and resources, and the silica

Table 1.1 Mixture proportions of UHPCC (kg/m^3)

Cement	Silica fume	Ultra-fine mineral admixture	Sand	Water	HRWR	Fiber content
700	140	110	1200	152	22.8	0 ~ 195

Table 1.2 Chemical composition of cement and silica fume

Composition	CaO	SiO	MgO	Fe_2O_3	Al_2O_3	SO_3	LOI
Cement	61.80	20.00	5.31	3.46	4.64	3.15	2.44
Silica fume	1.00	93.10	0.65	2.44	0.76	0.12	1.04

(a) (b) (c)

Fig. 1.1 Steel fibers **a** micro-straight **b** hooked **c** comparison, reprinted from Ren et al. (2018), copyright 2020, with permission from Elsevier fume is more expensive than fly ash and ultra-fine slag for its limited production, the present UHPCC is relative low-costs and energy saving.

1.2.2 Specimen Preparation and Curing

For the preparation of the UHPCC, a forced single-axis mixer with a capacity of 60L and a constant mixing speed (47 rpm) was used. To achieve good workability, particle distribution and packing density, mixing procedure for the present UHPCC was controlled rigorously. Firstly, since the small particles tended to agglomerate and it was easier to break these chunks when the particles were dry, the cementitious materials (cement, silica fume, ultra-fine mineral admixture) and sand were put together simultaneously and dry-mixed uniformly for 3 min. Secondly, the water pre-mixed with super-plasticizer was then added gradually and mixed for 3 ~ 5 min. Finally, steel fibers were dispersed carefully by hand into the mixture and mixed for another 3 ~ 5 min in order to guarantee the fibers well distributed throughout the matrix. The shear action of steel fibers helped to destroy any remaining agglomerates in the fresh mixture, thus improving the workability of the composites.

After the prepared matrix was poured into moulds, the freshly cast specimens were covered with plastic sheets to prevent moisture losing and kept at room temperature for 24 h. Then, they were demoulded and cured in a standard curing room with the temperature of 17.7 ~ 21.7 °C and relative humidity of 98.4% for another 28 days. After that, the specimens were taken out of the standard curing room and stored at room temperature until the time of tests. All the specimens were cast from the side face to ensure that the pressure-bearing surfaces are parallel to each other and the flatness tolerance is less than 0.0005d (d_s is the length of specimens) according to the Chinese standard GB-T50081-2002 (2003). Table 1.3 lists the arrangement of the present tests, in which three specimens were tested in each batch and total 165

Table 1.3 Test arrangement

Test	Steel fiber volume fraction (%)	Steel fiber type	Specimen size (mm × mm × mm)	Specimen number
Cubic compression	0 ~ 2.5	S, H	100 × 100 × 100	33
Axial compression	0 ~ 2.5	S, H	100 × 100 × 300	33
Direct tension	0 ~ 2.5	S, H	unnotched dog-bone	33
Four-point flexure	0 ~ 2.5	S, H	100 × 100 × 400	33
Three-point flexure	0 ~ 2.5	S, H	100 × 100 × 400	33

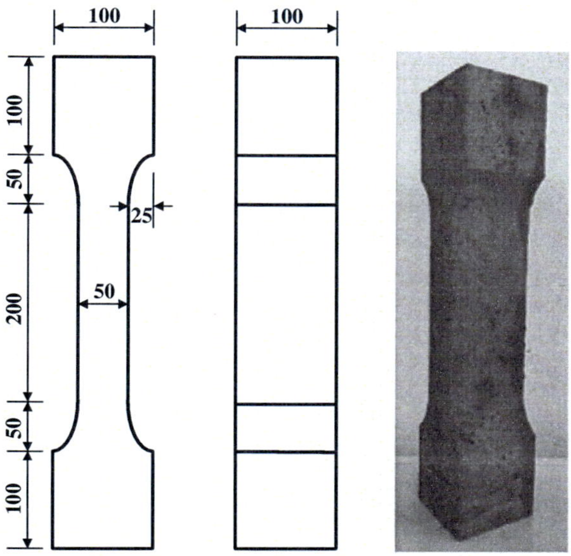

Fig. 1.2 Geometry of unnotched dog-bone specimen (*unit* mm), reprinted from Ren et al. (2018), copyright 2020, with permission from Elsevier

specimens were loaded. The capitals "S" and "H" represent the micro-straight and hooked steel fibers, respectively. The geometries of the specimens for cubic/axial compressive and four/three-point flexural tests were determined in accordance with the Chinese standard GB-T50081-2002 (2003). Due to the lack of relevant common standards, the unnotched dog-bone specimens with an overall length of 500 mm refering to Ref. (Kang and Kim 2011) were used in the direct tensile test, as shown in Fig. 1.2. The cross section of the specimens started with 100 mm × 100 mm and changed to a prismatic shape of 50 mm × 100 mm after 150 mm away from each ends of specimens, which enabled fracture to occur in the narrow cross section of the specimen. The dimension of narrow cross section was two times the maximal fiber length (25 mm), which guaranteed the random fiber orientation and distribution. Since the static tests were time consuming, the following tests were conducted between 28 and 84 days after the specimens were cast.

1.3 Instrumentation and Loading Scheme

1.3.1 Cubic Compressive Test

As shown in Fig. 1.3, the cubic compressive test was performed by using a digital display pressure testing machine with a maximum load capacity of 5000kN. A load

Fig. 1.3 Cubic compressive test setup, reprinted from Ren et al. (2018), copyright 2020, with permission from Elsevier

control at a rate of 5.0kN/s was applied to the UHPCC specimens. The cubic compressive strengths were measured and the averages of three parallel specimens for each batch was reported as the test results.

1.3.2 Axial Compressive Test

The axial compressive test through displacement control was conducted to measure the full compressive stress–strain curves. As shown in Fig. 1.4, a high rigidity material

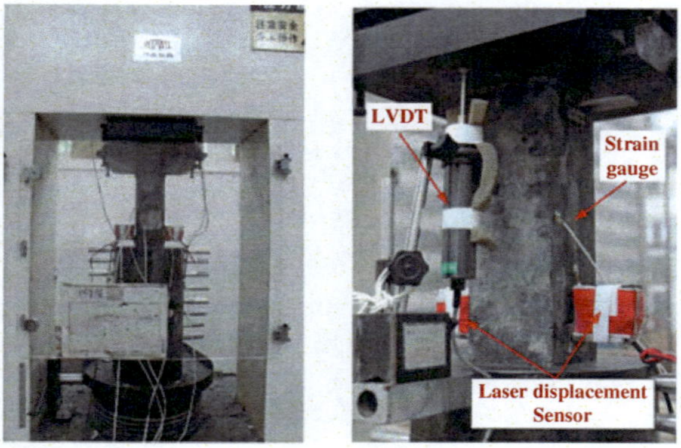

Fig. 1.4 Axial compressive test setup, reprinted from Ren et al. (2018), copyright 2020, with permission from Elsevier

testing machine was used, which consisted of a stiff loading frame with 3000kN load cell and the loading rate was set as 0.1 mm/min. One linear variable differential transformer (LVDT) and two laser displacement sensors positioned at opposite sides of the specimens were used to record the overall axial displacement of testing machine and the averaged overall axial displacement of specimens, respectively. Three sets of mutually perpendicular strain gauges attached on the surface of specimens were used to measure the averaged local axial and circumferential strains of specimens. Since the strain gauges attached on the surface of specimens would come off at the peak strength due to the spalling of concrete, the strain data recorded by the strain gauges and laser displacement sensors were respectively used for the pre-peak and post-cracking stages of the full compressive stress–strain curves, as shown in Sect. 1.4.1.2.

1.3.3 Direct Tensile Test

As shown in Fig. 1.5, the direct tensile test was performed by using an INSTRON-8803 testing machine with the maximum load capacity of 500 kN, and the displacement control rate was set as 0.05 mm/min. The axial displacement along the length of narrow cross section was measured by using two LVDTs positioned at opposite sides of the specimens. Averaged values from the two LVDTs were used in calcuating the tensile strain, and the full tensile stress–strain curves were acquired.

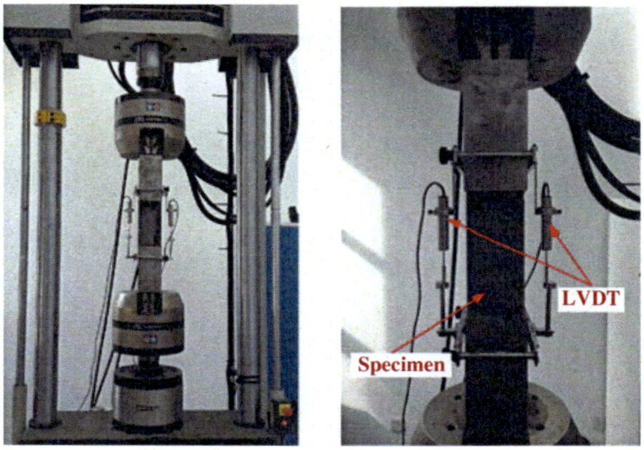

Fig. 1.5 Direct tensile test setup, Ren et al. (2018), copyright 2020, with permission from Elsevier

1.3.4 Four-Point Flexural Test

As shown in Fig. 1.6, in the four-point flexural test, the unnotched beam specimens with a span of 300 mm were loaded and the forces divided the span into thirds. The above INSTRON-8803 testing machine was used, and the displacement control rate was set as 0.1 mm/min. Two LVDTs attached at opposite sides of the specimens were used to measure the averaged mid-span deflection, and the flexural load–deflection curves were obtained.

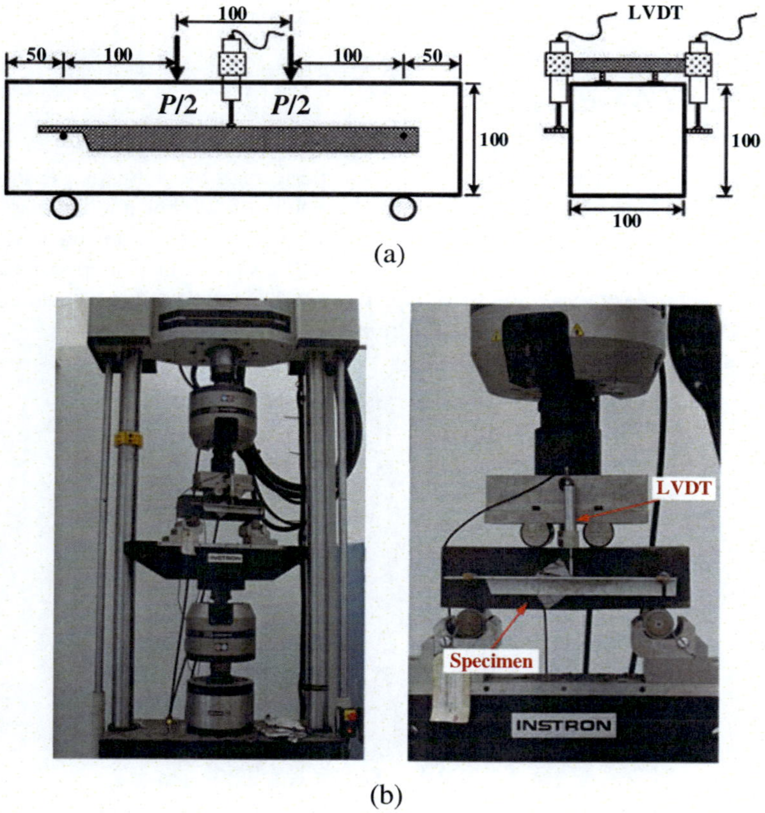

Fig. 1.6 Four-point flexural test with unnotched beam **a** dimensions of specimen (*unit* mm) **b** setup, Ren et al. (2018), copyright 2020, with permission from Elsevier

1.3.5 Three-Point Flexural Test

As shown in Fig. 1.7, the beam specimens with a 30 mm notch at mid-length were fabricated and loaded in the three-point flexural test. An electromechanical univerasl testing machine (MTS-810) with maximum load capacity of 250 kN was used, and the loading rate was set as 0.1 mm/min. The similar measuring method with the four-point flexural test was adopted to obtain the averaged mid-span deflection, and the CMOD was measured by using a clip gauge (with a maximum capacity of 5 mm) attached to the bottom of the specimens. Both the flexural load–deflection and the flexural load-CMOD curves were recorded in the test.

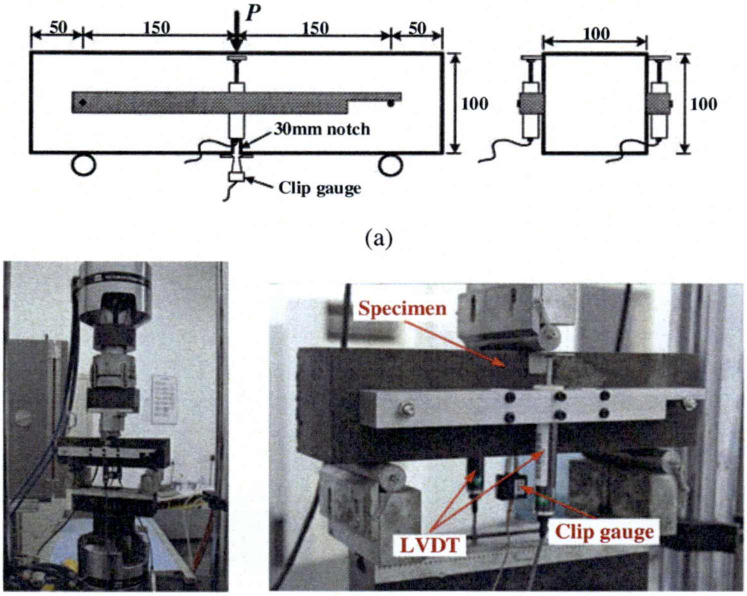

(a)

(b)

Fig. 1.7 Three-point flexural test with notched beam **a** dimensions of specimen (*unit* mm) **b** setup, Ren et al. (2018), copyright 2020, with permission from Elsevier

1.4 Test Results and Discussion

1.4.1 Compression Test

1.4.1.1 Cubic Compression Test

Table 1.4 shows the experimental cubic compressive strengths of UHPCC specimens with different steel fiber content and type at 28 days. "N100-0", "S100-0" and "H100-0.5" were denoted as the test No., where the capitals "N" "S" and "H" represent no mixing fiber, micro-straight and hooked steel fibers, "100" the length of specimen, "0" and "0.5" the steel fiber content. S1 ~ S3 represent the three specimens in each batch and "—" denotes that the data was not collected or discarded due to large deviation.

Figure 1.8 shows the effects of steel fiber content and type on the cubic compressive strength of UHPCC. It indicates that, the cubic compressive strength gradually increases with the increase of steel fiber content, either for the UHPCC with micro-straight or hooked steel fibers, and the influence of the steel fiber content on the

Table 1.4 Cubic compressive test results (*unit* MPa)

Test No.	S1	S2	S3	Average	Test No.	S1	S2	S3	Average
N100-0	110.3	110.9	111.5	110.9	–				
S100-0.5	118.5	120.3	124.5	121.1	H100-0.5	127.7	128.7	125.5	127.3
S100-1.0	128.3	125.0	130.0	127.8	H100-1.0	122.6	124.8	125.5	124.3
S100-1.5	–	136.2	130.4	133.3	H100-1.5	123.2	134.2	124.0	127.1
S100-2.0	132.5	140.0	132.8	135.1	H100-2.0	138.3	135.6	134.4	136.1
S100-2.5	141.3	144.2	146.3	143.9	H100-2.5	136.7	139.2	134.8	136.9

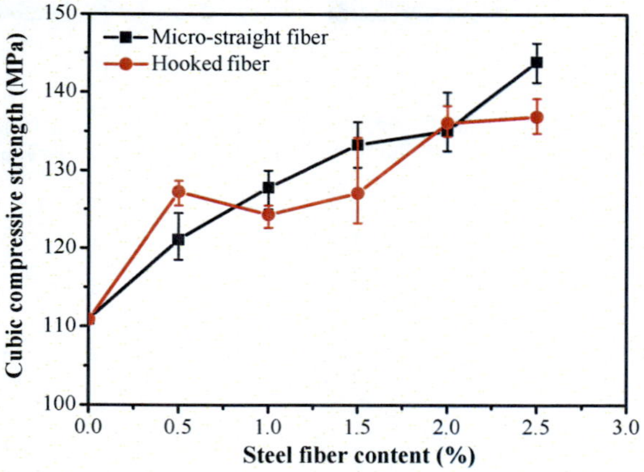

Fig. 1.8 Effects of steel fiber content and type on the cubic compressive strength of UHPCC, Ren et al. (2018), copyright 2020, with permission from Elsevier

former is more influential. For example, the cubic compressive strength of UHPCC with 2.5% micro-straight steel fibers is 143.9 MPa, which has increased by 29.8% and 5.1% compared to the plain UHPCC and UHPCC with 2.5% hooked steel fibers, respectively.

The reason lies in that, increasing steel fiber content could make more fibers to sustain the load and lead to multi-cracks. Meanwhile, the stress between fibers and matrix is reduced with increasing the fiber volume fraction, which delays the formation and propagation of cracks, and thus results in the increase of cubic compressive strengths. Besides, with the identical steel fiber content, due to the different size of present two typical fibers, relatively larger amount of micro-straight steel fibers endure the loading. Based on the compressive test on the UHPC specimens respectively reinforced by the straight and hooked steel fibers with the same length and diameter, Wu et al. (2016) obtained the same conclusion about the effect of steel fiber content on the cubic compressive strength, while the conclusion of the influence of steel fiber type is on the contrary.

1.4.1.2 Axial Compression Test

Figure 1.9 shows the representative compressive stress–strain curves of UHPCC. It indicates that, UHPCC shows almost linear behavior up to the peak strength, then more ductile behavior occurs in the post-cracking stage depending on the mixing ratio of steel fibers. The fibers restrict the lateral expansion of specimens and lead to higher tolerance for axial deformation, which also agrees with the experimental observations in Refs. (Graybeal 2006; Hassan et al. 2012). The stress–strain curve of C100 plain concrete suggested by the *fib* Model Code 2010 (2013) is also shown in Fig. 1.9a–b. It can be seen that the suggested curve could well predict the linear elastic stage of stress–strain curves of UHPCC. It should be noted that, for the relatively low steel fiber content, the post-cracking behaviors are not measured due to the addition

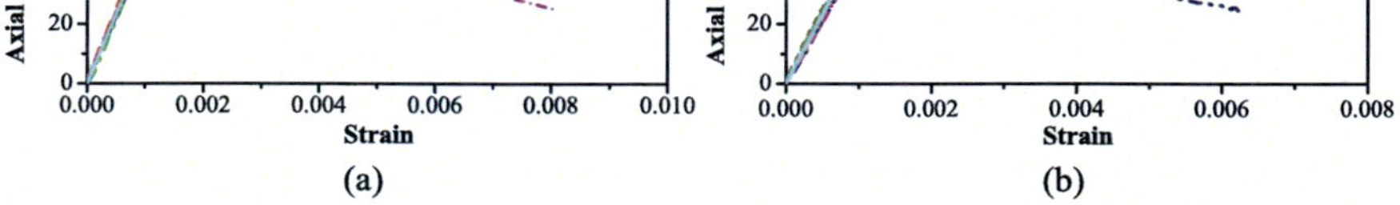

Fig. 1.9 Representative compressive stress–strain curves of UHPCC with **a** micro-straight fiber **b** hooked fiber, Ren et al. (2018), copyright 2020, with permission from Elsevier

Fig. 1.10 Failure patterns of UHPCC specimens with **a** no mixing fiber **b** micro-straight fiber **c** hooked fiber, Ren et al. (2018), copyright 2020, with permission from Elsevier

of silica fume which could make UHPCC more brittle. Compared with the plain UHPCC specimens, the interaction (i.e., bonding and slipping) between steel fiber and matrix makes the ductile failure of the UHPCC specimens, which remain nearly unseparated after the axial compression, as shown in Fig. 1.10.

Table 1.5 lists the test results of the axial compressive strength, compressive elastic modulus and Poisson's ratio from the axial compressive test. "—" denotes that the data is not collected due to the brittle characteristics of UHPCC, which makes the specimen collapse before reaching the peak stress. Figure 1.11 further illustrates the effects of steel fiber content and type. It can be derived that, (i) the steel fiber type has little influence on the axial compressive strength, compressive elastic modulus and Poisson's ratio of UHPCC; (ii) the axial compressive strengths of UHPCC obviously increase with rising the steel fiber content from 0% (~100 MPa) up to 1.0% (~120 MPa), while the values of that almost keep unchanged with the volumetric ratio further increasing. The reason may lie in that, increasing the steel fiber content could increase the bonding strength at fiber-matrix interface, which can restraint the lateral cracking of specimens to some extent. When the axial stress of the specimen reaches a certain value, i.e., nearly 120 MPa at present, the bonding strength is lower than the lateral stress, which leads to the formation of cracks and the specimens are broken. Thus the axial compressive strength could not further increase by increasing the steel fiber content; (iii) the steel fiber content has little effect on the compressive elastic modulus and Poisson's ratio of UHPCC, and the averaged values of which are 43.8GPa and 0.23. As we know, the compressive elastic modulus E is dominated by the matrix and could be predicted as a first approximation by Ref. (Wille et al. 2014).

$$E = (1 - V_f)E_m + V_f E_s \tag{1.1}$$

Table 1.5 Axial compressive test results

Test No.	Axial compressive strength (MPa)				Compressive elastic modulus (GPa)				Poisson's ratio			
	S1	S2	S3	Average	S1	S2	S3	Average	S1	S2	S3	Average
S300-0	104.7	99.7	–	102.2	42.1	45.2	–	43.7	0.19	0.25	–	0.22
S300-0.5	–	110.4	119.0	114.7	–	49.0	41.7	45.4	–	0.26	0.19	0.23
S300-1.0	117.4	–	125.6	121.5	46.5	–	50.1	48.3	0.21	–	0.26	0.24
S300-1.5	119.4	114.8	130.8	121.7	41.8	42.7	46.3	43.6	0.22	0.22	0.22	0.22
S300-2.0	–	106.5	131.9	119.2	–	39.6	44.3	42.0	–	0.18	0.22	0.20
S300-2.5	117.4	131.9	121.8	123.7	45.6	41.2	42.0	42.9	0.26	0.27	0.20	0.24
H300-0.5	114.6	111.3	114.7	113.5	43.5	42.1	42.3	42.6	0.22	0.23	0.23	0.23
H300-1.0	118.2	121.6	121.6	120.5	41.4	42.5	43.0	42.3	0.23	0.24	0.23	0.23
H300-1.5	121.0	–	115.6	118.3	43.0	–	43.6	43.3	0.23	–	0.22	0.23
H300-2.0	119.5	–	122.8	121.2	44.3	–	43.4	43.9	0.21	–	0.21	0.21
H300-2.5	122.2	120.8	123.2	122.1	43.1	44.5	–	43.8	0.23	0.21	0.26	0.23

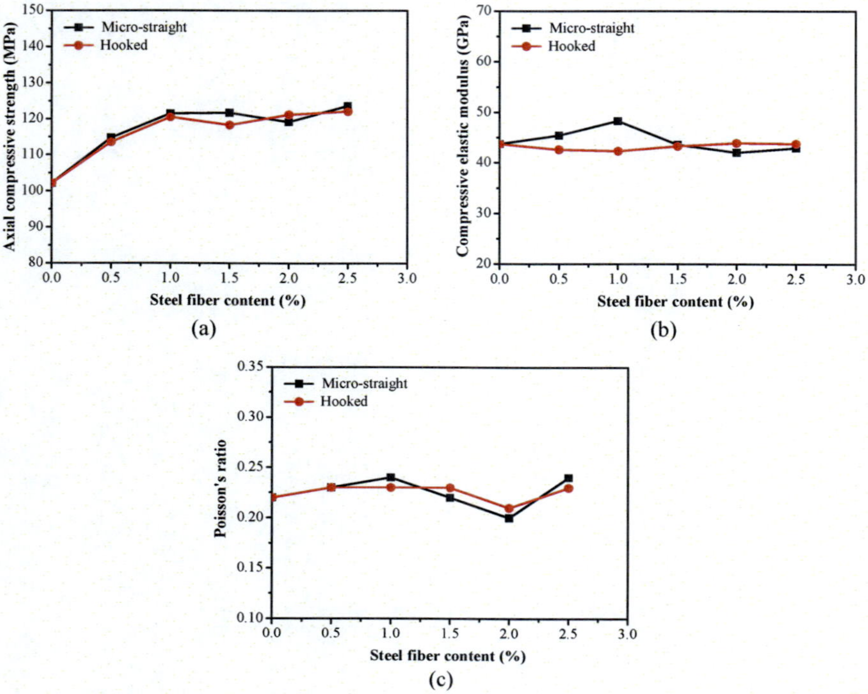

Fig. 1.11 Effects of steel fiber content and type on the **a** axial compressive strength **b** compressive elastic modulus **c** Poisson's ratio, Ren et al. (2018), copyright 2020, with permission from Elsevier

where E_m is the elastic modulus of matrix, V_f and E_s are the volumetric ratio and elastic modulus of steel fibers, respectively. For the present test, $0 \leq V_f \leq 2.5\%$ and $E_s = 210$ GPa, thus the compressive elastic modulus of UHPCC with different fiber volumetric ratios are almost identical, which conforms to the above experimental conclusion. Furthermore, the elastic modulus of C100 plain concrete suggested by the *fib* Model Code 2010 (2013) is 47.5 GPa, which is relatively larger than that obtained in this study.

1.4.2 Direct Tension Test

Figure 1.12 shows the measured tensile stress–strain curves of UHPCC with 2.0% hooked steel fibers. As can be seen, it is difficult to obtain the average curve based on these three curves, thus one of three curves which has the highest post-peak stress is chosen as the representative curve of each batch. Accordingly, Fig. 1.13 shows the representative tensile stress–strain curves of UHPCC with different steel fiber content and type. It can be seen that, all the UHPCC specimens behave elastically up to the

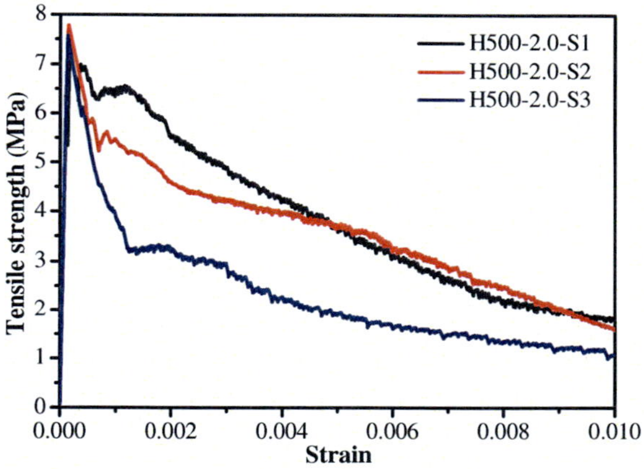

Fig. 1.12 Tensile stress–strain curves of UHPCC with 2.0% hooked steel fibers, Ren et al. (2018), copyright 2020, with permission from Elsevier

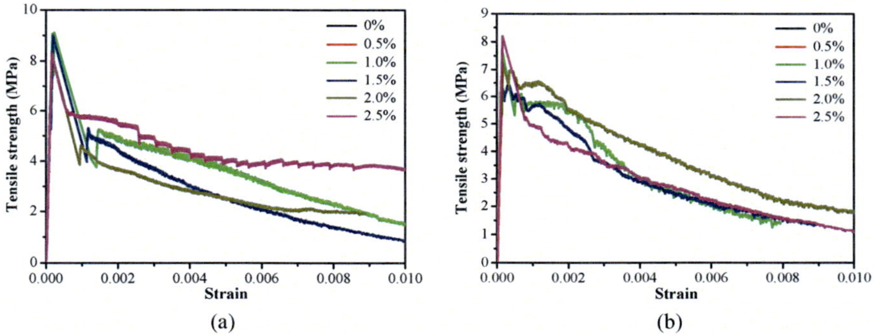

Fig. 1.13 Representative tensile stress–strain curves of UHPCC with **a** micro-straight fiber **b** hooked fiber, Ren et al. (2018), copyright 2020, with permission from Elsevier

first crack tensile strength, and at this instant, the concrete matrix is cracked. Then, since the first crack tensile strengths of these specimens are higher than the corresponding maximum fiber-bridging stresses, an abrupt drop in load-bearing capacity occurs, which indicates that the strain hardening and multiple cracking behaviors do not happen (Fig. 1.14). Furthermore, the tensile stress is completely carried by the fiber/matrix interactions, and the fibers are gradually pullout since the above tensile stress is larger than the bonding strength between the fiber and the matrix.

The test results of tensile strength are given in Table 1.6, in which "—" denotes that the data is not collected. For the test batch with three effective test data, the largest error between the maximum or minimum value and the intermediate value is 12.8%, which is located within the accepted range 15% according to the Chinese standard GB-T50081-2002 (2003). Figure 1.15 further shows the effects of steel fiber content

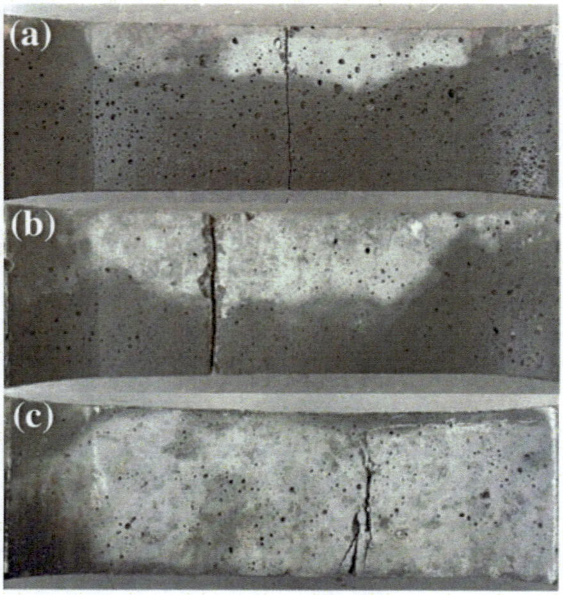

Fig. 1.14 Direct tensile cracking behavior of UHPCC with **a** no mixing fiber **b** micro-straight fiber **c** hooked fiber, Ren et al. (2018), copyright 2020, with permission from Elsevier

Table 1.6 Direct tensile test results (*unit* MPa)

Test No.	S1	S2	S3	Average	Test No.	S1	S2	S3	Average
N500-0	6.7	5.9	6.9	6.50	–				
S500-0.5	7.8	7.7	–	7.75	H500-0.5	–	6.0	7.5	6.75
S500-1.0	7.6	–	9.1	8.35	H500-1.0	7.0	7.6	8.0	7.53
S500-1.5	7.5	–	9.0	8.25	H500-1.5	–	6.5	8.3	7.40
S500-2.0	–	9.6	8.3	8.95	H500-2.0	7.2	7.6	7.8	7.53
S500-2.5	6.8	7.8	8.3	7.63	H500-2.5	8.2	8.5	–	8.35

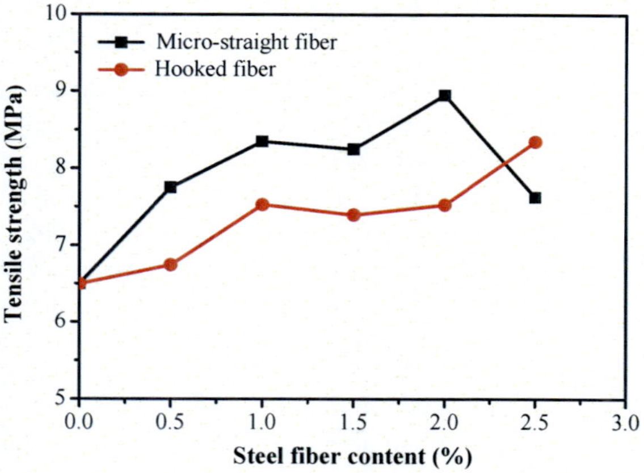

Fig. 1.15 Effects of steel fiber content and type on the tensile strength of UHPCC, Ren et al. (2018), copyright 2020, with permission from Elsevier

and type on the tensile strength of UHPCC. It can be found that, with the volume fraction increasing from 0.5 to 2.0%, the tensile strengths of UHPCC with micro-straight steel fiber increase by 19.2 ~ 37.7% compared to the tensile strength of plain UHPCC (N500-0, 6.50 MPa), and the corresponding values with hooked steel fiber only increase by 3.8 ~ 15.8%. Therefore, the tensile strength of UHPCC increases with the rising of the steel fiber mixing ratio, and the strengthening effect of micro-straight steel fiber is more remarkable due to the larger amount. As is not expected, for the UHPCC specimens with micro-straight steel fibers, the tensile strength for 2.5% volumetric ratio and post-peak stress for 2.0% volumetric ratio are relatively small, which may be caused by the casting quality and the measuring deviations.

1.4.3 Four-Point Flexure Test

1.4.3.1 Four-Point Flexural Load–Deflection Curves

Figure 1.16 illustrates the typical four-point flexural load–deflection response curves. Based on the variation of load carrying capacity after first cracking, the flexural behavior of UHPCC is categorized into deflection hardening and softening behaviors, respectively.

Figure 1.17 shows the representative four-point flexural load–deflection curves of UHPCC. As can be seen, due to the fiber-bridging effect, addition of steel fibers brings favourable effect on the ductility of UHPCC. The addition of steel fibers has little effect on the four-point flexural load–deflection curves before cracking, but significantly influences the post-cracking stage, which agrees well with the conclusions from Wu et al. (2016). When the steel fiber content is less than 1.0%, the four-point flexural load–deflection curves of UHPCC with micro-straight steel fiber show the behavior of deflection softening. As the steel fiber content increases, the behavior

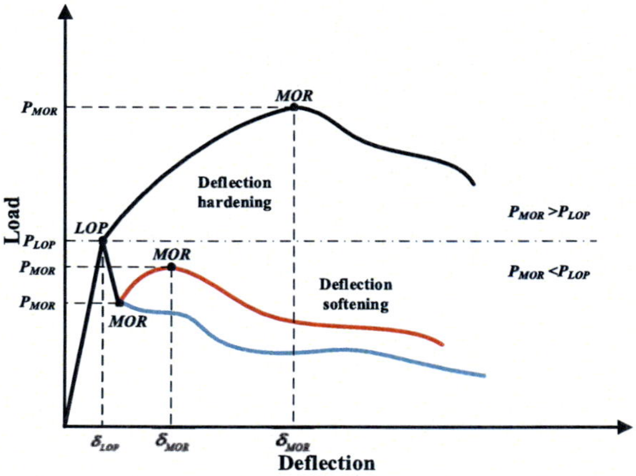

Fig. 1.16 Typical load–deflection response curves (Kim et al. 2008)

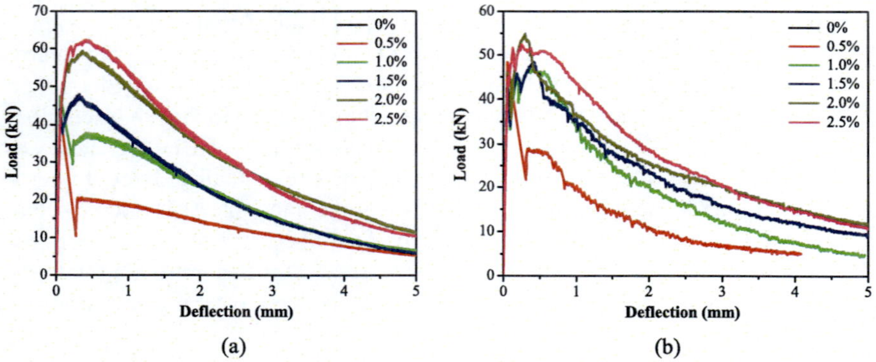

Fig. 1.17 Representative four-point flexural load–deflection curves of UHPCC with **a** micro-straight fiber **b** hooked fiber, Ren et al. (2018), copyright 2020, with permission from Elsevier

of four-point flexural load–deflection curves turns to be deflection hardening. While for the UHPCC with hooked steel fiber, the behavior of deflection hardening occurs when the steel fiber content is larger than 0.5%. The crack pattern of UHPCC specimens in the four-point flexural test is shown in Fig. 1.18, which indicates that only one major crack occurs, and the distributed cracking behavior is not emerged.

As shown in Fig. 1.16, several parameters are used to describe the flexural behavior of UHPCC. The first cracking point of UHPCC is defined as the Limit of proportionality (*LOP*) in ASTM C1018-97 (1998), and the point where deflection softening behavior starts to occur after the point *LOP* is defined as Modulus of rupture (*MOR*). The definitions of *LOP* and *MOR* are introduced by Kim et al. (2008) in detail. P_{LOP}, δ_{LOP} and P_{MOR}, δ_{MOR} are the loads and the corresponding deflections at *LOP* and *MOR*, respectively. Besides, the other two points *L*/600 and *L*/150 are also adopted

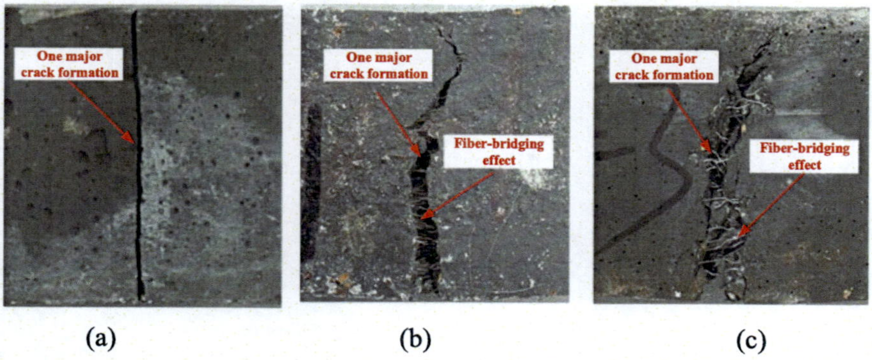

Fig. 1.18 Four-point flexural crack pattern of UHPCC specimens with **a** no mixing fiber **b** micro-straight fiber **c** hooked fiber, Ren et al. (2018), copyright 2020, with permission from Elsevier

to assess the flexural behavior of UHPCC according to ASTM C1609/C1609M-12 (2012).

Taking *MOR* for example, the energy which is equivalent to the area under the load–deflection curve up to *MOR* is denoted as the toughness $Toughness_{MOR}$. According to ASTM C1609/C1609M-12 (2012), the flexural strength at *MOR* can be derived as

$$f_{MOR} = P_{MOR}\frac{L}{bh^2} \tag{1.2}$$

where f_{MOR} is the flexural strength at *MOR*. *L* is the span length; *b* and *h* are the width and height of specimen, respectively.

Based on Fig. 1.17, Table 1.7 lists the corresponding deflection δ, flexural strength f and *Toughness* at four points, i.e., *LOP*, *MOR*, *L*/600 and *L*/150. "FN400-0", "FS400-0" and "FH400-0.5" denote the test No. and the capital "F" represents the four-point flexural test.

Figures 1.19 and 1.20 show the flexural strengths and toughness of UHPCC at the points *LOP*, *MOR*, *L*/600 and *L*/150, respectively. Based on the Table 1.7 and Figs. 1.19 and 1.20, it indicates that, (i) addition of steel fibers has little influence on the flexural strength, the corresponding deflection, and the toughness at the point *LOP*, which is consistent with the conclusion derived by Kazemi and Lubell (2012). This is because that the matrix of UHPCC mainly carries the load in the linear elastic stage before the point *LOP*, the contribution of steel fibers has not been developed; (ii) both the flexural strengths and toughness of UHPCC increase with the rising of steel fiber content at the points *MOR*, *L*/600 and *L*/150, thus adding steel fibers can effectively improve the load carrying capacity and energy absorption capacity of UHPCC. Additionally, the micro-straight steel fiber is more influential when the steel fiber content is larger than 1.5%.

1.4.3.2 Maximum Flexural Strength

Table 1.8 further shows the maximum flexural strengths of all the UHPCC specimens, in which "—" denotes that the data is not collected or discarded due to the large deviation. It should be noted that, for the "FH400-2.0" specimens, the deviation between the maximum value (16.5 MPa) and the intermediate value (14.2 MPa) is 16.2%, which exceeds the accepted range. Thus the intermediate value is regarded as the maximum flexural strength according to the Chinese standard GB-T50081-2002 (2003).

Figure 1.21 illustrates the effects of steel fiber content and type on the maximum flexural strength of UHPCC. It indicates that, addition of steel fibers has favourable effect on improving the maximum flexural strength. In detail, for the UHPCC with micro-straight steel fiber, when the volume fraction of steel fiber varies from 0.5 to 1.5%, the maximum flexural strengths almost keep unchanged, then the corresponding values gradually increase with the rising of steel fiber content. For the

Table 1.7 Flexural test results of UHPCC

Test No.	*LOP*			*MOR*			*L/600*			*L/150*		
	δ (mm)	*f* (MPa)	*Toughness* (N·m)	δ (mm)	*f* (MPa)	*Toughness* (N·m)	δ (mm)	*f* (MPa)	*Toughness* (N·m)	δ (mm)	*f* (MPa)	*Toughness* (N·m)
FN400-0	0.053	12.7	1.186	–	–	–	–	–	–	–	–	–
FS400-0.5	0.052	13.0	1.190	0.298	6.1	7.556	0.500	5.9	11.589	2.000	4.3	37.585
FS400-1.0	0.055	14.3	1.374	0.444	11.3	15.853	0.500	11.3	18.144	2.000	7.0	64.169
FS400-1.5	0.051	13.2	1.197	0.320	14.4	16.371	0.500	13.6	21.079	2.000	7.2	71.616
FS400-2.0	0.054	13.1	1.211	0.366	17.8	18.321	0.500	17.3	26.140	2.000	10.3	95.084
FS400-2.5	0.048	12.0	1.040	0.411	18.7	21.924	0.500	18.4	27.453	2.000	10.9	100.282
FH400-0.5	0.057	14.5	1.508	0.351	8.7	11.512	0.500	8.6	15.776	2.000	3.3	42.460
FH400-1.0	0.054	12.1	1.134	0.418	14.2	16.903	0.500	13.8	20.842	2.000	6.1	66.627
FH400-1.5	0.052	13.0	1.187	0.410	14.6	16.964	0.500	13.6	21.168	2.000	7.1	69.098
FH400-2.0	0.057	13.8	1.355	0.299	16.5	13.713	0.500	13.6	23.102	2.000	7.7	74.112
FH400-2.5	0.058	13.5	1.388	0.377	15.8	17.218	0.500	15.2	23.469	2.000	8.7	83.131

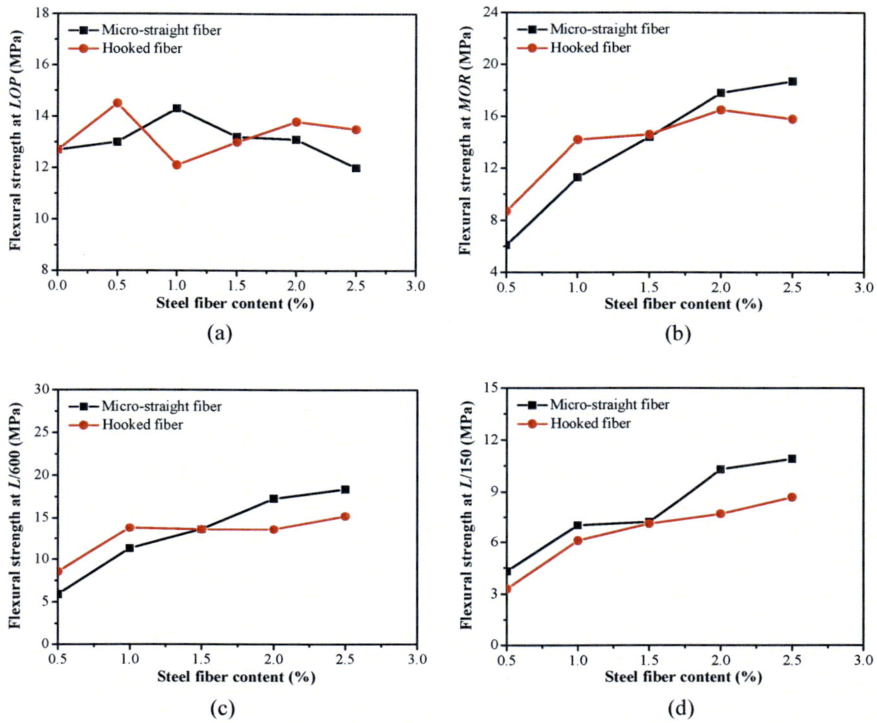

Fig. 1.19 Effects of steel fiber content and type on the load carrying capacity of UHPCC at **a** *LOP* **b** *MOR* **c** *L*/600 **d** *L*/150, Ren et al. (2018), copyright 2020, with permission from Elsevier

UHPCC with hooked steel fiber, when the steel fiber content is 1.5%, the flexural strength reaches the maximum value. With incorporating 2.5% micro-straight steel fibers and 1.5% hooked steel fibers, the maximum flexural strengths of UHPCC has increased by 48.4% and 28.6% compared to the corresponding value of plain UHPCC (12.6 MPa).

1.4.4 Three-Point Flexure Test

1.4.4.1 Three-Point Flexural Load-CMOD/deflection Curves

The representative three-point flexural load-CMOD and flexural load–deflection curves of UHPCC are shown in Figs. 1.22 and 1.23, respectively. As can be seen, except for the three-point flexural load–deflection curve of UHPCC with 0.5% micro-straight steel fibers, all the other curves exhibit the deflection hardening behavior. Figure 1.24 shows the crack pattern of UHPCC specimens, which is similar with that

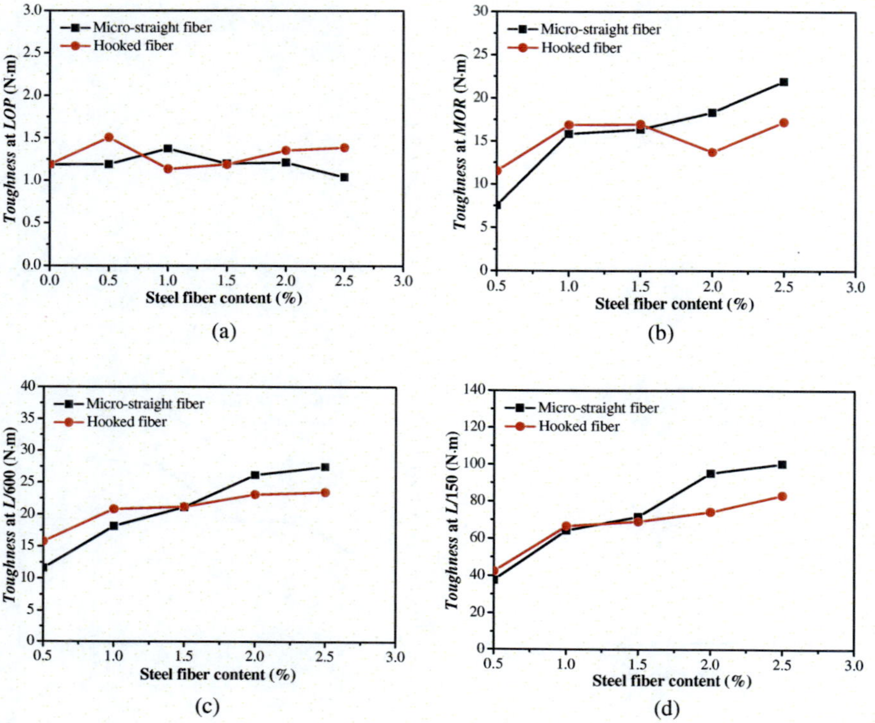

Fig. 1.20 Effects of steel fiber content and type on the toughness of UHPCC at **a** *LOP* **b** *MOR* **c** *L*/600 **d** *L*/150, Ren et al. (2018), copyright 2020, with permission from Elsevier

Table 1.8 Maximum flexural strengths of UHPCC (*unit* MPa)

Test No.	S1	S2	S3	Average	Test No.	S1	S2	S3	Average
FS400-0	12.7	12.7	12.4	12.6	–				
FS400-0.5	14.5	–	13.0	13.8	FH400-0.5	14.5	13.0	–	13.8
FS400-1.0	–	14.3	13.1	13.7	FH400-1.0	13.4	13.7	14.2	13.8
FS400-1.5	–	12.6	14.4	13.5	FH400-1.5	17.4	16.7	14.6	16.2
FS400-2.0	17.5	15.7	17.8	17.0	FH400-2.0	13.4	14.2	16.5	14.2
FS400-2.5	20.3	17.1	18.7	18.7	FH400-2.5	13.7	14.8	15.8	14.7

in the four-point flexural test. Based on Figs. 1.22 and 1.23, the fracture toughness and fracture energy of UHPCC can be derived and discussed at follows.

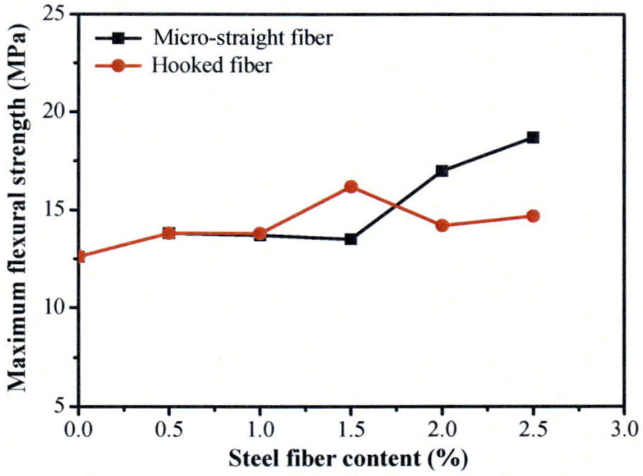

Fig. 1.21 Effects of steel fiber content and type on the maximum flexural strength of UHPCC, Ren et al. (2018), copyright 2020, with permission from Elsevier

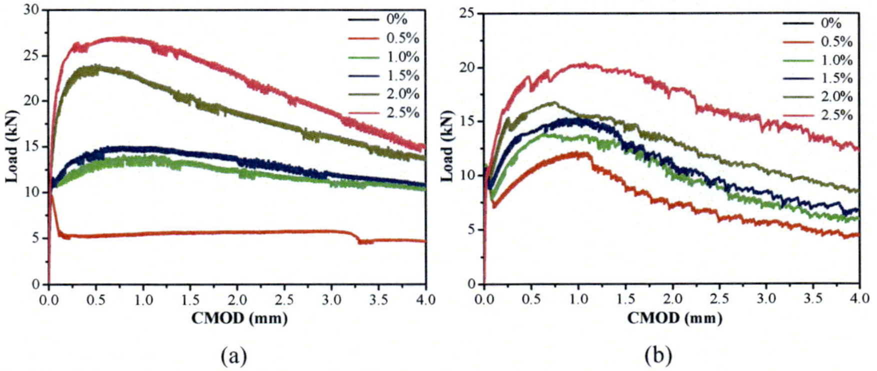

Fig. 1.22 Representative three-point flexural load-CMOD curves of UHPCC with **a** micro-straight fiber **b** hooked fiber, Ren et al. (2018), copyright 2020, with permission from Elsevier

1.4.4.2 Fracture Toughness

Fracture toughness K_{IC} represents the critical stress intensity value when the fracture failure of materials occurs. For the present non-standard three point bending concrete beam, a modified calculation formula suggested by Fan et al. (2012) is adopted to derive the fracture toughness of UHPCC, which is expressed as

$$K_{\mathrm{IC}}=\frac{3(F_{un}+0.5m_1g\times 10^{-3})L\times 10^{-3}}{2bh^2}\sqrt{h}\times k_{\beta}(\alpha) \tag{1.3}$$

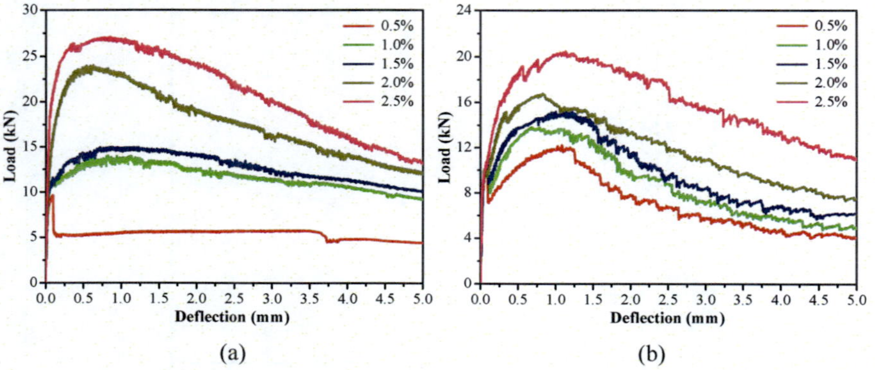

Fig. 1.23 Representative three-point flexural load–deflection curves of UHPCC with **a** micro-straight fiber **b** hooked fiber, Ren et al. (2018), copyright 2020, with permission from Elsevier

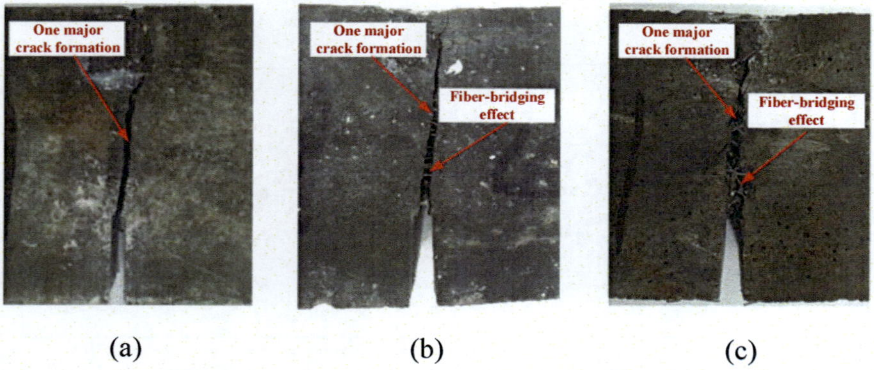

Fig. 1.24 Three-point flexural crack pattern of UHPCC specimens with **a** no mixing fiber **b** micro-straight fiber **c** hooked fiber, Ren et al. (2018), copyright 2020, with permission from Elsevier

where F_{un} and m_1 are the maximum load and the mass of specimen between the supports. g is the gravity. $\alpha{=}a_c/h$ and $\beta{=}L/h$ are the crack-to-depth ratio and span-to-depth ratio. a_c is the critical effective crack length at F_{un}, which can be determined as follows according to DL/T5332-2005 (2006).

$$a_c = \frac{2}{\pi}(h + h_0)\arctan\left(\frac{bEV_c}{32.6F_{un}} - 0.1135\right)^{1/2} - h_0 \tag{1.4}$$

where $h0 = 1$ mm is the thickness of the steel plate pasted on the specimen, E is the compressive elastic modulus and V_c is the critical value of CMOD at F_{un}.

When $\beta \geq 2.5$, $k_\beta(\alpha)$ can be determined by Guinea et al. (1998)

$$k_\beta(\alpha) = \frac{\alpha^{1/2}}{(1-\alpha)^{3/2}(1+3\alpha)}\left\{p_\infty(\alpha) + \frac{4}{\beta}\left[p_4(\alpha) - p_\infty(\alpha)\right]\right\} \tag{1.5}$$

where $p_4(\alpha)$ and $p_\infty(\alpha)$ are cubic polynomial and can be expressed as

$$p_4(\alpha) = 1.90 + 0.41\alpha + 0.51\alpha^2 - 0.17\alpha^3 \tag{1.6}$$

$$p_\infty(\alpha) = 1.99 + 0.83\alpha - 0.31\alpha^2 + 0.14\alpha^3 \tag{1.7}$$

Based on the Eqs. (1.3–1.7) and the three-point flexural load-CMOD curves of UHPCC shown in Fig. 1.22, the fracture toughness of UHPCC is derived and listed in Table 1.9. For example, "TN400-0" denotes the test No. and the capital "T" represents the three-point flexural test. The results show that, addition of steel fibers can significantly improve the fracture toughness of UHPCC and the steel fiber type has relatively minor effect. The fracture toughness of UHPCC with 2.5% micro-straight or hooked steel fibers (~15.8 MPa·m$^{1/2}$) is about 11 times compared to the corresponding value of plain UHPCC (~1.4 MPa·m$^{1/2}$).

1.4.4.3 Fracture Energy

Fracture energy G_F is defined as the energy required to generate unit fractured surface area, it is generally considered as a material property and determined by the following equation recommended by RILEM (1985).

$$G_F = \frac{W_0 + m_0 g \delta_0}{b(h - a_0)} \tag{1.8}$$

where W_0 is the area under the load–deflection curve. m_0 is the mass of specimen including the mass of measuring device attached to the specimen. δ_0 and a_0 are the deflection at finale fracture and the notch depth, respectively.

Based on the three-point flexural load–deflection curves of UHPCC given in Fig. 1.23, Table 1.10 lists the fracture energy of UHPCC, where "—" denotes that the data is not obtained or discarded due to the large deviation. Figure 1.25 further shows the effects of steel fiber content and type on the fracture energy. As can be seen, the fracture energy of UHPCC gradually increases with the rising of steel fiber content. When the steel fiber content is more than 1.0%, the micro-straight steel fiber has better effect on improving the fracture energy of UHPCC than the hooked steel fiber. For example, the maximal fracture energy of UHPCC with 2.5% micro-straight steel fibers (13583 N/m) can reach almost 1.2 and 3.5 times larger than the corresponding values of UHPCC with 2.5% hooked steel fibers (11308 N/m) and 0.5% micro-straight steel fibers (3879 N/m), respectively. Furthermore, the fracture energy of C100 plain concrete suggested by the *fib* Model Code 2010 (2013) is 169.6 N/m,

Table 1.9 Fracture toughness of UHPCC

Test No.	F_{un} (kN)	V_c (mm)	a_c (mm)	K_{IC} (MPa m$^{1/2}$)	Test No.	F_{un} (kN)	V_c (mm)	a_c (mm)	K_{IC} (MPa m$^{1/2}$)
TN400-0	10.1	0.029	30	1.430	–				
TS400-0.5	9.7	0.033	33	1.484	TH400-0.5	12.2	0.905	80	12.535
TS400-1.0	14.1	0.802	78	12.514	TH400-1.0	13.8	0.810	78	12.248
TS400-1.5	15.0	0.884	78	13.310	TH400-1.5	15.3	0.860	77	12.681
TS400-2.0	24.1	0.730	70	13.253	TH400-2.0	16.8	0.852	76	13.040
TS400-2.5	27.1	0.852	71	15.702	TH400-2.5	20.5	1.080	76	15.906

Table 1.10 Fracture energy of UHPCC (*unit* N/m)

Test No.	S1	S2	S3	Average	Test No.	S1	S2	S3	Average
TS400-0.5	3787	–	3970	3879	TH400-0.5	3925	6356	5213	5164
TS400-1.0	8152	8355	–	8254	TH400-1.0	6384	–	4070	5227
TS400-1.5	9466	9156	8213	8945	TH400-1.5	7127	8456	–	7792
TS400-2.0	12,570	12,638	10,027	11,745	TH400-2.0	8512	9170	–	8841
TS400-2.5	–	12,084	15,081	13,583	TH400-2.5	10,784	11,498	11,641	11,308

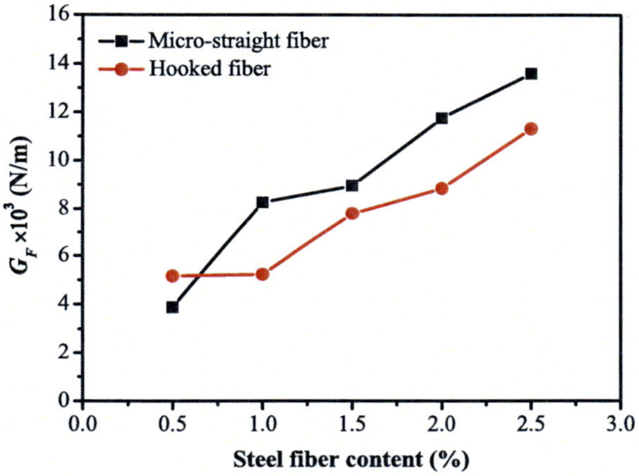

Fig. 1.25 Effects of steel fiber content and type on the fracture energy of UHPCC, Ren et al. (2018), copyright 2020, with permission from Elsevier

which is much smaller than those of UHPCC with steel fibers. Adding steel fiber could obviously improve the fracture energy of UHPCC. It should be pointed out that, limited by the test condition and time consumption, the present three-point load–deflection curves are measured with the deflections less than 5 mm, thus the fracture energy of UHPCC given in Table 1.10 would increase further if the experimental load reduces continuously to zero.

1.5 Summary

In this chapter, the static mechanical properties of UHPCC with six content (0 ~ 2.5%) and two types (micro-straight and hooked) of steel fibers are studied experimentally. By examining the influences of steel fiber content and type on the compressive, direct tensile and flexural behaviors of UHPCC, the following conclusions are obtained:

(1) For the cubic compressive strength of UHPCC, with the increase of steel fiber content, the cubic compressive strength gradually increases and the micro-straight steel fiber is more influential.

(2) For the axial compressive behavior of UHPCC, the axial compressive strength increases with the rising of the steel fiber content, but almost keeps unchanged when the steel fiber content is larger than 1.0%, and the steel fiber type almost has no influence. Addition of steel fibers has little influence on the compressive elastic modulus, Poisson's ratio and the pre-peak behavior of compressive stress–strain curves, but considerably influences the post-cracking ductile behavior.
(3) For the direct tensile strength of UHPCC, incorporation of steel fibers can effectively improve the tensile strength, and the present micro-straight steel fiber has a greater influence than the hooked steel fiber.
(4) For the four-point flexural properties of UHPCC, addition of steel fibers has little effect on the first crack flexural strength and the corresponding deflection, while it has favorable effect on the flexural strength, load carrying capacity and energy absorption capacity. When the steel fiber content is more than 1.5%, adding the present micro-straight steel fiber is more efficient than the hooked steel fiber.
(5) For the three-point flexural properties of UHPCC, the fracture toughness has significant improvement with the addition of steel fibers, while the steel fiber type has little influence on it. The fracture energy increases steadily with the increase of steel fiber content, and the effect of micro-straight steel fiber is more remarkable.

References

ABBAS S, SOLIMAN A M, NEHDI M L. Exploring mechanical and durability properties of ultra-high performance concrete incorporating various steel fiber lengths and dosages[J]. Construction and Building Materials, 2015, 75: 429-441.

ASTM C1018-97. Structural test method for flexural toughness and first crack strength of fiber reinforced concrete (using beam with third point loading)[S]. American Society for Testing and Materials, 1998: 514–551.

ASTM C1609/C1609M-12. Structural test method for flexural performance of fiber reinforced concrete (using beam with third point loading)[S]. American Society for Testing and Materials, 2012: 1–9.

Comite Euro-International du Beton. fib Model Code for Concrete Structures 2010[M]. Trowbridge, Wiltshire, UK: Redwood Books, 2013.

DL/T5332–2005. Norm for fracture test of hydraulic concrete[S]. Electric Power Industry Standard of the People's Republic of China, 2006. (in Chinese)

FAN X Q, HU S W, LU J. Experimental research on double-K fracture toughness of non-standard three point bending concrete beam[J]. Journal of Building Structures, 2012, 33(10): 152-157. (in Chinese)

GB-T50081–2002. Standard for test method of mechanical properties on ordinary concrete[S]. Beijing, China: 2003. (in Chinese)

GRAYBEAL B A. Material property characterization of ultra-high performance concrete: FHWA-HRT-06-103[R]. VA: U.S. Department of Transportation Federal highway Adminstration, 2006: 1-186.

GUINEA G V, PASTOR J Y, PLANAS J, ELICES M. Stress intensity factor, compliance and CMOD for a general three-point-bend beam[J]. International Journal of Fracture, 1998, 89: 103-116.

HASSAN A M T, JONES S W, MAHMUD G H. Experimental test methods to determine the uniaxial tensile and compressive behaviour of ultra high performance fiber reinforced concrete (UHPFRC)[J]. Construction and Building Materials, 2012, 37: 874-882.

KANG S T, KIM J K. The relation between fiber orientation and tensile behavior in an Ultra High Performance Fiber Reinforced Cementitious Composites (UHPFRCC)[J]. Cement and Concrete Research, 2011, 41(10): 1001-1014.

KAZEMI S, LUBELL A S. Influence of specimen size and fiber content on mechanical properties of ultra-high performance fiber-reinforced concrete[J]. ACI Materials Journal, 2012, 109(6): 675-684.

KIM D J, NAAMAN A E, EL-TAWIL S. Comparative flexural behavior of four fiber reinforced cementitious composites[J]. Cement & Concrete Composites, 2008, 30: 917-928.

LIU J Z, HAN F Y, CUI G, ZHANG Q Q, LV J, ZHANG L H, YANG Z Q. Combined effect of coarse aggregate and fiber on tensile behavior of ultra-high performance concrete[J]. Construction and Building Materials, 2016, 121: 310-318.

REN G M, WU H, FANG Q, LIU J Z. Effects of steel fiber content and type on static mechanical properties of UHPCC[J]. Construction and Building Materials, 2018, 163: 826-839.

RILEM TC 50-FMC. Determination of the fracture energy of mortar and concrete by means of three-point bending tests on notched beams[J]. Materials and Structures, 1985, 18: 285–290.

SCHMIDT M, FEHLING E. Ultra high performance concrete research development and application in Europe, in: Seventh International Symposium on the Utilization of High Strength/High Performance Concrete 51–78[C]. Washington, USA: 2005

WILLE K, TAWIL S E, NAAMAN A E. Properties of strain hardening ultra high performance fiber reinforced concrete (UHP-FRC) under direct tensile loading[J]. Cement & Concrete Composites, 2014, 48: 53-66.

WU Z M, SHI C J, HE W, WU L M. Effects of steel fiber content and shape on mechanical properties of ultra high performance concrete[J]. Construction and Building Materials, 2016, 103: 8-14.

YOO D Y, LEE J H, YOON Y S. Effect of fiber content on mechanical and fracture properties of ultra high performance fiber reinforced cementitious composites[J]. Composite Structures, 2013, 106: 742-753.

YOO D Y, SHIN H O, YANG J M, YOON Y S. Material and bond properties of ultra high performance fiber reinforced concrete with micro steel fibers[J]. Composites Part B, 2014, 58(mar): 122–133.

Chapter 2
Dynamic Compressive Mechanical Properties of UHPCC

2.1 Introduction

The mechanical properties of UHPCC could be affected by many factors (e.g., curing conditions, aggregate size, steel fiber content and type), in which the steel fiber plays an important role. In this Chapter, the dynamic compressive behavior of UHPCC with different steel fiber contents and types at high strain rate is mainly concerned.

In the existing studies on the influence of steel fiber reinforcement type on the dynamic compressive behavior of steel fiber reinforced concrete (SFRC), Wang et al. (2008) performed Split Hopkinson pressure bar (SHPB) compressive test on the SFRC specimens with 0, 3.0 and 6.0% ultrashort steel fibers at the strain rate up to 100 s^{-1}, and found that both the fiber volume fraction and strain rate exerted significant enhancing effect on the dynamic compressive strength. Lai and Sun (2009) studied the effect of steel fiber content (0, 3.0 and 4.0% by volume) on the dynamic compressive properties of UHPCC at different strain rates (23.7 ~ 99.2 s^{-1}) by using the SHPB device. It was concluded that UHPCC is strain rate dependent, and the elastic modulus, peak stress and the corresponding strain as well as the toughness of UHPCC increased with increasing the strain rate. While the enhancement effect of steel fiber content on the dynamic strength was reduced at higher strain rate. Hao and Hao (2013) conducted a series of SHPB test to study the effects of steel fiber content (0 ~ 1.5% by volume) and strain rate (50 ~ 200 s^{-1}) on the dynamic compressive behavior of spiral steel fiber reinforced concrete. The results showed that, the compressive strength and elastic modulus exhibited increasing sensitivity to strain rate with increasing the steel fiber content. Also, the energy absorption capacity had a strong dependency on the steel fiber volume fraction under high strain rate. Su et al. (2016) evaluated the influences of two types of steel fibers (micro-straight and twisted) with different lengths and aspect ratios on the dynamic compressive properties of ultra-high performance concrete (UHPC). It was found that the micro-straight steel fiber had better effect on increasing the dynamic compressive strength, and the increase of steel fiber length and aspect ratio had positive impact on it. Li et al. (2016) investigated the influence of steel fiber content (0 ~ 1.5% by volume) on the dynamic

Q. Fang et al., *UHPCC Under Impact and Blast*,
https://doi.org/10.1007/978-981-33-6842-2_2

compressive behavior of hybrid (steel fiber blended with 2% polyvinyl alcohol fiber) fiber reinforced ultra high toughness cementitious composites (UHTCC) at the strain rate ranging from 113.8 s^{-1} to 192.1 s^{-1}. It was concluded that, increasing the volume fraction of steel fibers had obvious strengthening effect on the dynamic compressive strength, while the above effect was weakened with increasing the strain rate. The energy absorption capacity significantly increased with increasing the strain rate and steel fiber content.

Generally, the existing experimental assessment for the effects of steel fiber content and type on the dynamic compressive mechanical properties of steel fiber reinforced concrete, particularly for UHPCC, is far from sufficient. And the variation range of strain rate for SHPB test in the existing studies is not wide enough. At present, the impact test is performed by using a 50-mm-diameter SHPB device at the strain rate ranging from 17.6 s^{-1} to 328.4 s^{-1}, in which two typical steel fibers (micro-straight and hooked) and three volume fractions (0%, 1.0% and 2.0%) are considered. The dynamic compressive stress–strain curves under different strain rates are derived, and the effects of steel fiber content and type as well as the strain rate on the dynamic compressive properties of UHPCC are assessed. Furthermore, the dynamic increase factor (DIF) for the dynamic compressive strength, as well as the visco-elastic damage model of UHPCC are discussed.

2.2 Specimen Preparation

For the preparation of UHPCC, a forced single-axis mixer with a capacity of 60L and a constant mixing speed (47 rpm) was used. To achieve good workability, particle distribution and packing density, mixing procedure for the present UHPCC was controlled rigorously. Firstly, since small particles tended to agglomerate and it was easier to break these chunks when the particles were dry. The cementitious materials (cement, silica fume, ultra-fine mineral admixture) and sand were put together simultaneously and dry-mixed uniformly for 3 min. Secondly, the water pre-mixed with super-plasticizer was then added gradually and mixed for 3 ~ 5 min. Finally, steel fibers were dispersed carefully by hand into the mixture and mixed for another 3 ~ 5 min in order to guarantee the fibers well distributed throughout the matrix. The shear action of steel fibers helped to destroy any remaining agglomerates in the fresh mixture, thus improving the workability of the composites.

After the prepared matrix was poured into the moulds, the freshly cast specimens were covered with plastic sheets to prevent moisture losing and kept at room temperature for 24 h. Then, they were demoulded and cured in a standard curing room with the temperature of 17.7 ~ 21.7 °C and relative humidity of 98.4% for another 28d. After that, the specimens were taken out of the standard curing room and stored at room temperature until the time of tests. For the present two steel fibers types and three volume fractions, 120 cylindrical specimens were cast for the dynamic SHPB test. As shown in Fig. 2.1a, the specimen size of Φ48mm × 35 mm was selected by considering the diameter of the SHPB device (50 mm) and the maximal fiber length

(a) Specimen size (b) Vertical cross-section of specimen (2% hooked steel fiber)

Fig. 2.1 SHPB specimen, Ren et al. (2018), copyright 2020, with permission from Elsevier

Table 2.1 Quasi-static cubic compressive strengths of UHPCC

Test No.	N-0	S-1	S-2	H-1	H-2
Quasi-static compressive strength (MPa)	116.8	142.7	160.6	126.6	141.7

(25 mm), which guarantees the random fiber orientation and distribution. Also, the vertical cross-section of UHPCC specimen with 2.0% hooked steel fiber is given in Fig. 2.1b, which indicates that the steel fibers were uniformly distributed in the matrix. In order to make the specimen-bar contact as perfect as possible, the end surfaces of these cylinders were polished, which were flattened to less than 0.1 mm, and the roughness on the surfaces was less than 0.05 mm. Besides, 15 cubic control specimens with the size of 100 mm × 100 mm × 100 mm were cast. Three specimens were tested for each batch and the average quasi-static compressive strengths are listed in Table 2.1. In which the capitals "N", "S" and "H" represent no mixing fiber, micro-straight steel fiber and hooked steel fiber, respectively. "0", "1" and "2" represent the volume fractions of steel fibers are 0%, 1.0% and 2.0%, respectively. The similar designation way is used in the following test.

2.3 SHPB Test

2.3.1 Test Device

The SHPB test was conducted on a 50-mm-diameter conic variable cross-sectional Hopkinson pressure bar, as shown in Fig. 2.2. The lengths of the striker bar, incident bar and transmitted bar are 400 mm, 2400 mm and 1400 mm, respectively. The smaller cross-sectional diameter of the incident bar is 37 mm, which is coincident with that of the striker bar. The specimens were grease-lubricated on both surfaces to minimize the effect of end friction confinement and then placed between the incident bar and transmitted bar. In order to eliminate the high-frequency oscillation and achieve equal stress distribution in the specimen before it was damaged, the pulse

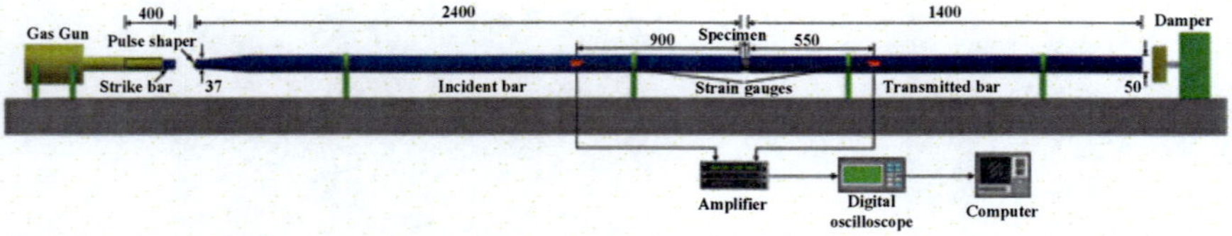

Fig. 2.2 SHPB setup (*unit* mm), Ren et al. (2018), copyright 2020, with permission from Elsevier

Fig. 2.3 Pulse shaper **a** rubber **b** brass, Ren et al. (2018), copyright 2020, with permission from Elsevier

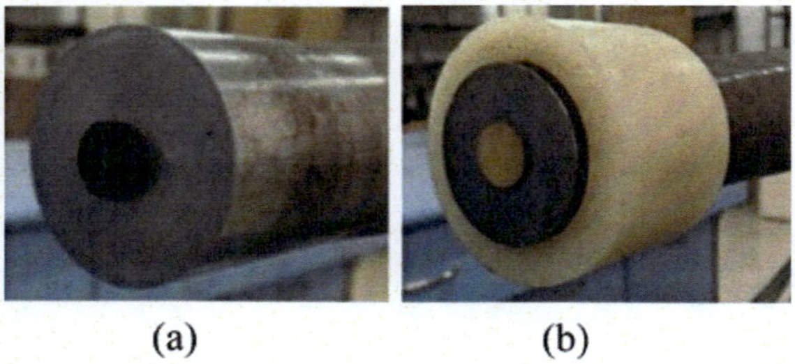

shaper was placed on the impact surface of the incident bar. As shown in Fig. 2.3, two different pulse shaping materials (rubber and brass) were considered at present to enlarge the range of strain rate. Total six strain rates ranging from 17.6 s^{-1} to 328.4 s^{-1} were achieved, in which the rubber disk (Φ10mm × 3 mm) and the brass disk (Φ12mm × 1 mm) were used for the two lower and the three medium strain rates, and no pulse shaper was adopted for the sixth strain rate in order to achieve the highest strain rate under the elastic limit state of this device.

2.3.2 Test Technique

The technique of SHPB test is based on the theory of one-dimensional wave propagation in an elastic bar. The striker bar was driven by air pressure with the values of 0.25 ~ 1.0 MPa, which led to the velocity of striker bar V_0 ranging from 11.6 ~ 30.1 m/s during the present test. The impact of striker bar on the incident bar generates an elastic compressive wave, which is referred to as the incident wave and propagates towards the specimen. When the incident wave reaches the incident bar/specimen interface, it is split into the reflected wave and transmitted wave due to the impedance mismatch between the incident/transmitted bars and specimen. The reflected wave is reflected away from the specimen and travels backwards to the impact end of the incident bar. While the transmitted wave travels through the specimen and into the transmitted bar, which causes plastic deformation in the specimen. As illustrated in Fig. 2.2, two strain gauges are respectively attached to the incident bar and the transmitted bar to measure the incident strain $\varepsilon_i(t)$, the reflected strain $\varepsilon_r(t)$ as well

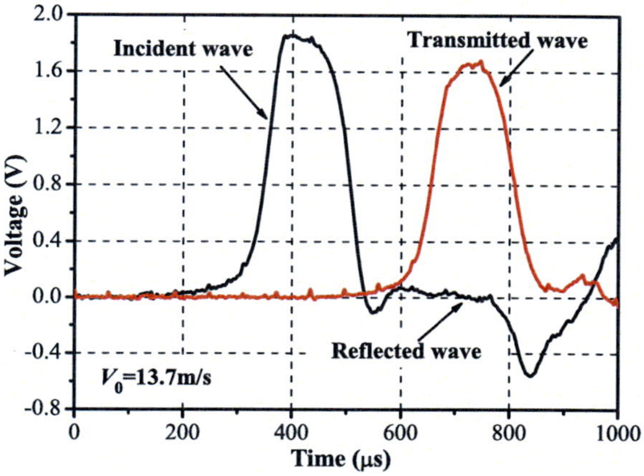

Fig. 2.4 Typical original pulse waves of SHPB test, Ren et al. (2018), copyright 2020, with permission from Elsevier

as the transmitted strain $\varepsilon_t(t)$ induced by the stress waves. The typical original pulse waves of SHPB test are shown in Fig. 2.4.

Based on the two-wave method, the dynamic stress $\sigma_s(t)$, the dynamic strain $\varepsilon_s(t)$ of the specimen and the strain rate $\dot{\varepsilon}_s(t)$ are derived by using the following equations (Lindholm 1964).

$$\begin{cases} \sigma_s(t) = \dfrac{E_b A_0}{A_s}\varepsilon_t(t) \\ \varepsilon_s(t) = \dfrac{2C_0}{l_s}\displaystyle\int_0^t [\varepsilon_i(t) - \varepsilon_t(t)]\,\mathrm{d}t \\ \dot{\varepsilon}_s(t) = \dfrac{2C_0}{l_s}[\varepsilon_i(t) - \varepsilon_t(t)] \end{cases} \tag{2.1}$$

where $E_b = 210\text{GPa}$, A_0 and $C_0 = 5172$ m/s are the elastic modulus, cross-sectional area and elastic wave velocity of the bars, respectively. A_s and l_s are the cross-sectional area and original length of the specimen, respectively. It should be noted that, the application condition of the above equations is that the longitudinal stress in the specimen should reach the equilibrium state, which will be discussed in Sect. 2.4.1.

At follows, an example will be taken to introduce the process of deriving the stress–strain curves and strain rate-strain curves of UHPCC. Figure 2.5a shows the strain signals of incident, reflected and transmitted waves (which has been shifted to $t = 0$) of four parallel UHPCC specimens (S1 ~ S4) with 2.0% micro-straight steel fiber at $V_0 = 20.1$ m/s. Based on Fig. 2.5a and Eq. (2.1), Figs. 2.5b–d show the calculation results of stress-, strain-, and strain rate-time histories. Then, the stress–strain curves and strain rate-strain curves of the above four UHPCC specimens as well as the corresponding average curves can be obtained, as shown in Figs. 2.6 and 2.7. The above strain rate-strain curves were further adopted to determine the strain rate at follows in Sect. 2.4.2. It can be seen that, the average curves are smoother

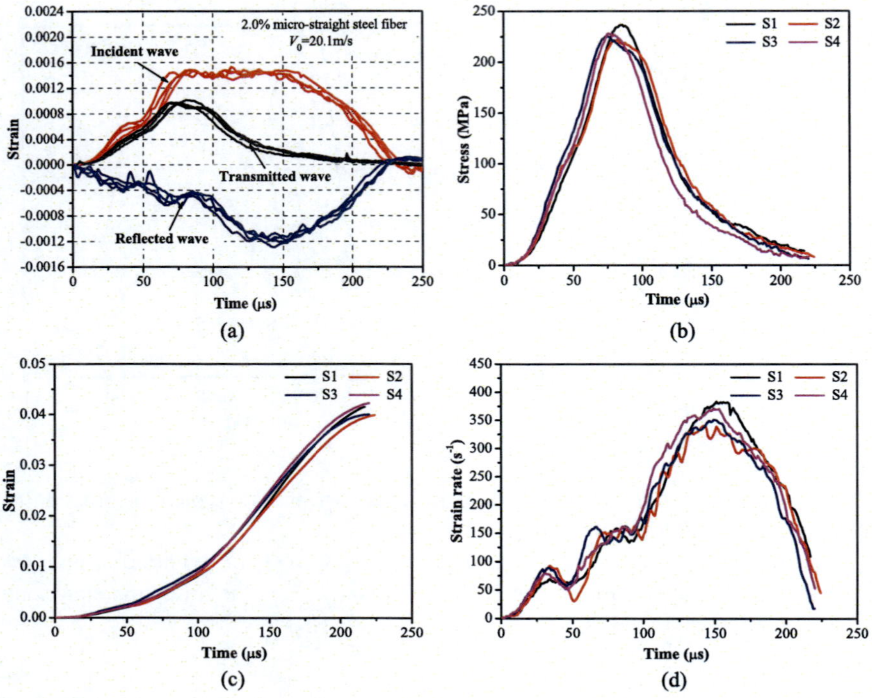

Fig. 2.5 **a** strain signals of the bars and the **b** stress time-history, **c** strain time-history, **d** strain rate time-history curves of specimens, Ren et al. (2018), copyright 2020, with permission from Elsevier

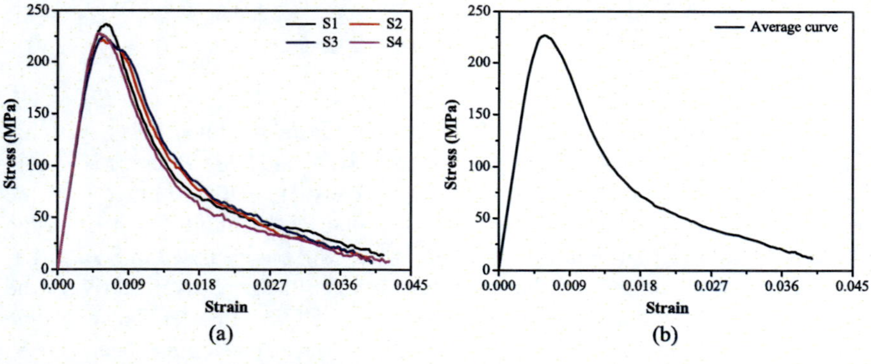

Fig. 2.6 Stress–strain curves of UHPCC specimens **a** before averaging **b** after averaging, Ren et al. (2018), copyright 2020, with permission from Elsevier

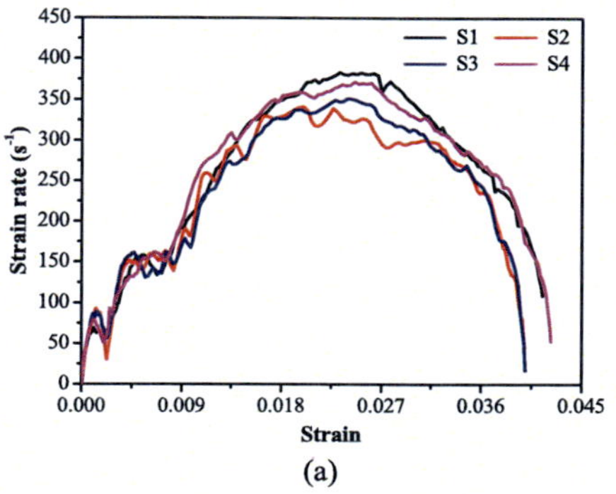

Fig. 2.7 Strain rate-strain curves of UHPCC specimens **a** before averaging **b** after averaging, Ren et al. (2018), copyright 2020, with permission from Elsevier

than the raw curves, which represent more robust and reliable results. In this test, four parallel specimens were carried out for each batch and the average values of which were reported as the representative results.

2.4 Test Results and Discussions

2.4.1 Stress Equilibrium

It is essential to conduct the longitudinal stress equilibrium check for a valid SHPB test (Lu and Li 2010). According to Eq. (2.2), the transmitted stress wave at the incident surface of the specimen in the SHPB test can be obtained, which will be used to check the stress equilibrium.

$$\sigma_t = \sigma_i + \sigma_r \tag{2.2}$$

where σ_t, σ_i and σ_r represent the transmitted, incident and reflected stress on the specimen, respectively.

For example, Fig. 2.8 shows the typical stress equilibrium check on UHPCC specimens with 2.0% micro-straight steel fiber at $V_0 = 20.1$ m/s. It can be seen that, the specimen reached the stress balance, which verified the validity of the SHPB test results in the present study.

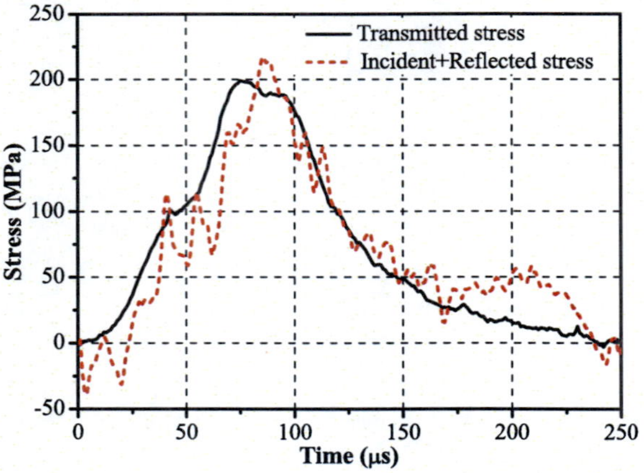

Fig. 2.8 Stress equilibrium check, Ren et al. (2018), copyright 2020, with permission from Elsevier

2.4.2 Strain Rate Determination

Based on the test technique introduced in Sect. 2.3.2, the two typical strain rate-strain curves can be obtained, as shown in Fig. 2.9. According to Eq. (2.1), the strain rate is linearly related to the reflected strain. Normally, the reflected wave could reach a plateau with relatively constant values, which corresponds to a nearly constant strain rate that the specimen has undergone, as shown in Fig. 2.9a. In this case, the average value of the strain rate plateau is used as the strain rate. However, as shown in Fig. 2.9b, the plateau of reflected wave could not always be achieved in SHPB test due to many uncertainties, e.g., the dynamic properties of wave shaper and tested materials. Under this circumstance, the value of strain rate corresponding to the peak stress is used as the strain rate.

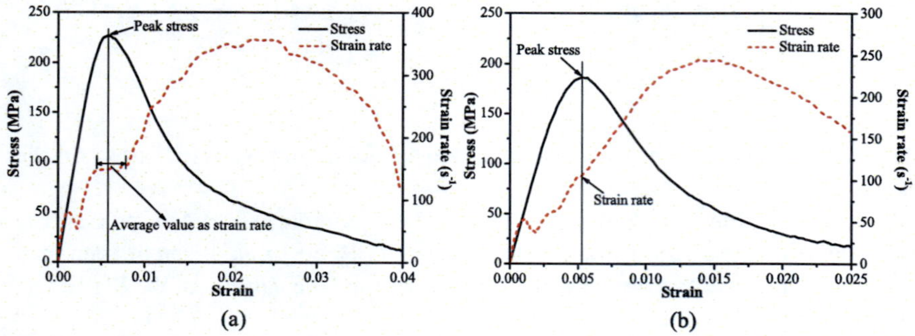

Fig. 2.9 Strain rate determination **a** with reflected plateau **b** without reflected plateau, Ren et al. (2018), copyright 2020, with permission from Elsevier

2.4.3 Dynamic Failure Pattern

The typical failure patterns of UHPCC specimens at different strain rates are shown in Fig. 2.10. It can be observed that the failure patterns were axial splitting and the damage degree increased with the rising of strain rate. At the lowest strain rate, the plain UHPCC (N-0) specimens broke into small pieces and the UHPCC specimens with hooked steel fibers only had several visible cracks, while the UHPCC specimens with micro-straight steel fibers almost remained intact. With the strain rate further increasing, the damage of UHPCC specimens became more severe and specimens crushed into smaller fragments.

Fig. 2.10 Failure patterns of UHPCC specimens **a** N-0 **b** S-1 **c** S-2 **d** H-1 **e** H-2, Ren et al. (2018), copyright 2020, with permission from Elsevier

Correspondingly, by comparing of the failure patterns of UHPCC specimens at a similar strain rate, the damage degree of UHPCC specimens weakened with increasing the steel fiber content, which indicates that adding steel fibers can reduce crack development and the micro-straight steel fiber has better crack confinement ability. For example, at the strain rate around 150 s^{-1}, UHPCC specimens with 2.0% micro-straight steel fibers broke into few larger parts (Fig. 2.10c), while those with 1.0% micro-straight steel fibers broke into much smaller pieces (Fig. 2.10b). Also, for the volume fraction of 2.0%, the UHPCC specimens with micro-straight steel fibers almost kept unseparated under the strain rate around 100 s^{-1}, but those with hooked steel fibers were fragmented into pieces.

2.4.4 Dynamic Stress–Strain Curve

Figure 2.11 shows the average stress–strain curves of the present UHPCC specimens at different strain rates. It can be seen that, the shapes of above curves are similar, which consist of the ascending and descending stages. In the ascending stage, the stress–strain relationship is initially linear elasticity and then progresses to the strain-hardening behavior, where micro-cracks begin to develop as the strain increases. After the stress reaches the peak value, the stress–strain relationship represents the strain-softening behavior with the continuous cracks further propagation.

Table 2.2 lists the corresponding dynamic compressive properties, where the 1 ~ 6 in the test No. represent the six different strain rates. It can be seen that, the dynamic compressive behaviors of UHPCC are significantly sensitive to the strain rate, the peak stress and the corresponding strain as well as the dynamic elastic modulus of UHPCC increase obviously with the increase of strain rate. As stated by Mindess et al. (2003), the strain rate sensitivity of concrete is usually caused by the time-dependent movement of free water in the matrix through voids and pores and the time-dependent nature of crack growth relative to the loading rate. Generally, the influence of moisture is not always considered. The crack propagating speed increases with increasing the strain rate, while it remains much lower than that of stress wave in concrete, which leads to the enhancement of dynamic compressive properties of concrete.

Figure 2.12 shows the effects of steel fiber content and type on the dynamic compressive properties of UHPCC. It should be noted that, based on Fig. 2.12, the average peak stress of "H-1–2" specimens (153.9 MPa) and average peak strain of "S-2-3" specimens (6.71×10^{-3}) are discarded due to the large deviation. It can be concluded that,

(i) The steel fiber content and type have little influence on the dynamic elastic modulus and peak strain of UHPCC at different strain rates.
(ii) Adding steel fibers could slightly improve the dynamic compressive strength of UHPCC, while there is an obvious effect of steel fiber reinforcement on the static compressive strength and it increases with increasing the volumetric

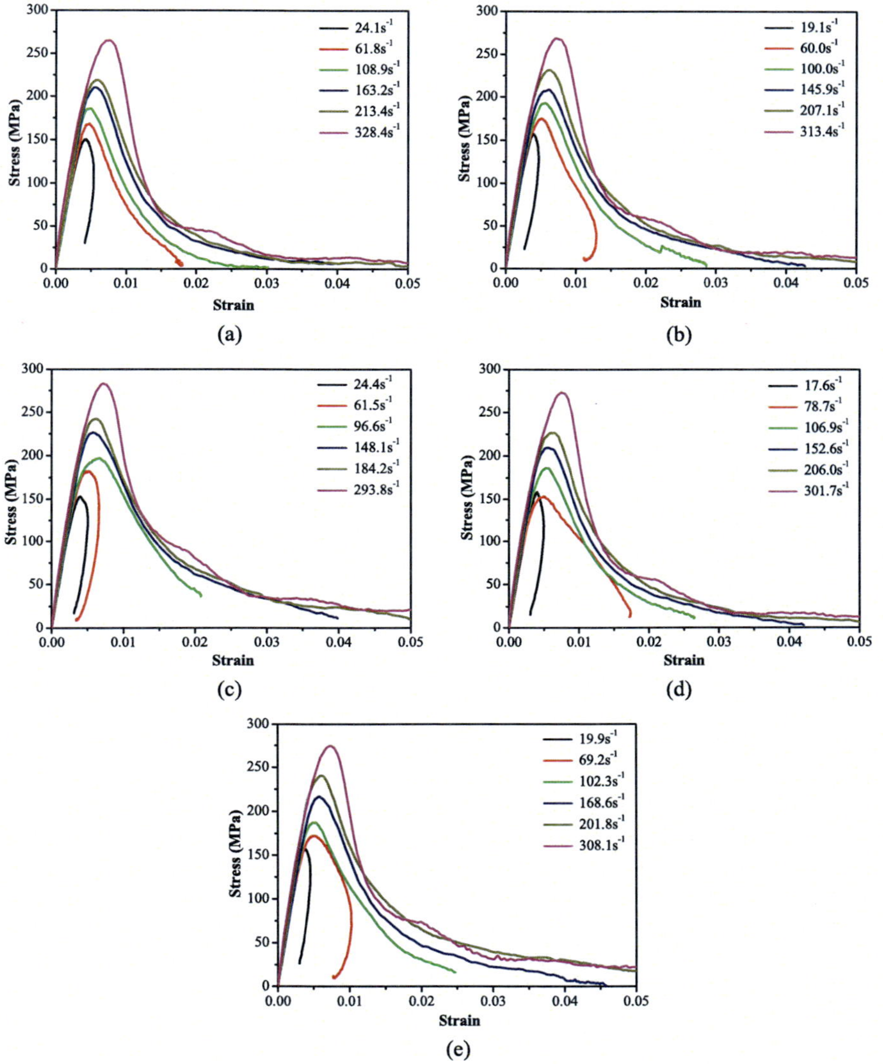

Fig. 2.11 Dynamic stress–strain curves of UHPCC specimens with different steel fiber content and type **a** N-0 **b** S-1 **c** S-2 **d** H-1 **e** H-2, Ren et al. (2018), copyright 2020, with permission from Elsevier

ratio of steel fiber. For example, at the strain rate around 309.1 s^{-1}, the dynamic compressive strength of UHPCC with mixing 2.0% micro-straight steel fibers (283.4 MPa) is only 6.8% larger than that of the plain UHPCC (265.3 MPa). While the quasi-static compressive strength of UHPCC with mixing 2.0% micro-straight steel fibers is 160.6 MPa, which increases by 37.5% compared to the plain UHPCC (116.8 MPa). The reason lies in that, the different failure

Table 2.2 Dynamic compressive properties of UHPCC

Test No.	Strain rate (s^{-1})	Peak stress (MPa)	Peak strain ($\times 10^{-3}$)	Dynamic elastic modulus (GPa)	Test No.	Strain rate (s^{-1})	Peak stress (MPa)	Peak strain ($\times 10^{-3}$)	Dynamic elastic modulus (GPa)
N-0-1	24.1	150.3	4.30	46.0	N-0-4	163.2	210.5	5.52	49.4
N-0-2	61.8	168.4	4.75	47.1	N-0-5	213.4	219.1	6.00	50.3
N-0-3	108.9	185.8	4.99	48.0	N-0-6	328.4	265.3	7.58	62.5
S-1-1	19.1	157.4	3.88	47.3	S-1-4	145.9	208.6	6.11	50.4
S-1-2	60.0	175.2	4.98	48.4	S-1-5	207.1	231.6	6.28	54.8
S-1-3	100.0	192.9	5.57	49.2	S-1-6	313.4	268.9	7.02	60.2
S-2-1	24.4	153.3	4.01	48.1	S-2-4	148.1	226.7	5.74	51.7
S-2-2	61.5	182.5	5.11	49.4	S-2-5	184.2	242.2	6.17	53.7
S-2-3	96.6	197.3	6.71	50.2	S-2-6	293.8	283.4	7.18	62.9
H-1-1	17.6	158.9	4.03	46.4	H-1-4	152.6	209.6	5.66	50.5
H-1-2	78.7	153.9	4.89	48.5	H-1-5	206.0	226.7	6.42	55.3
H-1-3	106.9	186.5	5.31	49.5	H-1-6	301.7	273.2	7.41	59.1
H-2-1	19.9	157.4	3.77	48.2	H-2-4	168.6	217.1	5.81	51.5
H-2-2	69.2	172.6	4.94	49.9	H-2-5	201.8	240.4	6.19	53.8
H-2-3	102.3	187.6	4.90	50.7	H-2-6	308.1	275.1	7.28	62.3

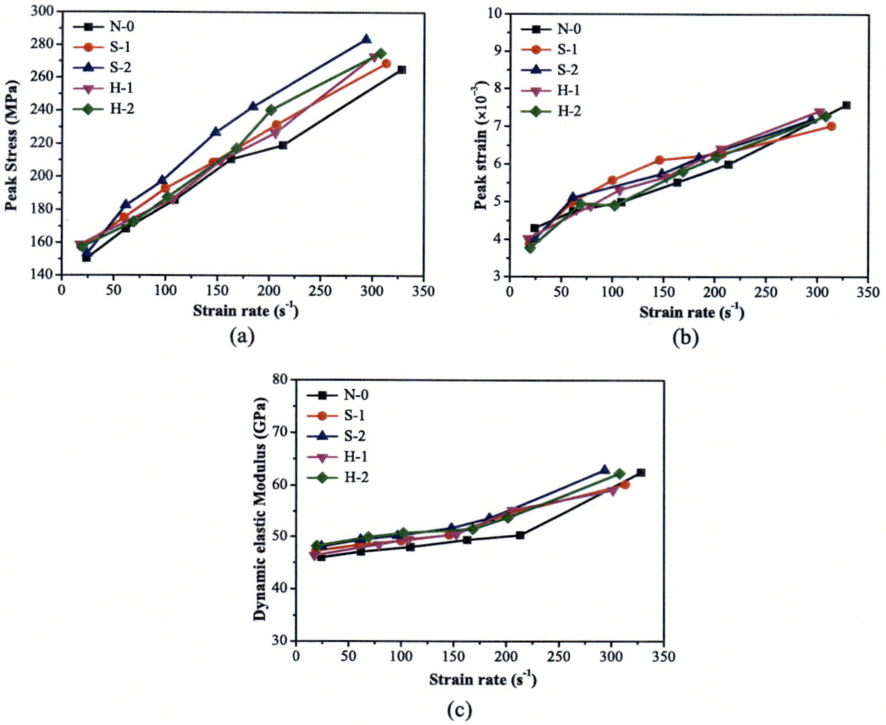

Fig. 2.12 Effects of steel fiber content and type on the dynamic compressive properties of UHPCC **a** peak stress **b** peak strain **c** dynamic elastic modulus, Ren et al. (2018), copyright 2020, with permission from Elsevier

patterns of UHPCC specimens under static and dynamic loadings lead to the different effects of steel fiber mixing ratio. Under the static loadings, the specimen only has one major crack across the weak zone and the fiber-bridging effect plays an important role to delay the formation and propagation of the cracks. Under the dynamic loadings, the specimen is fractured into fragments and the steel fibers are pulled out from the matrix, which makes the fiber-bridging effect weakening to some extent.

(iii) The micro-straight steel fiber has slightly better effect on improving the dynamic compressive strength of UHPCC with the volume fraction of 2.0% than the hooked steel fiber at a similar strain rate, while these two types of steel fibers have the similar enhancement effect on that when the steel fiber content is 1.0%. For example, at the strain rate around 155.7 s^{-1}, the dynamic compressive strength of UHPCC with 2.0% micro-straight steel fibers is 226.7 MPa, which is 7.7% and 4.4% larger than that of the plain UHPCC and UHPCC with 2.0% hooked steel fibers, respectively.

2.4.5 Dynamic Increase Factor

The dynamic increase factor (DIF) is defined as the ratio of the dynamic strength to the quasi-static strength in uniaxial compression (Li and Meng 2003), which is widely used to analyze and design of concrete protective structures under intensive loadings. The DIFs for compressive strength of UHPCC with different steel fiber contents and types are listed in Table 2.3 and illustrated in Fig. 2.13. It should be noted that, the stress–strain curves of UHPCC are hysteretic at the lower strain rate in Fig. 2.11, and the peak stresses are not the actual failure stresses, thus the DIFs for the UHPCC specimens with hysteretic stress–strain curves are not given in Fig. 2.13.

It is well accepted that there exists a transition strain rate for the relationship between the DIF and the logarithm of strain rate, beyond which the DIF increases

Table 2.3 DIFs for compressive strength

Test No.	Strain rate (s^{-1})	DIF	Test No.	Strain rate (s^{-1})	DIF	Test No.	Strain rate (s^{-1})	DIF
N-0-1	24.1	1.29	N-0-2	61.8	1.44	N-0-3	108.9	1.59
N-0-4	163.2	1.80	N-0-5	213.4	1.88	N-0-6	328.4	2.27
S-1-1	19.1	1.10	S-1-2	60.0	1.23	S-1-3	100.0	1.35
S-1-4	145.9	1.46	S-1-5	207.1	1.62	S-1-6	313.4	1.88
S-2-1	24.4	0.95	S-2-2	61.5	1.14	S-2-3	96.6	1.23
S-2-4	148.1	1.41	S-2-5	184.2	1.51	S-2-6	293.8	1.76
H-1-1	17.6	1.26	H-1-2	78.7	1.22	H-1-3	106.9	1.47
H-1-4	152.6	1.66	H-1-5	206.0	1.79	H-1-6	301.7	2.16
H-2-1	19.9	1.11	H-2-2	69.2	1.22	H-2-3	102.3	1.32
H-2-4	168.6	1.53	H-2-5	201.8	1.70	8H-2-6	308.1	1.94

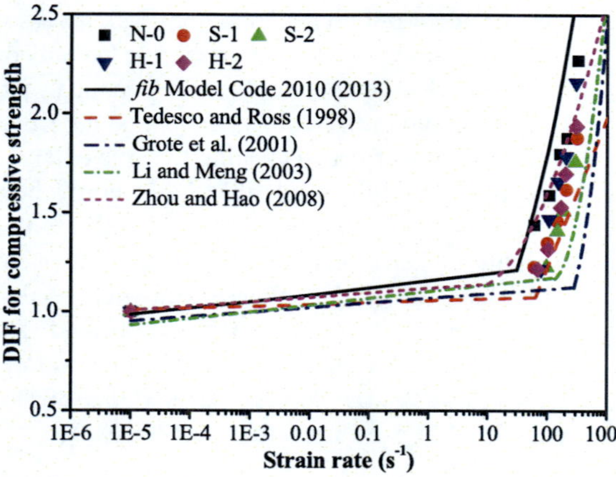

Fig. 2.13 Dynamic increase factor for compressive strength of UHPCC, Ren et al. (2018), copyright 2020, with permission from Elsevier

with strain rate rapidly and the material changes from low strain rate sensitivity to high strain rate sensitivity. As shown in Fig. 2.13, the DIFs for UHPCC material rapidly increase with the strain rate rising, and the highest DIF of 2.27 is obtained for the plain UHPCC at the strain rate of 328.4 s^{-1}. Hao et al. (2016) thought that the specimen inevitably undergoes lateral expansion under dynamic loading, which results in lateral stresses that will act as a form of confinement. Li and Meng (2003) derived that the influence of lateral inertia confinement on the apparent dynamic strength enchantment is significant and cannot be neglected for the strain rates exceeding 100 s^{-1} for mortar material. Thus, the DIF may not be a material property, but rather a physical result of lateral inertia confinement and many other influencing factors (e.g., specimen size, moisture) with increasing strain rate.

Besides, for the influences of the steel fiber content and type, it indicates that, (i) the plain UHPCC has constantly higher DIF values than UHPCC with different steel fiber content and type, which suggests that UHPCC with various steel fiber reinforcement is less sensitive to high strain rate than the plain UHPCC; (ii) for both fiber types, the values of DIF for UHPCC material decrease with the increase of steel fiber content; (iii) with the identical fiber mixing ratio, the DIFs for UHPCC with hooked steel fiber are relatively higher compared to those with micro-straight steel fiber within the experimental strain rate range. The reason may lie in that, addition of steel fibers could improve the concrete quality and the micro-straight steel fiber has relatively better effect on that, which leads to lower DIFs. The similar conclusion was summarized by Bischoff and Perry (1991). They discussed the influence of concrete quality on DIF by summarizing a large number of experimental data, and concluded that the poorer concrete exhibits a larger DIF for compressive strength.

For the concrete-like materials, several empirical formulae have been proposed to estimate the strain-rate effect on the compressive strength. The *fib* Model Code 2010 (2013) recommended

$$\mathrm{DIF} = \begin{cases} (\dot{\varepsilon}/\dot{\varepsilon}_{ts})^{0.014} & \text{for } \dot{\varepsilon} \le 30\mathrm{s}^{-1} \\ 0.012(\dot{\varepsilon}/\dot{\varepsilon}_{ts})^{1/3} & \text{for } \dot{\varepsilon} > 30\mathrm{s}^{-1} \end{cases} \tag{2.3}$$

where $\dot{\varepsilon}$ is the strain rate in the range of 30×10^{-6} to 300 s^{-1} and $\dot{\varepsilon}_{ts}$=$30 \times 10^{-6}s^{-1}$ is the static strain rate.

Based on a series of SHPB test conducted by Ross et al. (1989, 1995, 1996), Tedesco and Ross (1998) further proposed

$$\mathrm{DIF} = \begin{cases} 0.00965\log\dot{\varepsilon} + 1.058 \ge 1.0 & \text{for } \dot{\varepsilon} \le 63.1\mathrm{s}^{-1} \\ 0.758\log\dot{\varepsilon} - 0.289 \le 2.5 & \text{for } \dot{\varepsilon} > 63.1\mathrm{s}^{-1} \end{cases} \tag{2.4}$$

Grote et al. (2001) conducted the SHPB test on mortar with the strain rates between 250 s^{-1} and 1700s^{-1} and established

$$\mathrm{DIF}=\begin{cases}0.0235\log\dot{\varepsilon}+1.070.0235\log\dot{\varepsilon}+1.07 & \text{for } \dot{\varepsilon}\le 266.0\mathrm{s}^{-1}\\ 0.882(\log\dot{\varepsilon})^3-4.48(\log\dot{\varepsilon})^2+7.22\log\dot{\varepsilon}-2.64 & \text{for } \dot{\varepsilon}>266.0\mathrm{s}^{-1}\end{cases} \tag{2.5}$$

Based on the numerical results of SHPB test on mortar, Li and Meng (2003) formulated a relationship between DIF and the logarithm of strain rate as

$$\mathrm{DIF}=\begin{cases}1+0.03438(\log\dot{\varepsilon}+3) & \text{for } \dot{\varepsilon}\le 10^2\mathrm{s}^{-1}\\ 1.729(\log\dot{\varepsilon})^2-7.1372\log\dot{\varepsilon}+8.5303 & \text{for } \dot{\varepsilon}>10^2\mathrm{s}^{-1}\end{cases} \tag{2.6}$$

Zhou and Hao (2008) further proposed a DIF formula based on the existing SHPB test results of rock materials as

$$\mathrm{DIF}=\begin{cases}0.0225\log\dot{\varepsilon}+1.12 & \text{for } \dot{\varepsilon}\le 10.0\mathrm{s}^{-1}\\ 0.2713(\log\dot{\varepsilon})^2-0.3563\log\dot{\varepsilon}+1.2275 & \text{for } \dot{\varepsilon}>10.0\mathrm{s}^{-1}\end{cases} \tag{2.7}$$

The above five formulae are also plotted in Fig. 2.13, it indicates that, these formulae could not well describe the evolution of the DIF for UHPCC material and generally overestimate or underestimate the dynamic compressive strength.

Since there is no test data of UHPCC under quasi-dynamic loadings, it is difficult to determine the transition strain rate $\dot{\varepsilon}_{TR}$ exactly. As shown in Eq. (2.8), an empirical formula is proposed to describe the strain rate sensitivity of UHPCC, where the *fib* Model Code 2010 (2013) is adopted for the strain rates below the transition strain rate.

$$\mathrm{DIF}=\begin{cases}(\dot{\varepsilon}/\dot{\varepsilon}_{ts})^{0.014} & \text{for } \dot{\varepsilon}\le\dot{\varepsilon}_{TR}\\ A_f\cdot(\log\dot{\varepsilon})^2+B_f\cdot\log\dot{\varepsilon}+C_f & \text{for } \dot{\varepsilon}>\dot{\varepsilon}_{TR}\end{cases} \tag{2.8}$$

where A_f, B_f and C_f are the coefficients. By fitting the test data, the values of A_f, B_f, C_f and $\dot{\varepsilon}_{TR}$ for UHPCC are confirmed and given in Table 2.4, and the corresponding empirical DIF relations are shown in Fig. 2.14. These empirical formulae can be used to estimate the DIF in the numerical prediction of UHPCC structure responses under high strain rate loadings. It should be noted that, Eq. (2.8) is derived from test data

Table 2.4 Coefficients in proposed DIF formula for UHPCC

Test No.	A_f	B_f	C_f	$\dot{\varepsilon}_{TR}(\mathrm{s}^{-1})$
N-0	0.3714	−0.5622	1.2471	28
S-1	0.5835	−1.5905	2.1988	63
S-2	0.5103	−1.2301	1.6804	92
H-1	0.5065	−0.7046	0.7684	75
H-2	0.8152	−2.3884	2.8389	75

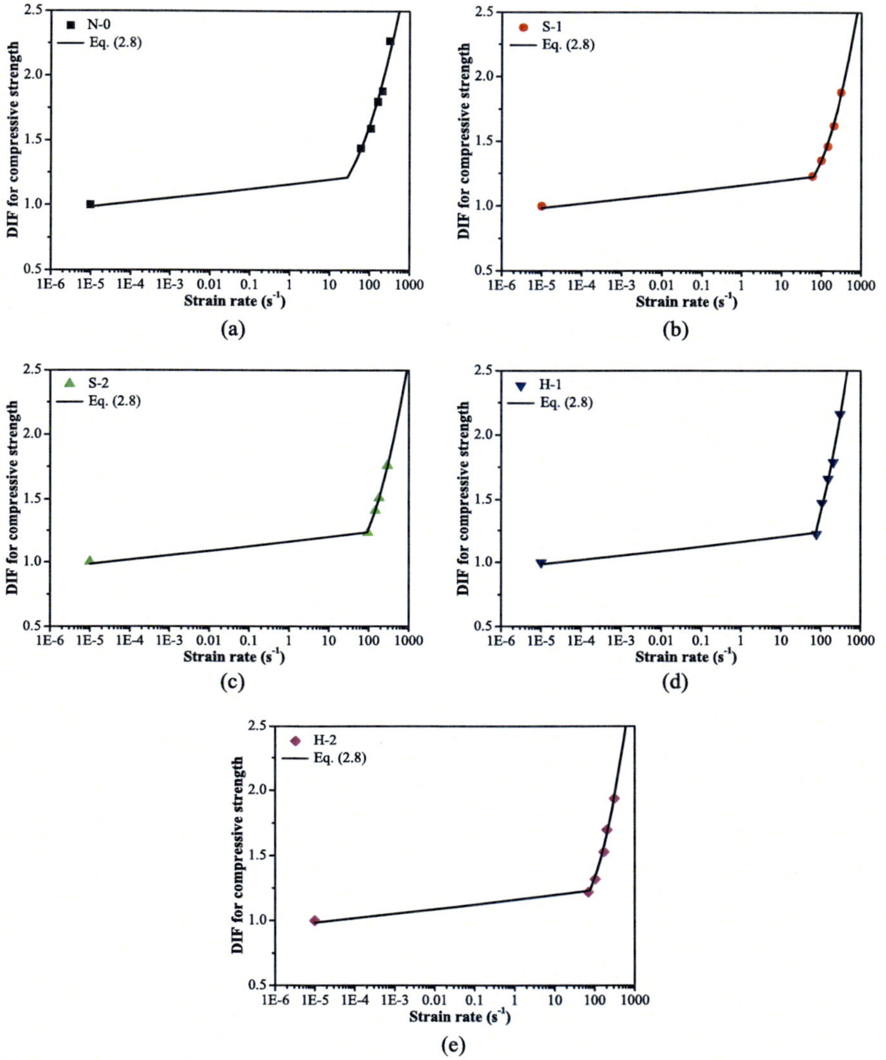

Fig. 2.14 Empirical DIF relations for UHPCC **a** N-0 **b** S-1 **c** S-2 **d** H-1 **e** H-2, Ren et al. (2018), copyright 2020, with permission from Elsevier

for the strain rates up to ~328.4 s^{-1}, beyond which there is no test data to substantiate the validity of the proposed formula.

2.4.6 Energy Absorption Capacity

The energy absorption capacity is always used to evaluate the dynamic mechanics performance of a material and usually expressed as toughness, which can be determined by calculating the area under the stress–strain curves according to $W = \int \sigma d\varepsilon$, where W denotes the toughness of a material, σ and ε are the stress and strain, respectively.

Based on the dynamic stress–strain curves of UHPCC given in Fig. 2.11, Table 2.5 lists the corresponding dynamic toughness of UHPCC under various strain rates. W_p and W_u represent the peak and ultimate dynamic toughness when the strain reaches the peak and ultimate value, respectively. As shown in Fig. 2.11, the stress is at a relative low level when the strain reaches 0.05. Thus, the ultimate strain is determined as 0.05. "—" denotes that the dynamic toughness is not given since the stress–strain curves are hysteretic and the peak stresses are not the actual failure stresses.

Figure 2.15 illustrates the comparison of the dynamic toughness of UHPCC with varying steel fiber reinforcement at different strain rates. It indicates that, (i) both the peak and ultimate dynamic toughness gradually increase with the increase of strain rate; (ii) the steel fiber content and type almost have no effects on the peak dynamic toughness of UHPCC; (iii) the ultimate dynamic toughness of UHPCC increases with increasing the volumetric ratio of steel fibers, either with the micro-straight or hooked steel fibers, and the influence of the former is more obvious. For example, at the strain rate around 150 s^{-1}, the ultimate dynamic toughness of UHPCC with mixing 2.0% micro-straight steel fibers (33590 J/m^3) is 20.7% and 12.4% larger than those of UHPCC with 1.0% micro-straight steel fibers (27820 J/m^3) and 2.0% hooked steel fibers (29890 J/m^3), respectively. The reason may lie in that, the matrix will be cracking firstly under the dynamic loadings and the steel fibers in the specimen resist the formation and propagation of cracks. Increasing steel fiber content could make more steel fibers to delay the crack propagation, which implies that more energy are

Table 2.5 Dynamic toughness of UHPCC

Test No.	W_p (J/m^3)	W_u (J/m^3)	Test No.	W_p (J/m^3)	W_u (J/m^3)	Test No.	W_p (J/m^3)	W_u (J/m^3)
N-0-1	–	–	N-0-2	4810	14,040	N-0-3	5620	17,610
N-0-4	6990	24,230	N-0-5	8270	27,200	N-0-6	12,250	31,980
S-1-1	–	–	S-1-2	5440	14,430	S-1-3	6740	21,530
S-1-4	8220	27,820	S-1-5	9400	31,540	S-1-6	12,220	34,980
S-2-1	–	–	S-2-2	–	–	S-2-3	8730	24,170
S-2-4	7880	33,590	S-2-5	9320	36,070	S-2-6	12,930	40,360
H-1-1	–	–	H-1-2	4860	16,200	H-1-3	6170	20,700
H-1-4	7400	26,320	H-1-5	9570	30,040	H-1-6	12,980	34,100
H-2-1	–	–	H-2-2	5330	12,890	H-2-3	5540	20,940
H-2-4	7680	29,890	H-2-5	8880	35,930	H-2-6	12,860	37,890

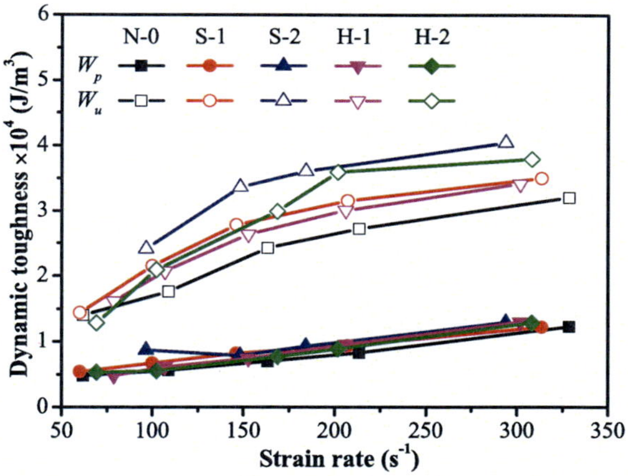

Fig. 2.15 Comparison of dynamic toughness at different strain rates, Ren et al. (2018), copyright 2020, with permission from Elsevier

consumed in the process of cracking resistance, thus the ultimate dynamic toughness of UHPCC is improved. Also, with the identical steel fiber content, due to the different size of two typical steel fibers, relatively larger amount of micro-straight steel fibers work to resist the cracks propagating.

2.5 Visco-Elastic Damage Model

2.5.1 Nonlinear Visco-Elastic Model

The dynamic stress–strain curves of UHPCC obtained from the SHPB test present a nonlinear visco-elastic behavior, and previous studies have reported that a nonlinear visco-elastic model (ZWT model) proposed by Zhu, Wang and Tang can well simulate the above behavior of concrete under dynamic loadings (Lai and Sun 2009; Zhang et al. 2016; Li et al. 2016). As shown in Fig. 2.16, the ZWT model is composed of a nonlinear elastic spring, a low frequency Maxwell element and a high frequency Maxwell element, which can be expressed as (Chu et al. 1985; Wang et al. 1995)

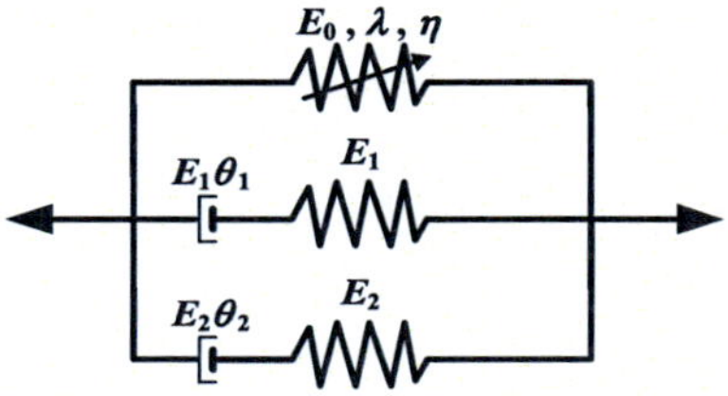

Fig. 2.16 Nonlinear visco-elastic model, Ren et al. (2018), copyright 2020, with permission from Elsevier

$$\sigma = f_e(\varepsilon) + E_1 \int_0^t \dot{\varepsilon} \exp\left(-\frac{t-\tau}{\theta_1}\right) d\tau + E_2 \int_0^t \dot{\varepsilon} \exp\left(-\frac{t-\tau}{\theta_2}\right) d\tau \tag{2.9}$$

where $f_e(\varepsilon) = E_0\varepsilon + \lambda\varepsilon^2 + \eta\varepsilon^3$ describes the nonlinear elastic behavior of the material. E_0, λ, and η are the elastic constants. t and τ denote the loading time and time variable, respectively. E_1 and θ_1 are the elastic modulus and relaxation time of the low frequency Maxwell element, which describes the material visco-elastic response at a low strain rate. E_2 and θ_2 are those of the high frequency Maxwell element, which describes the material visco-elastic response at a high strain rate.

For the dynamic loadings, the loading time is from 1 μs to 102 μs and the relaxation time (θ_1) is from 10 to 10^2 s, thus the low frequency Maxwell element is no enough time to relax (Wang 2005). When the strain of the material is very low, the elastic element is approximately linear, i.e., $f_e(\varepsilon) = E_0\varepsilon$. Therefore, the ZWT model in Eq. (2.9) can be simplified as

$$\sigma = E'\varepsilon + E_2 \int_0^t \dot{\varepsilon} \exp\left(-\frac{t-\tau}{\theta_2}\right) d\tau \tag{2.10}$$

where $E' = E_0 + E_1$.

Zhang et al. (2016) further modified the original ZWT model by considering the material damage as

$$\sigma_{\mathrm{M}} = \sigma(1 - D) + Dk \tag{2.11}$$

where σ_{M} is the modified stress, k is the bearing capacity of the steel fibers in the specimen damage zone. D is the damage factor and can be expressed as

$$D = \begin{cases} 0 & \varepsilon \le \varepsilon_{th} \\ 1 - e^{-(\varepsilon-\varepsilon_{th})^{m_d}/a_d} & \varepsilon > \varepsilon_{th} \end{cases} \tag{2.12}$$

where m_d and a_d are the damage evolution coefficients, which are determined by the cement strength, steel fiber content as well as the loading strain rate. ε_{th} is the ultimate elastic strain, which denotes the threshold strain for the damage evolution of material. $\varepsilon_{th} = (0.7 \sim 0.9)\varepsilon_u$ is suggested by Zhang et al. (2016), where ε_u is the peak strain. At present, $\varepsilon_{th} = 0.7\varepsilon_u$ is adopted and the modified ZWT model is expressed as

$$\sigma_{\mathrm{M}} = \begin{cases} E'\varepsilon + E_2\theta_2\dot{\varepsilon}\left[1 - e^{-\varepsilon/(\theta_2\dot{\varepsilon})}\right] & \varepsilon \le \varepsilon_{th} \\ e^{-(\varepsilon-\varepsilon_{th})^{m_d}/a_d}\left\{E'\varepsilon + E_2\theta_2\dot{\varepsilon}\left[1 - e^{-\varepsilon/(\theta_2\dot{\varepsilon})}\right]\right\} + \left[1 - e^{-(\varepsilon-\varepsilon_{th})^{m_d}/a_d}\right]k & \varepsilon > \varepsilon_{th} \end{cases} \tag{2.13}$$

Table 2.6 Values of the fitted parameters

Test No.	Strain rate (s^{-1})	$\varepsilon_{th}(\times 10^{-3})$	E'(GPa)	E_2 (GPa)	$\theta_2(\mu s)$	m_d	a_d ($\times 10^{-3}$)	k
N-0–5	213.4	4.2	12	48	30	1.35	1.0	15.2
S-1–5	207.1	4.4	13	50	30	1.28	1.5	15.4
S-2–5	184.2	4.3	14.5	51	30	1.24	2.2	23.4
H-1–5	206.0	4.5	12.5	49	30	1.34	1.0	24.0
H-2–5	201.8	4.3	13.2	51	30	1.33	1.1	35.9

To summarize, the modified ZWT model has seven parameters (ε_{th}, E', E_2, θ_2, m_d, a_d and k) to describe the dynamic mechanical properties of UHPCC.

2.5.2 *Model Calibration and Validation*

In this section, by choosing the five experimental stress–strain curves of UHPCC with different steel fiber reinforcement at the strain rate of ~ 202.5 s^{-1}, the modified ZWT model is calibrated and validated. Table 2.6 lists the corresponding fitted parameters of the modified ZWT model. As can be seen, the values of E' and E_2 gradually increase as the steel fiber content increases, while the relaxation time θ_2 keeps unchanged and is not sensitive to the steel fiber content. The parameters m decreases and a increases with increasing the volumetric ratio of steel fibers, which indicates that the damage factor D reduces as the steel fiber content increases, as illustrated in Eq. (2.12). In addition, the values of k increase with increasing the volume fraction of steel fibers, which means that the bearing capacity of the steel fibers is gradually improved.

Figure 2.17 illustrates the comparison of the experimental and the corresponding predicted stress–strain curves by Eq. (2.13). Good agreements are obtained and the validation of the modified ZWT model in describing the dynamic response of UHPCC is verified.

2.6 Summary

Investigations of the mechanical behaviors of UHPCC are important and essential to provide valuable information for the structural design and calibration/validation of the constitutive model. In this chapter, the dynamic compressive behavior of UHPCC with three volume fractions (0, 1.0 and 2.0%) and two types (micro-straight and hooked) of steel fibers is experimentally investigated by using a 50 mm diameter SHPB device. The effects of strain rate, steel fiber content and type on the dynamic

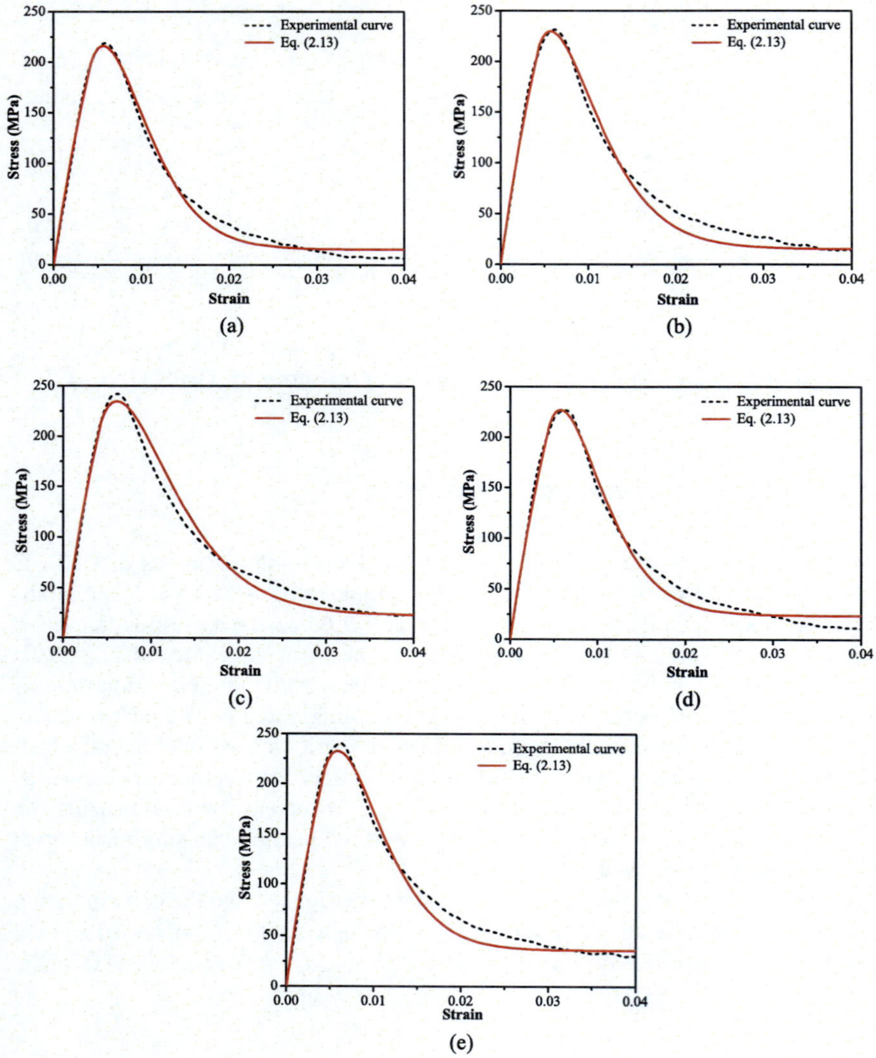

Fig. 2.17 Comparison of the experimental and predicted stress–strain curves of UHPCC **a** N-0-5 **b** S-1-5 **c** S-2-5 **d** H-1-5 **e** H-2-5, Ren et al. (2018), copyright 2020, with permission from Elsevier

stress–strain relationship and the toughness of UHPCC are studies. The following conclusions can be drawn:

(1) UHPCC is a strain rate dependent material, the dynamic elastic modulus, compressive strength and peak strain as well as the toughness increase obviously as the strain rate increases.

(2) Adding steel fibers has little effect on the dynamic elastic modulus, peak strain and peak toughness, and slightly improves the dynamic compressive strength of UHPCC. The micro-straight steel fiber has relatively better effect on improving the dynamic compressive strength. The ultimate toughness gradually increases with the volumetric ratio of steel fiber increasing, and the micro-straight steel fiber is more influential on that.

(3) The DIFs for UHPCC material decrease with the increase of steel fiber content for both typical steel fibers, and the reducing effect of micro-straight steel fiber is more obvious. Furthermore, the empirical DIF formulae for UHPCC with various steel fiber reinforcement are proposed.

(4) A modified visco-elastic damage model is established under dynamic loadings, which is calibrated and validated for UHPCC material.

References

BISCHOFF P H, PERRY S H. Compressive behavior of concrete at high strain rates[J]. Materials and Structures, 1991, 24: 425-450.

CHU C H, WANG L L, XU D B. A nonlinear thermo-viscoelastic constitutive equation for thermoset plastics at high strain rates, in: W.Z. Chien (Ed.), Proceedings of the international conference on nonlinear mechanics, Shanghai, China[C]. 1985: 92–97.

Comite Euro-International du Beton. Fib Model Code for Concrete Structures 2010[M]. Trowbridge, Wiltshire, UK: Redwood Books, 2013.

GROTE D L, PARK S W, ZHOU M. Dynamic behavior of concrete at high strain rates and pressures: I. experimental characterization[J]. International Journal of Impact Engineering, 2001, 25: 869–886.

HAO H, HAO Y F, LI J, CHEN W S. Review of the current practices in blast-resistant analysis and design of concrete structures[J]. Advances In Structural Engineering, 2016, 19(8): 1193-1223.

HAO Y, HAO H. Dynamic compressive behaviour of spiral steel fibre reinforced concrete in split Hopkinson pressure bar tests[J]. Construction and Building Materials, 2013, 48: 521-532.

LAI J Z, SUN W. Dynamic behaviour and visco-elastic damage model of ultra-high performance cementitious composite[J]. Cement and Concrete Research, 2009, 39: 1044-1051.

LINDHOLM U S. Some experiments with the split Hopkinson pressure bar[J]. Journal of the Mechanics and Physics of Solids, 1964, 12: 317-335.

LI Q H, ZHAO X, XU S L, GAO X. Influence of steel fiber on dynamic compressive behavior of hybrid fiber ultra high toughness cementitious composites at different strain rates[J]. Construction and Building Materials, 2016, 125: 490-500.

LI Q M, MENG H. About the dynamic strength enhancement of concrete-like materials in a split Hopkinson pressure bar test[J]. International Journal of Solids and Structures, 2003, 40: 343-360.

LU Y B, LI Q M. Appraisal of pulse-shaping technique in split Hopkinson pressure bar tests for brittle materials[J]. International Journal of Protective Structures, 2010, 1(3): 363-390.

MINDESS S, YOUNG J F, DARWIN D. Concrete[M]. New Jersey: Pearson Education Inc., 2003.

REN G M, WU H, FANG Q, LIU J Z. Effects of steel fiber content and type on dynamic compressive mechanical properties of UHPCC[J]. Construction and Building Materials, 2018, 164: 29-43.

ROSS C A, JEROME D M, TEDESCO J W, Hughes M L. Moisture and strain rate effects on concrete strength[J]. ACI Materials Journal, 1996, 93(3): 293-300.

ROSS C A, TEDESCO J W, KUENNEN S T. Effects of strain rate on concrete strength[J]. ACI Materials Journal, 1995, 92(1): 37-47.

ROSS C A, THOMPSON P Y, TEDESCO J W. Split-Hopkinson pressure-bar tests on concrete and mortar in tension and compression[J]. ACI Materials Journal, 1989, 86: 475-481.

SU Y, LI J, WU C Q, WU P T, LI Z X. Effects of steel fibres on dynamic strength of UHPC[J]. Construction and Building Materials, 2016, 114: 708-718.

TEDESCO J W, ROSS C A. Strain-rate-dependent constitutive equations for concrete[J]. Journal of Pressure Vessel Technology, 1998, 120(4): 398-405.

WANG L L. Foundation of stress waves[M]. Beijing: National defense industry press, 2005. (in Chinese)

WANG L L, HUANG D J, GAN S. Nonlinear viscoelastic constitutive relations and nonlinear viscoelastic wave propagation for polymers at high strain rates, in constitutive relation in high/very high strain rates, in: K. Kawata, J. Shioiri (Eds.), IUTAM Symposium, Noda, Japan[C].1995: 137–146.

WANG Z L, LIU Y S, SHEN R. Stress-strain relationship of steel fiber-reinforced concrete under dynamic compression[J]. Construction and Building Materials, 2008, 22: 811-819.

ZHOU X Q, HAO H. Modelling of compressive behaviour of concrete-like materials at high strain rate[J]. International Journal of Solids and Structures, 2008, 45: 4648-4661.

ZHANG H, LIU Y, SUN H, WU S F. Transient dynamic behavior of polypropylene fiber reinforced mortar under compressive impact loading[J]. Construction and Building Materials, 2016, 111: 30-42.

Chapter 3
Dynamic Tensile Mechanical Properties of UHPCC

3.1 Introduction

Spallation is a specific kind of fracture in solids, which is developed by tensile stress pulses produced by the reflection of compression waves at the interfaces adjacent to low impedance media. For example, in case of the projectile impacting on a target, the contact detonation of high explosives on the target or the sudden deposition of an intense pulse of energy on the target surface, the intensive compressive waves are generated and propagate through the structures. These compressive waves are reflected back at the free surface of structures and turn to the tensile waves, leading to the fracture of the concrete in the form of spallation due to its low tensile strength (Goldsmith et al. 1996; Watson and Sanderson 1979; Malvar and Crawford 1998). The study of dynamic tensile (spalling) behavior of UHPCC can provide helpful reference for exploring the dynamic mechanical properties and establishing the constitutive model of UHPCC.

In the existing studies about the dynamic tensile properties of concrete materials, Brara et al. (Brara et al. 2001; Klepaczko and Brara 2001) developed an efficient experimental technique to investigate the dynamic tensile behavior of concrete (45 MPa in compression) under the strain rates of 10 ~ 120 s^{-1} and found the high rate sensitivity of concrete strength in tension at high strain rates. Furthermore, a local damage cumulative criterion which incorporated the loading rate sensitivity was proposed and implemented in numerical analyses. By examining the strain wave profiles in a polymer buffer bar behind the slender concrete bar specimen, Zhang et al. (2009) proposed a new method to measure the dynamic spalling strength of C30, C60 and C80 concrete. The experimental results showed that the dynamic spalling strength is dependent on both the compressive strength of concrete and the strain rate. Wang et al. (2012) evaluated the influence of steel fiber content (0, 0.75 and 1.5% by volume) on the dynamic spalling strength of steel fiber reinforced normal strength concrete (SFRC) with the compressive strength of 62.7 ~ 68.6 MPa at medium strain rates (48.7–72.0 s^{-1}). It was concluded that SFRC is strain rate dependent, and the volume fraction of fibers exerted a significant influence on the dynamic spalling

Q. Fang et al., *UHPCC Under Impact and Blast*,
https://doi.org/10.1007/978-981-33-6842-2_3

strength. Rong and Sun (2012) studied the effect of steel fiber content (0, 3.0 and 4.0% by volume) on the dynamic tensile properties of UHPCC at different strain rates (21 ~ 66 s^{-1}) by conducting the dynamic spalling test and found that both the fiber volume fraction and strain rate exerted significant enhancing effect on the dynamic spalling strength. Thaomas and Sorensen (2017) collected the thirteen previously published studies about the strain rate sensitivity of Ultra-high performance concrete (UHPC) in tension. The comparisons were conducted between the reported dynamic increase factors (DIFs) for the dynamic tensile strength, first cracking tensile strength, modulus of elasticity, modulus of rupture, strain capacity and energy absorption of UHPC with the existing DIF formulae for conventional concrete or UHPC. The results showed that, (i) UHPC is much more strain rate-sensitive than the conventional concrete at high strain rate; (ii) the strain rate sensitivity of UHPC does not depend on the fiber volume fraction, fiber type, fiber orientation and compressive strength of the matrix; (iii) the existing DIF formulae could not well estimate the DIFs for UHPC and an improved DIF formulae are further proposed.

Generally, the existing experimental assessment for the steel fiber content and type on the dynamic tensile behavior of UHPCC is far from sufficient. There is no systematical assessment for the effects of steel fiber content and type on the dynamic spalling strength of UHPCC under a wider range of strain rate, and the corresponding predictions for the dynamic tensile strength of UHPCC are practically needed. At present, the dynamic spalling test on Φ48mm × 400 mm cylindrical UHPCC specimen was performed by using a 50 mm-diameter SHPB device at the strain rate ranging from 14.3 s^{-1} to 214.8 s^{-1}, in which two typical steel fibers (micro-straight and hooked) and three volume fractions (0%, 1.0% and 2.0%) were considered. The dynamic spalling strengths under different strain rates were derived and the effects of steel fiber content and type as well as the strain rate were experimentally assessed.

3.2 Specimen Preparation

For the present two steel fibers types and three volume fractions, 40 cylindrical specimens were cast for the dynamic spalling test. As shown in Fig. 3.1, the specimen size of Φ48mm × 400 mm was selected by considering the diameter of the SHPB device (50 mm) and the maximal fiber length (25 mm), which guaranteed the random

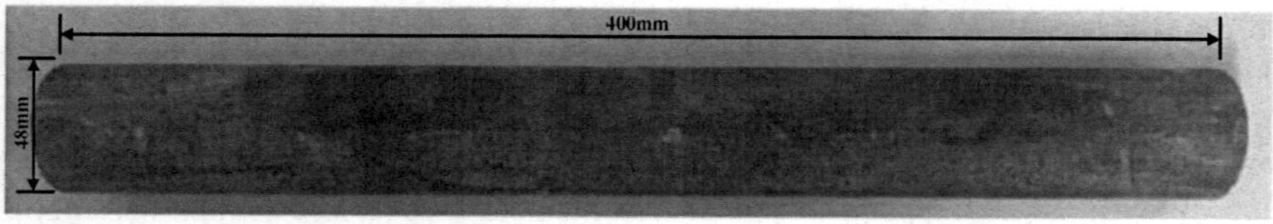

Fig. 3.1 Spalling test specimen, reprinted from Wu et al. (2018), copyright 2020, with permission from Elsevier

Table 3.1 Quasi-static cubic compressive strengths of UHPCC

Test no.	N-0	S-1	S-2	H-1	H-2
Quasi-static cubic compressive strength (MPa)	109	136	152	119	138

fiber orientation and distribution. In order to make the specimen-bar contact as perfect as possible, the end surfaces of all specimens were carefully polished, which were flatted to less than 0.1 mm, and the roughness on the surfaces was less than 0.05 mm. Besides, 15 cubic control specimens for each batch with the size of 100 × 100 × 100 mm were cast and tested, and the average quasi-static compressive strengths are listed in Table 3.1. In which the capitals "N", "S" and "H" represent no mixing fiber, micro-straight steel fiber and hooked steel fiber, and the numbers "0", "1" and "2" denote that the volume fractions of steel fibers are 0%, 1.0% and 2.0%, respectively. The similar designation way is adopted in the following spalling test.

3.3 Dynamic Spalling Test

3.3.1 Test Device

As shown in Fig. 3.2, the dynamic spalling test on UHPCC specimens was performed on the 50 mm-diameter conic variable cross-sectional Split Hopkinson Pressure Bar (SHPB), and the lengths of striker bar and incident bar were 70 and 2400 mm. The smaller cross-sectional diameter of the incident bar was 37 mm, which was coincident with the diameter of the striker bar. The gimbal device with the size of Φ50 × 55 mm was used to guarantee good contact between the incident bar and specimens, as shown in Fig. 3.3. In order to separate the compressive and tensile waves, the strain gauge should be positioned close to the incident end and adjusted according to the wave-length to ensure that the incident, transmit and reflected waves from the free end of specimen can be recorded separately. Four strain gauges were used in the present study, as illustrated in Fig. 3.2. The strain gauge 1 attached to the incident bar was used to provide the trigger signals for the data acquisition system,

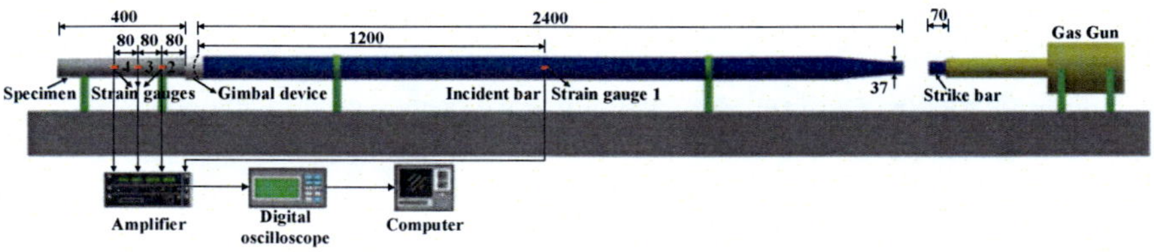

Fig. 3.2 Schematic diagram of SHPB setup for dynamic spalling test (unit: mm), reprinted from Wu et al. (2018), copyright 2020, with permission from Elsevier

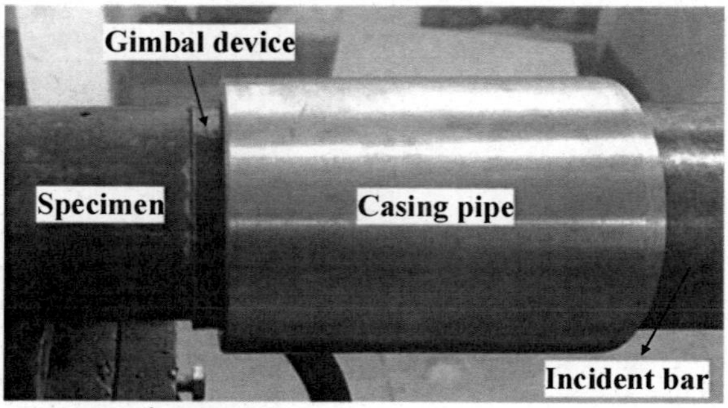

Fig. 3.3 Gimbal device, reprinted from Wu et al. (2018), copyright 2020, with permission from Elsevier

and the strain gauges 2 ~ 4 were attached onto the specimen to record the whole process of wave propagation. The distances from the strain gauges 2 ~ 4 to the incident end of specimen are 80 mm, 160 mm and 240 mm, respectively.

3.3.2 Test Technique

The technique of spalling test combines the principle of one-dimensional wave propagation in the Hopkinson bar and the phenomenon of spalling. In the present test, the striker bar is driven by air pressure and arrives at the velocity V_0 of 4.1 ~ 8.5 m/s. The impact of striker bar on the incident bar generates a compressive longitudinal incident wave, which then propagates towards the specimen. When the incident compressive wave reaches the incident bar/specimen interface, a small part of the incident compressive wave is reflected back to the incident bar due to the impedance mismatch between the incident bar and the specimen. The major part of the incident compressive wave transmits into the specimen and propagates towards to the free end, and it is then reflected at the free end of the specimen as a tensile wave. The superposition of the incident compressive wave and the reflected tensile wave generates a tensile stress, which grows rapidly along with the specimen. When the tensile stress reaches the critical value (dynamic tensile strength), the spalling fracture of the specimen occurs at a certain distance from the free end.

The cylindrical specimens in the spalling test can be regarded as the slender bars, based on the theory of one-dimensional wave propagation in an elastic bar, the one-dimensional wave equation can be formulated as (Díaz-Rubio et al. 2002)

$$\frac{\partial^2 u}{\partial t^2} - c^2 \frac{\partial^2 u}{\partial x^2} = 0 \tag{3.1}$$

where u, t and x are the particle displacement, time and the longitudinal coordinate. c is the elastic wave velocity of the specimen. By integrating Eq. (3.1), it derives

$$u(x,t) = f(x - ct) + g(x + ct) \tag{3.2}$$

where $f(x - ct)$ and $g(x + ct)$ represent the stress waves propagating in the positive and negative x-axis direction, respectively. Based on Eq. (3.2), the strain of specimen $\varepsilon(x,t)$ and the velocity of particle $v(x,t)$ can be obtained directly by the following expressions

$$\varepsilon(x,t) = \partial u(x,t)/\partial x = f_x'(x - ct) + g_x'(x + ct) \tag{3.3}$$

$$v(x,t) = \partial u(x,t)/\partial t = -cf_t'(x - ct) + cg_t'(x + ct) \tag{3.4}$$

where the subscript x and t are the derivation of coordinate and time. If linear elastic behavior is assumed, the stress can be calculated as follows:

$$\sigma(x,t) = E\varepsilon(x,t) = E\left[f_x'(x - ct) + g_x'(x + ct)\right] \tag{3.5}$$

By assuming an elastic compressive wave travelling along the positive x-axis direction of a semi-infinite bar, the function $f(x - ct)$ is the compressive wave and the function $g(x + ct)$ is zero. When this wave reaches the free end of the bar, the boundary condition is that the stress at the free end should be zero: $\sigma(x, t) = 0$. This means that the following relationship is satisfied at the boundary.

$$f_x'(x - ct) = -g_x'(x + ct) \tag{3.6}$$

Meanwhile, a reflected wave appears which has the same shape as the incident wave but has opposite sign (Fig. 3.4). During the reflection, the initial compressive wave is being reflected as a tensile pulse, thus the stress in the specimen is the

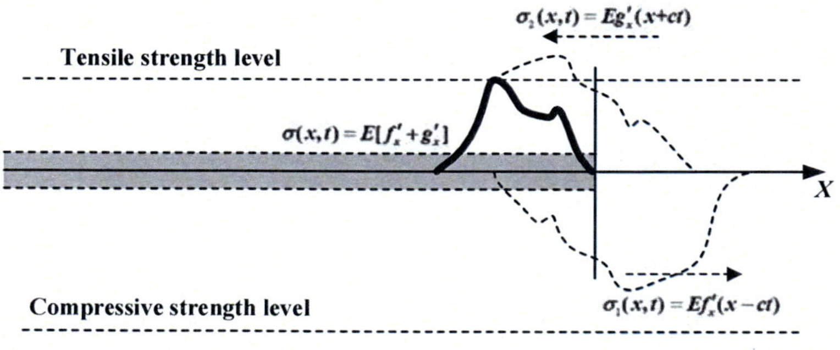

Fig. 3.4 Reflection of a wave at the free end (Díaz-Rubio et al. 2002)

superposition of the initial incident compressive wave and the reflected tensile wave. If the initial compressive wave is lower than the compressive strength but its absolute value exceeds the dynamic tensile strength of material, the specimen will fail during the reflection process. In actual experiments, the tensile stresses produced in the reflection will not emerge abruptly in the specimen. The compressive wave will decrease gradually while the tensile stress is growing up to the dynamic tensile strength of material. When this critical value is reached, the material is fractured.

In this study, the strain gauges 2 ~ 4 were used to measure the strains produced by the stress waves, as shown in Fig. 3.2. The satisfactory homogeneity of the specimens in the test resulted in the nearly consistent signals measured by the strain gauges. Without observing obvious stress-wave attenuation, we used the average signal values measured by the three gauges as the incident compressive waves in the specimens (Fig. 3.5). Based on the one-dimensional stress wave theory, the tensile stress waves propagating in the specimen at different time instants (1.0 μs time interval) can be derived by superposing the incident compressive wave and the reflected tensile wave at the free end of the specimen (Fig. 3.6). Furthermore, by connecting the peak points of the above stress waves, a straight line is drawn and intersects with the vertical line through the spalling location (first fracture point), and the intersection is the dynamic spalling strength σ_d.

The strain rate of the dynamic spalling process $\dot{\varepsilon}$ is described as (Rong and Sun 2012)

$$\dot{\varepsilon} = \frac{\sigma_d}{E_S \tau_t} \tag{3.7}$$

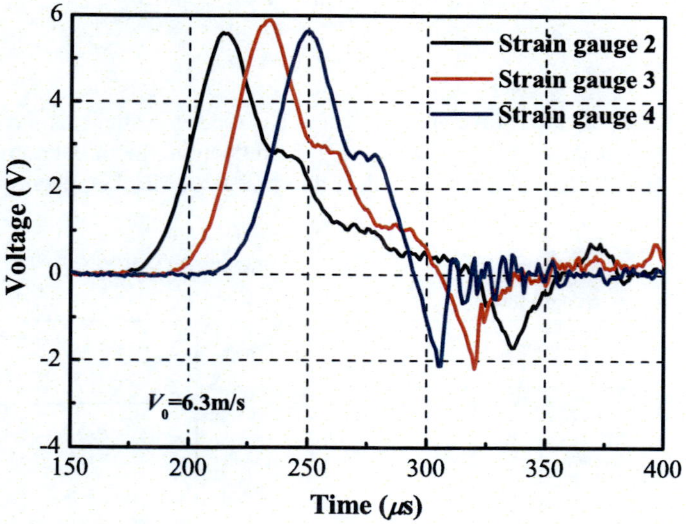

Fig. 3.5 Typical original pulse waves in specimen, reprinted from Wu et al. (2018), copyright 2020, with permission from Elsevier

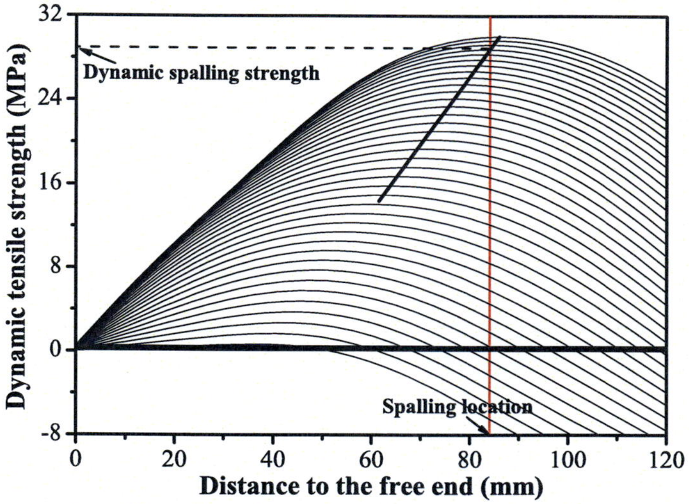

Fig. 3.6 Determination of the dynamic spalling strength (Rong and Sun 2012)

where τ_t is the time required from the tensile stress of the specimen reaching the rupture, which is obtained by multiplying the number of stress waves (the peak stress of which is lower than the dynamic spalling strength) and the time interval 1.0 μs, as shown in Fig. 3.6. E_S is the approximate tensile elastic modulus. Hao et al. (2016) has formulated the dynamic compressive/tensile elastic modulus of concrete, which is dependent on the strain rate and could not directly applied at present. In Ref. (Ren et al. 2018), we have experimentally studied the dynamic compressive mechanical behavior of UHPCC, by assuming the tension–compression symmetry of UHPCC, the E_S of the N-0, S-1, S-2, H-1 and H-2 UHPCC specimens are adopted as 46.0 GPa, 47.3 GPa and 48.1 GPa, 46.4 GPa and 48.2 GPa, respectively.

3.4 Test Results and Discussions

3.4.1 Dynamic Failure Patterns

The typical dynamic failure patterns of UHPCC specimens with different steel fiber contents and types are shown in Fig. 3.7. In which the white numbers represent the distance "*l*" between the first fractured section and the free end of specimens, and the corresponding tensile strain rates are given under the photos. Besides, the tiny cracks are highlightened by the solid red lines in Fig. 3.7(b–e).

It can be observed that the fractured sections of all specimens are relatively flat and perpendicular to the longitudinal axes of specimens, and the damage degree of UHPCC specimens without mixing fibers (N-0) is more severe than that of steel

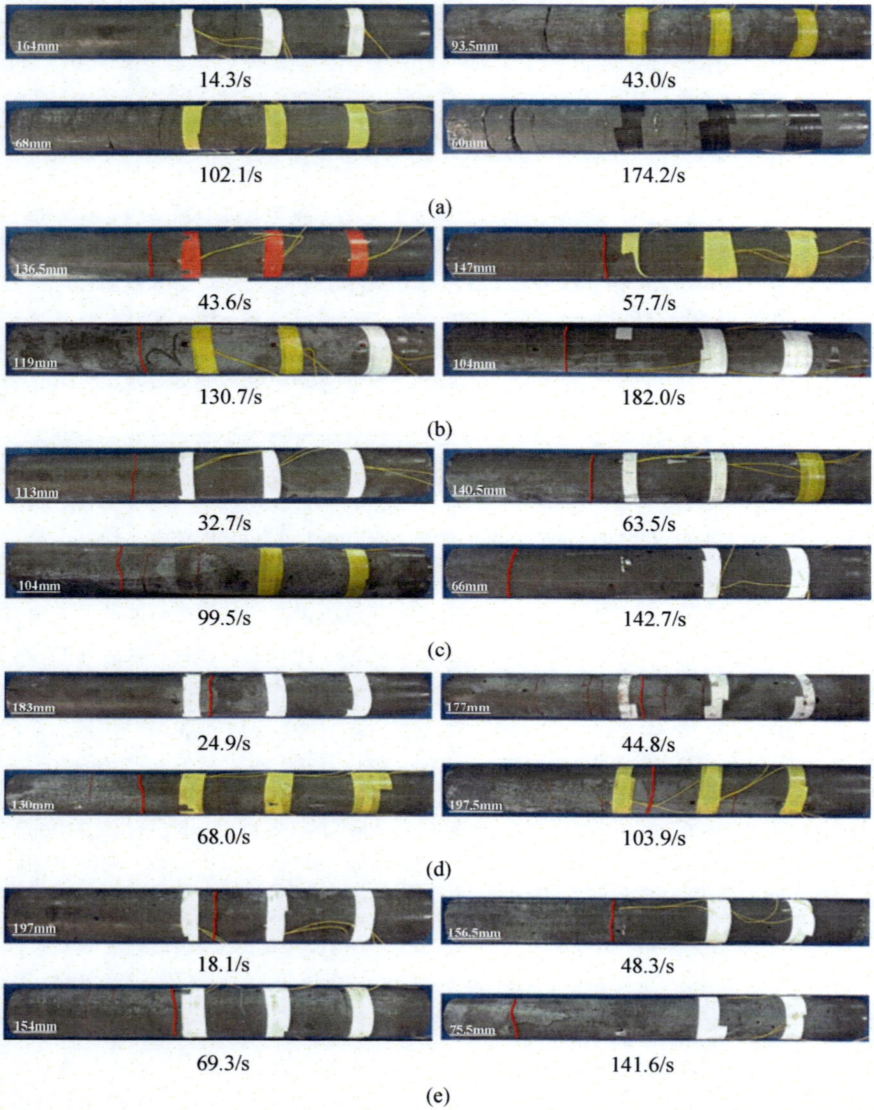

Fig. 3.7 Failure patterns of UHPCC specimens **a** N-0, **b** S-1, **c** S-2, **d** H-1, **e** H-2, reprinted from Wu et al. (2018), copyright 2020, with permission from Elsevier

fiber reinforced ones under the similar strain rates. With increasing the strain rate, the UHPCC matrix is fractured into two or more parts, while those reinforced with micro-straight or hooked steel fibers only have some small cracks. The UHPCC specimens with 1.0% or 2.0% steel fibers basically maintain the integrity due to the reinforcement of steel fibers. As well known, the matrix of UHPCC is a kind of brittle material. After the tensile stress exceeds its dynamic tensile strength, the formation and propagation of microcracks in brittle materials finally causes the concentration of these microcracks into a narrow zone, inducing the formation of a visible macrocrack which makes the fissure of specimens. Addition of steel fibers can effectively prevent the formation and propagation of cracks.

3.4.2 Dynamic Spalling Strength

Figure 3.8 shows the typical examples to determine the dynamic spalling strengths of UHPCC with different steel fiber contents and types, and the related test data are listed in Table 3.2. Furthermore, the influences of strain rate, steel fiber content and type on the dynamic spalling strength of UHPCC are shown in Fig. 3.9, respectively.

It can be seen that,

(i) As shown in Fig. 3.9, the dynamic spalling strength of UHPCC increases from 18.7 MPa to 112.1 MPa when the strain rate increases from 14.3 s^{-1} to about 110 s^{-1}. Its significant sensitivity to the strain rate may result from the insufficient time for the interior cracks to develop at high strain rates and under the impact loadings. However, when the input compressive stress to specimen exceeds the threshold required to trigger the compressive damage, the evolution and cumulation of compressive damage in the specimens result in the dynamic spalling strength decreases at the strain rate larger than 110 s^{-1}.

(ii) The dynamic spalling strengths of UHPCC reinforced with steel fibers are larger than those of the plain UHPCC at the strain rate higher than 60 s^{-1}, as shown in Fig. 3.9. Adding steel fibers could effectively improve the dynamic spalling strength. For example, the dynamic spalling strength of UHPCC with mixing 2.0% hooked steel fibers (112.1 MPa) is nearly two times larger than that of the plain UHPCC (58.7 MPa) at the strain rate around 110 s^{-1}. The increase in the dynamic spalling strength may result from the fiber-bridging effect which plays an important role to delay the formation and propagation of the cracks under the dynamic tensile loadings.

(iii) With the 1.0% volume fraction of steel fiber, the micro-straight steel fibers have slightly better effect on improving the dynamic spalling strength of UHPCC than the hooked steel fibers at the strain rate higher than 60 s^{-1}.

Besides, it is unreasonable that the dynamic spalling strengths of S-2 specimens are smaller than S-1 specimens with the strain rate larger than 60 s^{-1}. The reason may be that the specimens with 2.0% micro-straight steel fibers fail to reach complete fracture due to larger fiber-bridging stress than the other types of specimens under

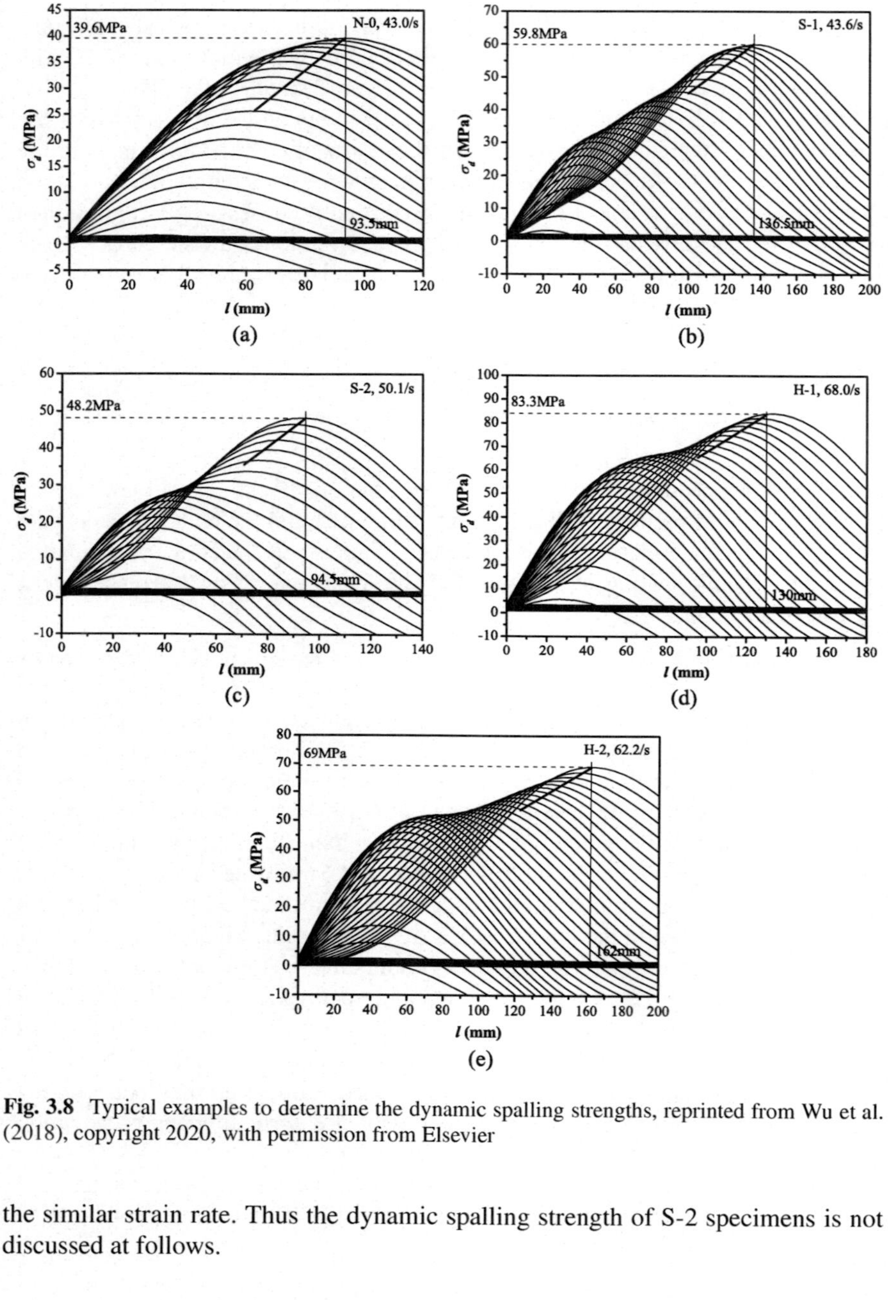

Fig. 3.8 Typical examples to determine the dynamic spalling strengths, reprinted from Wu et al. (2018), copyright 2020, with permission from Elsevier

the similar strain rate. Thus the dynamic spalling strength of S-2 specimens is not discussed at follows.

Table 3.2 Dynamic spalling test data

Test No.	E_S (GPa)	l (mm)	τ (μs)	$\dot{\varepsilon}$ (1/s)	σ_d (MPa)	l (mm)	τ (μs)	$\dot{\varepsilon}$ (1/s)	σ_d (MPa)
N-0	46.0	164	28.5	14.3	18.7	225.5	23.5	46.6	50.4
		214.5	29.5	29.2	39.6	114.5	24	55.5	61.3
		211	24	33.7	37.2	68	12.5	102.1	58.7
		93.5	20	43.0	39.6	60	9	174.2	72.1
S-1	47.3	170	27.5	35.7	46.5	188	26.5	82.7	103.7
		136.5	29	43.6	59.8	119	17	130.7	105.1
		103	21	48.7	48.4	104	12	182.0	103.3
		104.5	22	57.3	59.6	56.5	10	214.8	101.6
		147	26	57.7	71				
S-2	48.1	113	24	32.7	37.8	140.5	18.5	63.5	56.5
		130.5	29	35.6	49.6	104	20	99.5	95.7
		150.5	27	35.7	46.3	135.5	18.5	100.2	89.2
		94.5	20	50.1	48.2	66	13	142.7	89.2
H-1	46.4	183	30	24.9	34.7	130	26.5	68.0	83.3
		226	24.5	39.4	44.8	206.5	22	89.5	91
		177	31	44.8	64.5	197.5	20.5	103.9	98.4
H-2	48.2	197	42	18.1	36.7	154	23	69.3	76.8
		191	27.5	37.0	49.1	138.5	21	110.7	112.1
		156.5	33	48.3	76.8	113	15.5	141.5	105.7
		162	23	62.2	69	75.5	16	141.6	109.2

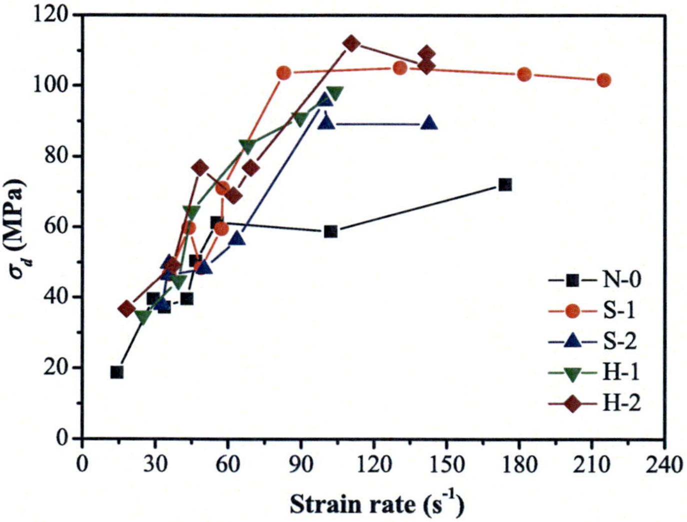

Fig. 3.9 Effects of strain rate, steel fiber content and type on dynamic spalling strength of UHPCC, reprinted from Wu et al. (2018), copyright 2020, with permission from Elsevier

3.4.3 Relations Between the Critical Time to Fracture and Dynamic Spalling Strength

Based on Table 3.2, Fig. 3.10 shows the correlation between dynamic spalling strengths σ_d and the critical time to fracture τ. As can be seen, the dynamic spalling strengths of UHPCC with mixing different steel fiber contents and types show close relation with the critical time to failure at different strain rates. Although the dispersion of points is relatively obvious, the dynamic spalling strengths generally change inversely proportional to the critical time and the dynamic spalling strength is reduced with increasing the critical time. It can be assumed that the asymptotic value of the dynamic spalling strength will approach the quasi-static tensile strength with further increasing the critical time.

As for metals and alloys, the effect of delayed spall was regarded as a fundamental characteristic by Tuler and Butcher (1968). While for concrete materials, the physical interpretation for the reason of increasing strength and delayed failure can be focused on the dynamics and interaction of cracks. During a short time of loading, the triggered micro or small cracks have no enough time to further propagate and the above intensive interaction of many cracks leads to the delayed cracking of specimen. Thus the higher dynamic spalling strength is reached.

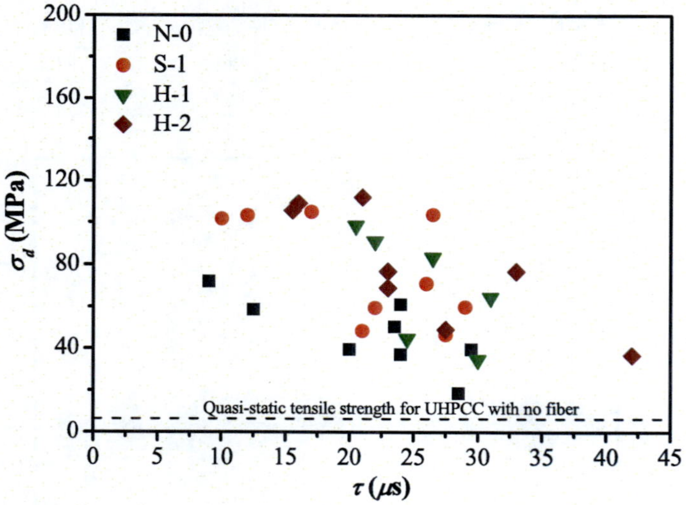

Fig. 3.10 Dependences of dynamic spalling strengths on the critical time to fracture, reprinted from Wu et al. (2018), copyright 2020, with permission from Elsevier

3.4.4 Dynamic Increase Factor

The dynamic increase factor (DIF) is defined as the ratio of the dynamic strength to the quasi-static strength in uniaxial tensile, which is widely used to analyze and design concrete protective structures under intensive loadings. Tran and Kim (2014), Su et al. (2016) and Millard et al. (2010) had experimentally investigated the dynamic tensile properties of UHPCC by using different test methods (direct tension, splitting tension and flexural tension). However, the corresponding strain rates were limited, i.e. 10 ~ 30 s^{-1}, 100 ~ 200 s^{-1} and 0.1 ~ 10 s^{-1}, and the empirical DIF formulae for UHPCC were not derived under a wide range of strain rate. At present, based on the spalling test, the DIFs for the dynamic tensile strength of UHPCC at the strain rate ranging from 14.3 s^{-1} to 214.8 s^{-1} are listed in Table 3.3 and illustrated in Fig. 3.11.

It is well accepted that there exists a transition strain rate for the relationship between the DIF and the logarithm of strain rate, beyond which the DIF increases rapidly with the strain rate and the material changes from low to high strain rate sensitivity. As shown in Fig. 3.11, the DIFs for dynamic tensile strength of UHPCC rapidly increase with rising the strain rate and the highest DIF of 14.89 is obtained for the UHPCC with 2.0% hooked steel fiber at the strain rate of 110.7 s^{-1}.

Besides, the experimental results indicate that, (i) UHPCC with different steel fiber contents and types has constantly higher DIF values than the plain UHPCC, suggesting that UHPCC with various steel fiber reinforcement is more sensitive to the strain rate than the plain UHPCC; (ii) for the hooked steel fibers, the values of DIF for UHPCC material increase with increasing the steel fiber content from 1 to 2%; (iii) with the mixing ratio of 1.0%, the DIFs for UHPCC with hooked steel fiber are slightly higher than those with micro-straight steel fibers within the experimental strain rate range.

Table 3.3 DIFs for dynamic tensile strength

Test No.	Strain rate (s^{-1})	DIF	Test No.	Strain rate (s^{-1})	DIF	Test No.	Strain rate (s^{-1})	DIF
N-0–1	14.3	2.88	N-0–2	29.2	6.09	N-0–3	33.7	5.72
N-0–4	43.0	6.09	N-0–5	46.6	7.75	N-0–6	55.5	9.43
N-0–7	102.1	9.03	N-0–8	174.2	11.09	—	—	—
S-1–1	35.7	5.57	S-1–2	43.6	7.16	S-1–3	48.7	5.80
S-1–4	57.3	7.14	S-1–5	57.7	8.50	S-1–6	82.7	12.42
S-1–7	130.7	12.59	S-1–8	182.0	12.37	S-1–9	214.8	12.17
H-1–1	24.9	4.61	H-1–2	39.4	5.95	H-1–3	44.8	8.57
H-1–4	68.0	11.06	H-1–5	89.5	12.08	H-1–6	103.9	13.07
H-2–1	18.1	4.87	H-2–2	37.0	6.52	H-2–3	48.3	10.20
H-2–4	62.2	9.16	H-2–5	69.3	10.20	H-2–6	110.7	14.89
H-2–7	141.5	14.04	H-2–8	141.6	14.50	—	—	—

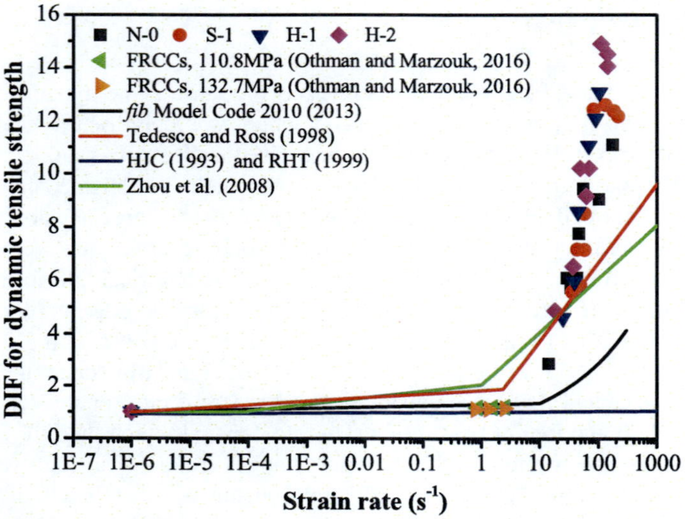

Fig. 3.11 DIFs for dynamic tensile strength of UHPCC, reprinted from Wu et al. (2018), copyright 2020, with permission from Elsevier

For the concrete-like materials, several empirical formulae have been proposed to estimate the strain-rate effect on the dynamic tensile strength. The *fib* model code 2010 (Comite Euro-International du Beton 2013) recommended

$$\mathrm{DIF} = \begin{cases} (\dot{\varepsilon}/\dot{\varepsilon}_{ts})^{0.018} & \text{for } \dot{\varepsilon} \le 10\mathrm{s}^{-1} \\ 0.0062(\dot{\varepsilon}/\dot{\varepsilon}_{ts})^{1/3} & \text{for } \dot{\varepsilon} > 10\mathrm{s}^{-1} \end{cases} \tag{3.8}$$

where $\dot{\varepsilon}$ is the actual strain rate in the range of $1.0 \times 10^{-6}\ \mathrm{s}^{-1}$ to $300\ \mathrm{s}^{-1}$ and $\dot{\varepsilon}_{ts}$=$1.0 \times 10^{-6}\mathrm{s}^{-1}$ is the static strain rate.

Based on a series of dynamic splitting tests on concrete specimens with different compressive strength (27.9 ~ 54.5 MPa), Tedesco and Ross (1998) established a bilinear tensile DIF regression formula as follows

$$\mathrm{DIF} = \begin{cases} 1 + 0.1425\left[\log\dot{\varepsilon} + 5.8456\right] \ge 1.0 & \text{for } \dot{\varepsilon} \le 2.32\mathrm{s}^{-1} \\ 1 + 2.929\left[\log\dot{\varepsilon} - 0.0635\right] \le 6.0 & \text{for } \dot{\varepsilon} > 2.32\mathrm{s}^{-1} \end{cases} \tag{3.9}$$

Both Holmquist-Johnson-Cook (Holmquist et al. 1993) and RHT (Riedel et al. 1999) concrete model adopted the same DIF formulations for dynamic tensile strength of concrete as

$$\mathrm{DIF} = 1 + c\ln(\dot{\varepsilon}/\dot{\varepsilon}_0) \tag{3.10}$$

where c is assumed to be 0.007 and $\dot{\varepsilon}_0$ is the reference strain rate and taken as $1.0\ \mathrm{s}^{-1}$.

Zhou et al. (2008) further proposed a DIF formula by curve fitting the existing experimental results of dynamic tensile tests on concrete materials (Schuler et al. 2006).

$$\mathrm{DIF} = \begin{cases} 1.0 & \text{for} \quad \dot{\varepsilon} \le 10^{-4}s^{-1} \\ 2.06 + 0.26\log\dot{\varepsilon} & \text{for} \quad 10^{-4}s^{-1} \le \dot{\varepsilon} \le 1s^{-1} \\ 2.06 + 2.0\log\dot{\varepsilon} & \text{for} \quad \dot{\varepsilon} > 1s^{-1} \end{cases} \tag{3.11}$$

The above four formulae are also plotted in Fig. 3.11. Because these formulae are mainly concentrated on normal strength concrete and not suitable for high strength concrete, they could not well describe the evolution of the DIFs for UHPCC material and generally underestimate the dynamic tensile strength.

Since the available test data of UHPCC under quasi-dynamic loadings is limited, it is difficult to determine the transition strain rate $\dot{\varepsilon}_{TR}$ exactly. Othman and Marzouk (2016) experimentally obtained the DIFs for two types of fiber-reinforced cementitious composites (FRCCs) with the compressive strengths of 110.8 and 132.7 MPa at low to medium strain rates (0.58 ~ 2.58 s^{-1}), as shown in Fig. 3.11. Due to the good consistence with the DIF for FRCCs, the empirical formula of *fib* Model Code 2010 (2013) is adopted for the strain rates below the transition strain rate. Thus an empirical formula is proposed to describe the strain rate effect of UHPCC as follows,

$$\mathrm{DIF} = \begin{cases} (\dot{\varepsilon}/\dot{\varepsilon}_0)^{0.018} & \text{for } \dot{\varepsilon} \le \dot{\varepsilon}_{TR} \\ A \cdot \log\dot{\varepsilon} + B & \text{for } \dot{\varepsilon} > \dot{\varepsilon}_{TR} \end{cases} \tag{3.12}$$

where A and B are the coefficients. By fitting the test data, the values of A, B and $\dot{\varepsilon}_{TR}$ for UHPCC are determined and listed in Table 3.4. Then the corresponding empirical DIF relations of Eq. (3.12) are given in Fig. 3.12. These empirical formulae can be used to predict the tensile DIFs in the numerical predictions of UHPCC structure responses under high strain rate loadings. It should be noted that Eq. (3.12) is derived from the test data for the strain rate up to ~214.8 s^{-1}, beyond which there is no available test data to substantiate its validity.

Table 3.4 Coefficients for the proposed DIF formula of UHPCC

Test No.	A	B	$\dot{\varepsilon}_{TR}(s^{-1})$
N-0	7.362	−5.167	7.8
S-1	9.637	−8.814	11.5
H-1	14.338	−15.695	15.5
H-2	11.561	−10.446	10.5

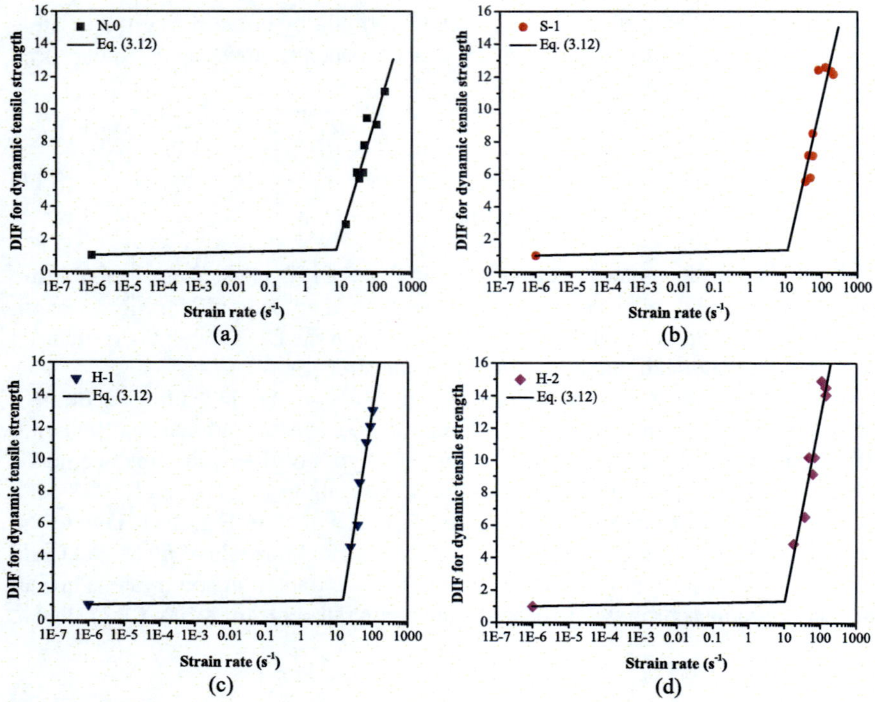

Fig. 3.12 Empirical DIF relations for UHPCC **a** N-0, **b** S-1, **c** H-1, **d** H-2, reprinted from Wu et al. (2018), copyright 2020, with permission from Elsevier

3.5 Summary

Investigations of the mechanical behaviors of UHPCC are important and essential to provide valuable information for the structural design and calibration/validation of its constitutive model. In this chapter, the dynamic tensile behavior of UHPCC with three volume fractions (0, 1.0 and 2.0%) and two types (micro-straight and hooked) of steel fibers is experimentally studied, respectively. The effects of strain rate, steel fiber content and type on the dynamic spalling strength of UHPCC are assessed. The following conclusions can be drawn:

1. The dynamic tensile behavior of UHPCC is significantly sensitive to the strain rate, and the dynamic spalling strength increases obviously with the strain rate increasing up to ~110 s^{-1}.
2. The damage degree of UHPCC matrix with no fibers is more severe than that of steel fiber reinforced ones under identical strain rates. The addition of steel fibers can effectively prevent the formation and propagation of cracks.
3. The dynamic spalling strength of UHPCC improves with increasing the content of both two typical steel fibers. With the fiber volume fraction of 1.0%, the

micro-straight steel fiber has slightly better effect on improving the dynamic spalling strength than the hooked one for the strain rate higher than 60 s^{-1}.
4. The dynamic spalling strength is reduced with increasing the critical time to fracture.
5. The DIFs for dynamic tensile strength of UHPCC increase with increasing the content for hooked steel fibers, and the enhancing effect of hooked steel fibers is more obvious than micro-straight steel fibers. Furthermore, the empirical DIF formulae for UHPCC with various steel fiber reinforcement are proposed.

References

BRARA A, CAMBORDE F, KLEPACZKO J R, MARIOTTI C. Experimental and numerical study of concrete at high strain rates in tension[J]. Mechanics of Materials, 2001, 33: 33–45.
Comite Euro-International du Beton. Fib Model Code for Concrete Structures 2010[M]. Trowbridge, Wiltshire, UK: Redwood Books, 2013.
DÍAZ-RUBIO F G, PÉREZ J R, GÁLVEZ V S. The spalling of long bars as a reliable method of measuring the dynamic tensile strength of ceramics[J]. International Journal of Impact Engineering, 2002, 27: 161–177.
GOLDSMITH W, POLIVKA M, YANG T. Dynamic behavior of concrete[J]. Experimental Mechanics, 1996, 6: 65–79.
HAO H, HAO Y F, LI J, CHEN W S. Review of the current practices in blast-resistant analysis and design of concrete structures[J]. Advances in Structural Engineering, 2016, 19(8): 1193–1223.
HOLMQUIST T J, JOHNSON G R, COOK W H. A computational constitutive model for concrete subjective to large strain, high strain rates, and high pressure. The 14th International Symposium on Ballistic, Quebec[C]. 1993: 591–600.
KLEPACZKO J, BRARA A. An experiment method for dynamic tensile testing of concrete by spalling[J]. International Journal of Impact Engineering, 2001, 25: 387–409.
MALVAR L, CRAWFORD J. Dynamic increase factors for concrete[R]. Orlando, FL: Twenty-Eighth DDESB Seminar, 1998.
MILLARD S G, MOLYNEAUX T C K, BARNETT S J, GAO X. Dynamic enhancement of blast-resistant ultra high performance fibre-reinforced concrete under flexural and shear loading[J]. International Journal of Impact Engineering, 2010, 37: 405–413.
OTHMAN H, MARZOUK H. Strain Rate Sensitivity of Fiber-Reinforced Cementitious Composites[J]. ACI Materials Journal, 2016, 113(2): 1–8.
RONG Z D, SUN W. Experimental and numerical investigation on the dynamic tensile behavior of ultra-high performance cement based composites[J]. Construction and Building Materials, 2012, 31: 168–173.
REN G M, WU H, FANG Q, LIU J Z. Effects of steel fiber content and type on dynamic compressive mechanical properties of UHPCC[J]. Construction and Building Materials, 2018, 164: 29–43.
RIEDEL W, THOMA K, HIERMAIER S, SCHMOLINSKE E. Penetration of reinforced concrete by BETA-B-500 numerical analysis using a new macroscopic concrete model for hydrocodes. The 9th International Symposium, Interaction of the Effects of Munitions with Structures[C]. Berlin-strausberg: IBMAC,1999: 315–22.
SU Y, LI J, WU C Q, WU P T, LI Z X. Effects of steel fibres on dynamic strength of UHPC[J]. Construction and Building Materials, 2016, 114: 708–718.
SCHULER H, MAYRHOFER C, THOMA K. Spall experiments for the measurement of the tensile strength and fracture energy of concrete at high strain rates[J]. International Journal of Impact Engineering, 2006, 32: 1635–1650.

THAOMAS R J, SORENSEN A D. Review of strain rate effects for UHPC in tension[J]. Construction and Building Materials, 2017, 1153: 846–856.

TULER F R, BUTCHER B M. A criterion for the time dependence of dynamic fracture[J]. International Journal of Fracture technology, 1968, 4(4): 431–437.

TRAN T K, KIM D J. High strain rate effects on direct tensile behavior of high performance fiber reinforced cementitious composites[J]. Cement & Concrete Composites, 2014, 45: 186–200.

TEDESCO J W, ROSS C A. Strain-rate-dependent constitutive equations for concrete[J]. Journal of Pressure Vessel Technology-Transactions of the ASME, 1998, 120: 398–405.

WATSON A, SANDERSON A. The resistance of concrete to shock. Proceedings of the Conference on Mechanics and Physics, Behaviour of Materials Under Dynamic Loading[C]. The Institute of Physics, Oxford, 1979.

WANG Z L, ZHOU N Q, WANG J G. Using Hopkinson pressure bar to perform dynamic tensile tests on SFRC at medium strain rates[J]. Magazine of Concrete Research, 2012, 64(8): 657–664.

WU H, REN G M, FANG Q, LIU J Z. Effects of steel fiber content and type on dynamic tensile mechanical properties of UHPCC[J]. Construction and Building Materials, 2018, 173:251–261.

ZHANG L, HU S S, CHEN D X, YU Z Q, LIU F. An experimental technique for spalling of concrete[J]. Experimental Mechanics, 2009, 49: 523–532.

ZHOU X Q, KUZNETSOV V A, HAO H, WASCHL J. Numerical prediction of concrete slab response to blast loading[J]. International Journal of Impact Engineering, 2008, 35: 1186–1200.

Chapter 4
Triaxial Compressive Behavior of UHPCC and Application in the Numerical Analyses of Projectile Impact

4.1 Introduction

Ultra-high performance cement based composite (UHPCC) is a relatively new type of composite material which possess very low water-to-binder ratio, high amount of high-range water reducer (HRWR), fine aggregates with the maximum size less than 1 mm and high-strength steel fibers with the diameter of 0.15 ~ 0.20 mm (Graybeal 2006). Compared with the newly developed engineered cementitious composites (ECC) (Maalej et al. 2005), multi-scale cement composite (MSCC) (Rossi et al. 2005), reactive powder concrete (RPC) (Richard and Cheyrezy 1995) and slurry infiltrated fiber concrete (SIFCON) (Naaman and Homrich 1989), the ordinary curing temperature and pressure preparation procedures, as well as the comprehensive prominent characteristics (e.g. high compressive strength and tensile strength, high fracture energy, self-consolidating workability and very low permeability), make UHPCC becoming the most prospective construction materials for both military and civil protective structures (fortifications, nuclear containment, defense shelter and etc.), which are designed to withstand the intentional or accidental impact and blast loadings caused by projectiles, fragments, aircrafts, explosives and etc. For above scenarios, the concrete material always undergoes multi-axial compressions with a high confinement, thus the investigations of triaxial compressive behavior of UHPCC are important and essential to provide valuable information (e.g. strength criteria, post-peak response) for the structural design and calibration/validation of the constitutive model.

The existing works on the mechanical behavior of UHPCC were mainly concentrated on the uniaxial compressive or tensile properties (Naaman and Reinhardt 1996; Petr et al. 2013; Wille et al. 2014), limited tests were carried out to investigate the triaxial compressive behavior of UHPCC with relatively high confinement ratio. In the present study, based on the previous works (Zhang et al. 2005a, b; Langberg and Markeset 1999; Wu et al. 2015a; Máca et al. 2010; Sovják et al. 2015; Peng et al. 2016) on impact resistance of cement composites, the optimal UHPCC (straight steel fiber with the volumetric ratio of 2%, high-strength basalt aggregates with the size

Q. Fang et al., *UHPCC Under Impact and Blast*,
https://doi.org/10.1007/978-981-33-6842-2_4

range of 5–10 mm, the designed uniaxial compressive strength of ~100 MPa) for protective structures against projectile high-speed impact was prepared under ordinary procedures. Triaxial compressive test on two batches of 50 × 100 mm cylindrical specimens with uniaxial compressive strength of 95 MPa and 129 MPa was conducted with the confinement ratio (the ratio of confining pressure to the uniaxial compressive strength) reached 0.84 and 0.78, respectively. The deviatoric stress–strain curves for various confinement pressure were derived, the failure criteria and toughness of UHPCC under triaxial compression were discussed. Finally, by using the present and existing triaxial compression tests data (Lu and Hsu 2006; Xie et al. 1995; Sovják et al. 2013; Farnam et al. 2010; Xiong 2009) on high-strength concrete, the dominant strength parameters of Holmquist- Johnson–Cook (HJC) constitutive model (Holmquist et al. 1993) were confirmed. Furthermore, aiming to verify the validity of the proposed model parameters, by utilizing the finite-element program LS-DYNA, the predicted projectile penetration depth and residual velocity were compared with the experimental data from eleven sets of projectile penetration or perforation tests on concrete and rock targets with the uniaxial compressive strength ranging from 67.5 to 157 MPa.

4.2 A Review of the Existing Works on Triaxial Behavior of Concrete

The existing related works are mainly on normal-strength concrete (NSC) and high-strength concrete (HSC). Wang et al. (1987) performed a series of triaxial compressive test on 100 mm cubic specimen, the unconfined compressive strength was around 10 MPa and the confinement ratio reached up to 3, the nearly linear relationship of normalized triaxial compressive strength and the confinement ratio was obtained. Sfer et al. (2002) conducted the triaxial compressive test on 150 mm × 300 mm cylindrical specimens with the unconfined compressive strength around 30 MPa and the confinement ratio varing within 0 ~ 2. It is observed that the failure envelope of which correlated well with the nonlinear Etse and Willam model (Etse and Willam 1994). Sirijaroonchai et al. (2010) experimentally studied the influences of fiber types (high-strength hooked steel fibers, ultra high molecular weight polyethylene fibers), fiber volumetric fraction (1 ~ 2%) and confining pressure (41 and 52 MPa) on the passive triaxial behavior of high performance fiber reinforced cement composites with the unconfined compressive strength of ~50 MPa. It indicated that, for relatively high confinement ratio, confining effect introduced by the fibers with various types and fractions became minor. Chern et al. (1992) experimentally investigated the influences of volumetric ratio of steel fibers (0 ~ 2%) and confinement pressure (0 ~ 70 MPa) on the triaxial strength of the NSC with the uniaxial compressive strength around 20 MPa. It was concluded that the addition of steel fibers has insignificant effect on the mechanical behavior of concrete at confining pressure up to 70 MPa.

Correspondingly, Lu and Hsu (2006) compared the triaxial compressive stress–strain relations and failure criteria of HSC and steel fiber reinforced HSC (SFHSC), and also found the slight reinforcing effect on the uniaxial and triaxial strengths of 100 × 150 mm cylindrical specimens, in which the unconfined compressive strengths of HSC and SFHSC were 67 and 69 MPa, and the maximum confinement ratios were 0.74 and 1, respectively. Xie et al. (1995) experimentally confirmed the three-parameter parabolic relationships between the confined pressure and the maximum as well as residual strengths of HSC containing silica fume specimens (55.5 × 110 mm cylinder), the variation ranges of unconfined compressive strengths and confinement ratio were 60.2 ~ 119 MPa and 0 ~ 0.504, respectively. Ansair and Li (1998), Li and Ansair (1997, 2010) experimentally studied the triaxial constitutive relationships of 100 × 200 mm and 75 × 150 mm cylindrical concrete specimens, with the uniaxial compressive strengths ranging from 47.7 MPa (maximum confinement ratio of 0.878) to 107.3 MPa (maximum confinement ratio of 0.7792). The less pronounced effect of confining pressure on the failure strength of HSC than NSC, and the distinct effect of confining pressure on the failure strain were found. Also the specimen size of HSC has slight effect on the triaxial compressive failure strength and failure surface. Furthermore, the empirical failure criterion based on Ottosen constitutive model (Ottosen 1977) was established for HSC. Attard and Setunge (1996) and Candappa et al. (2001) investigated the stress–strain relationships of concrete specimens (100 × 200 mm cylinders) with the unconfined compressive strength ranging from 60 ~ 130 MPa and 41.9 ~ 103.3 MPa, respectively. The empirical model for full triaxial stress–strain relationship was developed and the parameters for Ottosen constitutive model (Ottosen 1977) were suggested. By comprehensively assessing the validities of the existing nine empirical formulae for triaxial ultimate compressive strength of HSC, Girgin et al. (2007) proposed a new relationship to predict the ultimate strength of concrete with the unconfined strength ranging from 60 to 132 MPa. Sovják et al. (2013) experimentally studied the triaxial cylinder (100 × 200 mm) and cubic (100mm^3) compressive strengths of ultra high performance concrete, of which the unconfined compressive strengths were 123 MPa and 148 MPa and the confined ratio increased up to 0.25 and 0.6, respectively.

Additionally, Imran and Pantazopoulou (1996), Farnam et al. (2010), Vu et al. (2011) and Piotrowska et al. (2014) have systematically studied the influences of moisture content, loading path, steel fiber volumes, size, shape and composition of coarse aggregates as well as the cement paste volume on the triaxial compressive behavior of concrete.

However, for the widely applications of UHPCC in the on-site constructions of protective structures against projectile impact, limited works were concentrated on the composition optimization based on the balances of anti-strike ability and costs, preparation of optimal UHPCC under ordinary procedure, triaxial compressive behavior of optimal UHPCC as well as its applications in the projectile penetration or perforation analyses. The present chapter is aiming to fill in the above gaps.

4.3 Mixing Optimization of UHPCC and Triaxial Compression Test

4.3.1 Compositions

For the protective structures withstanding high-speed projectile impact, the resistance of concrete target depends mainly on the compressive strength of the target as well as the size and strength (hardness) of the coarse aggregates (Zhang et al. 2005a, b; Langberg and Markeset 1999; Wu et al. 2015a; Dancygier et al. 2007, 2014; Wu et al. 2015b). Therefore, adding the high-strength coarse aggregates into the concrete matrix, probably results in the decrease of compressive strength of the concrete to some extent, but enhancing the impact resistance.

As for the target strength, Zhang et al. (2005a, b), Langberg and Markeset (1999), Wu et al. (2015a) have experimentally found that the penetration depth of projectiles are no longer decreasing obviously when the compressive strength of the targets are larger than a certain value. Moreover, the compressive strengths of 90 ~ 150 MPa were recommended for protective structures by comprehensive considerations on the protective efficiency and production costs. Besides, as for the shapes (straight, hooked, twisted) and volumetric fractions of steel fibers, Wu et al. (2015a), Máca et al. (2010), Sovják et al. (2015) and Peng et al. (2016) have proposed that volume fraction of the straight steel fibers with 1.5 ~ 2% is the optimum choice with respect to the mechanical properties, workability, crack prohibiting ability as well as the cost of the matrix.

The optimal UHPCC was prepared by State Key Laboratory of High Performance Civil Engineering Materials, Jiangsu Research Institute of Building Science in China. The traditional high performance concrete is often costly and low production efficiency for its expensive raw materials (e.g. high content of silica fume, fine quartz sand and small-sized steel fiber). More importantly, the casting and curing of ultra-high strength concrete (e.g. RPC) needs high temperature (90 ~ 400 °C) and sufficient pressure to expel the excess liquids and air from the fresh mixture, which limits its large-scale on-site construction applications.

Table 4.1 lists the present mixing proportion of UHPCC, in which all the ingredients are commercially available in China. To reduce the porosity of binder and improve the interface zone between coarse aggregates and cement pastes, four cementitious materials were used in the present study, including the Chinese standard Graded 52.5 P.II type Portland cement (GB175-2007 2008) from Xiaoyetian corporation (20 ~ 30 μm in particle diameter), silica fume (particle size of 0.10 ~ 0.26 μm, density of 2.1 g/cm^3, specific surface area of 20,500 m^2/kg) from Elkem company, ultra-fine fly ash (density of 2.7 g/cm^3) from Nantong electric power plant and ultra-fine slag (particle size of 3 ~ 6 μm, density of 2.8 g/cm^3) from Jiangnan cement plant.

Compared with the traditional high performance concrete, the ground fine quartz sand (the size was less than 0.6 mm) was substituted by easily obtained natural river sand with maximal diameter of 2.5 mm and basalt aggregates with 5 ~ 10 mm

Table 4.1 Mixtures of UHPCC (kg/m^3)

Designing strength	Cement	Silica fume	Fly ash	Ultra-fine slag	Fine sand	Basalt aggregate	Water	HRWR (%)	Steel fibers (%)
CF90	300	50	50	100	900	700	150	1.2	2
CF130	366	61	61	122	794	700	140	1.6	2

in diameter. Also the partial silica fume was replaced by the ultra-fine industrial waste powder, e.g. fly ash and ultra-fine slag. Table 4.1 indicates that the ultra-fine industrial waste powder (silica fume, ultra-fine fly ash and slag) accounts for 40% of the binder, the ratio of expensive silica fume is only 10%, and the water-binder ratio is 0.3. Since the production of cement and fine quartz sands consumes much more energy and resources, and silica fume is more expensive than fly ash and ultra-fine slag for its limited production, the present UHPCC is low-costs and energy saving.

The volumetric ratio of the mixing fibers were designed as 2% and the straight brass-coated steel fiber was chosen since it provides a good trade-off between workability and mechanical properties of concrete. The equivalent diameter, length and tensile strength of steel fibers were 0.175 mm, 13 mm and 3000 MPa, respectively. Provided by the manufacture, Jiangsu SOBUTE new material Co., Ltd, the compressive strength and density of the basalt aggregates are 210 MPa and 2800 kg/m^3, respectively. The water-reducing ratio of the PCA®-I polycarboxylic type HRWR (high-range-water-reducer) was no less than 35%, which is also developed by Jiangsu SOBUTE new material Co., Ltd.

4.3.2 Mixing Procedure

To achieve good workability, particle distribution and packing density, mixing procedure for the present UHPCC was controlled rigorously. All the fine particles and coarse aggregates were blended before adding water and HRWR because small particles tend to agglomerate and it is easier to break these chunks when the particles are dry. At first, the dry cementitious materials (cement, fly ash, silica fume, slag) and sands were put together simultaneously and mixed uniformly for 3 min, then the coarse basalt aggregates were added and mixed uniformly for 1 min. Secondly, the water and super-plasticizer were added together and mixed for 3 ~ 5 min. Finally, steel fibers were incorporated into the mixture and mixed for another 3 ~ 5 min in order to guarantee the fibers well distributed throughout the matrix. The shear action of the fibers helps to destroy any remaining agglomerates in the mixture, and thus improvs the workability of the composites. A vertical pan mixer with a capacity of 60L and a constant mixing speed (47 rpm) was used. UHPCC is a self-consolidating concrete which does not require heat and high pressure preparation. The slump and spread of the present UHPCC were about 230 mm and 610 mm, respectively, which satisfies the workability requirements of on-site constructions.

The prepared matrix was poured into 50 × 100 mm cylindrical plastic moulds, the freshly cast specimens were kept in the moulds for 24 h. Then, they were demoulded and cured in a standard climate with temperature of 17.7 ~ 21.7ºC and relative humidity of 98.4% for 28 days. Afterwards, the specimens were taken out of the curing room and cured in air. To minimize the influences of the aging on the behavior, the triaxial compressive test was conducted at 120 days after the specimens were cast. Two batches of UHPCC specimens (CF90 and CF130) listed in Table 4.1 were cast, of which the densities were nearly 2570 kg/m^3, and the average 120d unconfined cylindrical (50 × 100 mm) compressive strengths were 95 MPa and 129 MPa, respectively.

4.3.3 *Triaxial Compression Test*

As shown in Fig. 4.1a, the triaxial compression test was performed by the MTS 815.03 material testing system, a very rigid frame MTS servo-hydraulic closed loop testing machine with a frame stiffness rated at 11.0×10^9 N/m. It has a load capacity of 4600 kN and the triaxial cell is able to apply confining pressure up to 140 MPa on a 50 × 100 mm cylindrical specimen. A confining pressure intensifier was used to fill and pressurize the triaxial cell with the confining fluid, and the confining pressure σ_3 was measured by an inside pressure transducer. In-vessel pressure and displacement transducers were employed for the control of the servo-valve through computer commands. The axial load was applied by the axial actuator and the in-vessel load cell recorded the axial deviatoric load F_0. The actual axial stress σ_1 can be obtained as $F_0/A_s + \sigma_3$, where A_s is the cross-sectional area of the cylindrical specimen. Before the test, the ends of specimens were flattened in a grinding machine. In order to keep the confining pressure constant, the concrete specimens were jacketed with a rubber membrane (2 mm in thick) to prevent penetration of fluid into the porous specimen during pressurization, shown in Fig. 4.1b.

The specimen surrounded by the rubber membrane was placed between upper and lower steel loading cap. Figure 4.1c shows the instrumentation of the specimen, including one linear variable differential transformer (LVDT) and one circumferential extensometer with the measuring capacity of ±2.5 mm and 5 mm, respectively.

(a) (b) (c)

Fig. 4.1 Test setup **a** MTS 815.03 system **b** specimen with rubber membrane **c** loading cell, reprinted from Ren et al. (2016), copyright 2020, with permission from Elsevier

Fig. 4.2 Loading path for triaxial tests, reprinted from Ren et al. (2016), copyright 2020, with permission from Elsevier

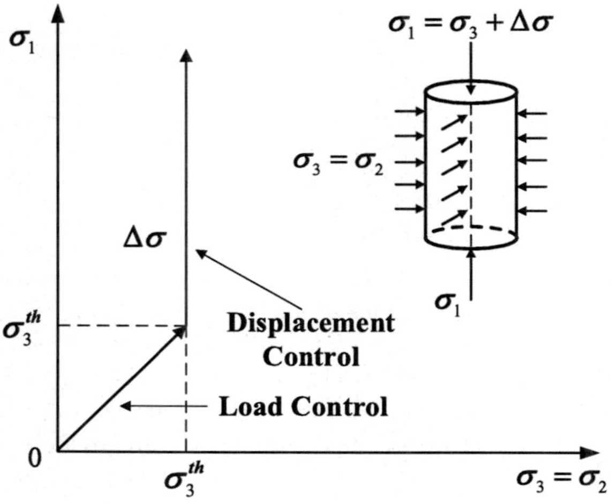

The LVDT was assembled at the side of a specimen to measure the overall axial displacement, and the axial strain ε_1 was obtained by subtracting the initial reading from the current reading of the LVDT and then divided by the length of specimen (100 mm). The circumferential extensometer was used at the middle of the height of the specimen for measuring the circumferential strain ε_3.

Under uniaxial compression, a displacement control with a rate of 0.001 mm/s was employed to obtain the complete stress–strain curve. While for triaxial compression, the loading path was shown in Fig. 4.2. Firstly, the triaxial compression tests were performed with a load control, in which equal magnititudes of axial and confining pressure were simultaneously applied to the specimens at the rate of 0.8 MPa/s until the confining pressure σ_3 reached the targeted values σ_3^{th}. And then, the confining pressure was kept constant at the target value, the axial control mode was switched to a displacement control with a rate of 0.001 mm/s and the axial deviatoric stress $\Delta\sigma = \sigma_1 - \sigma_3$ was increased until the strain reached the measuring capacity of the axial or circumferential extensometer.

4.4 Results and Analysis

4.4.1 Failure Pattern

Figure 4.3 shows the failure patterns of the recovered UHPCC specimens after the test. It could be seen obviously that all the specimens matained their intigrity after sustaining the traxial compressions. Under uniaxial compression, although relatively large cracks emerged, the major splitting failure that always occurred for HSC specimens (Xie et al. 1995; Lu and Hsu 2006) did not take place due to the addtion of steel

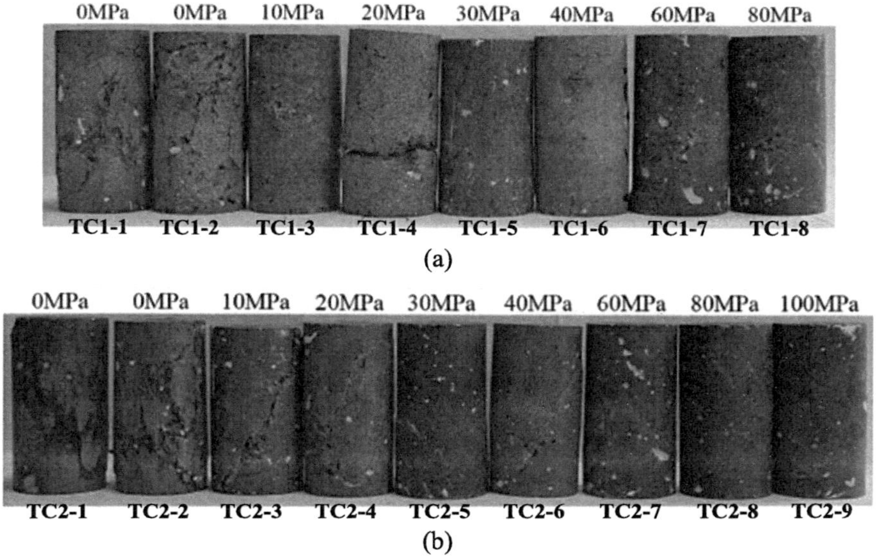

Fig. 4.3 Failure patterns of recovered UHPCC specimens **a** 95 MPa, **b** 129 MPa, reprinted from Ren et al. (2016), copyright 2020, with permission from Elsevier

fibers. For the specimens with relatively lower confining pressure, the shear failure induced by several minor cracks occurred. With further increasing the confining pressure, no visible cracks emerged and apparent swelling occurred around the specimens due to the axial contractions. Therefore, adding fibers can improve the failure mode of UHPCC when the confining pressure is relatively low. While for the high confining pressure, the fiber enhancing effect becomes marginal. It should be pointed out that the broken specimen TC1-4 was induced by the careless operations in taking off the rubber membrane after the test.

4.4.2 Stress–Strain Curve

Behavior of UHPCC specimens subjected to various levels of confining pressure is shown in Fig. 4.4. It can be seen obviously that the triaxial strength and the ductility of specimens increase with the rising of confining pressure. In the pre-peak phase, the slope of the curves also gradually increase. For the confining pressure $\sigma_3 \leq 60$ MPa, the deviatoric stress–strain curves show softening behavior. As the confining pressure increases, the behavior of the deviatoric stress–strain curves become more plastic, and the stress slightly drops after the peak stress.

Based on Figs. 4.4 and 4.5 further illustrates the relationships of volumetric strain

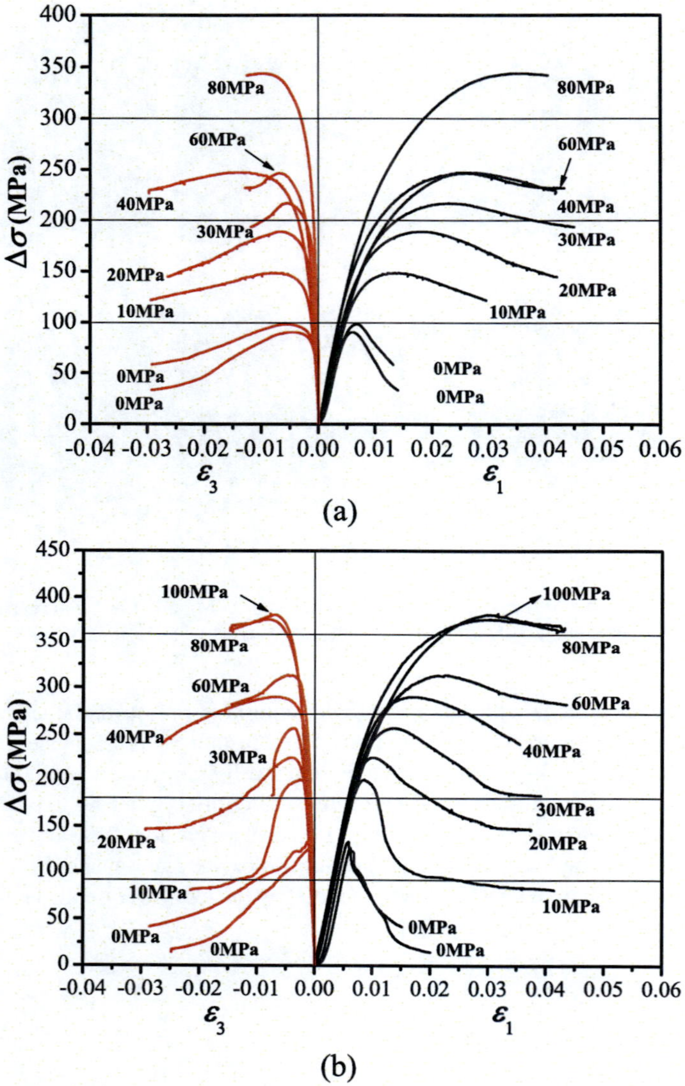

Fig. 4.4 Axial deviatoric stress–strain curves of two batches of UHPCC specimens **a** 95 MPa, **b** 129 MPa, reprinted from Ren et al. (2016), copyright 2020, with permission from Elsevier

ε_v to the axial and circumferential strains under various confining pressures, respectively. It can be found that, for specimens with $\sigma_3 \leq 60\,\text{MPa}$, as shown in Fig. 4.4, softening behavior is observed which is attributed to the development of more cracks. As the cracking develops, the volumetric strain changes from contraction to dilation mode. For confining pressures larger than 60 MPa, since the crack development is not

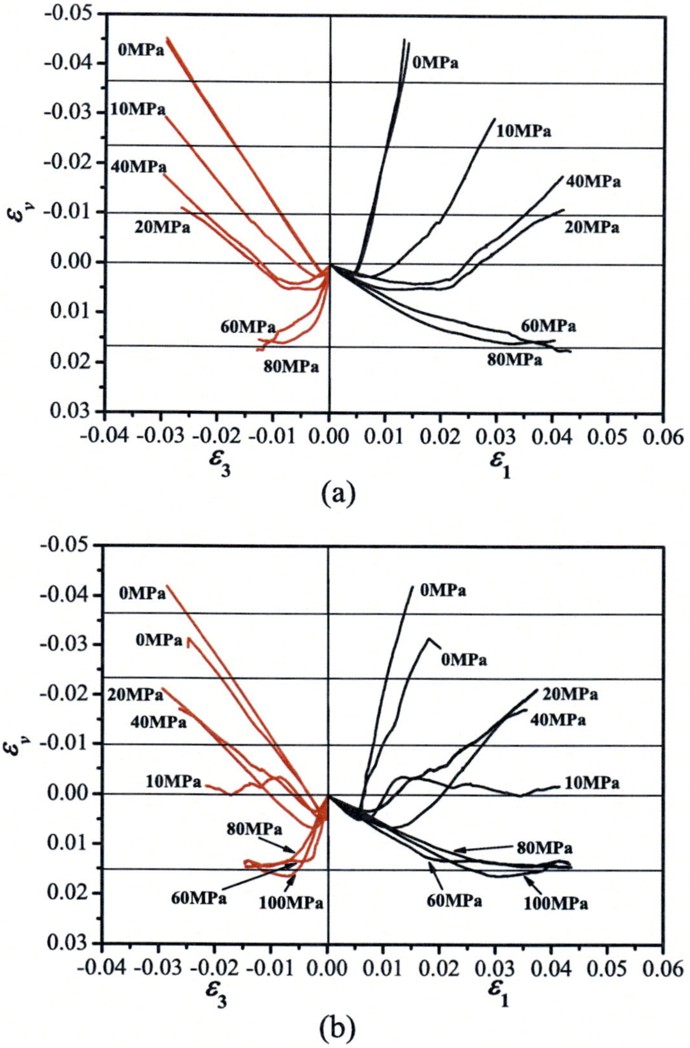

Fig. 4.5 Volumetric strain-strain curves of two batches of UHPCC specimens **a** 95 MPa, **b** 129 MPa, reprinted from Ren et al. (2016), copyright 2020, with permission from Elsevier

so pronounced, the specimens deform as a plastic material and the axial contraction is predominant.

From Fig. 4.4, the triaxial compression test results were listed in Table 4.2, where f_c' denotes the unconfined compressive strength of the specimens.

From Table 4.2, the test data of normalized peak axial strain ratio $\varepsilon_1^u/\varepsilon_c^u$ and confinement ratio σ_3/f_c' is shown in Fig. 4.6, in which $\varepsilon_c^u = 0.006$ and $\varepsilon_c^u = 0.0058$ are the peak axial strains of UHPCC under uniaxial compression for 95 MPa and

Table 4.2 Triaxial compression test results

No	f_c'(MPa)	Confinement ratio σ_3/f_c'	Peak axial strength ratio σ_1^u/f_c'	Peak axial strain ε_1^u	Peak lateral strain ε_3^u	Modulus of elasticity E (GPa)	Poisson's ratio ν
TC1-1	95	0	0.95	0.0060	−0.0045	24.0	0.25
TC1-2		0	1.04	0.0067	−0.0056	25.9	0.25
TC1-3		0.11	1.67	0.0141	−0.0075	24.2	0.38
TC1-4		0.21	2.20	0.0176	−0.0063	26.2	0.23
TC1-5		0.32	2.60	0.0229	−0.0055	24.9	0.15
TC1-6		0.42	3.02	0.0256	−0.0136	28.1	0.33
TC1-7		0.63	3.23	0.0262	−0.0068	24.9	0.21
TC1-8		0.84	4.46	0.0358	−0.0101	27.3	0.13
TC2-1	129	0	0.98	0.0062	−0.0009	30.8	0.07
TC2-2		0	1.03	0.0058	−0.0012	32.0	0.05
TC2-3		0.08	1.63	0.0086	−0.0026	31.8	0.16
TC2-4		0.16	1.90	0.0101	−0.0040	30.9	0.27
TC2-5		0.23	2.22	0.0135	−0.0036	32.7	0.28
TC2-6		0.31	2.56	0.0172	−0.0069	32.0	0.16
TC2-7		0.47	2.89	0.0226	−0.0046	30.1	0.15
TC2-8		0.62	3.54	0.0302	−0.0082	34.0	0.25
TC2-9		0.78	3.74	0.0318	−0.0078	28.0	0.22

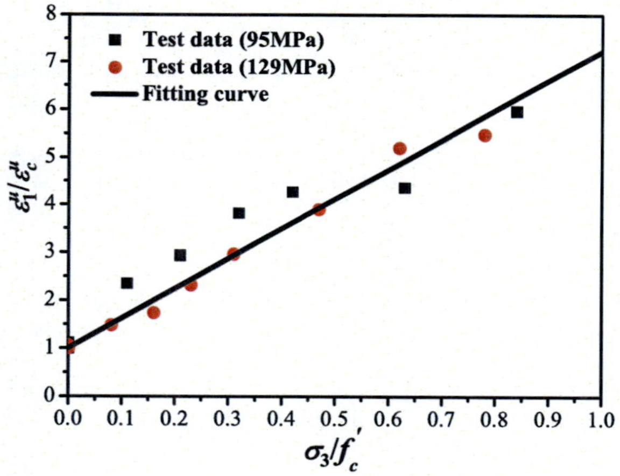

Fig. 4.6 Relationship of $\varepsilon_1^u/\varepsilon_c^u$ with σ_3/f_c' from present test on UHPCC, reprinted from Ren et al. (2016), copyright 2020, with permission from Elsevier

129 MPa specimens, respectively. It indicates that $\varepsilon_1^u/\varepsilon_c^u$ and σ_3/f_c' show nearly a linear relationship, and the function of the fitted straight line is

$$\varepsilon_1^u/\varepsilon_c^u = 1 + 6.24(\sigma_3/f_c') \tag{4.1}$$

The slope of the fitted straight line is 6.24 for UHPCC. Moreover, Lu and Hsu (2006), Ansari and Li (1998) and Candappa et al. (2001) did the same work and also found the above linear relationship, while the slopes are 19.21, 15.15 and 20 for concrete specimens with f_c' ranging from 32.6 MPa to 107.3 MPa, respectively. It indicates that, for UHPCC material, the relatively larger magnitude of the peak strain with the rising of the confinement ratio was not obvious compared with NSC and HSC.

4.4.3 *Failure Criteria*

In this section, the triaxial compression test results of UHPCC from Table 4.2 were used to calibrate the existing three most applied failure criteria.

4.4.3.1 Mohr–Coulomb Failure Criterion

Mohr–Coulomb failure criterion is a classical failure criterion for describing the failure envelop, and is still extensively used due to its simplicity and relatively good accuracy. The pressure-dependent Mohr–Coulomb model is expressed as

$$\sigma_1^u/f_c' = 1 + k_i\sigma_3/f_c' \tag{4.2}$$

where $k_i = (1 + \sin\varphi)/(1 - \sin\varphi)$ and φ represents the internal-friction angle of concrete. The previous works indicate that the value of k_i ranges from 2.6 to 5.3 for concrete specimens with diverse strengths (Lu and Hsu 2006; Xie et al. 1995; Ansari and Li 1998; Candappa et al. 2001; Lan and Guo 1997).

For the present test, shown in Fig. 4.7a, it is found that the best-fit value of k_i for the dimensionless peak axial stress-confining pressure relationship is 4.1. Figure 4.7b also gives the existing compressive tests data on high-strength concrete, including UHPCC, HPFRC and SIFCON as well as the curve of Eq. (4.2) with $k_i = 4.1$ obtained from Fig. 4.7a. The comparisons indicate that, (i) For HSC and the confining ratio $\sigma_3/f_c' \le 1$, the relationship of σ_1^u/f_c' versus σ_3/f_c' have nearly the same tendency, and the influences of strength, size, compositions of specimens were not pronounced; (ii) For low confinement levels ($\sigma_3/f_c \le 0.5$), $k_i = 4.1$ can better predict the peak axial stress values; (iii) For high levels of confinement ($\sigma_3/f_c > 0.5$), Mohr–Coulomb failure criterion overpredicts the peak triaxial stress of HSC.

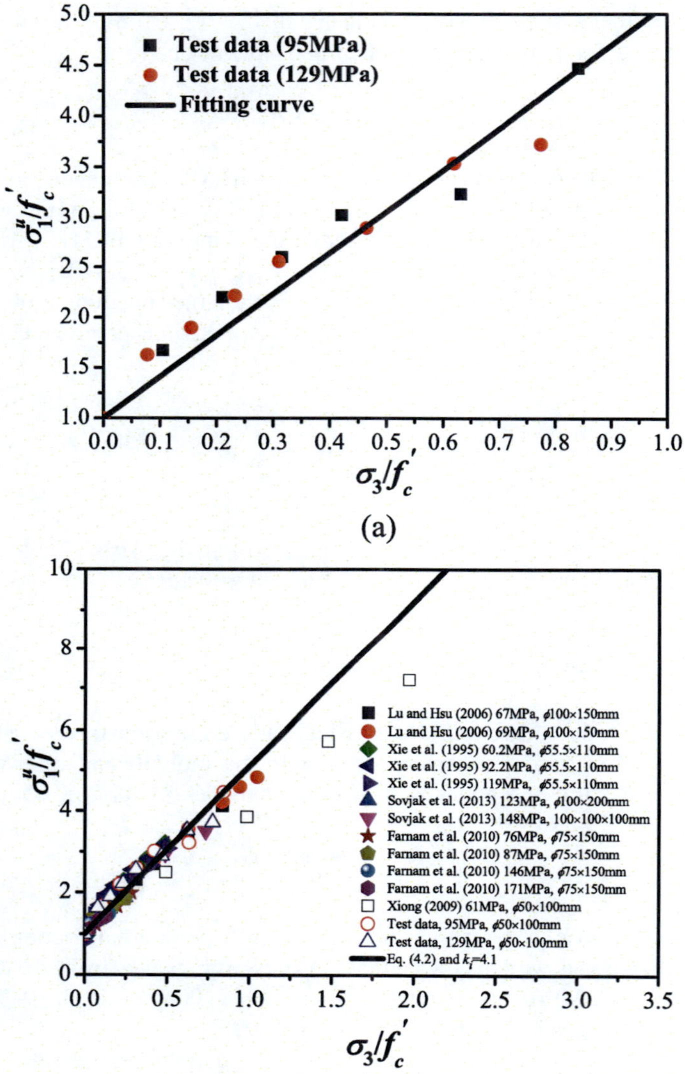

Fig. 4.7 Mohr–Coulomb failure criterion, reprinted from Ren et al. (2016), copyright 2020, with permission from Elsevier

4.4.3.2 Willam-Warnke Failure Criterion

Willam-Warnke failure criterion (Chen 1982) is a five-parameter model which has been adopted by the popular finite element software ANSYS. The compressive meridians equation is expressed as

$$\frac{\tau_m}{f_c'} = a_w + a_{w1}\frac{\sigma_m}{f_c'} - a_{w2}\left(\frac{\sigma_m}{f_c'}\right)^2 \tag{4.3}$$

where $\sigma_m = (\sigma_1 + \sigma_2 + \sigma_3)/3$ and $\tau_m = \frac{1}{\sqrt{15}}\left[(\sigma_1 - \sigma_2)^2 + (\sigma_2 - \sigma_3)^2 + (\sigma_3 - \sigma_1)^2\right]^{1/2}$ are the primary normal stress and shear stress, respectively. σ_1, σ_2 and σ_3 are the stresses in the three principal directions, a_w, a_{w1} and a_{w2} are the Willam-Warnke model coefficients, respectively.

Figure 4.8(a) shows the present test data from Table 4.2 as well as the fitting curves corresponding to Eq. (4.3), the compressive meridian of UHPCC in this study can be expressed as

$$\frac{\tau_m}{f_c'} = 0.177 + 0.656\frac{\sigma_m}{f_c'} - 0.062\left(\frac{\sigma_m}{f_c'}\right)^2 \tag{4.4}$$

Figure 4.8b also gives the existing compressive test data on high-strength concrete, including UHPCC, HPFRC and SIFCON as well as the curve of Eq. (4.4). The comparisons indicate that, compared with the Mohr–Coulomb failure criterion, Willam-Warnke failure criterion with the expression of Eq. (4.4) gives excellent predictions for high-strength concrete, and the influences of strength, size, and compositions of specimens are not pronounced.

4.4.3.3 Power-Law Failure Criterion

The Power-law failure criterion can be expressed as

$$\sigma_1^u/f_c' = 1 + a_p(\sigma_3/f_c')^{b_p} \tag{4.5}$$

where a_p and b_p are the coefficients.

Figure 4.9a shows the present test data on UHPCC as well as the power-law fitted curves with the form as

$$\sigma_1^u/f_c' = 1 + 3.50(\sigma_3/f_c')^{0.72} \tag{4.6}$$

Similarly, Fig. 4.9b gives the present and existing test data on high-strength concrete, as well as the curve of Eq. (4.6). It can be seen that, compared with the Mohr-Coulomb failure criterion, Power-law failure criterion with the expression of

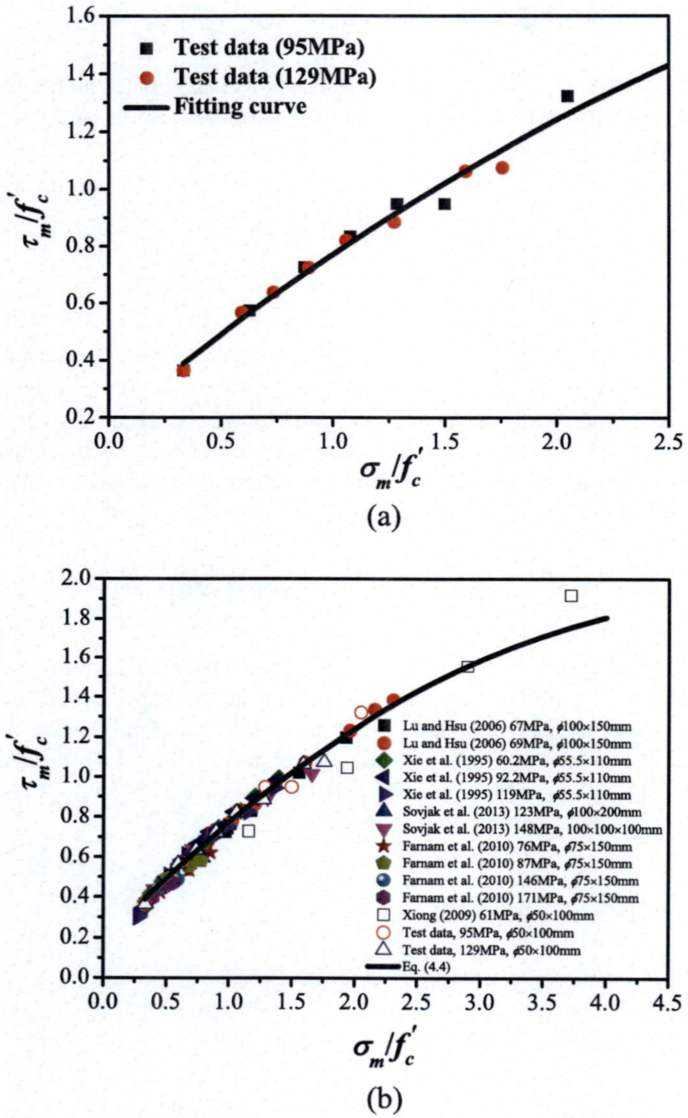

Fig. 4.8 Willam-Warnke failure criterion, reprinted from Ren et al. (2016), copyright 2020, with permission from Elsevier

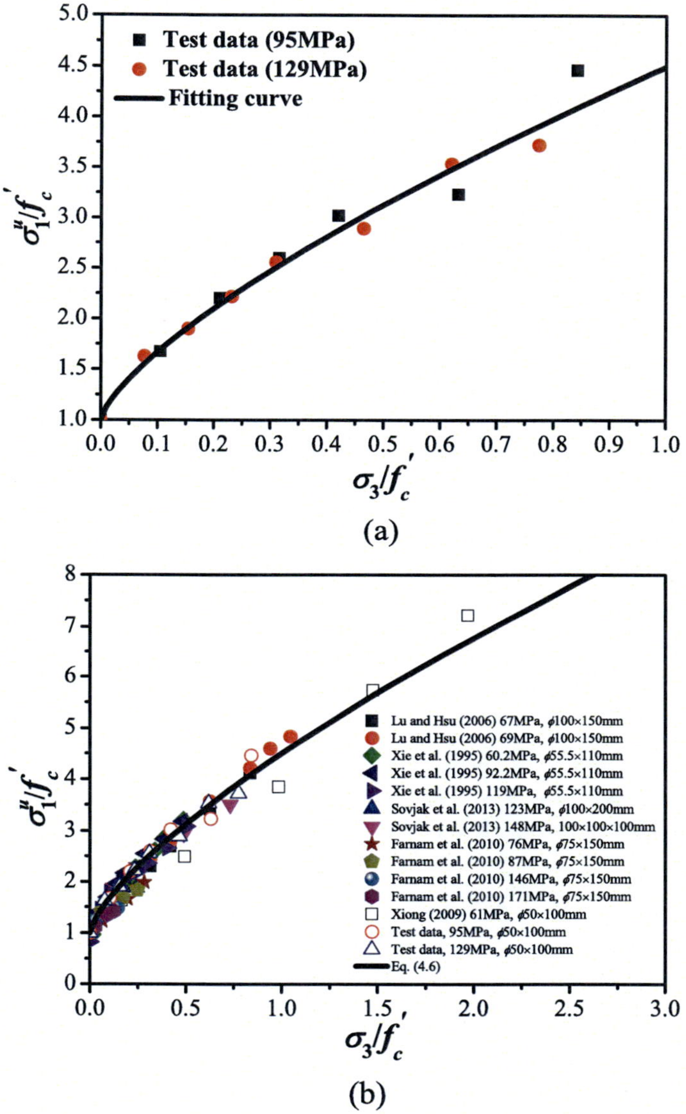

Fig. 4.9 Power-law failure criterion, reprinted from Ren et al. (2016), copyright 2020, with permission from Elsevier

Eq. (4.6) gives excellent predictions for high-strength concrete, and the influences of strength, size, and compositions of specimens are not pronounced.

4.4.4 Toughness

The toughness of concrete (R) generally refers to the energy absorbing performance of the material in compression and has been defined as the area under the stress–strain curve calculated up to a specified strain value (Taerwe 1992; Nataraja et al. 1999). Most researchers (Nataraja et al. 1999; Fanella and Naaman 1985; Poon et al. 2004) have obtained the toughness of fiber reinforced concrete in compression as the area under the stress–strain curve calculated up to a strain value of 0.015, while the specimens still have significant resistance left. In this study, in order to calculate the toughness of UHPCC under triaxial loading, a convenient method suggested by Farnam et al. (2010) was adopted.

This method was used to determine a ratio named toughness index (TI) that identify the toughness of material up to the selected strain criteria in compression to the unconfined concrete specimen. To calculate toughness, energy of all three directions is considered with the following equation:

$$R = \int (\sigma_1 - \sigma_3) d\varepsilon_1 - 2\int \sigma_3 d\varepsilon_3 \tag{4.7}$$

where σ_1, σ_3, ε_1 and ε_3 are the axial stress, confining stress, axial strain and lateral strain, respectively. $\int (\sigma_1 - \sigma_3) d\varepsilon_1$ is the area under the deviatoric stress-axial strain curve. Since the confining stress is constant, Eq. (4.7) can be further simplified as

$$R = \int (\sigma_1 - \sigma_3) d\varepsilon_1 - 2\sigma_3\varepsilon_3 \tag{4.8}$$

Since the maximum strain of the experimental deviatoric stress-axial strain curve is limited, in this chapter, four toughness indices (TI_1, TI_3, TI_5 and TI_7) in Eq. (4.9) introduced from Ref. (Farnam et al. 2010) were calculated for each triaxial test.

$$TI_1 = \frac{R_{cr}}{R_{cr_0}},\ TI_3 = \frac{R_{3cr}}{R_{cr_0}},\ TI_5 = \frac{R_{5cr}}{R_{cr_0}},\ TI_7 = \frac{R_{7cr}}{R_{cr_0}} \tag{4.9}$$

where R_{cr}, R_{3cr}, R_{5cr} and R_{7cr} are the toughness up to the corresponding first-crack strain (ε_{cr}), three times of ε_{cr}, five times of ε_{cr} and seven times of ε_{cr}, respectively. R_{cr_0} is the toughness of UHPCC with no confining pressure up to the first-crack strain. As shown in Fig. 4.10, ε_{cr} is calculated from the deviatoric stress-axial strain curve to the point where the elastic behavior ends and changes to nonlinearity. By this approach, the toughness of UHPCC can be easily calculated and the results are

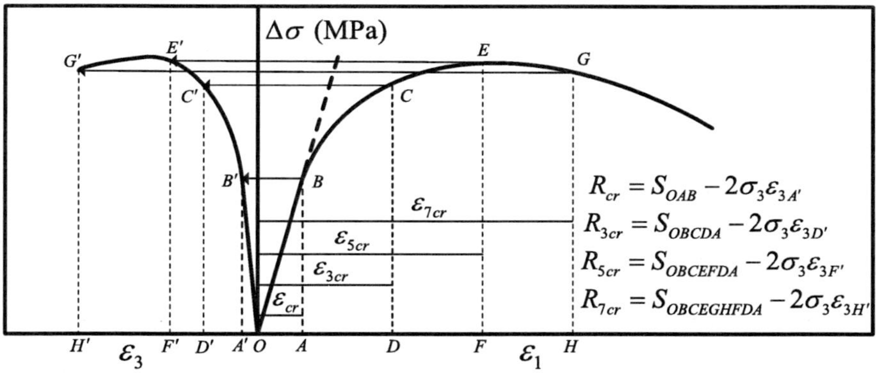

Fig. 4.10 Definition of toughness in terms of different multiples of first-crack deflection (Farnam et al. 2010)

listed in Table 4.3. Here, the lowest value of $R_{cr_0} = 0.123$ MPa for the specimen TC1-2 is adopted. "—" denotes that the maximum strain of the experimental deviatoric stress-axial strain curve could not reach the correspondingly specified strain. As can be seen in Table 4.3, the maximum triaxial toughness indices of UHPCC can reach the 82.52 times of the uniaxial toughness indices of UHPCC within elastic stage,

Table 4.3 Toughness indices of UHPCC under triaxial compression

No	σ_3 (MPa)	First-crack strain ε_{cr}	TI_1	TI_3	TI_5	TI_7
TC1-1	0	0.0040	1.05	5.89	—	—
TC1-2	0	0.0039	1.00	6.50	—	—
TC1-3	10	0.0045	1.40	10.27	19.05	—
TC1-4	20	0.0044	1.60	11.70	23.80	33.49
TC1-5	30	0.0053	2.13	15.95	32.75	49.31
TC1-6	40	0.0043	1.35	11.60	24.90	36.56
TC1-7	60	0.0047	1.24	12.10	26.63	41.94
TC1-8	80	0.0056	2.83	21.39	45.24	69.25
TC2-1	0	0.0053	1.62	7.29	—	—
TC2-2	0	0.0049	1.87	8.24	—	—
TC2-3	10	0.0057	3.22	15.94	23.74	30.56
TC2-4	20	0.0058	3.54	20.14	32.71	—
TC2-5	30	0.0057	3.07	23.15	43.08	59.07
TC2-6	40	0.0057	3.30	23.34	40.84	—
TC2-7	60	0.0058	3.28	19.46	49.71	71.94
TC2-8	80	0.0058	2.50	25.63	55.10	82.52
TC2-9	100	0.0058	1.78	22.88	51.92	75.81

and the confining pressure have good effect on improving the toughness indices. The same conclusion was obtained by Farnam et al. (2010).

4.5 Applications in the Numerical Analyses

In this section, based on the present and previous triaxial compression test data on high-strength concrete (Lu and Hsu 2006; Xie et al. 1995; Sovják et al. 2013; Farnam et al. 2010; Xiong 2009), the dominated strength parameters for Holmquist-Johnson-Cook (HJC) constitutive model were calibrated. Then, by utilizing the large commercial finite element program LS-DYNA, the HJC model and the corresponding parameters were utilized to predict the terminal ballistic parameters of projectile impacting high-strength concrete-like target.

4.5.1 Brief Introduction of HJC Constitutive Model

HJC model is an elastic-viscoplastic model and used to describe the dynamic compressive behaviors of concrete subjected to large strain, high strain and hydrostatic pressure. It was embedded into the commercial finite element program LS-DYNA (LS-DYNA 1997) as the *MAT_JOHNSON_HOLMQUIST_CONCRETE (111#) model, and widely applied into the numerical simulations of the dynamic responses of concrete-like material under projectile impact and explosive loadings. It mainly includes three parts: equation of yield surface, equation of state and equation of damage evolution, shown in Fig. 4.11.

Shown in Fig. 4.11a, the equation of yield surface of HJC model can be expressed as

$$\sigma^* = [A(1 - D) + BP^{*N}](1 + C \ln \dot{\varepsilon}^*) \leq S_{\max} \tag{4.10}$$

where $\sigma^* = \sigma / f_c'$ and $P^* = P / f_c'$ are the normalized equivalent stress and hydrostatic pressure, respectively. σ denotes the actual equivalent stress and P is the actual pressure. $T^* = T / f_c'$ is the normalized maximum tensile hydrostatic pressure, where T is the maximum tensile hydrostatic pressure the material can withstand. $\dot{\varepsilon}^* = \dot{\varepsilon} / \dot{\varepsilon}_0$ is the dimensionless strain rate, where $\dot{\varepsilon}$ and $\dot{\varepsilon}_0 = 1.0\,s^{-1}$ are the actual and reference stain rate, respectively. In addition, A, B, N and $S_{\max}$ are the material strength parameters obtained from experimental data, which represent the normalized cohesive strength, the normalized pressure hardening coefficient, the pressure hardening exponent and the normalized maximum strength that can be developed. D ($0 \leq D \leq 1$) is the damage parameter and C is the strain rate coefficient.

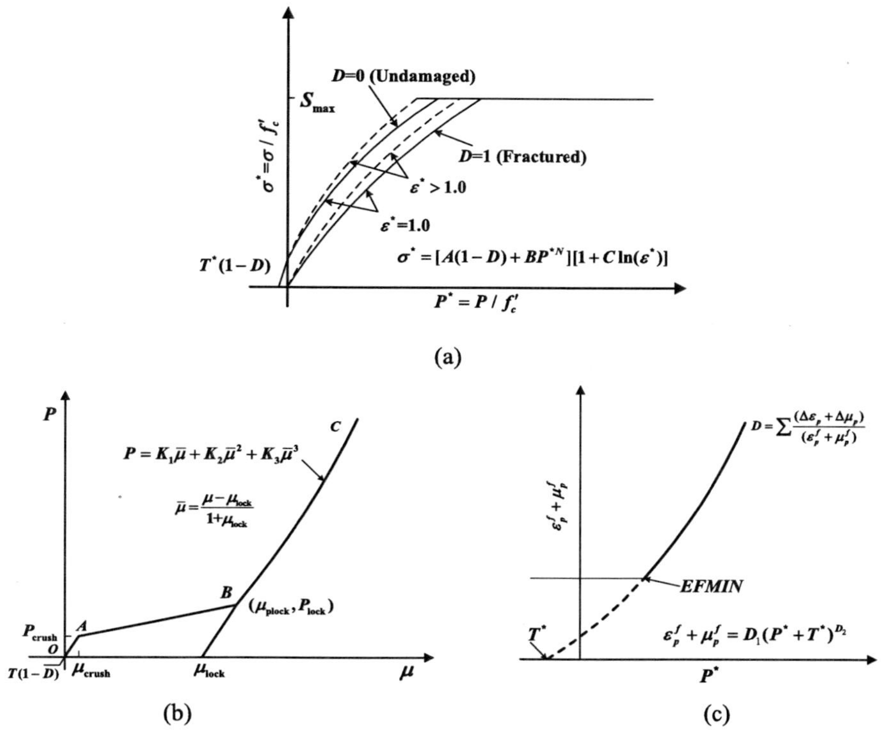

Fig. 4.11 HJC model **a** equation of yield surface, **b** equation of state **c** damage model, reprinted from Ren et al. (2016), copyright 2020, with permission from Elsevier

Shown in Fig. 4.11b, the pressure-volumetric strain relationship is separated into three response regions in the equation of state of HJC model. It is expressed as

$$\begin{cases} P = K\mu & P < P_{\text{crush}} \\ P = P_{\text{crush}} + K_{\text{lock}}(\mu - \mu_{\text{crush}}) & P_{\text{crush}} < P < P_{\text{lock}} \\ P = K_1\bar{\mu} + K_2\bar{\mu}^2 + K_3\bar{\mu}^3 & P > P_{\text{lock}} \end{cases} \tag{4.11}$$

where $\mu = \rho/\rho_0 - 1$ is the volumetric strain, ρ and ρ_0 are the current and initial density, respectively. The elastic bulk modulus is $K = P_{\text{crush}}/\mu_{\text{crush}}$, where P_{crush} and μ_{crush} are the pressure and volumetric strain at which the material begins to undergo plastic deformation. $K_{\text{lock}} = (P_{\text{lock}} - P_{\text{crush}})/(\mu_{\text{plock}} - \mu_{\text{crush}})$, and μ_{plock} is the volumetric strain at the compacting hydrostatic pressure P_{lock}. $\overline{\mu} = (\mu - \mu_{\text{lock}})/(1 + \mu_{\text{lock}})$ is the modified volumetric strain, and the constants K_1, K_2 and K_3 are equivalent to those used for material with no voids.

The damage for fracture is shown in Fig. 4.11c. The model discussed herein accumulates damage from both equivalent plastic strain and plastic volumetric strain, and is expressed as

$$D = \sum \frac{\Delta \varepsilon_p + \Delta \mu_p}{\varepsilon_p^f + \mu_p^f} \tag{4.12}$$

$$\varepsilon_p^f + \mu_p^f = D_1 (P^* + T^*)^{D_2} \geq EFMIN \tag{4.13}$$

where $\Delta \varepsilon_p$ and $\Delta \mu_p$ are the equivalent plastic strain and plastic volumetric strain during a cycle of integration, and $\varepsilon_p^f + \mu_p^f$ is the total plastic strain under a constant pressure until fracture. D_1 and D_2 are damage constants, and $EFMIN$ is a material constant used to suppress fracture from weak tensile waves.

4.5.2 HJC Model Parameters for HSC

Based on the projectile penetration test on Salem limestone targets as well as the symmetrical plate impact test on concrete panel, Fang et al. (2014) and Xiong et al. (2009) discussed the parameters sensitivity of HJC model. It has been found that ρ_0, f_c', B and N are the most sensitive parameters, in which ρ_0 and f_c' are the basic mechanical parameters, and should be given by the specific experimental data. B and N are the strength parameters determined by the triaxial compression tests. Besides, the normalized cohesive strength parameter A is assumed to be 0.30 in Refs. (Malvar et al. 1997, 2000), which is obtained from a large amount of experimental data. Therefore, Eq. (4.10) could be further simplified without considering the influences of damage and strain rate effect as follows:

$$\sigma^* = 0.30 + BP^{*N} \tag{4.14}$$

where $\sigma^* = (\sigma_1 - \sigma_3)/f_c'$, $P^* = (\sigma_1 + \sigma_2 + \sigma_3)/3f_c'$. Based on the discussions in Sect. 4.4.3, UHPCC has the uniform ultimate strength envelop with other types of HSC. Therefore, by fitting the previous and present triaxial compression test data on HSC specimens, the strength parameters $B = 1.73$ and $N = 0.79$ could be obtained as shown in Fig. 4.12.

As for the other parameters of HJC model, due to the weak sensitivity, they are still referred from the original documents (Holmquist et al. 1993). Table 4.4 gives the present parameters of HJC constitutive model for high-strength concrete.

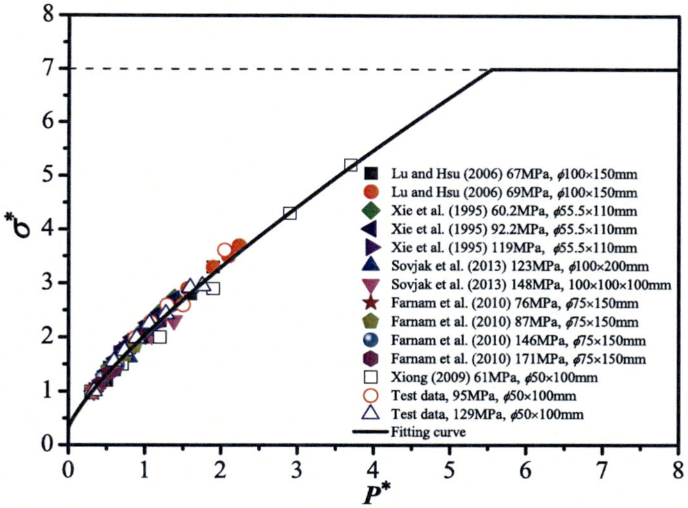

Fig. 4.12 Determination of strength parameters for HSC, reprinted from Ren et al. (2016), copyright 2020, with permission from Elsevier

4.5.3 *Validations*

In this section, by using the large commercial finite element program LS-DYNA (LS-DYNA 1997), a series of projectile impacting tests on HSC and high-strength rock targets conducted by Wu et al. (2015a), O'Neil et al. (1999), Yan (2001), Frew et al. (2000), Zhang et al. (2005a; b) (granite targets) and Hanchak et al. (1992) were numerically simulated to validate the proposed parameters of HJC constitutive model for HSC in Sect. 4.5.2.

Table 4.5 lists the related experimental parameters of total eleven sets of rigid projectile penetration and perforation tests on HSC and high-strength rock targets. Where d, CRH and V_0 are the diameter, caliber-radius-head (the ratio of curvature radius of the ogival nose to the projectile shank diameter) and striking velocity of projectile, respectively. D_t and H_t are the diameters and thickness of targets. "—" denotes the data that was not given in original references.

4.5.3.1 Finite Element Model

Since the projectiles and targets are axial symmetrical in the above experiments. The 2D Solid 162 element type was employed for modelling the projectile and concrete, and the 2D axisymmetric Lagrange algorithm was used. Local mesh refinement was deployed in the mesh division of concrete targets so as to guarantee the requirements of computational precision and save the computational cost. Minimum element size for target is considered as 1 mm near the impact location, and the

Table 4.4 Parameters of HJC constitutive model for HSC

Strength parameters				Equation of state parameters				Damage parameters			Strain rate parameter
A	B	N	S_{max}	P_{lock} (GPa)	K_1 (GPa)	K_2 (GPa)	K_3 (GPa)	D_1	D_2	$EFMIN$	C
0.30	1.73	0.79	7	0.80	85	−171	208	0.04	1.0	0.01	0.007

Table 4.5 Parameters of the projectile penetration and perforation tests

No	References	d (mm)	CRH	V_0 (m/s)	f_c' (MPa)	D (cm)	H (cm)	Erosion strain
1	Wu et al. (2015a)	25.3	3	510	67.5–125.2	75	50–65	0.60
2		25.3	3	850	67.5–125.2	75	60–100	0.60
3	O'Neil et al. (1999)	26.9	2	562–795	104	76.2	91.4	0.25
4		26.9	2	229–754	157	76.2	91.4	0.30
5	Yan (2001)	14.5	1	273–793	71.7	30	20	0.90
6	Frew et al. (2000)	7.1	3	497–1516	60	71	—	0.60
7		12.7	3	459–1502	60	71		0.80
8		25.4	3	407–1177	60	100		0.60
9	Zhang et al. (2005a, b)	15	3	332–839	154	44.3	60	0.60
10		20	3	298–615	154	44.3	60	0.50
11	Hanchak et al. (1992)	25.4	3	376–996	140	61	17.8	0.15

element size is gradually enlarged along the radius outward. The projectiles were modelled by the *MAT_RIGID (20#) model and the targets were modelled by the *MAT_JOHNSON_HOLMQUIST_CONCRETE (111#) model with the parameters given in Table 4.4, respectively. For the eleven sets tests listed in Table 4.5, the diameters of cylindrical targets were over 20 times larger than the projectiles diameters, and thus the influences of circumferential boundary was neglected, which was set as the free boundary. The contact between the projectile and concrete was modelled by using the keyword "CONTACT_2D_AUTOMATIC_SINGLE_ SURFACE" in LS-DYNA. In order to simulate the physical fracture or crushing of concrete numerically, the erosion algorithm *MAT_ADD_EROSION was brought in and the erosion criterion was defined by the maximum principal strain. During the high velocity impact analysis, when the instantaneous geometric strain of the element reach the above predefined value, the element was removed and would not further offer resistance. For each test, the value of erosion strain given in the last column was the one which give better predictions of the specific shot data (the first, middle or last), as shown in Table 4.5. Figures 4.13 and 4.14 show the finite element model of projectile penetration and perforation tests as well as the damage contour of targets, respectively.

4.5.3.2 Comparisons of Numerical Predictions with Test Data

The main terminal ballistic parameters during the impact of the projectile were the depth of penetration (DOP, the maximum normal depth of the projectile penetrates into the target) and the residual velocity (V_r, the remaining exit velocity of the projectile after perforating the target) of the projectile, shown in Figs. 4.13 and 4.14. Figure 4.15 shows the comparisons of test data and numerical predicted curves of

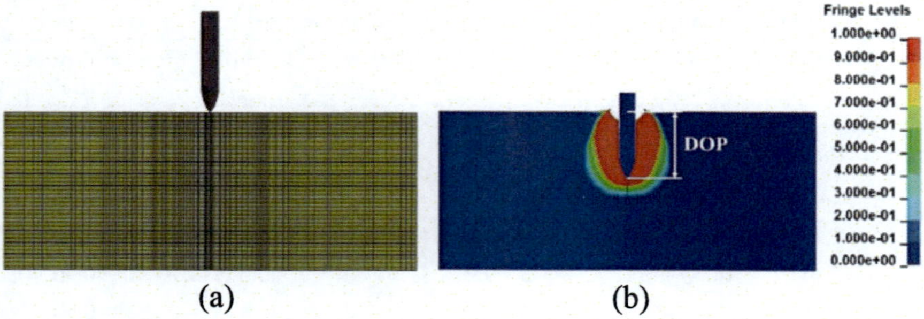

Fig. 4.13 Finite element model of projectile penetration test and the damage contour of target, reprinted from Ren et al. (2016), copyright 2020, with permission from Elsevier

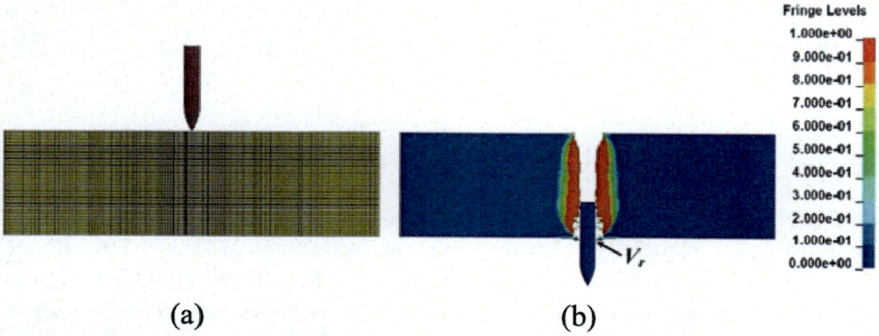

Fig. 4.14 Finite element model of projectile perforation test and the damage contour of target, reprinted from Ren et al. (2016), copyright 2020, with permission from Elsevier

dimensionless depth of penetration DOP/*d* and residual velocities V_r for penetration tests 1–10 and perforation test 11 in Table 4.5, respectively. From Fig. 4.15, it can be found that, for both DOP of projectile striking into semi-infinite target as well as residual velocity of projectile after perforating the finite-depth slab, the predicted results correlate well with the test data. Therefore, for predicting the terminal ballistic parameters of projectile impacting on high-strength concrete-like targets, the validations of HJC model as well as the proposed parameters given in Table 4.4 were verified.

4.6 Summary

For the construction and design of concrete protective structures, the triaxial compressive behavior of UHPCC and its applications into the analyses of high-speed projectile impact were investigated. The main works and conclusions are:

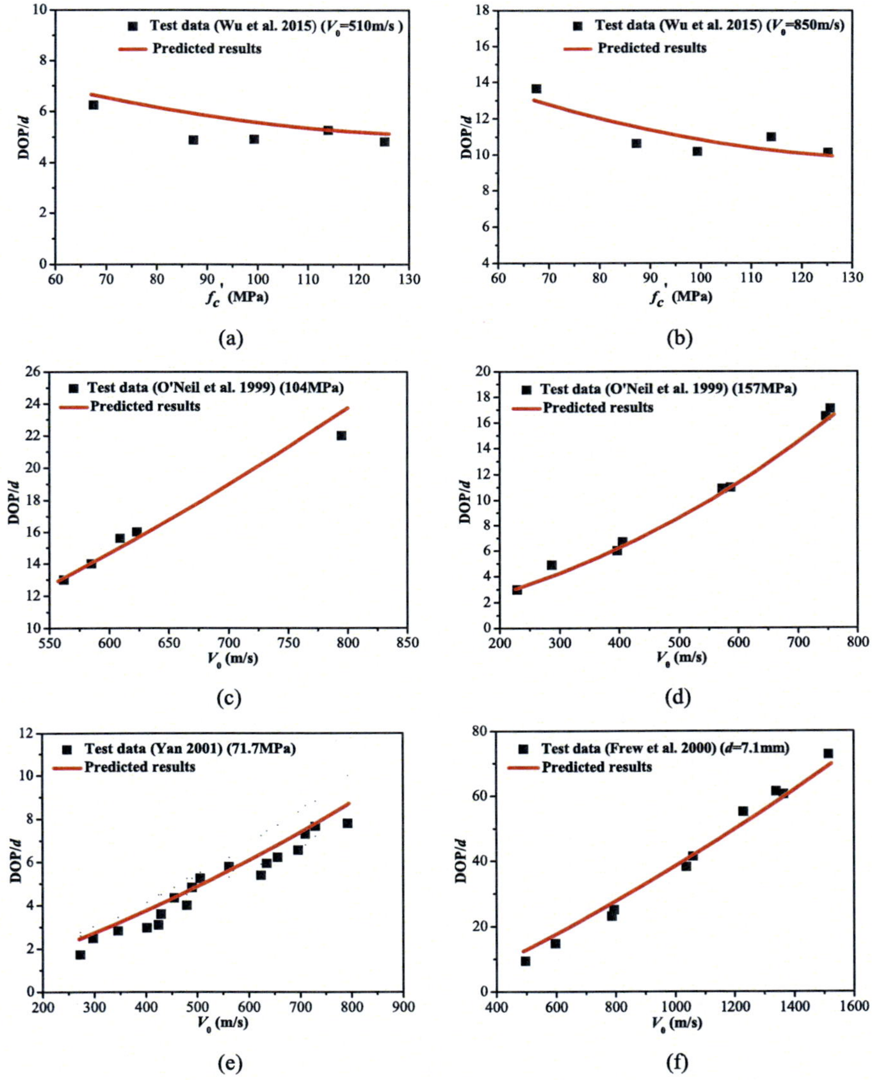

Fig. 4.15 Comparisons of test data and numerical results of **a** test 1, **b** test 2, **c** test 3, **d** test 4, **e** test 5, **f** test 6, **g** test 7, **h** test 8, **i** test 9, **j** test 10, **k** test 11, reprinted from Ren et al. (2016), copyright 2020, with permission from Elsevier

1. The optimal UHPCC with higher impact resistance, good workability, low-costs and energy saving was prepared under ordinary procedures without temperature and pressure curing;

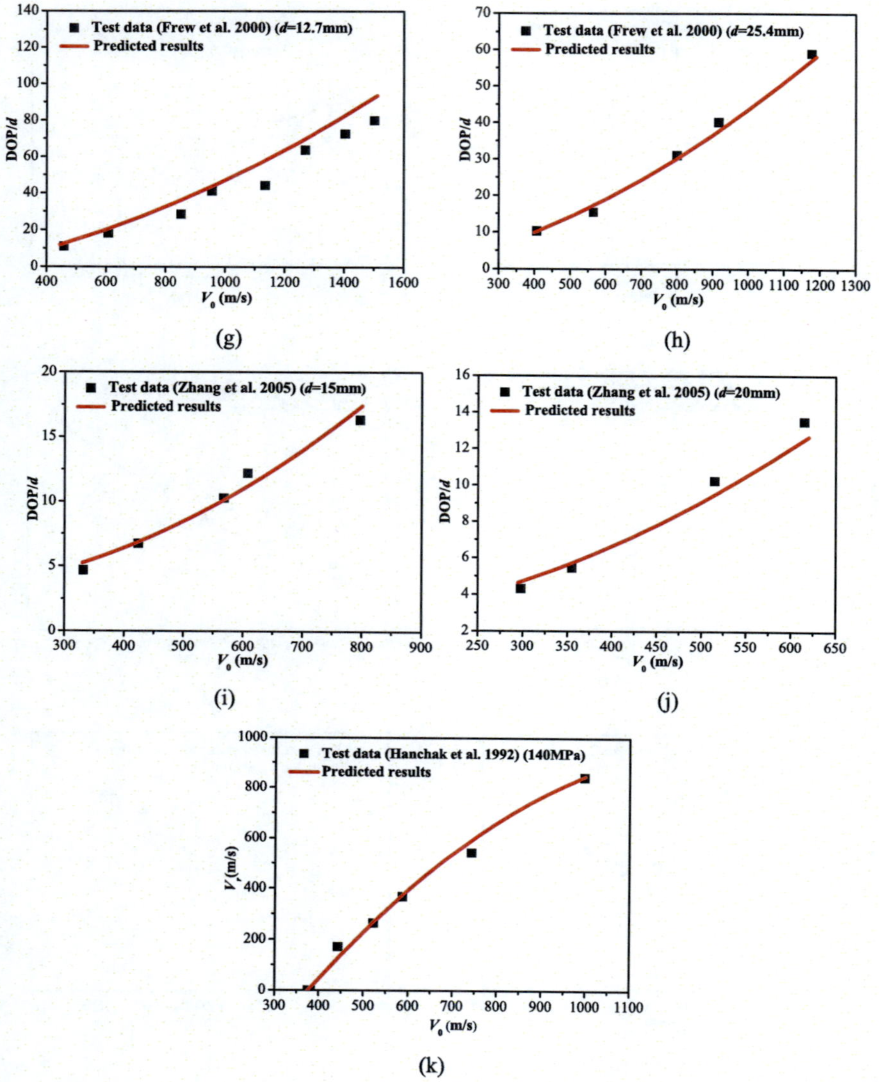

Fig. 4.15 (continued)

2. Triaxial compressive test on two batches UHPCC specimens (95 and 129 MPa) was conducted, and the deviatoric stress–strain curves with high confining pressure up to 100 MPa were obtained;
3. The triaxial compressive strength and ductility of UHPCC increases with rising the confining pressure, the shear failure and swelling damage modes occurred

successfully, and the normalized peak axial strain increases linearly with the confinement ratio;

4. Willam-Warnke and Power-law failure criteria show excellent predictions of triaxial compressive strength envelop of HSC, e.g. UHPCC, SIFCON, the influences of strength, size, and compositions of specimen were not pronounced;
5. The dominant strength parameters of HJC constitutive model for HSC were calibrated and validated by comparing the numerical results with eleven sets of high-speed projectile impact tests.

References

ANSARI F, LI Q. High-strength concrete subjected to triaxial compression[J]. ACI Materials Journal, 1998, 95(6): 747–755.

ATTARD M M, SETUNGE S. Stress-strain relationship of confined and unconfined concrete[J]. ACI Materials Journal, 1996, 93(5): 432–441.

CANDAPPA D C, SANJAYAN J G, SETUNGE S. Complete triaxial stress-strain curves of high-strength concrete[J]. Journal of Materials in Civil Engineering, 2001, 13(13): 209–215.

CHEN D C. Plasticity in reinforced concrete[M]. McGraw-Hill Book Company New York, USA, 1982.

CHERN J C, YANG H J, CHEN H W. Behavior of steel fiber reinforced concrete in multiaxial loading[J]. ACI Materials Journal, 1992, 89(1): 32–40.

DANCYGIER A N, YANKELEVSKY D Z, HAEGERMANN C. Response of high performance concrete plates to impact of non-deforming projectiles[J]. International Journal of Impact Engineering, 2007, 34(11): 1768–1779.

DANCYGIER A N, KATZ A, BENAMOU D, YANKELEVSKY D Z. Resistance of double-layer reinforced HPC barriers to projectile impact[J]. International Journal of Impact Engineering. 2014, 67: 39–51.

ETSE G, WILLAM K. Fracture energy formulation for inelastic behavior of plain concrete[J]. Journal of Engineering Mechanics, 1994, 120(9): 1983–2011.

FANELLA D A, NAAMAN A E. Stress-strain properties of fiber reinforced mortar in compression[J]. Journal of the American Concrete Institute, 1985, 82(4): 475–483.

FANG Q, KONG X Z, WU H, GONG Z M. Determination of Holmquist-Johnson-Cook constitutive model parameters of rock[J]. Engineering Mechanics, 2014, 31(3): 197–204. (in Chinese).

FARNAM Y, MOOSAVI M, SHEKARCHI M, BABANAJAD S K, BAGHERZADEH A. Behaviour of slurry infiltrated fibre concrete (SIFCON) under triaxial compression[J]. Cement and Concrete Research, 2010, 40: 1571–1581.

FREW D J, FORRESTAL M J, HANCHAK S J. Penetration experiments with limestone targets and ogive-nose steel projectiles[J]. Journal of Applied Mechanics, 2000, 67(4): 841–845.

GB175–2007. Common Portland cement[S]. National Standard of the People's Republic of China, 2008.

GIRGIN Z C, ANOGLU N, ANOGLU E. Evaluation of strength criteria for very-high-strength concretes under triaxial compression[J]. ACI Structural Journal, 2007, 104(3): 277–283.

GRAYBEAL B A. Material property characterization of ultra-high performance concrete: FHWA-HRT-06-103[R]. VA: U.S. Department of Transportation Federal highway Adminstration, 2006: 1–186.

HANCHAK S J, FORRESTAL M J, YOUNG E R, EHRGOTT J Q. Perforation of concrete slabs with 48MPa (7ksi) and 140MPa (20ksi) unconfined compressive strength[J]. International Journal of Impact Engineering, 1992, 12(1): 1–7.

HOLMQUIST T J, JOHNSON G R, COOK W H. A computational constitutive model for concrete subjective to large strain, high strain rates, and high pressure. The 14th International Symposium on Ballistic, Quebec[C]. 1993: 591–600.

IMRAN I, PANTAZOPOULOU S J. Experimental study of plain concrete under triaxial stress[J]. ACI Structural Journal, 1996, 93(6): 589–601.

LANGBERG H, MARKESET G. High performance concrete-penetration resistance and material development. Proceedings of the Ninth International Symposium on Interaction of the Effects of Munitions with Structures, Berlin-Strausberg[C]. 1999: 933–941.

LAN S, GUO Z. Experimental investigation of multiaxial compressive strength of concrete under different stress paths[J]. ACI Structural Journal, 1997, 94(5): 427–434.

LI Q, ANSARI F. Mechanics of damage and constitutive relationships for high-strength concrete in triaxial compression[J]. Journal oF Engineering Mechanics, 1997, 125(1): 1–10.

LI Q, ANSARI F. High-strength concrete in triaxial compression by different size of specimens[J]. ACI Structural Journal, 2010, 97(6): 684–689.

LS-DYNA. Theoretical Manual[M]. Livermore Software Technology Corporation, Livermore, USA, 1997.

LU X B, HSU C T T. Behavior of high strength concrete with and without steel fiber reinforcement in triaxial compression[J]. Cement and Concrete Research, 2006, 36: 1679–1685.

MAALEJ M, QUEK S T, ZHANG J. Behavior of hybrid-fiber engineered cementitious composites subjected to dynamic tensile loading and projectile impact[J]. Journal of Materials in Civil Engineering, 2005, 17(2): 143–152.

MÁCA P, SOVJÁK R, KONVALINKA P. Mix design of UHPFRC and its response to projectile impact[J]. International Journal of Impact Engineering, 2010, 63(5): 158–163.

MALVAR L J, CRAWFORD J E, WESEVICH J W, SIMONS D. A plasticity concrete material model for DYNA3D[J]. International Journal of Impact Engineering, 1997, 19: 847–873.

MALVAR L J, CRAWFORD J E, MORRILL K B. K&C concrete material model release III-automated generation of material model input[A]. Karagozian&Case Structural Engineers, 2000.

NAAMAN A E, HOMRICH J R. Tensile stress-strain properties of SIFCON[J]. ACI Materials Journal, 1989, 86(3): 244–251.

NAAMAN A E, REINHARDT H W. Characterization of high performance fiber reinforced cement composites- HPFRCC[M]. High performance fiber reinforced cement composites, 1996: vol. 2. Ann Arbor 1–24.

NATARAJA M C, DHANG N, GUPTA A P. Stress-strain curves for steel-fiber reinforced concrete under compression[J]. Cement and Concrete Research, 1999, 21: 383–390.

O'NEIL E F, NEELEY B D, CARGILE J D. Tensile properties of very-high-strength concrete for penetration- resistant structures[J]. Shock and Vibration, 1999, 6(5): 237–245.

OTTOSEN N S. A failure criterion for concrete[J]. Journal of Engineering Mechanics, 1977, 103(4): 527–535.

PENG Y, WU H, FANG Q, LIU J Z, GONG Z M. Impact resistance of basalt aggregated UHP-SFRC/fabric composite panel against small caliber arm[J]. International Journal of Impact Engineering, 2016, 88: 201–213.

PETR M, RADOSLAV S, TOMÁŠ V. Experimental investigation of Mechanical properties of UHPFRC[J]. Procedia Engineering, 2013, 65: 14–19.

PIOTROWSKA E, MALECOT Y, KE Y. Experimental investigation of the effect of coarse aggregate shape and composition on concrete triaxial behavior[J]. Mechanics of Materials, 2014, 79: 45–57.

POON C S, SHUI Z H, LAM L. Compressive behavior of fiber reinforced high-performance concrete subjected to elevated temperatures[J]. Cement and Concrete Research, 2004, 34(12): 2215–2222.

REN G M, WU H, FANG Q, LIU J Z, GONG Z M. Triaxial compressive behavior of UHPCC and applications in the projectile impact analyses[J]. Construction and Building Materials, 2016, 113: 1–14.

RICHARD P, CHEYREZY M. Composition of reactive powder concrete[J]. Cement and Concrete Research, 1995, 25(7): 1501–1511.

ROSSI P, ARCA A, PARANT E, FAKHRI P. Bending and compressive behaviors of a new cement composite[J]. Cement and Concrete Research, 2005, 35(1): 27–33.

SFER D, CAROL I, GETTU R, ETSE G. Study of the behavior of concrete under triaxial compression[J]. Journal of Engineering Mechanics, 2002, 128(2): 156–163.

SIRIJAROONCHAI K, EI-TAWIL S, PARRA-MONTESINOS. Behavior of high performance fiber reinforced cement composites under multi-axial compressive loading[J]. Cement & Concrete Composites, 2010, 32: 62–72.

SOVJÁK R, VAVŘINIK T, ZATLOUKAL J, MÁCA P, MIČUNEK T, FRYDRÝN M. Resistance of slim UHPFRC targets projectile impact using in-service bullet[J]. International Journal of Impact Engineering, 2015, 76: 166–177.

SOVJÁK R, VOGEL F, BECKMANN B. Triaxial compressive strength of ultra high performance concrete[J]. Acta Polytechnica Hungarica, 2013, 53(6): 901–905.

TAERWE L R. Influence of steel fibers on strain-softening of high-strength concrete[J]. ACI Materials Journal, 1992, 88(6): 54–60.

VU X H, DAUDEVILLE L, MALECOT YANN. Effect of coarse aggregate size and cement paste volume on concrete behavior under high triaxial compression loading[J]. Construction and Building Materials, 2011, 25: 3941–3949.

WANG C Z, GUO Z H, ZHANG X Q. Experimental investigation of biaxial and triaxial compressive concrete strength[J]. ACI Materials Journal, 1987, 84(2): 92–100.

WILLE K, TAWIL S-EL, NAAMAN A E. Properties of strain hardening ultra high performance fiber reinforced concrete (UHP-FRC) under direct tensile loading[J]. Cement & Concrete Composites, 2014, 48: 53–66.

WU H, FANG Q, CHEN X W, GONG Z M, LIU J Z. Projectile penetration of ultra-high performance cement based composites at 510m/s to 1320m/s[J]. Construction and Building Materials, 2015, 74(15): 188–200.

WU H, FANG Q, GONG J, LIU J Z, ZHANG J H, GONG Z M. Projectile impact resistance of corundum aggregated UHP-SFRC[J]. International Journal of Impact Engineering, 2015, 84: 38–53.

XIE J, ELWI A E, MACGREGOR J G. Mechanical properties of three high-strength concretes containing silica fume[J]. ACI Materials Journal, 1995, 92: 135–143.

XIONG Y B. Research on constitutive parameters of concrete based on the Johnson-Holmquist concrete model[D]. Xi'an: Northwest Institute Nuclear of Technology, 2009. (in Chinese).

XIONG Y B, CHEN J J, HU Y L. Preliminary identification of sensitive parameters in Johnson-Holmquist concrete constitutive model[J]. Acta Armamentarii, 2009, 30(2 suppl): 145–148. (in Chinese).

YAN S H. Study of penetration theory and experiments on high-strength fiber reinforced concrete[D]. Nanjing: PLA University of Science and Technology, 2001. (in Chinese).

ZHANG D Z, ZHANG X R, LIN J D, TANG R D. Penetration experiments for normal impact into granite targets with high-strength steel projectile[J]. Chinese Journal of Rock Mechanics and Engineering, 2005, 24(9): 1612–1618. (in Chinese).

ZHANG M H, SHIM V P W, LU G, CHEW C W. Resistance of high-strength concrete to projectile impact[J]. International Journal of Impact Engineering, 2005, 31(7): 825–841.

Chapter 5
Projectile Penetrations into Coarse Aggregated UHPCC Targets

5.1 Introduction

The cement based composites, such as high strength concrete (HSC) and the steel fiber reinforced HSC (SFRHSC), are generally considered to be the proper materials for both military and civil constructions, such as fortifications, nuclear containment and other protective structures, which are designed to withstand the intentional and accidental impact loadings caused by projectiles, fragments, etc. Investigations on the projectile impact resistances of HSC and SFRHSC have been drawn much attentions for both weapon designers and civil engineers in the past 20 years.

Based on the comparative projectile perforation tests on the normal strength concrete (NSC, 48 MPa) and HSC (140 MPa) slabs, Hanchak et al. (1992) found that, the three-fold increase in concrete compressive strength resulted in 20% decrease of the residual velocities of the projectile after perforating the concrete slabs. Dancygier et al. (1996, 1998) experimentally studied the influences of different ductility improvement approaches (steel fibers, small diameter steel wire mesh and woven steel fence mesh of various diameters) on the impact resistance of the HSC targets (95–100 MPa), where the projectile striking velocities ranged from 85 to 230 m/s. It was found that adding the steel fibers was the most effective approach in reducing the spalling and scabbing areas of both impacted and rear faces of the targets. Under the strike of the projectile at 229–754 m/s, O'Neil et al. (1999) also found that the DOP of the concrete target with a compressive strength of 157 MPa was approximately 50 and 30% less than that of the targets with a compressive strength of 35 MPa as well as 85–104 MPa, respectively. Additionally, the fibers did not significantly improve the penetration resistance, but prohibiting visible damage of the concrete surrounding the penetration crater. Based on the projectile impact test on SFRHSC at 364.9–378.3 m/s, Luo et al. (2000) found that the targets with the cubic compressive strengths of 72.4 MPa exhibited smash failure, while the targets with the cubic compressive strengths of 107.1–116.1 MPa remained intact with only several radial cracks on the front surface and some minor cracks on the side face. Langberg and Markeset (1999) derived that, although the penetration resistance of HSC was

Q. Fang et al., *UHPCC Under Impact and Blast*,
https://doi.org/10.1007/978-981-33-6842-2_5

much better than that of the NSC, the impact resistance was not improved obviously any more when the cylinder compressive strength was beyond 150 MPa. Liu et al. (2002) experimentally investigated the impact resistance of SFRHSC targets, where the fiber volumetric fraction ranged from 0% up to 12% and the projectile striking velocities were 314 m/s and 414 m/s, respectively. It was found that, the strength and toughness of the target increased tremendously with rising the fiber mixing ratio, especially when the fiber volume fraction was beyond 4%. The DOP of the projectile penetrating into 100 MPa concrete target decreased about 33% when the fiber volume fraction rised from 0 up to 10%. However, above-mentioned large addition ratios of fibers destroy the workability of the concrete, which cannot be realized in the field constructions. From the projectile penetration tests on concrete targets with the compressive strengths of 45–235 MPa and the impact velocities around 610–710 m/s, Zhang et al. (2005, 2007) discussed the effects of content, type and length of the incorporated fibers as well as the compressive and flexural tensile strengths of the concrete on the impact resistances. It indicated that, the compressive strength of the target and the coarse granite aggregates are both beneficial in decreasing penetration depth, while the incorporated fibers are mainly contributing to reducing crater diameter and prohibiting crack propagation. Zhang et al. (2005) further proposed that the SFRHSC with a compressive strength of 100 MPa was the most efficient choice for the anti-strike protective structures. From a series of comparative projectile impact tests on NSC (40 MPa) and HSC (93–119 MPa) targets with the striking velocities at 200–315 m/s, the enhancing effects of compressive strength of concrete as well as the type and size of coarse aggregates on the impact resistance of the target were also evaluated by Dancygier et al. (2007). Tai (2009) carried out the low-velocity (27–104 m/s) projectile impact tests on plain or fiber reinforced NSC (~25 MPa) and reactive powder concrete (RPC, 161.9–192.8 MPa) target, respectively. The results indicated that, adding steel fibers slightly increased the compressive strength of RPC, but remarkably increased the toughness and fracture energy of RPC, e.g. the scabbing area decreased by approximately 50% when 1% steel fiber was mixed into RPC.

The main conclusions from above works can be drawn as follows: (1) the impact resistance of the concrete target is enhanced with increasing the compressive strength, while the brittleness of the target also rises, which is detrimental to prevent the target from spalling and scabbing. (2) Mixing steel fibers in HSC can significantly improve the toughness and fracture energy, restrain the initiation and propagation of cracking due to the bridging effect, and thus reduce the damages of the impact crater as well as the scabbing of the rear face. However, too many fibers would impair the workability of the composite, and the allowable upper limit of the fiber volume ratio for construction is nearly 4%. (3) The type and size of coarse aggregates, especially the high-strength aggregates, are also very beneficial in decreasing DOP.

From the points of both engineering protection and weapon design, there are two main shortcomings in the existing experimental investigations. Firstly, the SFRHSC is recognized as the most efficient material until now in preventing projectile penetration and reducing the frontal and distal crater damages. However, the improvement of the strength needs decreasing or even eliminating the ingredients sizes in order to improve the homogeneity and density of the composite, and thus there are little coarse aggregates mixed into SFRHSC, especially when the compressive strength is larger than 100 MPa. In such a way, the anti-penetration capacity of coarse aggregates is neglected. Besides, the casting and curing of ultra-high strength concrete (>150 MPa, e.g. RPC) needs high temperature (90–400 °C) and sufficient pressure to expel the excess liquids and air from the fresh mixture, which prohibits its large-scale construction applications. Secondly, the projectile penetration efficiency as well as the damages of both projectile and concrete target are dependent on the striking velocity to a great extent. Most of the existing projectile penetration tests (Dancygier and Yankelevsky 1996; Dancygier 1998; O'Neil et al. 1999; Luo et al. 2000; Liu et al. 2002; Zhang et al. 2005, 2007; Dancygier et al. 2007; Tai 2009) conducted by weapon designer or protective engineers were mainly aiming to the conventional earth penetration weapons, of which the striking velocity were relatively low (<700 m/s). While for the high-speed advanced earth penetration weapons (the striking velocity could reach 1500 m/s), UHPCC is considered as the most proper construction materials for protective structures, especially for the military fortifications. However, the high-speed projectile penetration resistance of UHPCC as well as the influential factors are still not clear, and thus the related projectile penetration tests with broad striking velocity are urgently needed.

In Sect. 5.2, UHPCC with the additions of both steel fibers and basalt coarse aggregates, was prepared under ambient temperature and pressure by State Key Laboratory of High Performance Civil Engineering Materials, Jiangsu Research Institute of Building Science. For the applications of UHPCC in the anti-strike protective structures, the projectile penetration tests on UHPCC target was carried out with the striking velocities from 510 to 1320 m/s. The damages of both UHPCC target (impact crater dimensions, DOP) and the projectile (structural destruction and ballistic trajectory deviation), as well as the parametric influences (compressive strength of the target, the projectile striking kinetic energy, fiber mixing fraction and the coarse aggregates) were examined and discussed.

In Sect. 5.3, the corundum coarse aggregates (compressive strength exceeds 2000 MPa, Young's modulus 400 GPa, density 4000 kg/m^3, Mohs hardness 9.5, main chemical component is Al_2O_3) were further adopted and the corundum aggregated UHP-SFRC (UHP-CASFRC) was prepared. Based on a series of high-speed (510–850 m/s) projectile striking tests, the influences of the strength (hardness), size and volumetric ratio of coarse aggregate, incident velocity of the projectile as well as the repeated strikes on the impact resistance of UHP-CASFRC target were assessed. The outstanding impact resistance of UHP-CASFRC against the projectile penetration was verified. Then, by examining the structural integrity of the projectiles, the optimal corundum aggregate size of UHP-CASFRC target was suggested for the constructions of protective structures.

In Sect. 5.4, the corresponding numerical simulations aiming to examine the influences of coarse aggregates on the penetration resistance of coarse aggregated UHPCC were performed. Firstly, the generation algorithms of three-dimensional (3D) mesoscopic concrete models with the randomly distributed sphere and convex polyhedron coarse aggregates were proposed with the MATLAB compiler, respectively. The influences of two coarse aggregates shapes on the projectile impact resistance of concrete targets were further discussed, and the sphere coarse aggregates were adopted to guarantee both the calculation accuracy and efficiency. Then, by simulating the projectile impact tests on the CA-UHPC targets, the 3D mesoscopic concrete model with randomly distributed sphere coarse aggregates and the corresponding constitutive parameters were validated numerically by LS-DYNA. Finally, the influence of four coarse aggregates on the projectile penetration resistance of UHPC targets were discussed. The obtained conclusions are beneficial to the design and constructions of UHPCC in the protective engineering.

5.2 Basalt Aggregated UHPCC Target

5.2.1 Target

The compositions of UHPCC in the present penetration tests are listed in Table 5.1. The Portland cement with 28 days of compressive strength was 64.5 MPa, and 60% of cement in the conventional mixtures was replaced by ultra-fine industrial waste powders, such as silica fume (≤0.2 μm), ultra-fine fly ash and slag. The water-binder and sand-binder ratios were 0.15 and 1.2, respectively. The traditional fine quartz sand (≤600 μm) was substituted by natural sands with the maximum particle size of 2.5 mm. Thus, UHPCC is low-costs and energy savings because much more energy and resources are consumed during the production of the cement and fine quartz. The super-plasticizer is a kind of polycarboxylic type high performance water-reducer with the water-reducing ratio no less than 35%. The volume fractions of basalt aggregate V_b ranged up to 1042 kg/m^3, the steel fiber volumetric ratios V_f varied from 0% to 4%, and the equivalent diameter, length and tensile strength of the steel fibers were 0.175 mm, 13 mm and 1800 MPa, respectively (Rong et al. 2010; Lai et al. 2005). For the workability of the UHPCC, with increasing the fiber mixing ratio from 0 to 4%, the slump dropped from 260 to 170 mm (Zhang et al. 2012a, b). Compared with traditional high strength concrete, e.g. RPC (Zhang et al. 2008), steel fibers and basalt coarse aggregates were added successfully, and UHPCC can be prepared under ambient temperature and pressure. UHPCC exhibits high dynamic strength and impact toughness from SHPB tests (Rong et al. 2010), which has widely potential applications in the anti-impact protective structures.

Table 5.1 Compositions of UHPCC

Cement (kg/m^3)	Silica fume (kg/m^3)	Fly ash (kg/m^3)	Slag (kg/m^3)	Sand (kg/m^3)	V_b (kg/m^3)	w/cm ratio	Super-plasticizer	V_f
240–654	60–109	80–109	160–218	810–1090	0–1042	0.17–0.2	2%	0–4%

Fig. 5.1 UHPCC targets, reprinted from Wu et al. (2015a), copyright 2020, with permission from Elsevier

During the casting of UHPCC targets specimens, at first, the dry cementitious materials and sands were put into a compulsory mortar mixer simultaneously and mixed uniformly, and then the water and super-plasticizer were added together and mixed for 3–5 min. Finally, steel fibers were incorporated into the mixture and mixed for 3–5 min to guarantee the well distribution of fibers throughout the mortar (Rong et al. 2010). The mixture of UHPCC was cast into cylindrical steel culverts with the thickness of 3 mm, the diameter of the targets were 750 mm, which was nearly 30 times of the projectile diameter, and thus the circumferential boundary effects could be neglected. The target specimens were self-compacting and curing atmospherically for 28 days before the projectile penetration tests, shown in Fig. 5.1. The compressive strength and the maximal size of the basalt aggregate were 120 MPa and 10 mm, respectively. The volume fractions of basalt aggregate V_c (kg/m^3) and steel fibers V_f (%), unconfined cylindrical compressive strength f_c (MPa) and the thickness H_t (cm) of the UHPCC target in each shot are listed in Table 5.2.

5.2.2 *Projectile*

Two series of reduce-scaled ogive-nosed projectiles (A and B) were used in the penetration tests, of which the schemes of longitudinal section are given in Fig. 5.2 with the numbers in the brackets denoting the dimensions of projectiles B. As shown in Fig. 5.3a, projectiles A were machined from DT300 (SiMnCrNiMoV) steel rods with the yield strength of 1500 MPa. They were launched by the 25.3 mm-diameter, smooth-bore powder gun at the striking velocities 510 m/s and 850 m/s, respectively. The projectile cartridge thickness-to-diameter ratio was 0.14, and the average mass of the projectile was 342.2 g. Projectiles B shown in Fig. 5.3b were machined from D6A (45CrNiMoV) steel rods with the yield strength of 1420 MPa. Since the projectiles needed to be propelled to higher striking velocities, projectiles B were launched by a newly developed three-chambered 30 mm-caliber smooth-bore powder gun with the striking velocities at 1150–1320 m/s. The projectile cartridge was thickened and the cartridge thickness-to-diameter ratio was 0.2, and the average mass of the projectile

Table 5.2 Penetration tests data

Test	No.	M_0 (g)	V_0 (m/s)	f_c (MPa)	H_t (cm)	V_f (%)	V_d (kg/m^3)	DOP (mm)	Cratering	
									d_c (mm)	h_c (mm)
Test 1	A-1-1	333.1	510	67.5	65	1.5	978	148	189	50
	A-1-3	344.8		67.5	65	1.5	978	168	185	51
	A-2-1	336.7		87.3	55	0	1042	115	430	50
	A-2-3	341.9		87.3	55	0	1042	132	485	76
	A-3-1	345.6		99.3	55	1.5	870	133	335	65
	A-3-3	341.7		99.3	55	1.5	870	116	200	71
	A-4-1	341.0		125.2	50	2	800	124	172.5	42
	A-4-3	341.0		125.2	50	2	800	120	152.5	56
	A-5-1	341.1		114.0	50	3	0	134	153	52
	A-5-3	341.6		114.0	50	3	0	133	130	37
Test 2	A-1-2	350.1	850	67.5	100	1.5	978	345	216	62
	A-1-4	341.3		67.5	100	1.5	978	346	250	69
	A-2-2	340.6		87.3	65	0	1042	260	490	110
	A-2-4	341.1		87.3	65	0	1042	278	486	91
	A-3-2	342.3		99.3	65	1.5	870	255	314.5	58
	A-3-4	346.4		99.3	65	1.5	870	260	220.5	56
	A-4-2	342.7		125.2	60	2%	800	252	152.5	59
	A-4-4	341.5		125.2	60	2	800	261	212.5	60
	A-5-2	345.6		114.0	60	3	0	288	144	48
	A-5-4	344.1		114.0	60	3	0	268	189	43

(continued)

Table 5.2 (continued)

Test	No.	M_0 (g)	V_0 (m/s)	f_c (MPa)	H_t (cm)	V_f (%)	V_d (kg/m^3)	DOP (mm)	Cratering	
									d_c (mm)	h_c (mm)
Test 3	B-1-1	392.3	1150	35	150	0	750	–	516.75	120
	B-2-1	395.0		88	100	2	870	564	309	110
	B-3-1	394.5		128	100	3	800	402	252.5	115
	B-4-1	393.8		142	100	4	0	420	241.25	70
Test 4	B-1-2	392.3	1250	35	150	0	750	673	420	150
	B-2-2	393.3		88	100	2	870	540	250	–
	B-3-2	394.3		128	100	3	800	503	238.75	119
	B-4-2	394.3		142	100	4	0	473	167.5	75
Test 5	B-1-3	392.3	1320	35	150	0	750	748	403	95
	B-2-3	395.0		88	100	2	870	681	295.5	102
	B-3-3	393.6		128	100	3	800	130	448.75	130
	B-4-3	394.1		142	100	4	0	490	251.25	88

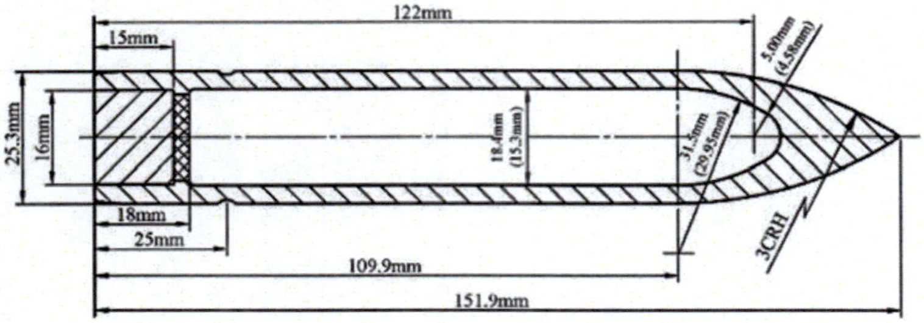

Fig. 5.2 Schemes of the projectiles, reprinted from Wu et al. (2015a), copyright 2020, with permission from Elsevier

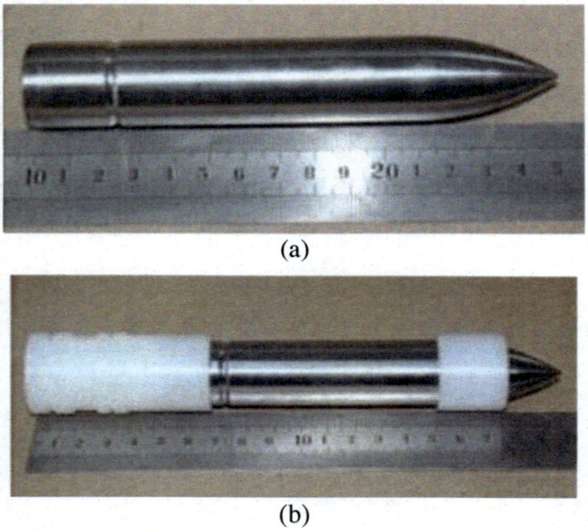

Fig. 5.3 Projectiles **a** series A, **b** series B, reprinted from Wu et al. (2015a), copyright 2020, with permission from Elsevier

was 393.7 g. The sub-caliber technology was used, projectiles B were fitted with the nylon sabots and obturators that fit snugly into the gun bore. After the projectiles were propelled out of the barrel, sabots and obturators can be stripped aerodynamically into debris prior to the projectile striking the target, which can be observed obviously from the penetration photograph recorded by a high-speed camera system given in Fig. 5.4.

Both projectiles A and B were filled by polymer inert materials with the density of (1.5 ± 0.05) g/cm^3, which was used to substitute the filled charge. During each shot, the initial mass of the projectile M_0 was measured and listed in Table 5.2, and the abrasions and deformations of the recovered projectiles were also recorded.

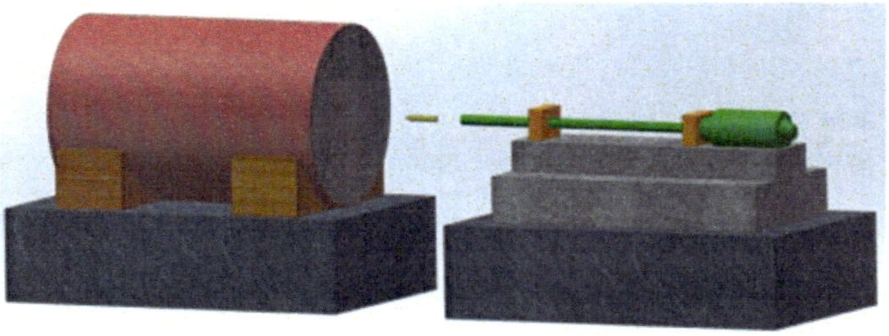

(a)

(b)

Fig. 5.4 Projectile penetrating into concrete target **a** schematic diagram, **b** photograph from high-speed camera system, reprinted from Wu et al. (2015a), copyright 2020, with permission from Elsevier

5.2.3 Test Setup

The schematic diagram of the projectile penetration tests is shown in Fig. 5.4a. During the tests, the cylindrical targets were seated on their sides, with the top face being placed perpendicular to and centered on the longitudinal axis of the barrel, and the wood-culvert was used to prevent the lateral rolling of the target. Striking velocities were measured from a pair of metal wire nets that were placed in the straight trajectory of the projectiles between the muzzle to the target, which formed part of the electrical circuits that were connected to an oscilloscope. Sequential breakage of the wire nets by the projectile generated voltage changes, thus dividing the distance between two metal wires nets by the time interval between the voltage changes, the impact velocity V_0 could be calculated. The flight process of the projectile was recorded by the high-speed camera system, and Fig. 5.4b shows the photograph at the instant of projectile impacting on the target. After each shot, the damages of the targets were recorded, including the maximum DOP of the projectile, as well as the depth h_c and the equivalent diameter d_c (average value of dimensions from four directions: horizontal, vertical, 45° and 135°) of the impact crater.

5.2.4 *Test Results*

In the present tests, 20 shots of projectiles A on UHPCC targets in tests 1–2, and 15 shots of projectiles B on both UHPCC and NSC targets in tests 3–5 were carried out, respectively. The penetration tests data are listed in Table 5.2 as follows, where – indicates that the data was not recorded.

5.2.4.1 Target Damage

There are no visible cracks emerged on the rear faces of the total 35 targets, the damage views of the impacted faces in each test are shown in Figs. 5.5, 5.6, 5.7, 5.8 and 5.9, respectively.

5.2.4.2 Projectile Abrasion and Deformation

During the penetration tests 1–2, for the relatively low striking velocity, projectiles A were not deformed or damaged, except for some minor abrasions on the projectile surfaces, as shown in Fig. 5.10a. While in the tests 3–5, for the increased initial striking velocities, the projectiles exhibited obvious nose-blunting and even bending or fractures, as shown in Fig. 5.10b–d.

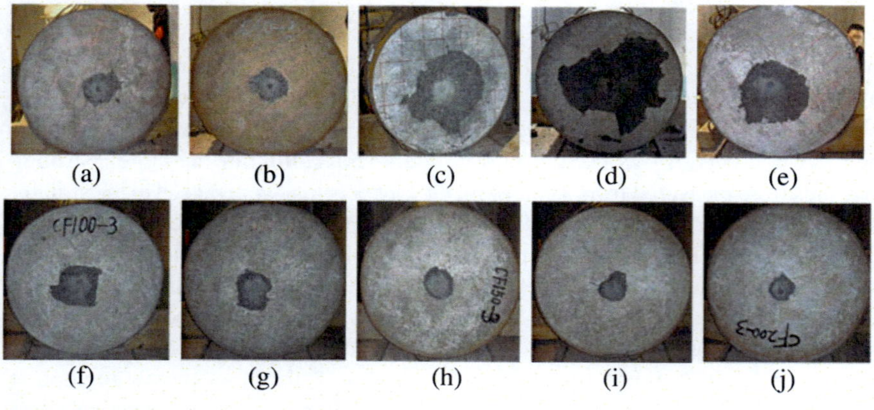

Fig. 5.5 Impact damages of the targets in test 1 **a** A-1-1, **b** A-1-3, **c** A-2-1, **d** A-2-3, **e** A-3-1, **f** A-3-3, **g** A-4-1, **h** A-4-3, **i** A-5-1, **j** A-5-3, reprinted from Wu et al. (2015a), copyright 2020, with permission from Elsevier

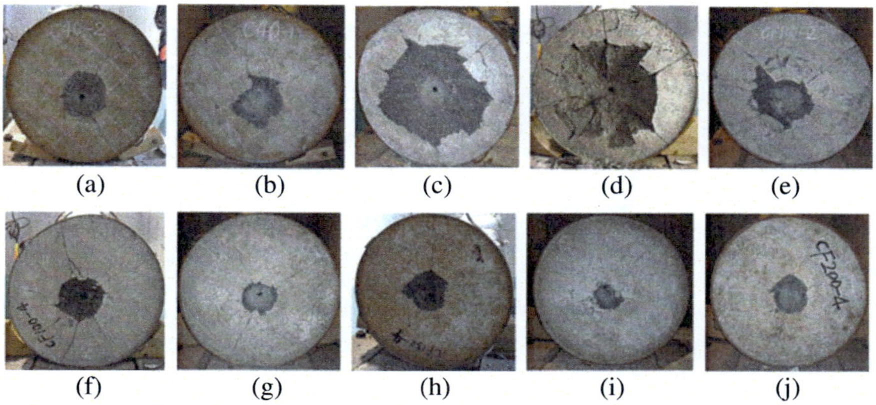

Fig. 5.6 Impact damages of the targets in test 2 **a** A-1-2, **b** A-1-4, **c** A-2-2, **d** A-2-4, **e** A-3-2, **f** A-3-4, **g** A-4-2, **h** A-4-4, **i** A-5-2, **j** A-5-4, reprinted from Wu et al. (2015a), copyright 2020, with permission from Elsevier

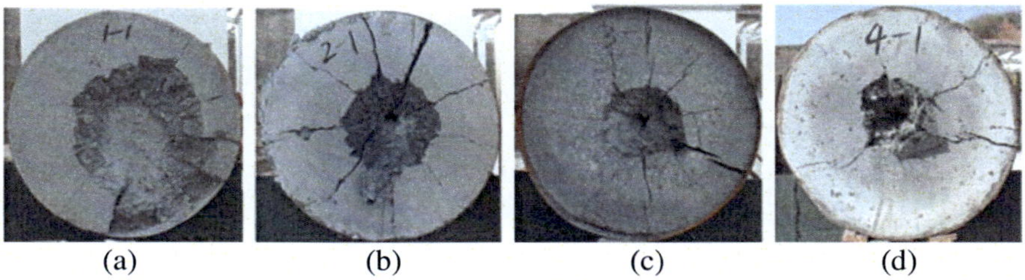

Fig. 5.7 Impact damages of the targets in test 3 **a** B-1-1, **b** B-2-1, **c** B-3-1, **d** B-4-1, reprinted from Wu et al. (2015a), copyright 2020, with permission from Elsevier

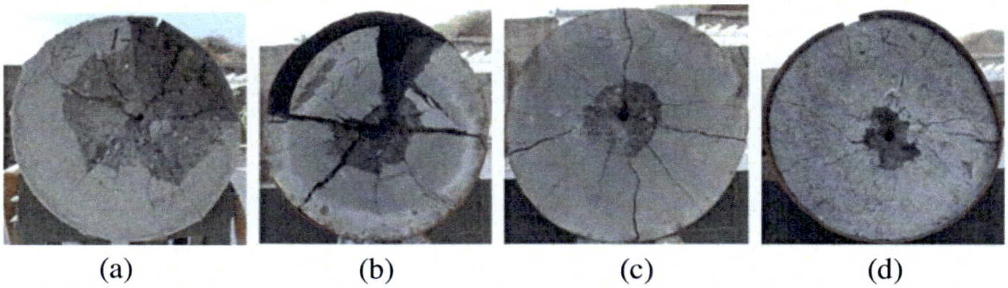

Fig. 5.8 Impact damages of the targets in test 4 **a** B-1-2, **b** B-2-2, **c** B-3-2, **d** B-4-2, reprinted from Wu et al. (2015a), copyright 2020, with permission from Elsevier

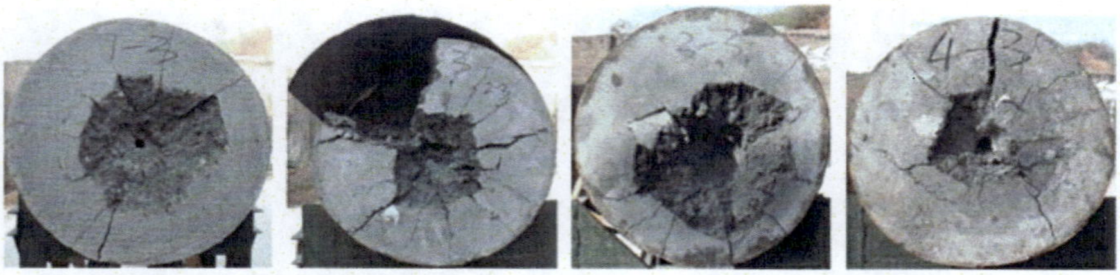

(a) (b) (c) (d)

Fig. 5.9 Impact damages of the targets in test 5 **a** B-1-3, **b** B-2-3, **c** B-3-3, **d** B-4-3, reprinted from Wu et al. (2015a), copyright 2020, with permission from Elsevier

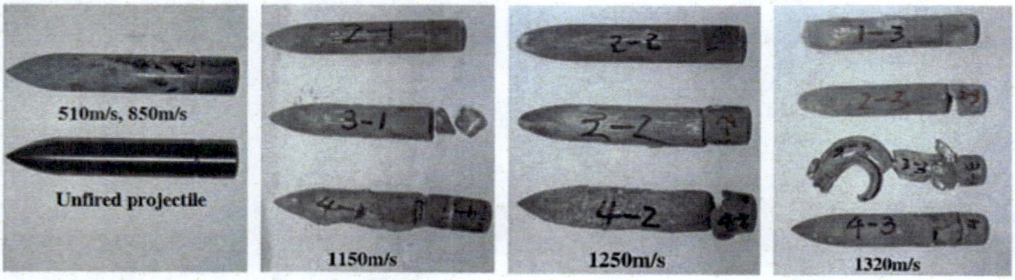

(a) (b) (c) (d)

Fig. 5.10 Abrasions and damages of the projectiles after penetration tests **a** tests 1–2, **b** test 3, **c** test 4, **d** test 5, reprinted from Wu et al. (2015a), copyright 2020, with permission from Elsevier

5.2.4.3 Terminal Ballistic Trajectory

During the penetration tests 1–2, the penetration paths of projectiles were all straight. During the penetration tests 3–5, the terminal ballistic trajectories of projectiles were curved to some extent, and some projectiles even exited the targets from the side surface (tests B-1-2 and B-4-3). Figure 5.11 illustrated the terminal ballistic trajectories in some tests obtained by cutting the targets after shots.

5.2.5 *Discussions*

Based on the works of Chen and Li (2004), Lundgren (1994), and Wu et al. (2014), we have proposed four regimes of penetration mechanics classified by the initial striking velocity of the projectile V_0 as follows: (i) The non-deformable penetration regime ($V_0 \leq 1$ km/s, such as the traditional Earth Penetration Weapon (EPW) and etc.), where the projectile can be treated as a rigid body; (ii) The semi-rigid penetration regime (1 km/s < $V_0 \leq 1.5$ km/s, such as the Advanced Earth Penetration Weapon (AEPW) and etc.), where the mass abrasion of the projectile should be considered; (iii) The semi-hydrodynamic penetration regime (1.5 km/s < $V_0 \leq 3$ km/s, such as the

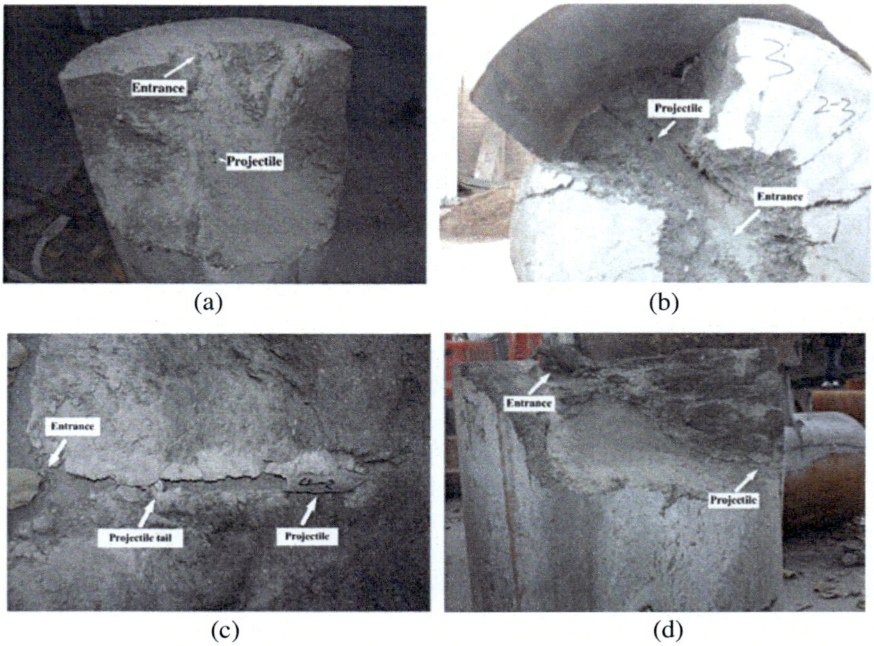

Fig. 5.11 Terminal ballistic trajectories obtained by cutting the targets **a** B-3-1, **b** B-2-3, **c** B-4-2, **d** B-4-3, reprinted from Wu et al. (2015a), copyright 2020, with permission from Elsevier

armor piercing projectile and etc.), where the erosions of both the nose and shank of projectile become obviously, the interface defeat and the failure wave phenomena are concerned occasionally, and the classical or improved Alekseevskii-Tate models can be applied; (iv) The hydrodynamic penetration regime ($V_0 > 3$ km/s, such as the metal jet, space debris and etc.), which can be characterized as a fluid–fluid interaction and governed by the steady-state fluid dynamics. In this case, the Bernoulli's equation can be used, and the strengths of both projectile and target can be neglected. It should be noted that, the above critical velocities are not absolute. Rather, they depend on the properties of both the targets and the projectiles, such as the strength, density, etc.

Therefore, in this section, rigid penetration analyses are conducted for tests 1–2 for $V_0 \leq 1$ km/s, the impact crater dimension and DOP of the rigid projectile as well as the parameters influential analyses are conducted in Sects. 5.3.1 and 5.3.2, respectively. In Sect. 5.3.3, our previously proposed model (Wu et al. 2012) for DOP of rigid projectile penetrating into fiber reinforced HSC are evaluated by comparing with the present test data.

5.2.5.1 Impact Crater Dimension

(1) The influence of target strength

In Table 5.2, under the same striking velocity and fiber mixing fraction, taking A-1-1, A-1-3, A-3-1 and A-3-3 in test 1 ($V_0 = 510$ m/s) for example, increasing the compressive strength of the target by 47% (from 67.5 to 99.3 MPa) increases the average impact cratering depth h_c and diameter d_c by 35% and 43% (h_c ranges from 50.5 to 68 mm, d_c ranges from 187 to 267 mm), respectively. Comparably, for A-1-2, A-1-4, A-3-2 and A-3-4 in tests 2 ($V_0 = 850$ m/s), with the same increment of the compressive strength of the target, h_c decreases by 13% (from 65.5 to 57 mm) and d_c increases by 15% (from 233 to 267 mm), respectively.

It indicates that, with the increase of the compressive strength of the target, the impact cratering depth and diameter both increases, while the above increasing degree tends to decrease with gradually rising the striking velocity of the projectile.

(2) The influence of fiber mixing fraction

Figure 5.12a, b illustrate the average impact cratering dimensions versus the compressive strength of target for tests 1–2, respectively. The compressive strength of the target and the steel fiber mixing ratio in each test are shown in the bracket. It indicates that, although the impact cratering dimensions enlarges with increasing the compressive strength, the addition of fibers can improve the fracture toughness and dynamic tensile strength of the composite, and thus help to decrease the cratering dimensions under projectile penetration with various striking velocities. The impact cratering dimensions reduces with the increase of fiber mixing fraction. Furthermore, in the variation ranges of the discussed parameters, the influential degree of mixing fibers on reducing impact cratering dimensions seems to be much greater than that of compressive strength on enlarging impact cratering dimensions.

Furthermore, as for two initial striking velocities in tests 1 and 2, when the compressive strength of the target varied from 87.3 MPa and 99.3 MPa, increasing

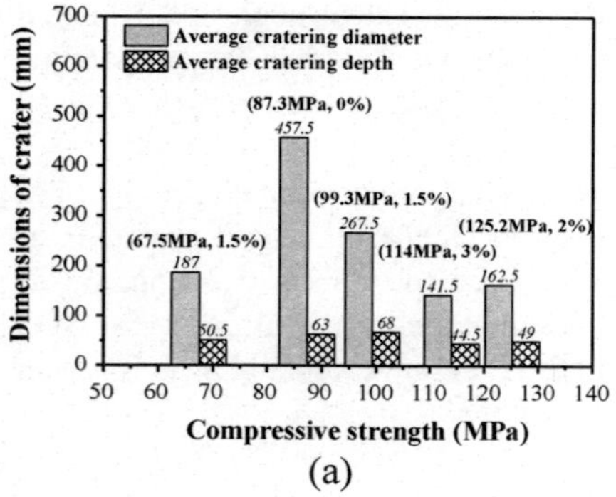

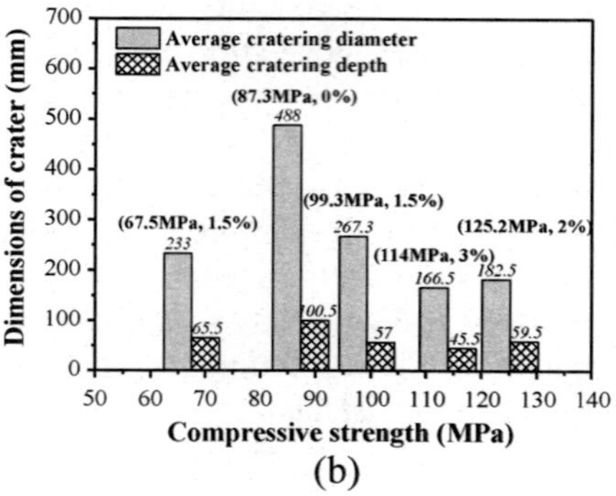

Fig. 5.12 Dimensions of the impact crater versus compressive strength of target **a** test 1, **b** test 2, reprinted from Wu et al. (2015a), copyright 2020, with permission from Elsevier

the fiber mixing ratio from 0% to 1.5% makes the corresponding averaging cratering diameter decrease from 457.5 mm to 267.5 mm (decreases by nearly 42%) and 488 mm to 267.5 mm (decreases by nearly 45%), respectively. In comparison, when the compressive strength of the target varied from 87.3 MPa to 114 MPa, increasing the fiber mixing ratio from 0% to 3% makes the corresponding averaging cratering diameter decrease from 457.5 mm to 141.5 mm (decreases by nearly 70%) and 488 mm to 166.5 mm (decreases by nearly 66%), respectively.

The above comparison indicates that the larger reduction of the cratering diameter is when the fiber mixing ratio increases from 0 to 1.5%. Furthermore, the cost of every cubic meter UHPCC (about 2.5t) with 3% steel fibers volumetric ratio is about 35% higher than that of UHPCC (about 2.5t) with 1.5% steel fibers volumetric ratio. By comprehensively considering the crack-prohibiting efficiency, costs and the workability of the composites, 1.5% of the fiber mixing ratio is the most efficient and economical for UHPCC in the construction of protective structures.

(3) The influence of projectile striking kinetic energy

By comparing with the tests 1 and 2, when the striking velocity of the projectile increases from 510 to 850 m/s, which means the impact kinetic energy increases about 2.7 times, for various compressive strength and fiber addition fractions, the corresponding average impact cratering diameters d_c increases about 6.7% (87.3 MPa), 0% (99.3 MPa), 25% (67.5 MPa), 12% (125.2 MPa) and 17.7% (114 MPa), respectively. Comparably, the cratering depths h_c increases about 60% (87.3 MPa), −16% (99.3 MPa), 30% (67.5 MPa), 21% (125.2 MPa) and 2% (114 MPa), respectively. It indicates that, under the same compressive strength and fiber mixing fraction, the impact cratering dimensions enlarges with rising the striking velocities.

Besides, taking tests 3–5 for considerations, for shots A-4-1 and A-4-3 (2% fiber addition) in test 1, shots A-4-2 and A-4-4 (2% fiber addition) in test 2, shot B-3-1 (3% fiber addition) in test 3, shot B-3-2 (3% fiber addition) in test 4, shot B-3-3 (3% fiber addition) in test 5, the compressive strengths of the targets are nearly equal (125.2 and 128 MPa, 2% deviation). Figure 5.13a shows the variations of impact crater dimensions with the impact energy of the projectiles under above shots.

Additionally, for shots A-2-1 and A-2-3 (no fiber addition) in test 1, shots A-2-2 and A-2-4 (no fiber addition) in test 2, shot B-2-1 (2% fiber addition) in test 3, shot B-2-2 (2% fiber addition) in test 4, shot B-2-3 (2% fiber addition) in test 5, compressive strengths (87.3 and 88 MPa, 1% deviation) are almost the same. Figure 5.13b shows the variations of impact crater dimensions with the impact energy of the projectiles under above shots.

From Fig. 5.13a, we can find that, under the same compressive strength, the dimensions of the impact crater increases with rising the impact kinetic energies of the projectile while the fiber addition ratio increases from 2 to 3%. As in Fig. 5.13b, the dimensions of the impact crater decreases with the rising of the impact kinetic energies of the projectile while the fiber addition ratio increases from 0 to 2%. Thus, we may draw a conclusion that, the influential degrees of the opposite contributions of the impact kinetic energy in enlarging the crater dimensions as well as the fibers in reducing the crater dimensions are difficult to determine. This is because the eventual

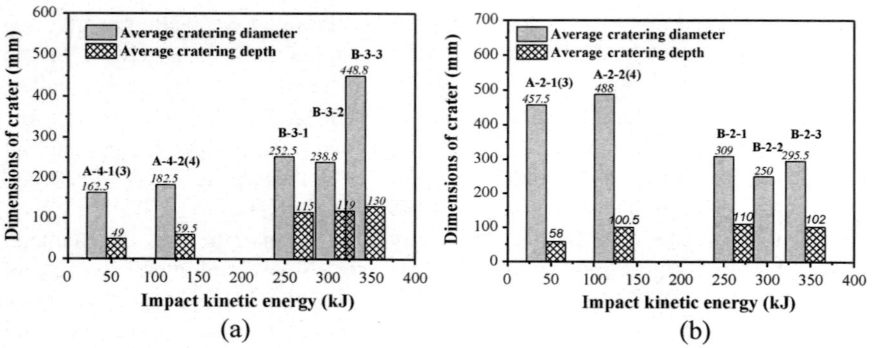

Fig. 5.13 Dimensions of the impact crater versus the impact kinetic energy of the projectile **a** target strength nearly 128 MPa, **b** target strength nearly 88 MPa, reprinted from Wu et al. (2015a), copyright 2020, with permission from Elsevier

results depend on the amounts of the fibers as well as the striking kinetic energy of the projectile. Additionally, we can find that, the cratering depth are limited into 2–5 times of the projectile diameter in the present tests.

5.2.5.2 DOP

(1) Influences of steel fiber volumetric ratio

By comparing the shots A-2-1, A-2-3, A-3-1 and A-3-3 in test 1, as well as the shots A-2-2, A-2-4, A-3-2 and A-3-4 in test 2, when the compressive strength increases from 87.3 MPa to 99.3 MPa, the corresponding average DOPs of the projectiles almost keep unchanged (from 123.5 to 124.5 mm, increases less than 1% in test 1, from 269 to 257.5 mm, decreases about 4% in test 2) with the fiber mixing fraction increasing from 0% to 1.5%, respectively.

It can be found that, considering that increasing both the compressive strength and the fiber mixing fraction can help to decrease DOP of the projectile, for the relatively low fiber addition ratio ($\leq$3%) discussed in this chapter, the contribution of fibers on strengthening the impact resistance against the projectile is very limited. The above findings coincide with the existing conclusions in O'Neil et al. (1999) and Zhang et al. (2005, 2007).

(2) Influences of concrete strength

Since that incorporating fibers cannot decrease the DOP of the projectile discussed above, Fig. 5.14 illustrates the average dimensionless penetration depth (DOP divided by the projectile shank diameter d) versus the compressive strength of target in tests 1–2. It indicates that the DOP of the projectile continuously diminishes with increasing the compressive strength of the concrete target. Additionally, for both two striking velocities, the DOP of the projectiles are no longer obviously decreasing when the

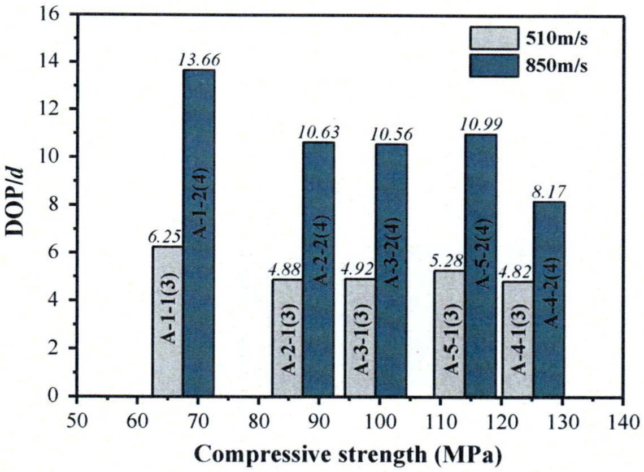

Fig. 5.14 Dimensionless DOP versus the concrete strength of the target, reprinted from Wu et al. (2015a), copyright 2020, with permission from Elsevier

compressive strength of the target is larger than nearly 90 MPa. By comprehensively considering the costs and the anti-strike performance, UHPCC with the compressive strength of 90 MPa and steel fiber mixing ratio of 1.5% is the optimal choice for protective structure constructions.

(3) Influences of coarse aggregate

For tests 1–2, Fig. 5.14 illustrates that the DOPs of projectiles slightly rise while increasing the compressive strength from 87.3 and 99.3 MPa to 114 MPa. The reason of such reverse rise lies in that the volumetric fractions of basalt coarse aggregates decrease gradually and no aggregates are added into the 114 MPa targets. Therefore, the coarse aggregates can also help to reduce the DOP of the projectile so as to enhance the impact resistance of UHPCC target. Comparably, the anti-strike contribution of coarse aggregate cannot be realized in the high strength RPC because coarse aggregates are difficult to be added under common curing conditions. Because of the experimental difficulties and random test data, the quantitative investigations of the contributions of the size and strength of the coarse aggregates on the impact resistance of the concrete target were not conducted in the present tests. Instead, the corresponding quantitative investigations were discussed numerically by Fang and Zhang (2013) based on the 3D meso-model of fiber reinforced concrete.

5.3 Corundum Aggregated UHPCC Target

5.3.1 Target and Projectile

Totally sixteen corundum aggregated UHPCC (UHP-CASFRC) targets and two comparative HSC targets with basalt coarse aggregates were cast, of which the compositions are listed in Table 5.3. The matrix of UHP-CASFRC was prepared

Table 5.3 Compositions of UHP-CASFRC and HSC targets (kg/m^3)

	Cement	Silica fume	Fly ash	Slag	Sand	V_c	w/cm ratio	HRWR	V_f
UHP-CASFRC	377–480	63–80	63–80	126–160	629–800	800–1700	0.26	1.5–16	160
HSC	266	–	38	76	720	1090	0.58	1.5	0

Fig. 5.15 UHP-CASFRC targets

and cast into cylindrical steel culverts with the thickness of 3 mm and the diameter of 750 mm, which is nearly 30 times of the projectile diameter. As a result, the circumferential boundary effects could be neglected. The targets were self-compacting and cured atmospherically for 28 days before the projectile penetration tests, as shown in Fig. 5.15. The unconfined cylindrical strength of the target f_c (MPa), the target thickness T_t (cm) in each shot as well as the corresponding weight V_c (kg/m^3) and the volumetric fractions V_{cp} (%) of corundum aggregates are given in Table 5.4. The steel fiber volumetric ratios V_f (%) were fixed as 2%. In the present test, three series of corundum sizes d_a (5–20, 35–45 and 65–75 mm, shown in Fig. 5.16) and two volumetric percentages V_{cp} (30 and 45%) were considered. Aiming to assess the performances of UHP-CASFRC target under multiple impacts, the repeated shots were conducted on the target 6-1 and 6-2, denoted by 6-1-1, 6-1-2 as well as 6-2-1, 6-2-2 in Table 5.4, respectively. Besides, the projectiles used in the present penetration tests were identical with which used in Sect. 5.2.2.

5.3.2 *Test Results*

5.3.2.1 Target Damage

In the present test, there were no visible cracks emerged on the rear faces of all targets, and the damage views of the front (impacted) faces for the UHP-CASFRC and HSC targets are shown in Fig. 5.17, respectively.

Table 5.4 Tests data

Test	No.	V_0 (m/s)	f_c (MPa)	T_t (cm)	V_f (%)	d_a (mm)	V_{cp} (%)	V_c (kg/m^3)	DOP (mm)	Crater[1]		Total volume (cm^3)	Crater[2]				
										Area[1] (cm^2)	Volume[1] (cm^3)		d_c (mm)	h_c (mm)	Area[2] (cm^2)	Volume[2] (cm^3)	α (°)
UHP-BAFRC [12]	1-1	510	67.5	65	1.5	5–15	36	978	148	242.1	–	–	189	50	280.4	538.3	58.6
	1-3			65	1.5	5–15	36	978	168	223.6	–	–	185	51	268.7	527.7	57.4
	2-1		87.3	55	0	5–15	39	1042	115	1260.5	–	–	430	50	1451.5	2569.8	76.1
	2-3			55	0	5–15	39	1042	132	1780.1	–	–	485	76	1846.5	4934.6	71.7
	3-1		99.3	55	1.5	5–15	32	870	133	780.3	–	–	335	65	881.0	2063.8	67.2
	3-3			55	1.5	5–15	32	870	116	370.1	–	–	200	71	314.0	849.0	50.9
	4-1		125.2	50	2	5–15	30	800	124	250.2	–	–	172.5	42	233.6	382.0	60.3
	4-3			50	2	5–15	30	800	120	178.7	–	–	152.5	56	182.6	406.7	48.6
	5-1		114.0	50	3	0	0	0	134	149.8	–	–	153	52	183.8	379.9	50.8
	5-3			50	3	0	0	0	133	114.9	–	–	130	37	132.7	201.7	54.7
	1-2	850	67.5	100	1.5	5–15	36	978	345	369.9	–	–	216	62	366.3	856.0	57.0
	1-4			100	1.5	5–15	36	978	346	392.1	–	–	250	69	490.6	1254.2	58.4
	2-2		87.3	65	0	5–15	39	1042	260	1789.8	–	–	490	110	1884.8	7286.1	64.7
	2-4			65	0	5–15	39	1042	278	1887.3	–	–	486	91	1854.1	5932.2	68.4
	3-2		99.3	65	1.5	5–15	32	870	255	525.4	–	–	314.5	58	776.5	1631.6	68.1
	3-4			65	1.5	5–15	32	870	260	363.6	–	–	220.5	56	381.7	803.6	60.2
	4-2		125.2	60	2	5–15	30	800	252	199.7	–	–	152.5	59	182.6	428.5	47.2
	4-4			60	2	5–15	30	800	261	296.7	–	–	212.5	60	354.5	803.4	57.3
	5-2		114.0	60	3	0	0	0	288	131.1	–	–	144	48	162.8	314.2	51.0
	5-4			60	3	0	0	0	268	266.0	–	–	189	43	280.4	462.9	62.3

(continued)

Table 5.4 (continued)

Test	No.	V_0 (m/s)	f_c (MPa)	T_t (cm)	V_f (%)	d_a (mm)	V_{cp} (%)	V_c (kg/m^3)	DOP (mm)	Crater[1]		Total volume (cm^3)	Crater[2]				
										Area[1] (cm^2)	Volume[1] (cm^3)		d_c (mm)	h_c (mm)	Area[2] (cm^2)	Volume[2] (cm^3)	α (°)
UHP-CAFRC	1-1	508	110.7	50	2	5–20	30	1126	98	237.4	387	410	174.5	54	239.0	501.7	54.1
	1-2	508		50	2	5–20	30	1126	101	372.5	676	700	218.8	55	375.8	777.9	60.4
	2-1	503	125.6	50	2	35–45	30	1126	21	135.8	126	126	138.8	21	151.2	128.7	73.2
	2-2	515		50	2	35–45	30	1126	68	302.9	548	548	191	68	286.4	746.5	54.5
	3-1	510	129.2	50	2	65–75	30	1126	59	197.8	214	214	161.5	59	204.8	475.6	53.8
	3-2	517		50	2	65–75	30	1126	37	136.1	121	121	132	37	136.8	207.2	60.7
	4-1	505	128.8	50	2	5–20	45	1700	145	212.0	299	341	161.5	54	204.8	435.3	51.6
	4-2	507		50	2	5–20	45	1700	106	278.7	479	492	181.8	75	259.5	751.5	46.2
	5-1	524	102.5	50	2	5–20	30	1126	106	253.4	453	468	179.3	56	252.2	546.7	54.0
	5-2	511		50	2	5–20	30	1126	88	258.3	512	521	178.5	51	250.1	425.2	56.3
	6-1-1	708	110.7	60	2	5–20	30	1126	171	549.3	995	1028	255.5	84	512.5	1591.0	53.9
	6-1-2	710		60	2	5–20	30	1126	310	645.4	1169	1288	279.5	84	613.2	1886.6	56.5
	6-2-1	722	110.7	60	2	5–20	30	1126	176	368.4	728	762	217.5	86	371.4	1202.8	48.2
	6-2-2	720		60	2	5–20	30	1126	323	435.9	943	1080	242.8	95	462.8	1634.1	48.9
	7-1	832	110.7	60	2	5–20	30	1126	185	244.5	409	450	179	82	251.5	798.4	43.1
	7-2	846		60	2	5–20	30	1126	191	352.4	596	639	206.5	84	334.7	1066.2	47.2
HSC	1-1	510	61.8	60	0	5–20	40	1090	151	818.5	1843	1859	310.8	93	758.3	2557.6	56.9
	1-2	516		60	0	5–20	40	1090	143	1204.7	2664	2686	376.5	85	1112.8	3378.9	64.2

Fig. 5.16 Corundum aggregates with various sizes, reprinted from Wu et al. (2015b), copyright 2020, with permission from Elsevier

5.3.2.2 Impact Crater Dimension

As shown in Fig. 5.17, projectile penetration into concrete target always forms the impact crater and followed by a cylindrical tunnel with the diameter nearly equal to the projectile shank diameter d. In this section, the pixel and equivalent diameter approaches are adopted to measure the areas of impact craters, respectively. The crater areas denoted by Area^1 in Table 5.4 were obtained by measuring the ratio of pixels for crater and target, and the details can be referred in Werner et al. (2013). Besides, the equivalent diameter d_c of the impact crater was obtained by averaging the maximal dimensions of crater from four directions: horizontal, vertical, 45° and 135°, respectively. The corresponding areas $\text{Area}^2 = \pi d_c^2/4$ were also given in Table 5.4.

5.3.2.3 Impact Crater Volume

Two approaches are adopted to obtain the volumes of the impact craters. Firstly, the Volume^1 in Table 5.4 was obtained by filling sands into the craters illustrated in Fig. 5.18a, and the Total Volume in Table 5.4 is the sum of both crater and tunnel volumes. Secondly, by measuring the crater depth h_c, the Volume^2 could be calculated by approximating the crater as a frustum-of-cone, graphed schematically in Fig. 5.18b. The slope angle $\alpha = \arctan[(d_c - d)/2h_c]$ of each target could be calculated and listed in Table 5.4.

5.3.2.4 DOP

After each shot, the penetration test data including the DOP of the projectile as well as the crater areas and volumes were recorded and listed in Table 5.4. To facilitate

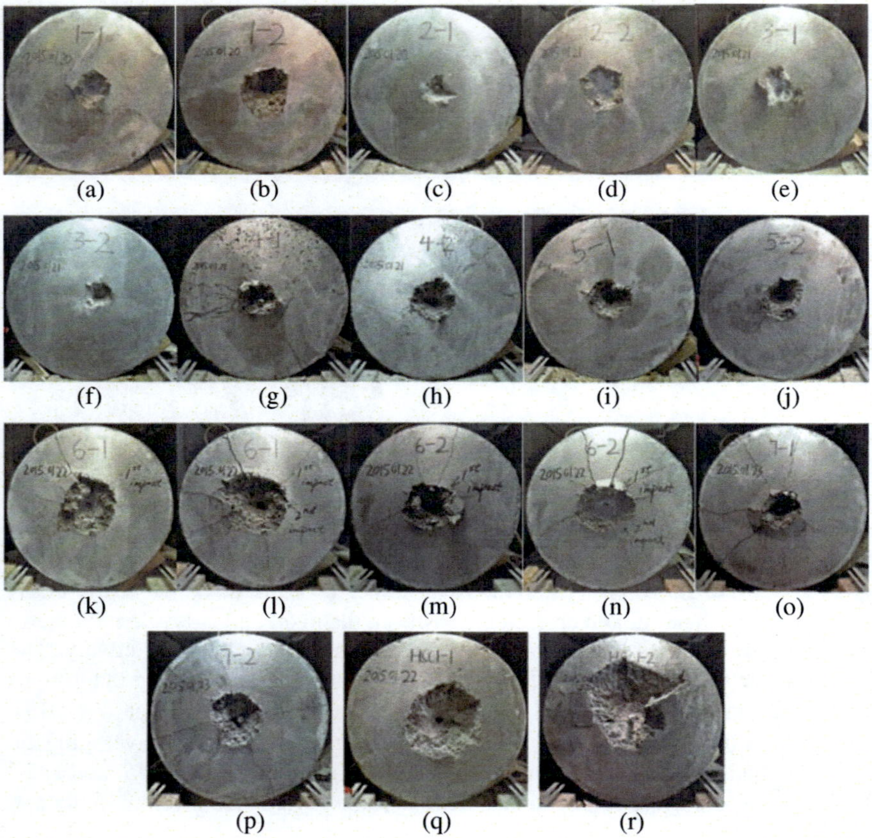

Fig. 5.17 Impact damages of the UHP-CASFRC targets **a** 1-1, **b** 1-2, **c** 2-1, **d** 2-2, **e** 3-1, **f** 3-2, **g** 4-1, **h** 4-2, **i** 5-1, **j** 5-2, **k** 6-1-1, **l** 6-1-2, **m** 6-2-1, **n** 6-2-2, **o** 7-1, **p** 7-2, **q** HSC1-1, **r** HSC1-2, reprinted from Wu et al. (2015b), copyright 2020, with permission from Elsevier

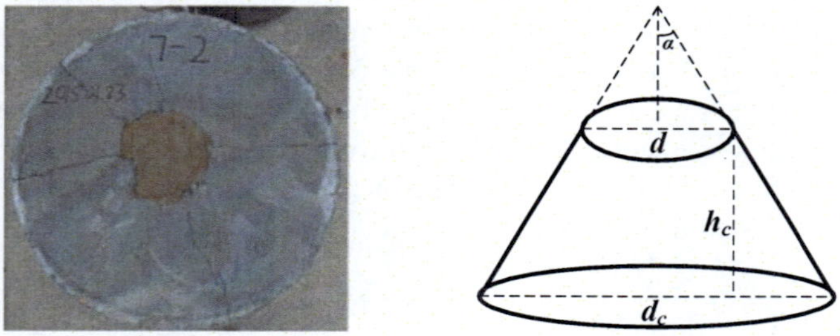

(a) Sand filling approach (b) Equivalent diameter approach

Fig. 5.18 Impact crater volumes, reprinted from Wu et al. (2015b), copyright 2020, with permission from Elsevier

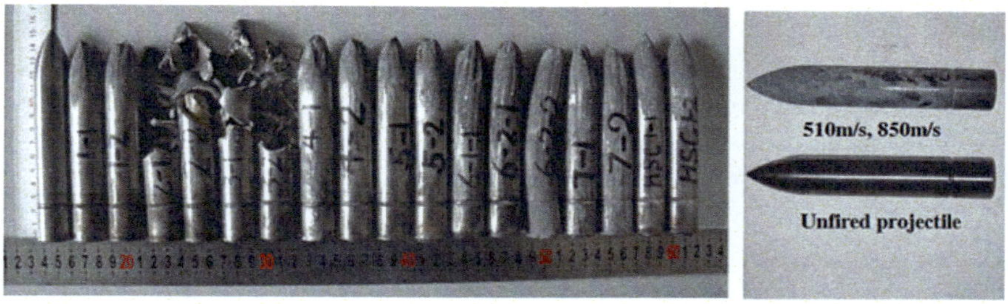

(a) (b)

Fig. 5.19 Unfired and recovered projectiles from the test on **a** UHP-CASFRC, **b** UHP-BASFRC, reprinted from Wu et al. (2015b), copyright 2020, with permission from Elsevier

comparison, the results of projectile penetration test in Sect. 5.2 (denoted as UHP-BASFRC targets) are also given, and "–" denotes that the data were not obtained.

5.3.2.5 Projectile Damage

Figure 5.19a shows the photographs of unfired projectile (far left one) and recovered projectiles (the projectile in 6-1-2 was missed for it penetrated deep in the target) in the present test. Table 5.5 lists the masses and lengths of unbroken projectiles during each shot before and after the penetrations. It can be found that the maximal relative mass losses and blunted lengths of the projectiles reached 8.4% and 9.9%, respectively. Figure 5.19b shows the typical unfired and recovered projectiles in the penetration test on UHP-BASFRC, in which the projectiles were not deformed or damaged, except for some minor abrasions on the surfaces of noses and shanks due to the high-speed frictions during the penetrations.

5.3.3 Discussions

5.3.3.1 DOP

The existing projectile penetrating tests (Zhang et al. 2005, 2007; O'Neil et al. 1999) indicate that, increasing the compressive strength and the fiber mixing fraction can both help decreasing DOP of the projectile. Furthermore, for the relatively low fiber addition ratio ($\leq$4%), the contribution of fibers on strengthening the impact resistance of the projectile is very limited. Also, the DOP of the projectile is no longer decreasing obviously when the compressive strength of the target increasing beyond a certain value, which was proposed as about 90 MPa (Wu et al. 2015a), 100 MPa (Zhang et al. 2005) and 150 MPa (Langberg and Markeset 1999), respectively. In this section,

Table 5.5 Comparisons of mass and length of unfired and recovered projectiles

Target	No.	Projectile mass			Projectile length			Projectile
		M_0 (g)	M_r (g)	Mass loss	L_0 (mm)	L_r (mm)	Length loss	
UHP-CASFRC	1-1	341.2	329.2	3.5%	150.7	140.5	6.8%	Bounced and moderately abraded
	1-2	341.2	329.5	3.4%	150.7	141.6	6.0%	Bounced and moderately abraded
	2-1	341.2	Broken	–	150.9	Broken	–	Bounced and broken
	2-2	341.2	Broken	–	151.9	Broken	–	Bounced and broken
	3-1	341.3	Broken	–	150.8	Broken	–	Bounced and broken
	3-2	341.2	Broken	–	151.1	Broken	–	Bounced and broken
	4-1	341.2	332.0	2.7%	150.9	143.6	4.8%	Embed and little abraded
	4-2	341.2	328.4	3.8%	151.9	139.2	8.4%	Bounced and moderately abraded
	5-1	341.2	327.7	4.0%	151.2	143.1	5.4%	Bounced and moderately abraded
	5-2	340.7	329.3	3.3%	150.8	142	5.8%	Bounced and moderately abraded
	6-1-1	341.3	318.6	6.7%	151.1	140.2	7.2%	Bounced and severely abraded
	6-1-2	341.3	–	–	150.7	–	–	Embed
	6-2-1	341.2	323.5	5.2%	150.4	139.7	7.1%	Bounced and severely abraded
	6-2-2	341.2	319.9	6.2%	150.8	135.8	9.9%	Bounced and bended

(continued)

Table 5.5 (continued)

Target	No.	Projectile mass			Projectile length			Projectile
		M_0 (g)	M_r (g)	Mass loss	L_0 (mm)	L_r (mm)	Length loss	
	7-1	341.2	312.7	8.4%	151.0	139.2	7.8%	Embed and severely abraded
	7-2	341.2	315.2	7.6%	150.9	136.4	9.6%	Embed and severely abraded
HSC	1-1	341.2	336.1	1.5%	151.2	149.2	1.3%	Bounced and nearly intact
	1-2	341.2	336.3	1.4%	150.8	150.3	0.3%	Bounced and nearly intact

based on Table 5.4, the influences of strength (hardness), size and the volumetric ratio of coarse aggregate as well as the striking velocity on the DOP were discussed.

(1) Influences of the coarse aggregate strength (hardness)

Figure 5.20 illustrates the DOPs of the UHP-BASFRC and UHP-CASFRC targets when the striking velocities are nearly 510 and 850 m/s, in which the size range and volumetric ratio of coarse aggregate are almost identical. It can be seen that, for both two striking velocities, although the compressive strengths of UHP-BASFRC targets are higher than those of the UHP-CASFRC targets, the DOP of the BASFRC targets are larger than that of the UHP-CASFRC targets inversely. For example, when the striking velocity is nearly 850 m/s, the average DOP of UHP-BASFRC targets (256.5 mm) is 36% larger than that of the UHP-CASFRC targets (188 mm). Therefore, we can derive that the strength (hardness) of the coarse aggregate can

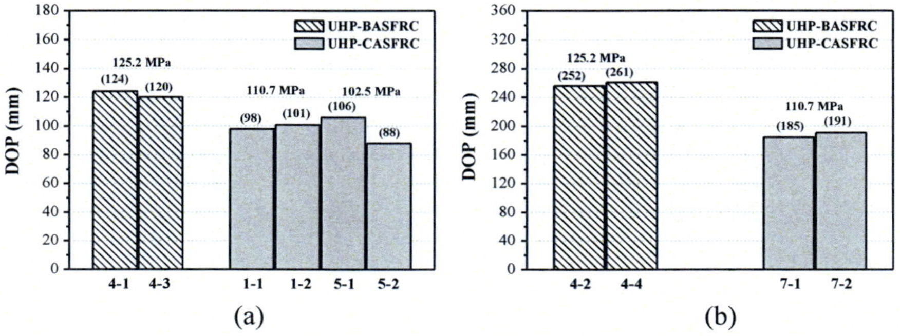

Fig. 5.20 Influences of the coarse aggregate strength (hardness) on DOP **a** $V_0 = 510$ m/s, **b** $V_0 = 850$ m/s, reprinted from Wu et al. (2015b), copyright 2020, with permission from Elsevier

help decreasing the DOP, and moreover, it seems that the decreasing degree of coarse aggregate was more influential than that of compressive strength, which agrees with the theoretical conclusions of Wu et al. (2014). Thus the harder coarse aggregate can reduce the DOP by aggravating the mass abrasion of the projectile.

(2) Influences of the coarse aggregate size

For UHP-CASFRC targets, Fig. 5.21a illustrates the DOPs of the eight targets with the nearly identical projectile striking velocities listed in Table 5.4. It can be easily obtained that, with the average coarse aggregates size increasing from nearly 0.5*d*, to 1.5*d* and 2.5*d*, the DOPs decrease gradually. For example, the average DOP of the targets 3-1 and 3-2 (48 mm) is 51% lower than that of the targets 1-1, 1-2, 5-1 and 5-2 (98.3 mm). Noted that the DOP in shot 2-1 is unexpectedly low, which was caused by the projectile directly hit the corundum aggregate and was broken into pieces, as shown in Figs. 5.19 and 5.21b. With enlarging the aggregate size, the probability that the projectile hit the coarse aggregate also increases. If high strength (hardness) coarse aggregates are used, the impact resistance of the target can be considerably enhanced. This is also the motivation we have been paying efforts into developing the basalt and corundum aggregated UHP-SFRC.

(3) Influences of the coarse aggregate volumetric ratio

Figure 5.22a illustrates the DOPs of the six targets with nearly identical projectile striking velocities listed in Table 5.4. It can be drawn that, the projectile penetrated the targets with higher strength and coarse aggregate volumetric ratio (4-1 and 4-2) inversely deeper than the targets with lower strength and coarse aggregate volumetric ratio (1-1 and 1-2, 5-1 and 5-2), which is opposite to our common sense that the probability that the projectile hits coarse aggregates becomes greater with rising the volumetric ratio. We suppose that such unexpected experimental results may be

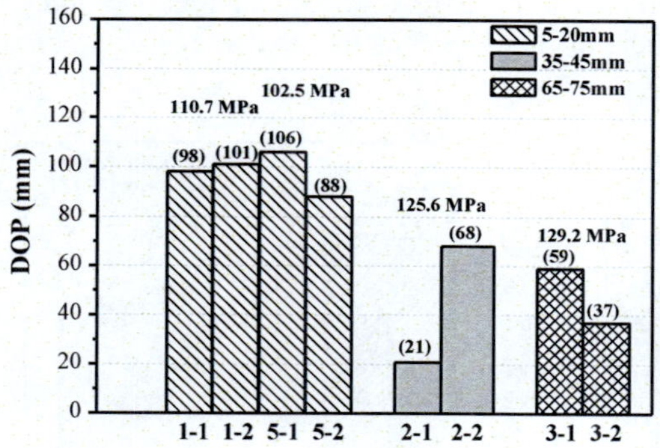

(a)

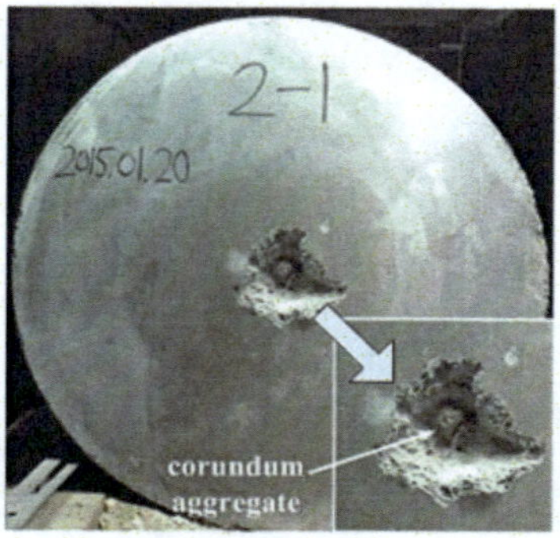

(b)

Fig. 5.21 Influences of the coarse aggregate size on DOP **a** DOP, **b** damage of 2-1 target, reprinted from Wu et al. (2015b), copyright 2020, with permission from Elsevier

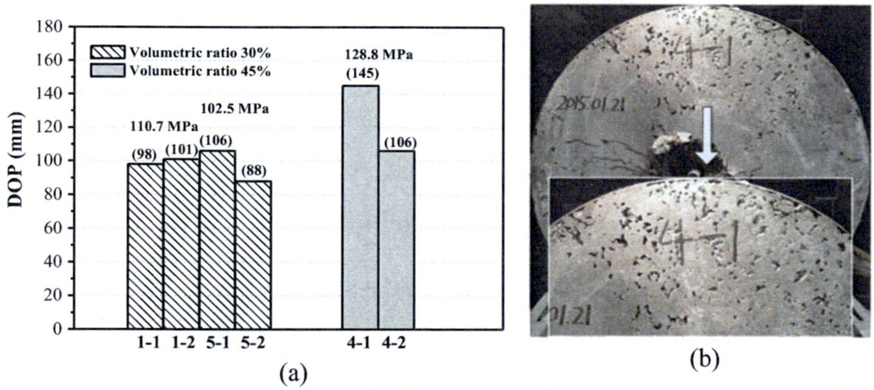

Fig. 5.22 Influences of the coarse aggregate volumetric ratio on DOP **a** DOP, **b** detail view of 4-1 target, reprinted from Wu et al. (2015b), copyright 2020, with permission from Elsevier

caused by the fact that the targets 4-1 and 4-2 were not self-densified enough with many voids forming on the impacted surfaces, as shown in Fig. 5.22b.

(4) Comparisons between UHP-CASFRC and HSC targets

Under nearly the same projectile penetrating velocities, Fig. 5.23 shows the DOPs of UHP-CASFRC and HSC targets of the present tests listed in Table 5.4. The average DOPs of UHP-CASFRC (98.3 mm) are 33% lower than those of the HSC targets (147 mm). Also, by considering the Fig. 5.20, the superior anti-penetration performance of UHP-CASFRC over UHP-BASFRC and HSC was validated.

(5) The influence of projectile striking velocity

Figure 5.24 shows the DOPs of UHP-CASFRC targets with the projectiles striking velocities are about 510, 700 and 850 m/s, in which the targets have nearly the same compressive strength and aggregate size range. It indicates that with rising

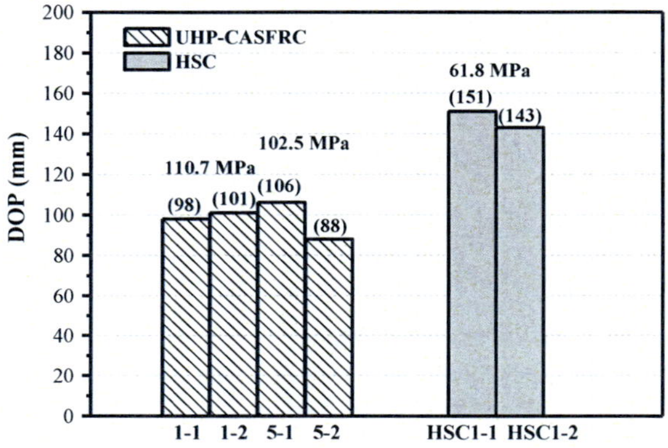

Fig. 5.23 Comparisons of DOPs between UHP-CASFRC and HSC targets, reprinted from Wu et al. (2015b), copyright 2020, with permission from Elsevier

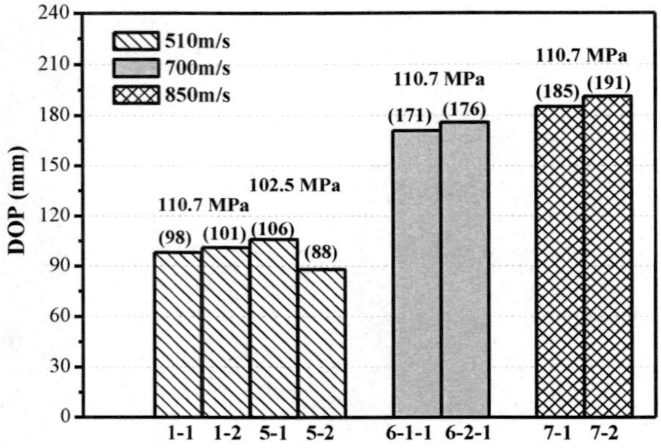

Fig. 5.24 DOPs of UHP-CASFRC target versus projectile striking velocity, reprinted from Wu et al. (2015b), copyright 2020, with permission from Elsevier

the projectile striking velocity, the DOP increases but the increasing magnitude is gradually weakened. The reasons lie in that, the mass abrasions of the projectiles become more and more serious with increasing the striking velocity, as seen in Fig. 5.19, which significantly reduces the efficiency of the projectile penetration.

5.3.3.2 Impact Crater Dimension

In Table 5.4, as for the two approaches proposed for measuring the area of the impact crater, the relative deviations of Area[1] and Area[2] for UHP-CASFRC are less than 7%, and the corresponding deviations for HSC are about 8%. Therefore, the above two approaches can both be utilized. The data of Area[1] were used in the following discussions.

(1) Influences of the coarse aggregate strength (hardness)

Figure 5.25 shows the comparisons of the impact crater areas of UHP-BASFRC and UHP-CASFRC targets under the projectile striking velocities are about 510 m/s and 850 m/s, respectively, in which the targets have the identical fiber mixing ratio and nearly the same aggregate size range and the compressive strength. It can be found that, the average crater areas of the UHP-CASFRC targets 280.4 cm^2 and 298.5 cm^2 under two striking velocities are 30.7% and 20.3% larger than those of the UHP-BASFRC targets 214.5 cm^2 and 248.2 cm^2, respectively. The reason lies in that, the surfaces of the corundum aggregates are smoother than basalt aggregates, leading to the weaker bonding strength of the interfacial transition zone (ITZ) between aggregate and matrix.

(2) Influences of the coarse aggregate size

Figure 5.26 illustrates the influences of corundum coarse aggregate size on the impact crater area. It indicates that, although the compressive strength has little increase (the brittleness of the concrete also increases) with the enlarging of the coarse aggregate

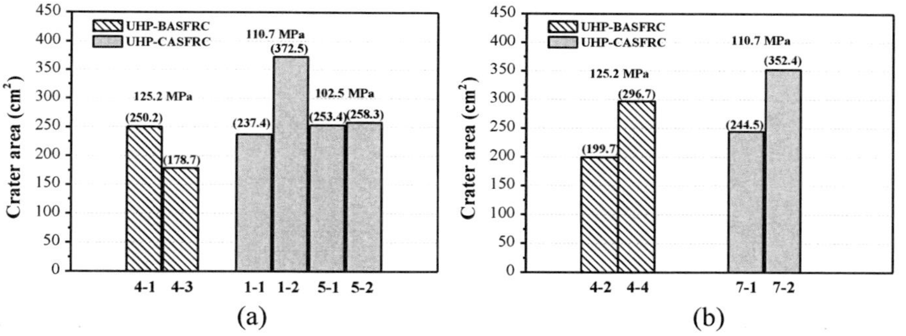

Fig. 5.25 Impact crater areas of UHP-BASFRC and UHP-CASFRC targets **a** $V_0 = 510$ m/s, **b** $V_0 = 850$ m/s, reprinted from Wu et al. (2015b), copyright 2020, with permission from Elsevier

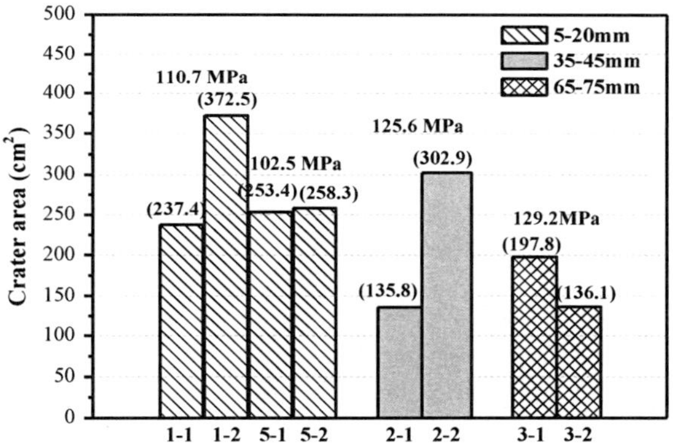

Fig. 5.26 Influences of aggregate sizes on impact crater areas, reprinted from Wu et al. (2015b), copyright 2020, with permission from Elsevier

size, the corresponding average crater areas decrease gradually from 280.4 to 219.4 and 167.0 cm^2. The reason may lie in that, the probability that the projectile hits the coarse aggregates also increases with the enlarging of the aggregate size, and the larger aggregates are difficult to be spalled.

(3) Influences of the coarse aggregate volumetric ratio

Figure 5.27 illustrates the influences of corundum coarse aggregate volumetric ratio on the impact crater area. It indicates that, the impact crater area decreases with the rising of the coarse aggregate volumetric ratio, even the compressive strength as well as the brittleness of the target also increases. The average crater area of the UHP-CASFRC with the aggregate volumetric ratio 45% (245.4 cm^2) is 12.5% lower than that of the UHP-CASFRC with the aggregate volumetric ratio 30% (280.4 cm^2).

(4) Comparisons between UHP-CASFRC and HSC targets

Figure 5.28 illustrates the comparisons of the crater areas between the UHP-CASFRC and HSC targets. It indicates that UHP-CASFRC targets has much smaller crater

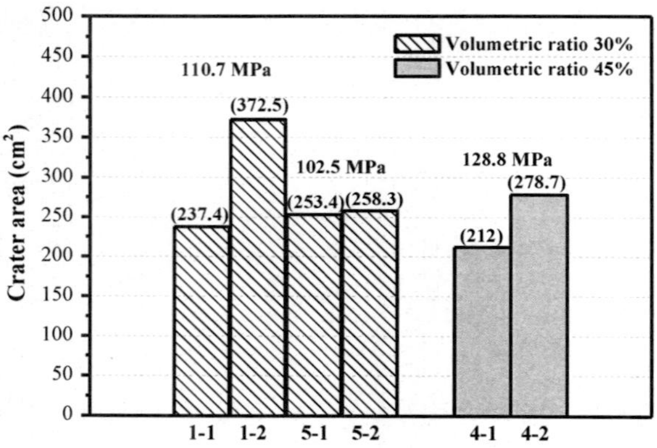

Fig. 5.27 Influences of aggregate volumetric ratio on impact crater areas, reprinted from Wu et al. (2015b), copyright 2020, with permission from Elsevier

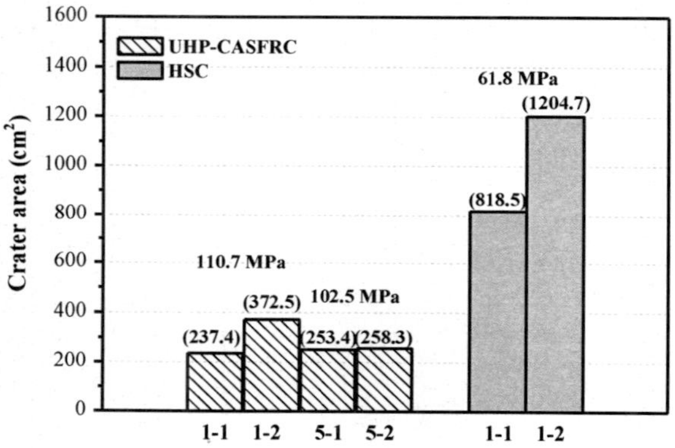

Fig. 5.28 Comparisons of the crater areas between the UHP-CASFRC and HSC targets, reprinted from Wu et al. (2015b), copyright 2020, with permission from Elsevier

area than the HSC targets. Under the projectiles striking velocities about 510 m/s, the average crater area of UHP-CASFRC targets (280.4 cm^2) is 72.3% smaller than that of the HSC targets (1011.6 cm^2). It indicates that, although the impact cratering area enlarges with increasing the compressive strength, the addition of fibers can improve the fracture toughness and dynamic tensile strength of the composite, so as to help decreasing the impact cratering dimensions by the bridging effect. Based on the discussions of the penetration test data on UHP-BASFRC targets listed in Table 5.4, Wu et al. (2015a) further found that the impact cratering dimensions reduces with the increase of fiber mixing fraction, and the influential degree of mixing fibers on reducing impact cratering dimensions seems to be much greater than that of compressive strength on enlarging impact cratering dimensions in the variation ranges of discussed parameters.

(5) The influence of projectile striking velocity

Figure 5.29 shows the variations of the crater areas with the striking velocities of the projectile, from which we cannot obtain the monotonic variation law of the

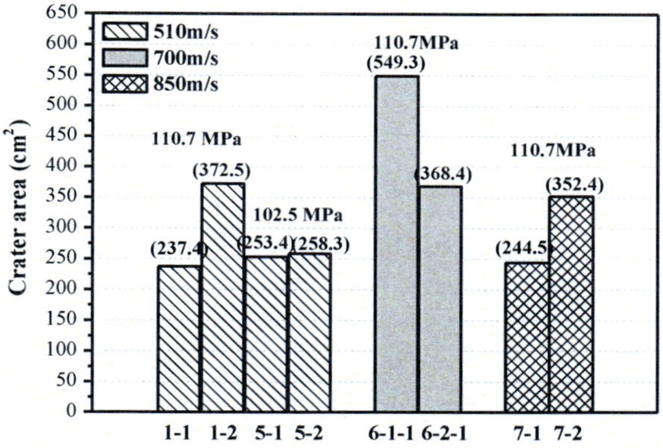

Fig. 5.29 Influences of striking velocity on impact crater areas, reprinted from Wu et al. (2015b), copyright 2020, with permission from Elsevier

crater area with the striking velocity. For the projectile penetration tests on UHP-BASFRC targets listed in Table 5.4, Wu et al. (2015a) concluded that, under the same compressive strength and fiber mixing fraction, the impact cratering areas enlarges with the rising of the striking velocities. While the influential degrees of the opposite contributions of the impact kinetic energy in enlarging the crater dimensions as well as the fibers in reducing the crater dimensions are difficult to determine, which mainly depends upon the amounts of the fibers as well as the striking kinetic energy of the projectile.

5.3.3.3 Impact Craters Volume

In Table 5.4, for the two approaches proposed for measuring the volume of the impact crater, the maximal relative deviations of Volume^1 and Volume^2 for UHP-CASFRC even reached 50%, thus the data of Volume^1 obtained by filling the sand into the craters are used for the following comparisons. As for the slope angle of the rear crater of the concrete panel, Dancygier (1998) proposed 66° and 76° for normal and high strength concrete target. From Table 5.4, since the steel fibers are incorporated, the average slope angles for UHP-BASFRC and UHP-CASFRC targets (aggregates sizes lower than 20 mm) are reduced as 60° and 51°, respectively. The slope angles are helpful in the future theoretical studies on the projectile perforations of the UHP-SFRC panels.

Since the data of Volume^1 were not obtained from the penetration test on UHP-BASFRC targets, the influences of the coarse aggregate strength (hardness) on the impact crater volumes is not discussed in this section.

(1) Influences of the coarse aggregate size

Figure 5.30 illustrates the influences of corundum coarse aggregate sizes on the impact crater volumes. It indicates that, although the compressive strength and the

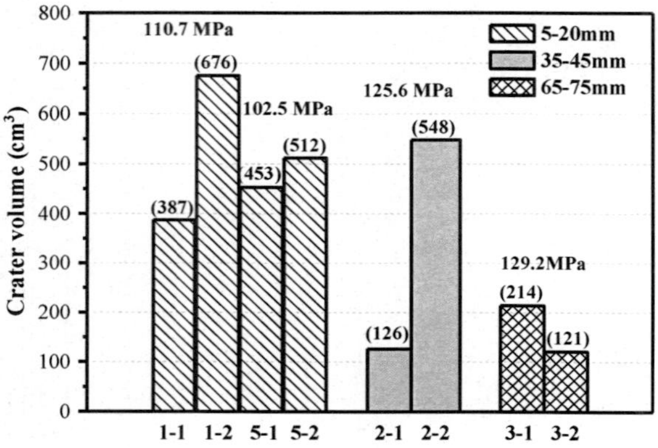

Fig. 5.30 Influences of aggregate size on impact crater volume, reprinted from Wu et al. (2015b), copyright 2020, with permission from Elsevier

corresponding brittleness have a little increase with the enlarging of the coarse aggregate size, the corresponding average crater volumes decrease gradually from 507 to 337 and 167.5 cm^2. The reason of which may be similar with the influences of coarse aggregate size on the crater areas discussed in Sect. 5.3.3.2.

(2) Influences of the coarse aggregate volumetric ratio

Figure 5.31 illustrates the influences of corundum coarse aggregate volumetric ratios on the impact crater volumes. It indicates that, the average impact crater volumes of UHP-CASFRC targets with the aggregate volumetric ratio 45% (389 cm^3) are 23.3% less than those of UHP-CASFRC targets with the aggregate volumetric ratio 30% (507 cm^3) under the strike velocity of about 510 m/s. The reason may be similar with the influences of the coarse aggregate size interpreted above.

(3) Comparisons between UHP-CASFRC and HSC targets

Figure 5.32 illustrates the comparisons of the crater volumes between the UHP-CASFRC and HSC targets. It indicates that, the average impact crater volumes

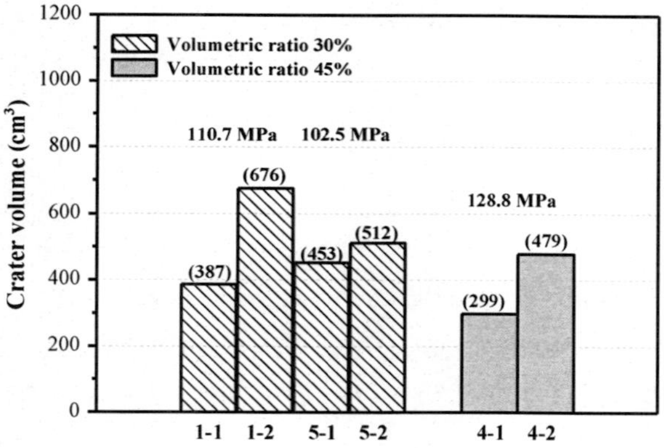

Fig. 5.31 Influences of aggregate volumetric ratio on impact crater volume, reprinted from Wu et al. (2015b), copyright 2020, with permission from Elsevier

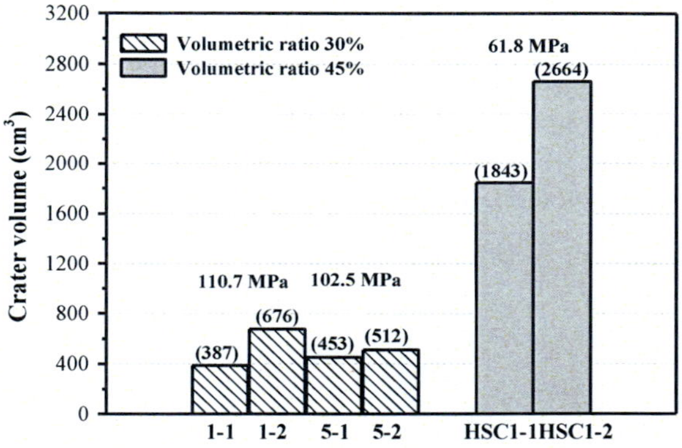

Fig. 5.32 Comparisons of the crater volumes between the UHP-CASFRC and HSC targets, reprinted from Wu et al. (2015b), copyright 2020, with permission from Elsevier

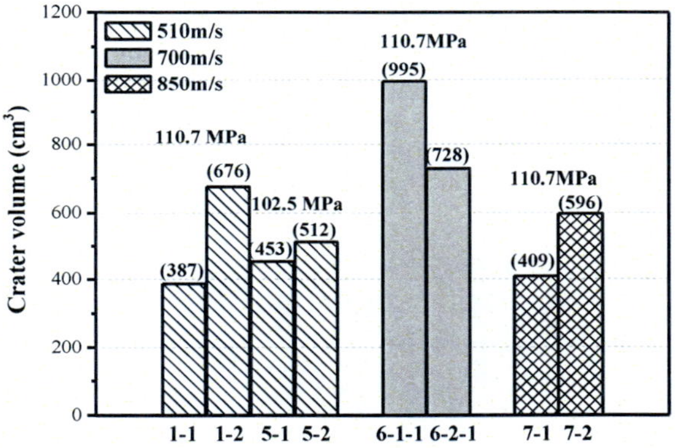

Fig. 5.33 Influences of striking velocity on impact crater volume, reprinted from Wu et al. (2015b), copyright 2020, with permission from Elsevier

of UHP-CASFRC targets (507 cm^3) is 77.5% less than those of HSC targets (2253.5 cm^3) under the strike velocity of 510 m/s. The results further verified the excellent functions of steel fibers on greatly improving the impact toughness of the concrete.

(4) The influence of projectile striking velocity

Figure 5.33 shows the variations of the crater volumes with the striking velocities of the projectile. Similar with the influences of striking velocity on the impact crater area discussed in Sect. 5.3.3.2, it is difficult to confirm a definite relationship between the projectile striking velocity and the impact crater volumes from the present test.

5.3.3.4 Structural Integrity of Projectile

From Fig. 5.19 and Tables 5.4 and 5.5, it can be drawn that:

(i) For shots UHP-CASFRC 1-1, 1-2, 2-1, 2-2, 3-1 and 3-2, where the striking velocity (~510 m/s), volumetric fractions of coarse aggregates (30%) and compressive strength (~129.3 MPa) were all nearly identical, with the increasing of the size range of corundum coarse aggregate, the projectiles were broken when the average corundum coarse aggregate size was nearly 1.5 times larger than the projectile shank diameter *d*. Besides, the increased coarse aggregate sizes are also beneficial to deviate the terminal ballistic trajectories.

(ii) For shots UHP-CASFRC 1-1 and 1-2 as well as UHP-BASFRC 4-1 and 4-3, with nearly the same coarse aggregate size range, compressive strength and striking velocity, the projectiles penetrating into UHP-CASFRC target endured much more abrasions than the projectiles penetrating into UHP-BASFRC target, while the structural integrity of projectiles could be maintained.

(iii) For shots HSC 1-1 and HSC 1-2 as well as UHP-BASFRC 2-1 and 2-3 with nearly the same basalt coarse aggregate size range and striking velocity, the mass abrasions of projectiles have no obvious distinctions although the compressive strength increases by about 35%.

(iv) For shots UHP-CASFRC 1-1 and 1-2, 6-1-1 and 6-2-1, 7-1 and 7-2 with nearly the same corundum coarse aggregate size range (5–20 mm) and compressive strength (129.3 MPa), the mass losses and blunted length gradually increase with the rising of the striking velocities.

5.4 Numerical Simulations Based on 3D Mesoscopic Concrete Model

5.4.1 3D Mesoscopic Concrete Model

In this section, the generation algorithms of 3D mesoscopic concrete model with the random distributed sphere and convex polyhedron coarse aggregates are given at first, respectively. Then, for obtaining the greater efficiency, the influences of above two coarse aggregates shapes on projectile impact resistance of concrete targets are discussed.

5.4.1.1 Modelling Approach

The coarse aggregates are randomly distributed within the concrete matrix, and thus the locations of coarse aggregates in the mesoscopic concrete model need to be determined by the random numbers which are generated by the Monte Carlo method. There are many widely used methods for generating random numbers to simulate the physical random process, e.g., displacement, iterative and linear congruential methods. However, the randomness of numbers generated by above methods is insufficient occasionally, which cannot satisfy the requirement of randomness of Monte Carlo method. At present, the Mersenne Twister algorithm (Matsumoto and Nishimura

1998) is adopted to generate the evenly distributed pseudo-random numbers to simulate the random distribution of coarse aggregates within the concrete matrix. The rand function which defaults to Mersenne Twister algorithm is invoked by the mathematics software MATLAB to generate the random variables evenly distributed on the interval [0, 1], which are then used to determine the location of coarse aggregates.

The distribution of coarse aggregates size is very important for establishing 3D mesoscopic concrete models, because it has a significant influence on the concrete performance. Therefore, it is necessary to consider the gradation of coarse aggregates. Fuller and Thompson (1907) proposed the continuous gradation curve to optimize the density and strength of concrete, namely the Fuller curve as follows

$$F = 100\left(\frac{d_a}{d_{a,\max}}\right)^{0.5} \tag{5.1}$$

where F is the cumulative percentage of coarse aggregates passing the sieve with diameter d_a; d_a and $d_{a,\max}$ are the size and maximum size of coarse aggregates, respectively.

Based on the information of actual gradation and particle size of coarse aggregates within the concrete, and combining the Fuller curve, the frequently-used gradation of coarse aggregates can be divided into three gradations, as given in Table 5.6. Comparing the different gradations curves shown in Fig. 5.34, it is found that three

Table 5.6 Distribution of coarse aggregates with different gradations

Gradation	Small: medium: big	Type	Size (mm)
1st gradation	1:0:0:0	Small stone	5–20
2nd gradation	5.5:4.5:0:0	Medium stone	20–40
3rd gradation	3:3:4:0	Big stone	40–80

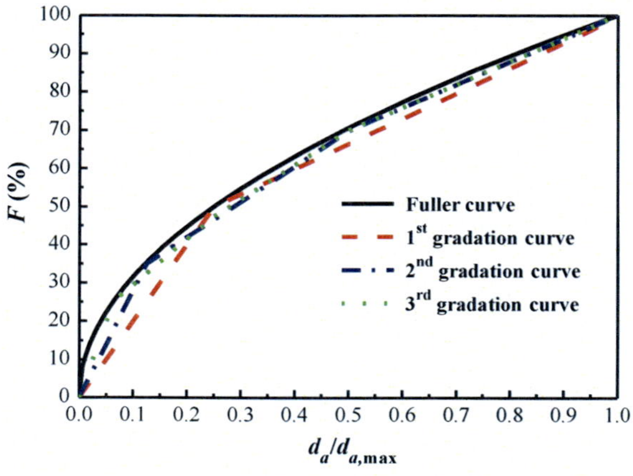

Fig. 5.34 Gradation curves, reprinted from Wu et al. (2019), copyright 2020, with permission from Elsevier

commonly used gradations of coarse aggregates are in good agreement with the Fuller curves. Thereafter, during establishing the 3D mesoscopic concrete model, the distribution of coarse aggregates sizes can be determined by different gradation curves.

The relatively simple shape of coarse aggregates, i.e., the sphere coarse aggregates, is discussed in this section. Firstly, generating the random numbers to determine the location of sphere coarse aggregates center (x_0, y_0, z_0), and the dropping area of coarse aggregates is limited in the first octant for convenient modelling. Then, the sphere coarse aggregates are dropped according to the distribution of coarse aggregates sizes of the actual concrete specimens and gradation curve. The diameter and radius of coarse aggregates to be dropped is denoted by D_0 and R_0, respectively. Before dropping the coarse aggregates, it is necessary to judge whether the following boundary conditions are satisfied

$$\begin{cases} X_{\min} \leq (x_0 + R_0) \leq X_{\max} \\ Y_{\min} \leq (y_0 + R_0) \leq Y_{\max} \\ Z_{\min} \leq (z_0 + R_0) \leq Z_{\max} \end{cases} \tag{5.2}$$

where $X_{\min}$, $Y_{\min}$, $Z_{\min}$ and $X_{\max}$, $Y_{\max}$, $Z_{\max}$ are the boundary for cuboid dropping area of the coarse aggregates. If the location of coarse aggregates does not satisfy above boundary conditions, the new random numbers will be regenerated until Eq. (5.2) is satisfied.

In order to avoid the invasion between the dropped coarse aggregates (x_0, y_0, z_0) and the previous ones (x_i, y_i, z_i), the distance between the new coarse aggregates and existing one must be greater than the sum of two spheres radius, i.e.,

$$R_0 + R_i \leq L_D \tag{5.3}$$

where R_i is the radius of any coarse aggregate which has been dropped, and L_D represents the distance between the two sphere coarse aggregates, which can be expressed as

$$L_D = \sqrt{(x_0 - x_i)^2 + (y_0 - y_i)^2 + (z_0 - z_i)^2} \tag{5.4}$$

When the new coarse aggregates meet boundary conditions and are not invaded by other existing coarse aggregates, they can be dropped until the required volume fraction is satisfied. When the concrete model is relatively large, the computational payload attributed to the invasion judgment will be increased significantly as more and more coarse aggregates are dropped, seriously slowing down the modelling speed. Aiming to improve the efficiency of dropping coarse aggregates and modelling speed, the methods of block modelling and dropping coarse aggregates according to their particle sizes are adopted at present. The block modelling breaks the whole dropping area into several parts, a larger concrete model is divided into several smaller concrete models, and dropping coarse aggregates is conducted within each smaller concrete model. When the volume fraction of coarse aggregates is satisfied for each small concrete model, then the smaller parts are gathered into a whole, which can reduce the computing cost and improve the modelling efficiency. Besides, in the process of dropping coarse aggregates, the larger ones are dropped first, then dropping the smaller ones according to the gradation curve of coarse aggregates. The above methods can be achieved by compiling the control conditions with MATLAB. The 3D mesoscopic concrete model with randomly distributed sphere coarse aggregates established by above methods is shown in Fig. 5.35, and the corresponding modelling program is compiled by software MATLAB and named Sphere_Generation.

The random convex polyhedron coarse aggregates are generated by extending the octahedron aggregates. There are two main methods: edge extension and surface extension. The edge extension method uses the longest edge as extension object, while the surface extension method uses maximum surface as extension object. At present, the former one is adopted herein. The process of establishing the 3D mesoscopic concrete model with random distributed convex polyhedron coarse aggregates is shown in Fig. 5.36, and it can be described by following four steps.

Step 1: Generating a quadrangle randomly in the circular plane

For the four vertices of the quadrangle A (x_A, y_A, z_A), B (x_B, y_B, z_B), C (x_C, y_C, z_C) and D (x_D, y_D, z_D), the 3D coordinates of A and C are ($-R_0$, 0, 0) and (R_0, 0, 0), respectively. In order to avoid generating malformed quadrangle, the random coordinates of B and D must satisfy the following conditions

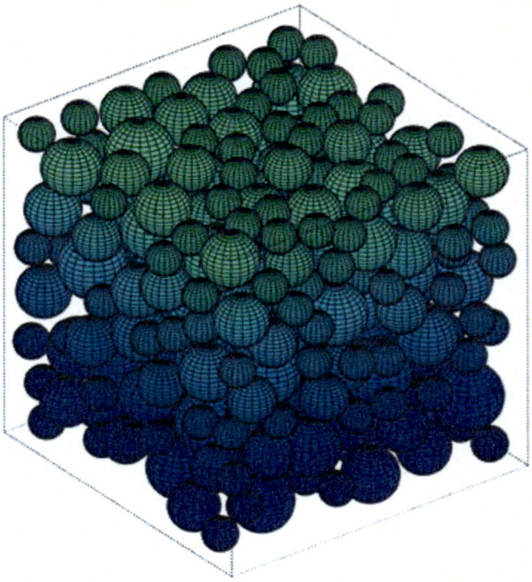

Fig. 5.35 3D mesoscopic concrete model with random distributed sphere coarse aggregates, reprinted from Wu et al. (2019), copyright 2020, with permission from Elsevier

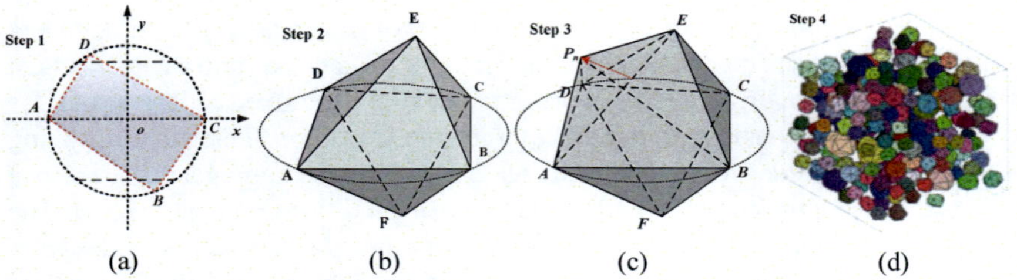

Fig. 5.36 Process of establishing the 3D mesoscopic concrete model with random distributed convex polyhedron coarse aggregates, reprinted from Wu et al. (2019), copyright 2020, with permission from Elsevier

$$\begin{cases} x_B^2 + y_B^2 = R^2 \\ -R_0 < x_B < R_0 \\ -R_0 \le y_B < -0.7R_0 \\ z_B = 0 \end{cases}, \quad \begin{cases} x_D^2 + y_D^2 = R^2 \\ -R_0 < x_D < R_0 \\ 0.7R_0 < y_D \le R_0 \\ z_D = 0 \end{cases} \tag{5.5}$$

The four vertices *A*, *B*, *C* and *D* are connected counter-clockwise to form a random quadrangle, and its schematic diagram in the *xoy* plane is shown in Fig. 5.36a.

Step 2: Forming an octahedron by generating a vertex on each side of the quadrangle

Along the *z* axis, two vertices *E* and *F* are randomly generated on the upper and lower sides of the quadrangle *ABCD*, respectively. Similarly, aiming to avoid generating malformed octahedron and generate the convex polyhedron coarse aggregates, the random coordinates of *E* and *F* must satisfy the following conditions

$$\begin{cases} x_E^2 + y_E^2 + z_E^2 = R_0^2 \\ -0.2R_0 < x_E < 0.2R_0 \\ -0.2R_0 < y_E < 0.2R_0 \\ 0 < z_E \le R_0 \end{cases}, \quad \begin{cases} x_F^2 + y_F^2 + z_F^2 = R_0^2 \\ -0.2R_0 < x_F < 0.2R_0 \\ -0.2R_0 < y_F < 0.2R_0 \\ -R_0 \le z_F < 0 \end{cases} \tag{5.6}$$

When the vertices of *E* and *F* are generated randomly, successively connecting *EA*, *EB*, *EC*, *ED*, *FA*, *FB*, *FC* and *FD* to generate the random octahedron *EABCDF*, as shown in Fig. 5.36b.

Step 3: Generating the random convex polyhedron by extending the octahedron outward

When the random octahedron is generated, it is necessary to calculate and store the length of each edge as well as the external normal vector of each surface. Comparing the lengths of each edge of the octahedron, the longest edge is determined as the extension object. The longest edge in octahedron, of which the vector are assumed as P_iP_j and $\overrightarrow{P_iP_j}$. Two adjacent surfaces of edge P_iP_j can be recorded as *i* and *j*, and the corresponding outward normal vector of each surface is denoted by $\overrightarrow{i}$ and $\overrightarrow{j}$, which are defined as $\vec{i} + \vec{j} = \vec{ij}$. Randomly generating a point $P_k\left(x_{P_k}, y_{P_k}, z_{P_k}\right)$ on the edge P_iP_j, in order to ensure that the newly generated coarse aggregates is the convex polyhedron, it is necessary to control the location of P_k, i.e.,

$$\left(x_{P_k}, y_{P_k}, z_{P_k}\right) = \left(x_{P_i}, y_{P_i}, z_{P_i}\right) + \alpha \cdot \overrightarrow{P_iP_j} \quad 0.3 < \alpha < 0.7 \tag{5.7}$$

Extending P_k outward along the vector $\overrightarrow{ij}$ until intersecting with the sphere which surrounds the octahedron, the intersection point is denoted by $P_l\left(x_{P_l}, y_{P_l}, z_{P_l}\right)$. Then, a point $P_n\left(x_{P_n}, y_{P_n}, z_{P_n}\right)$ is randomly generated from P_k to P_l along the $\overrightarrow{P_kP_l}$ as one of the vertex of the new polyhedron. In order to ensure that the newly generated coarse aggregates satisfy the requirements of convexity and particle size, the new vertex P_n needs to be generated under the following conditions

$$\left(x_{P_n}, y_{P_n}, z_{P_n}\right) = \left(x_{P_k}, y_{P_k}, z_{P_k}\right) + \beta \cdot \overrightarrow{P_kP_l} \quad 0.6 < \beta < 1.0 \tag{5.8}$$

Deleting the above longest edge P_iP_j and two adjacent surfaces i and j, and successively connecting the P_n with four vertices belong to the two surfaces i and j to generate four new surfaces. Thus a decahedron can be obtained, as shown in Fig. 5.36c. If more surfaces are required for the polyhedron, repeating above steps to extend the polyhedron until number of the surfaces is satisfied. In the formation of convex polyhedron, the information of newly generated edge length and outward normal vector of corresponding surfaces need to be continuously recorded and stored for easy invoking.

In order to further guarantee the distribution diversity of convex polyhedron coarse aggregates within the concrete model, the convex polyhedron needs to be rotated around the x, y and z axes with random rotation angle γ_x, γ_y and γ_z, respectively. Assuming the point $P_i\left(x_{P_i}, y_{P_i}, z_{P_i}\right)$ as one vertex of convex polyhedron, the coordinate after rotation can be denoted by $P_i'\left(x_{P_i'}, y_{P_i'}, z_{P_i'}\right)$, then the relationship between P_i and P_i' can be expressed as

$$H_x = \begin{bmatrix} 1 & 0 & 0 \\ 0 & \cos\gamma_x & \sin\gamma_x \\ 0 & -\sin\gamma_x & \cos\gamma_x \end{bmatrix}, \quad H_y = \begin{bmatrix} \cos\gamma_y & 0 & \sin\gamma_y \\ 0 & 1 & 0 \\ -\sin\gamma_y & 0 & \cos\gamma_y \end{bmatrix},$$
$$H_z = \begin{bmatrix} \cos\gamma_z & -\sin\gamma_z & 0 \\ \sin\gamma_z & \cos\gamma_z & 0 \\ 0 & 0 & 1 \end{bmatrix} \tag{5.9a}$$

$$\left(x_{P_i'}, y_{P_i'}, z_{P_i'}\right) = \left(x_{P_i}, y_{P_i}, z_{P_i}\right) \cdot H_x \cdot H_y \cdot H_z \tag{5.9b}$$

Step 4: Dropping the convex polyhedron coarse aggregates in the specified area

The convex polyhedron coarse aggregates will be dropped in the specified area after generated. Since the convex polyhedron aggregates are generated within the sphere, when the requirement of coarse aggregates volumetric fraction is not very high, the above criteria of invasion judgment for sphere can be used to save the modelling time. If the volume fraction of the coarse aggregates is required to be high, the criteria of invasion judgment based on the volume-index introduced in Liu and Gao (2003) can be adopted. In order to improve the modelling efficiency, similarly, the methods of block modelling and dropping the coarse aggregates in order of their particle sizes are adopted at present. The 3D mesoscopic concrete model with random distributed convex polyhedron coarse aggregates is shown in Fig. 5.36d, and the corresponding modelling program is written by MATLAB and named as Polyhedron_Generation.

5.4.1.2 3D Mesoscopic Finite Element Model

The commonly meshing methods of finite element model include mapping, grid-based, Delaunay triangulation and advancing front methods, etc. (Guan et al. 2003). Considering the element quality and meshing efficiency, the mapping method is adopted at present, and the hexahedral element was adopted to guarantee the calculation accuracy (Landertshamer and Steffan 2010). In order to obtain the mesoscopic concrete model, the program named 3D_RanMesh for meshing is written by MATLAB. When all the convex polyhedron coarse aggregates are completely dropped, the meshing process is also finished simultaneously, and then the finite element model is output.

Take the meshing of the 3D mesoscopic concrete model with random distributed convex polyhedron coarse aggregates for example, there are two meshing types for concrete with different components, one is the two-phase concrete composed of coarse aggregates and mortar, and the other is the three-phase concrete with three components: coarse aggregates, mortar and interfacial transition zone (ITZ).

(1) Meshing the two-phase concrete model

The elements generated by mapping method are all identical, and the coarse aggregates and mortar are not distinguished. Therefore, the generated elements must be identified with material properties. Assuming that the element center is G, the outward normal vector of any surface of coarse aggregates is $\vec{i}$ ($1 < i < \text{n}$, where n is the surface number of polyhedral convex coarse aggregates), and one vertex on this surface is P_i. Then, the vector from the P_i to G can be represented as $\overrightarrow{P_i G}$. The boundary conditions of the dropped coarse aggregates are identical with that in Sect. 5.4.4.1. In order to improve the calculation efficiency, only identify the material properties of elements which located around the coarse aggregates. The element sizes on the x, y and z axes are denoted by E_x, E_y, E_z, and n_x, n_y, n_z are the number of elements which needs to be identified in the corresponding direction. Then the element identification range can be expressed as

$$\begin{cases} \max\left(\left\lfloor \dfrac{x_0 - R_0}{E_x} \right\rfloor, 0\right) \le n_x \le \min\left(\left\lceil \dfrac{x_0 + R_0}{E_x} \right\rceil, \left\lceil \dfrac{X_{\max}}{E_x} - 1 \right\rceil\right) \\ \max\left(\left\lfloor \dfrac{y_0 - R_0}{E_y} \right\rfloor, 0\right) \le n_y \le \min\left(\left\lceil \dfrac{y_0 + R_0}{E_y} \right\rceil, \left\lceil \dfrac{Y_{\max}}{E_y} - 1 \right\rceil\right) \\ \max\left(\left\lfloor \dfrac{z_0 - R_0}{E_z} \right\rfloor, 0\right) \le n_z \le \min\left(\left\lceil \dfrac{z_0 + R_0}{E_z} \right\rceil, \left\lceil \dfrac{Z_{\max}}{E_z} - 1 \right\rceil\right) \end{cases} \tag{5.10}$$

Since the generated coarse aggregates are convex polyhedron, the finite element model of two-phase concrete can be obtained by the following identifying algorithm.

$$\begin{cases} T_i = 0 \text{ for } \overrightarrow{P_i G} \cdot \vec{i} \le 0 \\ T_i = 1 \text{ for } \overrightarrow{P_i G} \cdot \vec{i} > 0 \end{cases}, \quad \begin{cases} \text{Mortar} \quad \text{for } \sum_{i=1}^{n} T_i = 0 \\ \text{Aggregate for } \sum_{i=1}^{n} T_i > 0 \end{cases} \tag{5.11}$$

(2) Meshing the three-phase concrete model

The meshing method of three-phase concrete is similar to that of two-phase concrete, except that the ITZ is added within the finite element model of three-phase concrete. Denoting one of the element nodes by S_j, the element identification range and other parameters are identical with that in the process of meshing two-phase concrete model. The specific identifying algorithm is as follows

$$\begin{cases} T_i = 0 \text{ for } \forall j \in [1, 8], \overrightarrow{P_i S_j} \cdot \vec{i} \le 0 \\ T_i = 1 \text{ for } \exists j \in [1, 8], \overrightarrow{P_i S_j} \cdot \vec{i} > 0 \end{cases}, \quad \begin{cases} \text{Aggregate for } \sum_{i=1}^{\mathrm{n}} T_i = 0 \wedge \prod_{i=1}^{\mathrm{n}} T_i = 0 \\ \text{ITZ} \quad \text{for } \sum_{i=1}^{\mathrm{n}} T_i > 0 \wedge \prod_{i=1}^{\mathrm{n}} T_i = 0 \\ \text{Mortar} \quad \text{for } \sum_{i=1}^{\mathrm{n}} T_i > 0 \wedge \prod_{i=1}^{\mathrm{n}} T_i > 0 \end{cases} \tag{5.12}$$

The finite element model of three-phase concrete generated by above method is shown in Fig. 5.37.

The mesoscopic concrete is generally considered to be composed of three components: mortar, coarse aggregates and ITZ, where ITZ refers to the inhomogeneous composition which is formed between the surface of coarse aggregates and mortar. The thickness of ITZ is about 30–100 μm (Ollivier et al. 1995; Diamond and Huang 2001), if the finite element model of concrete is established by adopting the thickness of ITZ as the element size, the element number will be increased exponentially, which will greatly increase the calculation time. Meanwhile, the ITZ thickness is several orders of magnitude less than the sizes of large-diameter projectile and coarse aggregates, and there are few constitutive models that can better describe the performance of ITZ. Therefore, the 3D mesoscopic model of two-phase concrete is adopted in the following numerical simulations, and the ITZ is incorporated into the mortar.

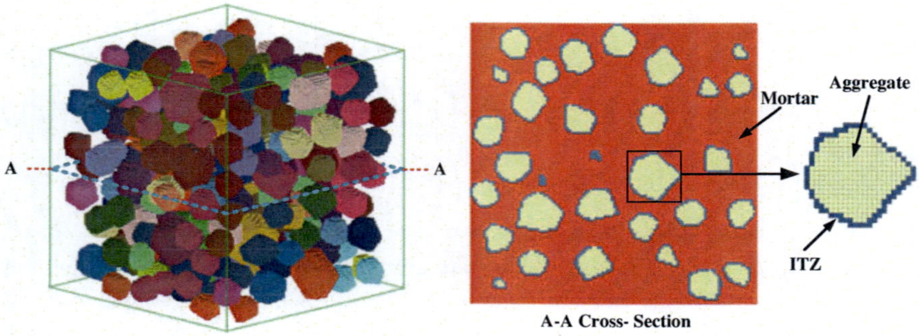

Fig. 5.37 Finite element model of three-phase concrete, reprinted from Wu et al. (2019), copyright 2020, with permission from Elsevier

Two types of 3D mesoscopic concrete models are established with randomly distributed sphere and convex polyhedron (the surface number ranges from 20 to 28) coarse aggregates, and the 2nd gradation of coarse aggregates and volume fraction of coarse aggregates is 20% are both adopted in two models. In order to examine the effect of coarse aggregates shapes on the impact resistance of concrete, by using the finite element program LS-DYNA, the numerical simulations of projectile penetration and perforation into concrete targets are performed by applying above two types mesoscopic models, respectively.

The projectile used in the test (Hanchak et al. 1992) is adopted. Because the mass abrasion, erosion and deformation of high-strength steel projectile are very limited during penetration into concrete target with striking velocity roughly less than 1 km/s, the projectile can be regarded as rigid body and described by the *MAT_RIGID material model (MAT#020). The target sizes in penetration and perforation scenarios are respectively 200 × 200 × 600 mm^3 and 200 × 200 × 200 mm^3, and the element sizes of targets are both 2 mm. For describing the mechanical behavior of concrete under the conditions of large deformation, high strain rate and high hydrostatic pressure, the *MAT_JOHNSON_HOLMQUIST_CONCRETE (HJC) model (MAT#111) (Hallquist 2007; Holmquist et al. 1993) is selected as the material model for mortar and coarse aggregates. The corresponding model parameters are listed in Table 5.7. The contact between the concrete and projectile is described by the keyword "CONTACT ERODING SURFACE TO SURFACE". In order to avoid the hourglass effect during numerical simulation, *Mat_ADD_EROSION is applied to remove the highly distorted elements, and the same maximum principal strain of 0.3 is adopted as the failure criterion of mortar and coarse aggregates for both two types concrete models (Tai 2009).

The calculated results are shown in Figs. 5.38 and 5.39 further shows the effective stress contours of projectile perforation into concrete target plate with the impact velocity of 800 m/s. By comparing the DOP, residual velocity (V_r) and the perforation process of projectile, it can be found that there is no obvious difference for the impact resistance of concrete targets with two shapes of coarse aggregates.

Table 5.7 Parameters of the HJC model for mortar and aggregates (Ren et al. 2016, 2017)

Parameter	Mortar	Aggregate	Parameter	Mortar	Aggregate
Density ρ (kg/m^3)	2300	2600	Pressure hardening exponent N	15	7.0
Shear modulus G (GPa)	12.5	30.0	Crushing pressure P_{crush} (MPa)	0.013	0.068
Normalized cohesive strength A	0.28	0.3	Crushing volumetric strain U_{crush}	8×10^{-4}	1.2×10^{-3}
Normalized pressure hardening B	1.85	1.73	Locking pressure P_{lock} (GPa)	1.21	3.47
Strain rate coefficient C	0.006	0.005	Locking volumetric strain U_{lock}	0.12	0.1
Uniaxial compressive strength f_c (MPa)	40	160	Damage constant D_1	0.04	0.04
Maximum tensile pressure T (MPa)	3.92	7.84	Damage constant D_2	1.0	1.0
Quasi-static threshold strain rate $EPS0$	1	1	Pressure constant K_1 (GPa)	12	116
Plastic strain before fracture $EFMIN$	0.01	0.01	Pressure constant K_2 (GPa)	135	−243
Normalized maximum strength $SFMAX$	15	7.0	Pressure constant K_3 (GPa)	698	506

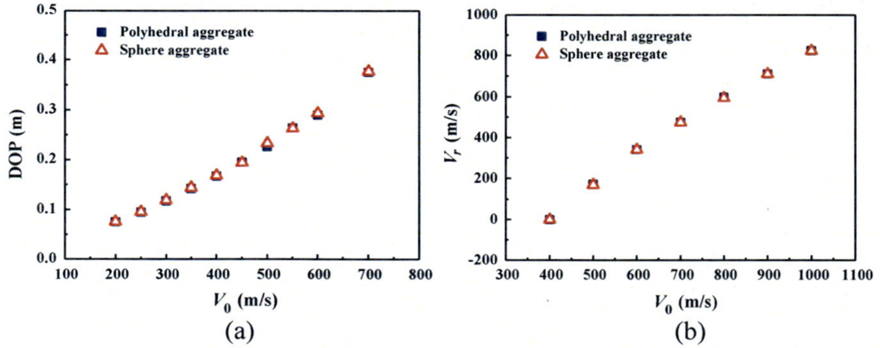

Fig. 5.38 Calculation results of two mesoscopic concrete models under penetration and perforation **a** DOP, **b** residual velocity, reprinted from Wu et al. (2019), copyright 2020, with permission from Elsevier

Generally, with regard to the 3D mesoscopic concrete model which is established with hexahedral elements, for both sphere or convex polyhedron coarse aggregates, it can be found that there are negligible differences on the impact resistance of concrete targets. Therefore, the sphere coarse aggregates are adopted in the following

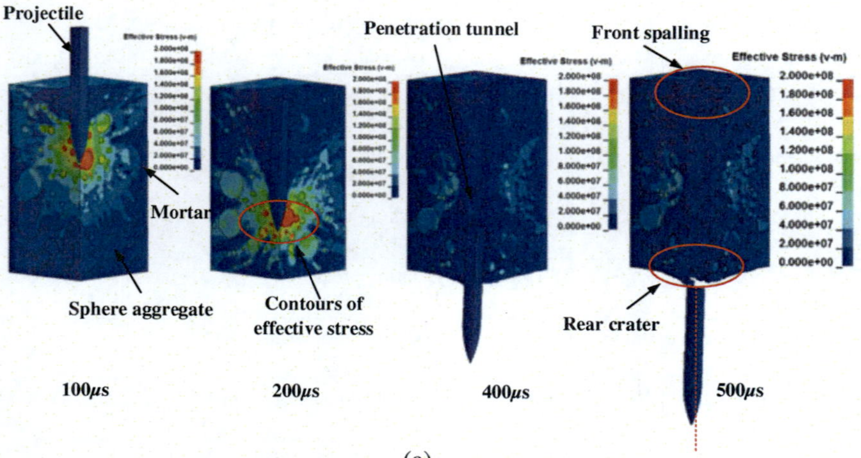

(a)

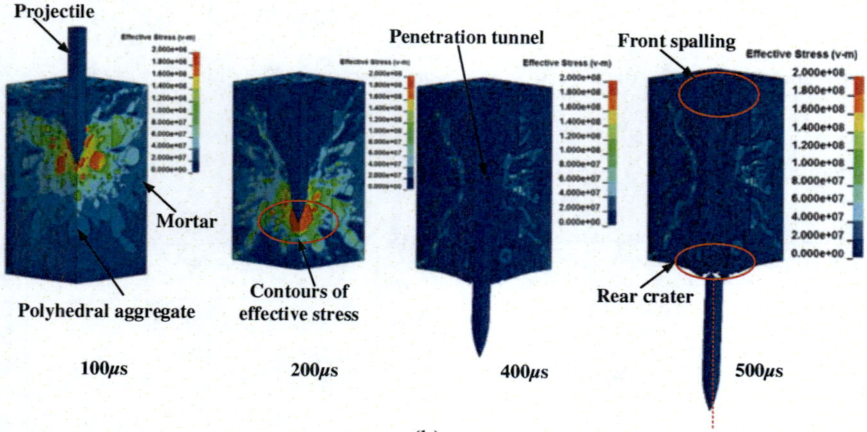

(b)

Fig. 5.39 Effective stress contours of projectile perforating into concrete target plates with **a** sphere, **b** polyhedral coarse aggregates, reprinted from Wu et al. (2019), copyright 2020, with permission from Elsevier

numerical simulations to guarantee both the modelling efficiency and calculation accuracy.

5.4.2 *Validations*

In this section, aiming to validate the 3D mesoscopic UHPC model with randomly distributed sphere coarse aggregates and the corresponding constitutive parameters, the numerical simulations of projectile penetration test in Sect. 5.3 (Wu et al. 2015b) are performed numerically by LS-DYNA.

5.4.2.1 Finite Element Model

The targets size of CA-UHPC are mainly Φ750 mm × 500 mm and Φ750 mm × 600 mm. Considering the actual damage of targets in the tests and calculation efficiency, the size of CA-UHPC target in the numerical simulation is determined as Φ400 mm × 400 mm. In order to reduce the calculation cost, the element size is 2 mm for the central impacted zone (200 mm × 200 mm) and 5 mm for the surrounding region, respectively. The coarse aggregates are only dropped in the central impacted zone, while volume fraction and size of coarse aggregates remain constant with test. The projectile used in the numerical simulation is identical with the test. It should be noted that, in order to improve the calculation efficiency and avoid the calculation errors, the solid projectile is adopted in the numerical simulation, which is different from the projectile filled by polymer inert materials in the test, and the element size of projectile is 3 mm. Figure 5.40 shows the half of finite element model for the test with coarse aggregate size is 35–45 mm and coarse aggregates volumetric fraction is 30%.

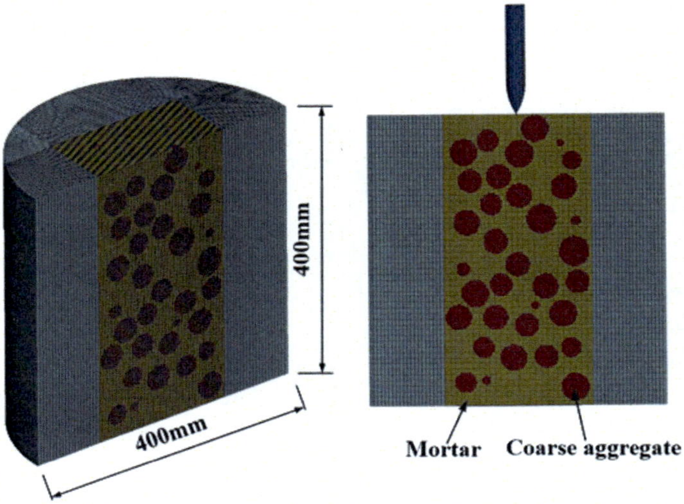

Fig. 5.40 Finite element model

5.4.2.2 Constitutive Model Parameters of Concrete and Projectile

The projectile deformed and damaged obviously in the test, and thus the *MAT_JOHNSON_COOK (JC) model (MAT#015) (Hallquist 2007; Johnson and Cook 1985) and Mie-Gruneisen equation of state are selected as the projectile material model, which can better describe the dynamic mechanics properties of metals under the large deformation, high strain rate and high temperature. Moreover, JC model is widely used in the numerical simulation of metal penetration. The corresponding parameters of JC model and Mie-Gruneisen equation of state are listed in Tables 5.8 and 5.9, respectively.

Similarly, the HJC model is selected as the material model for the mortar and corundum aggregates, and the corresponding parameters are listed in Table 5.10.

The keyword *Mat_ADD_EROSION is also applied to delete failed elements according to the maximum principal strain, and the corresponding value of 0.2 derived by trial and error is given as the failure criterion for the test.

5.4.2.3 Comparisons of Numerical Results with Test Data

The numerical simulation is performed by above 3D mesoscopic UHPC model with randomly distributed sphere coarse aggregates and the corresponding constitutive parameters. Meanwhile, the numerical results and test data are listed in Table 5.11,

Table 5.8 Parameters of JC model for projectile (Johnson and Cook 1985)

Parameter	Projectile	Parameter	Projectile
Density ρ (kg/m^3)	5208	Room temperature T_r (K)	298
Shear modulus G (GPa)	77	Specific heat C_v (J/(g*K))	469
Yield stress constant A' (MPa)	792	Quasi-static threshold strain rate *EPSO*	1
Strain hardening constant B' (MPa)	510	Failure constant D_1'	0.05
Strain hardening exponent N'	0.26	Failure constant D_2'	3.44
Strain rate correlation coefficient C'	0.014	Failure constant D_3'	−2.12
Temperature correlation index M	1.03	Failure constant D_4'	0.002
Melt temperature T_m (K)	1793	Failure constant D_5'	0.61

Table 5.9 Mie-Gruneisen equation of state (Johnson and Cook 1985)

Parameter	Projectile	Parameter	Projectile
Intercept of the curve C_1 (m/s)	4569	Slope coefficients of curve S_3	0
Slope coefficients of curve S_1	1.49	Grüneisen coefficient γ_0	2.17
Slope coefficients of curve S_2	0	First order volume correction coefficient α	0.46

Table 5.10 Parameters of the HJC model for mortar and corundum aggregates

Parameter	Mortar	Corundum	Parameter	Mortar	Corundum
ρ (kg/m^3)	2500	3800	N	0.79	0.79
G (GPa)	24.1	188	P_{crush} (MPa)	38.3	666.7
A	0.3	0.3	U_{crush}	0.0012	0.0026
B	1.73	1.73	P_{lock} (GPa)	3.47	3.47
C	0.005	0.005	U_{lock}	0.08	0.053
f_c (MPa)	115	2000	D_1	0.04	0.04
T (MPa)	6.65	27.7	D_2	1.0	1.0
*EPS*0	1.00	1.00	K_1 (GPa)	116	116
EFMIN	0.01	0.01	K_2 (GPa)	−243	−243
SFMAX	7	7	K_3 (GPa)	506	506

Table 5.11 Comparisons of test data and numerical results

Test No.	d_a (mm)	V_{agg} (%)	V_0 (m/s)	DOP (mm)		Deviation (%)
				Test	Simulation	
1-1	5–20	30	508	98	102.78	4.87
1-2	5–20	30	508	101	101.56	0.55
2-1	35–45	30	503	21	38.38	82.76
2-2	35–45	30	515	68	68.20	0.29
3-1	65–75	30	510	59	56.14	−4.85
3-2	65–75	30	517	37	34.1	−7.84
4-1	5–20	45	505	145	68.22	−52.95
4-2	5–20	45	507	106	80.64	−23.92
5-1	5–20	30	524	106	104.47	−1.44
5-2	5–20	30	511	88	102.16	16.09
6-1-1	5–20	30	708	171	148.27	−13.29
6 1-2	5–20	30	710	310	318.76	2.83
6-2-1	5–20	30	722	176	150.19	−14.66
6-2-2	5–20	30	720	323	336.22	4.09
7-1	5–20	30	832	185	182.13	−1.55
7-1	5–20	30	846	191	178.40	−6.60

in which "V_{agg}" denotes volume fraction of coarse aggregates. Figure 5.41 further shows the effective stress contours of projectile penetration into CA-UHPC target. It can be drawn from comparison that the numerical results are in relatively good agreement with the tests data and the errors are within the allowable range, except the tests 2-1, 4-1 and 4-2. For the test 2-1, projectile hits the corundum aggregates directly and thus causes a small DOP, which is shown in Fig. 5.41. While the large

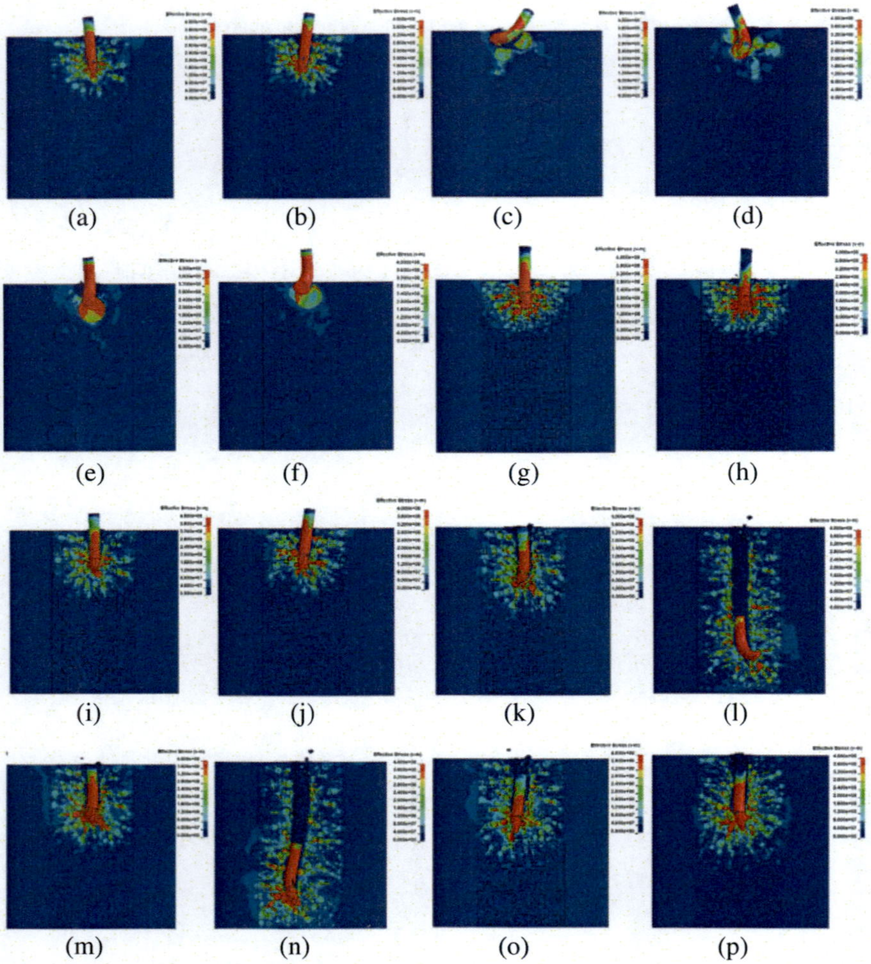

Fig. 5.41 Effective stress contours of projectile penetrating into CA-UHPC targets **a** 1-1, **b** 1-2, **c** 2-1, **d** 2-2, **e** 3-1, **f** 3-2, **g** 4-1, **h 4-2, i 5-1, j** 5-2, **k** 6-1-1, **l** 6-1-2, **m** 6-2-1, **n** 6-2-2, **o** 7-1, **p** 7-2

deflection is caused after the projectile hits the edge of corundum aggregates in numerical simulation, which gives rise to the deeper DOP than that in the test, the corresponding calculated result is shown in Fig. 5.41c. For the tests 4-1 and 4-2, the coarse aggregates volumetric fraction is higher than that in the tests 1-1 and 1-2, while the DOP is inversely deeper. Generally, the probability of projectile hitting the corundum aggregates will increase with the rising of the volumetric fraction, and the DOP should be reduced, which is in agreement with the results of numerical simulation. The unexpected experimental results may be caused by the fact that the targets 4-1 and 4-2 were not self-densified enough, as described in Sect. 5.3.2.

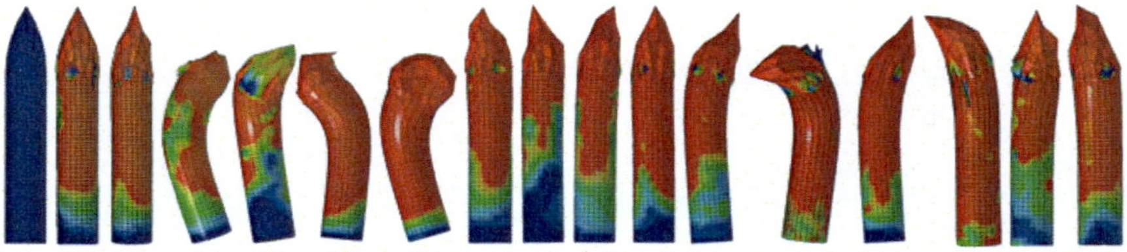

Fig. 5.42 Damages state of projectiles

Due to the high strength of corundum aggregate within the CA-UHPCC targets, the damage of projectile is relatively serious. Figure 5.42 further shows the damaged projectiles obtained from the numerical simulations. By comparing to Fig. 5.19, it indicates that the numerical simulation can reflect the deformation of projectile well. The projectiles were broken in tests 2-1, 2-2, 3-1 and 3-2, while the corresponding projectiles were only severely distorted in numerical simulation. The main reasons may lie in that, (i) the property of JC model makes the projectile is difficult to break; (ii) the projectiles were filled by polymer inert materials in test, while the solid projectile is selected to simply the calculation in numerical simulation.

In this section, based on the 3D mesoscopic concrete model with the randomly distributed sphere coarse aggregates, the projectile penetration test is numerically simulated. By comparisons of the DOP and projectiles damages, the above mesoscopic concrete model and corresponding constitutive parameters are validated.

5.4.3 *Impact Resistance of Different Aggregated UHPC*

In this section, based on above validated 3D mesoscopic concrete model and the constitutive parameters, the impact resistance of CA-UHPC (corundum aggregates), BA-UHPC (basalt aggregates), QA-UHPC (quartz aggregates) and LA-UHPC (limestone aggregates) targets are discussed.

The finite element model and the corresponding constitutive model parameters of projectile are identical with that in Sect. 5.4.2.1, respectively. For the four different aggregated UHPC targets, the volume fraction and size of coarse aggregates are the same, and the 2nd gradation of coarse aggregates (the size ranges from 5 to 40 mm) and volume fraction is of 30% are adopted. When analyzing the penetration and perforation resistance of four targets, the targets size is the same as that in Sect. 5.4.2.1, while analyzing the influence of coarse aggregate on projectile trajectory, the size of target is determined as 400 × 400 × 200 mm^3. The HJC model is selected as the material model for the mortar and four aggregates, and the corresponding model parameters of mortar and corundum aggregate are identical to those in Table 5.10. Meanwhile, the compressive strength of limestone, basalt and quartz aggregates are identified as 100 MPa, 200 MPa and 350 MPa based on experience, respectively. Referring to the method of determining the HJC model parameters for

Table 5.12 Parameters of the HJC model for limestone, basalt and quartz aggregates

Parameter	Limestone	Basalt	Quartz	Parameter	Limestone	Basalt	Quartz
ρ (kg/m^3)	2600	3000	2700	S_{max}	7.0	7.0	7.0
G (GPa)	31.7	41.6	47.4	P_{crush} (MPa)	33.3	66.7	117
A	0.3	0.3	0.3	U_{crush}	1.05×10^{-3}	1.20×10^{-3}	1.86×10^{-4}
B	1.73	1.73	1.73	P_{lock} (GPa)	3.47	3.47	3.47
C	0.005	0.005	0.005	U_{lock}	0.12	0.1	0.07
N	0.79	0.79	0.79	D_1	0.04	0.04	0.04
f_c (MPa)	100	200	350	D_2	1.0	1.0	1.0
T (MPa)	6.2	8.77	11.6	K_1 (GPa)	116	116	116
EPS0	1.00	1.00	1.00	K_2 (GPa)	−243	−243	−243
EFMIN	0.01	0.01	0.01	K_3 (GPa)	506	506	506

high strength concrete or rock provided by Ren et al. (2017), the main model parameters of limestone, basalt and quartz aggregates are summarized in Table 5.12. It should be noted that, the maximum principal strain for deleting failed elements is 0.2 which is identical to that in Sect. 5.4.2.2.

5.4.3.1 Projectile Penetration Resistance

In order to analyze anti-penetration performance of different aggregated UHPC targets, the numerical simulations of projectile penetrating into four targets with $V_0 = 400$, 600, 800 and 1000 m/s are performed, respectively. The numerically derived results are shown in Fig. 5.43. It indicates that the DOP decreases with the rise of the coarse aggregate strength (hardness), especially when the impact velocity

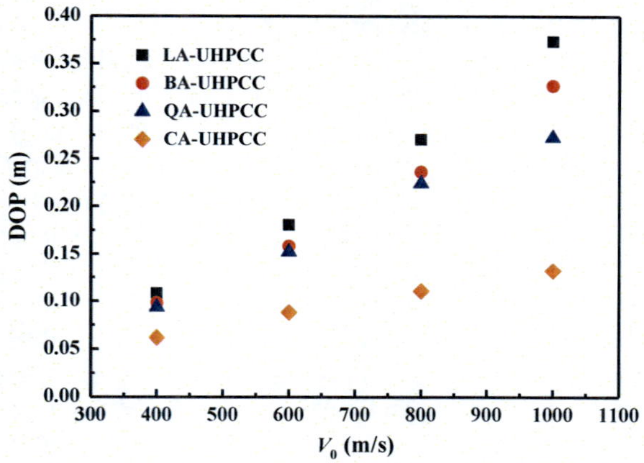

Fig. 5.43 Depth of penetration

of the projectile is relatively high. Meanwhile, compared with other three targets, the advantage of anti-penetration performance of CA-UHPC target becomes more obvious with the increase of impact velocity. For instance, when $V_0 = 400$ m/s, the DOP of projectile into CA-UHPC target is 0.66, 0.63 and 0.57 times that of projectile into QA-UHPC, BA-UHPC and LA-UHPC targets, respectively; when $V_0 = 1000$ m/s, the DOP of projectile into CA-UHPC target is 0.49, 0.41 and 0.35 times that of projectile into QA-UHPC, BA-UHPC and LA-UHPC targets, respectively. Figures 5.44 and 5.45 further show the damages state of projectiles with $V_0 = 1000$ m/s and the corresponding effective stress contours of projectiles penetration into different aggregated UHPC targets, respectively. It can be found that the projectile is severely deformed after penetrating the CA-UHPC target, which may be the reason for the corresponding small DOP. The projectile is also deformed after penetrating the QA-UHPC target, but the deformation is smaller than that of the former. For the other two projectiles penetrating into BA-UHPC and LA-UHPC targets, there are almost unchanged, and only projectile nose is slightly deformed.

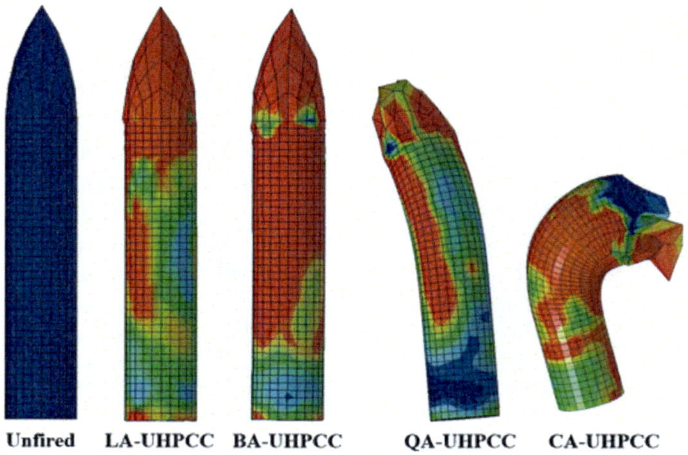

Fig. 5.44 Damages state of projectiles after perforation

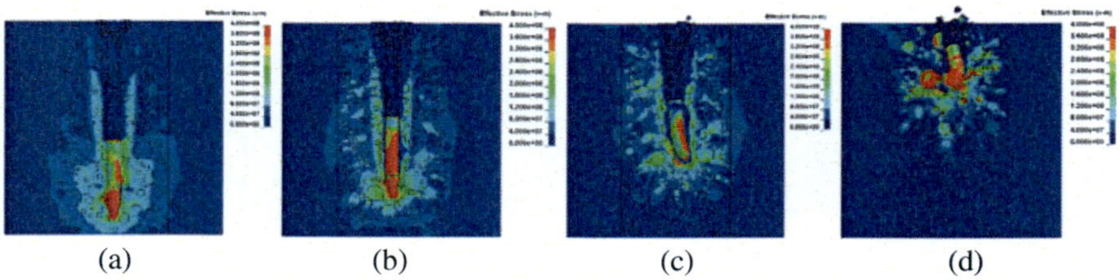

Fig. 5.45 Effective stress contours of projectile penetrating into different aggregated UHPC targets **a** LA-UHPC, **b** BA-UHPC, **c** QA-UHPC, **d** CA-UHPC

Therefore, it should be a feasible choice that rising the coarse aggregates strength (hardness) within the UHPC targets to improve the penetration resistance of targets.

5.4.3.2 Anti-perforation Performance

For investigating the influence of different aggregates on anti-perforation performance of UHPC targets, the numerical simulations of projectile perforating into four targets with $V_0 = 800$, 1000, 1200 and 1400 m/s are performed, respectively. Figure 5.46 shows the numerical simulation results. It should be noted that, when $V_0 = 1400$ m/s, the projectile still cannot perforate into CA-UHPC target, thus Fig. 5.46 only shows the calculated results of projectile perforating into the other three UHPC targets. Similarly, it can be drawn from Fig. 5.46 that the residual velocity of projectile will decrease with the rise of the coarse aggregate strength (hardness), but there is only slight difference in the residual velocity, and the difference decreases with the increase of projectile velocity. For instance, when $V_0 = 800$ m/s, the residual velocity of projectile perforating into the QA-UHPC target is 48 m/s and 121 m/s lower than that of projectile perforating into the BA-UHPC and LA-UHPC targets, respectively; when $V_0 = 1400$ m/s, the residual velocity of projectile perforating into the QA-UHPC target is 30 m/s and 82 m/s lower than that of projectile perforating into the BA-UHPC and LA-UHPC targets, respectively. Figures 5.47 and 5.48 further show the damages state of projectiles with $V_0 = 1400$ m/s and the corresponding effective stress contours of projectiles perforation into different aggregated UHPC targets, respectively. It indicates that the projectiles deform in varying degrees after perforating into different targets, and the deformation increases with the rise of the coarse aggregate strength (hardness). Moreover, for the condition of projectile impacting the CA-UHPC target, it can be drawn from Fig. 5.47 that the projectile is severely deformed, and Fig. 5.48d further shows the projectile deflects seriously within the

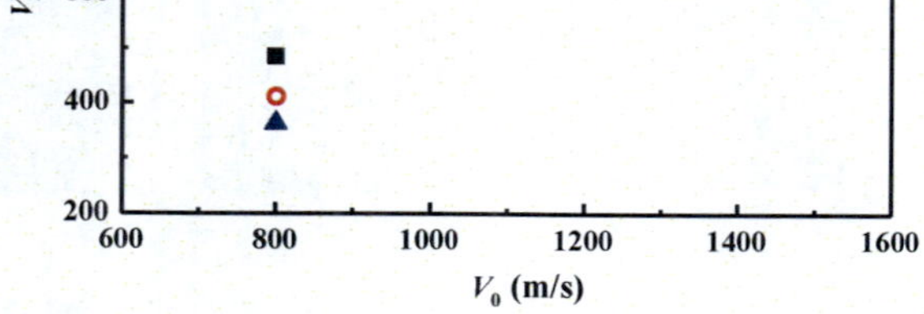

Fig. 5.46 Depth of penetration

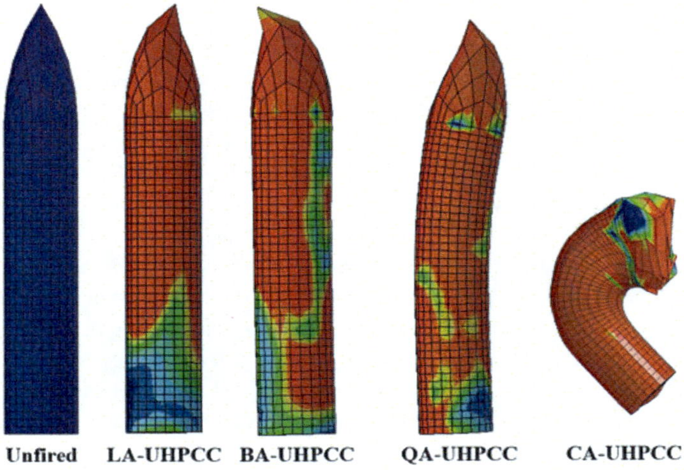

Fig. 5.47 Damages state of projectiles after perforation

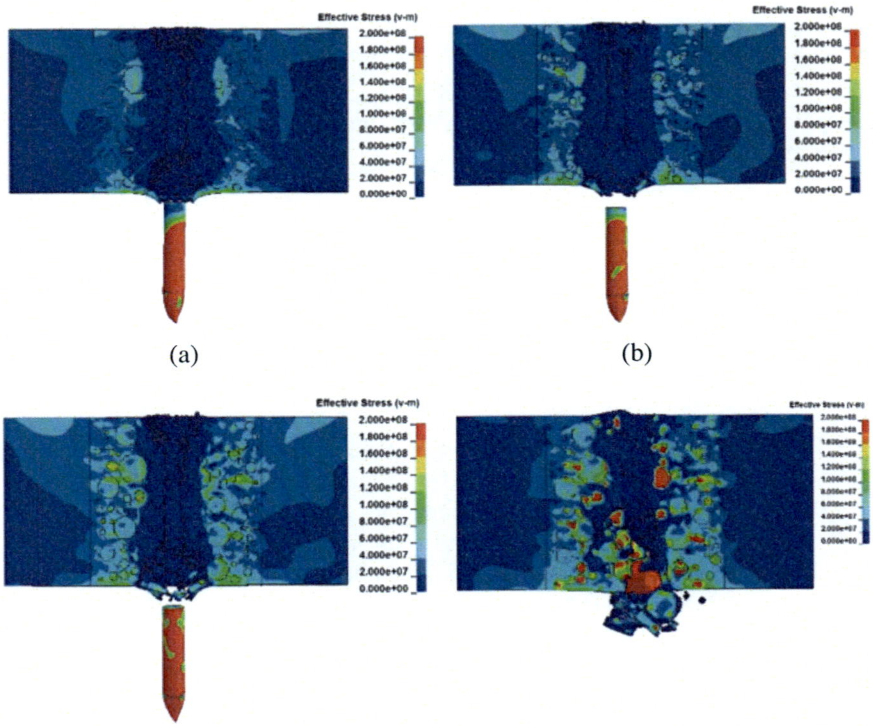

Fig. 5.48 Effective stress contours of projectile perforating into different aggregated UHPC targets **a** LA-UHPC, **b** BA-UHPC, **c** QA-UHPC, **d** CA-UHPC

target. This should be the reason why the projectile failes to perforate the CA-UHPC target. Therefore, the anti-perforation performance of UHPC targets is improved with the rise of the coarse aggregates strength (hardness), and the penetration resistance of CA-UHPC is much better than that of the other three targets.

5.4.3.3 Projectile Trajectory

Considering that the projectiles sometimes impact the target with a certain oblique angle, and in order to analyze the influence of different aggregates on projectile trajectory, the numerical simulations of projectiles impact aggregated UHPC targets by different oblique angles (5, 10 and 15°) are performed in the present section. The impact velocity of projectile is 1000 m/s. Figures 5.49 and 5.50 show the effective stress contours of projectiles penetration into different aggregated UHPC targets and the corresponding damage states of projectiles, respectively. It indicates that although the size and distribution of four kinds of coarse aggregates within the UHPC targets are the same, but the ballistic trajectories of projectiles penetrating different aggregated UHPC targets are obviously different. Furthermore, it can be seen from the comparison that the deflection and deformation of projectiles penetrating the CA-UHPC and QA-UHPC targets are more serious than that of projectiles penetrating the other two UHPC targets. Therefore, the UHPC targets with high strength (hardness) aggregates can increase the deflection and deformation of projectile to improve the impact resistance of the targets.

5.5 Summary

In this chapter, the impact resistance of coarse aggregated UHPCC target against the medium caliber projectile is studied experimentally and numerically, the main work and conclusions are as follows:

(1) In Sect. 5.2, the high-speed projectile penetration tests on UHPCC targets with the striking velocities at 510–1320 m/s were carried out. It derives that, the projectile impact crater dimensions enlarges with the increase of the compressive strength of the target and the projectile striking kinetic energy, and decreases with the increase of fiber mixing fraction. Within the variation ranges of the discussed parameters, the influential degree of mixing fibers seems to be much greater than that of compressive strength. The influential degrees of the impact kinetic energy in enlarging the crater dimensions as well as the fibers in reducing the crater dimensions are dependent on the amounts of the fibers as well as the projectile striking velocity; The compressive strength, the fiber mixing fraction as well as coarse aggregates can help to decrease the DOP of the projectile and thus enhance the impact resistance of the target,

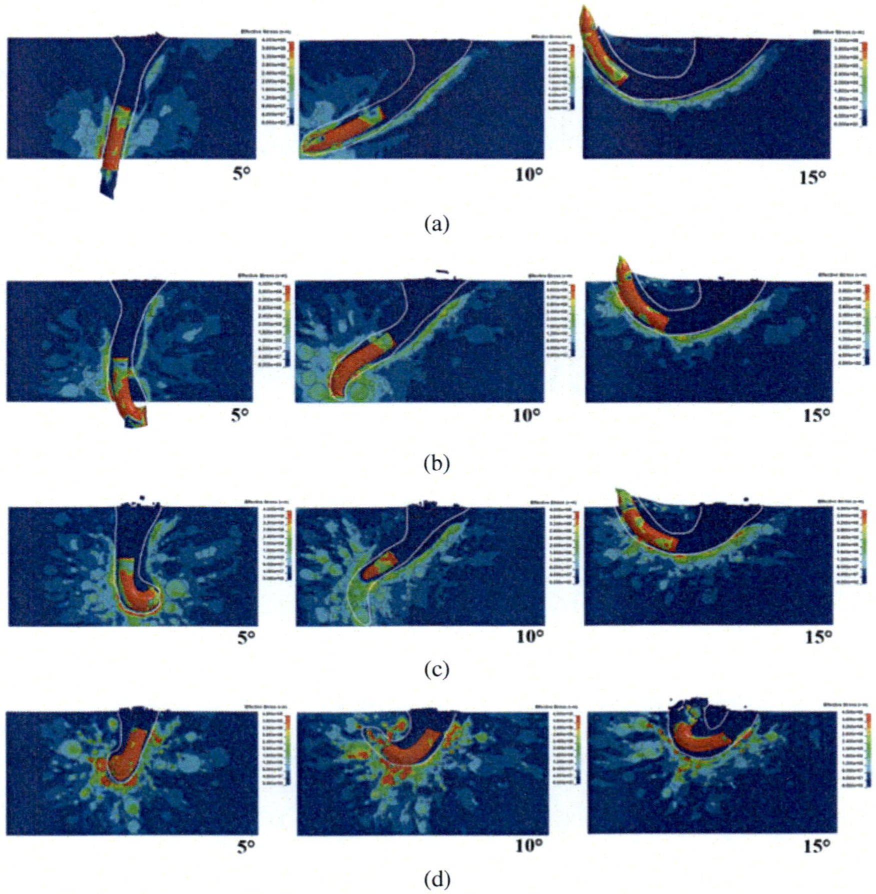

Fig. 5.49 Effective stress contours of projectile penetrating into different aggregated UHPC targets by different obliquity angles **a** LA-UHPC, **b** BA-UHPC, **c** QA-UHPC, **d** CA-UHPC

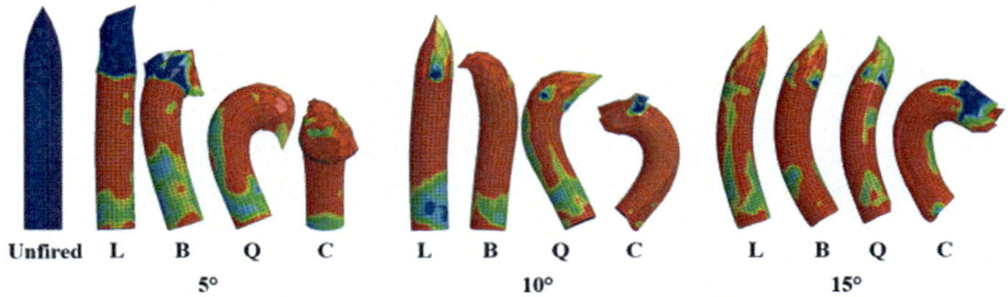

Fig. 5.50 Damage of projectiles after oblique impact

while the contribution of fibers is very limited for the relatively low fiber addition ratio ($\leq$3%); UHPCC with the compressive strength of 90 MPa and steel fiber mixing ratio of 1.5% is the most optimum choice for protective structure constructions.

(2) In Sect. 5.3, by conducting the projectile penetration test on UHP-CASFRC and comparative HSC targets, it derives that, larger sized coarse aggregates help to reduce the DOP, impact crater area and volume, and the harder coarse aggregates decrease the DOP by aggravating the mass abrasions of the projectile. For the coarse aggregates with the size less than projectile shank, the harder coarse aggregates only aggravate the mass abrasions of the projectile, and the structural stability of the projectile could be maintained; For the structural integrity of the projectile penetrating into UHP-CASFRC target, the structural destruction of the penetrator (bend even fracture) occurs with enlarging the corundum size. By comprehensively considering the anti-strike performance and the workability of UHP-CASFRC, the corundum aggregate sizes $\geq 1.5d$ is suggested for the constructions of protective structures.

(3) In Sect. 5.4, based on the 3D mesoscopic concrete model, the influence of coarse aggregates strength (hardness) on the impact resistance of UHPC targets is analyzed and discussed. The generation algorithms of 3D mesoscopic concrete model with randomly distributed sphere and convex polyhedron coarse aggregates are proposed. Regarding the 3D mesoscopic concrete model meshed by hexahedral elements, the sphere and convex polyhedron coarse aggregates have negligible differences for impact resistance of concrete target; By simulating the experiments of projectile penetration into the CA-UHPC targets, the 3D mesoscopic UHPC model with the randomly distributed sphere coarse aggregates and corresponding constitutive parameters are validated; Based on the verified 3D mesoscopic UHPC model, the numerical studies of projectile penetration into CA-UHPC, QA-UHPC, BA-UHPC and LA-UHPC targets under different impact conditions are performed, respectively. According to the simulated results of DOP, residual velocity and ballistic trajectory of projectile, it is found that the UHPC targets with high strength (hardness) aggregates can increase the deflection, deformation and abrasion of projectiles to improve the impact resistance of target. Furthermore, the relationship of impact resistance of four targets can be determined as: CA-UHPC > QA-UHPC > BA-UHPC > LA-UHPC, and the impact resistance of CA-UHPCC is much better than the others; for design concrete structures with high protection standard, it is a good choice to adopt the CA-UHPC. In addition, under the premise of construction conditions, the particle size of corundum aggregate is as large as possible, and it is best to be consistent with the diameter of the projectile.

References

CHEN X W, LI Q M. Transition from non-deformable projectile penetration to semi-hydrodynamic penetration [J]. Journal of Engineering Mechanics, 2004, 130(1): 123–127.

DANCYGIER A N, YANKELEVSKY D Z. High strength concrete response to hard projectile impact [J]. International Journal of Impact Engineering, 1996, 18(6): 583–599.

DANCYGIER A N. Rear face damage of normal and high-strength concrete elements caused by hard projectile impact [J]. ACI Structural Journal, 1998, 5(3): 291–304.

DANCYGIER A N, YANKELEVSKY D Z, HAEGERMANN C. Response of high performance concrete plates to impact of non-deforming projectiles [J]. International Journal of Impact Engineering, 2007 34(11): 1768–1779.

DIAMOND S, HUANG J. The ITZ in concrete-a different view based on image analysis and SEM observations [J]. Cement & Concrete Composites, 2001, 23(2–3): 179–188.

FANG Q, ZHANG J H. Three-dimensional modelling of steel fiber reinforced concrete material under intense dynamic loading [J]. Construction and Building Materials, 2013, 44: 1318–1323.

FULLER W B, THOMPSON S E. The laws of proportioning concrete[J]. Transactions of the American Society of Civil Engineers, 1907, lix: 67–143.

GUAN Z Q, SONG C, GU Y X. Recent Advances of Research on Finite Element Mesh Generation Methods [J]. Journal of Computer-Aided Design & Computer Graphics, 2003, 15(1): 1–14. (in Chinese).

HANCHAK S J, FORRESTAL M J, YOUNG E R, EHRGOTT J Q. Perforation of concrete slab with 48MPa (7ksi) and 140MPa (20ksi) unconfined compressive strength [J]. International Journal of Impact Engineering, 1992, 12(1): 1–7.

HALLQUIST J O. LS-DYNA keyword user's manual[M]. Livermore Software Technology Corporation, 2007.

HOLMQUIST T J, JOHNSON G R, COOK W H. A computational constitutive model for concrete subjective to large strain, high strain rates, and high pressure. The 14th International Symposium on Ballistic, Quebec[C]. 1993: 591–600.

JOHNSON G R, COOK W H. Fracture characteristics of three metals subjected to various strains, strain rates, temperatures and pressures [J]. Engineering Fracture Mechanics, 1985, 21(1): 31–48.

LAI J Z, SUN W, LIN W, JIN Z Q. Static and dynamic mechanical behavior of ECO-RPC (Eng Ed) [J]. Journal of Southeast University, 2005, 21(2): 197–202.

LANDERTSHAMER F, STEFFAN H. Method to generate complex computational meshes efficiently [J]. International Journal for Numerical Methods in Biomedical Engineering, 2010, 10(5): 373–384.

LANGBERG H, MARKESET G. High performance concrete-penetration resistance and material development. Proceedings of the Ninth International Symposium on Interaction of the Effects of Munitions with Structures, Berlin-Strausberg[C].1999: 933–941.

LIU R C, WU B, ZHANG X Z, SHEN G S, NIU X L. Tests on resisting projectiles penetration of high strength volume steel fiber concrete [J]. Chinese Journal of Explosion and Shock Waves, 2002, 22(4): 368–372. (in Chinese).

LIU G T, GAO Z G. Random 3-D aggregate structure for concrete [J]. Journal of Tsinghua University (Science and Technology), 2003, 43(8): 1120–1123. (in Chinese).

LUNDGREN R G. High-velocity penetrators, the American Institute of Aeronautics and Astronautics Missile Sciences Conference, Monterey, CA, 7–9 Nov. 1994[C]. 1994. SAND94–2724C Conf-9411142–1.

LUO X, SUN W, SAMMY CHAN Y N. Characteristics of high-performance steel fiber-reinforced concrete subject to high velocity impact[J]. Cement and Concrete Research, 2000, 30(6): 907–914.

MATSUMOTO M, NISHIMURA T. Mersenne twister: a 623-dimensionally equidistributed uniform pseudo-random number generator [J]. ACM Transactions on Modeling & Computer Simulation, 1998, 8(1): 3–30.

OLLIVIER J P, MASO J C, BOURDETTE B. Interfacial transition zone in concrete [J]. Advanced Cement Based Materials, 1995, 2(1): 30–38.

O'NEIL E F, NEELEY B D, CARGILE J D. Tensile properties of very-high-strength concrete for penetration-resistant structures [J]. Shock and Vibration, 1999, 6(5–6): 237–245.

REN G M, WU H, FANG Q. Determinations of HJC constitutive model parameters for normal strength concrete [J]. Journal of Vibration shock, 2016, 35(18): 9–16. (in Chinese).

REN G M, WU H, FANG Q. Parameters of HJC model for high-strength concrete-like materials under projectile impact [J]. International Journal Protection Structures, 2017, 8(3): 352–367.

RONG Z D, SUN W, ZHANG Y S. Dynamic compression behavior of ultra-high performance cement based composites [J]. International Journal of Impact Engineering, 2010, 37(5): 515–520.

TAI Y S. Flat ended projectile penetrating ultra-high strength concrete plate target [J]. Theoretical and Applied Fracture Mechanics, 2009, 51(2): 117–128.

WERNER S, THIENEL K-C, KUSTERMANN A. Study of fractured surfaces of concrete caused by projectile impact[J]. International Journal of Impact Engineering, 2013, 52: 23–27.

WU H, FANG Q, GONG Z M. Experimental and engineering analyses on DOP of HSFRC target under the impact of rigid projectile [J]. Journal of Ballistics, 2012, 24(3): 19–24. (in Chinese).

WU H, CHEN X W, HE L L, FANG Q. Stability analyses of the mass abrasive projectile high-speed penetrating into concrete target Part I: Engineering model for the mass loss and nose-blunting of the ogive-nosed projectile [J]. Acta Mechanica Sinica, 2014, 30(6): 933–942.

WU H, FANG Q, CHEN X W, GONG Z M, LIU J Z. Projectile penetration of ultra-high performance cement based composites at 510m/s to 1320m/s [J]. Construction and Building Materials, 2015, 74(15): 188–200.

WU H, FANG Q, GONG J, LIU J Z, ZHANG J H, GONG Z M. Projectile impact resistance of corundum aggregated UHP-SFRC [J]. International Journal of Impact Engineering, 2015, 84: 38–53.

WU H, LI Y C, FANG Q, PENG Y. Scaling effect of rigid projectile penetration into concrete target: 3D mesoscopic analyses [J]. Construction and Building Materials, 2019, 208: 506–524.

ZHANG M H, SHIM V P W, LU G, CHEW C W. Resistance of high-strength concrete to projectile impact [J]. International Journal of Impact Engineering, 2005, 31(7): 825–841.

ZHANG M H, SHARIF M S H, LU G. Impact resistance of high-strength fiber-reinforced concrete [J]. Magazine of Concrete Research, 2007, 59(3): 199–210.

ZHANG Y S, SUN W, LIU S F, JIAO C J, LAI J Z. Preparation of C200 Green Reactive Powder Concrete and its static-dynamic behaviors [J]. Cement & Concrete Composites, 2008, 30(9): 831–838.

ZHANG Y S, ZHANG W H, SHE W, MA L G, ZHU W W. Ultrasound monitoring of setting and hardening process of ultra-high performance cementitious materials [J]. Ndt & E International, 2012, 47: 177–184.

ZHANG W H, ZHANG Y S, ZHANG G R. Static. Dynamic mechanical properties and microstructure characteristics of ultra-high performance cementitious composite [J]. Science and Engineering of Composite Materials, 2012, 19(3): 237–245.

Chapter 6
Impact Resistance of Basalt Aggregated UHP-SFRC/Fabric Composite Panels Against Small Caliber Arm

6.1 Introduction

Persons and valuable equipment in protective structures (e.g. buildings with military and strategic importance, ground emergency shelters) are threatened by the perforated small caliber arms and rear ejecting fragments of the concrete wall during intentionally attacks. Therefore, the concrete materials with good on-site workability, distinguished protective performance under high-speed impact, as well as the related design approach are needed urgently.

For the prominent static mechanical properties (compressive strength about 150–200 MPa, tensile strength nearly 7–15 MPa, and fracture energy about 20,000–40,000 J/m^2) and the dynamical performances (Farnam et al. 2010; Habel and Gauvreau 2008), UHP-SFRC becomes the most potential construction material to resist the intensive loadings, such as impact and blast.

Aiming to the protection against earth penetration weapon (EPW), most of the existing high-speed impact tests on plain or steel fiber (bar) reinforced high-strength concrete (HSC) were mainly focused on middle caliber projectile (12.6–37 mm) (Hanchak et al. 1992; Dancygier and Yankelevsky 1996; Dancygier 1998; O'Neil et al. 1999; Luo et al. 2000; Liu et al. 2002; Zhang et al. 2007; Tai 2009; Wu et al. 2015a), the detailed review of these tests can be referred to Wu et al. (2015b), few studies were conducted on the impact resistance of concrete targets against small caliber arms.

For the bare concrete slabs, Almansa and Cánovas (1999) fired three types of small arms (5.56, 7.62, 12.7 mm in diameter) impacting on normal and steel fiber reinforced concrete panels (compressive strength of ~40 MPa) with thickness of 40–200 mm. The ratio of kinetic energy loss of the bullet to the plate thickness was studied and a model was proposed to predict the minimal thickness to avoid the perforation or scabbing. They also qualitatively pointed out that adding fibers tends to reduce the crater size while it has no significant effect on the target thickness necessary to avoid perforation. Bludau et al. (2006) experimentally studied the influences of coarse aggregate type (basalt, corundum, quartzite, recycled and boron glass), the

Q. Fang et al., *UHPCC Under Impact and Blast*,
https://doi.org/10.1007/978-981-33-6842-2_6

maximum aggregate diameter (5–16 mm), the grading curve of the coarse aggregates as well as binder content on the impact resistances of the HSC panels (90–130 MPa) impacted by the 7.62 mm hardcore bullet. Recently, Sovják et al. (2013, 2015) and Máca et al. (2014) conducted the deformable (Pb core) and non-deformable (steel core) 7.92 mm bullet impacting tests on the traditional no coarse aggregated UHP-SFRC (148–164 MPa), HSC (69–130 MPa), fiber reinforced concrete (FRC, 37–38 MPa), ultra-high performance concrete (UHPC, 110–141.9 MPa) and normal strength concrete (NSC, 43 MPa) slabs with the striking velocity at 691–720 m/s. The excellent impact resistance of UHP-SFRC was validated and the influences of steel fiber volumetric fractions (1–3%) on the depth of penetration (DOP), the impact crater size and debris fragment mass were studied, respectively.

Furthermore, aiming to reduce the damage degree of the concrete panels and block the high-speed ejecting fragments from the rear face of the concrete slab, compared with the rear attached steel liner (Kojima 1991; Tetsuo et al. 1997; Siddiqui et al. 2014; Wu et al. 2015c), pasting fabric on the rear face of the concrete panel is more practical for both operation and interior decoration. Vossoughi et al. (2007) conducted 12.7 mm conical-nosed projectile impact test on the normal strength concrete (30–43 MPa) panels with the front and/or rear Polypropylene or Zylon fabric liner. It is found that the rear scabbing of the fabric protected concrete panel was considerably reduced and the debris was contained by the fabric. Recently, Almusallam et al. (2015) presented the responses of rear CFRP strengthened RC panels (49 MPa, 90 mm thick) impacted by the hemispherical nosed projectile (40 mm in diameter, striking velocity at 92–158 m/s). It indicated that the rear strengthened CFRP fabric could increase the ballistic limit (the minimum initial impact velocity of the projectile to perforate the target with given thickness) by about 18%, reduce the front crater damage and block the fragments from the rear face of the panel.

The resistance of concrete panel under projectile impact depends mainly on the compressive strength of the target as well as the size and strength (hardness) of the coarse aggregate (Zhang et al. 2005; Dancygier et al. 2007, 2014; Wu et al. 2015a). Because the improvement of the strength needs decreasing or even eliminating the ingredients sizes in order to improve the homogeneity and density of the composite, there are few coarse aggregates mixed into UHP-SFRC (Graybeal 2006; Farnam et al. 2010; Habel and Gauvreau 2008; Sovják et al. 2013, 2015; Máca et al. 2014), and the impact resistance of coarse aggregates are neglected. For the optimization design of the mixing proportions of concrete target, (i) compressive strength of the target, Wu et al. (2015a) and Zhang et al. (2005) have found that DOPs of the projectiles are no longer decreasing obviously when the compressive strength of the target are larger than ~100 MPa; (ii) sizes of the high strength coarse aggregates, Bludau et al. (2006) proposed that the existence of coarse aggregates with size equal to or larger than the projectile's diameter would obtain a larger impact resistance, and basalt as well as quartzite aggregates give a good ratio between costs of production and anti-strike resistance. Also, Wu et al. (2015a) suggested that the average size of the coarse aggregates should be larger than 1.5 times of the projectile shank diameter for the constructions of protective structures; (iii) type, dimension and volumetric ratio of incorporating fibers, Wu et al. (2015a), Sovják et al. (2013, 2015) proposed that the straight steel fibers with volumetric ratio of 1.5–2% is the optimum choice by comprehensively considering the mechanical properties, workability, crack prohibiting ability as well as the costs.

In this chapter, based on the above suggestions of existing works (Wu et al. 2015a, b; Bludau et al. 2006; Sovják et al. 2013, 2015; Zhang et al. 2005), by mixing the high strength basalt aggregates with the size range of 5–10 mm and the straight steel fibers with the volumetric ratio of 2% into matrix of UHPC, the optimal basalt aggregated UHP-SFRC (UHP-BASFRC) with the designed uniaxial compressive strength of about 100 MPa was prepared under common curing temperature and pressure procedures. Furthermore, the API bullet impact test on above bare or rear fabric strengthened UHP-BASFRC panels with the thickness of 40–130 mm were conducted. The impact resistances of the bare and composite UHP-BASFRC panels, as well as the effect of externally protected fabric (CFRP or UHMWPE fabrics) were evaluated. Finally, a practical approach to predict the terminal ballistic parameters of bullet impacting on both traditional no coarse aggregated UHP-SFRC and present UHP-BASFRC slabs was suggested. The experimental conclusions as well as the suggested practical approach could provide references for the design and retrofit of ground structures threatened by high speed small caliber arms or fragments.

6.2 Bullet Perforation Test

6.2.1 Bullet

As shown in Fig. 6.1, the in-service armor-piercing incendiary (API) bullet was used in the present test, which was supposed to cover the most loads caused by small caliber arms (Bludau et al. 2006). The weight of the API bullet with 7.62 mm in diameter and 37.88 mm in length was 10.4 g. It consists of a brass jacket, a lead spacer, incendiary agent filler and a 5.15 g, ogive-nosed hard steel core. The hard steel core (6.12 mm in diameter and 27.5 mm in length) was machined by T12A steel with a hardness of HRC 62–67 and the yield strength of 980 MPa, respectively. Muzzle velocity of the in-service 7.62 mm API bullet fired from Chinese sniper rifle 85 was usually in the range of 800–815 m/s, while the actual impact velocity would be measured during test.

6.2.2 UHP-BASFRC Panels

The optimal UHP-BASFRC was prepared by State Key Laboratory of High Performance Civil Engineering Materials, Jiangsu Research Institute of Building Science in China. The mix proportion of UHP-BASFRC in the present impact test is listed in Table 6.1, in which all the ingredients are commercially available in China. Based

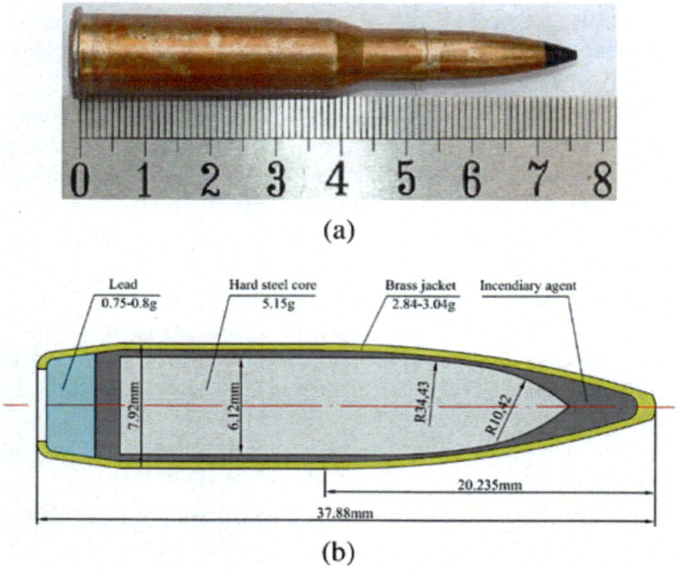

Fig. 6.1 7.62 mm API bullet **a** photograph (with cartridge case), **b** dimensions, reprinted from Peng et al. (2016), copyright 2020, with permission from Elsevier

Table 6.1 Mix proportions of UHP-BASFRC (kg/m^3)

Cement	Water	Fly ash	Silica fume	Ultra-fine slag	Fine sand	Basalt	Fibers	HRWR
450	115	75	75	150	750	750	150	15

on the discussions in Sect. 6.1, the size range of the basalt aggregates and the volumetric ratio of the mixing straight steel fiber were designed as 5–10 mm and 2%, respectively. Provided by the manufacturer, the compressive strength and density of the basalt aggregates were 210 MPa and 2800 kg/m^3, respectively. The diameter, length and tensile strength of the straight brass coated steel fibers were 0.175 mm, 13 mm and 3000 MPa, respectively.

To achieve good workability, particle distribution and packing density, mixing procedure for UHP-BASFRC was controlled rigorously. At first, the dry cementitious materials (cement, fly ash, silica fume, slag) and sands (average size 0.8 mm) were put together simultaneously and mixed uniformly for 30 s, and then the basalt coarse aggregates were added and mixed uniformly for 1 min. Secondly, the water and superplasticizer were added together and mixed for 3–5 min. Finally, steel fibers were incorporated into the mixture and mixed for another 3–5 min in order to guarantee the fibers well distributed throughout the matrix. A vertical pan mixer with a capacity of 600 L and a constant mixing speed (60 rpm) was used. UHP-BASFRC was a self-consolidating concrete which did not require heat and high pressure preparation. The slump and spread of UHP-BASFRC were respectively 250 mm and 630 mm, which guarantees the workability requirements of on-site constructions.

Nineteen square UHP-BASFRC panels with dimensions of 400 mm × 400 mm and nominal thicknesses of 40–130 mm as well as the control specimens were cast simultaneously. Nominal thickness is the height of the mould and the actual thicknesses of panels T_t are listed in Table 6.2. For such a small caliber bullet, only local impact responses occurred and low steel bar reinforcing ratio had no obvious effect on enhancing the impact resistance when the inner rebar mesh was not hit (Dancygier 1997). Therefore, the steel meshes were not considered from the viewpoint of conservation. The specimens were demoulded after 24 h and cured in a standard climate with temperature of 17.7–21.7 °C and 98.4% relative humidity. Concrete cylinders with length of 150 mm and diameter of 75 mm as well as the specimens with the dimension of 100 mm × 100 mm × 500 mm (loading procedure given in Wu et al. (2015a)) were cast for uniaxial compressive and direct tensile strength. The typical compressive and tensile curves of the present UHP-BASFRC specimen were shown in Fig. 6.2. The average 28d unconfined cylindrical compressive strength and the density of UHP-BASFRC were 106.2 MPa and 2530 kg/m^3, respectively.

6.2.3 Fabric Strengthening

Attributed to low density and high tensile strength, carbon fiber reinforced polymer (CFRP) and ultra-high molecular weight polyethylene fiber (UHMWPE) fabrics have been widely used to strengthen/retrofit the structural members under/after blast and impact loads (Wu et al. 2009; Rodriguez-Nikl et al. 2012; Azrul 2012; Jena et al. 2009; Ong et al. 2003). Currently, in order to reduce the ejected fragments from the rear face of the concrete panels, as shown in Fig. 6.3, an integral piece of squared CFRP or UHMWPE fabric (single-layered and bidirectional) with the dimensions of 400 mm × 400 mm was adhered onto the rear face of the 10 of the total 19 UHP-BASFRC panels (Table 6.2).

Thickness of the CRFP and UHMWPE fabrics were 0.35 mm and 0.12 mm, respectively. Provided by the manufacturer, the density, tensile strength, Young's modulus and elongation at break were 1850 kg/m^3, 2.3 GPa, 390 GPa, 1.5% and 970 kg/m^3, 3.1 GPa, 100 GPa, 3.7% for CRFP and UHMWPE fabrics respectively. Before adhering the fabric, the surface of the panel was polished and then a primer layer was evenly brushed on the rear side. After the primer layer was dried, the fabric was then adhered and solidified with epoxy resins. The basic mechanic properties of the epoxy resins were given by the National Research Center of Testing Techniques for Building Materials, Beijing, as tensile strength of 42 MPa, Young's modulus of 2672 MPa and elongation of 1.6%.

Table 6.2 Bullets impact test data

Group no.	Shot no.	T_t (mm)	M_{p0} (kg)	M_{pr} (kg)	Response	DOP (mm)	Front crater			Rear crater			
							$d_{c,f}$ (mm)	$h_{c,f}$ (mm)	$V_{c,f}$ (cm^3)	$d_{c,r}$ (mm)	$h_{c,r}$ (mm)	$V_{c,r}$ (cm^3)	α (°)
1 (Bare panels)	1-4	41.5	/	17.40	P	–	63.5	18	18	79.0	23	40	57.7
	1-5	47.5	18.65	18.50	P_p	–	68.5	19	24	86.0	27	31	55.9
	1-6	60.9	24.10	24.00	P_p	–	57.0	26	20	66.2	22	31	53.8
	1-7	67.9	26.95	26.65	P_p	–	69.5	24	25	94.5	40	70	47.8
	1-8	77.2	31.05	30.90	PL	–	70.2	27	33	72.7	19	18	60.3
	1-9	90.7	36.00	35.90	UP	43*	91.2	29	43	–	–	–	–
	1-10	95.9	/	37.65	UP	65.5*	71.2	28	25	–	–	–	–
	1-11	107.0	42.55	42.50	UP	33	81.7	31	43	–	–	–	–
	1-13	128.2	50.90	50.85	UP	78.5*	50.5	33	18	–	–	–	–
2 (UHMWPE protected)	2-4	43.2	18.00	17.90	P_p	–	54.5	21	12	61.7	32	25	40.2
	2-7	68.7	28.35	28.20	P	–	82.0	29	24	30.7	21	4	28.8
	2-8	78.7	31.3	31.25	UP	70.5	69.2	31	32	–	–	–	–
	2-9	83.6	34.2	34.15	UP	70	59.5	27	25	–	–	–	–
	2-10	98.5	39.75	39.65	UP	79.5*	79.0	32	25	–	–	–	–
3 (CFRP protected)	3-4	42.4	17.35	17.30	P_p	–	50.0	24	15	–	–	–	–
	3-7	67.2	27.15	27.00	P_p	–	71.2	45	28	–	–	–	–
	3-8	77.7	30.90	30.80	UP	58.5*	79.2	31	33	–	–	–	–
	3-9	84.5	33.95	33.90	UP	55.5	57.7	28	16	–	–	–	–
	3-10	97.5	38.85	38.70	UP	66.5	64.5	25	25	–	–	–	–

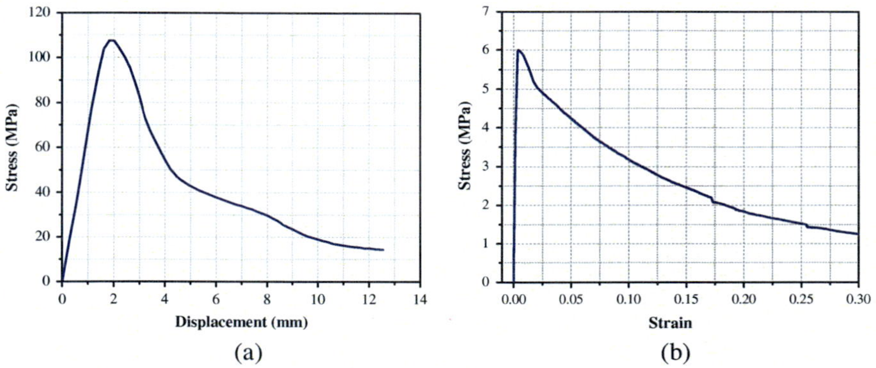

Fig. 6.2 Typical mechanical behavior of UHP-BASFRC **a** compressive curve, **b** tensile curve, reprinted from Peng et al. (2016), copyright 2020, with permission from Elsevier

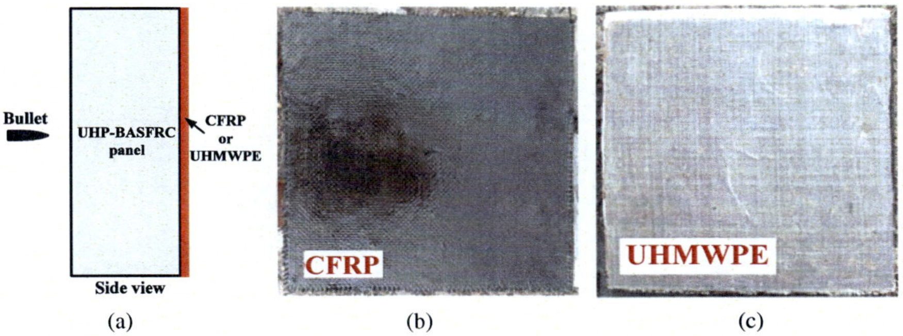

Fig. 6.3 Strengthened panels **a** schematic, **b** with rear CFRP, **c** with rear UHMWPE, reprinted from Peng et al. (2016), copyright 2020, with permission from Elsevier

6.2.4 *Test Setup*

Figure 6.4 illustrates the sketch of the present impact test, where the API bullets were launched by a fixed sniper rifle 85 and impact normally on the center of UHP-BASFRC panels. UHP-BASFRC panel was mounted in a position perpendicular to the rifle by a special steel frame and constrained by four adjustable angle irons on four corners of the panel. A piece of aluminum plate with the thickness of 2 mm was placed 0.6 m behind the rear face of the UHP-BASFRC panel to evaluate the residual impact potential and record the trajectory of the bullet qualitatively. The arrangement of the panel and witness sheet is shown in Fig. 6.5. By using a pair of metal wires nets placed both on the rifle muzzle and the front face of the target connected with a digital timer, the impact velocity could be calculated with the tested passing time and a fixed distance between the rifle muzzle to the target. For the present test, the

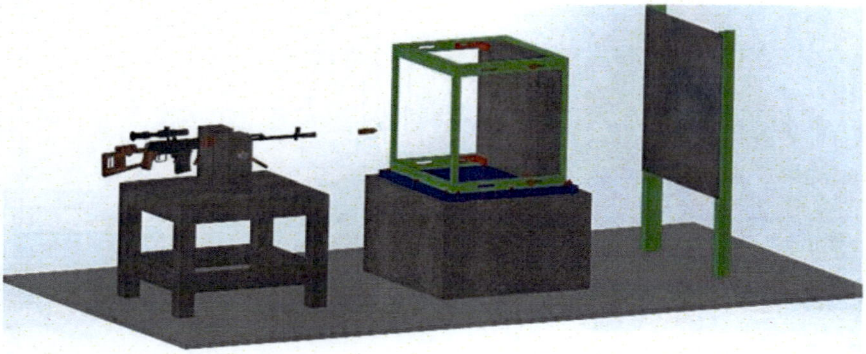

Fig. 6.4 Sketch of the test setup, reprinted from Peng et al. (2016), copyright 2020, with permission from Elsevier

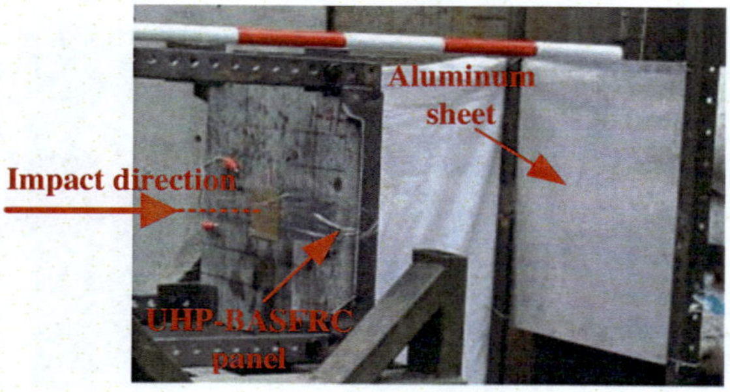

Fig. 6.5 Test setup, reprinted from Peng et al. (2016), copyright 2020, with permission from Elsevier

above distance was about 1.6 m and the average impact velocity of the bullets was 810 m/s.

The total 19 panels were divided into three groups, as listed in Table 6.2, in which Group 1–3 denote the bare UHP-BASFRC panels, UHMWPE and CFRP strengthened UHP-BASFRC panels, respectively. The first and the second numbers of the Shot No. in Table 6.2 denote the Group and nominal thickness of the slab in centimeter, respectively. For example, 2–8 means UHP-BASFRC panel with nominal thickness 8 cm and strengthened by UHMWPE fabric in Group 2.

6.3 Test Results

6.3.1 Damage of Target

Visible damage, including the front and the rear faces of the panels were recorded after each shot as shown in Figs. 6.6, 6.7 and 6.8. Figure 6.7 shows the rear faces of

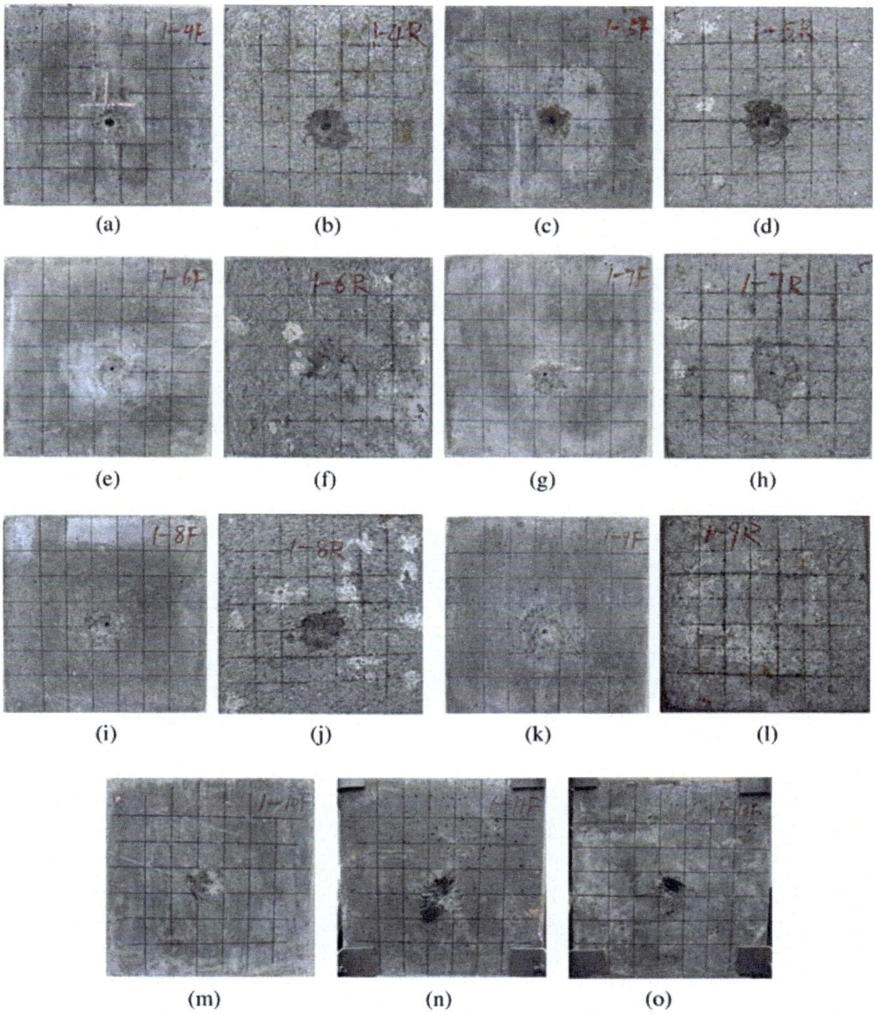

Fig. 6.6 Damage of the bare UHP-BASFRC panels **a** 1-4F, **b** 1-4R, **c** 1-5F, **d** 1-5R, **e** 1-6F, **f** 1-6R, **g** 1-7F, **h** 1-7R, **i** 1-8F, **j** 1-8R, **k** 1-9F, **l** 1-9R, **m** 1-10F, **n** 1-11F, **o** 1-13F, reprinted from Peng et al. (2016), copyright 2020, with permission from Elsevier

the concrete panels in Group 2 after the UHMWPE fabric were removed, while they were not given for Group 3 since the rear CFRP fabrics were adhered too tight to be removed.

The symbols written on the targets, e.g. the labels 1-4F and 1-4R, represent the front and rear faces of the panel in Shot 1-4, respectively. The labels 2-4F, 2-4R and 2-4R0 denote the front face, the rear face with fabric and the rear face with UHMWPE removed, respectively. Panels in Shots 1-9, 1-10, 1-11 and 1-13 did not

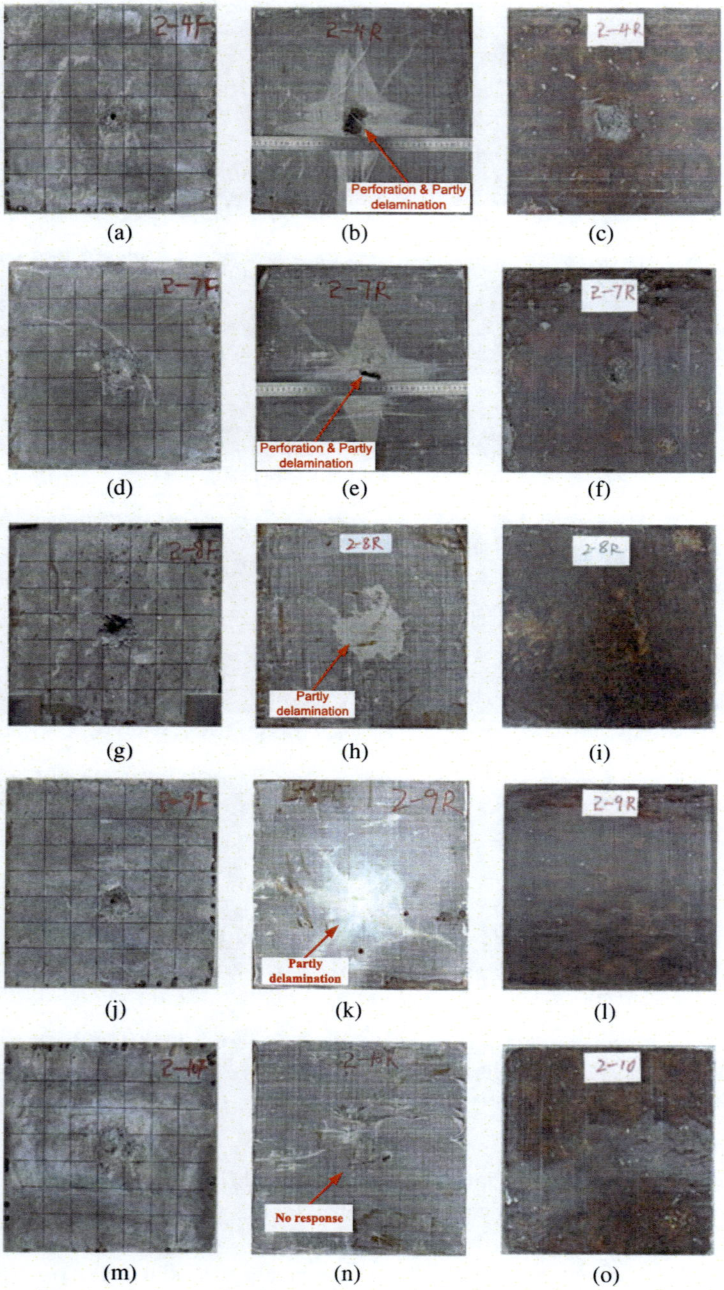

Fig. 6.7 Damage of the panels protected by UHMWPE fabric **a** 2-4F, **b** 2-4R, **c** 2-4R0, **d** 2-7F, **e** 2-7R, **f** 2-7R0, **g** 2-8F, **h** 2-8R, **i** 2-8R0, **j** 2-9F, **k** 2-9R, **l** 2-9R0, **m** 2-10F, **n** 2-10R, **o** 2-10R0, reprinted from Peng et al. (2016), copyright 2020, with permission from Elsevier

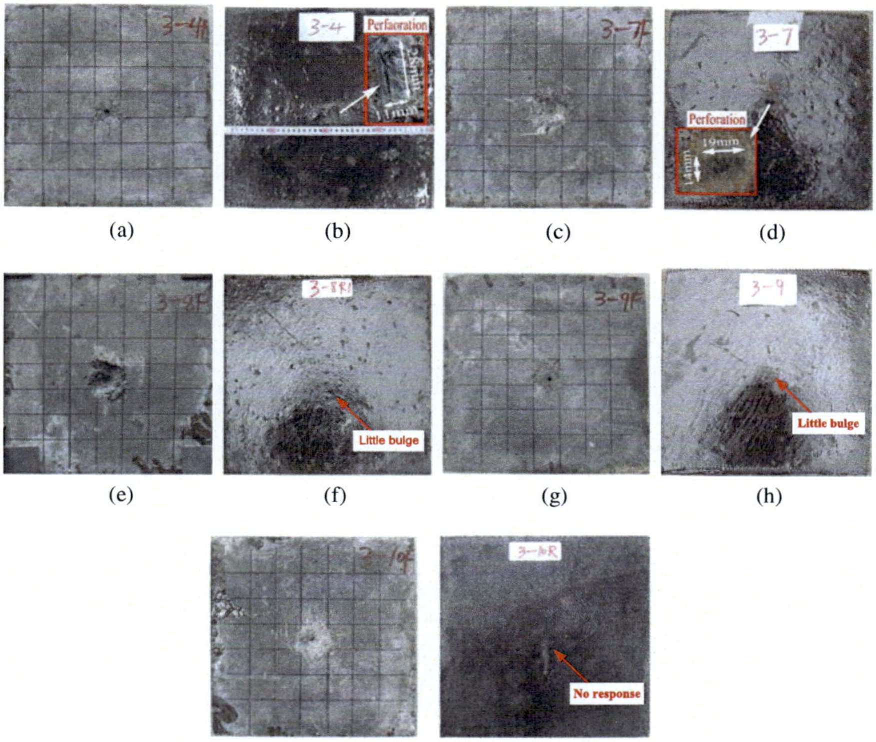

(a) (b) (c) (d)

(e) (f) (g) (h)

(i) (j)

Fig. 6.8 Damage of the panels protected by CFRP fabric **a** 3-4F, **b** 3-4R, **c** 3-7F, **d** 3-7R, **e** 3-8F, **f** 3-8R, **g** 3-9F, **h** 3-9R, **i** 3-10F, **j** 3-10R, reprinted from Peng et al. (2016), copyright 2020, with permission from Elsevier

show any visible cracks and deformation on the rear face, and thus only the rear face photograph of thinner slab 1-9 was given.

From Figs. 6.6, 6.7 and 6.8, for the relative thick panels, the impact of the bullet only induced a conical crater followed by a tunnel with the diameter nearly equal to that of the bullet core. With decreasing the slab thickness, the extruding of the bullet induced the perforation process which consisted of impact cratering, tunneling and rear and shear plugging. The scabbing did not occur in the present test, which mainly lies in that the magnitude of the tensile stress wave occurred at the rear face due to the reflections did not exceed the tensile strength of the target, and moreover, the larger increment of target thickness (10 mm) for the present test might be another reason.

Furthermore, by comparing the rear face damage of the bare, CFRP and UHMWPE fabric strengthened slabs with the identical concrete slab thickness, it can be derived that, (i) the rear fabrics can reduce the rear crater sizes obviously (e.g. Shots 1-7, 2-7 and 3-7); (ii) the rear delamination areas of the CFRP fabric protected

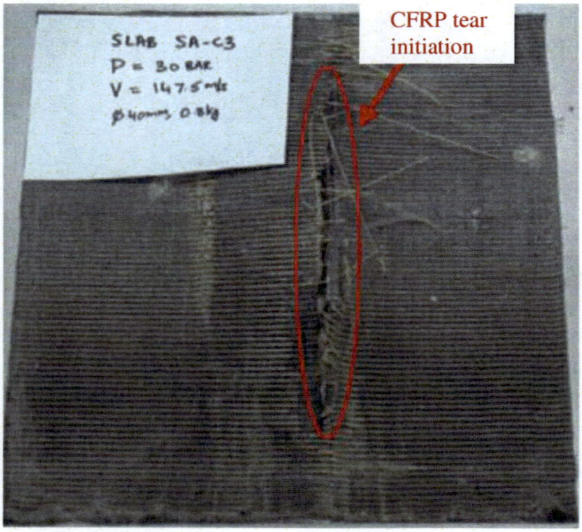

Fig. 6.9 Rear face damage of the panel protected by one-directional CFRP fabric (Almusallam et al. 2015)

slabs were considerably less than that of the UHMWPE fabric strengthened slabs (e.g. Shots 2-4 and 3-4); (iii) as introduced in Sect. 6.1, Almusallam et al. (2015) conducted projectile perforation test on unidirectional CFRP (single layer, thickness 1 mm, longitudinal tensile strength 846 MPa) rear strengthened RC panels, and the typical damage of CRFP fabric was shown in Figs. 6.6, 6.7, 6.8 and 6.9. It indicates that, the bidirectional CFRP fabric protected slabs in the present test (Figs. 6.6, 6.7 and 6.8) suffered less damage after the projectile perforations. Therefore, the bidirectional CFRP fabric should be the first choice in the retrofitting of protective structures.

6.3.2 *Dimension of Crater*

The response types of the panels listed in Table 6.2 were classified as perforated—P, perforation limit—PL, non-perforated—UP, and P_p denotes that the bullets perforated both the target and the rear aluminum plate. Illustrated in Fig. 6.10a, the impact of the bullet induces the front or rear frustum-of-cone shaped craters approximately, the dimensions of which includes the height $h_{c,f}$ (or $h_{c,r}$) and equivalent diameter $d_{c,f}$ (or $d_{c,r}$), where $d_{c,f}$ (or $d_{c,r}$) was obtained by averaging the maximal dimensions of crater from four directions as shown in Fig. 6.10b: i.e., horizontal, vertical, 45° and 135°, respectively. The slope angle α_s as shown in Fig. 6.10a was obtained by $\alpha_s = \tan^{-1}[(d_{c,r} - d_s)/2h_{c,r}]$, where d_s is the diameter of the hard steel core. The volume of the crater $V_{c,f}$ (or $V_{c,r}$) was measured by filling fine sand as shown in Fig. 6.10c, the tunneling volumes were not included in both front and rear crater

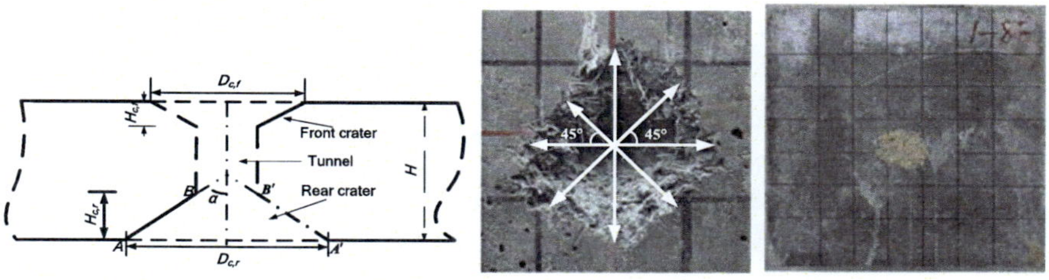

(a) (b) (c)

Fig. 6.10 Impact crater **a** schematic, **b** equivalent diameter determination, **c** crater volume determination, reprinted from Peng et al. (2016), copyright 2020, with permission from Elsevier

volumes by blocking the tunnel entrances. Symbol "–" denotes that the data are not obtained.

Besides, due to the material lost in both craters and tunnel, the initial and residual masses of the panel in each shot were weighed and listed as M_{p0} and M_{pr} in Table 6.2. For the unperforated panels, DOPs were obtained by measuring the distance of the projectile penetrating trajectory, where the DOP* denotes the penetrations deviated from the normal direction to some degree. It indicates that, the existence of high-strength basalt coarse aggregates made the terminal ballistic trajectories in most shots deviated.

However, it should be noted that, aiming to study the influence of UHP-BASFRC panel thickness on the impact resistance against the small caliber arms, and considering the costs, only a single shooting test was carried out for each combination of target thickness and fabric.

6.3.3 *Perforation Limit*

For the given ammunition, like the present API bullet, it is important to determine the perforation limit (the minimum thickness of the target required to prevent perforation) for safety design. Figure 6.11 shows the local view of the target rear face in Shot 1-8, in which the hard steel core of the bullet was stuck with 21 mm-length body exposed and 6.5 mm-length shank left embedded in the target. However, the composite panels 2-8 and 3-8 with nominal thickness of 80 mm prevented the bullet and had little damage on rear face except for partly delamination of UHMWPE fabric and little bulge of CFRP fabric as shown in Figs. 6.6, 6.7 and 6.8.

Therefore, the perforation limit of the present bare UHP-BASFRC panel was experimentally determined as 77.2 mm, which could provide a certain basis for the design of UHP-BASFRC walls/panels in protective structures. It also must be pointed that, more systematic bullet impact tests should be carried out on the UHP-BASFRC slab with the thickness around 77.2 mm in the future.

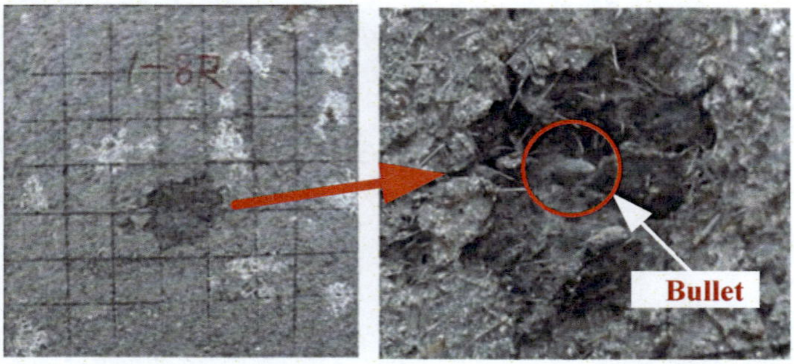

Fig. 6.11 Critical perforation in Shot 1-8, reprinted from Peng et al. (2016), copyright 2020, with permission from Elsevier

6.3.4 Recovered Bullet

During the penetrations, the brass jackets of the API bullets were all stripped from the hard steel core by targets but only one steel core was found clung with a half jacket in Shot 2-4. Figure 6.12a shows partly recovered hard steel core of the API bullets. All steel cores suffered mass abrasions to some degree, and some of them were even fractured. Figure 6.12b shows the typical recovered truncated-ogive nosed steel core of bullet (the initial yield strength was 550 MPa) after the perforation test on no coarse aggregated UHP-SFRC slab conducted by Sovják et al. (2015). It can be seen that the steel core was not deformed or damaged except for some minor abrasions on the surfaces of noses due to the high-speed frictions during the impact. Comparisons between Fig. 6.12a, b indicate that, adding the high strength basalt aggregates into UHP-SFRC material can obviously destruct the structural integrity of the hard core of API bullets, and thus decrease the penetration efficiency.

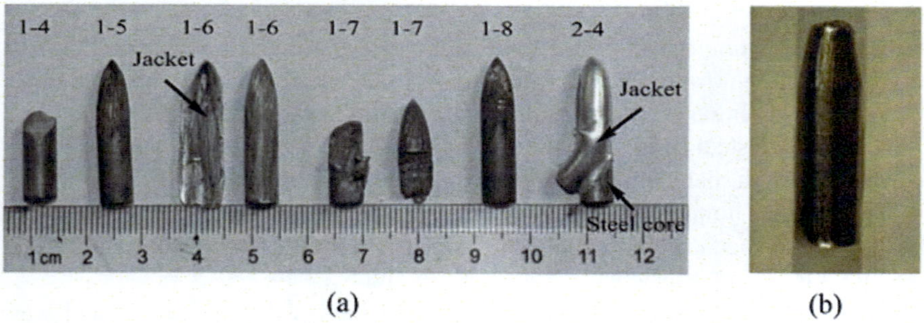

Fig. 6.12 Recovered bullets from **a** present test, **b** test in Sovják et al. (2015)

6.3.5 Damage of Aluminum Plate

Commonly, a piece of thin metallic plate was set behind the target to qualitatively evaluate the residual kinetic energy and trajectory of the perforated projectile. For instance, Bludau et al. (2006) considered that the perforation of 1.5 mm steel sheets is equivalent to a fatal hit of the human body. In Table 6.2, 6 of 8 perforated bullets impacted and perforated the aluminum plate.

In the present test, the fight trajectories of the steel core after perforating the concrete slab were examined. Since the poor visibility of the field testing site, the legible perforated steel core was not recorded from the high-speed camera. Figure 6.13a shows the locations of the steel cores impacted on the rear aluminum plate. It can be seen that, the ballistic trajectory of the steel cores after perforating the concrete slab were deviated off the center obviously. Additionally, the gestures of the perforated steel core were rotated obviously, e.g. Shot 1-5 in Fig. 6.13a. Figure 6.13b shows the high-speed photograph of the bullet steel core after perforating UHP-SFRC panel, obtained by Sovják et al. (2015). It indicates that, although the trajectory of the steel core was also deviated to some extent, the gesture of perforated hard core was maintained well and the steel core still had high penetration ability. Therefore, we can derive that, adding the high strength basalt aggregates into UHP-SFRC material can both deviate the ballistic trajectory and the gesture of the perforated core of API bullet, and correspondingly decrease the penetration efficiency obviously.

Besides, for the present test, under the impact of the small caliber arms, the ejected concrete fragments did not perforate the metallic plate arranged 0.6 m behind the concrete panels. However, the ejected coarse aggregates and steel fibers might be threatening in terms of the human safety. Therefore, attaching the fabric on the rear face of concrete panels to prevent the majority of ejected fragments is an efficient approach.

6.4 Discussions

6.4.1 Crater Dimensions

For the frontal impact craters, Wu et al. (2015a) proposed that the frontal crater depths are reduced with the decreasing of the target thickness. In the present test, due to the existence of the coarse aggregate with the average size nearly equal to the diameter of the bullet, the definite influential laws of target thickness and rear fabric on the frontal cratering dimensions, depth and volume were difficult to be derived.

For the rear impact craters, Fig. 6.14 shows the dimensionless rear crater area (the ratio of the actual crater area to the cross section area of the hard steel core) and rear crater volumes of three types targets, respectively. Since the rear craters of panels with rear CFRP fabric were very slight, some small values are given in Fig. 6.14b for the convenience of comparison. It can be seen from Fig. 6.14 that,

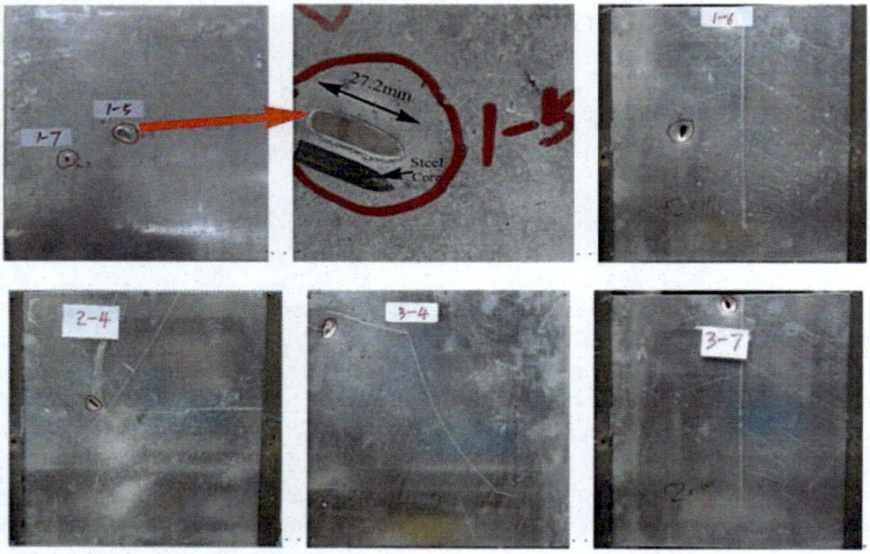

(a)

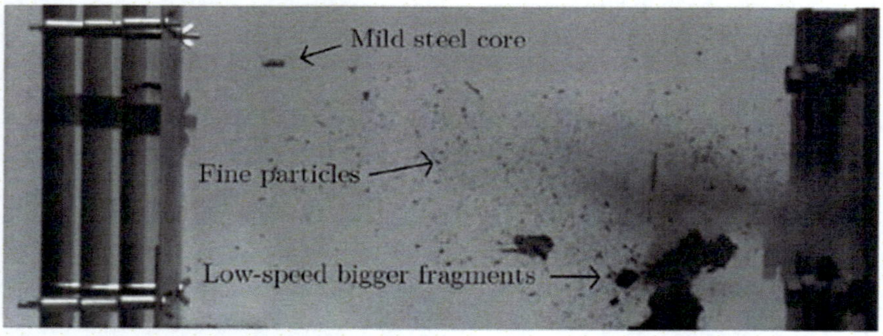

(b)

Fig. 6.13 After perforation effects **a** damage of the aluminum plate in the present test, **b** high-speed photograph from Sovják et al. (2015)

the rear crater of the CFRP strengthened panels are the smallest and the one of the UHMWPE fabric protected panels is much lower than that of bare panels. As for the rear cone slope angle α, based on Sect. 6.2, it can be obtained from Table 6.2 that $\alpha = 55°$ and $\alpha = 34.5°$ for the bare and UHMWPE strengthened UHP-BASFRC panels, respectively. While Dancygier (1998) obtained $\alpha = 76.3°$ for high-strength bare reinforced concrete slabs. It indicates that the adding fibers and attaching rear fabrics can both considerably reduce the rear fragments of the concrete target.

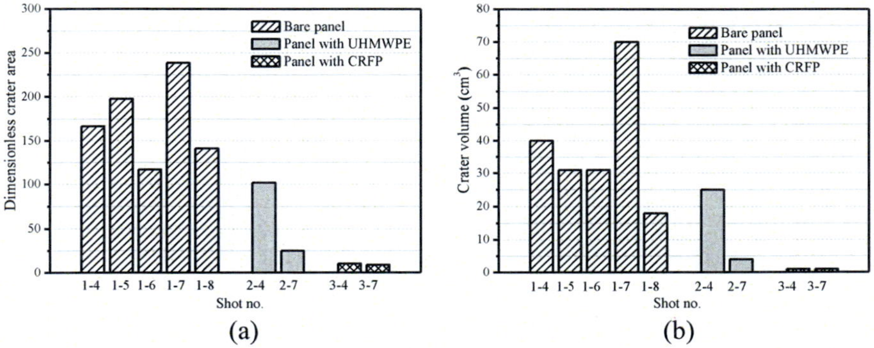

Fig. 6.14 Comparisons of the rear crater damage **a** dimensionless crater area, **b** crater volume, reprinted from Peng et al. (2016), copyright 2020, with permission from Elsevier

6.4.2 Terminal Ballistic Parameter

Forrestal et al. (2010) pointed out that the brass jacket and the filler of bullet had a slight effect on the perforation process, and the impact resistance of the hard steel core dominated the bullet perforation. Thus only the hard core of the API bullet is discussed in this section. There are no proper models to predict the terminal ballistic parameters of small caliber arms impacting on the traditional no coarse aggregated UHP-SFRC and the present UHP-BASFRC target. In order to find a practical approach, four existing models (Wu et al. 2015a; Li et al. 2005; Kar 1978; Chen et al. 2008) are adopted to estimate the terminal ballistic parameters of the bullet for a trial.

(i) The modified National Defense Research Committee (NDRC) formula (Li et al. 2005) is widely used and the ratio of DOP x to the projectile's diameter d is given as

$$\begin{cases} \frac{x}{d} = 2G^{0.5} \text{ for } G \le 1 \\ \frac{x}{d} = G + 1 \text{ for } G > 1 \end{cases} \tag{6.1a}$$

where

$$G = 3.8 \times 10^{-5} \frac{N^* M_0}{d\sqrt{f_c}} \left(\frac{V_0}{d}\right)^{1.8} \tag{6.1b}$$

where N^* is the projectile nose geometry factor, which equals to 0.72, 0.84, 1.0 and 1.14 for flat, hemispherical, blunt, and very sharp noses, respectively. The parameters M_0 and V_0 are the mass and the striking velocity of the projectile, while f_c is the compressive strength of the concrete target. Perforation limit h_{per} is given as

$$\begin{cases} \frac{h_{per}}{d} = 3.19\left(\frac{x}{d}\right) - 0.718\left(\frac{x}{d}\right)^2 & \text{for } \frac{x}{d} \le 1.35 \text{ or } \frac{h_{per}}{d} \le 3 \\ \frac{h_{per}}{d} = 1.32\left(\frac{x}{d}\right) + 1.24\left(\frac{x}{d}\right)^2 & \text{for } 1.35 < \frac{x}{d} < 13.5 \text{ or } 3 < \frac{h_{per}}{d} < 18 \end{cases} \tag{6.1c}$$

(ii) Kar (1978) further proposed the equations for predicting perforation limit of steel projectile

$$\begin{cases} \frac{h_{per}-a}{d} = 3.19\left(\frac{x}{d}\right) - 0.718\left(\frac{x}{d}\right)^2 & \text{for } \frac{x}{d} \le 1.35 \\ \frac{h_{per}-a}{d} = 1.32\left(\frac{x}{d}\right) + 1.24\left(\frac{x}{d}\right)^2 & \text{for } 1.35 < \frac{x}{d} \le 13.5 \end{cases} \tag{6.2}$$

where a is half of the aggregate size in concrete and x is derived from Eq. (6.1a).

(iii) Chen et al. (2008) proposed the formula to predict the perforation limit as follows

$$\begin{cases} \frac{x}{d} = \sqrt{\frac{(4k/\pi)(1+k\pi/4N)}{(1/I_0+1/N)}} & \text{for } \frac{x}{d} \le k \\ \frac{x}{d} = \frac{2}{\pi}\ln\left[\frac{(1+I_0/N)}{1+k\pi/4N}\right] + k & \text{for } \frac{x}{d} > k \end{cases} \tag{6.3a}$$

$$I_0 = \frac{M V_0^2}{Sf_c d^3};\ N = \frac{M}{N_1 \rho d^3} \tag{6.3b}$$

$$\begin{cases} \frac{h_{per}}{d} = \frac{x}{d} + \frac{\sqrt{1+\sqrt{3}S(x+kd)\tan\alpha} - 1}{2\tan\alpha} & \text{for } \frac{x}{d} \le k \\ \frac{h_{per}}{d} = \frac{x}{d} + \frac{\sqrt{1+\sqrt{3}S\tan\alpha} - 1}{2\tan\alpha} & \text{for } \frac{x}{d} > k \end{cases} \tag{6.3c}$$

where $k = 0.707 + h_c/d$ is a dimensionless parameter and h_c is the length of the hard core nose; ρ_t is the density of the target; N_1 is the nose shape factor and its value for different nose shaped projectile can be referred in Whiffen (1943) and Wu et al. (2012); and $\alpha = 55°$ is adopted as discussed in Sect. 6.4.1. Equations (6.3a–c) are also widely used to predict the ballistic limit V_{BL} and residual velocity V_r of the projectile impacting on the concrete target. For the relatively thick target in the present test where $x > kd$, it can be obtained reversely from Eqs. (3a–c) that

$$V_{BL} = \sqrt{\frac{Sf_c d^3 I_{BL}}{M}};\ I_{BL} = \left(N + \frac{k\pi}{4}\right)e^{\frac{\pi}{2N}\left(\frac{H}{d} - \frac{H_{plug}}{d} - k\right)} - N \tag{6.3d}$$

$$V_r = \left(V_0^2 - V_{BL}^2\right)^{0.5} \tag{6.3e}$$

where H_t is the thickness of the given target plate, and H_{plug} is the thickness of the rear crater (Chen et al. 2008).

(iv) In Wu et al. (2015a), we have proposed an empirical approach for predicting the terminal ballistic parameters of projectile perforating finite-thickness concrete panel. The residual impact function I_r is expressed as follows

Table 6.3 Residual velocities of the bullet in Sovják et al. (2015)

f_c (MPa)	V_0 (m/s)	V_r (m/s)	Predicted V_r (m/s)	
			Chen et al. (2008)	Wu et al. (2015a)
132	691	260	223.5 (−14.0%)	300.8 (15.7%)
154	705	226	201.1 (−11.0%)	303.1 (34.1%)
149	703	228	209.7 (−8.0%)	304.1 (33.4%)

$$V_r = \sqrt{\frac{Sf_c d^3 I_r}{M}};\ I_r = I_0 - K(I_0)H/d;$$
$$K(I_0) = 1.99 - 1.52\exp(-I_0/15.11);\ S = 127.7(f_c/10^6)^{-0.675} \tag{6.4}$$

Also the perforation limit h_{per} could be solved through Eq. (6.4) by setting $I_r = 0$. Equation (6.3) can be applied for different projectile nosed shapes, including (truncated-) ogive nose, hemi-spherical nose, conical nose and flat nose. However, Eq. (6.4) is only suitable for ogive-nosed projectile.

The available small caliber projectile impacting tests on UHP-SFRC are limited. Sovják et al. (2015) have conducted a set of truncated-ogive nosed bullet impacting tests on no coarse aggregated UHP-SFRC panels with the thickness of 45 mm, and the experimental and predicted residual velocities are listed in Table 6.3.

Máca et al. (2014) also obtained the perforation limit of the bullet impacting on no coarse aggregated UHP-SFRC panels with the striking velocity of 710 m/s, and the experimental and predicted perforation limits for experiments in Máca et al. (2014) and present test are listed in the first two rows of Table 6.4.

For predicting the bullet impact resistance of UHP-BASFRC, compared with the no coarse aggregate UHP-SFRC, it is difficult to study the enhancing degree of the adding coarse aggregates in tests. By considering the effect of the coarse aggregates, Whiffen formula (Whiffen 1943) for dimensionless DOP is suggested by British Road Research Laboratory as

$$\frac{x}{d} = \left(\frac{2.61}{\sqrt{f_c}}\right)\left(\frac{M}{d^3}\right)\left(\frac{d}{d_{a,\max}}\right)^{0.1}\left(\frac{V_0}{533.4}\right)^{\frac{97.51}{f_c^{0.25}}} \tag{6.5}$$

where $d_{a,max}$ is the maximum aggregate size. The application ranges are 5.52 MPa < f_c < 68.95 MPa, 0.136 kg < M < 9979.2 kg, 12.7 mm < d < 965.2 mm, 0 < V_0 < 1127.8 m/s and 0.5 < $d/d_{a,max}$ < 50 for ogival projectile nose shapes with caliber radius between 0.8 and 3.5. At follows, the influence of coarse aggregates on the impact resistance is discussed according to Eq. (6.5). It could be derived from Eq. (6.5) that, with the identical other parameters of both projectile and targets, the projectile penetration depth of the UHP-BASFRC with basalt coarse aggregates ($d_{a,max}$ = 10 mm) is nearly 77.7% of that of the UHP-SFRC with only fine sands ($d_{a,max}$ = 0.8 mm). By considering the above reduction, the corresponding predicted perforation limits of the present test are listed in the 3rd row of Table 6.4, and the

Table 6.4 Perforation limits of the UHP-SFRC and UHP-BASFRC panels

Targets	f_c (MPa)	M (g)	V_0 (m/s)	Nose shape	Perforation limit (mm)				
					Exp	Chen et al. (2008)	Wu et al. (2015a)	NDRC (Li et al. 2005)	Kar (1978)
UHP-SFRC (Máca et al. 2014)	151.7	3.65	710	Truncated-ogive	50	47.6 (−4.9%)	55.8 (11.6%)	33.4 (−33.2%)	33.8 (−32.4%)
UHP-BASFRC	106.2	5.15	810	Ogive	77.2	99.7 (29%)	90.9 (17.7%)	60.2 (−22%)	65.2 (−15.5%)
UHP-BASFRC	106.2	5.15	810	Ogive	77.2	77.5* (0.4%)	70.6* (−8.5%)	46.8* (−39%)	50.7* (−34%)
UHP-SFRC	151.7	5.15	810	Ogive	–	90.41	–	–	–

superscript * is added to distinguish the data of no aggregated UHP-SFRC in the 2nd row. It should be noted that, two assumptions were adopted during the above calculations: (i) the enhancing degree of adding coarse aggregates given by Eq. (6.5) for NSC was also suitable for UHP-SFRC; (ii) the enhancing degree of adding coarse aggregates on DOP was identical to the one on perforation limit.

Besides, aiming to further validate the distinguished impact resistance of the present optimal UHP-BASFRC, another impact scenario was assumed in the last row of Table 6.4, in which the UHP-SFRC (151.7 MPa) panels without coarse aggregate in Máca et al. (2014) was struck by the 7.62 mm API bullet used in the present test.

From Tables 6.3 and 6.4, it is derived that:

(1) As for no coarse aggregated UHP-SFRC in Sovják et al. (2015) and Máca et al. (2014), Chen's model (Chen et al. 2008) shows high degree of accuracy with the test data; Wu's model (Wu et al. 2015a) gives over predictions, and the reason lies in that Eq. (6.4) was obtained by fitting the ogive-nosed projectile perforation tests data and overestimate the perforation capacity of the truncated-ogive-nosed projectile; the modified NDRC (Li et al. 2005) and Kar (1978) formulae both show poor prediction accuracies.
(2) As for UHP-BASFRC, the existing four models all give poor predictions. When the decreasing effect on perforation limit of the coarse aggregate was considered, Chen's model (Chen et al. 2008) also shows good predictions (the 3rd row in Table 6.4). Thus, Chen's model (Chen et al. 2008) given in Eq. (6.3) with a reduction coefficient of 77.7% can be used as a practical approach to predict the terminal ballistic parameters of bullet impacting on UHP-BASFRC targets. However, the validation of above approach should be further verified by more tests.
(3) In the last row of Table 6.4, the predicted perforation limit 90.41 mm was obtained from Chen's model (Chen et al. 2008), and it was considered correct based on the above discussions. By comparing the predicted results in rows 2 and 4, for the same bullet and strike velocity, the perforation limit of the present UHP-BASFRC (77.2 mm) panel is nearly 14.6% lower than that of the UHP-SFRC (90.41 mm) panel even though the former compressive strength (106.2 MPa) is only 70% of the latter (151.7 MPa). The excellent impact resistance of the designed UHP-BASFRC was validated.

6.4.3 Fabric Effect

Comparisons among the damage of targets in Shots 1-8, 2-7, 2-8, 3-7 and 3-8 show that the perforation limits of the UHP-BASFRC/UHMWPE and UHP-BASFRC/CFRP panel located within the range of 68.7–77.2 mm and 67.2–77.2 mm, respectively. It can be derived that rear fabrics help to decrease the perforation limit of the UHP-BASFRC panels impacted by the small caliber arms, and the maximal reducing magnitudes are $(77.2 - 68.7)/77.2 = 11\%$ and $(77.2 - 67.2)/77.2 = 13\%$ for UHMWPE and CFRP, respectively. Almusallam et al. (2015) obtained that the ballistic limit of projectile was increased about 18% by attaching CFRP on the rear face of 49 MPa normal strength concrete slab.

As for reducing rear crater and blocking ejecting fragments, from the figures in Sect. 6.3.1 and discussions in Sect. 6.4.1, CFRP performed better than UHMWPE in the present test, which might be caused by the poor bonding forces at the interfaces between UHMWPE and epoxy resins induced by the surface chemical inertness of the UHMWPE (Chukov et al. 2014).

6.5 Summary

Aiming to protect persons and valuable equipment in the building structures threatened by the perforated small caliber arms and scabbing fragments, the optimal UHP-BASFRC was prepared and 7.62 mm API bullet impacting tests on bare and rear fabric (CFRP or UHMWPE) strengthened UHP-BASFRC panels were conducted. The prominent protective performance of bare and composite UHP-BASFRC panels was validated, and the main conclusions are as follows.

(1) Perforation limit of bare UHP-BASFRC panel was experimentally determined as 77.2 mm, which provides reference for the design of protective structures.
(2) Mixing the coarse aggregates could considerably improve the resistance of the UHP-SFRC panel and even deviate the terminal trajectory or destruct the bullet.
(3) Chen's model (Chen et al. 2008) show high degree of accuracy with the bullet impact test data of UHP-SFRC panel and a practical reduction coefficient 77.7% is suggested for UHP-BASFRC target.
(4) The rear strengthened two fabrics could reduce the perforation limit and considerably restrain the rear crater dimensions as well as block the ejected fragments, and moreover, CFRP performed better than UHMWPE.

References

ALMANSA E M, CÁNOVAS M F. Behaviour of normal and steel fiber-reinforced concrete under impact of small projectiles [J]. Cement and Concrete Research, 1999, 29: 1807–1814.

ALMUSALLAM T, AL-SALLOUM Y, ALSAYED S, IQBAL R, ABBAS H. Effect of CFRP strengthening on the response of RC slabs to hard projectile impact [J]. Nuclear Engineering and Design, 2015, 286: 211–226.

AZRUL A. Damage assessment and prediction of FRP strengthened RC structures subjected to blast and impact loads [D]. Perth: Univ. of West Australia, 2012.

BLUDAU C, KEUSER M, KUSTERMANN A. Perforation resistance of high-strength concrete panels [J]. ACI Structural Journal, 2006, 103(2): 188–195.

CHEN X W, LI X L, HUANG F L, WU H J, CHEN Y Z. Normal perforation of reinforced concrete target by rigid projectile [J]. International Journal of Impact Engineering, 2008, 35(10): 1119–1129.

CHUKOV D I, STEPASHKIN A A, GORSHENKOV M V, TCHERDYNTSEV V V, KALOSHKIN S D. Surface modification of carbon fibers and its effect on the fiber-matrix interaction of UHMWPE based composites [J]. Journal of Alloys and Compounds, 2014, 586(4): S459–S463.
DANCYGIER A N. Effect of reinforcement ratio on the resistance of reinforced concrete to hard projectile impact [J]. Nuclear Engineering and Design, 1997, 172(1–2): 233–245.
DANCYGIER A N. Rear face damage of normal and high-strength concrete elements caused by hard projectile impact [J]. ACI Structural Journal, 1998, 5(3): 291–304.
DANCYGIER A N, KATZ A, BENAMOU D, YANKELEVSKY D Z. Resistance of double-layer reinforced HPC barriers to projectile impact [J]. International Journal of Impact Engineering, 2014, 67: 39–51.
DANCYGIER A N, YANKELEVSKY D Z. High strength concrete response to hard projectile impact [J]. International Journal of Impact Engineering, 1996, 18(6): 583–599.
DANCYGIER A N, YANKELEVSKY D Z, HAEGERMANN C. Response of high performance concrete plates to impact of non-deforming projectiles [J]. International Journal of Impact Engineering, 2007, 34(11): 1768–1779.
FARNAM Y, MOHAMMADI S, SHEKARCHI M. Experimental and numerical investigations of low velocity impact behavior of high-performance fiber-reinforced cement based composite [J]. International Journal of Impact Engineering, 2010, 37(2): 220–229.
FORRESTAL M J, BØRVIK T, WARREN T L. Perforation of 7075-T651 aluminum armor plates with 7.62 mm APM2 bullets [J]. Experimental Mechanics, 2010, 50(8): 1245–1251.
GRAYBEAL B A. Material property characterization of ultra-high performance concrete: FHWA-HRT-06-103 [R]. VA: U.S. Department of Transportation Federal highway Adminstration, 2006: 1–186.
HABEL K, GAUVREAU P. Response of ultra-high performance fiber reinforced concrete (UHPFRC) to impact and static loading [J]. Cement & Concrete Composites, 2008, 30(10): 938–946.
HANCHAK S J, FORRESTAL M J, YOUNG E R, EHRGOTT J Q. Perforation of concrete slab with 48 MPa (7 ksi) and 140 MPa (20 ksi) unconfined compressive strength [J]. International Journal of Impact Engineering, 1992, 12(1): 1–7.
JENA P K, RAMANJENEYULU, SIVA K K, BALAKRISHNA B. Ballistic studies on layered structures [J]. Materials & Design, 2009, 30: 1922–1929.
KAR A K. Local effects of tornado generated missiles [J]. Journal of the Structural Division, 1978, 104(ST5): 809–816.
KOJIMA I. An experimental study on local behavior of reinforced concrete slabs to missile impact [J]. Nuclear Engineering and Design, 1991, 130(2):121–132.
LI Q M, REID S R, WEN H M, TELFORD A R. Local impact effects of hard missiles on concrete targets [J]. International Journal of Impact Engineering, 2005, 32(1): 224–284.
LIU R C, WU B, ZHANG X Z, SHEN G S, NIU X L. Tests on resisting projectiles penetration of high strength volume steel fiber concrete [J]. Chinese Journal Explosion and Shock Waves, 2002, 22(4): 368–372. (in Chinese).
LUO X, SUN W, SAMMY CHAN Y N. Characteristics of high-performance steel fiber-reinforced concrete subject to high velocity impact [J]. Cement and Concrete Research, 2000, 30(6): 907–914.
MÁCA P, SOVJÁK R, KONVALINKA P. Mix design of UHPSFRC and its response to projectile impact [J]. International Journal of Impact Engineering, 2014, 63: 158–163.
O'NEIL E F, NEELEY B D, CARGILE J D. Tensile properties of very-high-strength concrete for penetration-resistant structures [J]. Shock and Vibration, 1999, 6(5–6): 237–245.
ONG C W, BOEY C W, HIXSON R S, SINIBALDI J O. Advanced layered personnel armor [J]. International Journal of Impact Engineering, 2003, 38: 369–383.
PENG Y, WU H, FANG Q, LIU J Z, GONG Z M. Impact resistance of basalt aggregated UHP-SFRC/fabric composite panel against small caliber arm [J]. International Journal of Impact Engineering, 2016, 88: 201–213.

RODRIGUEZ-NIKL T, LEE C S, HEGEMIER G A, SEIBLE F. Experimental performance of concrete columns with composite jackets under blast loading [J]. Journal of Structural Engineering, 2012, 138: 81–89.
SIDDIQUI N A, KHATEEB B M A, ALMUSALLAM T H, AL-SALLOUM Y A, IQBAL R A, ABBAS H. Reliability of RC shielded steel plates against the impact of sharp nose projectiles [J]. International Journal of Impact Engineering, 2014, 69: 122–135.
SOVJÁK R, VAVŘINÍK T, MÁCA P, ZATLOUKAL J, KONVALINKA P, SONG Y. Experimental investigation of ultra-high performance fiber reinforced concrete slabs subjected to deformable projectile impact [J]. Procedia Engineering, 2013, 65: 120–125.
SOVJÁK R, VAVŘINÍK T, ZATLOUKAL J, MÁCA P, MIČUNEK T, FRYDRÝN M. Resistance of slim UHPSFRC targets to projectile impact using in-service bullets [J]. International Journal of Impact Engineering, 2015, 76: 166–177.
TAI Y S. Flat ended projectile penetrating ultra-high strength concrete plate target [J]. Theoretical and Applied Fracture Mechanics, 2009, 51(2): 117–128.
TETSUO S, ATSUSHI K, TOMONORI O, HAJIME T, MASATOSHI U, NOBUTAKA I. Experimental and numerical simulation of double-layered RC plates under impact loadings [J]. Nuclear Engineering and Design, 1997, 176(3): 195–205.
VOSSOUGHI F, OSTERTAG C P, MONTEIRO P J M, JOHNSONB G C. Resistance of concrete protected by fabric to projectile impact [J]. Cement and Concrete Research, 2007, 37(1): 96–106.
WHIFFEN P. UK Road Research Laboratory Note: No. MOS/311 [M]. 1943.
WU C, OEHLERS D J, REBENTROST M, LEACH J, WHITTAKER A S. Blast testing of ultra-high performance fibre and FRP-retrofitted concrete slabs [J]. Engineering Structures, 2009, 31: 2060–2069.
WU H, FANG Q, CHEN X W, GONG Z M, LIU J Z. Projectile penetration of ultra-high performance cement based composites at 510 m/s to 1320 m/s [J]. Construction and Building Materials, 2015, 74(15): 188–200.
WU H, FANG Q, GONG J, LIU J Z, ZHANG J H, GONG Z M. Projectile impact resistance of corundum aggregated UHP-SFRC [J]. International Journal of Impact Engineering, 2015, 84: 38–53.
WU H, FANG Q, PENG Y, GONG Z M, KONG X Z. Hard projectile perforation on the monolithic and segmented RC panels with a rear steel liner [J]. International Journal of Impact Engineering, 2015, 76: 232–250.
WU H, FANG Q, ZHANG Y D, GONG Z M. Semi-theoretical analyses of the concrete plate perforated by a rigid projectile [J]. Acta Mechanica Sinica, 2012, 28(6): 1630–1643.
ZHANG M H, SHARIF M S H, LU G. Impact resistance of high-strength fibre-reinforced concrete [J]. Magazine of Concrete Research, 2007, 59(3): 199–210.
ZHANG M H, SHIM V P W, LU G, CHEW C W. Resistance of high-strength concrete to projectile impact [J]. International Journal of Impact Engineering, 2005, 31(7): 825–841.

Chapter 7
Impact Resistance of Armsector Steel/Ceramic/UHPCC Layered Composite Targets Against 30CrMnSiNi2A Steel Projectiles

7.1 Introduction

Aiming to resist the military and accidental high-speed missiles, the developments of material and structure with high impact resistance have been mainly concerned for both weapon designers and protective structure engineers. Considering the excellent impact resistance and ductility, steel has been widely used in tanks, armored vehicles, and other equipment to resist the high-speed penetrations of projectiles with different shapes, diameters, and striking velocities. Besides, concrete, one of the most widely used construction materials, has been extensively applied in protective civil structures such as nuclear power plant containment, air-raid shelter, etc.

Amounts of experimental and numerical simulation studies on the penetration resistance of steel targets have been conducted. Silsby (1984), Keele et al. (1990), Magness and Farrand (1990), and Woolsey et al. (1989a, b, 1990) have conducted the systematic long-rod uranium or tungsten alloy projectiles penetration tests on RHA steel (yield strength about 880 ~ 900 MPa) targets, in which the corresponding striking velocities and length-to-diameter ratios of projectiles are 750 ~ 4500 m/s and 10 ~ 23, respectively. With the improvement of forging and heat treatment technology, the strength and ductility of steel are also improved. For instances, Fras et al. (2018, 2019) conducted a series of projectile penetration tests to investigate the fracture and dynamic perforation modes of high strength armor steel (yield strength of 1380 MPa) using experimental and numerical simulation methods. Choudhary et al. (2020) experimentally assessed the ballistic behavior of newly developed armor grade steel (yield strength of 1290 MPa) and numerically evaluated its performance by adopting four different ductile fracture criteria.

Besides, as for the impact resistance of concrete, Forrestal et al. (1994, 1996, 2003) conducted the systematic ogive-nosed steel projectile penetration tests with different diameters and striking velocities on normal strength concrete targets, in which the corresponding empirical prediction equations for the depth of penetration (DOP) were proposed and validated. Furthermore, Hanchak et al. (1992) performed the projectile perforation tests on finite depth concrete slabs with striking velocities of

Q. Fang et al., *UHPCC Under Impact and Blast*,
https://doi.org/10.1007/978-981-33-6842-2_7

300 ~ 1100 m/s and concrete compressive strength of 48 MPa and 140 MPa, and it was found that the projectiles after perforating the 140 MPa concrete targets have lower residual velocities. Ultra-high performance cementitious composites (UHPCC) is a relatively new type of cementitious material, and it has very low water-to-binder ratio, high amount of high-range water reducer (HRWR), fine aggregates and high-strength steel fibers. With the prominent mechanical properties, i.e., high compressive and tensile strengths, high ductility as well as the high fracture energy, UHPCC has been becoming the most prospective construction material for both civil and military protective structures to resist the impact loadings. Aiming to protect the persons and equipment in the protective structures against high-speed small caliber arms, Sovják et al. (2013, 2015) and Máca et al. (2014) conducted a series of bullets impact tests on traditional fiber reinforced concrete (FRC) and UHPC targets with striking velocity in the range of 691 ~ 720 m/s, and it was verified that UHPC has much greater impact resistance compared to the traditional FRC. As for the medium caliber projectiles, Wu et al. (2015a, b) carried out the 25.3 mm-caliber ogive-nosed projectile penetration tests on the basalt and corundum aggregated ultra-high performance steel fiber reinforced concrete, and the contribution of coarse aggregates against the projectile is discussed. In addition, for the numerical simulations of UHPC against projectile impact, the new constitutive model and the corresponding model parameters of UHPC were also calibrated and validated in Refs. (Yang et al. 2019; Ren et al. 2017; Liu et al. 2017, 2018).

Attributed to its high hardness and intrinsic strength, the lightweight ceramic target can induce the interface defeat response (the projectile is forced to flow radially on the surface of ceramic target for a period of time without significant penetration), which has a prominent effect on the penetration capacity of the armor piecing projectile (Lundberg et al. 2000; Lundberg and Lundberg 2005). Therefore, both the lightweight ceramic and high strength homogeneous metallic material have great potential applications in the military protective structures. For example, the ceramic/metal composite plates were successfully equipped to the military vehicles and helicopters to against the armor-piercing projectile and were verified to be effective to improve the motility simultaneously. The corresponding experimental and theoretical investigations on the ballistic limit velocities and the optimal design of the ceramic/metal composite plates with the constant thickness or areal density (the mass per unit area of a material with certain thickness) were further carried out (Serjouei et al. 2015; Fawaz et al. 2006; Bendor et al. 2000). Recently, several penetration tests into composite structures consisting of concrete targets as well as armor steel or ceramic plates were also performed (Feng et al. 2016; Shao et al. 2019; Shi et al. 2006;). By carrying out 9 shots of penetration test with small caliber projectiles (6 and 10.8 mm in diameter) at striking velocities of about 800 and 1300 m/s, Feng et al. (2016) quantitatively analyzed the impact resistance of UHPC (compressive strength of about 300 MPa)/armor steel composite targets. Compared with the spaced configuration, the layered configuration of UHPC/armor steel composite target was experimentally and numerically verified to have superior impact resistance due to the confinement at the interface between the front and rear plate. Aiming to evaluate the effectiveness of the alumina ceramic balls (30 mm in diameter) in resisting the high-speed projectile impact, 3 shots of penetration test of ogive-nosed projectiles (25.3 mm in diameter and 328.6 g in mass) impacting on ceramic balls/UHPC (110 MPa) composite targets were conducted by Shao et al. (2019). The experimental observations indicated that the layer of ceramic balls with thickness of about 150 mm can effectively prevent the UHPC target from destruction of the impact load with striking velocity up to 810 m/s. Furthermore, Shi et al. (2006) performed total 19 shots of ogive-nosed projectiles (9.9 mm in diameter and 64 g in mass) penetration test on armor steel/concrete (47.5 MPa) and Al_2O_3 ceramic/concrete composite targets, respectively. Based on the comparisons of DOP and damage models of projectiles, the ceramic/concrete composite target was proved to possess higher resistance than the armor steel/concrete configuration.

The aforementioned experimental studies were mainly focused on the plain concrete (NSC and UHPC) targets or ceramic/metal composite targets, while the comparative discussions on the ballistic performances of the armor steel/UHPC, ceramic/UHPC and armor steel/ceramic/UHPC composite targets were relatively limited. Moreover, compared with the high-speed penetration of small caliber armor-piercing projectiles, few impact tests on the composite targets with the medium caliber projectiles were carried out.

At present, in Sect. 7.2, totally 14 shots of 30 mm-caliber ogive-nosed projectile penetration test with the striking velocity of about 500 m/s into seven configurations of 10CrNi3MoV21A armor steel/SiC ceramic/concrete composite targets were carried out. The impact resistance of the composite targets was assessed quantitively by discussing the penetration depth as well as the ballistic efficiency factor. In addition, the 2D axisymmetric numerical simulations with ALE algorithm applied to the projectile were performed to reproduce the impact scenarios, in which the detailed Johnson–Cook constitutive model (Johnson and Cook 1983) parameters for the armor steel plate were calibrated from the static and dynamic mechanical tests. Furthermore, based on the validated numerical algorithm as well as the constitutive models and the corresponding model parameters, the energy evolutions of the projectiles and targets during the penetration process were examined. In Sect. 7.3, twelve shots of ogive nosed 30CrMnSiNi2A steel projectile with the diameter of 30 mm penetrating into UHPCC target and armor steel/UHPCC composite targets were firstly conducted with the striking velocities ranging from 372 m/s to 646 m/s. For the composite targets, two new types of armor steels developed by Nanjing Nangang Iron and Steel United Co., Ltd, i.e., NP450 and NP500, with different thicknesses are combined with UHPCC. The experimental DOPs, damages of armor steel (damage mode), UHPCC (impact crater area), and projectile (residual length, residual mass) are recorded and discussed. Then, the major parameters of Johnson–Cook (J-C) model (Johnson and Cook 1983) of the present two armor steels are calibrated experimentally based on a series of fundamental tests. The corresponding numerical simulations for the present penetration test are conducted and compared with the test data. Finally, the ballistic characterization of armor steel/UHPCC composite target is assessed.

7.2 Impact Test on 10CrNi3MoV21A Armor Steel/SiC Ceramic/UHPCC Composite Targets and Numerical Simulations

7.2.1 Impact Test

7.2.1.1 Projectile and Target

(1) Projectile

As depicted in Fig. 7.1a, the ogive-nosed projectile machined from 30CrMnSiNi2A steel with the yield strength of about 1260 MPa was designed and used in the test. The geometric dimensions of the projectile are illustrated in Fig. 7.1b, where the caliber-radius-head (CRH, the ratio of the ogival nose curvature radius to the shank diameter) of projectile nose is 4, and the length and diameter are 180 mm and 30 mm, respectively. A polymer obturator with the mass of about 30 g was attached to the projectile before the test, and the corresponding total mass of the projectile was about 682 g.

(2) Armor steel plate

As shown in Fig. 7.2a, the 10CrNi3MoV21A armor steel plate with the width of 200 mm × 200 mm and the thickness of 5 mm and 10 mm were adopted, respectively. The chemical components of the 10CrNi3MoV21A armor steel are given in Table 7.1, and the corresponding static yield strength and ultimate tensile strength were measured as about 665 MPa and 709.5 MPa through the quasi-static tensile test introduced in Sect. 7.2.2.2.

(3) SiC ceramic plate

The impact resistance of common ceramic materials, i.e., Al_2O_3, SiC, B_4C and AlN were assessed quantitatively by Reaugh et al. (1999). Besides, by further considering the physical property and sintering cost, SiC ceramic plate with the width of 150 mm × 150 mm and the thickness of 6 mm and 11 mm were adopted at present,

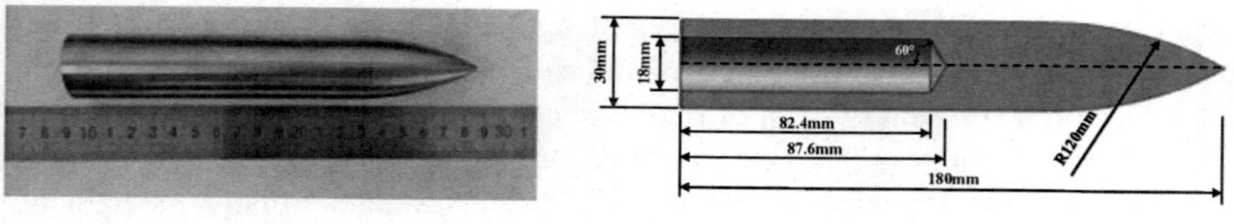

(a) (b)

Fig. 7.1 Ogive-nosed projectile **a** photograph **b** dimensions

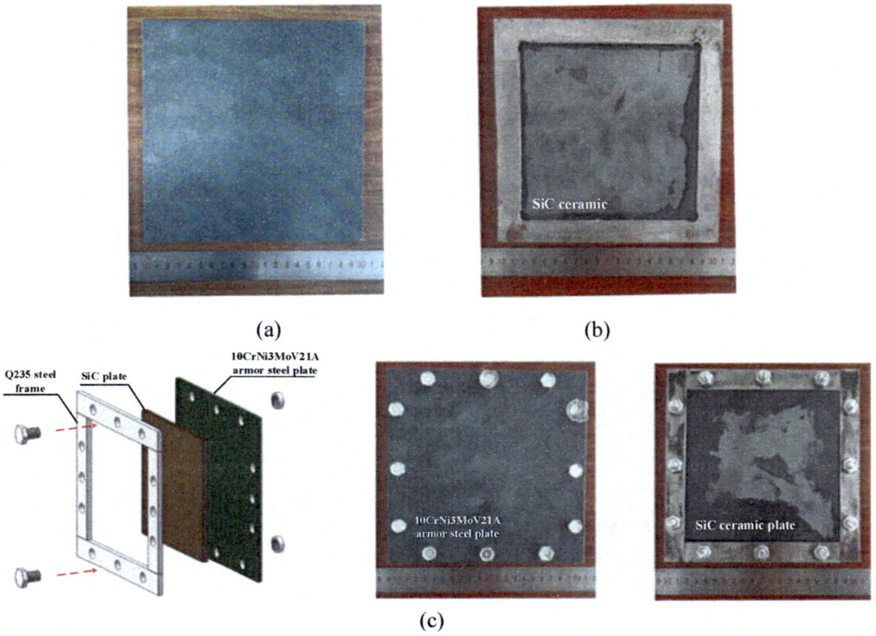

Fig. 7.2 Composite target **a** 10CrNi3MoV21A armor steel plate **b** SiC ceramic plate with steel frame **c** assembling diagram and photograph

Table 7.1 Chemical components of 10CrNi3MoV21A armor steel

Smelting analysis (‰)								
C	Si	Mn	P	S	V	Cr	Ni	Mo
0.9	2.7	5.6	0.07	0.04	0.5	9.3	26.6	2.4

respectively. The chemical components as well as the distinguished physical properties, e.g., high hardness and compressive strength, low density and small thermal expansion, are listed in Tables 7.2 and 7.3, respectively.

As shown in Fig. 7.2b, a Q235 steel frame was bonded to the SiC plate with epoxy resin for the convenience of assembling the composite plate. The fixation of the composite plate is further illustrated in Fig. 7.2c in which the SiC ceramic plate with steel frame was fixed to the 10CrNi3MoV21A armor steel plate with twelve bolts.

Table 7.2 Chemical components of SiC ceramic

Chemical components (%)						
SiC	SiO_2	CaO	Free Si	Na_2O	V	MgO
99.11	0.25	0.09	0.07	0.02	50	0.01

Table 7.3 Physical properties of SiC ceramic

Density (g/cm^3)	Vickers hardness	Compressive strength (MPa)	Thermal conductivity (Cal/cm·s·°C)	Coefficient of thermal expansion (1/K)
3.2	2500	>3000	0.3	4.3×10^{-6}

The corresponding front and back faces of the 10CrNi3MoV21A armor steel/SiC ceramic composite plate are also shown in Fig. 7.2c.

(4) UHPCC target

UHPCC is developed by Jiangsu Research Institute of Building Science in China, which is prefabricated by mixing the Chinese standard Graded 52.5 P.II type Portland cement, silica fume, ultra-fine mineral admixture which consists of fly ash and ultra-fine slag, the natural river sand with a maximal diameter of 2.5 mm, polycarboxylic type high-range-water-reducer (HRWR) and 2% (volume fraction) straight brass-coated steel fiber. The corresponding mixing proportions of the UHPCC and material properties of steel fiber are listed in Tables 7.4 and 7.5, respectively.

The casting process as well as the UHPCC target with 10CrNi3MoV21A armor steel/SiC ceramic composite plate are shown in Fig. 7.3. The cylindrical UHPCC target was firstly cast into a 5 mm-thick steel hoop with the height and diameter of 500 mm and 700 mm, respectively. The fluidity of UHPCC was then significantly weakened after curing atmospherically for about 4 ~ 5 h, after which, the composite plates were embedded into the UHPCC targets at the centre of front faces (the front surface of the plate was guaranteed in the same plane with the UHPCC target, as depicted in Fig. 7.3b). It should be noted that, all the UHPCC targets were cured in the air for 30 days before the impact tests. The average 30-d unconfined compressive strength of the UHPCC was about 135 MPa, which was tested on cylinder specimens with the dimensions of $\Phi 75 \times 150$ mm.

7.2.1.2 Test Setup

Seven different target configurations were designed in the present tests, i.e., NSC (35 MPa), UHPCC, 5 mm armor steel/UHPCC, 10 mm armor steel/UHPCC,

Table 7.4 Mix proportions of UHPCC (kg/m^3)

Cement	Silica fume	Viscosity reducing admixture	River sand	Water	Superplasticizer	Steel fiber
700	140	110	1200	152	22.8	156

Table 7.5 Material properties of steel fiber

Fiber type	Length (mm)	Diameter (mm)	Aspect ratio	Density (g/cm^3)	Tensile strength (MPa)	Elastic modulus (GPa)
Micro-straight	13	0.2	65	7.8	2800	210

(a) (b)

Fig. 7.3 UHPCC target **a** casting process **b** composite plate embedded in the front

6 mm ceramic/UHPCC, 11 mm ceramic/UHPCC and 5 mm armor steel/6 mm ceramic/UHPCC composite target. As shown in Fig. 7.4, the cylindrical target was placed horizontally on the support platform with its front face perpendicular to the barrel of a 30-mm-caliber smooth bore cannon, which was used to launch the projectiles with initial velocities V_0 of about 500 m/s. Besides, two layers of tinfoil connected to an oscilloscope were placed in the straight trajectory of the projectile to measure the initial striking velocities. The typical impact instants of projectile were captured by a high-speed camera system and the perpendicular impact can be identified, as shown in Fig. 7.5. After the test, the projectiles were recovered and the residual length and mass were measured and recorded in Table 7.5.

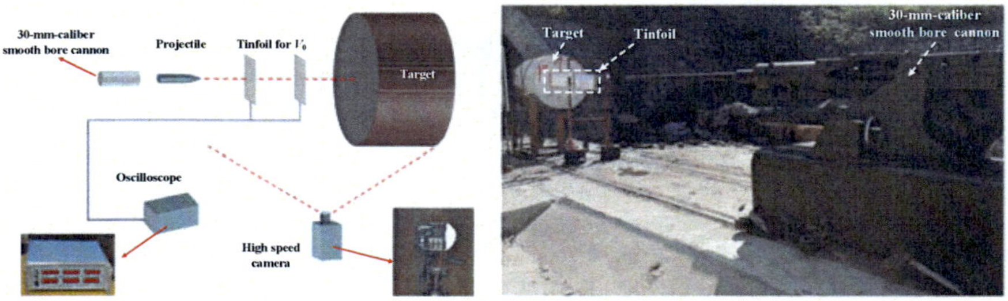

(a) (b)

Fig. 7.4 Test setup **a** sketch map **b** test site

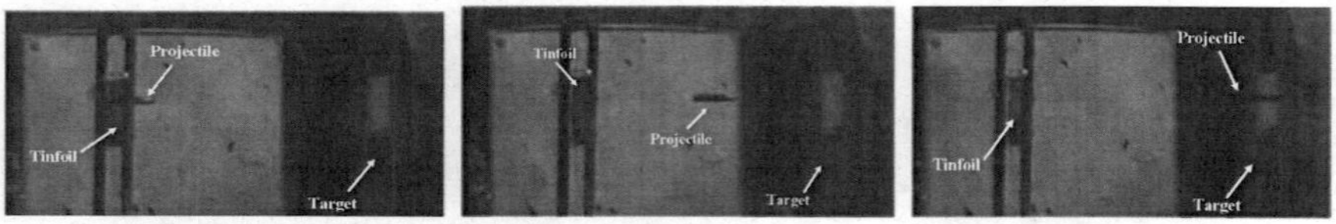

Fig. 7.5 Impact process captured by the high-speed camera

7.2.1.3 Test Results and Discussions

The target configurations are denoted by the target labels and listed in Table 7.6 in which the capitals N, U, S and C stand for NSC, UHPCC, 10CrNi3MoV21A armor steel and SiC ceramic, respectively. Besides, the number behind the labels denotes the thickness of the plate. For example, S5/C6 indicates that the UHPCC target embedded with a 5 mm 10CrNi3MoV21A armor steel/6 mm SiC ceramic composite plate. Repeated tests were carried out to eliminate the accidental deviations, and thus 14 shots of penetration test in total were conducted.

The test data is summarized in Table 7.6, in which L_0 and M_0 are the initial length and mass of the projectiles, and L_r and M_r are the residual length and mass of the recovered projectiles, respectively. As illustrated in Fig. 7.6, P and P_r are the total penetration depth and the residual penetration depth into the UHPCC target, i.e., P equal to the sum of P_r and the thickness of the front armor steel and ceramic plates. $\overline{L_r}$ and $\overline{P}$ are the averaged values of two repeated shots, which are used to compare with the following numerical simulation results in Sect. 7.2.2.3. Besides, it should be pointed out that, the symbol "a" in Table 7.6 denotes that the projectiles were not recovered as penetrating deeply into NSC targets, the symbol "b" represents that the projectile was not recovered as bouncing back, and "c" means that the projectile was fractured and only the nose was recovered.

(1) Penetration depth

The DOPs of each shot in the present test are depicted in Fig. 7.7, and the average values are also given in the brackets. Taking the average penetration depth (287.5 mm) of projectile penetrating into NSC target as a reference, it can be seen from Table 7.6 and Fig. 7.7 that the penetration depth into the UHPCC target decreases by 37.2%, while the S5 and S10 composite targets reduce the total penetration depth by 42% and 59%, respectively. Meanwhile, 50% and up to about 77.4% reductions of the corresponding total penetration depth are achieved by the C6 and C11 composite targets. Furthermore, there is 58.1% of the total penetration depth into the NSC target is reduced by the S5/C6 composite target. Therefore, it can be drawn that, (i) both the 10CrNi3MoV21A armor steel and the SiC ceramic plate have prominent effects on reducing the penetration depth and the SiC ceramic plates play a more effective role when they have almost the equal thickness; (ii) the thickness as well as the ballistic performance of S5/C6 composite plates are almost identical with the S10 plate, while about 25.4% of the areal density can be reduced by the S5/C6 composite plate. Thus,

Table 7.6 Test data

Target	Configuration	V_0 (m/s)	L_0 (mm)	L_r (mm)	$\overline{L}_r$ (mm)	M_0 (g)	M_r (g)	P_r (mm)	P (mm)	$\overline{P}$ (mm)
N-1	NSC	516	180	a	–	682.8	a	–	295	287.5
N-2	NSC	521	180	a	–	681.8	a	–	280	
U-1	UHPCC	504	180	179	178.5	683.8	675	–	181	180.5
U-2	UHPCC	520	180	178		681.8	673	–	180	
S5-1	Steel/UHPCC	509	180	177	177	682.8	674	163	168	166.5
S5-2	Steel/UHPCC	509	180	177		681.8	672	160	165	
S10-1	Steel/UHPCC	502	180	158	161.5	681.8	677	96	106	118.5
S10-2	Steel/UHPCC	503	180	165		681.8	675	121	131	
C6-1	SiC/UHPCC	508	180	b	165	683.8	b	144	150	143.5
C6-2	SiC/UHPCC	507	180	165		682.8	664	131	137	
C11-1	SiC/UHPCC	507	180	140	136	681.8	603	62	73	65
C11-2	SiC/UHPCC	513	180	132		680.8	547	46	57	
S5/C6-1	Steel/SiC/UHPCC	508	180	168	168	681.8	673	124	135	120.5
S5/C6-2	Steel/SiC/UHPCC	513	180	c		683.8	c	95	106	

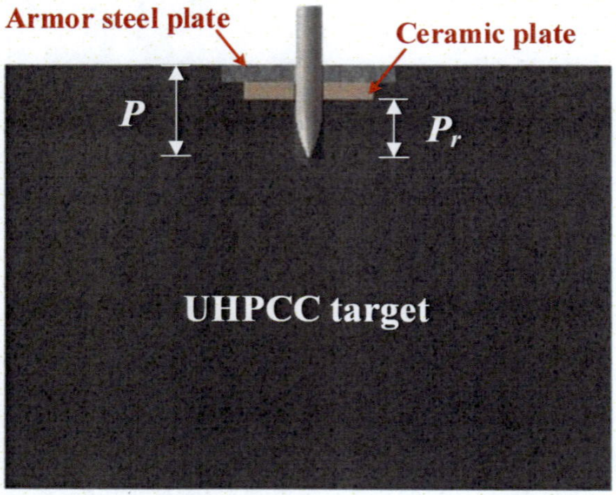

Fig. 7.6 Diagram of total DOP P and the residual penetration depth P_r

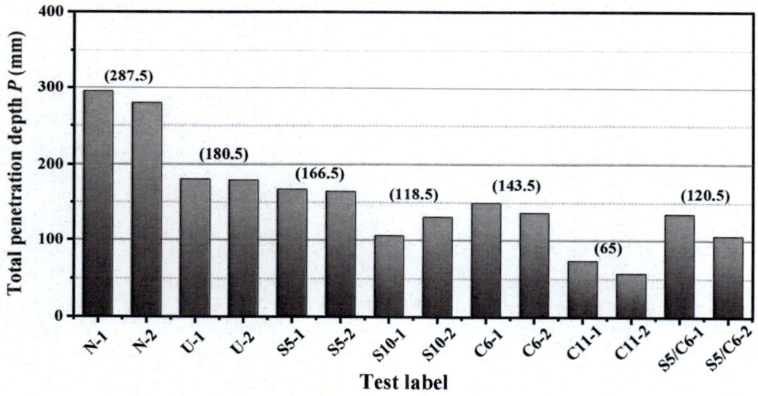

Fig. 7.7 DOPs of the test

the feasibility and applicability of the armor steel/ceramic/UHPCC composite target are verified.

(2) Damage of projectiles and targets

The comparisons between the unfired and recovered projectiles are shown in Fig. 7.8. In addition, the damage of the post-test NSC and UHPCC targets as well as the armor steel plates and armor steel/ceramic composite plates are further shown in Figs. 7.9 and 7.10, respectively. The ceramic plates in the present test are incapable of being recovered due to the serious fracture.

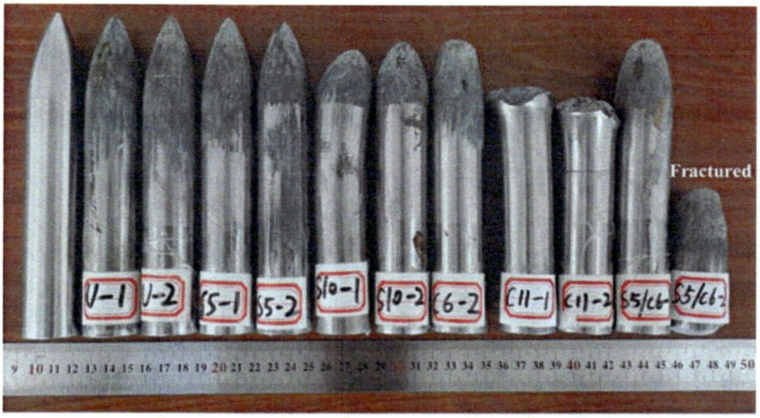

Fig. 7.8 Comparisons between the unfired and recovered projectiles

As for the deformations of projectiles, several damage model can be found. At first, for projectiles impacting on UHPCC and S5 composite targets, the maximal length reduction and mass loss of projectiles are less than 1.67% and 1.44%, respectively. In addition, only slight abrasions and no significant erosion of projectiles can be seen in Fig. 7.8, which indicates that the 30CrMnSiNi2A steel projectiles can be treated as rigid bodies when impacting on the above two targets with the present impact velocities. Secondly, the shortened projectiles (except for the broken projectile of S5/C6-2) with residual length of about 160 mm and residual mass of about 670 g can be identified after impacting the S10, C6 and S5/C6 composite targets, from which the eroding penetration can be identified consequently. Thirdly, projectiles with obvious erosion and mushrooming deformation after impacting C11 composite targets can be seen in Fig. 7.8. Overall, the following conclusion can be derived that the abrasion of projectiles can be enhanced with increasing the thickness of both the 10CrNi3MoV21A armor steel plates and SiC ceramic plates, while the abrasive effects of SiC ceramic plates to the projectiles are relatively more significant.

As for the deformations of the armor plates, it can be found firstly from Fig. 7.10 that all the 10CrNi3MoV21A armor steel plates are bulged towards to the opposite direction of projectiles impact, with the S5/C6 composite plates leading to the largest bulge height of about 23 mm, followed by S5 and S10 plates. Besides, all the perforating holes existing on the 10CrNi3MoV21A armor steel plates are close to circular with the diameters slightly larger than the projectiles. Meanwhile, obvious petal deformations around the holes can be found on both the S10 and S5/C6 composite plates, whereas a regular hole is perforated on the S5 plate, which indicates the punching failure of the relatively thin armor steel plate. Moreover, as shown in Fig. 7.11, it is worth noting that, even if the SiC ceramic plates possess the superior ballistic performance as discussed above, the ceramic plate is not recovered due to the splashing of the fractured ceramic fragments, and thus the damage of the ceramic plate is not discussed herein. In addition, the test results are also consistent with

Fig. 7.9 Targets damage

the general perspective that the ceramic is not suitable to be the single armor plate especially at the front face.

(3) Ballistic efficiency factor

Generally, the ballistic performance of a given plate with the thickness of T_t and density of ρ_t can be assessed quantitatively by the dimensionless thickness efficiency factor E_t and mass efficiency factor E_m as given in Eqs. (7.1) and (7.2) (Madhu et al. 2005), respectively. E_t is defined as the ratio of the penetration depth reduced by the composite target to the thickness of the test plate. E_m is defined as the ratio of the areal density reduced by the composite target to the areal density of the test plate.

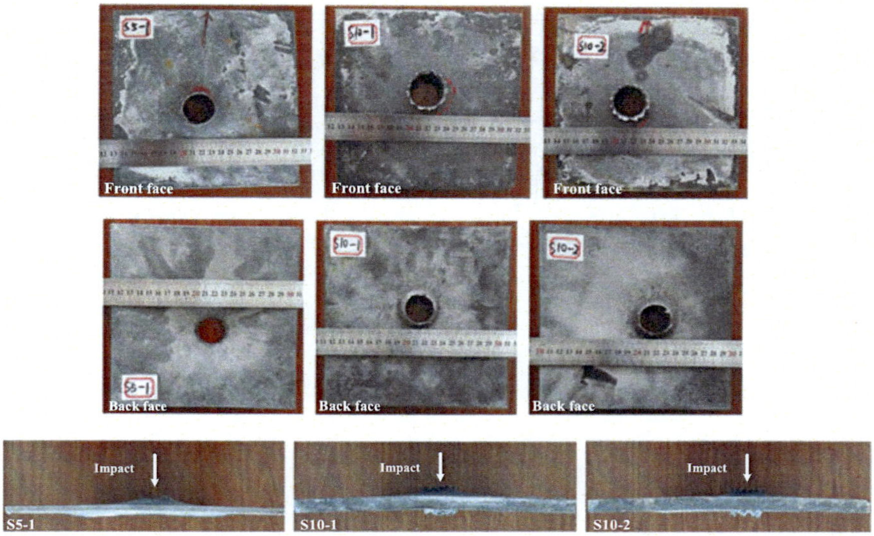

(a)

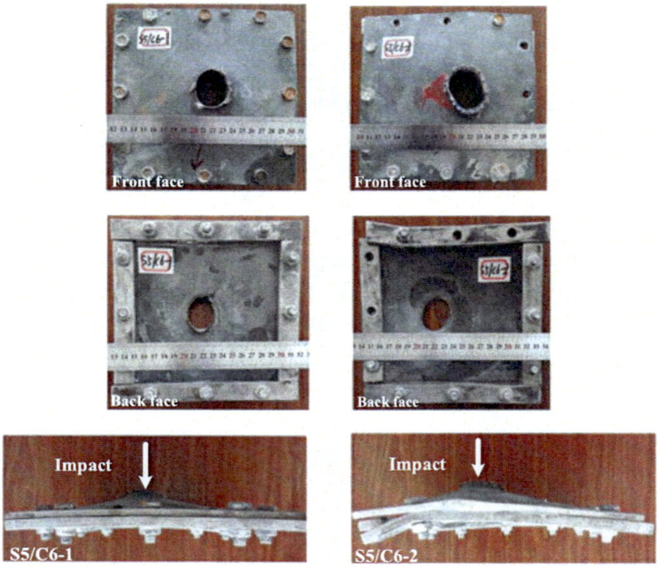

(b)

Fig. 7.10 Recovered target plate **a** 10CrNi3MoV21A armor steel plates **b** 10CrNi3MoV21A armor steel/SiC ceramic composite plates

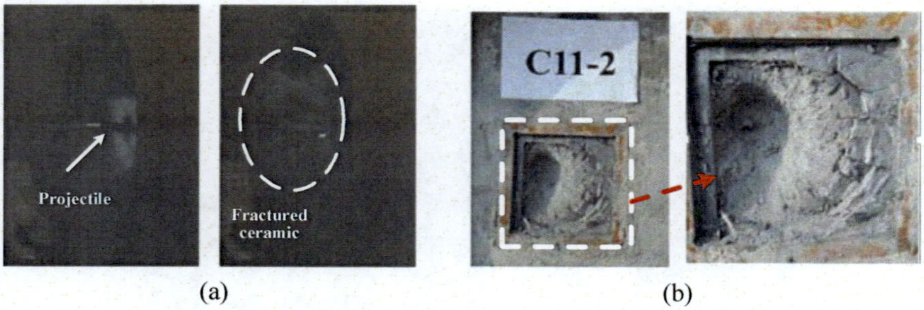

Fig. 7.11 Damage of ceramic plate **a** photographs of projectile impacting C11 target **b** damage of fractured ceramic with radial cracks

$$E_t = \frac{P_{ref} - P_r}{T_t} \tag{7.1}$$

$$E_m = \frac{\left(P_{ref} - P_r\right)\rho_{ref}}{T_t \rho_t} \tag{7.2}$$

where P_{ref} and ρ_{ref} are the total penetration depth into reference target (UHPCC at present) and the density of reference target, respectively. Obviously, when the values of E_t and E_m are larger than 1, it indicates that, for the specified threat, the test target is thinner and lighter than the reference target and is indicative of superior ballistic performance.

Based on Eqs. (7.1) and (7.2), the thickness and mass efficiency factors of present composite target configurations are illustrated in Fig. 7.12. The following conclusions, consequently, can be drawn that, (i) with almost the same thickness, SiC ceramic has nearly twice the thickness efficiency factor and more than four times the mass efficiency factor than 10CrNi3MoV21A armor steel; (ii) both the thickness and

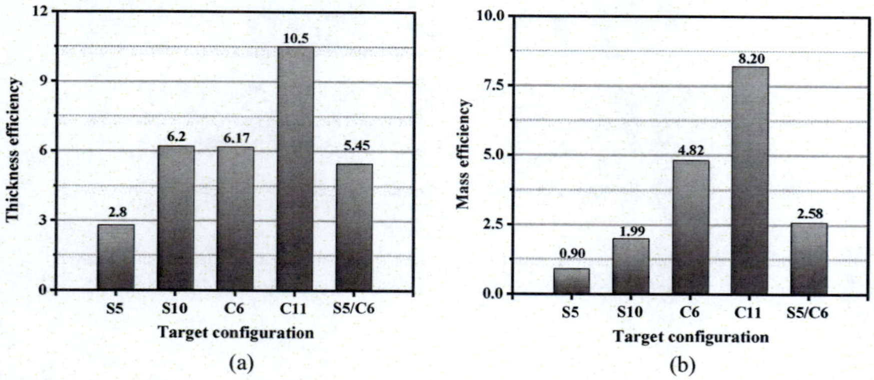

Fig. 7.12 Ballistic efficiency factors of composite target **a** thickness efficiency **b** mass efficiency

mass efficiency factors of S5/C6 configuration are almost twice higher than those of the S5 configuration, which confirms the contribution of ceramic plate to the light weight optimal design of the composite target.

7.2.2 *Numerical Simulations*

7.2.2.1 Numerical Model and Method

Generally, Lagrange, Euler and ALE (Arbitrary Lagrange Euler) are the three widely applied numerical algorithms to deal with the problem of continuum in solid mechanics. Based on the material coordinates, the Lagrange algorithm is often adopted to analyze the response of solid structures in which the change of the structures is completely consistent with the change of the finite elements. Besides, the spatial-coordinate-based Euler algorithm is usually suitable for the numerical analysis of fluid and fluid–solid coupling interaction. Therefore, the Lagrange algorithm with finite elements has difficulty in handling large distortions subjected to extreme loadings, such as impact or explosion. While for the Euler algorithm, in addition to the relatively low computation efficiency, it is also difficult to capture the material boundary during the simulation process. In order to overcome the aforementioned drawbacks, ALE algorithm which has the advantages of both the Lagrange and Euler algorithms is proposed (Benson 1989). The ALE algorithm introduces the characteristics of Lagrange algorithm in dealing with the movement of the material boundary, and more importantly, the finite element in ALE algorithm is independent with the material entity, and thus the elements can be adjusted properly during the numerical process and will not appear serious distortions.

At present, the suitable numerical algorithm for reproducing the eroded projectile with mushrooming deformation in test C11 was examined firstly. The commonly employed 3D numerical simulation was conducted by adopting the commercial finite element program LS-DYNA (2001) with Lagrange algorithm. The double symmetries of the problem were exploited and only a quarter of the geometry was discretized into eight-node hexahedral finite elements with the size of 1 mm. Besides, the smoothed particle hydrodynamics (SPH), a special Lagrange and meshless algorithm, was further adopted to simulate the response of fracture and splash of the brittle ceramic material. A total of 61,875 particles were distributed at the volume center of the original finite elements of ceramic material, so that the corresponding particle spacing was also 1 mm. The above two 3D numerical models as well as the finite elements and SPH particles domain of the ceramic plate are depicted in Fig. 7.13. The eroding-contact based algorithm was defined between the projectile and the target in which surface-to-surface algorithm was adopted for the finite elements and nodes-to-surface algorithm was implemented for the SPH particles and finite elements. The 3D numerical simulation results are shown in Fig. 7.14 where the acceptable penetration depth and residual length of the projectile are predicted. The fracture and splash response of the ceramic material are reasonably reproduced by

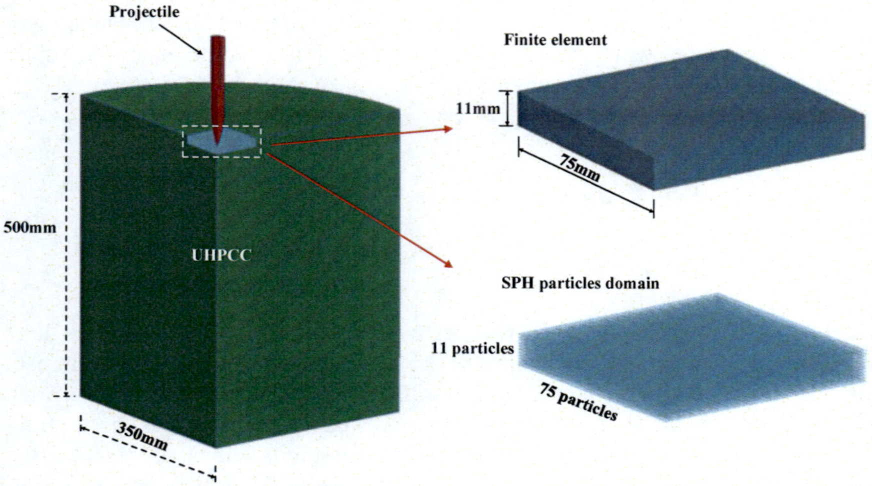

Fig. 7.13 3D numerical models with ceramic plate discretized into **a** finite elements and **b** SPH particles

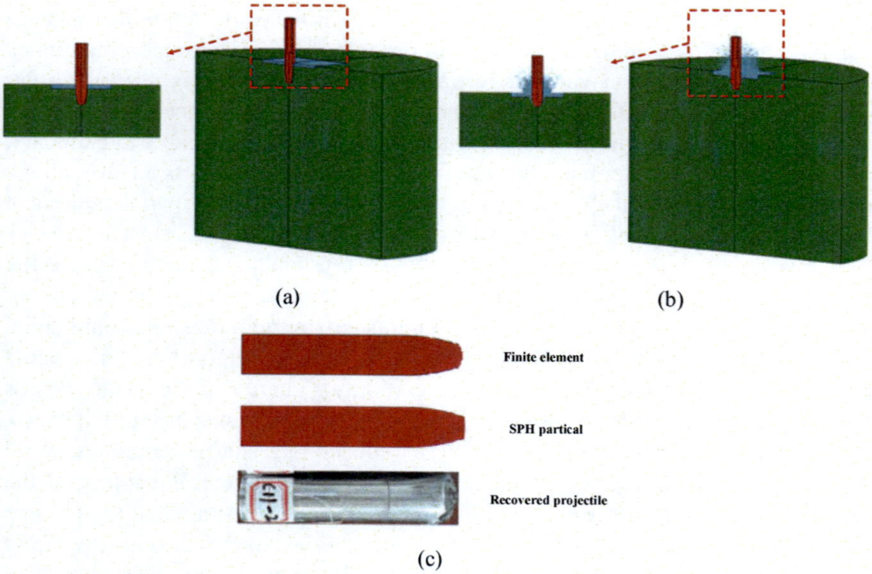

Fig. 7.14 3D numerical simulation results with ceramic plate discretized into **a** finite elements and **b** SPH particles **c** corresponding comparisons of recovered projectiles

the SPH algorithm whereas the mushrooming deformation of projectile is not reproduced well. However, the defects of the 3D Lagrange-algorithm based finite element and SPH algorithm can also be found as follows: (i) attributed to the relatively low computational efficiency, the time cost is correspondingly increased; (ii) the erosion criterion has to be added on the finite elements to obtain the eroded residual projectile. However, unreasonable deletions and distortions of the finite elements are usually occurred when a small and large values of erosion criterion are adopted, respectively. In other words, the erosion criterion in Lagrange algorithm is in contradiction with the reproduction of the mushrooming deformation of the projectile in the present simulation.

By taking the axisymmetric characteristic of the projectile and target (except the armor steel and ceramic plate) into consideration, the applicability of 2D axisymmetric model is further assessed at present. Besides, the ALE algorithm and adaptive remeshing method are the two practical numerical techniques to overcome the severe element distortion scenarios. Therefore, both the ALE algorithm and adaptive remeshing method (the element remeshing frequency was 2 μs and the minimum mesh size was 0.5 mm) were applied to the projectile and coupled with 2D axisymmetric model to solve the problem of mesh distortion with high computational efficiency. The corresponding simulation results of the ALE algorithm and adaptive remeshing method are summarized by the plots in Fig. 7.15.

It can be found from Fig. 7.15 that the mushrooming deformation of projectile is reasonably reproduced with the ALE algorithm while the adaptive remeshing method leads to a relatively smoothed nose of the projectile. Moreover, based on the 2D axisymmetric model, ALE algorithm holds the advantage of high computational efficiency against the adaptive remeshing method in which the computational efficiency and accuracy are gradually reduced on account of the frequent element remeshing. In conclusion, the 2D axisymmetric model together with ALE algorithm applied to the projectile are adopted to improve the computational efficiency and reproduce the large deformation of the projectile simultaneously.

Furthermore, taking the S5/C6 composite target for example, the details of the 2D axisymmetric FE model are depicted in Fig. 7.16, in which the projectile, armor steel and ceramic plates are all discretized into solid 162 finite elements with the size of

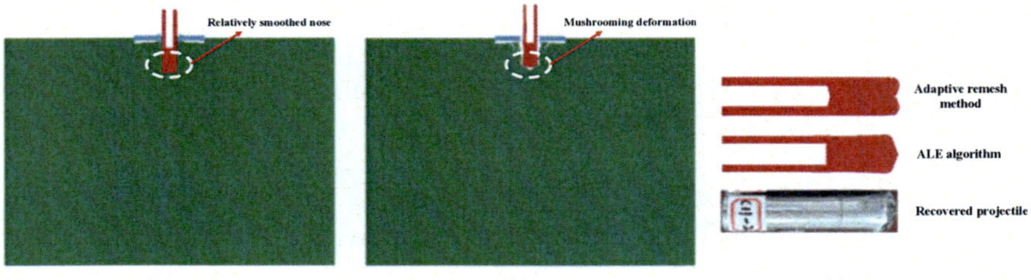

Fig. 7.15 2D axisymmetric numerical simulation results of **a** adaptive remeshing method **b** ALE algorithm and **c** comparisons of recovered projectiles

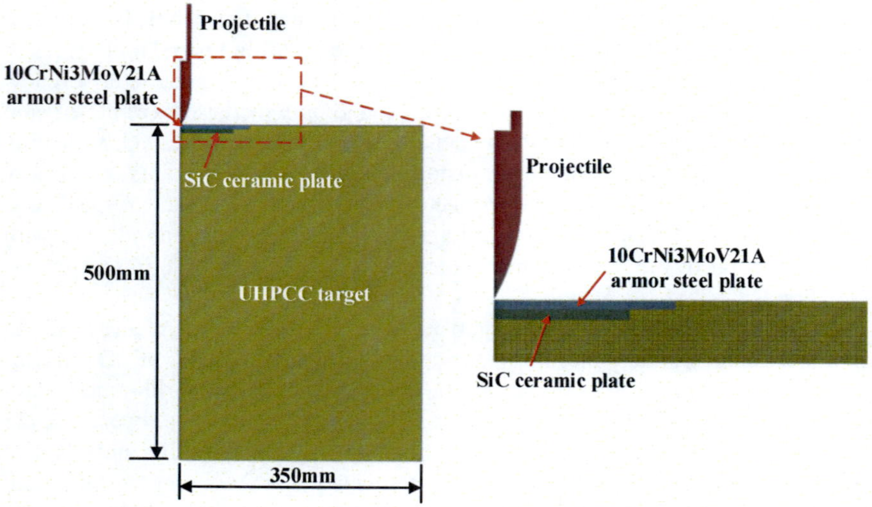

Fig. 7.16 2D axisymmetric FE model of S5/C6 scenario

1 mm to ensure the calculation efficiency as well as the numerical convergence. ALE algorithm is applied to the projectile for sake of the analysis of large deformation. Besides, the *CONTACT_2D_AUTOMATIC_SINGLE_SURFACE is implemented to reproduce the interactions between the projectile and target. The 10CrNi3MoV21A armor steel and SiC ceramic are therefore regarded as circle plates with diameters equaling to the side lengths of the actual target plates. It is worth pointing out that, the boundary effect on penetration depth in the present axisymmetric model is ignored since the side lengths of the armor steel and ceramic plates (200 and 150 mm) are already more than five times than that of the projectile shank diameter (30 mm).

7.2.2.2 Constitutive Model and Parameters

The appropriate constitutive models as well as the detailed parameters are of vital importance to the accuracy of numerical simulations. As for the projectile, according to the recovered projectiles shown in Fig. 7.8, the material models of *MAT_JOHNSON_COOK (denoted as JC model) with equation of state EOS_GRUNEISEN and *MAT_RIGID are adopted to represent the abrasive and rigid response of projectiles in the current study, respectively. Based on the static and dynamic mechanical tests on 30CrMnSiNi2A steel with Vickers hardness of 45 carried out by Li et al. (2017), the rigid and JC constitutive model parameters for the projectile are listed in Tables 7.7 and 7.8.

For the ceramic plate, a widely used constitutive model *MAT_JOHNSON_HOLMQUIST_CERAMICS, namely, the JH-2 model (Johnson and Holmquist 1994) is adopted to simulate the brittle behavior of the present SiC

Table 7.7 Constitutive model parameters for 30CrMnSiNi2A (Li et al. 2017) and 10CrNi3MoV21A steel

Material model	Parameters			
Rigid model (30CrMnSiNi2A steel)	Density (g/cm^3)	7.72	Young's modulus (GPa)	210
	Poisson's ratio	0.3		
	30CrMnSiNi2A steel		10CrNi3MoV21A steel	
JC model	Yield stress A (MPa)	1269	Yield stress A (MPa)	665
	Hardening constant B (MPa)	810	Hardening constant B (MPa)	950
	Hardening exponent n	0.479	Hardening exponent n	0.83
	Strain rate constant C	0.04	Strain rate constant C	0.028
	Thermal Softening m	1	Thermal Softening m	1.188

Table 7.8 EOS parameters for 30CrMnSiNi2A and 10CrNi3MoV21A steel (Steinberg 1996)

C (m/s)	S_1	S_2	S_3	γ_0	α
4569	1.49	0	0	2.17	0.46

ceramic plates subjected to the projectile impacts. Limited to the test condition, the present study do not carry out the mechanical tests on SiC ceramic material. Cronin et al. (2003) has proposed the JH-2 constitutive model parameters for four common ceramic materials, i.e., Al_2O_3, SiC, B_4C and AlN. Therefore, the corresponding model parameters for SiC ceramic from Ref. (Cronin et al. 2003) are listed in Table 7.9.

As for concrete targets, the constitutive model *MAT_JOHNSON_HOLMQUIST _CONCRETE (HJC model) (Holmquist et al. 1993), which is verified to be effective if the material failure is mainly controlled by compression, is applied to simulate the response of present NSC and UHPCC materials subjected to the projectile impact

Table 7.9 Parameters of JH-2 model for SiC ceramic (Johnson and Holmquist 1994)

Parameter	SiC	Parameter	SiC
Density (g/cm^3)	3.215	Hugoniot elastic limit (HEL) (GPa)	14.567
Shear modulus (GPa)	183	Pressure component at HELP_{HEL} (GPa)	5.9
Intact strength constant A	0.96	Damage constant D_1	0.48
Fracture strength constant B	0.35	Damage constant D_2	0.48
Strain rate constant C	0	Bulk modulus K_1 (GPa)	204.785
Fracture strength exponent M	1.0	Pressure coefficient K_2 (GPa)	0
Intact strength constant N	0.65	Pressure coefficient K_3 (GPa)	0

loadings. Based on a number of static and dynamic mechanical tests, the HJC constitutive model parameters for NSC and UHPCC materials were calibrated by Ren et al. (2016, 2017), respectively. In the present study, parameters from Refs. (Woolsey 1990; LSTC, LS-DYNA 2001) are adopted due to the close mixing proportions and summarized in Table 7.10.

Moreover, JC constitutive model is also employed to represent the response of 10CrNi3MoV21A armor steel plate. At follows, the detailed parameters are calibrated comprehensively by performing the systematic static and dynamic mechanical tests.

(1) Brief introduction of JC constitutive model

Johnson and Cook (1983) proposed an empirically based constitutive model to describe the response of metallic material subjected to large strains, high strain rates and high temperatures. In JC constitutive model, the effects of the strain, strain rate, and temperature on the flow stress σ of the metallic material are incorporated as a multiplication form as

$$\sigma=\left(A+B\varepsilon^{n}\right)\left(1+C\ln\dot{\varepsilon}^{*}\right)\left(1-T^{*m}\right) \tag{7.3}$$

where ε is the equivalent plastic strain; $\dot{\varepsilon}^{*}=\dot{\varepsilon}/\dot{\varepsilon}_0$ represents the dimensionless strain rate with the current strain rate $\dot{\varepsilon}$ and the reference strain rate $\dot{\varepsilon}_0$. A is the yield stress at reference strain rate and reference temperature, B and n are the coefficient and exponent of strain hardening effect, respectively. C and m are the constants which represent the coefficient of strain rate hardening and exponent of thermal softening, respectively. Besides, $T^{*}=\left(T-T_{ref}\right)/\left(T_m-T_{ref}\right)$ is the homologous temperature

Table 7.10 Parameters of HJC model for NSC and UHPCC (Ren et al. 2016, 2017)

Parameter	NSC	UHPCC	Parameter	NSC	UHPCC
Density (g/cm^3)	2.32	2.5	Maximum tensile hydrostatic pressure (MPa)	6	6
Shear modulus (GPa)	15.3	17.4	Normalized maximum strength S_{max}	15	7
Normalized cohesive strength A	0.28	0.30	Pressure constant K_1 (GPa)	12	116
Normalized pressure hardening B	1.85	1.73	Pressure constant K_2 (GPa)	135	−243
Pressure hardening exponent N	0.84	0.79	Pressure constant K_3 (GPa)	698	506
Strain rate coefficient C	0.006	0.005	Damage constant D_1 (GPa)	0.04	0.04
Uniaxial compressive strength (MPa)	35	135.1	Damage constant D_2 (GPa)	1.0	1.0

in which T, T_{ref} and Tm are the current temperature, reference temperature and melting temperature of the metallic material, respectively.

(2) Mechanical tests on armor steel

A series of static and dynamic mechanical tests were carried out to obtain the detailed parameters (A, B, n, C, m) of JC constitutive model for 10CrNi3MoV21A armor steel.

Quasi-Static Tensile Test at Room Temperature.

The quasi-static tensile test at room temperature was conducted to obtain the stress–strain relationship of the 10CrNi3MoV21A armor steel. As depicted in Fig. 7.17, the rectangular cross-section specimen was designed and three repeated tests were carried out under displacement control regime with an identical loading rate of 4.2 mm/min, i.e., the nominal strain rate is 1×10^{-3} s^{-1}. Figure 7.18 illustrates the corresponding true quasi-static tensile stress–strain curves of three specimens. It is found that the good repeatability and the tensile fracture of specimens were obtained. It can be derived that, the elastic modulus and static yield strength of 10CrNi3MoV21A armor steel are about 209 GPa and 665 MPa, respectively.

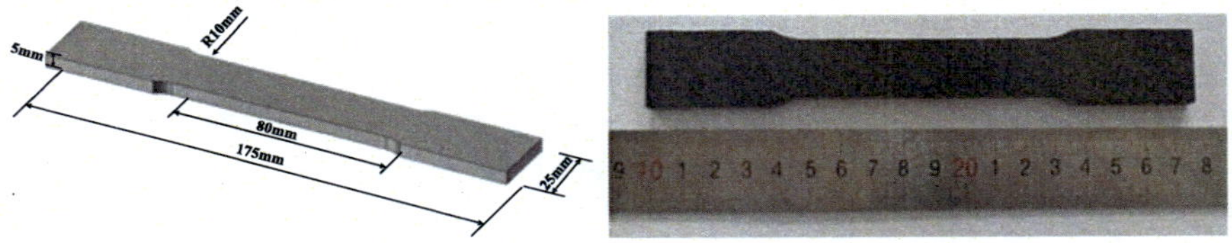

Fig. 7.17 Geometric dimensions of armor steel specimen for quasi-static tensile test at room temperature

Fig. 7.18 True quasi-static tensile stress–strain curves and tensile fracture of armor steel

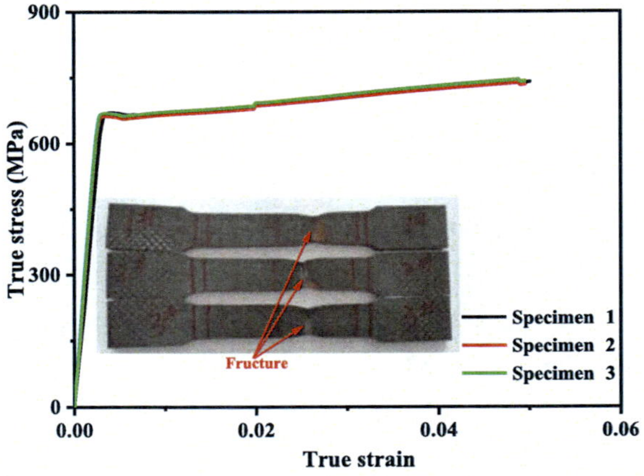

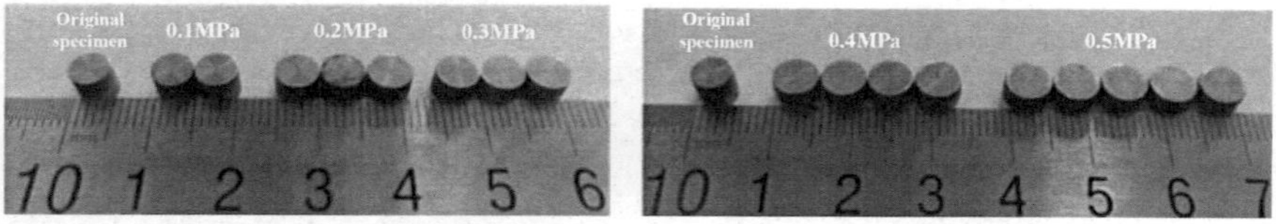

Fig. 7.19 Comparisons between the original and recovered specimen

Split Hopkinson Pressure Bar (SHPB) Test.

The dynamic response of the metal material subjected to extreme loadings such as explosion and impact is often accompanied with large deformation and high strain rate, which is significantly different from the mechanical behavior under the quasi-static loadings. Therefore, SHPB test was conducted to examine the dynamic mechanical properties of 10CrNi3MoV21A armor steel. Total 17 shots of SHPB test were carried out under the incident air pressure of 0.1 ~ 0.5 MPa, and the corresponding strain rates of the specimens were ranging from 1140 to 2820 s^{-1}. The comparisons between the original (Φ5 × 5 mm) and recovered specimens are shown in Fig. 7.19, and the dynamic compressive stress–strain curves of the tests under incremental incident air pressure with different strain rates are further given in Fig. 7.20.

Quasi-Static Isothermal Hot Tensile Tests

In order to investigate the temperature softening effect on the 10CrNi3MoV21A armor steel, the quasi-static isothermal hot tensile test of totally 13 specimens was carried out with the temperature ranging from 100 to 700 °C, and the temperature gradient was 50 °C. The rectangular cross-sectional specimen satisfying the isothermal hot tensile test was designed, and the corresponding geometric dimensions were shown in Fig. 7.21. The tests were conducted under displacement control regime with the loading rate of 6 mm/min, i.e., the nominal strain rate of 1×10^{-3} s^{-1}.

The tensile fracture of the specimens together with a failed specimen immediately after the test with temperature of 700 °C are shown in Fig. 7.22, respectively. The corresponding true tensile stress–strain curves are further given in Fig. 7.23, in which obvious decrease of both the elastic modulus and yield strength of 10CrNi3MoV21A armor steel can be found when the temperature exceeds 450 °C.

(3) Calibration of JC constitutive model parameters for 10CrNi3MoV21A armor steel

The parameters of JC constitutive model for 10CrNi3MoV21A armor steel are fitted and calibrated based on the above quasi-static and dynamic mechanical tests. At reference strain rate and temperature, Eq. (7.3) can be simplified as:

$$\sigma = \left(A + B\varepsilon^{n}\right) \tag{7.4}$$

Taking the natural logarithm of Eq. (7.4), it can be obtained that

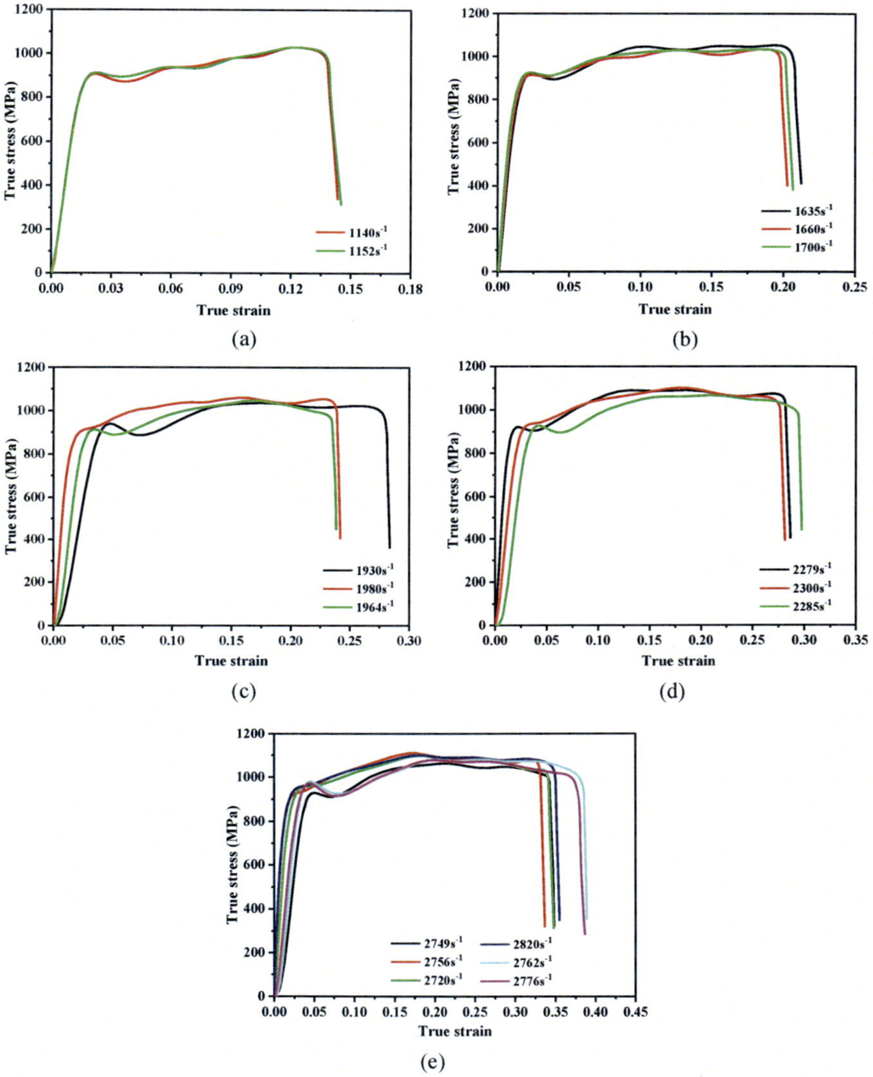

Fig. 7.20 Dynamic compressive stress–strain curves of armor steel under different incident air pressures **a** 0.1 MPa **b** 0.2 MPa **c** 0.3 MPa **d** 0.4 MPa **e** 0.5 MPa

$$\ln(\sigma - A) = \ln\ B + n\ \ln\ \varepsilon \tag{7.5}$$

The value of constant A can be determined as 665 MPa from the quasi-static tension tests. Then, based on the data within the plastic stage of the quasi-static tensile test, and substituting the value of A into Eq. (7.5), the values of $B = 950$ MPa and $n = 0.83$ are further obtained by fitting the relationship between ln(σ-A) and lnε. The

Fig. 7.21 Geometric dimensions of armor steel specimen for quasi-static isothermal hot tensile test

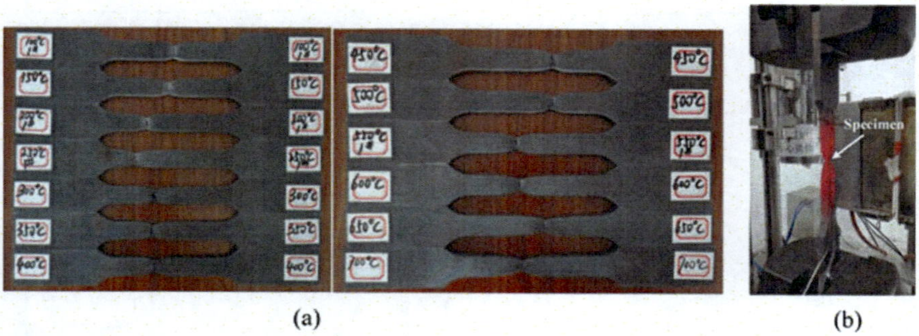

Fig. 7.22 Tensile fracture of specimens **a** at 100 ~ 700 °C **b** failed specimen immediately after the test at 700 °C

Fig. 7.23 True tensile stress–strain curves of armor steel at different temperatures

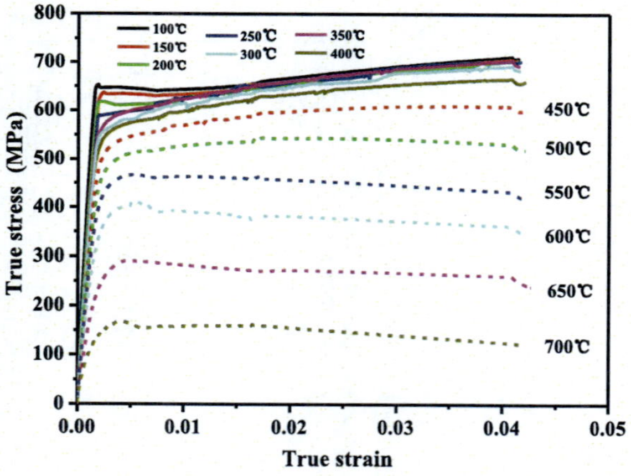

comparison between the fitting curve of JC constitutive model for 10CrNi3MoV21A armor steel based on the confirmed parameters (A, B and n) and the test data is given in Fig. 7.24.

The analysis of strain rate hardening effect on the 10CrNi3MoV21A armor steel was carried out based on the above SHPB test data. At reference temperature ($T^* = 1$, $\varepsilon = 0$), the expression of the dynamic yield strength of JC constitutive model can be simplified as

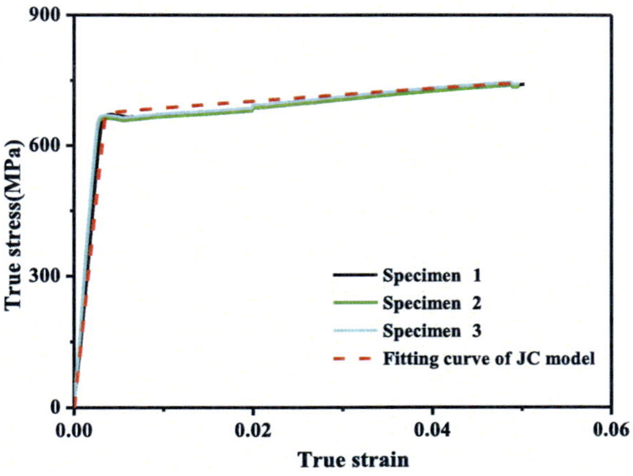

Fig. 7.24 Comparisons between the fitting curve of JC constitutive model and test data

$$\sigma = A(1 + C \ln \dot{\varepsilon}^*) \tag{7.6}$$

By substituting the value of A into Eq. (7.6), the relationship between the dynamic yield strength obtained from the above SHPB test and $\ln \dot{\varepsilon}^*$ is shown in Fig. 7.25. The strain rate effect parameter $C = 0.028$ is therefore obtained by calculating the slope of the linear fitting curve.

Furthermore, the thermal softening effect was investigated by carrying out the quasi-static isothermal hot tensile tests. The expression of JC constitutive model under reference strain rate ($10^{-3}\ s^{-1}$, $\varepsilon = 0$) can be simplified as

$$\sigma = A[1 - (T^*)^m] \tag{7.7}$$

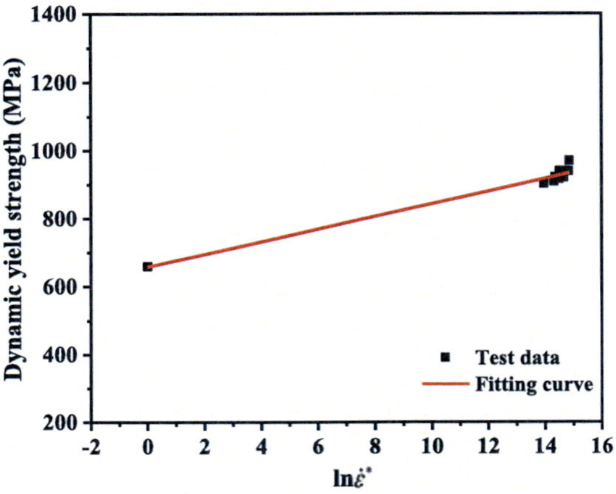

Fig. 7.25 Linear fitting of the relationship between dynamic yield strength and $\ln \dot{\varepsilon}^*$

Equation (7.7) can be further transformed into the following expression by taking the natural logarithm on both side to obtain the thermal softening effect parameter m, as

$$m \ln T^{*} = \ln(1 - \sigma/A) \tag{7.8}$$

Substituting the values of $A = 665$ MPa, $T_0 = 298$ K and $T_m = 1765$ K into Eq. (7.8), the relationship between $\ln(1-\sigma/A)$ and $\ln T^{*}$, as well as the linear fitting curve (across the origin point) are shown in Fig. 7.26. The thermal softening effect parameter m is therefore calculated as 1.188 based on the slope of the fitting curve.

Until now, the JC constitutive model parameters A, B, n, C and m for 10CrNi3MoV21A armor steel are all calibrated based on the above quasi-static and dynamic mechanical tests and summarized in Table 7.7. Besides, due to the lack of relevant test data, the EOS parameters of 4340 steel (Steinberg, 1996), as listed in Table 7.8, are adopted for both the present 30CrMnSiNi2A and 10CrNi3MoV21A armor steel.

7.2.2.3 Numerical Results and Discussions

(1) Simulation results

Based on the calibrated constitutive model parameters, the 2D axisymmetric numerical simulations of the present impact test were carried out. The numerical erosion algorithm *MAT ADD EROSION with the maximum principal strain was employed for the concrete, and the failure parameters D_1 and FS in JC and JH-2 damage model were also utilized to avoid the element distortions. The value of the maximum principal strain as well as the failure parameters were determined by trial and error method and assumed as the one which gave the best prediction to the experimental

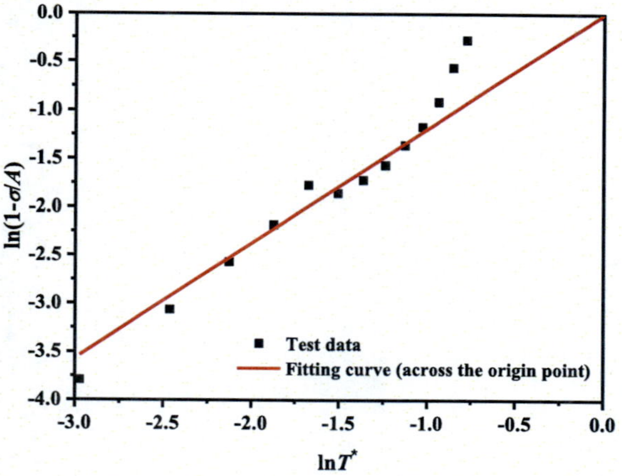

Fig. 7.26 Linear fitting of relationship between $\ln(1-\sigma/A)$ and $\ln T^{*}$ across the origin point

penetration depth. Firstly, as for rigid penetration scenarios of N, U and S5 tests, the maximum principal strain for NSC and UHPCC and failure parameter D_1 for 10CrNi3MoV21A armor steel were determined as 0.22, 0.2 and 0.58, respectively. It should be noted that, only D_1 in JC damage model was adopted at present following the method proposed by Rosenberg and Dekel (2003), and all the failure parameters would be adopted in the remaining simulations once they were validated. Then, the failure parameter $D_1 = 0.8$ in JC damage model for the ALE-algorithm-based projectile and FS = 1.2 in JH-2 damage model for the SiC ceramic were determined based on the tests of S10 and C6, respectively. Lastly, the simulations of C11 and S5/C6 scenarios were conducted based on the determined numerical parameters.

Figure 7.27 shows the numerically simulated penetration depths as well as the damage contour of concrete targets. The comparisons between the predicted abrasion and damage of projectiles and 10CrNi3MoV21A armor steel plates, as well as the corresponding experimental observations are further depicted in Fig. 7.28, in which the plastic strain ranging from 0 to 0.2 is applied to express the corresponding damage extent. The abrasion of projectiles, especially the mushrooming deformation of the projectile in test C11 and the deflection of the 5 mm-thick armor steel plate are reproduced well numerically. Therefore, the validations of the adopted numerical algorithm as well as the constutive models and the corresponding material parameters are preliminarily verified.

Moreover, the numerical and experimental results regarding $\overline{P}$ and $\overline{L}_r$ are shown in Fig. 7.29 and summarized in Table 7.11, in which the "Deviation" columns denote the relative deviations of numerical results compared with the test data. It can be found from Table 7.11 that the numerical predicted penetration depths and the residual lengths of the projectiles are in good agreement with the test data, and the absolute relative deviations are all in a reasonable range which is less than 10% except for $\overline{P}$ in test C11 (the penetration depth in test C11-2 is relatively small). Therefore, the proposed numerical algorithm as well as the constutive models and the corresponding material parameters are further verfied and can be adopted to conduct the extented assessments such as energy evolution analysis.

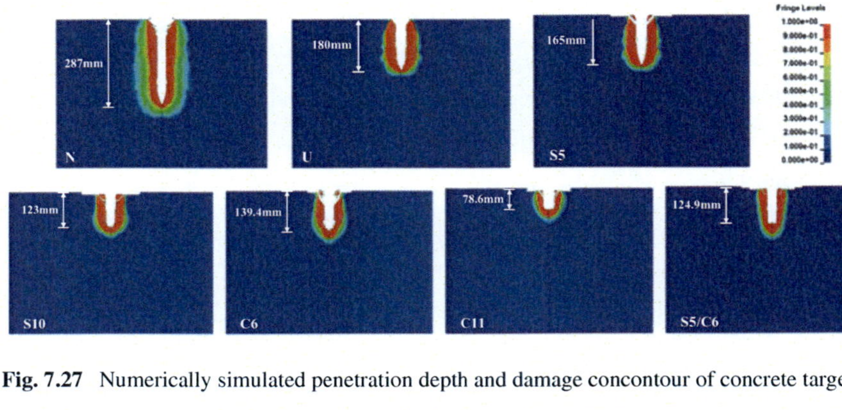

Fig. 7.27 Numerically simulated penetration depth and damage concontour of concrete targets

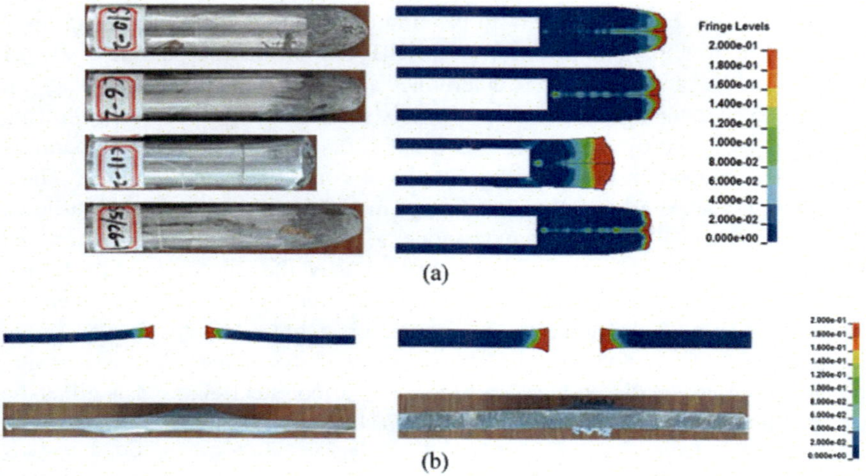

Fig. 7.28 Comparisons of the abrasion and damage between numerically simulated and test results **a** projectiles **b** 10CrNi3MoV21A armor steel plates

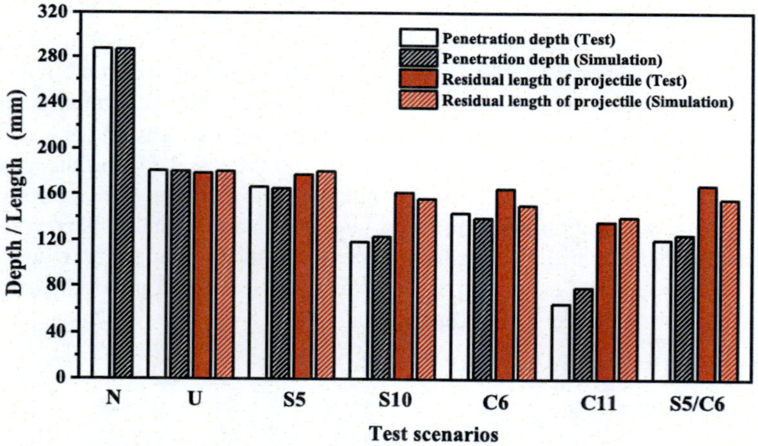

Fig. 7.29 Comparisons between the numerical simulation results and test data

(2) Discussions

Based on the above validated numerical algorithm as well as the constitutive models and the corresponding model parameters, the energy evolutions of both projectiles and targets during the penetration process are further discussed.

During the penetration process, the initial kinetic energy (KE) of the projectile is mainly dissipated by the following aspects. Firstly, a large proportion of the initial kinetic energy is dissipated by the fracture of the target subjected to the impact loading

Table 7.11 Deviations of the experimental and numerically simulated results of $\overline{P}$ and $\overline{L}_r$

Target	$\overline{P}$(mm)			$\overline{L}_r$(mm)		
	Test	Simulation	Deviation (%)	Test	Simulation	Deviation (%)
N	287.5	287	–	–	–	–
U	180.5	180	–	178.5	180	0.84
S5	166.5	165	–	177	180	1.69
C6	143.5	139.4	−2.86	165	150.5	−8.79
S10	118.5	123	3.80	161.5	156.2	−3.28
C11	65	78.6	20.92	136	139.9	2.87
S5/C6	120.5	124.9	3.65	168	155.8	−7.26

(Liu et al. 2018). Then, the increase of internal energies of the projectile and target is another important aspect of the energy transformation. Lastly, the rest of the initial kinetic energy can also be consumed by the friction between the projectile and target. The normalized energy time-histories (ratio of the instantaneous energy to the initial KE of projectile) of the projectiles and targets are obtained numerically and shown in Fig. 7.30, in which the internal energy time-histories of the armor steel and ceramic plates are displayed together for the comparative purpose. It can be concluded from the figures that the internal energy of the UHPCC target accounts for about 23% of the initial KE of projectile, and this proportion is also close to the numerical simulation result of the test conducted by Liu et al. (2018), in which the 120 MPa UHPCC target is impacted by the ogive-nosed projectile (25.3 mm in diameter) with striking velocity of 549 m/s. More importantly, although the penetration depth into the UHPCC target is about 100 mm less than that of the NSC target, the internal energy of UHPCC is almost identical to the NSC target, which indicates higher resistance and better energy absorption capacity of UHPCC under projectile impact. As for the internal energies of the armor plates, only a slight amount of internal energy is increased

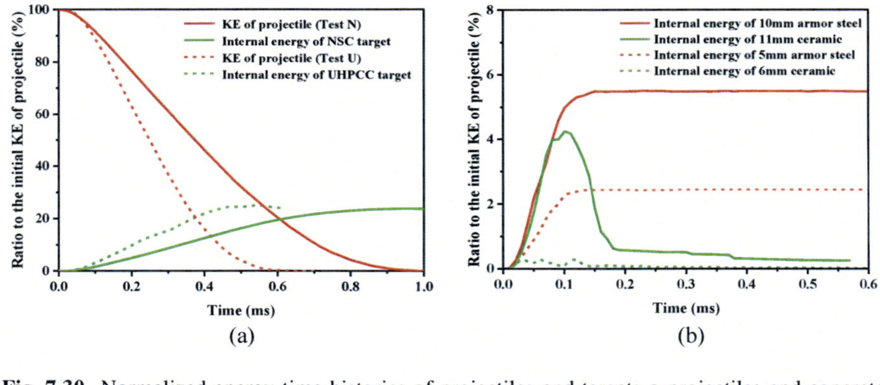

Fig. 7.30 Normalized energy time-histories of projectiles and targets **a** projectiles and concrete targets **b** armor steel and ceramic plates

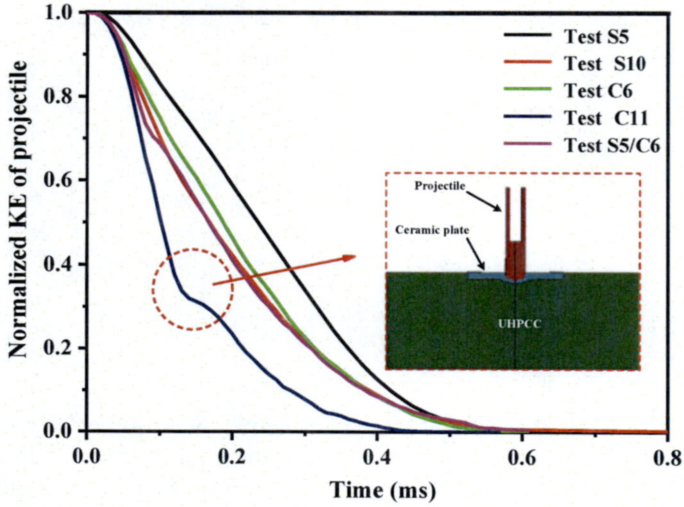

Fig. 7.31 Normalized KE time-histories of projectiles

in the 6 mm ceramic plate, and the internal energy of the 11 mm ceramic plate is also significantly reduced compared with the 10 mm armor steel plate. Moreover, the internal energy in 11 mm ceramic plate is not sustained for a period of time like amor steel but begins to attenuate at about 0.1 ms.

The time-histories of normalized KE of projectiles together with the numerical simulation image of scenario C11 at 0.1 ms are shown in Fig. 7.31 to further discuss the impact resistance of the armor plates. It can be noted that the 11 mm SiC ceramic plate is just perforated and fractured at about 0.1 ms. Moreover, obvious abrasion and mushrooming deformation of the projectile are occurred, and more than 60% of the KE of projectile is dissipated simultaneously. As for S5/C6 composite target, the KE time-history of the projectile is almost identical to the S10 scenario, while it should be noted that about 25% of the target areal density is reduced in the S5/C6 scenario. Therefore, the mechanism and advantages of the SiC ceramic plate to resist the impact loadings are observed as follows: during the penetration and perforation of the projectile into SiC ceramic plate, the decreases of both the velocity and residual length of projectile occur on account of the high intrinsic strength and hardness of the ceramic. Consequently, the penetration capacity of the blunt-nosed projectile is weakened obviously due to the significant drop of the kinetic energy. Simultaneously, attributed to the low thermal conductivity, the internal energy generated in the SiC ceramic plate is limited and will not last for a long time during the penetration process. Therefore, SiC ceramic can be considered as a potential high-performance protective material to resist the projectile impact. Correspondingly, the armor steel/ceramic/UHPCC composite target is also verified to be a promising configuration with high impact resistance, light weight and low thermal conductivity, which has wide potential applications in military facilities, nuclear power plants and etc.

7.3 Impact Test on NP450 Armor Steel/UHPCC and NP500 Armor Steel/UHPCC Composite Targets and Numerical Simulations

7.3.1 Impact Test

7.3.1.1 Projectile and Target

(1) Projectiles

In the present impact test, the adopted projectiles are the same as those in Sect. 7.2.1.1. The same cannon in Sect. 7.2.1.2 was utilized to launch the projectiles to reach the striking velocities between 372 m/s and 646 m/s by adjusting the charge weight.

(2) Composite targets

Ballistic performance of the armor steel/UHPCC composite target was examined by conducting projectile penetration test with different combinations of armor steel types and thicknesses, as shown in Fig. 7.32. The armor steel plate and the backed UHPCC cylindrical were pasted by binder.

Two types of newly developed armor steels named NP450 and NP500 with the plane dimensions of 150 mm × 150 mm and thicknesses of 5, 8, 10, and 13 mm were adopted and provided by Nanjing Nangang Iron and Steel United Co., Ltd., as shown in Fig. 7.33. The corresponding chemical compositions of those two new steels are listed in Table 7.12, and the yield strength of NP450 and NP500 armor steels are 1136 MPa and 1323 MPa, respectively.

The mix proportions of UHPCC and size of UHPCC targets are both same as those in Sect. 7.2.1.1.

Fig. 7.32 Armor steel/UHPCC composite target

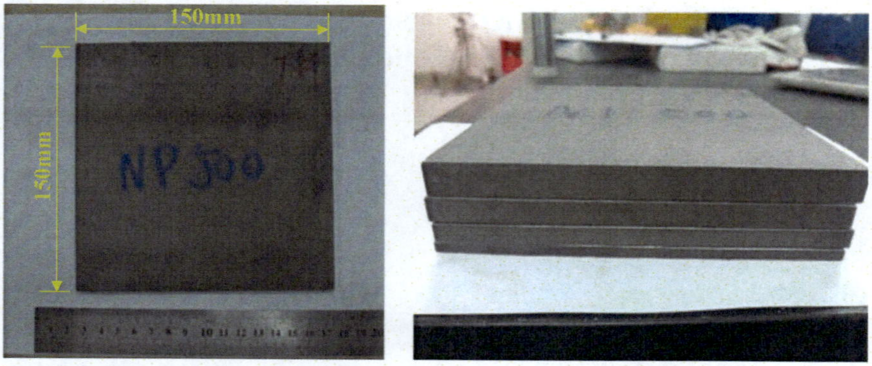

Fig. 7.33 Dimensions of armor steel plate

Table 7.12 Compositions of NP450 and NP500 armor steels (mass fraction %)

Component	C	Si	Mn	P	S	Alt	Ti	Ni	Cr	Mo	B	Ceq
NP450	0.20	0.51	1.29	0.01	0.008	0.041	0.02	0.02	0.66	0.15	0.0018	0.58
NP500	0.28	0.64	0.74	0.005	0.001	0.034	0.015	0.69	0.24	0.42	0.0016	0.60

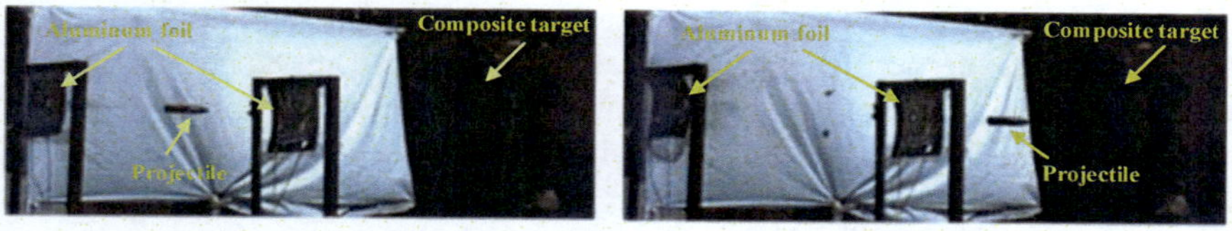

Fig. 7.34 Typical projectile impacting photographs

7.3.1.2 Test Setup

The setup of projectile penetration test is same as that in Sect. 7.2.1.2. The aluminum foils were placed in front of the launch cannon to record the transit time and derive the projectile striking velocities. The high-speed camera arranged aside was used to record the flight path of the projectile. The typical projectile impacting photographs are shown in Fig. 7.34, indicating horizontal flight of the projectile.

7.3.1.3 Test Results and Discussions

Twelve penetration shots on NP450/UHPCC and NP500/UHPCC composite targets, and the corresponding test results are listed in Table 7.13. As for the target, for example, NP450_5mm represents that the armor steel type is NP450 with a thickness of 5 mm in the composite target. The asterisk "*" denotes that the projectile was

Table 7.13 Test data

Target	V_0 (m/s)	L_r (mm)	M_r (g)	DOP (mm)	d_1 (mm)	d_2 (mm)	d_3 (mm)	d_4 (mm)	d_c (mm)	H_c (mm)	A_c (cm^2)
NP450_5mm	474	168	673	126	250	275	260	250	259	85	493
NP450_5mm	581	165	682	159	260	255	240	265	255	87	515
NP450_5mm	639	169	680	175	210	250	245	230	234	125	419
NP450_8mm	481	*	*	32	115	125	125	120	121	24	478
NP450_10mm	399	135	495	–	–	–	–	–	–	–	–
NP450_13mm	394	144	588	–	–	–	–	–	–	–	–
NP500_5mm	484	155	668	81	160	170	205	180	179	76	248
NP500_5mm	584	160	676	130	220	290	280	270	265	82	478
NP500_5mm	646	162	675	166	210	190	200	180	195	96	285
NP500_8mm	643	*	*	71	190	255	250	205	225	63	378
NP500_10mm	422	129	476	–	–	–	–	–	–	–	–
NP500_13mm	372	134	520	–	–	–	–	–	–	–	–
UHPCC	520	180	675	181	340	280	300	286	302	88	577

broken and the residual mass and length were not obtained. "–" represents that the armor steel plate was not perforated and the damage of UHPCC was ignored. In Table 7.13, the test data of striking velocity V_0 and DOP are listed. It should be noted that DOP contains the thickness of penetrated armor steel plate. Besides, the residual length and mass of the recovered projectiles denoted respectively by L_r and M_r are also listed in Table 7.13. The corresponding damage and failure of recovered projectiles, armor steel plates and UHPCC targets in each shot are shown in Fig. 7.35.

Fig. 7.35 Damage of projectiles, armor steel plates and UHPCC targets

Projectile penetration into concrete target always forms the frustum-of-cone shaped crater and is followed by a cylindrical tunnel with the diameter nearly equal to the projectile shank diameter. The average front crater diameter d_c of UHPCC target can be derived by averaging the dimensions of crater surface from four directions, i.e., horizontal, vertical, 45° and 135°. The crater height H_c, and the surface area A_c are also listed in Table 7.13. As suggested in Ref. (Wu et al. 2015a, b), the pixel approach as illustrated in Fig. 7.36 was adopted to measure the area of the crater surface A_c.

Figure 7.37 illustrates the photograph of unfired projectile (far left one) and recovered projectiles in composite targets after penetration test. It can be found from Fig. 7.37 and Table 7.13 that for the present two types of armor steels with a thickness of 5 mm, the maximal relative mass losses of the projectiles were less than 4.3%, but the blunted lengths reached 13.9%. Comparably, for the two types of armor steel plates with thicknesses of 8 mm, 10 mm, and 13 mm, since the interface pressure between projectile and target was far greater than the dynamic yield strength of the projectile, the projectiles endured obvious erosions or fractures, as shown in Fig. 7.37. For the NP450_8mm/UHPCC composite target, the steel plate was critically perforated with a striking velocity of 481 m/s.

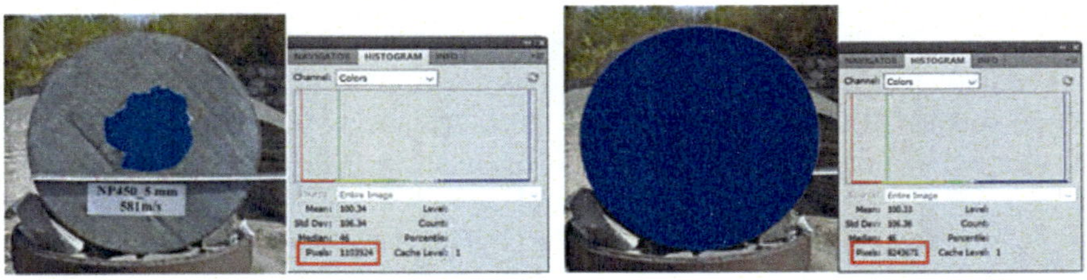

(a) (b)

Fig. 7.36 Pixel approach for measuring the area of impact crater surface **a** crater surface **b** target face

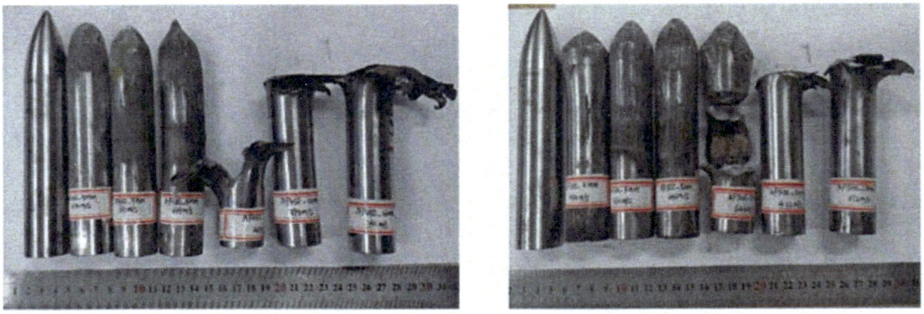

(a) (b)

Fig. 7.37 Photographs of unfired and recovered projectiles **a** NP450 series **b** NP500 series

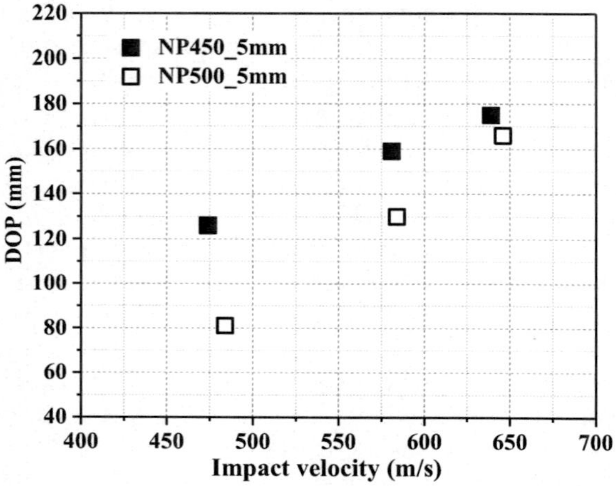

Fig. 7.38 Influences of striking velocity on DOP

For the composite targets with 5 mm-thick armor steel plate, DOP gradually increases with the increase of projectile striking velocity, as shown in Fig. 7.38. It indicates that, for the armor steel plate with a thickness of 5 mm and nearly identical projectile striking velocities, NP500/UHPCC composite targets lead to smaller DOPs than NP450/UHPCC composite targets. Especially, as illustrated in Fig. 7.35, for the NP450_5mm/UHPCC and NP500_5mm/UHPCC composite targets with the striking velocities of 474 m/s and 484 m/s, nearly no obvious tunneling region occurred for the rear UHPCC target of the NP500_5mm/UHPCC composite target. Therefore, it can be concluded that NP500 armor steel exhibits better penetration resistance.

7.3.2 *Numerical Simulations*

7.3.2.1 Numerical Model and Method

The penetration test is numerically simulated with the finite element program LS-DYNA (2001). The 2D axisymmetric Lagrange algorithm is adopted considering its high computational efficiency and 2D solid 162 element type is employed for modeling both the projectile and target. Figure 7.39 shows the finite element model, in which the geometry of projectile is identical with that in the test. The radius of armor steel plate is 75 mm. The radius and height of the UHPCC target are 350 mm and 500 mm, respectively. The boundary of the composite target is set as non-reflect. Considering the dimension of projectile and thickness of armor steel plate, the element size for the projectile and target is selected to be 1 mm, and there are 1935 elements in the projectile and 175,000 elements in the UHPCC target. The mesh convergence analyses are performed in the following Sect. 7.3.2.3. The element number of armor steel plate varies with the thickness,

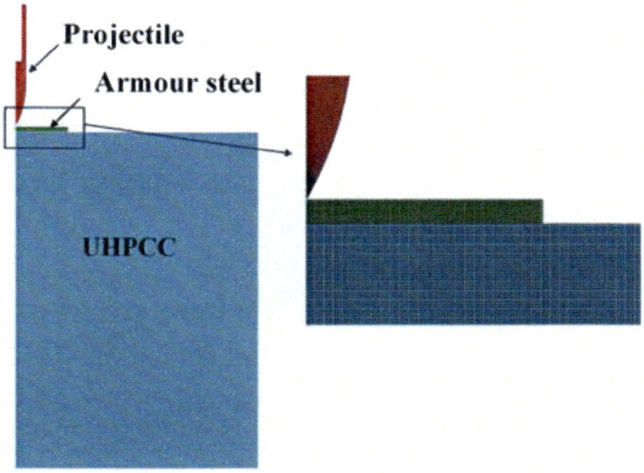

Fig. 7.39 2D axisymmetric finite element model

in which the element numbers are 375, 600, 750, and 975 corresponding to the armor steel plate thicknesses of 5 mm, 8 mm, 10 mm, and 13 mm, respectively. The 2D_AUTOMATIC_SINGLE_SURFACE contact is employed to reflect both the contact between the projectile and target, as well as the contact between armor steel and UHPCC.

7.3.2.2 Constitutive Models and Parameters

(1) Projectile

For the armor steel/UHPCC composite targets, the projectiles were eroded with obvious mass losses. Therefore, J-C constitutive model *MAT_JOHNSON_COOK (MAT_015) embedded in LS-DYNA (2001) is adopted. J-C constitutive model (Johnson and Cook 1983) given in Eq. (7.3) has been widely used for the metallic materials subjected to large strains, high strain rates, and high temperatures, which considers the coupled influences of strain hardening, strain rate hardening and thermal softening. Grüneisen equation of state (EOS) (Steinberg 1996) is employed corresponding to the J-C constitutive model.

The partial J-C constitutive model parameters of 30CrMnSiNi2A steel have been determined in Ref. (Li et al. 2017) and listed in Table 7.7. Besides, the EOS parameters of 30CrMnSiNi2A steel are listed in Table 7.8.

(2) Armor steel

Similarly, J-C constitutive model is also employed for describing the present NP450 and NP500 armor steels. At follows, the constitutive parameters of J–C model are experimentally determined by conducting the quasi-static tension test under varying temperatures and SHPB dynamic compression test under room temperature.

As shown in Fig. 7.40, the quasi-static tension test for NP450 and NP500 armor steels was conducted on a WAW-300B testing machine with the displacement loading speed of 6.3 mm/min. The corresponding strain rate is 0.001 s^{-1}, and the room temperature is 298 K, i.e., 25 °C, which is taken as the reference strain rate and temperature. Thus, J–C model given in Eq. (7.3) can be simplified as Eq. (7.4).

The quasi-static tensile stress–strain curves of NP450 and NP500 armor steel materials are shown in Fig. 7.40c, d. By fitting the stress–strain curves with Eq. (7.4), the strain hardening parameters $A = 1230$ MPa, $B = 1674$ MPa and $n = 0.4985$ of NP450 armor steel, as well as $A = 1340$ MPa, $B = 3825$ MPa and $n = 0.634$ of NP500 armor steel are determined. The fitting curves of J-C constitutive model are also shown in Fig. 7.40c, d.

Under the room temperature, the SHPB dynamic compression test on sixteen cylindrical armor steel specimens was conducted with the strain rates from 810 s^{-1} to 1623 s^{-1}. Figure 7.41 shows the intact specimens in the SHPB test, in which the initial height and diameter of the cylindrical specimen are both 5 mm. In the SHPB test, with increasing the axial strain rate, the specimens expanded radially and the corresponding heights were reduced gradually. Figure 7.41 further shows the dynamic stress–strain curves under different strain rates. For the dynamic yield strength under the reference temperature ($\varepsilon_P = 0$ and $T^* = 1$), the J-C constitutive model given in Eq. (7.3) can be simplified as Eq. (7.6).

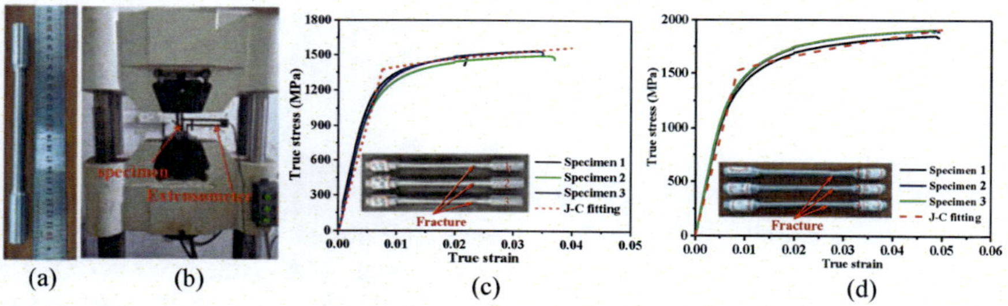

Fig. 7.40 Quasi-static tension test of armor steels under room temperature **a** specimen **b** WAW-300B testing machine **c** true stress–strain curves of NP450 and **d** NP500 armor steels

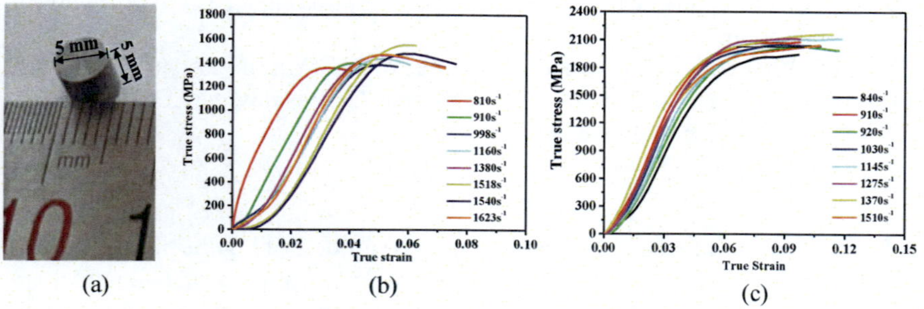

Fig. 7.41 SHPB dynamic compression test of armor steel materials **a** specimen **b** dynamic stress–strain curves of NP450 and **c** NP500 armor steels

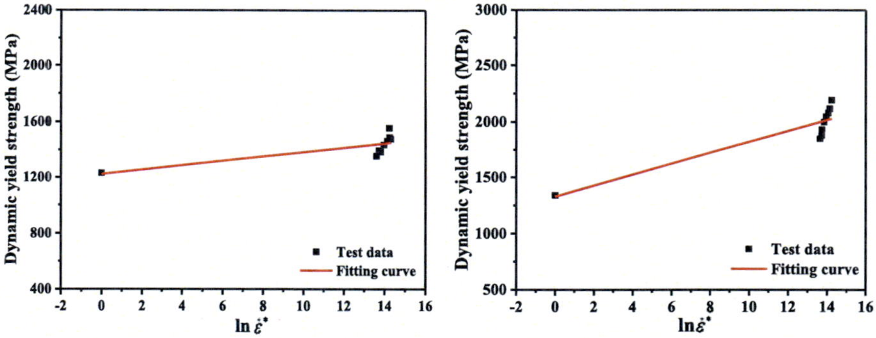

(a) (b)

Fig. 7.42 Dynamic yield strength data and the fitting curves of **a** NP450 and **b** NP500 armor steels

Figure 7.42 further shows the experimental dynamic yield strength data under different strain rates, as well as the fitting curves by Eq. (7.6). $C = 0.013$ for NP450 armor steel and $C = 0.0368$ for NP500 armor steel are determined.

As shown in Figs. 7.43 and 7.44, the quasi-static tension test for NP450 and NP500 armor steels under varying temperatures were conducted on 300 kN high temperature electronic material test system with the displacement loading speed

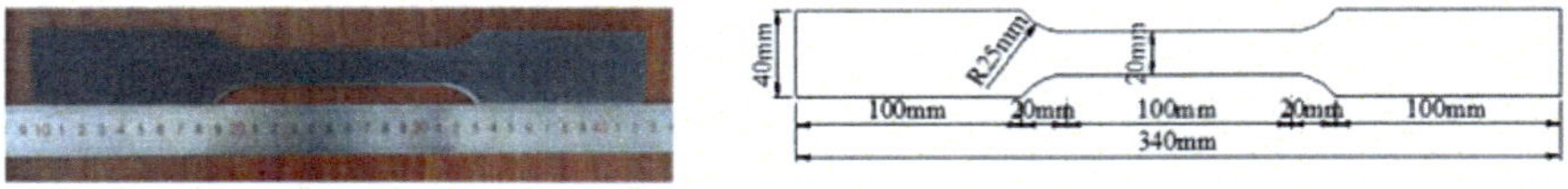

(a) (b)

Fig. 7.43 Test specimen **a** photograph **b** dimensions

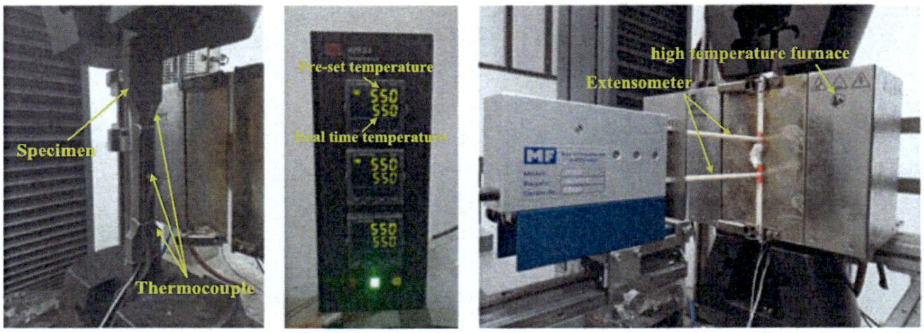

(a) (b) (c)

Fig. 7.44 Quasi-static tension test of armor steel materials under high temperature **a** arrangement of thermocouples **b** temperature control system **c** high temperature furnace and arrangement of extensometers of 6 mm/min, corresponding to the strain rate of 0.001 s^{-1}. Three thermocouples were pre-fixed on the specimen to record the temperature changes at the upper, middle and lower positions of the test specimens. The pre-set temperatures and real-time temperatures detected by thermocouples were explicitly displayed in the temperature control system to ensure that the temperature of specimen reached the pre-set temperature.

Considering the high yield strength of two newly developed armor steels and limited loading capability of the testing machine, the quasi-static tension test was conducted under the temperature of 700 ~ 900 °C. Figure 7.45 shows the damaged specimens and the corresponding dynamic stress–strain curves under different temperatures. For the dynamic yield strength under the reference strain rate ($\varepsilon_P = 0$ and $\dot{\varepsilon}/\dot{\varepsilon}_0 = 1$), J-C constitutive model given in Eq. (7.3) can be simplified as Eq. (7.7).

Figure 7.46 further shows the dimensionless yield strength under different temperatures. Due to the high yield strength of the two types of armor steels, the thermal

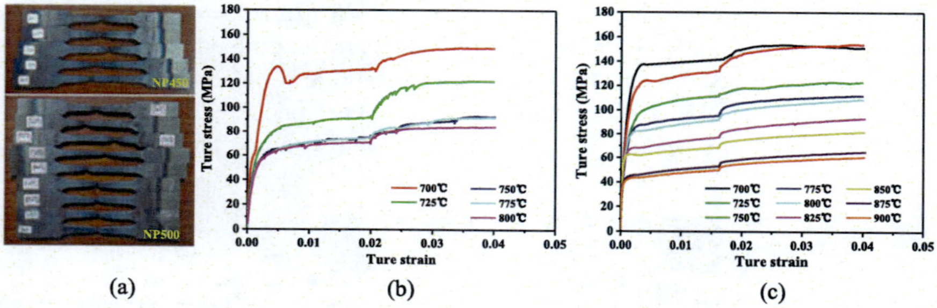

Fig. 7.45 Quasi-static tension test results of armor steels under high temperature **a** damaged specimens **b** true stress–strain curves of NP450 and **c** NP500 armor steels

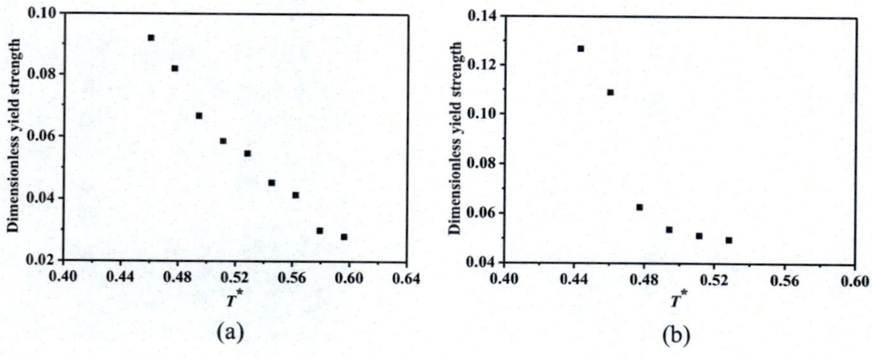

Fig. 7.46 Dimensionless dynamic yield strength data under high temperature of **a** NP450 and **b** NP500 armor steels

Table 7.14 J–C model parameters of NP450 and NP500 armor steels

Basic parameters			Constitutive parameters			EOS parameters (Steinberg 1996)		
	NP450	NP500		NP450	NP500		NP450	NP500
ρ (kg/m^3)	7860	7860	A (MPa)	1230	1340	C (m/s)	4569	4569
E (GPa)	210	210	B (MPa)	1647	3825	s_1	1.49	1.49
G (GPa)	80.8	80.8	N	0.4985	0.634	s_2	0	0
T_m (K)	1765	1765	C	0.013	0.0368	s_3	0	0
T_0 (K)	298	298	m	1.0	1.0	γ_0	1.93	1.93
C_v (J/(g · K))	477	477	$\dot{\varepsilon}_0(\mathrm{s}^{-1})$	1×10^{-3}	1×10^{-3}	a	0.5	0.5

softening coefficient m cannot be accurately determined by fitting the obtained experimental data using Eq. (7.8). Therefore, $m = 1$ is adopted for simplicity.

Up to now, the major J-C constitutive model parameters for the present NP450 and NP500 armor steels are calibrated and listed in Table 7.14. Besides, Grüneisen equation of state is employed and the corresponding parameters are referred from Ref. (Steinberg 1996).

(3) UHPCC

The MAT_JOHNSON_HOLMQUIST_CONCRETE (MAT_111) embedded in LS-DYNA (2001) is selected as the material model for UHPCC, which is suitable to describe the mechanical behavior of cementitious materials under the conditions of large deformation, high strain rate, and high hydrostatic pressure. The main material parameters of HJC model are listed in Table 7.10. To avoid the hourglass effect during numerical simulation, the MAT_ADD_EROSION is applied to remove the highly distorted elements and reproduce the physical fracture or crushing of concrete numerically. The maximum principal strain is adopted as the failure criterion of UHPCC.

7.3.2.3 Test Results and Discussions

(1) Comparisons

The erosion strain of UHPCC has been determined in Sect. 7.2.2.3. Then, adopting the above determined erosion strain of UHPCC, $D_1 = 1.6$, 0.54, and 0.46 for projectile, NP450 armor steel, and NP500 armor steel were determined by satisfying that the simulated DOP and L_r are close to the corresponding experimental values for the shots of NP450_5mm and NP500_5mm composite targets with the striking velocities of 474 m/s and 484 m/s, respectively.

Additionally, it is widely known that the mesh size affects the accuracy of calculation results and computational efficiency in simulations. Refining the mesh size can further obtain the accurate calculation results but increase the computation amount and computation cycle. Therefore, it is necessary to find a balance between the calculation accuracy and cost. One shot of NP450_5mm/UHPCC composite target with a striking velocity of 474 m/s is selected and the mesh sizes of 2 mm, 1.5 mm, 1 mm, and 0.8 mm of target are adopted to study the convergence of mesh size. The results of DOP with various mesh sizes are presented in Fig. 7.47. Therefore, considering both the calculation accuracy and cost, 1 mm mesh size is sufficiently fine and adopted in the present simulations.

Figure 7.48 shows the typical damage processes for the composite targets with perforated and non-perforated armor steel plates, respectively. The corresponding velocity–time histories of projectiles are further illustrated in Fig. 7.49. It can be seen that the projectile velocity drops more rapidly when impacting on the armor steel than that of UHPCC, which demonstrates that the armor steel provides greater impact resistance than UHPCC in composite target.

Figure 7.50 further shows the simulated projectiles in NP450_8mm/UHPCC and NP500_8mm/UHPCC composite targets. The test projectiles in those two scenarios are broken. It can be seen that severe deformation was occurred in the position where the projectile fractured in the test. The possible reason lies in that only the damage parameter D_1 is adopted in the present simulations to describe the damage and failure of projectile, while actually the failure of metallic materials may be influenced by the stress state, strain rate and temperature, etc.

For the shots with the armor steel plate thicknesses of 10 mm and 13 mm, the plates were not penetrated, i.e., NP450_10mm/UHPCC, NP450_13mm/UHPCC, NP500_10mm/UHPCC, and NP500_13mm/UHPCC composite targets. The comparisons between the experimental and simulated indentations on front impact face and bulges on rear face of armor steel plate are shown in Fig. 7.51. Besides, the corresponding comparisons of the erosive projectiles are also illustrated.

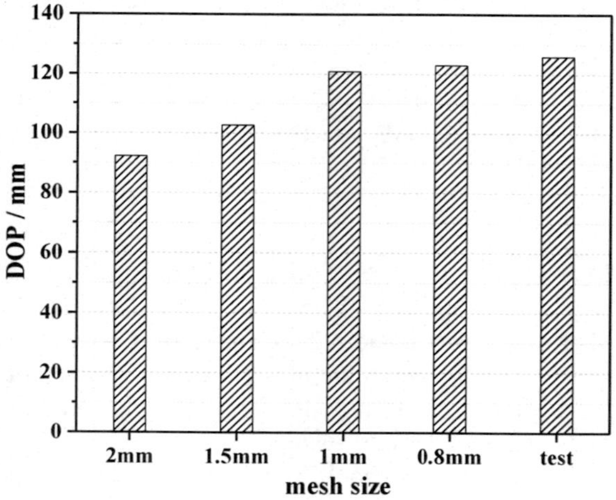

Fig. 7.47 Mesh convergence study

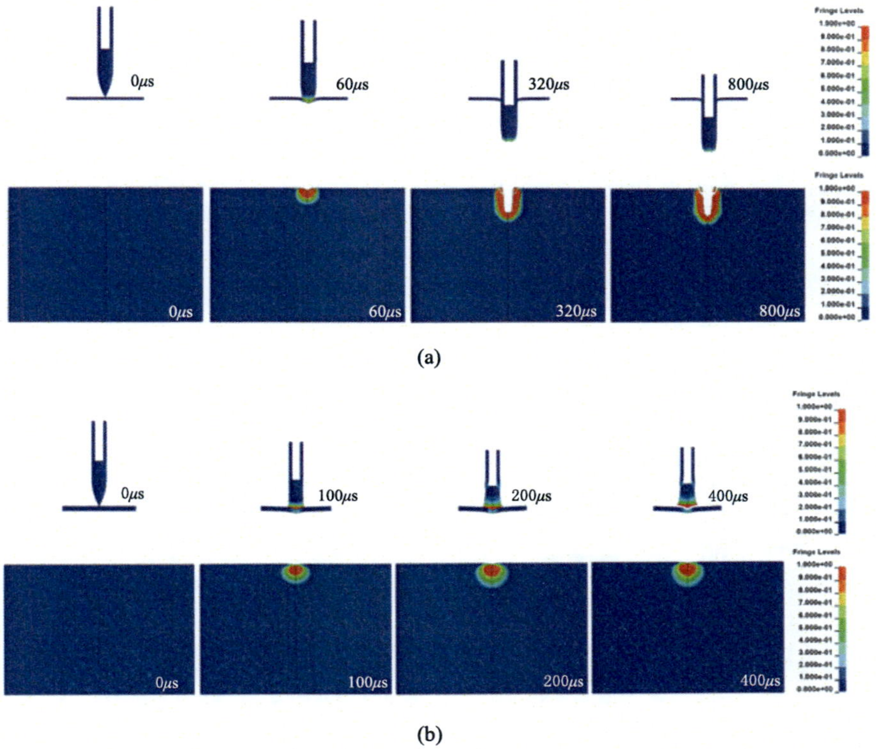

Fig. 7.48 Typical damage processes of **a** NP450_5mm/UHPCC composite target, 474 m/s **b** NP500_10mm/UHPCC composite target, 422 m/s

Fig. 7.49 Velocity–time histories of the projectiles for UHPCC, NP450_5mm/UHPCC and NP500_10mm/UHPCC composite targets

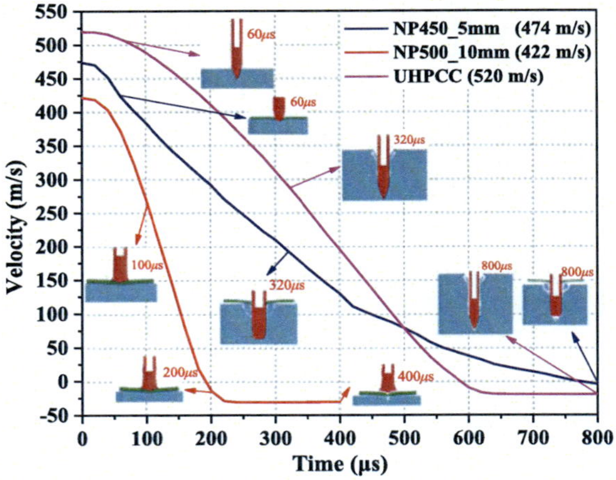

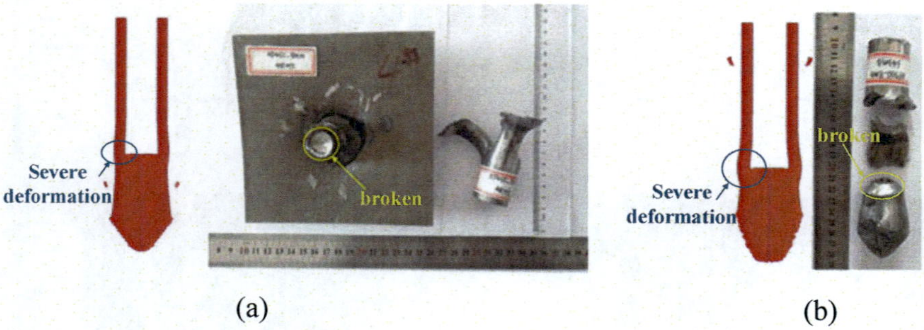

Fig. 7.50 Experimental and simulated broken projectiles for **a** NP450_8mm/UHPCC and **b** NP500_8mm/UHPCC composite targets

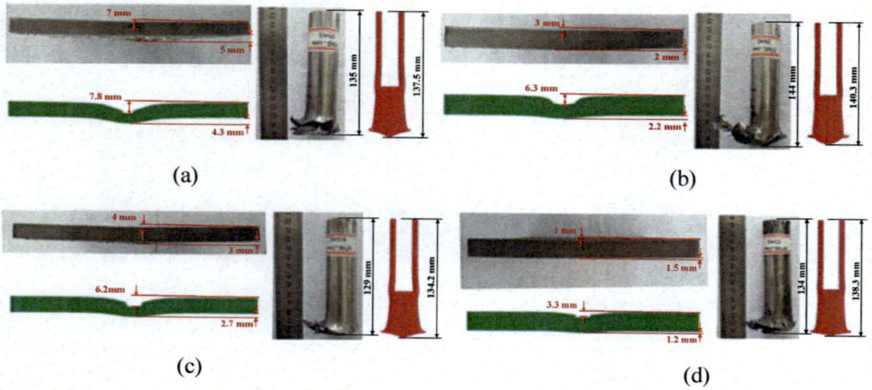

Fig. 7.51 Comparisons between numerical and experiment results **a** NP450_10mm/UHPCC **b** NP450_13mm/UHPCC **c** NP500_10mm/UHPCC **d** NP500_13mm/UHPCC

Finally, the comparisons between the simulated and experimental results of DOP and L_r are recorded in Table 7.15, and good agreements are obtained. "/" denotes that the deviations between experimental and simulated results cannot be derived due to the projectiles fractured in composite targets or projectile scratched slightly in UHPCC target. It should be noted that the composite targets with 10 mm and 13 mm armor steels are not listed in Table 7.15 since those armor steel plates were not penetrated.

(2) Ballistic characterization

The mass efficiency factor E_t, thickness efficiency factor E_{m}, and ballistic efficiency factor q^2 are commonly used to assess the impact resistance of the composite target (Madhu et al. 2005), which were defined in Eqs. (7.1, 7.2 and 7.9). Under various striking velocities, the ballistic performance of armor steel/UHPCC composite target is calculated by comparing the DOP of the projectile in a semi-infinite UHPCC

Table 7.15 Comparisons between simulated and experimental results of DOP and L_r

Target	V_0 (m/s)	DOP (mm)			L_r (mm)		
		Experiment	Simulation	Deviation (%)	Experiment	Simulation	Deviation (%)
NP450_5mm	474	126	120.9	−4.0	168	154.1	−8.3
NP450_5mm	581	159	162.6	2.3	165	151.9	−7.9
NP450_5mm	639	175	195.5	11.7	169	149.5	−11.5
NP450_8mm	481	32	30.3	−5.3	*	141.1	/
NP500_5mm	484	81	83.1	2.6	155	144.2	−7.0
NP500_5mm	584	130	135.2	−4.0	160	144.7	−9.6
NP500_5mm	646	166	164.3	1.0	162	144.1	−11.0
NP500_8mm	643	71	67.6	−4.8	*	130.0	/
UHPCC	520	181	174.1	−3.8	180	180	/

target (P_{REF}) to the residual penetration in the UHPCC when backing against armor steel plates with various thicknesses (P_{RES}). The DOPs in semi-infinite UHPCC target at different striking velocities (P_{REF}) are obtained by the numerical simulations. In Table 7.16, the bulge on rear face of armor steel plate was adopted as the P_{RES} in the UHPCC for shots in which the front steel plate not penetrated, i.e., NP450_10mm/UHPCC, NP450_13mm/UHPCC, NP500_10mm/UHPCC, and NP500_ 13 mm/UHPCC composite targets. T_t, ρ_{REF}, and ρ_t in Eqs. (7.1, 7.2) are the thickness of armor steel plate, densities of UHPCC and armor steel, respectively.

$$q^2 = E_m \times E_t \tag{7.9}$$

Table 7.16 Ballistic characterization results

Target	V_0 (m/s)	P_{REF} (mm)	P_{RES} (mm)	E_t	E_m	q^2
NP450_5mm	474	155.5	121	6.90	2.22	15.32
NP450_5mm	581	200.8	154	9.36	3.01	28.20
NP450_5mm	639	228.1	170	11.62	3.74	43.46
NP450_8mm	481	158.7	24	16.84	5.42	91.28
NP450_10mm	399	129.7	5	12.47	4.01	50.05
NP450_13mm	394	126.6	2	9.58	3.08	29.54
NP500_5mm	484	160.7	76	16.94	5.45	92.37
NP500_5mm	584	202.8	125	15.56	5.01	77.93
NP500_5mm	646	231.7	161	14.14	4.55	64.36
NP500_8mm	643	231.6	63	21.08	6.79	143.03
NP500_10mm	422	137.2	3	13.42	4.32	57.97
NP500_13mm	372	119.6	1.5	9.08	2.92	26.54

The ballistic efficiency factor has a great significance for designers since it connects both the mass and thickness efficiency factors. Values of 1.0 or above for E_t and E_m indicate that the armor steel under evaluation is thinner and lighter than the equivalent reference material, i.e., UHPCC at present, under the identical specified threat. Therefore, it can be drawn that, the armor steel/UHPCC in the protective structures of civil engineering can more effectively reduce the weight or improve the impact resistance of the structures.

In view of the fact that the composite target with 8 mm-thick armor steel plate obtained the maximum ballistic efficiency factor, the NP500_8mm/UHPCC composite target with a striking velocity of 481 m/s is further simulated to quantitatively compare the ballistic performance of the current two kinds of composite targets. The simulated P_{RES} of NP500_8mm/UHPCC composite target (481 m/s) is 6 mm, which is only 25% of that of NP450_8mm/UHPCC composite target. In this scenario, the calculated E_t, E_m, and q^2 are 19.09, 6.14 and 117.79 while those are 16.84, 5.42 and 91.28 for NP450_8mm/UHPCC composite target, respectively.

7.4 Summary

The medium caliber projectile impact resistance of armor steel/ceramic/UHPCC layered composite targets against 30CrMnSiNi2A steel projectiles are studied both experimentally and numerically. By calibrating the JC constitutive model parameters of the present 10CrNi3MoV21A, NP450 and NP500 armor steel, the numerically predicted penetration depth, residual length of the projectile, as well as the abrasion and damage of projectiles and armor steel plates were compared with the experimental observations. The main conclusions are summarized as follows:

(1) Compared with the NSC target, UHPCC and the present 10CrNi3MoV21A armor steel/SiC ceramic/UHPCC composite targets were experimentally and numerically verified to be effective in resisting the projectile penetration. For example, up to about 77.4% of the DOP into the NSC target was reduced by the 11 mm SiC ceramic/UHPCC composite target.

(2) Rigid and eroding penetration regimes were revealed experimentally, and the most severe abrasion of projectile occurred in scenario C11. Meanwhile, with almost the same thickness, SiC ceramic plate was verified to possess superior impact resistance to the 10CrNi3MoV21A armor steel plate considering both thickness and mass efficiency factors.

(3) The armor steel/ceramic/UHPCC composite structure was verified to be a promising protective configuration with excellent ballistic performance including high impact resistance, light weight and low thermal conductivity.

(4) NP450/UHPCC and NP500/UHPCC composite targets have a much more outstanding impact resistance regarding DOP and crater diameter than UHPCC target, and NP500/UHPCC composite target exhibits the best impact resistance.

(5) For two types armor steel/UHPCC composite targets, the mass efficiency factor E_t, thickness efficiency factor E_m, and ballistic efficiency factor q^2 of composite target are the largest when the thickness of armor steel plate is 8 mm. For instance, for the projectile with a striking velocity of 481 m/s, the residual penetration depths in UHPCC when backing against 8 mm-thick NP4500 and NP500 armor steel plate are 3.78% and 15.12% of DOP in the UHPCC target, respectively.

References

BENDOR G, DUBINSKY A, ELPERIN T, FRAGE N. Optimization of two component ceramic armor for a given impact velocity[J]. Theoretical and Applied Fracture Mechanics, 2000, 33(3): 185-190.

BENSON D J. An efficient, accurate, simple ALE method for nonlinear finite element programs[J]. Computer Methods in Applied Mechanics and Engineering, 1989, 72(3): 305-350.

CHOUDHARY S, SINGH P K, KHARE S, KUMAR K, VERMA R K. Ballistic impact behaviour of newly developed armour grade steel: An experimental and numerical study[J]. International Journal of Impact Engineering, 2020, 103557.

CRONIN D S, BUI K, KAUFMANN C, BERSTAD T. Implementation and Validation of the Johnson-Holmquist Ceramic Material Model in LS-DYNA. 4th Europe LS-DYNA Users Conference[C]. 2003.

FAWAZ Z, BEHDINAN K, XU Y G. Optimum design of two-component composite armors against high-speed impact[J]. Composite Structures, 2006, 73(3): 253-262.

FENG J, SUN W W, LIU Z L, CUI C, WANG X M. An armor-piercing projectile penetration in a double-layered target of ultrahigh-performance fiber reinforced concrete and armor steel: Experimental and numerical analyses[J]. Materials & Design, 2016, 102: 131-141.

FORRESTAL M J, ALTMAN B S, CARGILE J D, HANCHAK S J. An empirical equation for penetration depth of ogive-nose projectiles into concrete targets[J]. International Journal of Impact Engineering, 1994, 15(4): 395-405.

FORRESTAL M J, FREW D J, HANCHAK S J, BRAR N S. Penetration of grout and concrete targets with ogive-nose steel projectiles[J]. International Journal of Impact Engineering, 1996, 18(5): 465-476.

FORRESTAL M J, FREW D J, HICKERSON J P, ROHWER T A. Penetration of concrete targets with deceleration-time measurements[J]. International Journal of Impact Engineering, 2003, 28(5): 479-497.

FRAS T, ROTH C C, MOHR D. Fracture of high-strength armor steel under impact loading[J]. International Journal of Impact Engineering, 2018, 111: 147-164.

FRAS T, ROTH C C, MOHR D. Dynamic perforation of ultra-hard high-strength armor steel: Impact experiments and modeling[J]. International Journal of Impact Engineering, 2019, 131: 256-271.

HANCHAK S J, FORRESTAL M J, YOUNG E R, EHRGOTT J Q. Perforation of concrete slabs with 48 MPa (7 ksi) and 140 MPa (20 ksi) unconfined compressive strengths[J]. International Journal of Impact Engineering, 1992, 12(1): 1-7.

HOLMQUIST T J, JOHNSON G R, COOK W H. A computational constitutive model for concrete subjected to large strains, high strain rates, and high pressures. 14th International Symposium on Ballistics[C]. 1993.

JOHNSON G R, COOK W H. A constitutive model and data for metals subjected to large strains, high strain rates, and high temperatures. 7th International Symposium on Ballistics[C]. 1983.

JOHNSON G R, HOLMQUIST T J. An improved computational constitutive model for brittle materials. AIP Conference, American Institute of Physics AIP[C]. 1994. 309(1): 981–984

KEELE M, RAPACKI E J, BRUCHEY W J. High velocity ballistic performance of a uranium alloy long rod penetrator. 12th International Symposium on Ballistics[C]. 1990.

LI L, ZHANG X F, WU X, GAO F, LIU C. Dynamic constitutive and damage parameters of 30CrMnSiNi2A steel with different hardness[J]. Chinese Journal of High Pressure Physics, 2017, 31(3): 239-248. (in Chinese)

LIU J, WU C Q, CHEN X W. Numerical study of ultra-high performance concrete under non-deformable projectile penetration[J]. Construction and Building Materials, 2017, 135: 447-458.

LIU J, WU C Q, SU Y, LI J, SHAO R Z, CHEN G, LIU Z X. Experimental and numerical studies of ultra-high performance concrete targets against high-velocity projectile impacts[J]. Engineering Structures, 2018, 173: 166-179.

LSTC, LS-DYNA Keywords User's Manual[M]. Livermore Software Technology Corporation, 2001.

LUNDBERG P, RENSTRÖM R, LUNDBERG B. Impact of metallic projectiles on ceramic targets: transition between interface defeat and penetration[J]. International Journal of Impact Engineering, 2000, 24(3): 259-275.

LUNDBERG P, LUNDBERG B. Transition between interface defeat and penetration for tungsten projectiles and four silicon carbide materials[J]. International Journal of Impact Engineering, 2005, 31(7): 781-792.

MÁCA P, SOVJÁK R, KONVALINKA P. Mix design of UHPFRC and its response to projectile impact[J]. International Journal of Impact Engineering, 2014, 63: 158-163.

MADHU V, RAMANJANEYULU K, BHAT T B, GUPTA N K. An experimental study of penetration resistance of ceramic armor subjected to projectile impact[J]. International Journal of Impact Engineering, 2005, 32(1): 337-350.

MAGNESS L S, FARRAND T G. Deformation behavior and its relationship to the penetration performance of high-density KE penetrator materials. 1990 Army Science Conference Durham, NC[C]. 1990.

REAUGH J E, HOLT A C, WILKINS M L, CUNNINGHAM B J, HORD B L, KUSUBOV A S. Impact studies of five ceramic materials and pyres[J]. International Journal of Impact Engineering, 1999, 23: 771-782.

REN G M, WU H, FANG Q, ZHOU J W, GONG Z M. Determinations of HJC constitutive model parameters for normal strength concrete[J]. Journal of Vibration and Shock, 2016, 35(18): 9-16. (in Chinese)

REN G M, WU H, FANG Q, KONG X Z. Parameters of Holmquist–Johnson–Cook model for high-strength concrete-like materials under projectile impact[J]. International Journal Protection Structures, 2017, 8(3): 352-367.

ROSENBERG Z, DEKEL E. Numerical study of the transition from rigid to eroding-rod penetration[J]. Journal De Physique IV, 2003, 110: 681-686.

SERJOUEI A, CHI R Q, SRIDHAR I, TAN G E B. Empirical ballistic limit velocity model for bi-layer ceramic-metal armor[J]. International Journal Protection Structures, 2015, 6(3): 509-527.

SHAO R Z, WU C Q, SU Y, LIU Z X, LIU J, CHEN G, XU S C. Experimental and numerical investigations of penetration resistance of ultra-high strength concrete protected with ceramic balls subjected to projectile impact[J]. Ceramics International, 2019, 45(6): 7961-7975.

SHI Z Y, TANG W H, WU K G, ZHANG R Q. Experimental study of the penetration into the steel fiber reinforced concrete targets within armor plates and ceramic plates[J]. Journal of Experimental Mechanics, 2006, 21(2): 172-176. (in Chinese)

SILSBY G F. Penetration of semi-infinite steel targets by tungsten rods at 1.3 to 4.5 km/s. 8th International Symposium on Ballistics[C]. 1984.

SOVJÁK R, VAVŘINÍK T, MÁCA P, ZATLOUKAL J, KONVALINKA P, SONG Y P. Experimental investigation of ultra-high performance fiber reinforced concrete slabs subjected to deformable projectile impact[J]. Procedia Engineering, 2013, 65: 120-125.

SOVJÁK R, VAVŘINÍK T, ZATLOUKAL J, MACA P, MICUNEK T, FRYDRYN M. Resistance of slim UHPFRC targets to projectile impact using in-service bullets[J]. International Journal of Impact Engineering, 2015, 76: 166-177.

STEINBERG D J. Equation of state and strength properties of selected materials[R]. Livermore: Lawrence Livermore Natl Lab, 1996.

WOOLSEY P, MARIANO S A, KOKIDKO D. Alternative test methodology for ballistic performance ranking of armor ceramics. 5th Annu TACOM Armor Coord Conf Monterey, CA[C]. 1989a.

WOOLSEY P, KOKIDKO D, MARIANO S A. Alternative test methodology for ballistic performance ranking of armor ceramics[R], Watertown, Massachusetts, U.S: US Army Laboratory Command, 1989b. MTL TR 89–43.

WOOLSEY P, KOKIDKO D, MARIANO S A. Progress report on ballistic test methodology for armor ceramics. TACOM Combat Vehicle Survivability Symposium, Gaithersburg, MD, US[C]. 1990.

WU H, FANG Q, CHEN X W, GONG Z M, LIU J Z. Projectile penetration of ultra-high performance cement based composites at 510–1320 m/s[J]. Construction and Building Materials, 2015a, 74: 188-200.

WU H, FANG Q, GONG J, LIU J Z, ZHANG J H, GONG Z M. Projectile impact resistance of corundum aggregated UHP-SFRC[J]. International Journal of Impact Engineering, 2015b, 84: 38-53.

YANG L, LIN X S, LI H Y, GRAVINA R J. A new constitutive model for steel fibre reinforced concrete subjected to dynamic loads[J]. Composite Structures, 2019, 221: 110849.

Chapter 8
Response of UHPCC-FST Subjected to Low-Velocity Impact

8.1 Introduction

Concrete filled steel tube (CFST) has been extensively applied as the load-bearing members in critical infrastructures, e.g., the high-rise buildings, bridges and electricity transmission towers due to their enhanced compressive and flexural performances, improved ductility and economic advantage (Zhao et al. 2010; Uy 1998). One important load condition that must be considered in some structural design situations is the impact load which can cause severe damage to structures. For example, bridge piers located close to a navigable waterway may be vulnerable to vessel collisions (Sha and Hao 2013), bridge piers in mountain areas may be hit by falling rocks (Lu and Zhang 2012) and the viaduct piers may be impacted by heavy vehicles (Thilakarathna et al. 2010). Once the above CFST members are damaged, it may lead to serious consequences of infrastructure collapse. Several design codes offer general rules for the design of structures under impact load. Eurocode 1 (Draft prEN 1991-1-7, 2005) provides the formulas for calculating an equivalent static force on structures in an impact event based on the type of traffic (i.e., vehicle or ship). However, the real impact process is ignored in the current design guidance. For more precise design purposes, the impact performance of specific structural type should be studied in detail. Hence, it is of great importance to evaluate the performance of CFST subjected to transverse impact.

Most existing works (Madurapperuma and Wijeyewickrema 2013; Adhikary et al. 1997; Zhan et al. 2015; Yoo et al. 2016) are focused on the reinforced concrete (RC) column, which showed that the RC columns may suffer localized damage (e.g., spalling of concrete cover and diagonal shear cracks) or global damage (e.g., flexure). Comparably, the outer steel tube of CFST column could effectively prevent the spall damage of concrete and reducing the localized damage of the CFST column under transverse impact load. Bambach et al. (2008) performed a series of drop hammer tests to study the fully clamped hollow and concrete (compressive strength of 88.4 MPa) filled steel square beams subjected to low-velocity and large mass

Q. Fang et al., *UHPCC Under Impact and Blast*,
https://doi.org/10.1007/978-981-33-6842-2_8

transverse impacts. It was shown that the concrete filled steel square beams demonstrate significant reduction in transverse failure deflection and much superior energy absorption capacity compared to the hollow steel square beams. Remennikov et al. (2011) experimentally and numerically (using the program LS-DYNA) investigated the impact behavior of square hollow section tubes filled with the rigid polyurethane foam (RPF) and the normal weight concrete (compressive strength of 41 MPa). It was indicated that the square concrete-filled tube has the highest impact resistance and energy absorption capacity, followed by the square RPF-filled tube and then the square hollow tube. Deng et al. (2012) conducted drop hammer impact test on simply supported circular CFST, circular post-tensioned concrete-filled steel tube (PTCFST) and circular steel fiber reinforced concrete-filled steel tube (FRCFST) specimens with the compressive strength of concrete less than 60 MPa. It was derived that, the PTCFST and the FRCFST specimens demonstrate superior impact resistance than the regular CFST specimens, and using the pre-stressed strands and steel fibers can effectively restrain the cracks in the tension area of the core concrete. Yousuf et al. (2012) performed an experimental study to investigate the behavior of hollow and concrete-filled mild steel tubular columns under static and impact loading. The dynamic moment capacity of hollow and concrete filled columns under impact loading indicated a significant increase, compared to the capacity associated with static loading. Han et al. (2014) experimentally and numerically (using the program ABAQUS) investigated the response of high strength concrete (compressive strength up to 75 MPa) filled steel tubular specimens subjected to transverse impact by conducting the drop hammer test. The results showed that the CFST specimens deform in a ductile manner and have good resistance under transverse impact load, and the ductile damage model provided by ABAQUS/Explicit (Karlsson Sorensen Inc 2005) package can accurately predict the impact behavior of CFST specimens. Wang et al. (2013, 2015, 2016) performed an investigation into the impact behavior of CFST members and FRP-concrete-steel double skin tubular members (compressive strength of concrete is 48.7 ~ 60.2 MPa). The influences of the axial load level, confinement factor (related to the yield strength of steel tube and compressive strength of core concrete, as well as the cross-sectional area of both) and hollow ratio of section were systematically discussed. The results showed that the axial load level has a significant effect on the lateral deflections and the impact force–time histories of CFST members, and the failure mode of specimens changed from brittle manner to ductile manner with increasing the confinement factor. The main influencing factor of peak impact force was hollow ratio of section for the FRP-concrete-steel double skin tubular members. Furthermore, a finite element (FE) analysis model was developed by using ABAQUS and verified. Du et al. (2018) experimentally and numerically studied on the behavior of concrete (compressive strength of 31.2 MPa) filled circular steel tubular members under lateral impact loading. The results showed that, with increasing the outer diameter or the thickness of steel tube, the flexural deformation and mid-span deflection of specimens became less obviously but the peak impact force increased. Then a three-dimensional FE model was established by using ABAQUS, and the accuracy of the proposed model was verified by comparing with the test data.

The existing works are mainly concentrated on the normal strength concrete (NSC) and high-strength concrete (HSC) filled steel tubes (Bambach et al. 2008; Remennikov et al. 2011; Deng et al. 2012; Yousuff et al. 2012; Han et al. 2014; Wang et al. 2013, 2015, 2016; Du et al. 2018), relatively limited experimental and numerical simulation works were conducted to investigate the structural performance and impact resistance of UHPCC-FSTs subjected to transverse impact.

At present, a comprehensive study on the impact behavior of circular UHPCC-FST under transverse impact load is conducted. Firstly, the axial compression test was conducted to analyze the axial load capacities and obtain the constraining factor of UHPCC-FSTs. Secondly, three UHPCC-FSTs are tested horizontally by using a drop-hammer impact device, and the dynamic response of the specimens under transverse impact load, i.e., the impact force- and deflection-time histories, are recorded and discussed. Thirdly, the plasticity concrete material model (K&C model) (Malvar et al. 1997) is adopted and the constitutive model parameters for UHPCC are calibrated based on a series of test data. Finally, by using the calibrated K&C model, a FE analysis model is established to predict the dynamic behavior of UHPCC-FSTs under transverse impact load.

8.2 Test Program

8.2.1 UHPCC-FST Specimens

UHPCC is prepared in the State Key Laboratory of High Performance Civil Engineering Materials, Jiangsu Research Institute of Building Science in China, which is used as the core concrete and the mixture proportions are given in Table 8.1. It should be pointed out that, the above mix proportion is identical with that used in the previous works of the authors (Ren et al. 2016, 2018a).

Three cementitious materials are used at present, including Chinese standard Graded 52.5 P.II type Portland cement (20 ~ 30 μm in particle diameter), silica fume (particle size of 0.10 ~ 0.26 μm, density of 2.1 g/cm^3, specific surface area of 20500m^2/kg) and ultra-fine mineral admixture (density of 2.45 g/cm^3, specific surface area of 8500m^2/kg, 28d activity index of 115%). The chemical compositions of cement and silica fume are listed in Table 8.2. The water-to- cementitious materials ratio and HRWR-to-cementitious materials ratio are 0.16 and 0.024, respectively. The water-reducing ratio of the PCA®-I polycarboxylic type HRWR is no less than 35%, which is developed by Jiangsu SOBUTE new material Co., Ltd.. The

Table 8.1 Mixture proportions of UHPCC (kg/m^3)

Cement	Silica fume	Ultra-fine mineral admixture	Sand	Water	HRWR	Fiber content
700	140	110	1200	152	22.8	156

Table 8.2 Chemical composition of cement and silica fume (%)

Composition	CaO	SiO	MgO	Fe_2O_3	Al_2O_3	SO_3	LOI
Cement	61.80	20.00	5.31	3.46	4.64	3.15	2.44
Silica fume	1.00	93.10	0.65	2.44	0.76	0.12	1.04

Table 8.3 Material properties of steel fiber

Fiber type	Length (mm)	Diameter (mm)	Aspect ratio	Density (g/cm^3)	Tensile strength (MPa)	Elastic modulus (GPa)
Micro-straight	13	0.2	65	7.8	2800	210

micro-straight steel fiber is incorporated with the volume fraction of 2.0% and the corresponding material properties are given in Table 8.3. The slump and spread of the present UHPCC were about 230 mm and 610 mm respectively, which guarantees the workability requirements of on-site constructions.

The control specimens of UHPCC including three cubes of 100 mm × 100 mm × 100 mm and three prisms of 100 mm × 100 mm × 300 mm were cast and tested according to the Chinese standard GB/T 50081–2002 (2003). The average quasi-static compressive strength, elastic modulus and Poisson's ratio of UHPCC are 141.5 MPa, 43.8GPa and 0.23, respectively.

The outer steel tubes are made from 20# steel, and the material properties of which are determined from the standard static tensile coupon test. The coupons were cut from the longitudinal direction of the tube and a total of five standard coupons were tested according to the Chinese standard GB/T 228.1–2010 (2003). Figure 8.1 illustrates the experimental tensile stress–strain curves of 20# steel coupons 1 ~ 5. The engineering yield stress, ultimate tensile stress and rupture strain of the coupons can be derived and listed in Table 8.4. Besides, the elastic modulus and Poisson's ratio of 20# steel are taken as 206GPa and 0.3 according to the Chinese standard GB 50,017–2017 (2017).

Five large-sized circular UHPCC-FSTs were fabricated. As shown in Fig. 8.2, the thickness of steel tube is 6 mm, the length and outer diameters of the specimen are 2000 mm and 203 mm, respectively. A 20 mm thick and 400 mm × 400 mm sized square steel plate was welded on the bottom of the specimen to guarantee the preparation and stability of the column.

8.2.2 *Axial Compression Test*

As shown in Fig. 8.3, the axial compression test was conducted by using a 1000t hydraulic testing machine and two UHPCC-FSTs were tested to analyze their axial load bearing capacities. The load was applied at the top of each specimen with the

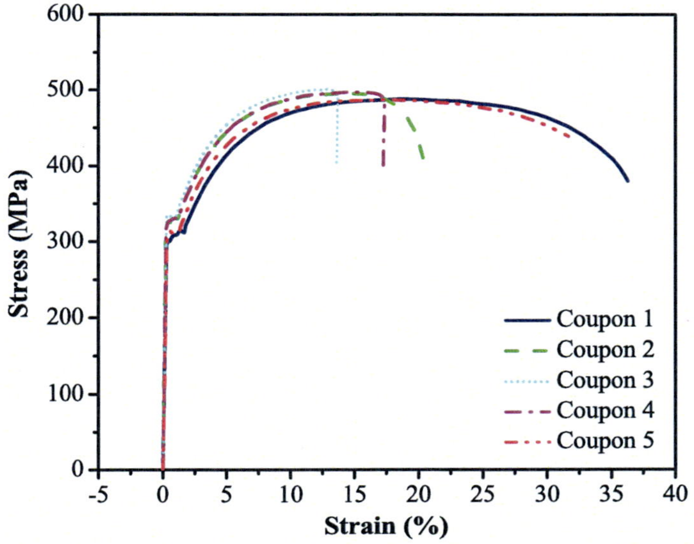

Fig. 8.1 Tensile stress–strain curves of 20# steel, reprinted from Wu et al. (2019), copyright 2020, with permission from Elsevier

Table 8.4 Material properties of 20# steel

Coupons	Width × Thickness (mm)	Yield stress (MPa)	Ultimate stress (MPa)	Rupture strain
1	9.81 × 3.04	302.6	503.6	0.14
2	9.85 × 3.02	318.1	508.5	0.17
3	9.85 × 3.02	338.3	511.8	0.17
4	9.87 × 3.03	320.2	509.9	0.31
5	9.86 × 3.03	313.4	501.6	0.35
Average		318.5	507.1	0.23

displacement rate of 1 mm/min. Several Linear variable differential transformers (LVDTs) were employed in the axial compression test. The middle lateral deformation of each specimen was measured by one LVDT with the measuring capacity of 500 mm installed at the middle height of the column. Four LVDTs with the measuring capacity of 100 mm were symmetrically arranged around the top of the column to obtain the vertical displacement at the top of the column, while two LVDTs with the maximum range of 50 mm were symmetrically placed at the steel bottom plate to obtain the vertical displacement of the bottom plate. Hence, the net vertical deformation of the specimen could be obtained by subtracting the displacement of the bottom plate from the top displacement.

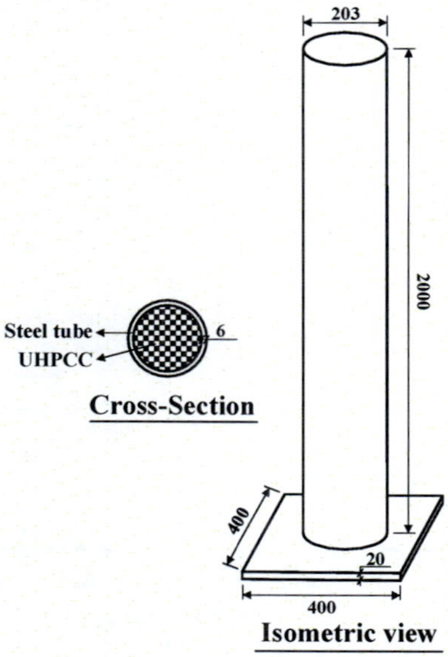

Fig. 8.2 Geometry of the UHPCC-FST (*unit* mm), reprinted from Wu et al. (2019), copyright 2020, with permission from Elsevier

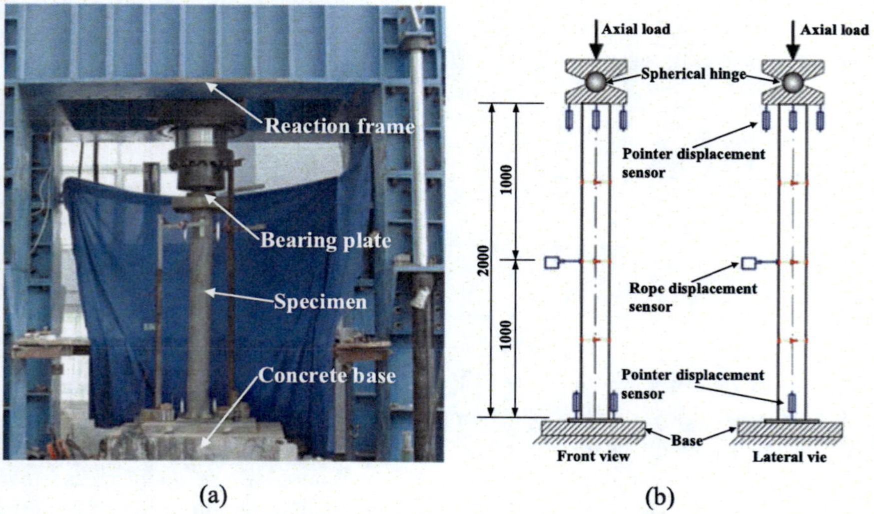

Fig. 8.3 Axial compression test setup **a** 1000t hydraulic testing machine **b** displacement meter layout (unit: mm), reprinted from Wu et al. (2019), copyright 2020, with permission from Elsevier

8.2.3 Drop-Hammer Impact Test

The drop-hammer impact test was conducted by using a drop-hammer impact device, as shown in Fig. 8.4. In this test, the release heights of the drop hammer were 5.0, 8.0 and 11.0 m, and the drop hammer fell as a free-falling body along the guide rails once the electromagnet release mechanism was manually triggered. The total mass of the drop hammer is 530 kg, which is composed of the drop hammer frame (200 kg), weight blocks (6×50 kg) and indenter (30 kg), as shown in Fig. 8.5. The indenter is a cylindrical rigid body with the thickness of 8 cm and the diameter of 20 cm. In the test, the specimen was pinned supported by the supporting frames with the clear span length of 1600 mm. Four linear variable differential transformers (LVDTs) were used

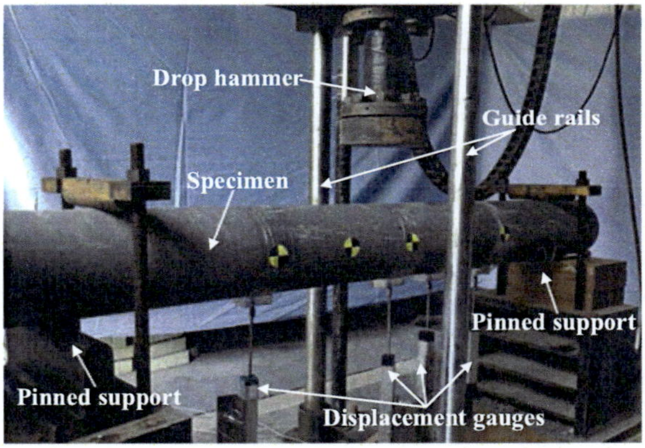

Fig. 8.4 Drop-hammer impact device, reprinted from Wu et al. (2019), copyright 2020, with permission from Elsevier

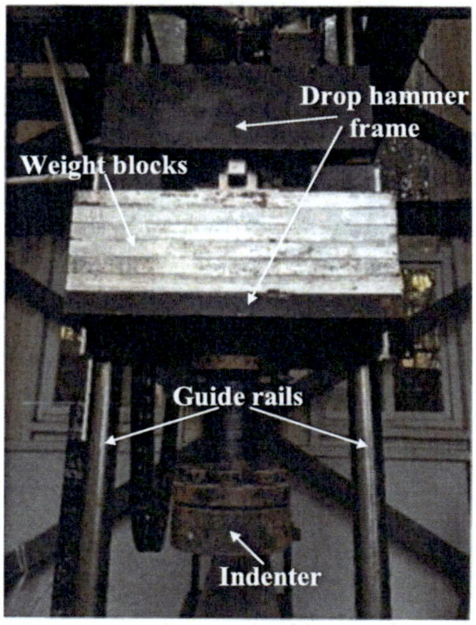

Fig. 8.5 Details of drop hammer, reprinted from Wu et al. (2019), copyright 2020, with permission from Elsevier

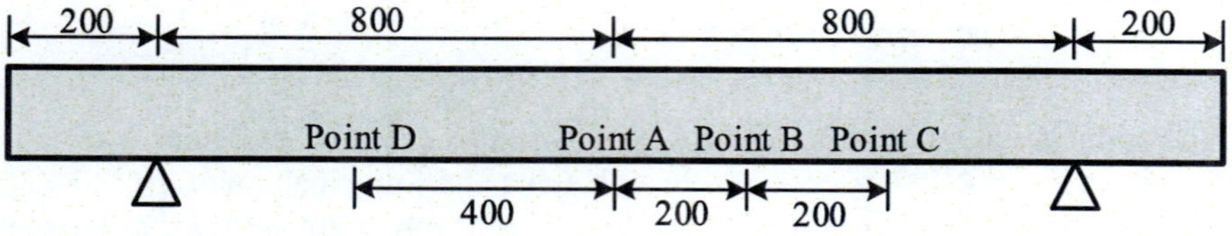

Fig. 8.6 Specific locations of measurement points (unit: mm), reprinted from Wu et al. (2019), copyright 2020, with permission from Elsevier

for measuring the deflections of the specimens, the corresponding specific locations of measurement points are illustrated in Fig. 8.6.

8.3 Test Results and Discussions

8.3.1 Axial Compression

8.3.1.1 Axial Load Bearing Capacity

Figure 8.7 shows the axial load-axial displacement and -lateral displacement curves of the specimens, all of which include the linear ascending stage, short platform stage and fast descending stage. As can be seen, the difference in the peak load is quite small and the average peak load is 4650 kN, suggesting that the axial load capacity of the specimens was consistent. A certain deviation of initial ascending stage exists in Fig. 8.7b, which probably attributes to the initial eccentricity and

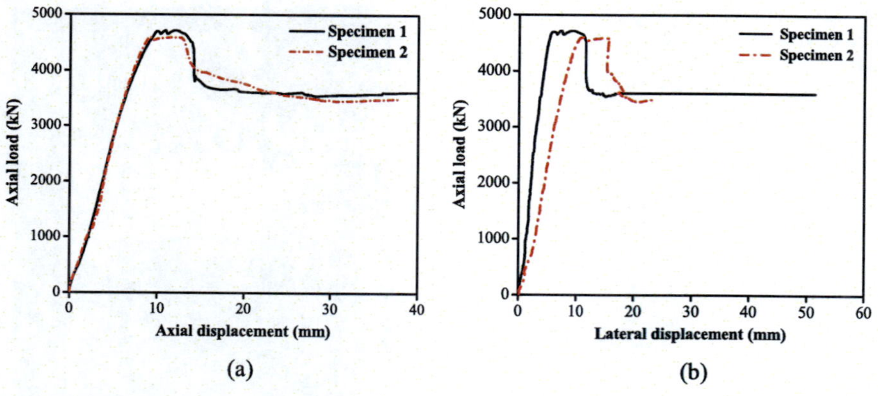

Fig. 8.7 Axial compressive test results **a** Axial load-axial displacement curves **b** Axial load-lateral displacement curves, reprinted from Wu et al. (2019), copyright 2020, with permission from Elsevier

material inhomogeneity of the specimens. However, in terms of theirs axial load capacities, such discrepancy can be neglected.

8.3.1.2 Constraining Factor

The confining effect is essential for the CFST members under different loading conditions. The constraining factor ξ proposed by the Chinese Standard GB50936 (2014) is used to describe the composite action between the steel tube and the filled concrete, which is expressed as

$$\xi = \frac{A_s f_y}{A_c f_c} = \alpha \frac{f_y}{f_c} \tag{8.1}$$

where A_s and A_c are the cross-sectional areas of the steel tube and core concrete, respectively. α_r is the steel ratio of CFST members, f_c is the axial compressive strength of core concrete and f_y is the yield strength of steel tube.

According to the Chinese Standard GB50936 (2014), the constraining factor for CFST members is suggested to be 0.5 ~ 2.0. For the present axial compressive strength (119.2 MPa) and prominent ductility of UHPCC, the corresponding constraining factor ξ is 0.35, which is smaller than the above recommended values. The reason may be that the maximum compressive strength of concrete suggested in the Chinese Standard GB50936 (2014) is 80 MPa, which is much smaller than the present UHPCC.

8.3.2 Lateral Impact Resistance

Figure 8.8 shows the recovered specimens after impact, in which "C-5", "C-8" and "C-11" are denoted as the test No., where the capital "C" represents the UHPCC filled steel tube; "5", "8" and "11" represent the release heights (unit: m) of drop hammer. It can be seen that all specimens displayed flexural deformation and the plastic hinges appeared in the impact area, and other parts of the specimens remained almost straight. Local buckling deformations are observed in the compression zone around the top of the mid-span section, but are not very noticeable. The specimens have excellent ductile deformation capacity due to the enhancement of the core UHPCC. The impact energy is dissipated effectively with the formation of plastic hinges, and there is no severe local buckling in the steel tube.

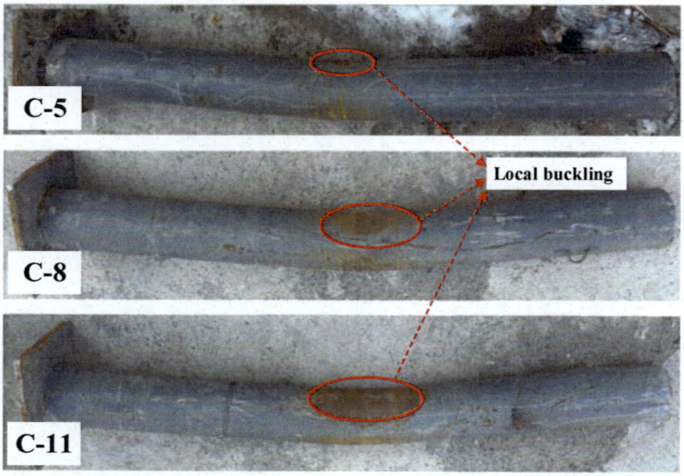

Fig. 8.8 Failure mode of the specimens, reprinted from Wu et al. (2019), copyright 2020, with permission from Elsevier

8.3.3 *Impact Force–Time History*

The most important data recorded during the impact procedure is the impact force–time history between the drop hammer and the specimen, as shown in Fig. 8.9. As can be seen, the peak impact force gradually increases with the rising of release

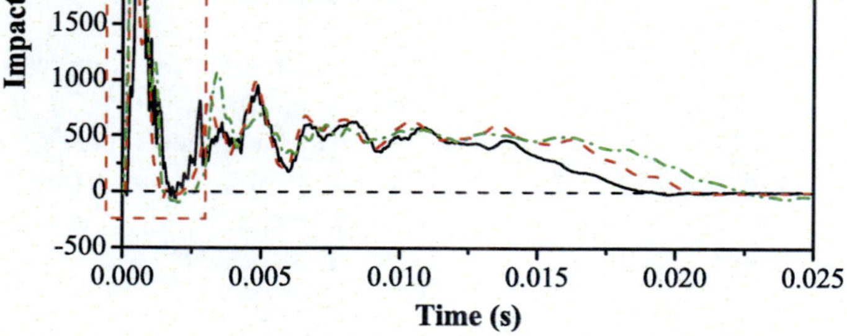

Fig. 8.9 Impact force–time history between the drop hammer and specimen, reprinted from Wu et al. (2019), copyright 2020, with permission from Elsevier

height, and the impact force versus time curves can be generally divided into three stages (i.e. the peak value stage, the plateau stage and the unloading stage). Firstly, the impact force increases sharply to a peak value when the hammer is in contact with the specimen, and then the specimen accelerates from a zero velocity to a speed approaching that of the drop hammer. This impact force could induce the drastic vibrations of the drop hammer, leading to the fast change of the contact area between the specimen and the drop hammer, reflected as fluctuations in the impact force history. The other reason for the fluctuations in the impact force is the sequential crushing and cracking of core UHPCC. Subsequently, the specimen and the drop hammer move downwards together and remain in contact, which reflects the actual structural capacity under the lateral impact. The average value of the impact force keeps almost constant for a relatively long period, which is regarded as the "plateau" stage. Finally, when the specimen reaches its maximum mid-span deflection, the specimen and the drop hammer start to rebound upwards, accompanied by an unloading on the specimen. The curve goes into the descending stage and the impact force gradually decreases to zero due to the complete separation of the specimen and the drop hammer.

8.3.4 Deflection-Time History

The deflection-time histories of specimens under three release heights are shown in Fig. 8.10. It can be seen that all curves show the following similar trends. The deflection increases rapidly after the impact, and the gradient of the curve decreases continuously with the impact energy is dissipated. The curve attains the maximum value when the velocity of the specimen declines to zero. Then, the specimen rebounds due to the release of elastic energy which leads to the deflection undergoes a descending stage. Finally, the specimens oscillate at a new equilibrium position and approach to a certain value.

Table 8.5 lists the experimental maximum and residual deflections of specimens, where E_K represents the impact energy and can be defined as:

$$E_K = mgH \tag{8.2}$$

where m and H are the mass and release height of drop hammer, respectively; g is the gravity acceleration and equal to 9.8 kg·m/s^2. "—" denotes that the data is not collected due to the accidental damage of LVDTs. As can be seen, both the maximum and residual deflections increase obviously with the increase of release heights.

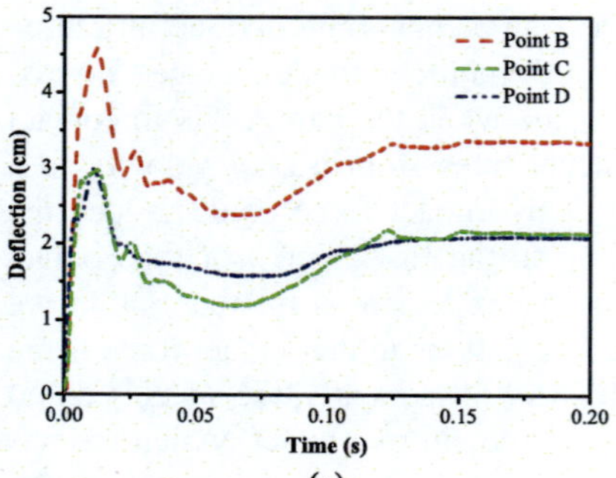

(a)

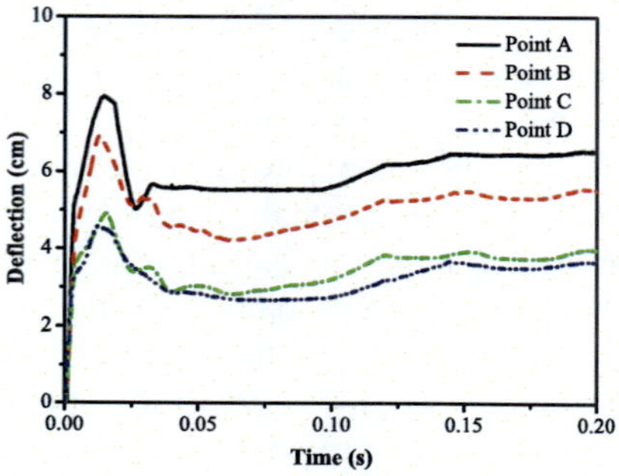

(b)

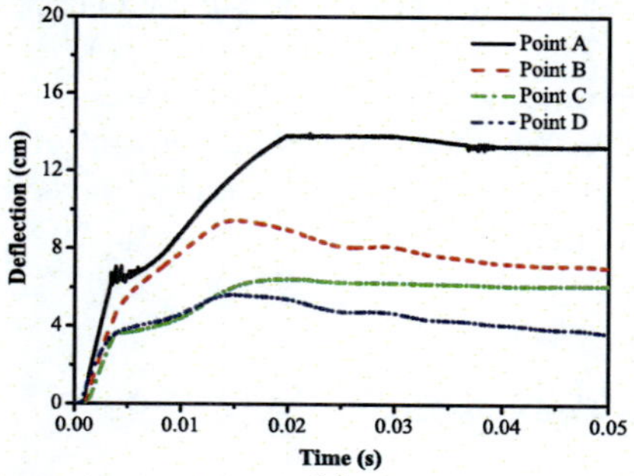

(c)

Fig. 8.10 Deflection-time histories of specimens under different release heights **a** C-5 **b** C-8 **c** C-11, reprinted from Wu et al. (2019), copyright 2020, with permission from Elsevier

Table 8.5 Maximum and residual deflections of the tested specimens

Test No.	H (m)	E_K (kJ)	Maximum deflection (cm)				Residual deflection (cm)			
			A	B	C	D	A	B	C	D
C-5	5.0	25.97	–	4.58	2.97	2.98	–	3.36	2.14	2.09
C-8	8.0	41.55	7.92	6.91	4.89	4.56	6.46	5.48	3.91	3.63
C-11	11.0	57.13	13.8	9.46	6.44	5.62	–	–	–	–

8.4 Calibration of K&C Model Parameters for UHPCC

The material model *MAT_CONCRETE_DAMAGE_REL3 (*MAT_72R3) named K&C model (Malvar et al. 1997, LS-DYNA Keyword User's Manual 2014) is used for UHPCC at present. K&C model is embedded in LS-DYNA and originally developed to describe the responses of concrete structures subjected to blast loads, and covers a number of important factors that are pertinent to the dynamic behaviors of concrete (Tu and Lu, 2009). The most significant user improvement provided by *MAT_72R3 is a model parameter auto-generated capability, based solely on the unconfined compression strength of the concrete. Guo et al. (2018) derived that, based on the automated generated parameters, K&C model has low accuracy in predicting the behavior of UHPCC under low-velocity impact. In this section, after brief introduction of K&C model, the constitutive model parameters of which for UHPCC are further calibrated by a series of static and dynamic tests of UHPCC mainly conducted by authors (Ren et al. 2016, 2018a, b; Wu et al. 2018).

8.4.1 Brief Introduction of K&C Model

Three independent strength surfaces of compressive meridians, namely, the initial yield strength surface $\Delta\sigma_y$, the maximum strength surface $\Delta\sigma_m$ and the residual strength surface $\Delta\sigma_r$ are applied in K&C model. The formations of above three surfaces can be expressed as (Malvar et al. 1997):

$$\Delta\sigma_y = \begin{cases} a_{0y} + p/(a_{1y} + a_{2y}p) & p \geq 0.15f_c \\ 1.35f_t + 3p/(1 - 3f_t/f_c) & 0 \leq p \leq 0.15f_c \\ 1.35(p + f_t) & p \leq 0 \end{cases} \tag{8.3}$$

$$\Delta\sigma_m = \begin{cases} a_0 + p/(a_1 + a_2p) & p \geq f_c/3 \\ 1.35/\psi(p + f_t) & 0 \leq p \leq f_c/3 \\ 3(p/\eta + f_t) & p \leq 0 \end{cases} \tag{8.4}$$

$$\Delta\sigma_r = a_{0f} + p/(a_{1f} + a_{2f}p) \tag{8.5}$$

where a_i, a_{iy} and a_{if} $(i = 0, 1, 2)$ are strength constants, which can be determined by suitable triaxial compression data. f_c and f_t are the unconfined uniaxial compressive and tensile strengths, respectively. p is the pressure and ψ denotes the tensile-to-compressive meridian ratio. Tu and Lu (2009) introduced the advantages of above strength surfaces represented by a segmented form in detail, especially at $p \leq f_c/3$. Figure 8.11 shows the typical normalized strength surfaces.

Current failure surface γ is determined as follows:

$$\gamma = \sqrt{3J_2} = \begin{cases} r'[\eta(\Delta\sigma_m - \Delta\sigma_y) + \Delta\sigma_y] & \text{Strain hardening} \\ r'[\eta(\Delta\sigma_m - \Delta\sigma_r) + \Delta\sigma_r] & \text{Strain softening} \end{cases} \tag{8.6}$$

where J_2 is the second deviatoric stress invariant. r' is the ratio of the current meridian to the compressive meridian. η is the yield scale factor referring to the damage function λ, which is defined as

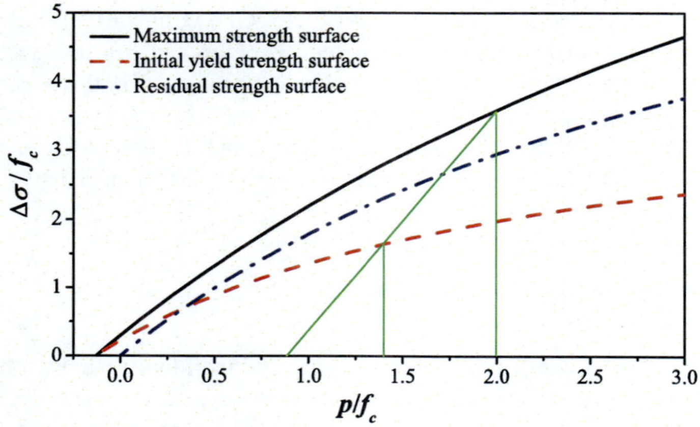

Fig. 8.11 Normalized independent strength surfaces, reprinted from Wu et al. (2019), copyright 2020, with permission from Elsevier

$$\lambda = \begin{cases} \int_0^t d\bar{\varepsilon}_p/(1+p/f_t)^{b1} & p > 0 \\ \int_0^t d\bar{\varepsilon}_p/(1+p/f_t)^{b2} & p \leq 0 \end{cases} \tag{8.7}$$

where $d\bar{\varepsilon}_p = \sqrt{2/3 d\varepsilon_{ij}^p d\varepsilon_{ij}^p}$ is the effective plastic strain increment and $d\varepsilon_{ij}^p$ is the plastic strain increment tensor. The damage constants b_1 and b_2 are usually different by considering different damage evolution of concrete in compression and tension. With the increase of λ, the yield scale factor η varies from zero to unity when the current failure surface moves from the initial yield strength surface to the maximum strength surface, in corresponding to the strain-hardening branch. As for the strain-softening branch, the yield scale factor η varies from unity back to zero when the current failure surface moves from the maximum strength surface to the residual strength surface.

To account for the strain rate effect, the current failure surface γ and damage function λ are determined as

$$\gamma = r_f \gamma(p/r_f) \tag{8.8}$$

$$d\lambda = d\lambda(p/r_f)/r_f = \begin{cases} d\bar{\varepsilon}_p/r_f(1+p/f_t)^{b_1} & p > 0 \\ d\bar{\varepsilon}_p/r_f(1+p/f_t)^{b_2} & p \leq 0 \end{cases} \tag{8.9}$$

where r_f is the strength dynamic increase factor (DIF), which could be defined separately for compression and tension to describe the much higher DIF in tension than it in compression (Ren et al. 2018b; Wu et al. 2018).

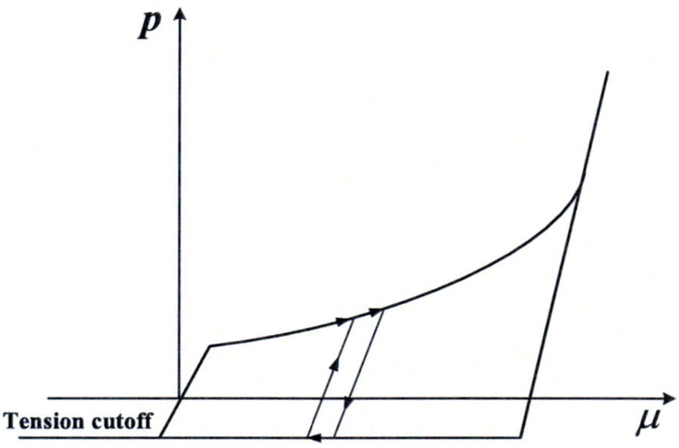

Fig. 8.12 Equation of state, reprinted from Wu et al. (2019), copyright 2020, with permission from Elsevier

To complete the description of mechanical behaviors in concrete material model, a proper equation of state (EOS) is necessary to correlate the pressure to the volumetric strain. K&C model uses the tabulated compaction EOS (*EOS #8 in LS-DYNA (LS-DYNA Keyword User's Manual, 2014)), which gives the current pressure as a function of volumetric strain μ by

$$p = C(\mu) + \gamma_i T(\mu) E_i \tag{8.10}$$

where E_i is the internal energy per initial volume and γ_i is the ratio of specific heat. $C(\mu)$ and $T(\mu)$ are the tabulated pressure and temperature as functions of volumetric strain, respectively. In the loading phase, pressure is determined by extrapolating the tabulated function given in Eq. (8.10). As shown in Fig. 8.12, unloading occurs along the unloading bulk modulus to the pressure cutoff. Reloading follows the unloading path to the point where unloading begins, and continues following the loading path.

8.4.2 Calibration

In this section, the model parameters of K&C model for UHPCC are calibrated systematically. For UHPCC with another different compressive strength, the following approach can be adopted as a reference.

(1) Strength parameters

The uniaxial compressive and conventional triaxial compressive tests on UHPCC previously conducted by authors (Ren et al. 2016) were used to determine the strength parameters, i.e., a_0, a_1, a_2, a_{1f}, a_{2f}, a_{0y}, a_{1y} and a_{2y}. Figure 8.13 illustrates the axial deviatoric stress–strain curves of UHPCC at different confining pressures.

$\Delta\sigma = \sigma_1 - \sigma_3$ is the axial deviatoric stress in which σ_1 and σ_3 are the actual axial stress and circumferential stress, respectively. ε_1 is the axial strain. Since the

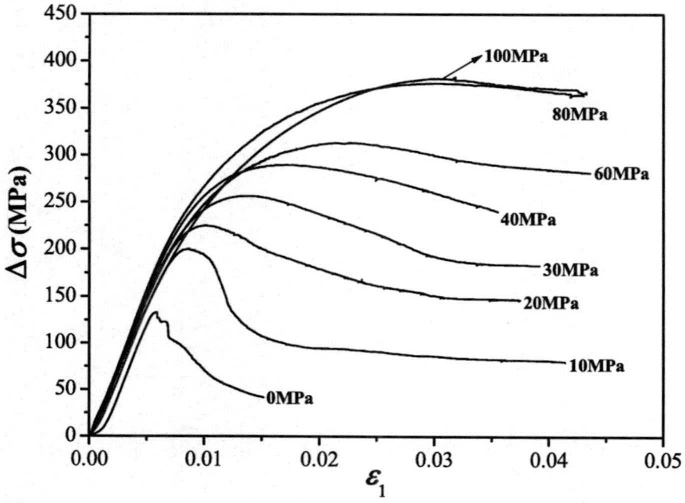

Fig. 8.13 Axial deviatoric stress–strain curves of UHPCC at different confining pressures (Ren et al. 2016)

circumferential and radial principal stresses are identical for cylinder specimen, i.e., $\sigma_2 = \sigma_3$, thus the equations for calculating stress difference $\Delta\sigma$ and mean stress p can be given as follows:

$$\Delta\sigma = \sqrt{3J_2} = \sqrt{\frac{(\sigma_1 - \sigma_2)^2 + (\sigma_1 - \sigma_3)^2 + (\sigma_2 - \sigma_3)^2}{2}} = |\sigma_1 - \sigma_2| \tag{8.11}$$

$$p = \frac{1}{3}(\sigma_1 + \sigma_2 + \sigma_3) = \frac{1}{3}(\sigma_1 + 2\sigma_2) \tag{8.12}$$

Based on Fig. 8.13, Fig. 8.14 further illustrates the experimental data of $\Delta\sigma \sim p$ from the triaxial compressive test. By curve fittings of Eqs. (8.3 ~ 8.5) for the three meridian planes, the strength parameters of K&C model for UHPCC can be determined and listed in Table 8.6.

(2) Strain rate effect parameters

Dynamic increase factor (DIF) is defined as the ratio of the dynamic strength to the quasi-static strength, which is proposed to quantify the magnitude increase of material strength under high strain rate loadings. In K&C model, the DIF versus strain rate is given by a curve (*DEFINE_CURVE in LS-DYNA (LS-DYNA Keyword User's Manual, 2014)), in which the negative (tensile) values of strain rate are input firstly, followed by the positive (compressive) values of strain rate. DIF should be positive and equals to 1.0 at the strain rate of zero. In Refs. (Ren et al. 2018b; Wu et al. 2018),

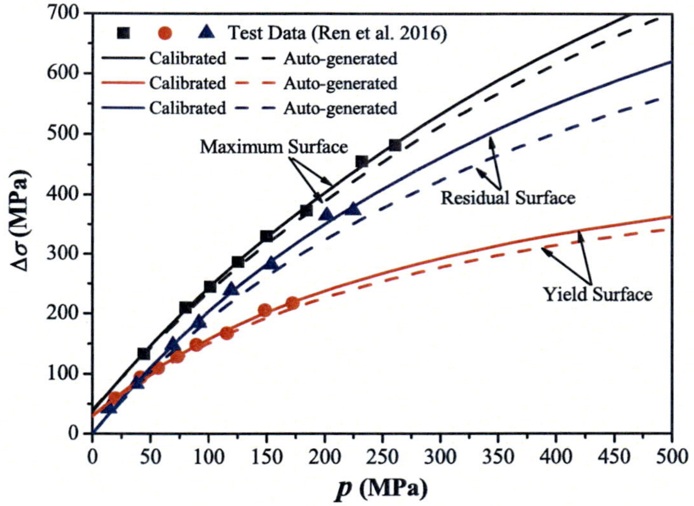

Fig. 8.14 Meridian plane, reprinted from Wu et al. (2019), copyright 2020, with permission from Elsevier

Table 8.6 Strength parameters of K&C model for UHPCC

Parameters	a_0	a_1	a_2	a_{1f}	a_{2f}	a_{0y}	a_{1y}	a_{2y}
Calibrated	37.50	0.43	5.87e-04	0.41	7.82e-04	30.26	0.60	1.80e-03
Auto-generated	39.31	0.45	6.08e-04	0.44	8.90e-04	29.69	0.63	1.94e-03

by conducting the split Hopkinson pressure bar (SHPB) compression and spalling tests, we have obtained the empirical DIF relations for dynamic compressive (DIF_c) and tensile (DIF_t) strengths of UHPCC, as given in Eqs. (8.13, 8.14) and illustrated in Fig. 8.15.

$$\text{DIF}_c = \begin{cases} (\dot{\varepsilon}/\dot{\varepsilon}_0)^{0.014} & \text{for } \dot{\varepsilon} \leq 62\text{s}^{-1} \\ 0.587(\log \dot{\varepsilon})^2 - 1.435\log \dot{\varepsilon} + 1.911 & \text{for } \dot{\varepsilon} > 62\text{s}^{-1} \end{cases} \tag{8.13}$$

$$\text{DIF}_t = \begin{cases} (\dot{\varepsilon}/\dot{\varepsilon}_0)^{0.018} & \text{for } \dot{\varepsilon} \leq 10.3\text{s}^{-1} \\ 9.959\log \dot{\varepsilon} - 8.718 & \text{for } \dot{\varepsilon} > 10.3\text{s}^{-1} \end{cases} \tag{8.14}$$

(3) Equation of state parameters

The relationship between pressure and volumetric strain can be described as the equation of state, and the change in the volumetric strain will induced the change in pressure. As shown in Sect. 8.4.1, the *EOS #8 (LS-DYNA Keyword User's Manual, 2014) which can be determined by the triaxial compressive test and fly plate impact test is used in K&C model. The auto-generated EOS parameters of K&C model are

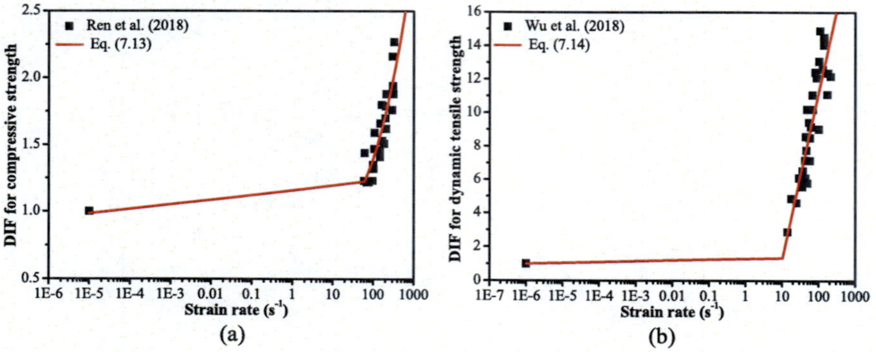

Fig. 8.15 Empirical DIF relations for UHPCC **a** Compression (Ren et al. 2018b) **b** Tension (Wu et al. 2018)

adopted based on Table 8.7 (Malvar et al. 2000), in which the elastic bulk modulus $K = E/[3(1-2\nu)]$, the elastic modulus $E = 43.8\text{GPa}$, Poisson's ratio $\nu = 0.23$ and $f_c = 141.5$ MPa at present.

Based on the Hugoniot tests (the plate impact which generates extremely high strain rate and high pressure in a state of uniaxial strain) conducted by Yan et al. (2000) and Marsh (1980), Fig. 8.16 presents the above pressure-volumetric strain test data along with the predictions of auto-generated and calibrated EOS parameters of K&C model. It is observed that the auto-generated K&C model underestimate the test data. While the calibrated K&C model gives a satisfactory agreement with the test data. The above EOS parameters of K&C model are listed in Table 8.8.

(4) Damage factor parameters

In K&C model, the relationship between η and λ is defined as a piece-wise linear curve and an automated generation procedure is provided for normal strength concrete. The automated generation procedure is not suitable to UHPCC and 13 pairs of $(\lambda,\ \eta)$ must be determined by using the uniaxial compression stress–strain relationship of UHPCC. The parameter η and λ could be derived from the current stress and the current strain, respectively. The current failure surface is interpolated from the maximum failure surface and either the yield or the residual failure surface. It was supposed that $\Delta\sigma_y = 0.45\Delta\sigma_m$ (Malvar et al. 1997) and the residual strength under unconfined uniaxial compression is zero. Thus, η could be solved from the current stress as

$$\eta = \begin{cases} \frac{\Delta\sigma - \Delta\sigma_y}{\Delta\sigma_m - \Delta\sigma_y} = \frac{\sigma - 0.45 f_c}{0.55 f_c} & \varepsilon \le \varepsilon_0 \\ \frac{\Delta\sigma}{\Delta\sigma_m} = \frac{\sigma}{f_c} & \varepsilon > \varepsilon_0 \end{cases} \tag{8.15}$$

The plastic potential function of fractionally associated flow rule adopted in K&C model could be written as (Malvar et al. 1997, Malvar and Simons 1996)

Table 8.7 Auto-generated EOS parameters of K&C model (Malvar et al. 2000)

Volumetric strain	μ_1	μ_2	μ_3	μ_4	μ_5	μ_6	μ_7	μ_8	μ_9	μ_{10}
Values	0	$f_c/(3K)$	0.0043	0.0101	0.0305	0.0513	0.0726	0.0943	0.1740	0.2080
Bulk unloading modulus	K_1	K_2	K_3	K_4	K_5	K_6	K_7	K_8	K_9	K_{10}
Values	K	K	1.014 K	1.065 K	1.267 K	1.470 K	1.672 K	1.825 K	4.106 K	5 K
Pressure	p_1	p_2	p_3	p_4	p_5	p_6	p_7	p_8	p_9	p_{10}
Values	0	$f_c/3$	$2.18p_2$	$3.50p_2$	$6.65p_2$	$10p_2$	$14.2p_2$	$21.8p_2$	$127p_2$	$194p_2$

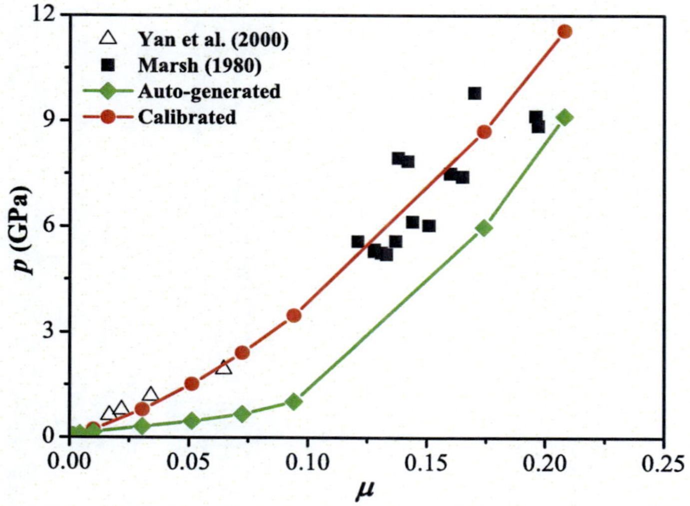

Fig. 8.16 Equation of state, reprinted from Wu et al. (2019), copyright 2020, with permission from Elsevier

Table 8.8 EOS input data in K&C model

i	μ_i	K_i (GPa)	p_i (GPa)	
			Auto-generated	Calibrated
1	0	27	0.000	0.000
2	0.0017	27	0.047	0.028
3	0.0043	27.39	0.102	0.094
4	0.0101	28.76	0.164	0.310
5	0.0305	34.21	0.313	0.800
6	0.0513	39.69	0.470	1.522
7	0.0726	45.14	0.667	2.411
8	0.0943	49.28	1.025	3.470
9	0.1740	110.86	5.969	8.698
10	0.2080	135	9.118	11.568

$$g_p = \sqrt{3J_2} - \omega\gamma \tag{8.16}$$

where $\omega = 0$ refers to the non-associative flow rule (J_2 flow rule), while $\omega = 1$ refers to the full associative flow rule. The value of ω usually ranges from 0 to 1, and the definition of plastic flow rule (Malvar and Simons 1996) is

$$d\varepsilon_{ij}^{p} = \frac{\partial g}{\partial \sigma_{ij}} d\tilde{\mu} = \left(\frac{3\sigma_{ij}^{'}}{2\sqrt{3J_2}} + \frac{\omega \delta_{ij}}{3} \frac{d\gamma}{dp} \right) d\tilde{\mu} \tag{8.17}$$

where $d\tilde{\mu}$ is a proportionality constant and the effective plastic strain increment can be derived from Eq. (8.17) as follows

$$d\overline{\varepsilon}_p = \sqrt{\frac{2}{3} d\varepsilon_{ij}^{p} d\varepsilon_{ij}^{p}} = \sqrt{1 + \frac{2}{27}\left(\omega \frac{d\gamma}{dp}\right)^2} d\tilde{\mu} \tag{8.18}$$

It could also be obtained from Eq. (8.17) in the uniaxial stress state that

$$d\varepsilon_{11}^{p} = \left(1 + \frac{1}{3}\omega \frac{d\gamma}{dp}\right) d\tilde{\mu} \tag{8.19}$$

where $d\varepsilon_{11}^{p}$ denotes the axial plastic strain increment. Substituting Eq. (8.19) into Eq. (8.18), the effective plastic strain increment could be expressed by the axial plastic strain increment as

$$d\overline{\varepsilon}_p = \frac{\sqrt{1 + \frac{2}{27}\left(\omega \frac{d\gamma}{dp}\right)^2}}{1 + \frac{1}{3}\omega \frac{d\gamma}{dp}} d\varepsilon_{11}^{p} \tag{8.20}$$

Substituting Eq. (8.20) into Eq. (8.9), the increment of damage function could be obtained as

$$d\lambda = \frac{\sqrt{1 + \frac{2}{27}\left(\omega \frac{d\gamma}{dp}\right)^2}}{1 + \frac{1}{3}\omega \frac{d\gamma}{dp}} \frac{d\varepsilon_{11}^{p}}{r_f (1 + p/f_t)^{b1}} \tag{8.21}$$

In K&C model (Malvar et al. 1997), the value of b_1 is 1.6, $\omega = 0.5$ represents the partially associative flow rule, and $d\gamma/dp$ could be solved from the yield surface as the function of pressure shown in Eqs. (8.4 ~ 8.6). Without considering the influences of strain rate effect, r_f is equal to 1.0. $d\varepsilon_{11}^{p}$ denotes the axial plastic strain increment which could be obtained from the uniaxial compressive stress–strain curve. Therefore, the dependence of η ~λ could be determined through a given uniaxial compressive stress–strain curve.

Attard and Setunge (1996) proposed an empirical formula to predict the complete compressive stress–strain curve of concrete. Based on the test data of UHPCC (Ren et al. 2018a), the specific form of formula is given in Eq. (8.22) and plotted in Fig. 8.17. Therefore, the relationship between η and λ for UHPCC could be obtained, as shown in Fig. 8.18. Then the 13 pairs of (λ, η) input in K&C model for UHPCC are listed in Table 8.9.

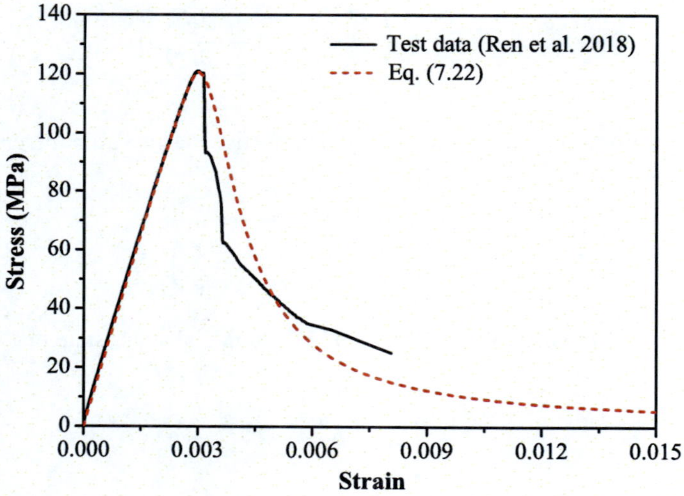

Fig. 8.17 Comparison of complete compressive stress–strain curves, reprinted from Wu et al. (2019), copyright 2020, with permission from Elsevier

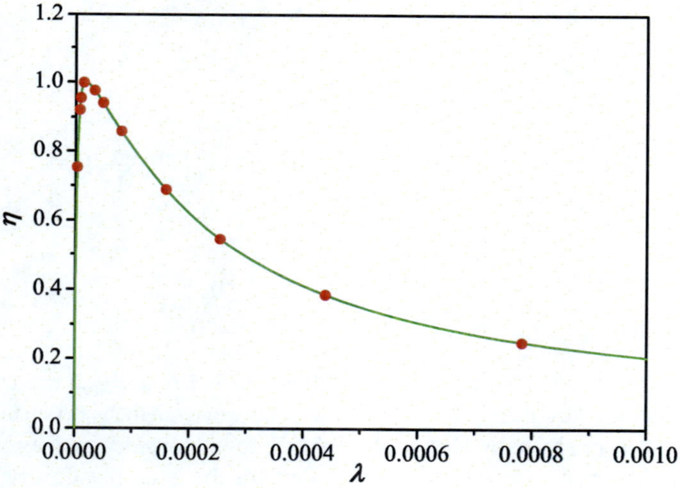

Fig. 8.18 η~λ relationship for UHPCC, reprinted from Wu et al. (2019), copyright 2020, with permission from Elsevier

$$Y = \begin{cases} \frac{1.089X - 0.986X^2}{1 - 0.911X + 0.014X^2} & \text{Ascending segment} \\ \frac{0.152X}{1 - 1.848X + X^2} & \text{Dscending segment} \end{cases} \tag{8.22}$$

Table 8.9 13 pairs of (λ, η) input in K&C model for UHPCC

No.	1	2	3	4	5	6	7
λ	0	3.13E-06	6.98E-06	8.67E-06	1.43E-05	3.32E-05	4.79E-05
η	0	0.75277	0.92074	0.95452	0.99837	0.97529	0.94049
No.	8	9	10	11	12	13	
λ	8.04E-05	1.59E-04	2.54E-04	4.38E-04	7.82E-04	1000	
η	0.85889	0.68833	0.54628	0.38554	0.25174	0	

where $Y = f/f_0$ and $X = \varepsilon/\varepsilon_0$ are the normalized equivalent stress and strain, in which f_0 and ε_0 are the peak stress and peak strain, respectively.

(5) Softening parameters b_1 and b_2

The softening parameters b_1 and b_2 control the concrete softening behavior after peak stress in uniaxial compression and uniaxial tension, respectively. These two parameters can be obtained through single element simulations of unconfined uniaxial compressive and tensile tests. It should be noted that, for different element sizes, the values of b_1 and b_2 should be changed accordingly in order to obtain the identical stress–strain relationship. The reason lies in that the stress–strain relationship will be changed with the variation of element size. The corresponding relationship is shown as follows (Tu and Lu 2009, Magallanes et al. 2010):

$$\int \sigma d\varepsilon = G/h_e \tag{8.23}$$

where G and h_e are the strain energy and element size, respectively.

Single element simulation under unconfined uniaxial compressive test was firstly performed and the element size of 10 mm was selected in this study. By changing the value of b_1 from 1.6 to 1.2 and re-running the analysis, the ratio G_c/h_e coincides with the area under the stress–strain curve of UHPCC for the uniaxial compressive test (Ren et al. 2018a), where $G_c = 4.3 \times 10^{-3}$MPa · m is the compressive strain energy. Hence, the b_1 value for a 10 mm mesh size was determined as 1.2 at present.

The parameter b_2 is further determined by assigning tensile strain energy G_t in the use of single element simulation (10 mm mesh size) of the uniaxial tensile test. b_2 is obtained via an iterative procedure until the ratio G_t/h_e coincides with the area under the uniaxial tensile stress–strain curve of UHPCC (Ren et al. 2018a). After updating the b_2 from 1.35 to -2, $G_t = 2.7 \times 10^{-4}$MPa · m and the material softening behavior was better modelled despite some discrepancy was observed, as shown in Fig. 8.19. Hence, the b_2 value for a 10 mm mesh size was determined as −2.0.

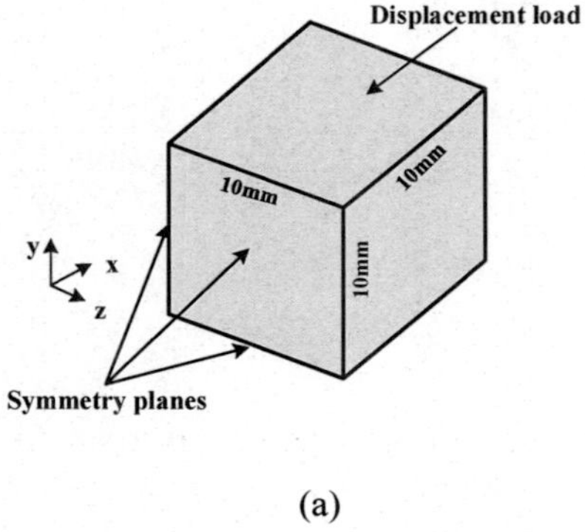

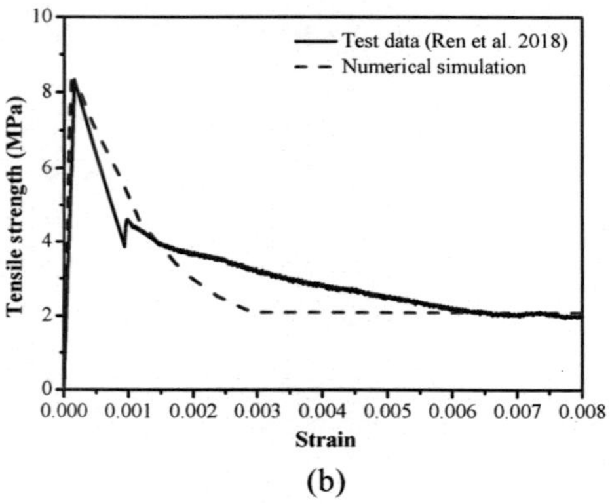

Fig. 8.19 Single element simulation **a** FE model **b** Comparison of uniaxial tensile stress–strain curve, reprinted from Wu et al. (2019), copyright 2020, with permission from Elsevier

8.5 Numerical Simulation

8.5.1 Present Test

8.5.1.1 FE Model

To further understand the performance of UHPCC filled steel tubes under transverse impact load, in this section, a FE model is established by using the commercial FE program Hypermesh/LS-DYNA, as shown in Fig. 8.20. The 3D solid element is

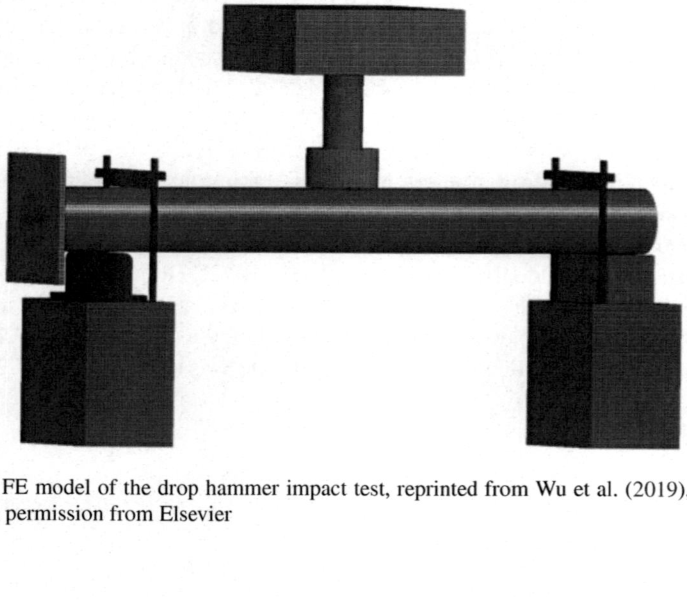

Fig. 8.20 FE model of the drop hammer impact test, reprinted from Wu et al. (2019), copyright 2020, with permission from Elsevier

employed for modelling the steel tube, UHPCC, drop hammer, supporting frames as well as the steel plate welded at the end of the column.

For the numerical simulation of the interface between the steel tube and core concrete, by adding a slip element between a steel tube element and a core concrete element, Hajjar (1998) and Bangash (1989) derived that, considering the slip property between the interfaces has little influence on the dynamic behavior of CFST. Thus the slip is not considered at present.

The interface between the steel tube and UHPCC were completely bonded by using a common-node approach, which is consistent with the fact that the steel tube and core concrete are tightly bonded. The contact between the drop hammer and steel tube, as well as the contact between the steel tube and supporting frames, were described by using the keyword "*CONTACT_AUTOMATIC_SURFACE_TO_SURFACE" implemented in LS-DYNA (LS-DYNA Keyword User's Manual 2014), respectively. As for the two deformable surfaces in contact, the master surface refers usually to the stiffer body or the surface with a coarser mesh if the two surfaces have comparable stiffness (LS-DYNA Keyword User's Manual, 2014). Therefore, the FE procedure selects the supporting frame as the master surface in contact with the steel tube (slave surface), and defines the steel tube as the slave surface in contact with the drop hammer (master surface).

8.5.1.2 Steel Tube Material Model

The Cowper-Symonds (CS) model (1957) and Johnson–Cook (JC) model (1983) are the common constitutive models for steel material. JC model (*MAT_JOHNSON_COOK in LS-DYNA (LS-DYNA Keyword User's Manual 2014)) is specially developed for materials subjected to large strain, high strain rate and high temperature, it is less popular than the former one in modeling the behavior of steel under low-velocity impact. Therefore, CS model (*MAT_PLASTIC_KINEMATIC in LS-DYNA (LS-DYNA Keyword User's Manual 2014)) with linear isotropic hardening and strain rate effect is adopted in the simulation, which has the reputation of minimizing the duration of analysis.

In CS model, the yield stress scaled by a strain rate dependent factor is expressed by:

$$\frac{\sigma_d}{\sigma_s} = 1 + \left(\frac{\dot{\varepsilon}}{C_c}\right)^{1/P_c} \tag{8.24}$$

where σ_d and σ_s denote the dynamic flow stress and the associated static flow stress, respectively. $\dot{\varepsilon}$ represents the strain rate. C_c and P_c are the strain rate parameters, and the values of 40 and 5 are adopted according to the findings in Stouffer and Dame (1996). Table 8.10 lists the material parameters of CS model for the steel tube, in which the tangent modulus E_t is obtained by using the following equation.

Table 8.10 CS model parameters of steel tube

Density (g/cm^3)	Elastic modulus (GPa)	Poisson's ratio	Yield stress (MPa)	Tangent modulus (GPa)	Rupture strain
7.85	206	0.3	318.5	0.8	0.23

$$E_t = \frac{\sigma_u - \sigma_y}{\varepsilon_u - \varepsilon_y} \tag{8.25}$$

where σ_u and σ_y are the ultimate stress and yield stress, respectively. ε_u and ε_y are the ultimate strain and yield strain, respectively. Based on Fig. 8.1 and Table 8.4, the values of σ_u, σ_y, ε_u and ε_y are taken as 507 MPa, 318.5 MPa, 0.23 and 0.0036, respectively.

8.5.1.3 Mesh Size Sensitivity

Element size has a great influence on the results of numerical simulation. In order to analyze the mesh size sensitivity, the C-5 case was taken as an example and the numerical test of mesh convergence was performed firstly. Figure 8.21 shows the deflection-time histories of C-5 case at Point B obtained by using four different specimen mesh sizes, i.e., 6 mm, 10 mm, 16 mm and 20 mm. It can be seen that 16 mm and 20 mm mesh predicts relatively larger deflections than 6 mm and 10 mm

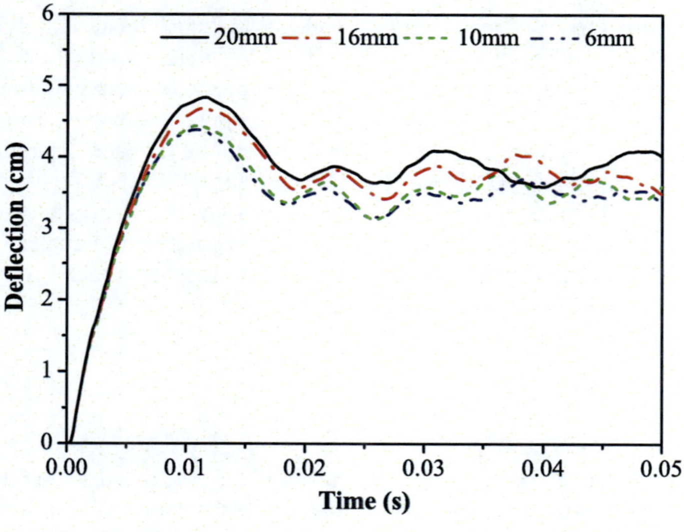

Fig. 8.21 Deflection-time histories of C-5 case at point B for various mesh sizes, reprinted from Wu et al. (2019), copyright 2020, with permission from Elsevier

mesh. Besides, using mesh sizes of 6 mm and 10 mm yields very close numerical results. Considering the computational efficiency, 10 mm mesh was used for the present numerical simulation.

8.5.1.4 Comparisons of Numerical Results with Test Data

This section presents the comparisons between the experimental and predicted dynamic responses of three UHPCC filled steel tubes under transverse impact load, as shown in Figs. 8.22 and 8.23. The main objective of this section is to assess whether the calibrated input parameters of K&C model are able to predict the response of UHPCC filled steel tubes with different release heights of the drop hammer. As can be seen, the numerically predicted impact force- and deflection-time histories are in good agreement with the test data in terms of peak values, response time and overall shape. The mean numerical simulation-to-test ratio of peak values for impact force and deflection is found to be in the range of 0.89 to 1.04 with a coefficient of variation ranging from 1.5 to 5.6%. It should be pointed out that, the calculation time is adopted as 0.05 s by considering the time consumption.

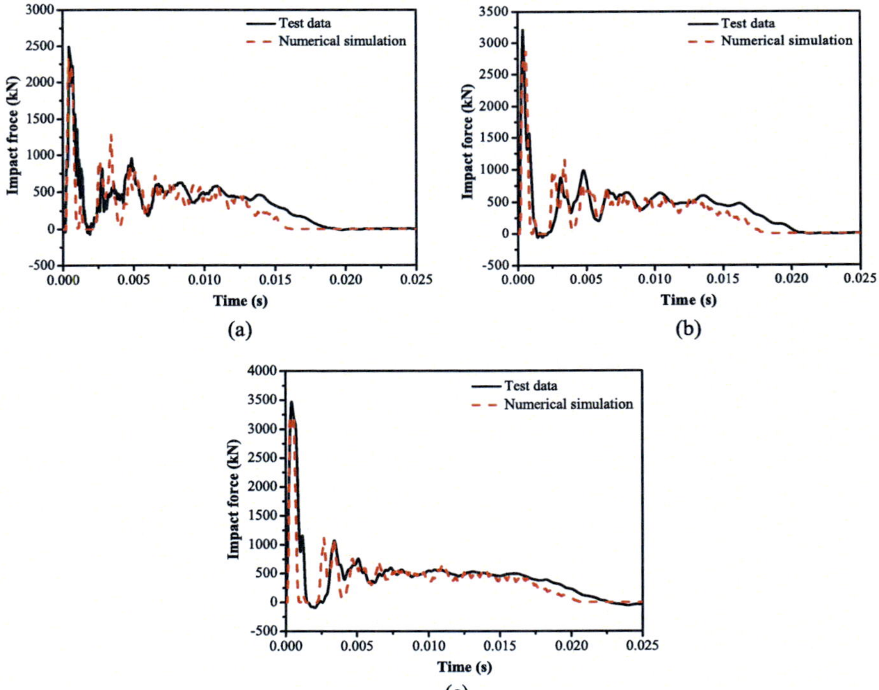

Fig. 8.22 Comparisons of experimental and predicted impact force–time histories **a** C-5 **b** C-8 **c** C-11, reprinted from Wu et al. (2019), copyright 2020, with permission from Elsevier

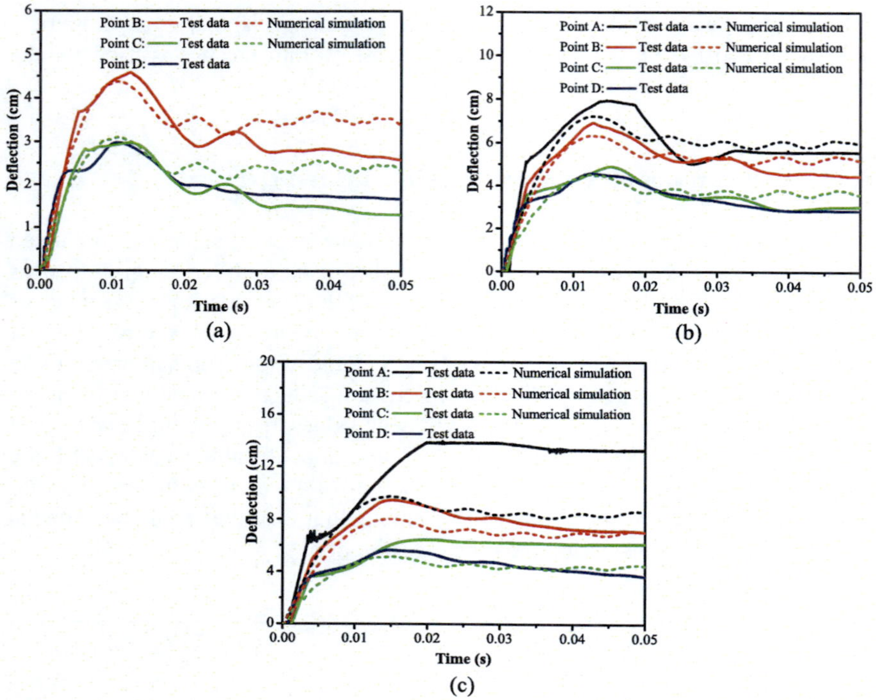

Fig. 8.23 Comparisons of experimental and predicted deflection-time histories **a** C-5 **b** C-8 **c** C-11, reprinted from Wu et al. (2019), copyright 2020, with permission from Elsevier

8.5.2 *Yoo et al. (2015) Test*

In this section, by using the finite element program Hypermesh/LS-DYNA, the drop-hammer impact test on the ultra-high performance fiber reinforced concrete (UHPFRC) beams are numerically simulated to further validate the proposed K&C model parameters.

Yoo et al. (2015) conducted a series of drop-hammer impact tests on the ultra-high performance fiber reinforced concrete (UHPFRC) beams, as shown in Fig. 8.24. The total mass and release heights of the drop hammer are 270 kg and 1.6 m, respectively. A total of four large-sized (200 × 270 × 2900 mm) beams were fabricated and the detailed geometry and arrangement of reinforcements (diameter of 12.7 mm) are shown in Fig. 8.25. "UH-N" represents a beam series without steel rebar and "UH-S-ρ_r" represents a beam series reinforced with steel rebar, where the letters ρ_r indicates the reinforcement ratio. The compressive strength and elastic modulus of UHPFRC are 152.5 MPa and 44GPa, respectively. The yield strength, ultimate strength and elastic modulus of rebar are 522.7 MPa, 627.6 MPa and 200GPa, respectively.

Figure 8.26 shows the corresponding numerical and experimental mid-span deflection-time histories of UHPFRC beams. As can be seen, the deflection-time

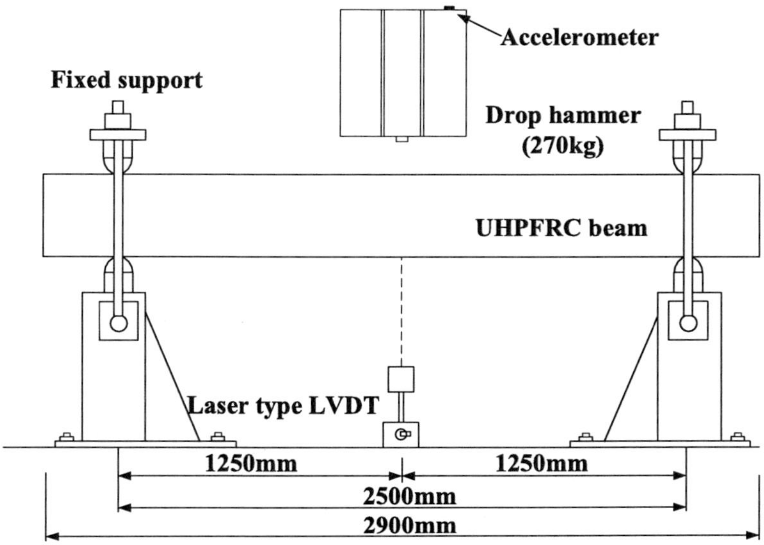

Fig. 8.24 Drop-hammer impact test setup, reprinted from Wu et al. (2019), copyright 2020, with permission from Elsevier

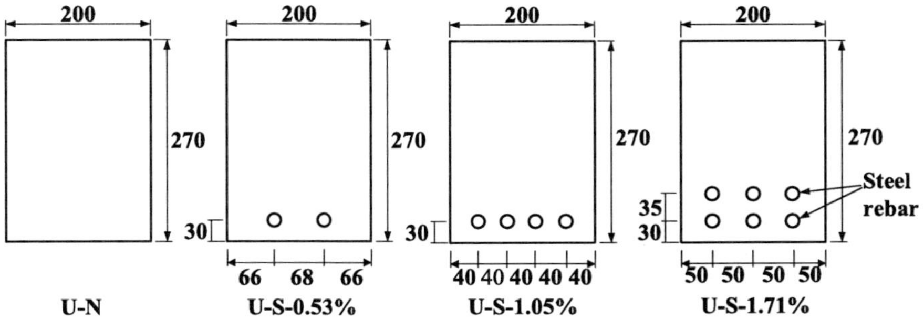

Fig. 8.25 Section details of impact test program (unit: mm), reprinted from Wu et al. (2019), copyright 2020, with permission from Elsevier

responses of UHPFRC beams are predicted fairly well by using the calibrated K&C model parameters, especially in terms of maximum deflections. Also. It should be pointed out that, the time instant when the UHPFRC beams begin to deform obtained numerically are slightly slower than the test data.

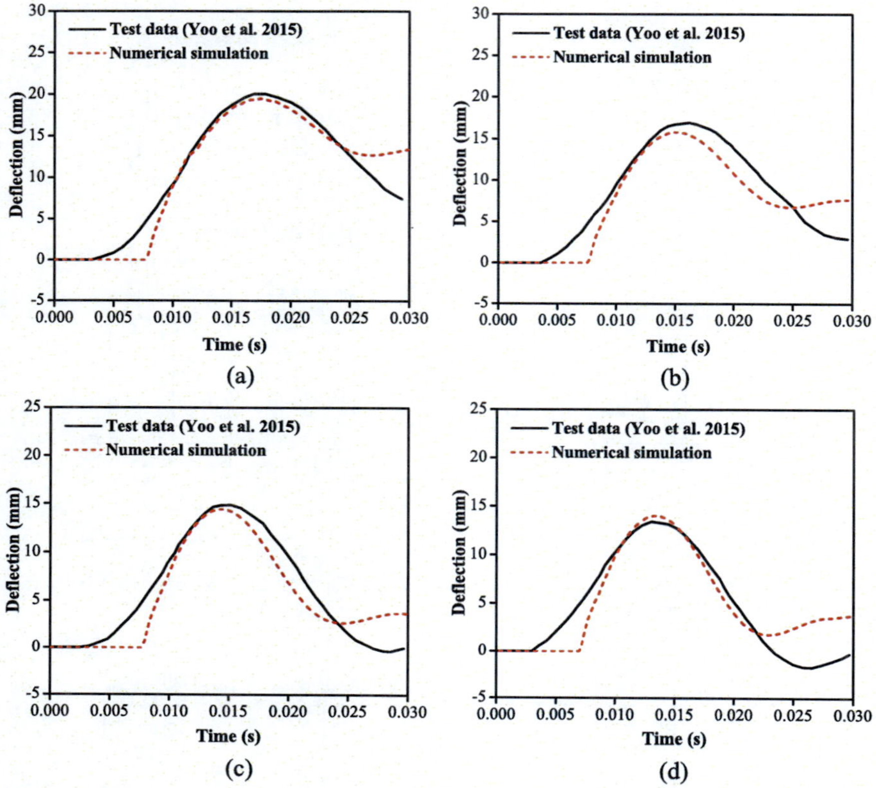

Fig. 8.26 Comparisons of numerical and experimental mid-span deflection-time histories **a** U-N **b** U-S-0.53% **c** U-S-1.05% **d** U-S-1.71%, reprinted from Wu et al. (2019), copyright 2020, with permission from Elsevier

8.6 Summary

In this chapter, the impact behavior of UHPCC-FSTs under transverse impact load is investigated experimentally and numerically. A set of three UHPCC-FSTs are fabricated and tested subjected to low-velocity impact by using a drop-hammer impact device. The model parameters of K&C model for UHPCC are calibrated by using the existing static and dynamic experimental data. Then a FE analysis model is established to predict the dynamic responses of UHPCC-FSTs under transverse impact load. The main work and conclusions are drawn:

(1) The axial compression test on UHPCC-FST is conducted, the axial load-axial displacement and -lateral displacement curves are obtained, and the constraining factor of the present UHPCC-FST is confirmed.
(2) UHPCC-FSTs deform in a ductile manner and there is no severe local buckling in the steel tube. UHPCC core restricts effectively the indentation within

a highly localized region around the impact point and limits the deformation of steel tube. UHPCC-FSTs demonstrate superior impact resistance under transverse impact load.

(3) The impact behavior of UHPCC-FSTs under transverse impact load can be generally divided into three stages: peak value stage, plateau stage and unloading stage. The peak impact force, maximum and residual deflections gradually increase with increasing the release height.

(4) The K&C model parameters (strength parameters, strain rate effect parameters, equation of state parameters, damage factor parameters as well as softening parameters) for the present UHPCC are calibrated based on a series of static and dynamic experimental data.

(5) By comparing with the present and existing drop hammer test on UHPCC beams, the validation of the calibrated K&C model parameters for UHPCC are verified.

References

ADHIKARY S D, LI B, FUJIKAKE K. Dynamic behavior of reinforced concrete beams under varying rates of concentrated loading[J]. International Journal of Impact Engineering, 1997, 19(9): 847-873.

ATTARD M M, SETUNGE S. Stress-strain relationship of confined and unconfined concrete[J]. ACI Materials Journal, 1996, 93: 432-442.

BAMBACH M R, JAMA H, ZHAO X L, GRZEBIETA R H. Hollow and concrete filled steel hollow sections under transverse impact loads[J]. Engineering Structures, 2008, 30(10): 2859-2870.

BANGASH M Y H. Concrete and concrete structure: numerical modeling and applications[M]. London and New York: Elsevier Applied Science, 1989.

COWPER G R, SYMONDS P S. Strain hardening and strain rate effects in the impact loading of cantilever beams[R]. Brown University, 1957. Technical Report No. 28.

DENG Y, TUAN C Y, XIAO Y. Flexural behavior of concrete-filled circular steel tubes under high-strain rate impact loading[J]. Journal of Structural Engineering, 2012, 138(3): 449-456.

Draft prEN 1991-1-7. Eurocode 1-actions on structures[M]. Brussels: European Committee for Standardization, 2005: part 1–7: general actions-accidental actions.

DU G F, BABIC M, WU F H, ZENG X, BIE X M. Experimental and numerical studies on concrete filled circular steel tubular (CFCST) members under impact loads[J]. International Journal of Civil Engineering, 2018, 17(8): 1211-1226.

GB/T 50081–2002. Standard for test method of mechanical properties on ordinary concrete[S]. Beijing: Standards Press of China, 2003.

GB/T 228.1–2010. Metallic materials-Tensile testing, Part I: Method of Test at Room Temperature[S]. Beijing: Standards Press of China, 2010.

GB 50017–2017. Code for design of steel structure[S]. Beijing: Standards Press of China, 2017.

GB50936–2014. Design Code for Composite Structure[S]. National Standard of China, 2014.

GUO W, FAN W, SHAO X D, SHEN D J, CHEN B S. Constitutive model of ultra-high-performance fiber-reinforced concrete for low-velocity impact simulations[J]. Composite Structures, 2018, 185: 307-326.

HAJJAR J F. A distributed plasticity model for concrete filled steel tube beam-columns with interlayer slip[J]. Engineering Structures, 1998, 20(8): 663-676.

HAN L H, HOU C C, ZHAO X L, RASMUSSEN K J R. Behaviour of high-strength concrete filled steel tubes under transverse impact loading[J]. Journal of Constructional Steel Research, 2014, 92: 25-39.

JOHNSON G R, COOK W H. A constitutive model and data for metals subjected to large strains, high strain rates and high temperatures. Proceedings of the 7th international symposium on ballistics, Hague, Netherlands[C]. 1983: 541–547.

KARLSSON SORENSEN INC. HIBBITT. ABAQUS/Explicit User's Manual[M]. New York: Rhode Island, 2005.

LSTC.LS-DYNA Keyword User's Manual[M]. Livermore Software Technology Corporation, 2014.

LU Y, ZHANG L. Analysis of failure of a bridge foundation under rock impact[J]. Acta Geotechnica, 2012, 7: 57-68.

MADURAPPERUMA M A K M, WIJEYEWICKREMA A C. Response of reinforced concrete columns impacted by tsunami dispersed 20′ and 40′ shipping containers[J]. Engineering Structures, 2013, 56: 1631-1644.

MAGALLANES J M, WU Y, MALVAR L J, CRAWFORD J E. Recent developments to release III of the K&C concrete model. In proceedings of the 11th international LS-DYNA user's conference 2010. Dearborn, MI[C]. 2010.

MALVAR L J, CRAWFORD J E, WESEVICH J W, SIMONS D. A plasticity concrete material model for DYNA3D[J]. International Journal of Impact Engineering, 1997, 19(9): 847-873.

MALVAR L J, CRAWFORD J E, MORRILL K B. K&C concrete material model release III-automated generation of material model input[R]. Karagozian and Case Structural Engineers, 2000. Technical Report TR-99–24.3.

MALVAR L J, SIMONS D. Concrete material modeling in explicit computations. Proceedings, Workshop on Recent Advances in Computational Structural Dynamics and High Performance Computing, USAE Waterways Experiment Station, Vicksburg, MS[C]. 1996: 165–194.

MARSH S P. LASL shock Hugoniot data[M]. California: University of California Press, 1980.

REMENNIKOV A M, KONG S Y, UY B. Response of foam- and concrete-filled square steel tubes under low-velocity impact loading[J]. Journal of Performance of Constructed Facilities, 2011, 25(5): 373-381.

REN G M, WU H, FANG Q, LIU J Z, GONG Z M. Triaxial compressive behavior of UHPCC and applications in the projectile impact analyses[J]. Construction and Building Materials, 2016, 113: 1-14.

REN G M, WU H, FANG Q, LIU J Z. Effects of steel fiber content and type on static mechanical properties of UHPCC[J]. Construction and Building Materials, 2018, 163: 826-839.

REN G M, WU H, FANG Q, LIU J Z. Effects of steel fiber content and type on dynamic compressive mechanical properties of UHPCC[J]. Construction and Building Materials, 2018, 164: 29-43.

SHA Y, HAO H. Laboratory tests and numerical simulations of barge impact on circular reinforced concrete piers[J]. Engineering Structures, 2013, 46: 593-605.

STOUFFER D C, DAME L T. Inelastic deformation of metals: models, mechanical properties, and metallurgy[M]. New York: John Wiley and Sons, 1996: 520.

THILAKARATHNA H M I, THAMBIRATNAM D P, DHANASEKAR M, PERERA N. Numerical simulation of axially loaded concrete columns under transverse impact and vulnerability assessment[J]. International Journal of Impact Engineering, 2010, 37: 1100-1112.

TU Z, LU Y. Evaluation of typical concrete material models used in hydrocodes for high dynamic response simulations[J]. International Journal of Impact Engineering, 2009, 36(1): 132-146.

UY B. Concrete-filled fabricated steel box columns for multistory buildings: behaviour and design[J]. Progress in Structural Engineering and Materials, 1998, 1(2): 150-158.

WANG R, HAN L H, HOU C C. Behavior of concrete filled steel tubular (CFST) members under lateral impact: Experiment and FEA model[J]. Journal of Constructional Steel Research, 2013, 80: 188-201.

WANG R, HAN L H, TAO Z. Behavior of FRP-concrete-steel double skin tubular members under lateral impact: Experimental study[J]. Thin-Walled Structures, 2015, 95: 363-373.

WANG R, HAN L H, ZHAO X, RASMUSSEN K J R. Analytical behavior of concrete filled double steel tubular (CFDST) members under lateral impact[J]. Thin-Walled Structures, 2016, 101: 129-140.

WU H, REN G M, FANG Q, LIU J Z. Effects of steel fiber content and type on dynamic tensile mechanical properties of UHPCC[J]. Construction and Building Materials, 2018, 173: 251-261.

WU H, REN G M, FANG Q, LIU J Z. Response of ultra-high performance cementitious composites filled steel tube (UHPCC-FST) subjected to low-velocity impact[J]. Thin-Walled Structures, 2019, 144: 106341.

YAN S H, QIAN Q H, ZHOU Z S, LU Y S, YIN F L. Experimental study of equation of state for high-strength concrete and high-strength fiber concrete[J]. Journal of PLA University of Science and Technology, 2000, 1(6): 49-53. (in Chinese).

YOO D Y, BANTHIA N, KIM S W, YOON Y S. Response of ultra-high-performance fiber-reinforced concrete beams with continuous steel reinforcement subjected to low-velocity impact loading[J]. Composite Structures, 2015, 126: 233-245.

YOO D Y, BANTHIA N, YOON Y S. Flexural behavior of ultra-high-performance fiber-reinforced concrete beams reinforced with GFRP and steel rebars[J]. Engineering Structures, 2016, 11: 246-262.

YOUSUFF M, UY B, TAO Z, REMENNIKOV A M, LIEW J Y R. Behaviour and resistance of hollow and concrete-filled mild steel columns due to transverse impact loading[J]. Journal of Constructional Steel Research, 2012, 13(1): 65-80.

ZHAN T B, WANG Z H, NING J G. Failure behaviors of reinforced concrete beams subjected to high impact loading[J]. Engineering Failure Analysis, 2015, 56: 233-243.

ZHAO X L, HAN L H, LU H. Concrete-filled tubular members and connections[M]. Oxford: Taylor & Francis, 2010.

Chapter 9
Dynamic Responses of Reinforced UHPCC Members Under Low-Velocity Lateral Impact

9.1 Introduction

In modern infrastructures, reinforced concrete (RC) members, e.g., beams, columns and piers, may undergo various kinds of lateral impact loading scenarios, e.g., vehicle and ship collisions, rockfall rolling impact, etc. The local damage on members caused by impact loading will spread in the entire structure and even leads to the whole structural collapse (Institute 2006; Wardhana and Hadipriono 2003). For more precise and adequate robustness design purposes, the lateral impact behaviors of structural members should be studied in-depth.

The low velocity impact resistance of structural members is generally studied by drop hammer impact tests and numerical simulations. Kishi et al. (2002) experimentally derived that, when the static shear-bending capacity ratio was less than 1.0, the diagonal cracks would emerge on the NSC beam without stirrups under the drop hammer impact loadings. Fujikake et al. (2009) carried out the drop hammer test on NSC (42 MPa) beams with longitudinal reinforcement ratio ranging from 0.34 to 2.06%, and derived that NSC beams with lower amounts of longitudinal steel reinforcement exhibit overall flexural failure comparatively. Saatci and Vecchio (2009) conducted a series of drop hammer tests on NSC beams with different shear reinforcement ratios. It was found that the severe diagonal shear cracks with a shear-plug occurred under the impact point of the specimens, and the shear resistance of NSC members played an important role in their overall performance. Based on the numerical simulations, Ozbolt and Sharma (2011) reported that the loading rate was a significant factor in the cracking pattern of RC beams, and the influence of shear reinforcement was reduced as the load rate increasing continuously. Wang et al. (2011) preformed the lateral rapid loading tests on NSC columns (25 MPa), and it was stated that the shear failure usually occurred with the low stirrup ratio, and the ultimate bearing capacity of NSC columns can be enhanced with increasing the load rate or longitudinal reinforcement ratio. Zhao et al. (2017, 2018) experimentally and numerically examined the dynamic responses of NSC beams under different

Q. Fang et al., *UHPCC Under Impact and Blast*,
https://doi.org/10.1007/978-981-33-6842-2_9

drop hammer impact energies. It was found that the additional transverse reinforcement could inhibit the development of shear cracks, and the effect of stress wave propagation on impact response was more pronounced with increasing the beam span. Jiang and Zhao (2015) put forward a method to determine the major parameters of the elasto-plastic damage cap (EPDC) model (#MAT145 in LS-DYNA), which was suitable for low-velocity impact analysis of NSC members. It indicates that NSC members may suffer severe localized damage (e.g., spalling of concrete cover, diagonal shear cracks) and global damage (e.g., flexure) (Adhikary et al. 1997, Madurapperuma and Wijeyewickrema 2013, Zhan et al. 2015).

Comparably, UHPCC is an advanced type of cementitious material with the low water-to-binder ratio, high amount of high-range water reducer (HRWR), fine aggregates and high-strength steel fibers. Considering its excellent mechanical properties, e.g., high compressive and tensile strength, high ductility, high fracture energy and very low permeability, UHPCC has been becoming the most prospective construction cement-based material of protective civil structures to resist high-speed penetration (Dancygier et al. 2007; Farnam et al. 2010; Máca et al. 2010; Wu et al. 2015; Sovják et al. 2015), low-velocity impact (Fujikake et al. 2006, Fan et al. 2019, Yoo et al. 2015, 2017, Wang et al. 2019b, Wei et al. 2019, Wu et al. 2019) and blast loadings (Li et al. 2015, 2016, Mao et al. 2014, Xu et al. 2016, Wu et al. 2009).

The low-velocity impact resistance of UHPCC members is mainly concerned in the current study. Fujikake et al. (2006) performed the 400 kg drop hammer impact test on simply supported I cross-sectional UHPC (214.7 MPa) beams with the longitudinal reinforcement ratio of 2.60% and five different drop heights from 0.8 m to 1.6 m. The distributed flexural cracks were detected on the mid-span bottom of UHPC beams and an effective theoretical model of two degrees of freedom mass-spring-damper system was further proposed. Yoo et al. (2015, 2017) conducted the 270 kg drop hammer test on the fixed supported rectangular sectional UHPC beams with different longitudinal reinforcement ratios (0, 0.53, 1.05 and 1.71%) and steel fiber types (long straight, short straight and twisted). It was proved that increasing the reinforcement ratio and adding 2% long straight steel fibers were effective to improve the impact resistance of UHPC beams. Fan et al. (2019) experimentally examined the drop hammer impact resistance of end-sliding supported UHPC (155 MPa) columns, as well as the influences of axial force and three different types of UHPC jackets. They concluded that the axial force has a significant enhanced effect on the impact resistance of UHPC columns, and the UHPC jackets on both the impact location and two ends were the worst configuration. Wei et al. (2019) carried out the comparable drop hammer impact test on both the UHPC (136 MPa) and NSC (40 MPa) columns with relatively low axial force (200 kN), square and circle cross section and releasing height from 1.0 m to 1.75 m, and proposed a formula for assessing the residual loading capacity of UHPC column after impact. Wu et al. (2019) and Wang et al. (2019b) respectively conducted a series of drop-hammer tests on UHPC (141.5 MPa and 136 MPa) filled steel tubular (UHPC-FST) and double-skin steel tubular (UHPC-FDST) with simply and fixed supports, various impact energy, steel tubular thickness and axial force, respectively. They concluded that UHPC core restricted effectively the indentation and limited the deformation of steel tube, and the impact energy and the thickness of outer steel tube can significantly affect the lateral impact behavior than the thickness of inner steel tube and axial force. Generally, most existing impact tests were carried out on UHPC specimens with small or without axial force, i.e., the low axial compressive ratio, the influences of impact energy and the axial force on the impact resistance of UHPC specimens needs to be assessed systematically.

Comparably, numerical simulation is an economical and efficient approach to study the low-velocity impact resistance of structural members. However, most of the numerical constitutive models for concrete which were embedded in the commercial finite element program LS-DYNA were developed based on NSC. For example, the auto-generation algorithm of K&C model (#MAT 72 in LS-DYNA) is established based on 41 MPa concrete, and the corresponding application range of concrete strength for Continuous Surface Cap (CSC) model (#MAT 159 in LS-DYNA) is 20-58 MPa. Therefore, those material models should be calibrated for UHPC before applied in the structural dynamic analyses. For instances, in our previous work (Wu et al. 2015), K&C model parameters of UHPC were calibrated and validated by comparing with the drop-hammer impact test on simply supported UHPC-FST members with different impact height from 5.0 m to 11.0 m. Similarly, partial CSC model parameters were respectively calibrated by Guo et al. (2018), Wei et al. (2019) and Wang et al. (2019b), and the validities of which were verified by comparisons with the low-velocity drop hammer impacting tests on simply supported reinforced UHPC and UHPC-FDST beams, respectively. Guo et al. (2018) pointed out that the numerically predicted results of UHPC beam with K&C model overpredicted the mid-span displacement than experimental data. However, some critical constitutive model parameters were not involved in above work (Fan et al. 2019, Guo et al. 2018, Wang et al. 2019b, Wei et al. 2019, Wu et al. 2019), e.g., fracture energy of the damage parameter and hardening parameters.

In this chapter, the dynamic responses of UHPCC members under low-velocity lateral impact are examined experimentally and numerically. Firstly, in Sect. 9.2, by fabricating the UHPCC beams with the compressive strength of 148 MPa, the drop hammer impact test on eleven UHPCC specimens and two NSC control specimens were conducted, in which different levels of impact energy and axial force are considered. Then, in Sect. 9.4, the CSC model parameters for UHPCC were calibrated comprehensively based on the systematic tests data, e.g., static compressive and direct tensile tests, dynamic compressive and tensile tests, triaxial compressive test and hydrostatic compression test. The validations of the proposed CSC model parameters generation method for UHPCC are discussed by comparing with the present tests data. Furthermore, in Sect. 9.5, the applicability of the generation method for above parameters was further assessed by the existing six sets of low-velocity lateral impact tests on UHPC and UHPC-FST members systematically.

9.2 Test Program

9.2.1 Specimen Fabrication

Total eleven reinforcement UHPCC specimens and two reinforcement NSC specimens with the identical reinforcement arrangement were fabricated, the specimen dimensions and the symmetric reinforcing details are illustrated in Fig. 9.1. The concrete cover thickness was 30 mm. The longitudinal and transverse reinforcement were deformed bars with 22 mm diameter and plain round bars with 6 mm diameter, respectively. The corresponding elastic modulus, yield strength and ultimate strength of reinforcing bars were derived according to the Chinese Standard GB/T 228.1-2010 (2010), as listed in Table 9.1.

The conventional concrete was used in NSC specimens, whose compressive strength and elastic modulus were 30.4 MPa and 30 GPa according to the Chinese Standard GB/T 50081-2002 (2003). The UHPCC specimens were cast with the material developed by Jiangsu SOBUTE new material Co., Ltd., the mixture proportions of UHPCC are given in Table 9.2, and the detailed properties of the mixed cementitious materials can be found in our previous work (Wu et al. 2015). Especially, the micro-straight steel fiber was incorporated with the volume fraction of 2.0% and the corresponding material properties are given in Table 9.3. The corresponding compressive strength, elastic modulus and Poisson's ratio of UHPCC were 148 MPa, 48.6 GPa and 0.23, respectively.

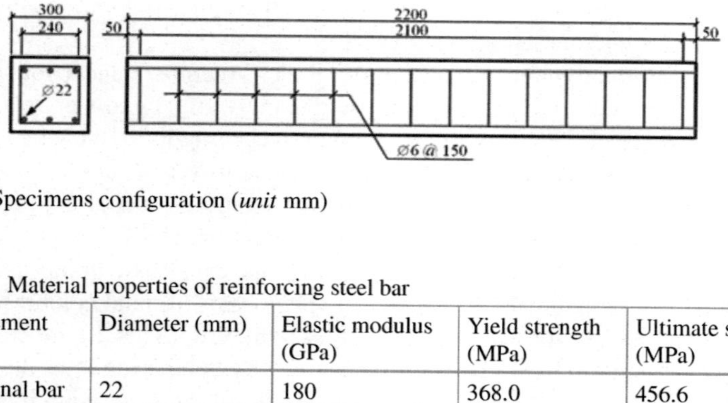

Fig. 9.1 Specimens configuration (*unit* mm)

Table 9.1 Material properties of reinforcing steel bar

Reinforcement	Diameter (mm)	Elastic modulus (GPa)	Yield strength (MPa)	Ultimate strength (MPa)
Longitudinal bar	22	180	368.0	456.6
Transverse bar	6	200	420.9	530.0

Table 9.2 Mixture proportions of UHPCC (kg/m^3)

Cement	Silica fume	Ultra-fine mineral admixture	Sand	Water	HRWR	Fiber content
700	140	110	1200	152	22.8	156

Table 9.3 Material properties of steel fiber

Fiber type	Length (mm)	Diameter (mm)	Aspect ratio	Density (g/cm^3)	Tensile strength (MPa)	Elastic modulus (GPa)
Micro-straight	13	0.2	65	7.8	2800	210

9.2.2 Drop-Hammer Impact Test

As shown in Fig. 9.2a, the lateral impact test was conducted by a drop-hammer impact system, which consists of the drop hammer, fix-sliding support and axial forcing system. The mass of drop hammer was 850 kg, which was composed of three weight blocks (3 × 250 kg) and a wedge-shape indenter (100 kg). The impact interface between the hammer and specimens was a flat subface with the dimension of 80 mm × 300 mm. Additionally, the indenter-nose was geometrically aligned at the middle span center of specimens through targeting laser. The maximum releasing height of the experimental system was 5 m, and the drop hammer was released freely

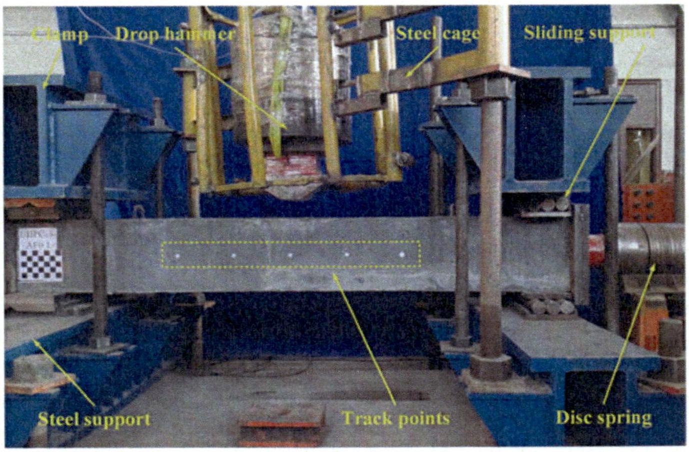

(a)

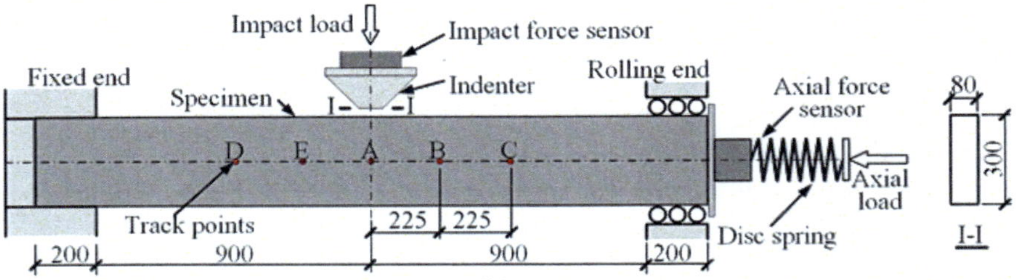

(b)

Fig. 9.2 Test setup **a** photograph **b** schematic view (*unit* mm)

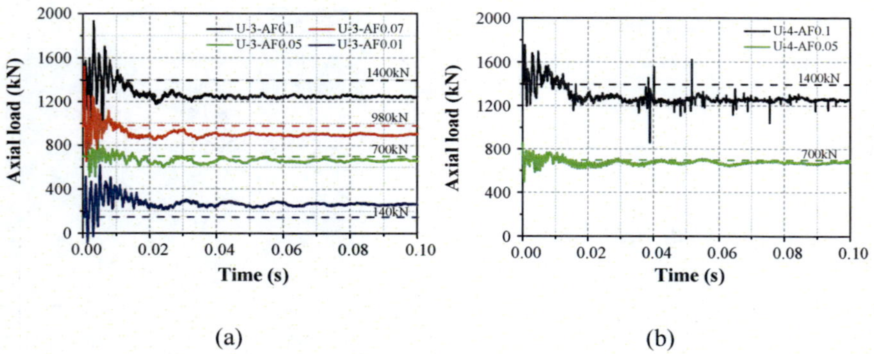

Fig. 9.3 Axial force–time history of specimens **a** 3 m drop-height **b** 4 m drop-height

from 2 to 4 m to achieve different levels of impact energy in this study. A steel cage was designed and fixed on the base of supports to ensure the safety if the drop hammer rotated and rebounded after impact.

A special support system was designed, in which the specimens were supported by a fixed support at one end (left end in Fig. 9.2) and an axially sliding one at the other end (right end in Fig. 9.2). The clear span for the specimen was 1800 mm. As shown in Fig. 9.2b, the steel clamps were used to restrain 200 mm of both ends of the specimen to prevent rotation. For the specimens with axial force, the axial force was applied at the sliding end according to the hydraulic jack with a disc spring, which was used to prevent the great loss of axial forces during the whole impact process. The various levels of axial forces can be realized steadily by adjusting the stiffness of disc spring, and the detailed information of disc springs can be referred from Wang et al. (2013). In the present test, the axial forces were measured by a force sensor during the whole impact process, as illustrated in Fig. 9.3. It can be found that, the axial force on the specimens after impact were slightly lower than initial value, due to the damage and shortening of specimens. The deviations are acceptable because the axial force reductions are less than 10% except for the specimen U-3-AF0.01, in which the abnormal response is mainly attributed to the too strong stiffness of disc springs. It should be noted that, the drop hammer would continue to impact the specimen repeatedly after the first impact. Considering that the remaining kinetic energy of drop hammer was much smaller than the original energy, and thus only the first impact process was concerned in this study.

As for the experimental measurements, two force sensors were embedded into the hammer and mounted between the sliding end of specimen and disc spring to record the impact force- and axial force–time histories, respectively. The corresponding sampling frequency of the data processing system was 100,000 Hz. Besides, a high-speed camera with 3000 Hz sampling frequency was arranged in front of test system to record the whole impact process. Five video capture points were marked along the axis of specimen, as presented in Fig. 9.2. During impact, the above tracking points were also captured by a high-speed camera to access the deflection-time histories

Table 9.4 Test matrix

Specimen	Concrete type	Drop height (m)	Impact velocity (m/s)	Axial force (kN)	Axial compressive ratio
N-2-AF0	NSC	2	6.26	0	–
U-2-AF0	UHPCC	2	6.26	0	–
U-2.5-AF0	UHPCC	2.5	7.00	0	–
N-3-AF0	NSC	3	7.67	0	–
U-3-AF0	UHPCC	3	7.67	0	–
U-3.5-AF0	UHPCC	3.5	8.28	0	–
U-4-AF0	UHPCC	4	8.85	0	–
U-3-AF0.01	UHPCC	3	7.67	140	0.01
U-3-AF0.05	UHPCC	3	7.67	700	0.05
U-3-AF0.07	UHPCC	3	7.67	980	0.07
U-3-AF0.1	UHPCC	3	7.67	1400	0.1
U-4-AF0.05	UHPCC	4	8.85	700	0.05
U-4-AF0.1	UHPCC	4	8.85	1400	0.1

by extracting from video footage using the Olympus iSpeed 3 Viewer software. The residual deflection at the mid-span was measured by height gauge to rectify the results obtained by high-speed camera.

In the present test, the influences of concrete type, impact energy (releasing height) and the axial force on the damage and dynamic responses of specimens were examined, and the corresponding test matrix was given in Table 9.4. For easy reference, the specimens' notation consists of three parts as follow: (i) "U-" and "N-" denote the concrete type, i.e., UHPCC and NSC; (ii) the middle number indicates the releasing height of hammer; (iii) "AF" and the afterward number indicate the axial force and axial compressive ratio (the ratio of the axial force to the designed axial bearing capacity of specimens). For example, the specimen "U-3-AF0.1" represents that the reinforced UHPCC specimen with the axial compressive ratio of 0.1 was impacted by the drop hammer released from 3 m height.

9.3 Test Results and Discussions

9.3.1 Failure Mode

During the impact test, the crack development process was recorded by the high-speed camera. Figure 9.4 shows the typical cracking process of specimens with different impact energy, axial force, and concrete type, i.e., N-2-AF0, N-3-AF0, U-3-AF0, U-4-AF0 and U-4-AF0.1, respectively.

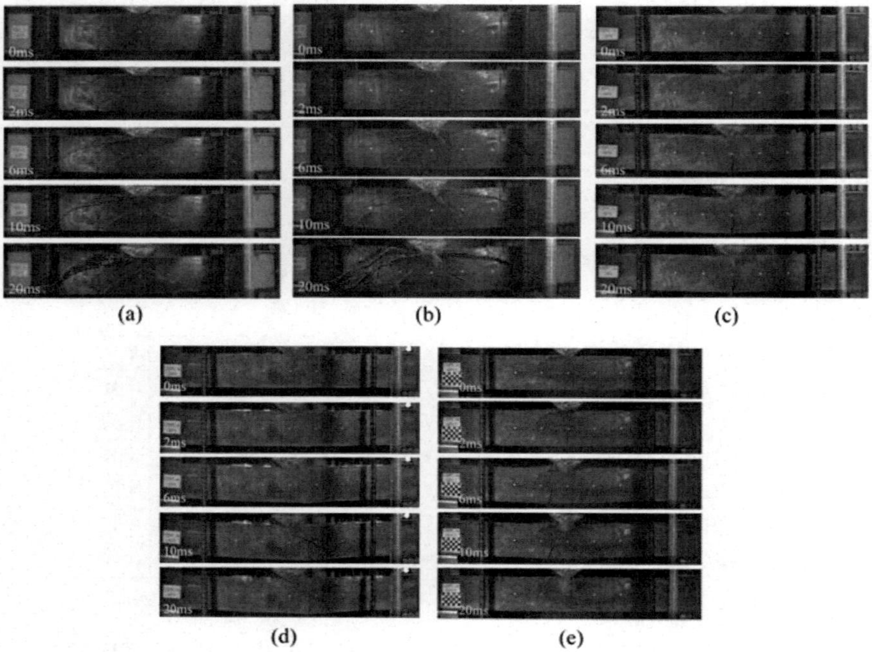

Fig. 9.4 Cracking process of specimen **a** N-2-AF0 **b** N-3-AF0 **c** U-3-AF0 **d** U-4-AF0 **e** U-4-AF0.1

Figure 9.4a, b show the progressive failure process of NSC specimens. It can be found that, the first two inclined cracks were observed at the very early stage (2 ms) on NSC specimens, which formed the shape of impact plug. At 6 ms, new diagonal crack was developed from impact location to the edge of support and formed the main crack. Then, the shear failure pattern of NSC specimens formed basically at 20 ms. Comparably, Fig. 9.4c–e present the crack development process of UHPCC specimens. Only some tiny vertical cracks were observed on the middle span at 2 ms, which continuously propagated to be the main crack. Especially, the crack pattern of U-4-AF0.1 almost disappeared at 20 ms, which is different from other specimens.

Through the above crack propagation process, the specimens finally formed different damage patterns. Two characteristic points on the crack propagation process need to be noted, i.e., the initial impact damage and final damage. (i) The initial impact damage was exhibited at 2 ms, since the peak impact stage on all specimens terminated before 2 ms (as shown in following Fig. 9.8). The initial impact damage on NSC specimens was inclined cracks (around 45°), while that on UHPCC specimens was a few vertical cracks at the same time instant. This phenomenon shows that, UHPCC plays a positive role against the initial impact damage and avoiding punching shear damage; (ii) The final damage of UHPCC specimens was formed from the propagation of initial impact damage, while the diagonal shear fractures on NSC specimens

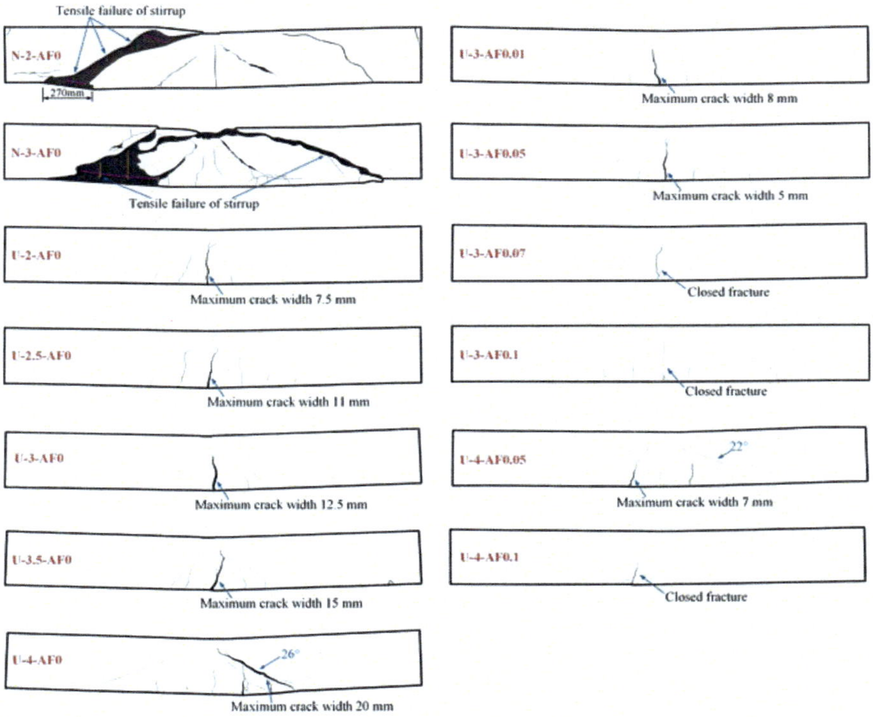

Fig. 9.5 Final failure patterns of specimens after impact

developed to be main cracks instead of the initial impact damage. It indicates that the impact resistance of NSC specimens was much weaker than UHPCC specimens during the whole impact process.

Furthermore, Fig. 9.5 presents the final damage pattern of all specimens after impact. It can be seen that, the diagonal shear fractures of NSC specimens were observed, and the main cracks propagated from the impact location to the supports with the obvious spalling of concrete cover. The damage of NSC specimens was not symmetrical due to asymmetric boundary supports, and more severe cracks was observed in the vicinity of the fixed support end. The tensile failure of stirrup occurred in the severe cracking zone of concrete. This brittle failure is similar to that of bridge pier under vehicle collisions, i.e., the direct shear damage at the bottom of pier-column. In addition, with the increase of impact velocity from 6.26 m/s (N-2-AF0) to 7.78 m/s (N-3-AF0), the damage at the sliding end of NSC specimen was extended obviously.

Comparably, all UHPCC specimens exhibited the flexural damage mode with a main flexural crack developing under the impact position. Especially, the punching shear cracks (about 25° from the horizontal) occurred on the specimens U-4-AF0 and U-4-AF0.05, while a few flexural cracks were also observed in the mid-span of

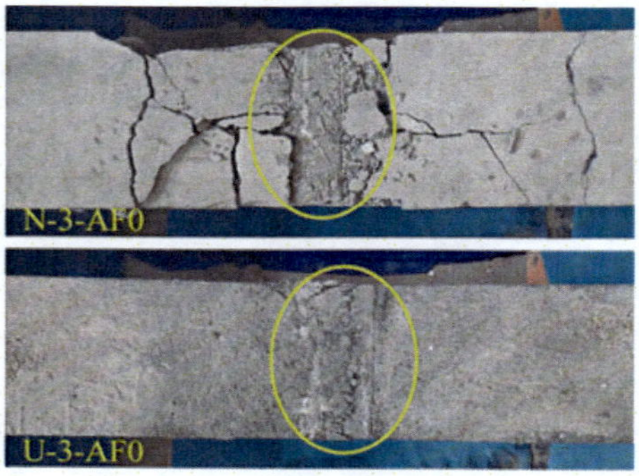

Fig. 9.6 Local damage of the impact

these two UHPCC specimens. The change of crack pattern from typical shear (NSC specimens) to a flexural-type (UHPCC specimens) could be attributed to the mix of steel fibers which could enhance both strength (tensile and shear) and fracture toughness of specimens (Biolzi and Cattaneo 2017; Mészöly and Randl 2018). It could be claimed that UHPCC specimens were more superior than NSC specimens in terms of prohibiting shear damage. In addition, the typical local damage of the impact location on the top surface of the specimens is presented in Fig. 9.6. The impact zone of specimen N-3-AF0 was severely fractured, while the local damage of specimen U-3-AF0 was relatively slight. The reason can be interpreted by the peak impact force. The corresponding peak compressive stress on specimen N-3-AF0 was 124.4 MPa which was more than the concrete strength, while the peak stress on specimen U-3-AF0 was 144.5 MPa which was less than strength of UHPCC. Additionally, the crack width increased continuously as the impact velocity increasing, while no cracks occurred on two ends of the specimens, due to the strong stiffness of UHPCC specimens. Under the identical impact velocity, from Fig. 9.5, the main crack width on the specimens decreased continuously with the increase of axial force. Especially, the punching shear cracks generally disappeared from U-4-AF0 to U-4-AF0.1. Therefore, it can be derived that the axial force could enhance both the shear capacity and impact-resistance of UHPCC specimens under impact loadings effectively. In conclusion, within the discussed parametric ranges, the employment of UHPCC and axial force can both improve the impact-resistance of concrete members.

9.3.2 *Impact Force–Time History*

Figure 9.7 presents the impact force–time history curves of all the specimens, which can be divided into three stages, i.e., peak, plateau and attenuation, as shown in the Fig. 9.8. At the first stage, the sudden lateral impact of the drop hammer induces severe vibrations of the specimen, which leads to the momentary variations of the contact area between the indenter and specimen (increasing and decreasing). Thus, the impact force immediately increases from zero to the peak value and then instantly

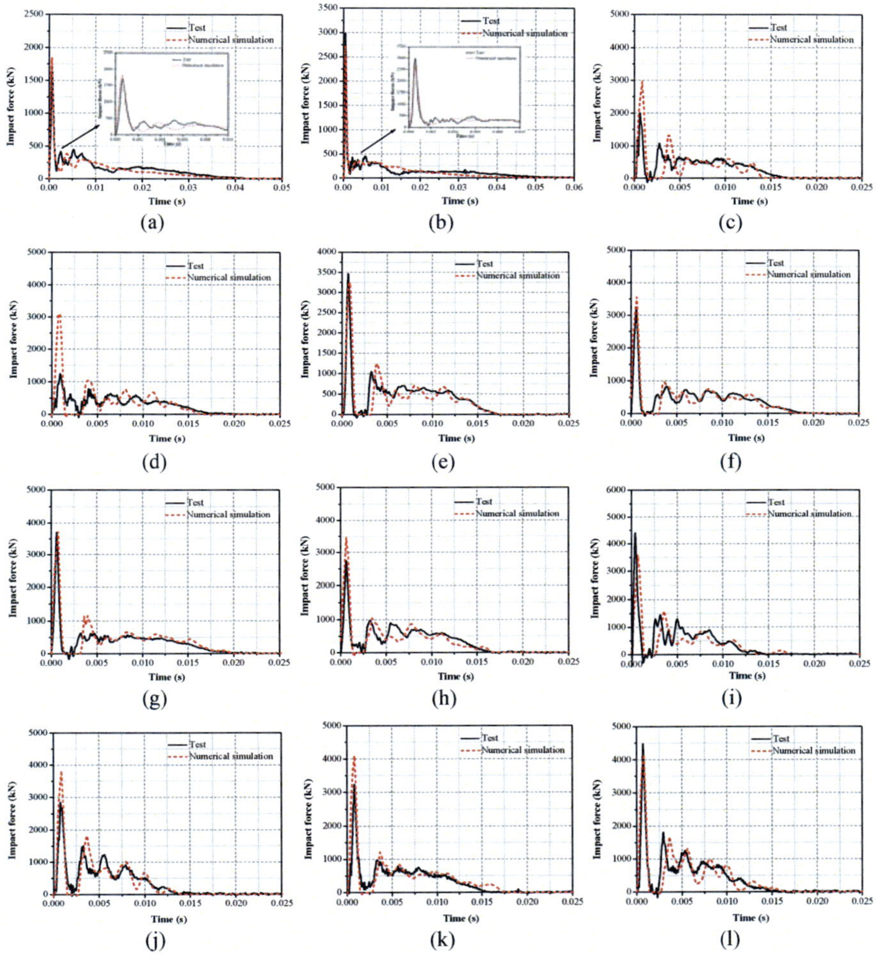

Fig. 9.7 Impact force–time histories of specimens **a** N-2-AF0 **b** N-3-AF0 **c** U-2-AF0 **d** U-2.5-AF0 **e** U-3-AF0 **f** U-3.5-AF0 **g** U-4-AF0 **h** U-3-AF0.01 **i** U-3-AF0.07 **j** U-3-AF0.1 **k** U-4-AF0.05 **l** U-4-AF0.1

decreases to the minimum value. At the second stage, the indenter and the specimen fall together and remain in contact, and the impact force fluctuates around a constant value, i.e., plateau value. At the third stage, the indenter and the specimen bounce up together after the maximum deflection, and the impact force decreases with the reduction of contact area until they separate to each other. Herein, the plateau impact force is calculated by the following equation (Wang et al. 2019b):

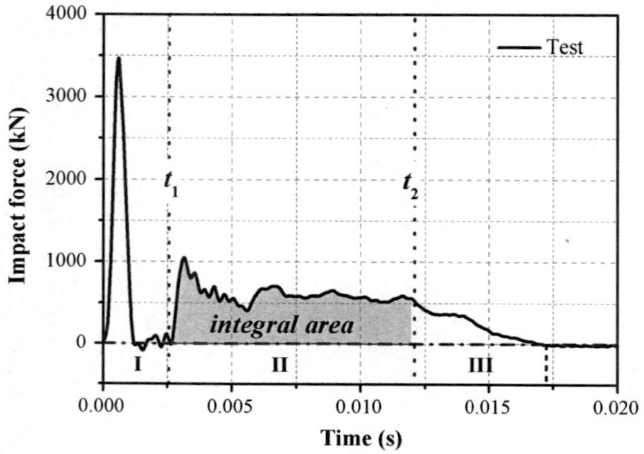

Fig. 9.8 Typical impact force–time history

$$F_s = \int_{t_1}^{t_2} F(t)dt/T \tag{9.1}$$

where t_1 and t_2 are the time instants at the start and the end of the plateau impact stage; $F(t)$ is the instantaneous impact force and T is the duration of plateau impact stage.

The present test results of specimens are listed in Table 9.5, including the impact velocity, the peak impact force F_p, the plateau impact force F_s, the maximum mid-span deflection Δ_1, the residual mid-span deflection Δ_2, the impact duration t_0, the axial force before impact F_{ab} and axial force after impact F_{aa}. It should be mentioned that the data of F_p and F_s of specimen U-3-AF0.05 is blank in Table 9.5 and Fig. 9.7 since the corresponding impact force–time history was not recorded during the test.

It is obvious that the responses of NSC and UHPCC specimens have some differences. The duration of impact force on NSC specimens is approximately twice than that on UHPCC specimens under the same impact energy. NSC specimens provided less impact resistance, and thus it takes longer time to dissipate the impulse of drop hammer. Focusing on the characteristic values (F_p and F_s) of impact force–time histories, as given in Table 9.5, these values on UHPCC specimens are obviously lager than that on NSC specimens. The peak impact force and plateau impact force could indicate the magnitude of the local stiffness and global bending stiffness, respectively (Pham et al. 2018). Thus, with the identical reinforcement arrangement, the local stiffness and global bending stiffness of UHPCC specimens are all larger than NSC specimens under impact loading. In addition, the peak impact force on NSC specimens increased with increasing the impact energy, while no regularity of the peak impact force on UHPCC specimens can be found with different impact energy. This phenomenon may be caused by the present experimental apparatus. The impact inclination angle of indenter can’t be determined in all impact scenarios since the slide rail of the hammer was absent, while the inclination angle has an obvious influence on the peak impact force (Li et al. 2019).

Table 9.5 Test data

Specimen	V_0 (m/s)	F_p (kN)	F_s (kN)	Δ_1(mm)	Δ_2(mm)	t_0 (s)	F_{ab} (kN)	F_{aa} (kN)	E_{ab} (kJ)
N-2-AF0	6.26	1703.4	303.9	50.6	35.2	28.6	–	–	14.3
N-3-AF0	7.67	2985.9	285.8	72.7	50.0	50.9	–	–	20.0
U-2-AF0	6.26	2000.7	510.1	24.8	16.2	15.2	–	–	14.2
U-2.5-AF0	7.00	1245.9	430.4	29.0	21.4	23.2	–	–	12.9
U-3-AF0	7.67	3467.0	557.9	35.5	25.8	20.0	–	–	23.5
U-3.5-AF0	8.25	3210.5	520.7	39.8	31.2	23.3	–	–	26.0
U-4-AF0	8.85	3704.6	455.5	48.4	35.6	27.6	–	–	32.4
U3-AF0.01	7.67	2746.4	595.2	31.1	19.5	17.9	140	267	20.0
U-3-AF0.05	7.67	–	–	28.0	15.9	18.7	698	674	–
U-3-AF0.07	7.67	5439.2	763.0	26.1	9.8	15.6	977	915	22.5
U-3-AF0.1	7.67	2836.7	780.0	24.7	4.7	15.1	1402	1264	22.4
U-4-AF0.05	8.85	3233.2	604.9	36.2	18.4	20.7	703	674	28.2
U-4-AF0.1	8.85	4476.3	805.0	28.8	8.6	16.8	1401	1277	31.8

As shown in Table 9.5, under the same impact energy, the plateau impact force on UHPCC specimens increases as axial force increasing. Thus, the global bending stiffness can be strengthened with a larger axial force. Typically, the confinement effect induced by the presences of axial force and spiral stirrups can improve the impact resistance of RC columns. Considering that the stirrup of present specimens is constructional reinforcement, which is deficient for confinement effect, the compressive membrane action (or called arching effect) may be the mechanism of axial force (Fan et al. 2019). With the increase of axial force, the horizontal force of the arch increases, and the dynamic bending stiffness of the specimen is improved. It is generally considered that the impact resistance of specimens can be reflected by the values of the plateau impact force, and thus the values of impact force platform are related to the ultimate bearing capacity of static loading. According to the Chinese Standard GB 50010-2010 (2010), the ultimate shear capacity of NSC and flexure capacity UHPCC specimens are 273 and 446 kN, and the corresponding plateau impact forces listed in Table 9.5 are slightly larger due to the strain rate effect.

9.3.3 Deflection-Time History

The deflection-time histories of all specimens are shown in Fig. 9.9. Generally, the deflection-time history curve can be divided into two stages in Fig. 9.9c. (i) Forced vibration stage under impact, in which deflection of tracing points goes through the maximum value and decreases until the end of impact process; (ii) Free vibration stage after impact, in which the specimen vibrates freely and then maintains a stable position. Besides, considering that the tracing points are symmetrical about middle span of specimens, i.e., points B and D, points C and E, and thus only the deflection-time histories of tracing points A, B and C are shown in Fig. 9.9. In addition, there are two characteristic values on the deflection-time history curve, i.e., the maximum and residual mid-span deflections given in Table 9.5, which are usually used to describe the impact response of the specimen.

Figure 9.10 illustrates the instantaneous deformation shape of four typical specimens without the axial force. As for the NSC specimens, due to the larger fragmentation close to fixed end, both the specimens “N-2-AF0” and “N-3-AF0” exhibit larger deflections at the fixed support side. With the increase of impact energy from 16.7 to 25.0 kJ, the maximum mid-span deflections of NSC specimens increase to 72.6 from 50.6 mm correspondingly. The relatively slight influence of support on specimen deformation can be found with larger impact energy, and the deflection of UHPCC specimens are comparably symmetric.

Figures 9.11 and 9.12 further show the maximum and residual mid-span deflections of specimens according to various concrete type and axial force levels, respectively. It can be concluded that, (i) the deflection of UHPCC specimen is almost half that of NSC specimen under the same impact energy; (ii) with the impact energy increasing, the deflection of specimens with two concrete types increases and the ratio of the maximum to the residual mid-span deflection (Δ_1/Δ_2) is almost kept

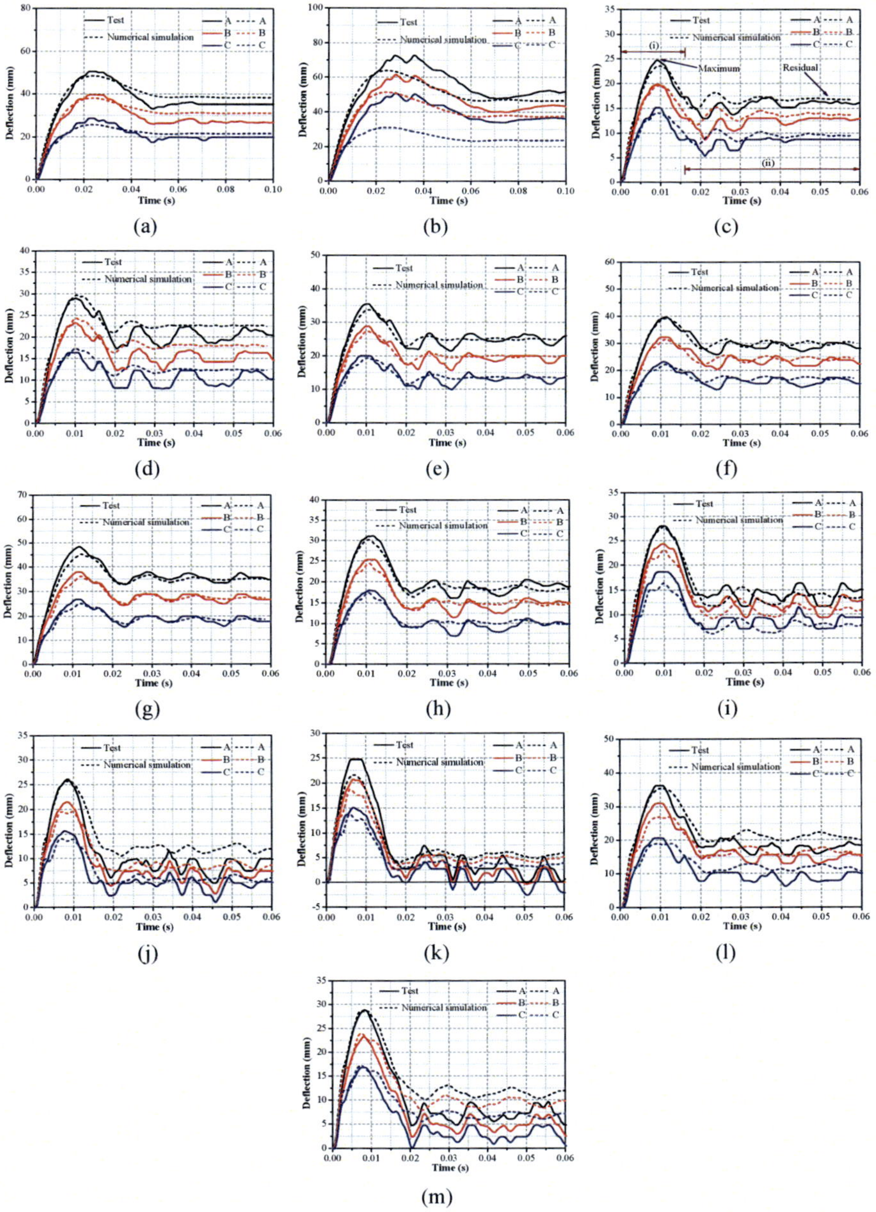

Fig. 9.9 Deflection-time histories of specimens **a** N-2-AF0 **b** N-3-AF0 **c** U-2-AF0 **d** U-2.5-AF0 e U-3-AF0 **f** U-3.5-AF0 **g** U-4-AF0 **h** U-3-AF0.01 **i** U-3-AF0.05 **j** U-3-AF0.07 **k** U-3-AF0.1 **l** U-4-AF0.05 **m** U-4-AF0.1

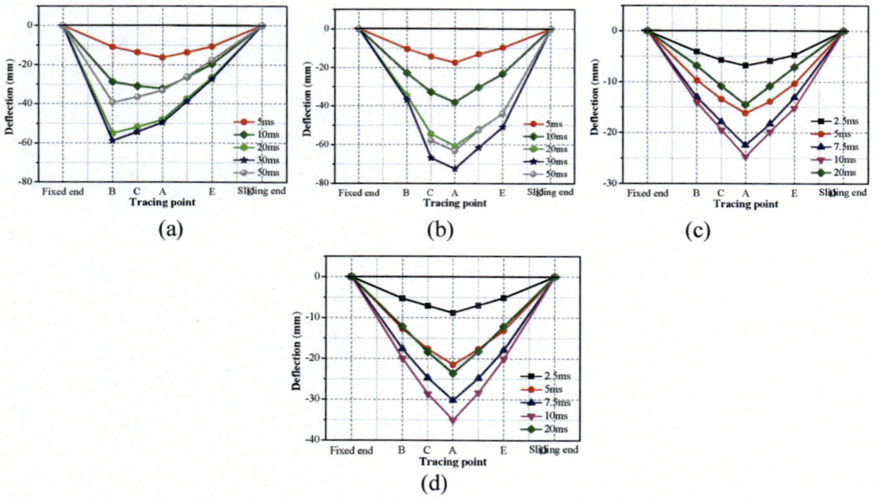

Fig. 9.10 Instantaneous deformation shape of specimens **a** N-2-AF0 **b** N-3-AF0 **c** U-2-AF0 **d** U-3-AF0

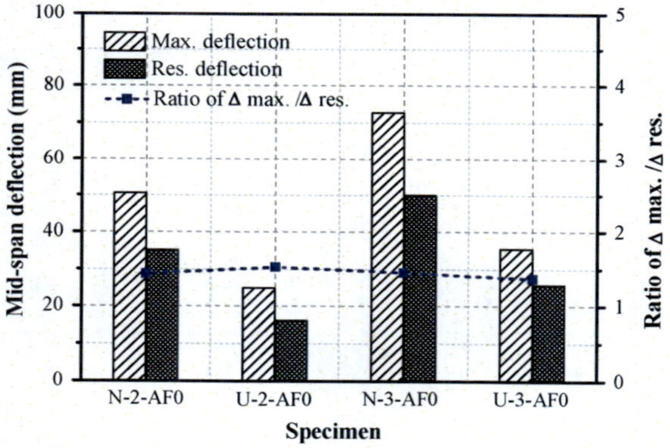

Fig. 9.11 Comparisons of specimens' deflections with different concrete type

constant; (iii) with increasing the axial force, the deflection of all UHPCC specimens decreases steadily while the Δ_1/Δ_2 ascends increasingly. Therefore, axial force is more effective for reducing the residual deflection of UHPCC specimens than the maximum deflection. In conclusion, the lateral impacted deformations of RC members can be reduced by adopting the UHPCC and applying the larger axial force.

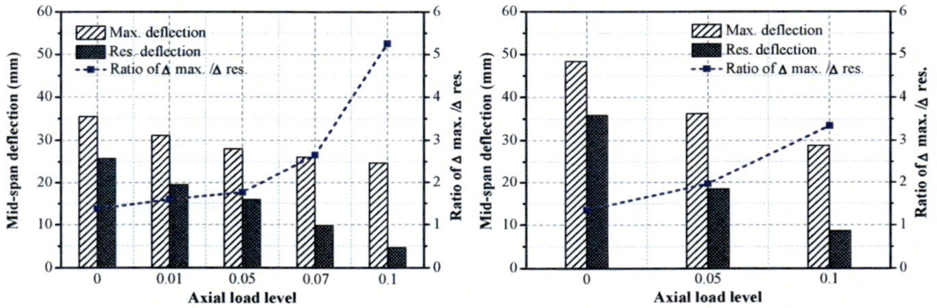

(a) (b)

Fig. 9.12 Comparisons of maximum and residual specimens' deflections according to axial force level **a** 3 m drop-height **b** 4 m drop-height

9.3.4 Energy Dissipation

Based on the experimental impact force- and mid-span deflection-time histories given in Figs. 9.7 and 9.9, the impact force versus displacement curve can be derived as schemed in Fig. 9.13a. A characteristic value E_{ab} is defined as the energy absorbed by the specimen at the maximum deflection position and given in Table 9.5, i.e., the area under the impact force–displacement curve illustrated in Fig. 9.13a. E_k is the initial kinetic energy of hammer at the beginning of impact. As shown in Fig. 9.13b, the value of the ratio of E_{ab}/E_k is varied from 62.0 to 97.3%, due to the energy dissipations such as frictional energy and sound energy during the impact process. In addition, except for the most deviated value of specimen U-2.5-AF0, it can be drawn that the energy dissipation ratio E_{ab}/E_k of all specimens is almost kept around 87.4%. Therefore, there are no influence regularities of concrete type, impact energy and axial force for dissipation of specimens' deformation.

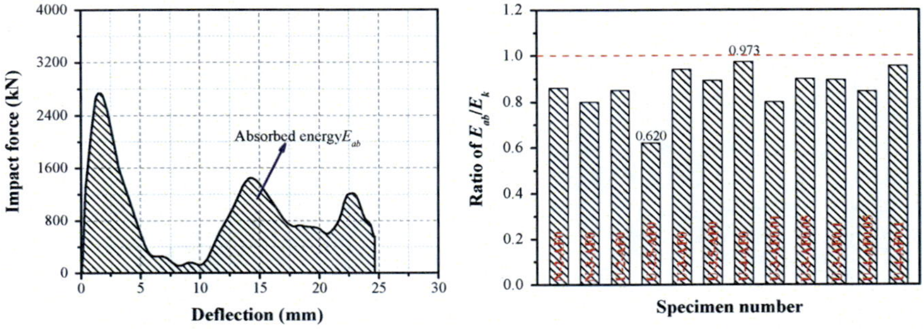

(a) (b)

Fig. 9.13 Energy dissipations of specimens' deformation **a** impact force–displacement curve **b** ratio of E_{ab}/E_k

9.4 Numerical Simulation

9.4.1 FE Model

To further study the impact resistance of specimens under low-velocity lateral impact, the numerical simulations are performed by using the commercial FE program LS-DYNA in this section. The FE model of drop hammer impact test was established based on the program Hypermesh, including the numerical models of test setup, axial force system and drop hammer, as shown in Fig. 9.14. The eight-node solid elements were employed for modelling the concrete, drop hammer, supporting frames, rollers and the disc spring. The longitudinal and stirrup bars were modeled by Hughes-Liu beam elements with 2 × 2 Gauss quadrature integration. According to the mesh convergency analysis in Sect. 9.4.3.1, the mesh size of both the solid and beam elements were determined as 10 mm. In total, the detailed FE models contained 472,607 solid and 2640 beam elements.

Considering the perfect bond between the reinforcement bars and concrete, the concrete and steel bars in specimens were modelled with shearing nodes. Besides, the interactions between different parts were realized by using the *CONTACT_AUTO_SURFACE_TO_SURFACE algorithm, and the contact parameters values are referred from Ref. (Pham et al. 2018). The contact stiffness parameters SLSFAC, SFS and SFM is 0.01, 1 and 1 respectively to reproduce the contact in test. The bottom layer elements of model were all fixed to represent the boundary conditions of test. The axial force was applied prior to some impact scenarios through compressing disc spring, and it was applied slowly enough to eliminate the axial vibration and instabilities. The initial impact velocity was adopted through *INITIAL_VELOCITY_GENERATION function to simplify the hammer free-falling process.

CSC model was adopted for both NSC and UHPCC. For NSC, only the mass density, compressive strength and maximum aggregate size are needed to input for concrete strength from 20 to 58 MPa, they are 2400 kg/m^3, 30 MPa, and 20 mm

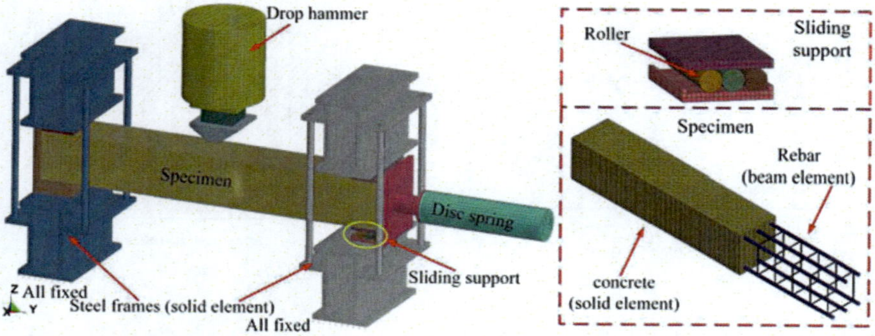

Fig. 9.14 FE model of drop hammer impact test

respectively in this study. The other model parameters can be generated automatically based on the above three parameters. For UHPCC, since the concrete compressive strength of which is beyond the default application range of CSC model, the constitutive model parameters of CSC model for UHPCC are calibrated in Sect. 9.4.2.

The plastic kinematic model (#MAT3 in LS-DYNA) is a bilinear model and usually applied to describe the behavior of steel, where the yield stress scaled by a strain rate dependent factor was expressed as

$$\sigma_d/\sigma_s = 1 + (\dot{\varepsilon}/C)^{1/P} \tag{9.2}$$

where σ_d and σ_s denote the dynamic flow stress and the associated static flow stress, respectively. $\dot{\varepsilon}$ represents the strain rate. C and P are the strain rate parameters and the values of which are adopted as 40 s^{-1} and 5, respectively (Stouffer and Dame 1996). The disc spring was also simulated with the plastic kinematic model, and its elastic modulus, i.e., 500 and 700 MPa, was calculated by the stiffness of disc spring.

9.4.2 *Calibration*

9.4.2.1 Brief Introduction of CSC Model

CSC model is originally developed to predict the dynamic performance of concrete used in roadside safety structures when involved in a vehicle collision (Murray 2007), in which the strain rate effect, material hardening and softening, damage and modulus reduction, etc. are considered comprehensively. Two versions of CSC model are available in the program LS-DYNA, i.e. MAT_CSCM_CONCRETE (automatic parameter generation with the minimum three parameters) and MAT_CSCM (full parameters input). The parameters generation algorithm in MAT_CSCM_CONCRETE is appropriate for concrete with the unconfined compressive strength ranging from 20 to 58 MPa and the maximum aggregate size ranging from 8 to 32 mm, respectively (LS-DYNA Keyword User's Manual 2014). For *MAT_CSCM, there are in total 45 parameters with a minimum 19 parameters need to be determined from basic material tests.

9.4.2.2 Calibration of FE Modelling

In this section, based on the static uniaxial tensile/compression tests, dynamic compressive and tensile tests (Hopkinson tests) and triaxial compressive tests we have conducted previously (Ren et al. 2018a; Ren et al. 2016; Wu et al. 2018), as well as the hydrostatic tests conducted by Williams et al. (2009), the parameters for modulus, strength, cap, damage, strain rate effect and hardening of CSC model were calibrated comprehensively.

(1) Modulus parameters

The bulk modulus K and shear modulus G of UHPCC can be determined by the elastic modulus E and Poisson's ratio ν in Eq. (9.3). Generally, the Poisson's ratio and elastic modulus E are derived from the static compression test on prism specimens.

$$K = \frac{E}{3(1-2\nu)}, \quad G = \frac{E}{2(1+\nu)} \tag{9.3}$$

(2) Strength parameters

The material strength in CSC model is controlled by the shear failure surface in the tensile and low confining pressure regimes (Murray 2007). There are three meridians for shear failure surface, i.e., compression, tensile and torsion, which can be determined by the triaxial compression test (TXC), triaxial tensile test (TXE) and torsion test (TOR). The compression meridian given in Eq. (9.4) is the foundation for defining the shear failure surface. Besides, the states on torsion and tensile meridians can be converted from compression meridian by the Rubin scaling function $\Re$. In Eq. (9.5), Q_1 and Q_2 are the Rubin scaling factors for torsion and extension meridians, respectively. The strength on torsion and tensile meridians are modeled as $Q_1 F_f$ and $Q_2 F_f$, respectively. The compression meridian, tension meridian and the control coordinates are illustrated in Fig. 9.15.

$$F_f = \alpha - \lambda \exp(-\beta I_1) + \theta I_1; \; I_1 = \sigma_1 + \sigma_2 + \sigma_3 \tag{9.4}$$

$$Q_1 = \alpha_1 - \lambda_1 \exp(-\beta_1 I_1) + \theta_1 I_1, \; Q_2 = \alpha_2 - \lambda_2 \exp(-\beta_2 I_1) + \theta_2 I_1 \tag{9.5}$$

where F_f is the strength in TXC;α, λ, β, θ are four strength parameters which can be determined from the TXC tests, and the parameters α_1, λ_1, β_1, θ_1 and α_2, λ_2, β_2, θ_2 respectively correspond to the torsion and tensile meridians; σ_i is the principal stress ($\sigma_3 \leq \sigma_2 \leq \sigma_1$); σ_1 and $\sigma_3(\sigma_2)$ are the axial and confining stresses in the confined compression test, respectively.

The key points on compression meridian in Fig. 9.15a are specified as follow: I, triaxial tensile state $(-3f'_{tt}, 0)$; II, biaxial tensile state $(-2f'_{bt} f'_{bt}/\sqrt{3})$; III, uniaxial

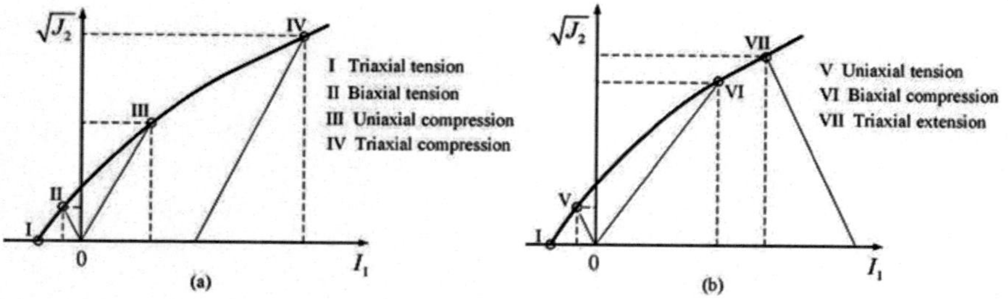

Fig. 9.15 Meridians of tensile and compression shear failure surfaces **a** compression **b** tension

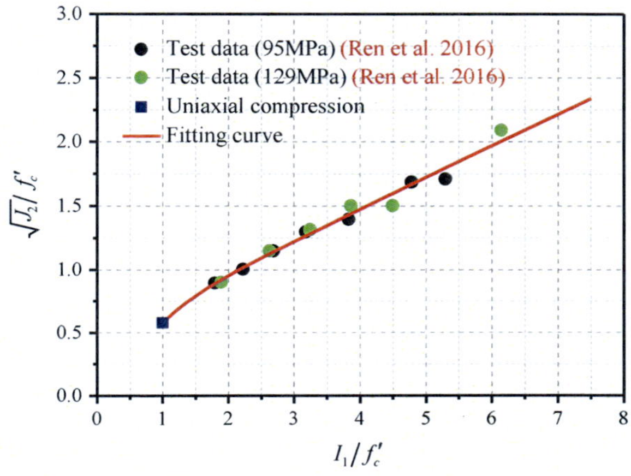

Fig. 9.16 Triaxial compression test data and the fitting curve

compression state (f_c',$f_c'/\sqrt{3}$); IV, triaxial compression state ($\sigma_1+2\sigma_3$, $(\sigma_1-\sigma_3)/\sqrt{3}$). The f_c' is the uniaxial compression strength of UHPCC. The triaxial tensile strength f_{tt}' and the biaxial tensile strength f_{bt}' of concrete are regarded approximately equaling to the uniaxial tensile f_t' (Chen 2007). The uniaxial tensile strength of UHPCC in different curing conditions can be estimated by the following empirical formula proposed by Graybeal (2006).

$$f_t' = \begin{cases} 0.55\sqrt{f_c'} & \text{for untreated} \\ 0.65\sqrt{f_c'} & \text{for underwent steam treatment} \\ 0.69\sqrt{f_c'} & \text{for tempered and delayed steam treated regimes} \end{cases} \tag{9.6}$$

Previously, we have conducted the triaxial compressive test on UHPCC specimens with the uniaxial compressive strengths of 95 and 129 MPa (Ren et al. 2016). By fitting the test data as shown in Fig. 9.16, we can obtain the formula as

$$\sqrt{J_2}/f_c' = 0.491 - 0.886 \times \exp\left(-1.71 \times I_1/f_c'\right) + 0.247 \times I_1/f_c' \tag{9.7}$$

where J_2 is the invariant of the deviatoric stress tensor, and the square root of that represents the strength on the shear failure surface. The key points at triaxial compression states can be determined by Eq. (9.7), which passes through the uniaxial compression point. In order to determine the parameters of compression meridian, the minimum five stress states are requisite. Apart from the points III and IV, another three stress states at different pressure, e.g., $I_1 = 1.2f_c'$, $3f_c'$ and $5f_c'$, computed by Eq. (9.7) are selected for the triaxial compression states. Based on the control coordinate points mentioned above, the parameters (α, λ, β, θ) of compression meridian can be obtained by fitting the Eq. (9.4).

The key points on tensile meridian in Fig. 9.15b can also be specified as follow: I, triaxial tensile state ($-3f_{tt}'$, 0); V, uniaxial tensile state ($-f_t'$, $f_t'/\sqrt{3}$); VI, biaxial compression state ($2f_{bc}'$,$f_{bc}'/\sqrt{3}$); VII, triaxial extension state ($\sigma_1+2\sigma_3$,

$(\sigma_1 - \sigma_3)/\sqrt{3}$). By performing the biaxial compressive test on UHPCC specimens, Manfred and Speck (2008) found that the biaxial compression strength f'_{bc} is approximately 1.1 times as large as the uniaxial compression strength f'_c. Moreover, considering that there are few available TXE tests on UHPCC and the ratio f'_t/f'_c of UHPCC is similar to that of NSC, the TXE test data of NSC and high strength concrete (HSC) are adopted at present. Figure 9.17 presents the tests data from Mills and Zimmerman (1970), Kotaovos and Newmen (1979) and Dupray et al. (2010), the corresponding linear fitting equation can be derived as

$$\sqrt{J_2}/f'_c = 0.261 \times \left(I_1/f'_c\right) + 0.0396 \tag{9.8}$$

Similar with the compression meridian, three another stress states at different pressure, e.g., $I_1 = 1.2f'_c$, $3f'_c$ and $5f'_c$, are selected as the critical points for tensile meridian. It is also worth noting that $Q_1 = 0.5774$ and $Q_2 = 0.5$ at zero pressure need to be considered for a smooth transition between the tensile and compressive pressure regions (Murray 2007; Murray et al. 2007). The parameters ($\alpha_2, \lambda_2, \beta_2, \theta_2$) of tensile meridian can be obtained by fitting the Eq. (9.5). In addition, the Rubin scaling factor Q_1 for torsion meridian can be determined by Q_2 according to Willians-Warnke formulation given in Eq. (9.9), and the parameters $\alpha_1, \lambda_1, \beta_1, \theta_1$ of torsion meridian can be acquired through fitting Eq. (9.5).

$$Q_1 = \frac{\sqrt{3}\left(1 - Q_2^2\right) + (2Q_2 - 1)\sqrt{3\left(1 - Q_2^2\right) + 5Q_2^2 - 4Q_2}}{3\left(1 - Q_2^2\right) + (1 - 2Q_2)^2} \tag{9.9}$$

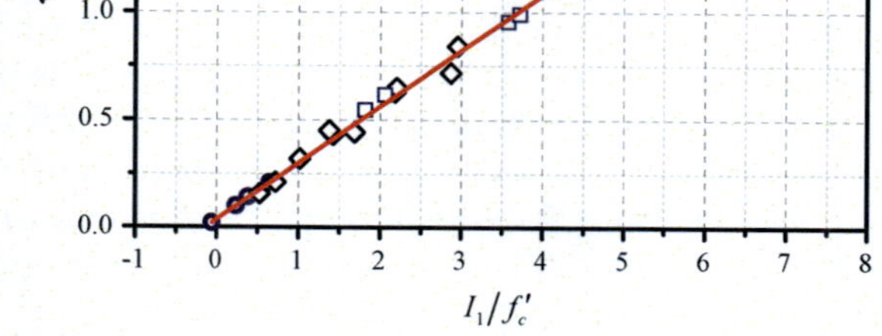

Fig. 9.17 Triaxial tensile test data and the fitting curve

(3) Cap parameters

UHPCC is a type of cement-based composite with internal defect like micro-void, etc., which is generally compacted and collapsed with increasing the hydrostatic pressure. Thus, the plastic deformation occurs on UHPCC when the pressure reaches some threshold value, which is described by the cap hardening surface given in Eq. (9.10) in CSC model. The cap hardening parameters consists of R, X_0, W, D_1 and D_2, which can be fitted by pressure-volumetric strain curves of hydrostatic compression tests. The parameter X_0 is the initial location of the cap where the compaction initiates in isotropic compression. Figure 9.18 presents the test data (ANSYS Autodyn User's Manual 2013, Schwer 2001; Williams et al 2009) concerning to the parameter X_0, the linear fitted equation is listed in Eq. (9.11).

$$\varepsilon_v^p = W(1 - e^{-D_1(X - X_0) - D_2(X - X_0)^2}) \tag{9.10}$$

$$X_0 = 3.474 f_c' - 62.24 \tag{9.11}$$

where ε_v^p is the plastic volumetric strain and W is the maximum plastic volume compaction. In addition, the cap curve usually passes through uniaxial compression state III $(f_c', f_c'/\sqrt{3})$ (Jiang and Zhao 2015). Thus, the cap aspect ratio R can be obtained by the following formula.

$$R = (X_0 - f_c')/(f_c'/\sqrt{3}) \tag{9.12}$$

$$R = (2.474 f_c' - 62.24)/(f_c'/\sqrt{3}) \tag{9.13}$$

Williams et al. (2009) had conducted a series of hydrostatic compression tests on UHPCC and obtained the pressure-volumetric strain curve. Figure 9.19 presents the corresponding test data, and the parameters of cap surface can be derived by fitting

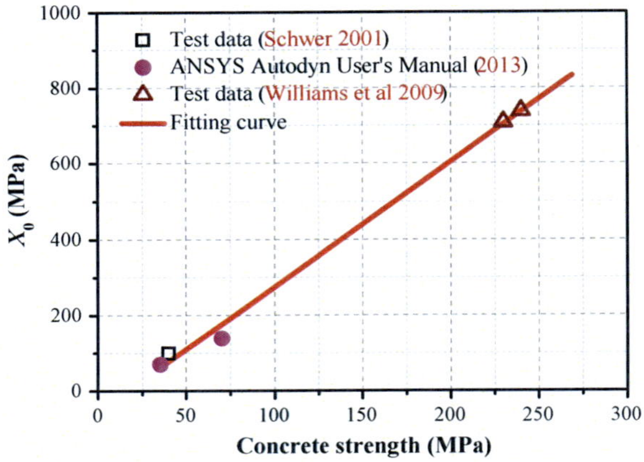

Fig. 9.18 Cap initial location parameter X0

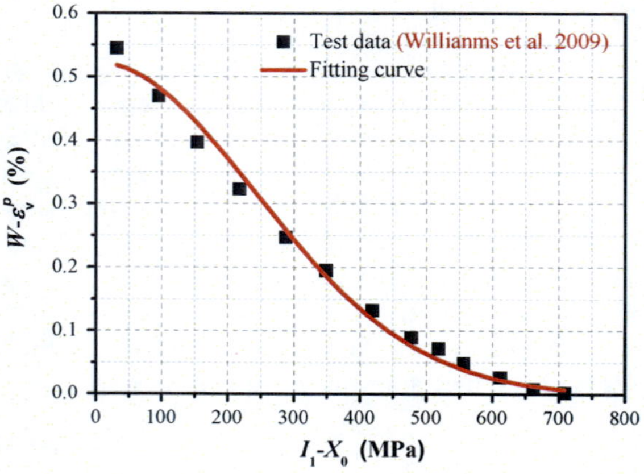

Fig. 9.19 Cap shape parameters

Eq. (9.14) rewritten from Eq. (9.10), i.e., the maximum plastic volume W is 0.00522, the other parameters D_1 and D_2 are 8.461 × 10^{-6} and 2.254 × 10^{-10}, respectively.

$$D_1(\mathrm{X}-X_0)+D_2\,(\mathrm{X}-X_0)^2=-\ln\left(\frac{W-\varepsilon_v^p}{W}\right) \tag{9.14}$$

(4) Damage parameters

The damage formulation of CSC model includes the strain softening and modulus reduction. Strain softening refers to the decreasing of strength after the initial yielding, and modulus reduction refers to the decreasing of loading or unloading slopes in cyclic loading or unloading tests. With the accumulation of damage, the damage-level parameter d increases from zero to one. The damage is cumulative according to the following Eqs. (9.15, 9.16) with the parameters A, B, C and D, in which B and D are user-specified input parameters, while the parameters A and C are determined by the model's formulation. Besides, the parameter *pmod* can reduce A only when A is less than 0.999 (Murray 2007), as shown in Eq. (9.17).

$$\sigma_{ij}^d=(1-d)\sigma_{ij}^{vp} \tag{9.15}$$

$$d(\tau_b)=\frac{0.999}{D}\left\{\frac{1+D}{1+D\exp[-C(\tau_b-r_{0b})]}-1\right\} \tag{9.16a}$$

$$d(\tau_d)=\frac{d_{\max}}{B}\left\{\frac{1+B}{1+B\exp[-A(\tau_d-r_{0d})]}-1\right\} \tag{9.16b}$$

$$A=A(d_{\max}+0.001)^{p\ \mathrm{mod}} \tag{9.17}$$

where σ_{ij}^{vp} and σ_{ij}^{d} are the stress tensors without and with damage; τ_b and τ_d are the brittle damage threshold and ductile damage threshold, r_{0b} and r_{0d} are the corresponding initial threshold. In addition, the brittle fracture energy G_f^B and ductile fracture energy G_f^D are adopted to dominate the damage, which can be expressed as

$$G_f^B = GFS + trans_B(GFT - GFS) \tag{9.18a}$$

$$G_f^D = GFS + trans_D(GFC - GFS) \tag{9.18b}$$

$$trans_B = \left(-I_1/\sqrt{3J_2}\right)^{pwrt},\ trans_D = \left(I_1/\sqrt{3J_2}\right)^{pwrc} \tag{9.18c}$$

where parameters *GFC*, *GFT* and *GFS* are the fracture energy under uniaxial compression, uniaxial tension and pure shear, respectively. The parameters *pwrt* and *pwrc* are the transition parameter for shear-to-compression and shear-to-tension, respectively. To make that transition smooth, the default values 5.0 and 1.0 are adopted for the parameters *pwrt* and *pwrc* (Murray 2007). The fracture energy parameters (*GFC*, *GFT* and *GFS*) are related to the critical length of the element l_e in the simulation, as given in Eq. (9.19). Considering that the 10 mm mesh size is determined by the mesh convergency analyses in the following numerical simulations, the critical length of element is adopted as 10 mm at present.

$$\int \sigma d\varepsilon = G_f/l_e \tag{9.19}$$

The complete stress–strain curves of uniaxial compression tests on UHPCC specimens with different compressive strength are obtained by Ren et al. (2018a), Hassan et al (2012) and Prabha et al (2010), respectively. The test data are shown in Fig. 9.20, and the formulation of the fitting curve can be derived as

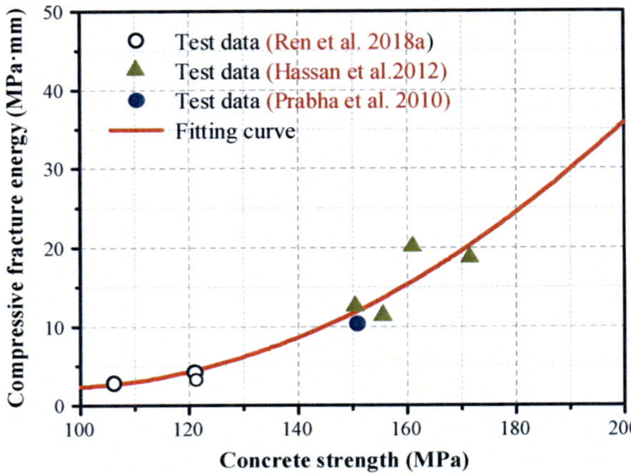

Fig. 9.20 Compression fracture energy of UHPCC

$$\int \sigma_c d\varepsilon = 0.000296 \times (f_c')^2 - 0.0553 \times f_c' + 2.812 \tag{9.20}$$

The default values for the compressive fracture energy are set as 100 times of tensile fracture energy (Murray 2007). Actually, the tensile fracture energy of UHPCC is assumed to be 1/20 of the compressive fracture energy based on the uniaxial tensile-compression test data (Ren et al. 2018a). The above assumption is based on the UHPCC without obvious tensile hardening effect. The shear fracture energy of CSC model is the same as that of uniaxial tensile fracture energy by default (Murray 2007). The parameter *repow* is recommended as 1.0 to make the fracture energy to increase with increasing the strain rate effect (Murray et al. 2007).

In addition, the softening parameters B and D can be calibrated through single element numerical simulation. The parameter B controls the ductile softening behavior. Figure 9.21a shows the influence of the parameter B on the compression softening behavior when the parameter GFC is kept constant, i.e., $GFC =$ 3.385 MPa·mm. Comparing the descending trend of compression stress in Fig. 9.21a, the parameter $B = 0.1$ is adopted. Besides, the parameter D controls the ductile softening behavior. Similarly, as illustrated in Fig. 9.21b, the fastest descending trend of uniaxial tensile curve approaching to the test curve when the parameter D is set as 0.001.

(5) Strain rate parameters

The viscoplastic algorithm is used in CSC model to describe the strain rate effect of concrete material on the yield surface. As a user-specified input parameter, the rate effect parameter η is applied on the Simo and Ju (1987) extension of the Duvaut-Lions formulation to realize viscoplastic algorithm. At each time step, the viscoplastic algorithm interpolates between the elastic trial stress σ_{ij}^T and the inviscid stress without rate effect σ_{ij}^P to set the viscoplastic stress with rate effect σ_{ij}^{vp}, which is expressed as (Murray 2007):

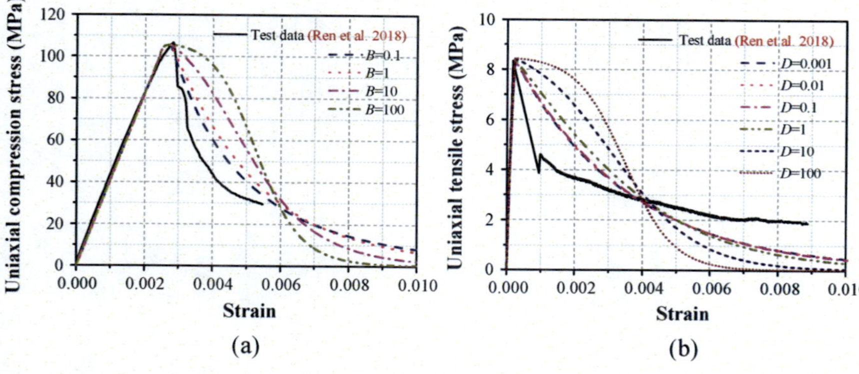

Fig. 9.21 Strain softening parameters **a** uniaxial compression **b** uniaxial tensile

$$\sigma_{ij}^{vp} = (1-\gamma)\times\sigma_{ij}^{T} + \sigma_{ij}^{P},\ \gamma = (\Delta t/\eta)/(1+\Delta t/\eta) \tag{9.21}$$

This interpolation γ depends on the effective fluidity coefficient η and the time step Δt. In addition, the coefficient η can be estimated by four user input parameters given in Eq. (9.22), such as the fluidity parameters (η_t, η_s, η_c) in uniaxial tensile stress, shear stress, and uniaxial compressive stress. The dynamic tensile and compressive strength $f'_{t,d}$ and $f'_{c,d}$ can be obtained according to Eq. (9.23).

$$\eta_t = \eta_{0t}/\dot{\varepsilon}^{N_t},\ \eta_c = \eta_{0c}/\dot{\varepsilon}^{N_c},\ \eta_s = Srate\times\eta_t \tag{9.22}$$

$$f'_{c,d} = f'_c + E\dot{\varepsilon}\eta_D,\ f'_{t,d} = f'_t + E\dot{\varepsilon}\eta_B \tag{9.23a}$$

$$\eta_D = \eta_s + trans_D(\eta_c-\eta_s),\ \eta_B = \eta_s + trans_B(\eta_t-\eta_s) \tag{9.23b}$$

where the parameter $\dot{\varepsilon}$ is the strain rate, η_{0t} and η_{0c} are the strain rate effect parameters for uniaxial tension and compression, respectively. N_t and N_c are the strain rate effects power for uniaxial tension and compression, respectively. *Srate* is the ratio of effective shear stress to tensile stress fluidity parameters. Therefore, the dynamic increasing factors (DIF_t and DIF_c) for direct tension and unconfined compression can be expressed as

$$DIF_t = f'_{t,d}/f'_t = 1 + E\dot{\varepsilon}\eta_{0t}/(f'_t\dot{\varepsilon}^{N_t}) \tag{9.24a}$$

$$DIF_c = f'_{c,d}/f'_c = 1 + E\dot{\varepsilon}\eta_{0c}/(f'_c\dot{\varepsilon}^{N_c}) \tag{9.24b}$$

Previously, we have examined the strain rate effect of UHPCC under uniaxial tension and compression based on the Hopkinson bars impact test (Ren et al. 2018b; Wu et al. 2018), and presented the following Eq. (9.25) to estimate the DIF_t and DIF_c of UHPCC.

$$DIF_t = \begin{cases} (\dot{\varepsilon}/\dot{\varepsilon}_{0t})^{0.018}\ \dot{\varepsilon} \le \dot{\varepsilon}_{TR} \\ 11.561\times\log\dot{\varepsilon} - 10.446\,\dot{\varepsilon} > \dot{\varepsilon}_{TR} \end{cases} \tag{9.25a}$$

$$DIF_c = \begin{cases} (\dot{\varepsilon}/\dot{\varepsilon}_{0c})^{0.014}\ \dot{\varepsilon} \le \dot{\varepsilon}_{TR} \\ 0.5103\times(\log\dot{\varepsilon})^2 - 1.2301\times\log\dot{\varepsilon} + 1.6804\,\dot{\varepsilon} > \dot{\varepsilon}_{TR} \end{cases} \tag{9.25b}$$

where $\dot{\varepsilon}_{0t}$ is the strain rate for static tensile loading (1.0×10^{-6} s^{-1}) and $\dot{\varepsilon}_{0c}$ is the strain rate for static compression loading (3.0×10^{-5} s^{-1}); and the inflection points $\dot{\varepsilon}_{TR}$ in Eq. (9.25a) and (9.25b) are 10.5 s^{-1} and 90 s^{-1}, respectively. Therefore, the strain rate formula Eq. (9.24) of CSC model can be determined based on the data from Eq. (9.25), as shown in Fig. 9.22. The corresponding strain rate parameters can be obtained as

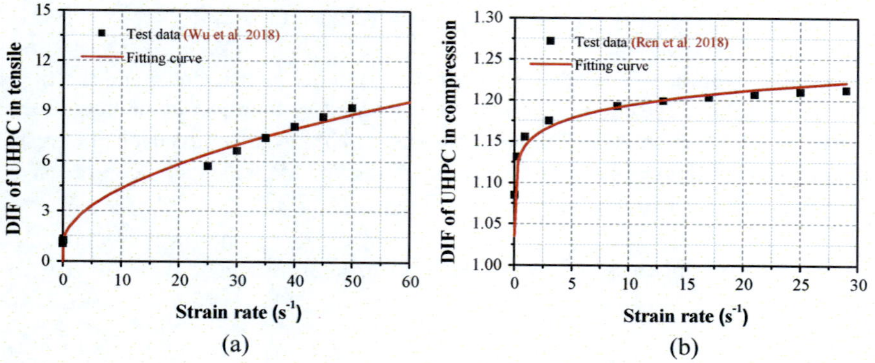

Fig. 9.22 Strain rate effect parameters **a** tension **b** compression

$$N_t = 0.4783,\ N_c = 0.8703,\ \eta_{0t} = 1.013f_t'/E,\ \eta_{0c} = 0.1430f_c'/E \tag{9.26}$$

(6) Hardening parameters

The dynamic hardening effect of UHPCC at high strain rate is more obvious than that under static loadings. Furthermore, based on the dynamic stress–strain curve we have previously obtained from the Hopkinson test under different strain rates (Ren et al. 2018b), e.g., 60 and 100 s^{-1} are chosen at present, the numerical simulation of single element is carried out to obtain the hardening parameters by trial and error. As shown in Fig. 9.23, by comparing with the test data, the hardening parameters N_H and D_H are determined as 0.98 and 150, respectively.

Until now, we can derive that, for different strength grades of UHPCC, only the basic mechanical properties such as compressive strength, Poisson's ratio and elastic modulus are necessary for the full parameters of CSC model for UHPCC. The parameters generation method is given in Table 9.6. As presented in

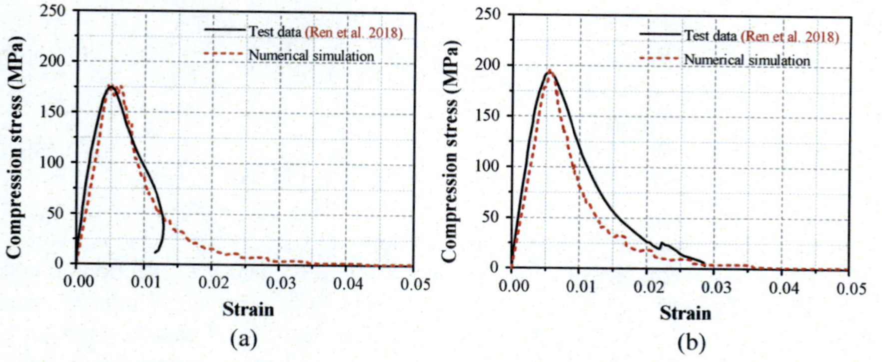

Fig. 9.23 Kinematic hardening parameters **a** strain rate of 60 s^{-1} **b** strain rate of 100 s^{-1}

Table 9.6 Parameters generation method for CSC model parameters of UHPCC

Shear failure surface parameters							
$\alpha, \lambda, \beta, \theta, \alpha_1, \lambda_1, \beta_1, \theta_1, \alpha_2, \lambda_2, \beta_2, \theta_2$				Calculated by fitting the meridian curve in Fig. 9.15			
Cap surface parameters		Damage parameters				Strain rate parameters	
X_0	$3.474 \times f_c' - 62.24$	B	0.1			N_c	0.8703
R	$(2.474f_c' - 62.24)/(f_c'/\sqrt{3})$	D	0.001			η_{0c}	$0.1430f_c'/E$
W	0.00522	GFC	$(0.000296f_c' - 0.0553 \times f_c' + 2.812)l_e$			N_t	0.4783
D_1	2.254×10^{-10}	GFT	$GFC/20$			η_{0t}	$1.013f_t'/E$
D_2	8.461×10^{-6}	GFS	$GFC/20$			$Srate$	1.0
		$pmod$	0			$Repow$	1.0
Modulus parameters		Harden parameters		Transition parameters		Over stress parameters	
G	$G = E/2(1+\nu)$	C_H	0.98	$pwrc$	5.0	$overc$	1000
K	$K = E/3(1-2\nu)$	N_H	150	$pwrt$	1.0	$overt$	50

Table 9.7 Shear failure surface parameters for verification tests

Test	α	λ	β	θ	α_1	λ_1	β_1	θ_1	α_2	λ_2	β_2	θ_2
This study	85.64	72.24	0.00516	0.2262	1.0	0.4226	0.00122	0	1.0	0.5	0.00098	0
Fujikake et al. (2006)	116.18	97.19	0.00405	0.2328	1.0	0.4226	0.00109	0	1.0	0.5	0.00086	0
Yoo et al. (2015)	94.93	79.42	0.00458	0.2250	1.0	0.4226	0.00116	0	1.0	0.5	0.00093	0
Fan et al. (2019)	95.09	78.92	0.00448	0.2233	1.0	0.4226	0.00135	0	1.0	0.5	0.00106	0
Wei et al. (2019)	84.65	77.07	0.00537	0.2235	1.0	0.4226	0.00132	0	1.0	0.5	0.00041	0
Wu et al. (2019)	85.64	72.24	0.00516	0.2262	1.0	0.4226	0.00122	0	1.0	0.5	0.00098	0
Wang et al. (2019a)	83.35	69.71	0.00521	0.2250	1.0	0.4226	0.00132	0	1.0	0.5	0.00041	0

Table. 9.7, the shear failure surface parameters of UHPCC for the present and existing low-velocity impact tests are listed.

9.4.3 *Comparisons of Numerical Results with Test Data*

9.4.3.1 Mesh Convergency Analyses

To accurately capture the deformation and damage of specimens against lateral impact, sufficiently small element is required in the numerical simulation, while the corresponding computation cost will be increased correspondingly. The computational solution could be converged by refining the mesh, and the specimen U-2-AF0 was taken as an example for mesh convergence analyses at present. Figure 9.24 shows the mid-span deflection time-histories and damage contours of specimen U-2-AF0 with four different element sizes, i.e., 5, 10, 20 and 30 mm. The "Fringe Levels" presented the scalar damage parameter ranged from 0 (no damage) to 1 (complete damage). It is noted that maximum deflections for different element sizes are similar, but larger residual deflection is caused by the refiner mesh. Regarding both the computing efficiency and numerical accuracy, the 10 mm element was adopted for the present numerical simulation. The hourglass energy is the negative effect of using the reduced integration in explicit analysis to speed up the calculation. The hourglass control do not affect the accuracy of the simulation results when hourglass

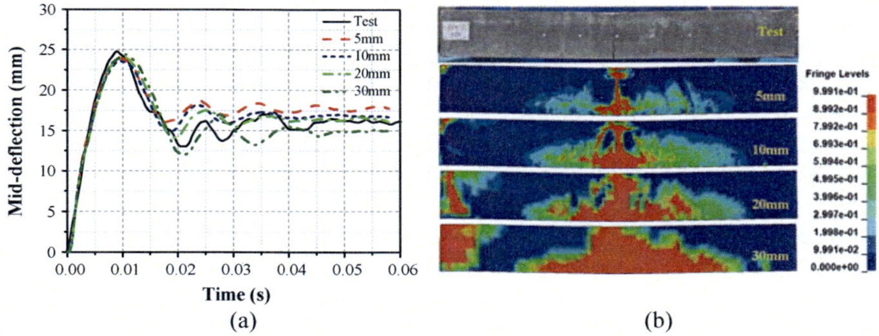

Fig. 9.24 Mesh sensitivity tests on U-2-AF0 **a** time-histories of mid-deflection **b** failure modes

Fig. 9.25 Energy-time histories of specimen U-2-AF0

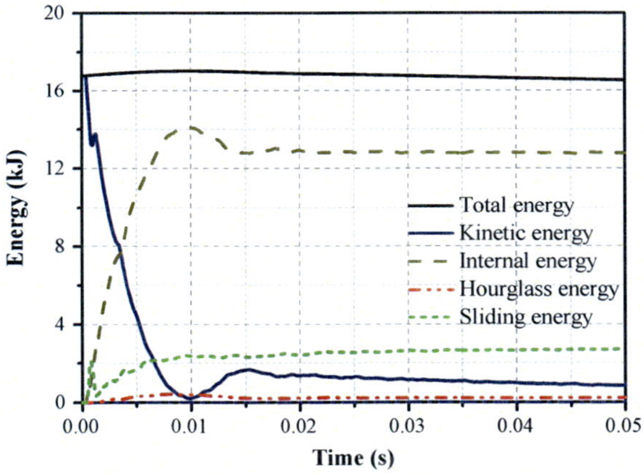

energy is below 5% of the total energy (Wang et al. 2019b). The energy time-history of the test on specimen U-2-AF0 is shown in Fig. 9.25. The initial kinetic energy of drop hammer is transformed into the internal energy, sliding energy, residual kinetic energy and hourglass energy during the whole impact process. The hourglass energy is 1.4% of the total energy, and thus the hourglass control in this FE model is reasonable. In addition, the dissipation energy of specimen U-2-AF0 when the maximum deflection is reached accounts for 84.6% of the total energy, while the corresponding value is 85% in the test. In conclusion, the present established FE model can be proven to be effective.

9.4.3.2 Validation of FE Model

The comparisons between the experimental and numerical impact force- and deflection-time histories of the present NSC and UHPCC specimens are presented in Figs. 9.7 and 9.9, respectively. It indicates that, the numerically simulated impact force- and deflection-time histories are in good agreement with the test data except for the peak value of impact force, which may be caused by the change of contact stiffness, as well as the impact inclination of indenter. It is worthily noted that the contact stiffness would significantly affect the peak impact force, and the contact stiffness is sensitive to the actual impact inclination and the contact condition (Li et al. 2019). In addition, the contact stiffness has little effect on the impact impulse and the deformation response of specimens.

Furthermore, the comparisons between the experimental and numerical damage of specimens are shown in Fig. 9.26. It can be drawn that, the numerical model can capture the identical diagonal shear fractures of NSC specimens as exhibited experimentally in Fig. 9.26a. As shown in Fig. 9.26b, the numerical results of UHPCC specimens with the developed CSC model exhibits the same ductile bending damage as the experimental results. The shear damage on UHPCC specimens without axial force exaggerated generally with increasing the impact energy in the simulation. Additionally, as shown in Fig. 9.27, the simulated instantaneous damage contour agrees well with the experimental photograph recorded by the high-speed camera of specimen U-3-AF0.

Generally, by comparisons of the experimental and numerically predicted damage and dynamic response of present specimens, the validations of the calibrated CSC model parameters for UHPCC are verified.

9.4.3.3 Discussion on Axial Force

The axial force has a great degree of inhibition to the deflection response of the UHPCC specimens based on the above experimental and numerical analyses, which can be attributed to the following two reasons. The first one is that the core UHPCC is restrained by the axial force, and the pre-load axial compressive stress could help postpone the propagation of tensile cracks to enhance the impact-resistance of specimen. The second reason is compressive membrane action (or called arching effect) of column under impact, which might be more important (Fan et al.2019). From the numerical results, as shown in Fig. 9.28, the mechanism of axial force is an energy transfer system based on compressive membrane action. As the deflection of UHPCC specimens increasing, the axial force absorbs a part of impact energy by doing the negative work. Comparably, the axial force provides the energy needed for the recovery of UHPCC specimens by doing positive work during the rebound stage of specimen. The internal energy in disc spring and mid-span deflection-time histories are presented in Fig. 9.28. The axial compressive length of the disc spring is 5.35 mm, which is calculated from spring internal energy.

Aiming to clearly demonstrate the mechanism of compressive membrane action, a simplified arch model has been provided in Fig. 9.29. The h' in simplified model is vertical distance between mid-span and supported plastic hinges, which is 140 mm in present test. According to the geometric relation of the arch model, the axial elongation of specimen U-3-AF0.1 at the maximum deflection is 5.26 mm. The elongation

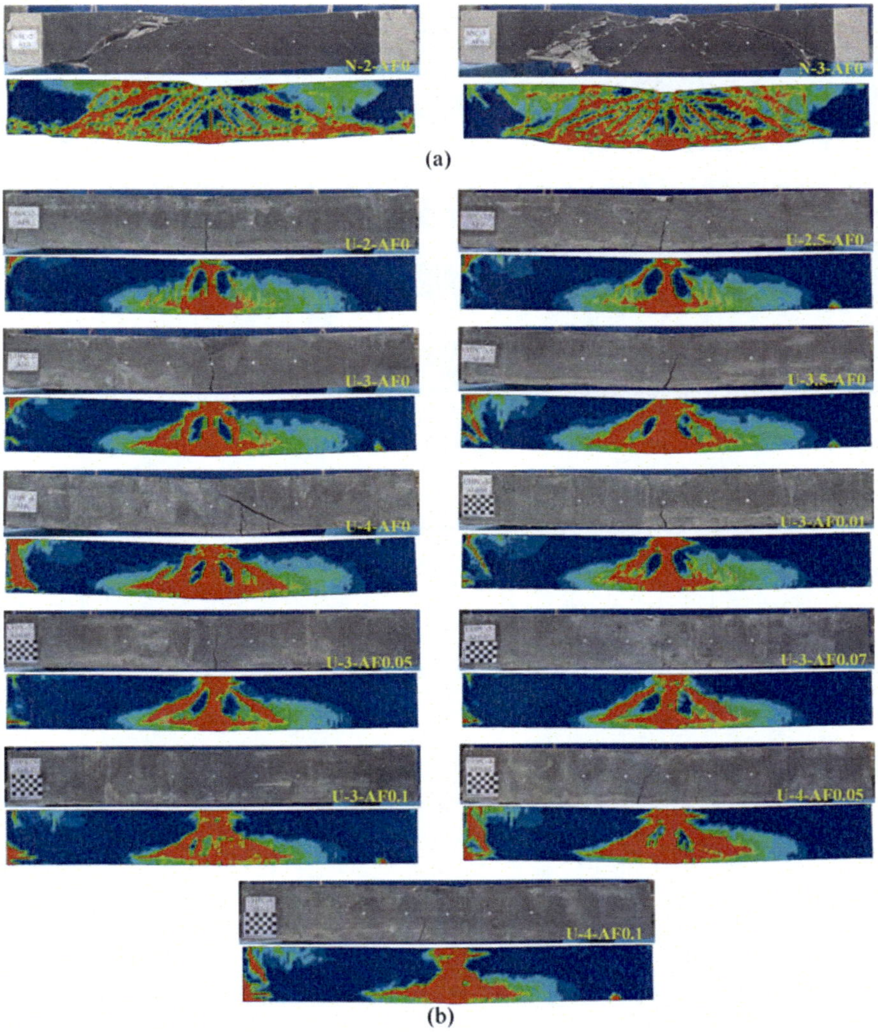

Fig. 9.26 Final damage of NSC/UHPCC specimens after impact **a** NSC **b** UHPC

calculated by the simplified arch model is basically same as axial compressive length of disc spring in numerical simulation. Therefore, the arching effect and simplified arch model can be verified.

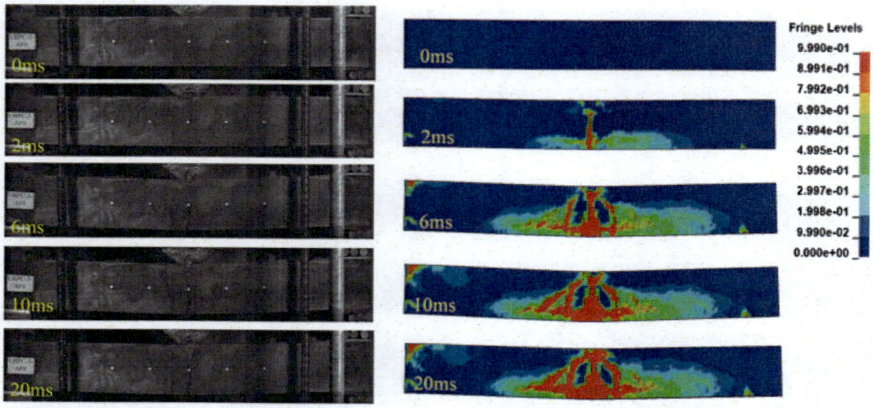

Fig. 9.27 Instantaneous damage process of specimen U-3-AF0

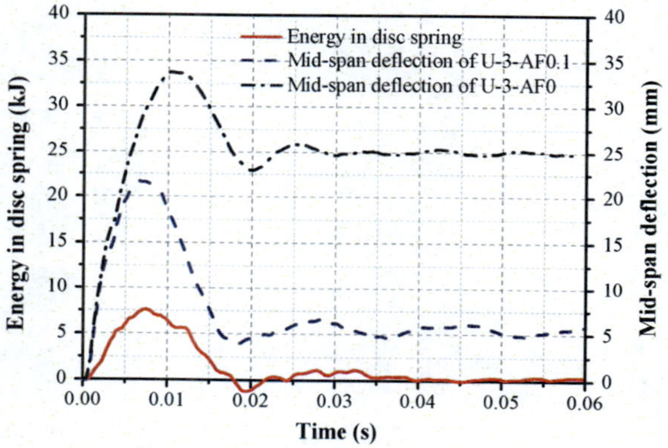

Fig. 9.28 Energy in disc spring and mid-span deflection-time histories (releasing height 3 m)

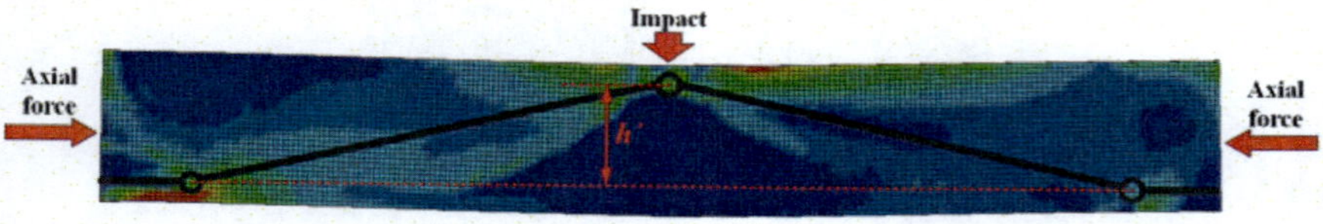

Fig. 9.29 Mechanism of axial force

9.5 Further Validations

In this section, based on the existing drop hammer impact tests on UHPC and UHPC-FST members, the applicability of the proposed determination approach of CSC model parameters for UHPC is further examined. Table 9.8 lists the corresponding details of the test setup.

Table 9.8 Experiments of further validation

References	Specimen type	f_c'(MPa)	Length (mm)	Boundary	Axial force (kN)
Fujikake et al. (2006)	UHPC beam	214.7	1700	Simply supported	0
Yoo et al. (2015)	UHPC beam	152.5	2900	Fixed supported	0
Fan et al. (2019)	UHPC column	155	2200	Sliding supported	0/400
Wu et al. (2019)	UHPC-FST	141.5	2000	Simply supported	0
Wang et al. (2019b)	UHPC-FST UHPC-FDST	136	2000	Fixed supported	0/200
Wei et al. (2019)	UHPC column	136	2000	Fixed supported	200

9.5.1 Reinforced UHPC Members

9.5.1.1 Fujikake et al. (2006) Test

Fujikake et al. (2006) conducted the drop hammer test on I-type UHPC beams with different impact energy. The experimental device and cross section of specimens are shown in Fig. 9.30. The specimen is simple supported, with a total length of 1700 mm and a net span of 1200 mm. The weight of the drop hammer is 400 kg, the diameter of the hemispheric indenter is 90 mm, and the impact position is in the middle span of the specimen. The releasing heights of the drop hammer were ranged from 0.8 to1.6 m. The diameter, elastic modulus and yield strength of longitudinal reinforcement in specimens are 13 mm, 200 GPa and 295 MPa, respectively. The comparisons between the experimental and simulated maximal deflection of middle span and the impact force–time histories are given in Table 9.9 and Fig. 9.31, in which the CSC model parameters for UHPC are determined from Tables 9.6 and

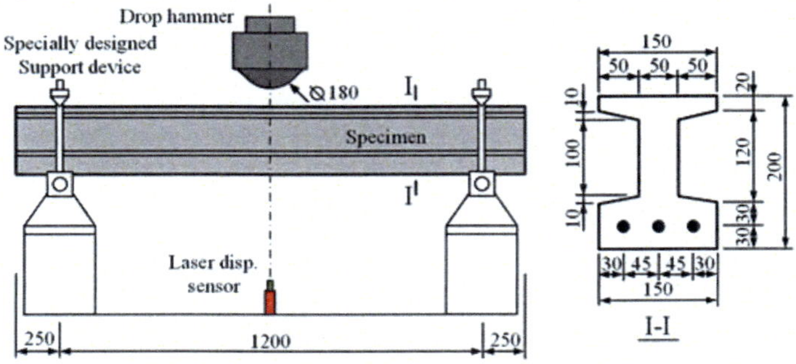

Fig. 9.30 Test layout and sectional reinforcement of specimen (*unit* mm)

Table 9.9 Comparisons of experimental and numerical mid-span deflection

Drop height (m)	Max-deflection in test (mm)	Max-deflection in simulation (mm)	Error (%)
0.8	14.1	14.5	2.8
1.0	14.5	18.6	28.2
1.2	24.7	23.6	−4.5
1.4	29.4	28.7	−2.4
1.6	36.1	34.2	−5.3

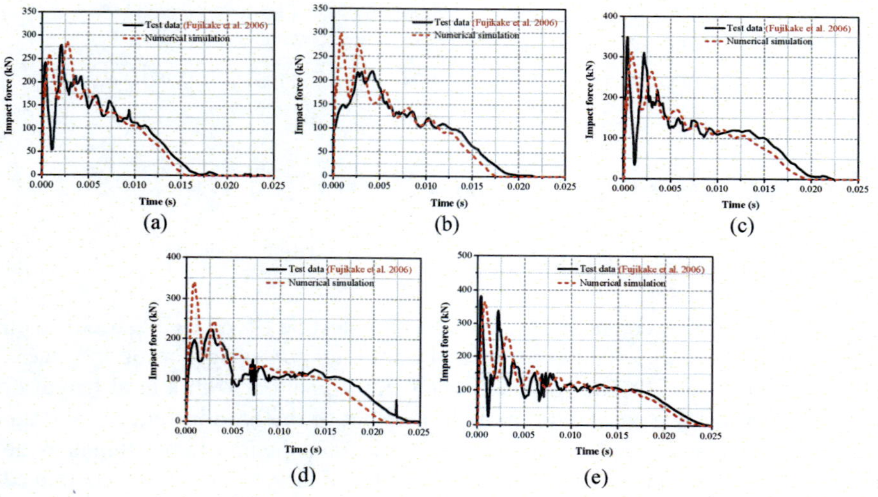

Fig. 9.31 Impact force–time histories for the dropping height of **a** 0.8 m **b** 1.0 m **c** 1.2 m **d** 1.4 m **e** 1.6 m

9.7. Good agreements are derived concerning the maximal deflections, peak impact force and impact durations. In addition, it should be pointed out that the first peak value of impact force at the height of 0.8, 1.2 and 1.6 m is closed to the second peak value, while the first peak impact force is not obvious for another two specimens, i.e., releasing heights of 1.0 and 1.4 m. This abnormal fluctuation of impact force could be attributed to the experimental deviations.

9.5.1.2 Yoo et al. (2015) Test

UHPC beams with different reinforcement arrangements subjected to a series of drop hammer impact test was performed by Yoo et al. (2015), as shown in Fig. 9.32a. The mass of the drop hammer was 270 kg, and the release height was 1.6 m. The cross-sectional dimension of specimens and steel bars are shown in Fig. 9.32b. The label

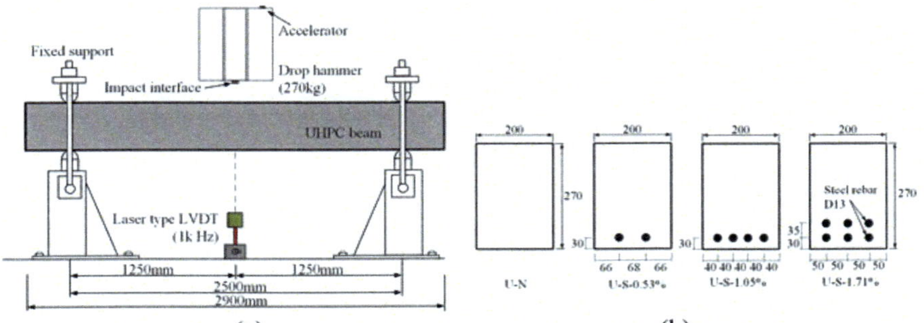

Fig. 9.32 Experimental layout **a** test setup **b** sectional details of beams (*unit* mm)

"UH-N" stands for the non-reinforced beam series, "UH-S-*n*" for the reinforced beam series, and *n* denotes the reinforcement ratio. The compressive strength and elastic modulus of UHPC are 152.5 MPa and 44 GPa respectively. Besides, the yield strength, ultimate strength and the modulus of elasticity of steel bars are 522.7 MPa, 627.6 MPa and 200 GPa, respectively. The FE model of this test is established based on the experimental layout, and the CSC model parameters of UHPC are generated according to Tables 9.6 and 9.7. The comparisons between the experimental and numerically simulated mid-span deflection-time histories of specimens, as well as the simulation results derived by Guo et al. (2018) are shown in Fig. 9.33. It can be seen that in the deflection rebound stage, the numerical results based on the calibrated CSC model in this study agree much better with the test results. The main reason lies

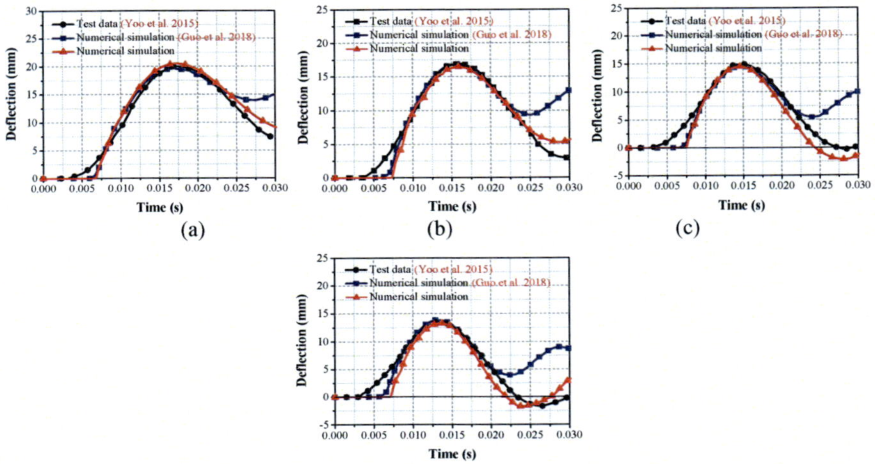

Fig. 9.33 Mid-span deflection-time histories of beam with the reinforcement ratio of **a** 0% **b** 0.53% **c** 1.05% **d** 1.71%

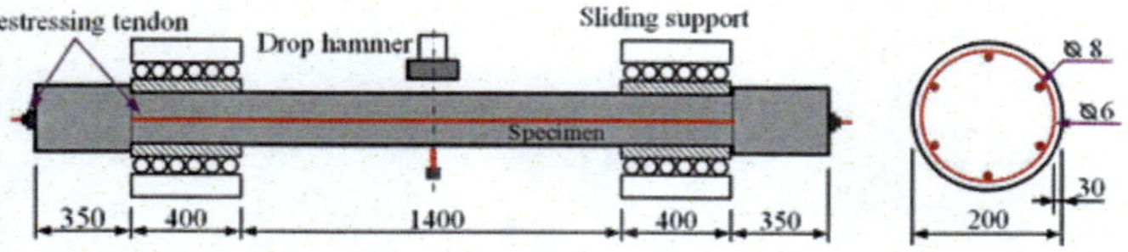

Fig. 9.34 Test layout and cross-sectional reinforcement of column (*unit* mm)

in that the fracture energy and softening parameters are calibrated at present, and the same specimen boundary with the fixed support in test is also crucial.

9.5.1.3 Fan et al. (2019) Test

Fan et al. (2019) conducted the comparative drop hammer impact test on reinforced UHPC (155 MPa) and NSC (~30 MPa) columns, respectively. The schematic of the experimental system is shown in Fig. 9.34. The disc spring and external prestressing tendon were used to realize the constant axial force, and the sliding support was applied to achieve the longitudinal free translation of specimens. The mass of the drop hammer for UHPC specimens was 586 kg and the release height was 2.4 m. The compressive strength and elastic modulus of UHPC were 155 MPa and 47 GPa, respectively. For the longitudinal reinforcing bars and stirrups, the diameters of which were 8 mm and 6 mm, the yield strength of which were 456.7 MPa and 333.4 MPa, and the elastic modulus of which were both 200 GPa. The axial forces for specimens U1 and U2 are 400 kN and 0 KN, respectively. The corresponding FE model is established based on test layout, and the CSC model parameters of UHPC are determined according to Tables 9.6 and 9.7. In addition, the axial force in this model is realized by *LOAD_SEGMENT in LS-DYNA. Figure 9.35 show the comparisons between the experimental and simulated impact force- and mid-span deflection-time histories of two specimens. It can be derived that the simulated impact force- (−F) and mid-span deflection- (−D) time histories agree well with the experimental data. A few of deviations can be found in the results of specimen U2, which is attributed to the excessive specimen damage. The predicted secondary peak impact force is relatively larger, because there is no additional damping in FE model.

9.5.1.4 Wei et al. (2019) Test

Wei et al. (2019) conducted a series of drop hammer impact tests on the UHPC columns, as shown in Fig. 9.36. The total mass of the drop hammer for hemispherical and wedge-shaped indenter are 411 kg and 427 kg, respectively. A total of three UHPC columns (2000 mm at length) were fabricated and the detailed cross-section and reinforcements (diameter of 12 mm) are shown in Fig. 9.36. The diameter and spacing of stirrup were 6 mm and 200 mm, respectively. The axial forces for all specimens are kept identical as 200 kN. The label "U-S/C-*h*" represented a UHPC

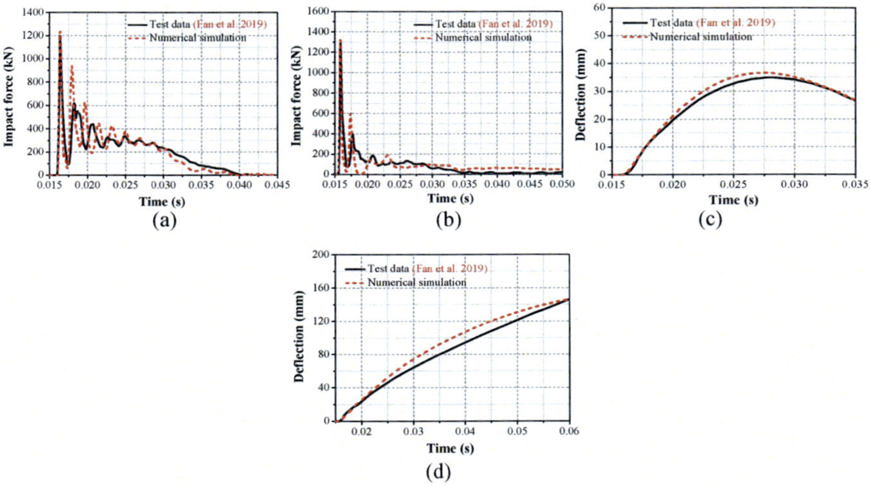

Fig. 9.35 Time histories for specimen **a** U1-F **b** U2-F **c** U1-D **d** U2-D

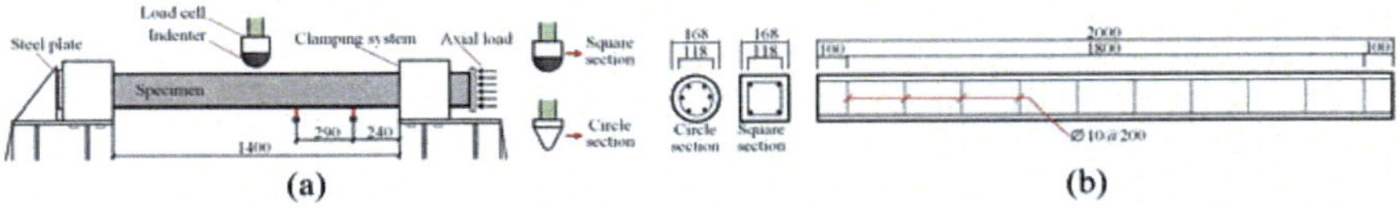

Fig. 9.36 Experimental layout **a** test setup **b** details of specimen (*unit*: mm)

column with square or circular sections (S- is square section, C— is circular section), where the letters *h* denotes the releasing height of dropping hammer. The compressive strength and elastic modulus of UHPC were 136 MPa and 45.5 GPa, respectively. The yield strength and elastic modulus of all reinforcing bars were 500 MPa and 200 GPa, respectively. The FE model of this test is established based on test layout in Fig. 9.36, the CSC model parameters of UHPC are determined according to Tables 9.6 and 9.7, and the axial force is also realized by the keyword *LOAD_SEGMENT. Figures 9.37,

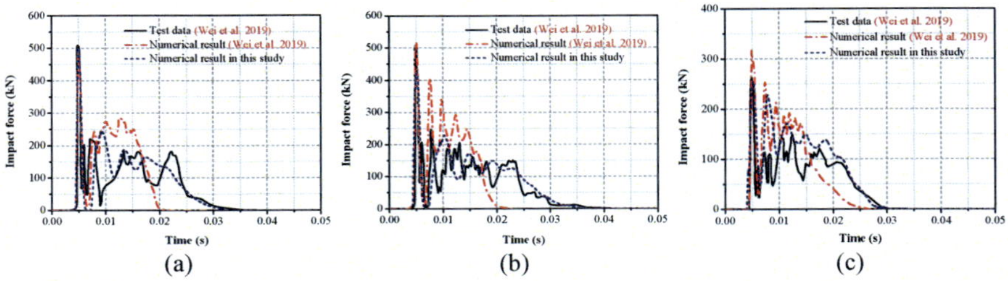

Fig. 9.37 Impact force time-histories for specimen **a** U-C-1.5 **b** U-C-1.75 **c** U-S-1

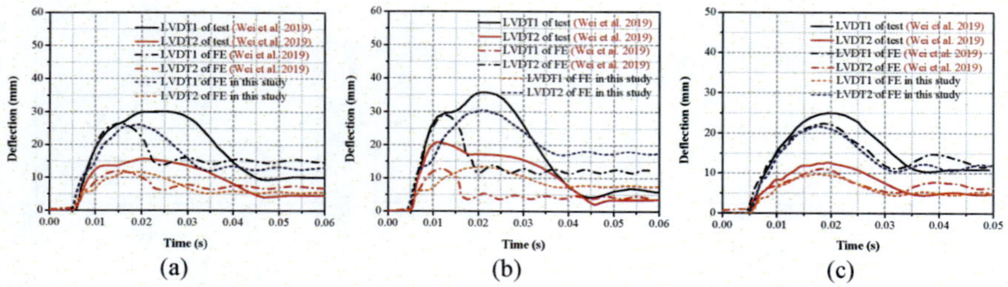

Fig. 9.38 Deflection-time histories for specimen **a** U-C-1.5 **b** U-C-1.75 **c** U-S-1

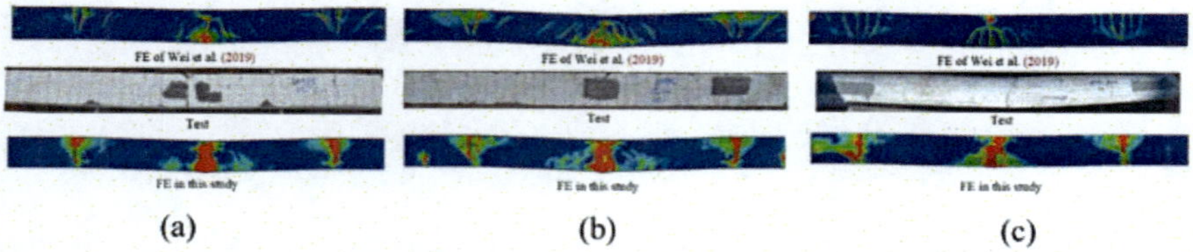

Fig. 9.39 Damage contour of specimen **a** U-C-1.5 **b** U-C-1.75 **c** U-S-1

9.38 and 9.39 show the comparisons between the experimental and numerical simulated impact force- and deflection-time histories, as well as the damage contour of UHPC columns, respectively. As can be seen obviously that, the predicted instantaneous impact forces are consistent fairly well with the experimental data, which is also better than the numerical results of Wei et al. (2019), especially for the impact time durations. It should be pointed out that, the present final damage level of specimens is not reproduced well than that of Wei et al. (2019), which is caused by the larger element size, i.e., 10 mm mesh, adopted in this study.

9.5.2 UHPC-FST Members

9.5.2.1 Wu et al. (2019) Test

The effect of impact energy on the dynamic response of simply supported UHPC-FST beams with circular cross-section was studied according to the drop weight test on the middle span of specimens with different release heights, i.e., 5, 8 and 11 m (Wu et al. 2019). As shown in Fig. 9.40, the length and net span of specimen are 2000 and 1600 mm, the outer diameter of cross-section is 203 mm, and outer steel tube thickness is 6 mm. respectively. The compressive strength, elastic modulus and Poisson's ratio of UHPC were 141.5 MPa, 43.8 GPa and 0.23, respectively. The yield strength and elastic modulus of steel were 318.5 MPa and 206 GPa, respectively. The mass of drop hammer is 530 kg, and its indenter diameter of flat head type drop

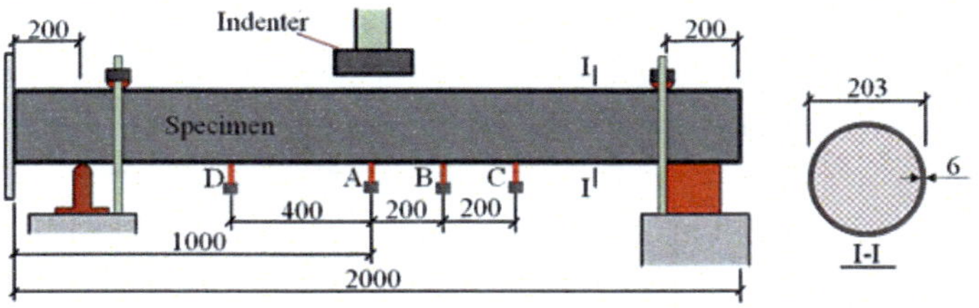

Fig. 9.40 Experimental apparatus (*unit* mm)

hammer is 220 mm. The FE model has been established referring the arrangement of test, and the CSC model parameters of UHPC are determined according to Tables 9.6 and 9.7. The comparisons between the numerical results and experimental data are show in Table 9.10, Figs. 9.41 and 9.42, including the deflection-, impact force–time histories and surface damage of steel tube. It can be seen that the maximum and residual deflections of each measuring point, variation trend and impact time duration, as well as the surface damage of the specimen can be precisely predicted. It should be pointed out that, the deviations of peak impact forces are attributed to the inadequate capability to describe the dilatancy effect of CSC model, which have been also mentioned in several literatures (Saini and Shafei 2019, Wu et al. 2012).

9.5.2.2 Wang et al. (2019b) Test

As shown in Fig. 9.43, Wang et al. (2019b) conducted a series of drop-hammer impact test on the UHPC-FST and UHPC-FDST beams with fixed supports, and the mainly concerned parameters are the impact energy, outer and inner steel tube thicknesses, as well as the axial force, respectively. The total mass of the drop hammer was 430 kg, the releasing heights were 2, 3 and 4 m, and the clear span of specimens was 1400 mm. A total of seven specimens were fabricated and the arrangement of steel tubes were shown in Fig. 9.43. In their test, "U-" indicated that the filling concrete of the specimen is UHPC. For UHPC-FDST beams, "To" and the number afterwards indicated the outer steel tube thickness, but "Ti" and the number afterwards indicated the inner steel tube thickness. In addition, "-F" indicates that 200 kN axial force was applied on specimen and "H-" and the number afterwards denotes the releasing height of hammer. The compressive strength of UHPC was 136 MPa. The yield strength, ultimate strength and elastic modulus of inner steel tube were 286 MPa, 352 MPa and 206 GPa, respectively. Additionally, the corresponding values for the outer steel tube were 310 MPa, 385 MPa, and 206 GPa, respectively. The FE model has been established referring the arrangement of test. And the CSC model parameters of UHPC are determined according to Tables 9.6 and 9.7, while the plastic kinematic model is applied on steel bars with various parameters. Figures 9.44 and 9.45 show the corresponding comparisons of the impact force- and midspan deflection-time histories of UHPC-FST and UHPC-FDST specimens, respectively. As can be seen, the

Table 9.10 Comparisons of experimental and numerical mid-span deflections of UHPC-FST

Measure point		Releasing height 5 m			Releasing height 8 m			Releasing height 11 m		
		Test	Simulation	Error (%)	Test	Simulation	Error (%)	Test	Simulation	Error (%)
Maximum deflection (mm)	A	–	52.5	–	79.2	80.3	0.76	138	108.2	−21.59
	B	45.8	45.3	−1.65	69.1	69.8	1.01	94.6	94.9	0.32
	C	29.7	31.3	5.39	48.9	48.1	1.63	64.4	65.2	1.24
	D	29.8	31.5	5.70	45.6	47.6	4.38	56.2	63.7	13.34
Residual deflection (mm)	A	–	42.1	–	64.6	68.9	6.65	–	90.9	–
	B	33.6	35.3	5.06	54.8	59.8	9.12	–	79.3	–
	C	21.4	24.6	14.95	39.1	40.4	3.32	–	52.7	–
	D	20.9	23.9	9.57	36.3	37.0	1.93	–	49.2	–

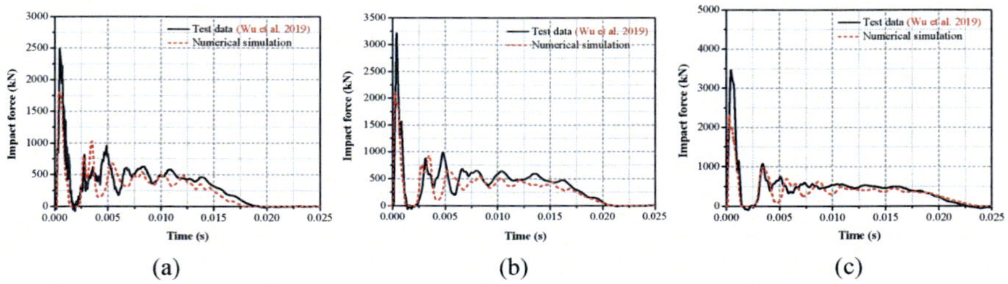

Fig. 9.41 Impact force–time histories under different drop heights **a** 5 m **b** 8 m **c** 11 m

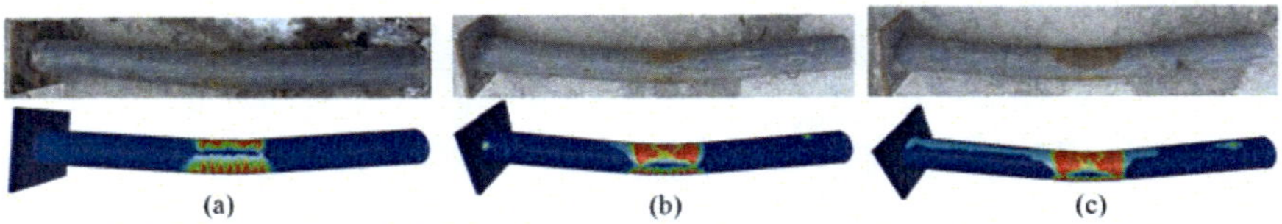

Fig. 9.42 Damage of specimens under different drop heights **a** 5 m **b** 8 m **c** 11 m

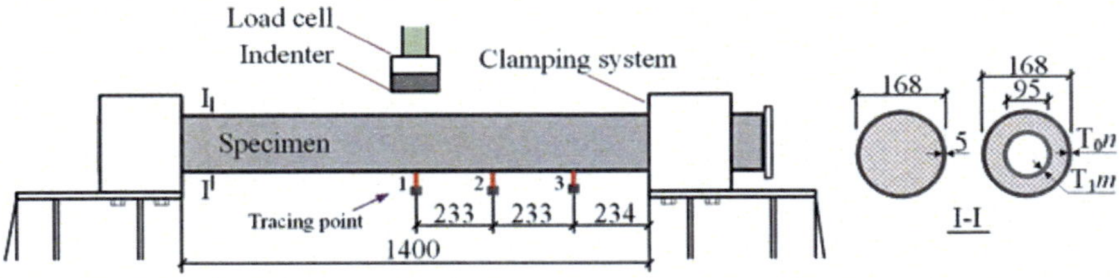

Fig. 9.43 Test layout (*unit* mm)

instantaneous impact force and deflection responses of the specimens are predicted fairly well, especially in terms of maximum deflection. It should be pointed out that, the deviations about the peak impact force are also induced by the essential defect of CSC model as stated in Sect. 9.5.2.1.

9.6 Summary

In this chapter, the lateral impact behavior of UHPCC members with different impact energy and axial force is investigated experimentally and numerically. A set of eleven UHPC and two NSC members are fabricated and tested under drop-hammer impact. The model parameters of CSC model for UHPCC are calibrated and the FE models are established to predict the damage and dynamic responses of UHPCC specimens. The main work and conclusions are drawn as follows:

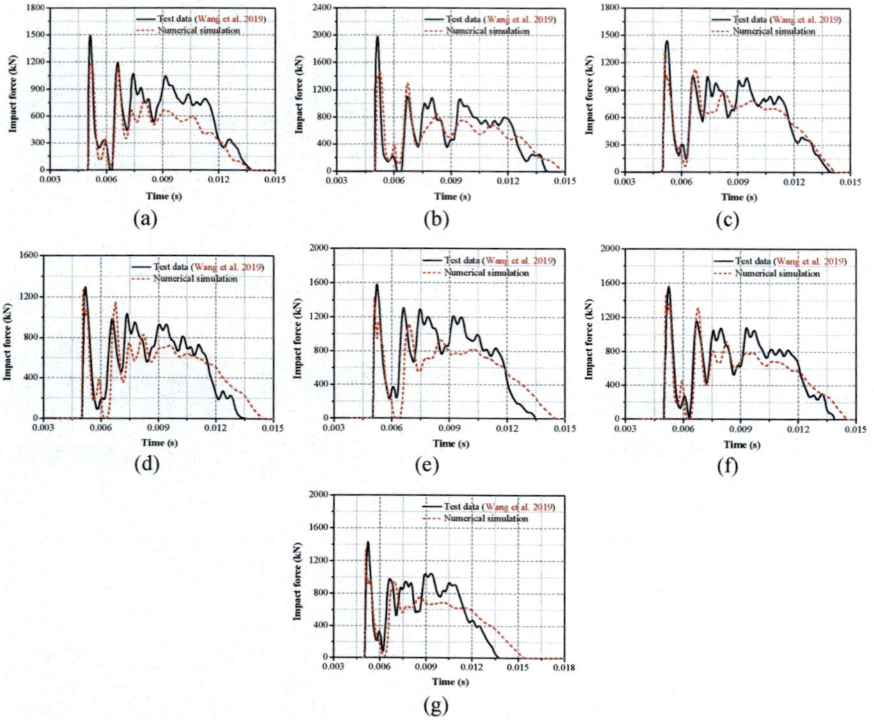

Fig. 9.44 Impact force–time histories **a** U-FST-H2 **b** U-FST-H4 **c** U-To5-Ti4-H4 **d** U-To5-Ti4-H3 **e** U-To8-Ti4-H4 **f** U-To5-Ti6-H4 **g** F-U-To5-Ti4-H4

(1) Reinforced UHPCC members exhibit outstanding impact resistance. From experimental observation, the NSC specimens experienced brittle shear failure with severe concrete fragmentation, while UHPCC specimens only showed minor flexural damage and deformation under identical impact scenario.
(2) Based on the experimental observations, the impact energy and axial force can significantly influence the lateral impact behavior of reinforced UHPCC members. With the increase of axial force, the residual deflection of UHPCC specimens decreases significantly, and the residual deflection almost decreases to zero when the axial force ratio is 0.1. Thus, the effect of axial force should be taken into account when the impact resistance of UHPCC load-bearing columns is studied.
(3) The simplified arch model is proposed and validated to expound the mechanism of axial force on UHPCC members during the impact process. Within the discussed parameters variation ranges, as the deflection of UHPCC specimens increasing, axial force does negative work to dissipate partial impact energy, while the axial force provides the energy for the recovery of UHPC specimen by doing the positive work during the deflection rebound.

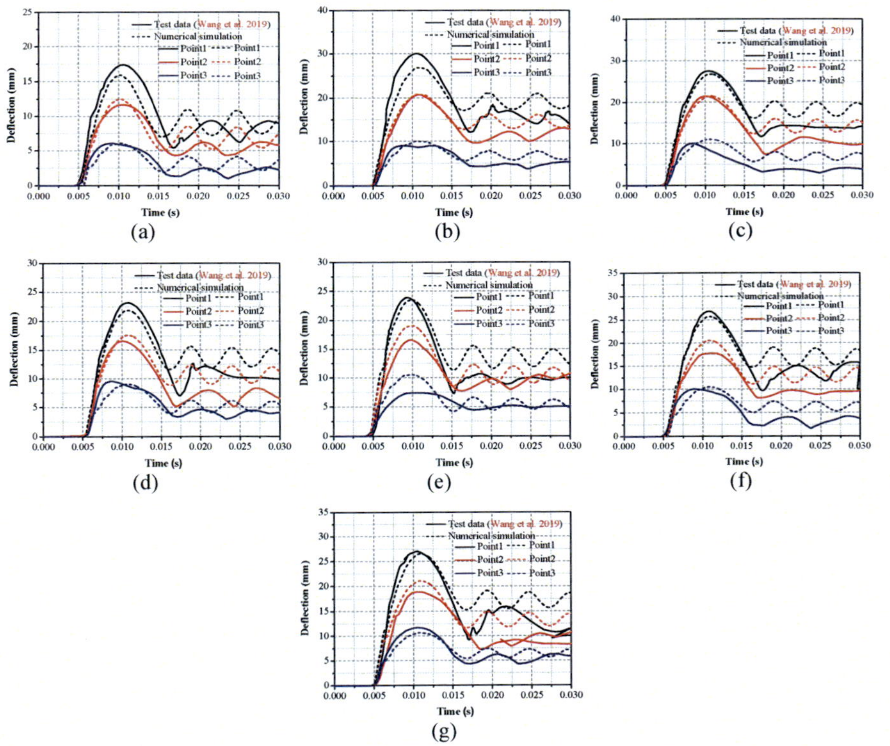

Fig. 9.45 Deflection-time histories **a** U-FST-H2 **b** U-FST-H4 **c** U-To5-Ti4-H4 **d** U-To5-Ti4-H3 **e** U-To8-Ti4-H4 **f** U-To5-Ti6-H4 **g** F-U-To5-Ti4-H4

(4) CSC model parameters (strength parameters, cap parameters, damage and softening parameters, strain rate effect parameters as well as hardening parameters) for the present UHPCC are calibrated based on a series of static and dynamic experimental data. Furthermore, the parameters generation method of CSC model for UHPCC is proposed, which was fully verified by the present and existing drop hammer tests on reinforced UHPC and UHPC-FST members.

References

ADHIKARY S D, LI B, FUJIKAKE K. Dynamic behavior of reinforced concrete beams under varying rates of concentrated loading[J]. International Journal of Impact Engineering, 1997, 19(9): 847-873.

ANSYS Autodyn User's Manual, Release 15.0[M]. ANSYS, Inc, 2013.

BIOLZI L, CATTANEO S. Response of steel fiber reinforced high strength concrete beams: Experiments and code predictions[J]. Cement & Concrete Composites, 2017, 77:1-13.

CHEN W F. Plasticity in reinforced concrete[M]. New York: J. Ross Publishing, 2007.
DANCYGIER A N, YANKELEVSKY D Z, HAEGERMANN C. Response of high performance concrete plates to impact of non- deforming projectiles[J]. International Journal of Impact Engineering, 2007 34(11): 1768-1779.
DUPRAY F, MALECOT Y, DAUDEVILLE L. Experimental behaviour of high-performance concrete in confined tension[J]. Materials and Structures, 2010, 43(5): 699-707.
FARNAM Y, MOHAMMADI S, SHEKARCHI M. Experimental and numerical investigations of low velocity impact behavior of high-performance fiber-reinforced cement based composite[J]. International Journal of Impact Engineering, 2010, 37(2): 220-229.
FAN W, SHEN D, YANG T, SHAO X. Experimental and numerical study on low-velocity lateral impact behaviors of RC, UHPFRC and UHPFRC-strengthened columns[J]. Engineering Structures, 2019, 191:509-525.
FUJIKAKE K, LI B, SOEUN S. Impact Response of Reinforced Concrete Beam and Its Analytical Evaluation[J]. Journal of Structural Engineering, 2009, 135(8): 938-950.
FUJIKAKE K, SENGA T, UEDA N, OHNO T, KATAGIRI M. Study on Impact Response of Reactive Powder Concrete Beam and Its Analytical Model[J]. Journal of Advanced Concrete Technology, 2006, 4(1): 99-108.
GB/T 228.1-2010. Metallic Materials-Tensile Testing, Part I: Method of Test at Room Temperature[S]. Beijing: Standards Press of China, 2010.
GB/T 50081-2002. Standard for Test Method of Mechanical Properties on Ordinary Concrete [S]. Beijing: Standards Press of China, 2003.
GB 50010-2010. Code for Design of Concrete Structures[S]. Beijing: Standards Press of China, 2010.
GUO W, FAN W, SHAO X, SHEN D, CHEN B. Constitutive model of ultra-high-performance fiber-reinforced concrete for low-velocity impact simulations[J]. Composite Structures, 2018, 185: 307-326.
GRAYBEAL B A. Material property characterization of ultra-high performance concrete: FHWA-HRT-06-103[R]. VA: U.S. Department of Transportation Federal highway Adminstration, 2006: 1-186.
HASSAN A M T, JONES S W, MAHMUD G H. Experimental test methods to determine the uniaxial tensile and compressive behaviour of ultra high performance fibre reinforced concrete (UHPFRC)[J]. Construction and Building Materials, 2012, 37: 874-882.
Institute SE. Minimum design loads for buildings and other structures[S]. Amer Society of Civil Engineers, 2006.
JIANG H, ZHAO J. Calibration of the continuous surface cap model for concrete[J]. Finite Elements in Analysis and Design, 2015, 97: 1-19.
KISHI N, MIKAMI H, MATSUOKA K G, ANDO T. Impact behavior of shear-failure-type RC beams without shear rebar[J]. International Journal of Impact Engineering, 2002, 27(9): 955-968.
KOTSOVOS M D, NEWMEN J B. A mathematical description of the deformational behaviour of concrete under complex loading[J]. Magazine of Concrete Research, 1979, 31(107): 77-90.
LSTC. LS-DYNA Keyword User's Manual[M]. Livermore Software Technology Corporation, 2014.
LI J, WU C Q, HAO H. An experimental and numerical study of reinforced ultra-high performance concrete slabs under blast loads[J]. Materials & Design, 2015, 82: 64-76.
LI J, WU C Q, HAO H, WANG Z Q, SU Y. Experimental investigation of ultra-high performance concrete slabs under contact explosions[J]. International Journal of Impact Engineering, 2016, 93: 62-75.
LI H, CHEN W, HAO H. Influence of drop weight geometry and interlayer on impact behavior of RC beams[J]. International Journal of Impact Engineering, 2019, 131: 222-237.
MADURAPPERUMA M A K M, WIJEYEWICKREMA A C. Response of reinforced concrete columns impacted by tsunami dispersed 20′ and 40′ shipping containers[J]. Engineering Structures, 2013, 56: 1631-1644.
MÁCA P, SOVJÁK R, KONVALINKA P. Mix design of UHPFRC and its response to projectile impact[J]. International Journal of Impact Engineering, 2010, 63(5): 158-163.

MAO L, BARNETT S, BEGG D, SCHLEYER G, WIGHT G. Numerical simulation of ultra high performance fibre reinforced concrete panel subjected to blast loading[J]. International Journal of Impact Engineering, 2014, 64: 91-100.

MÉSZÖLY T, RANDL N. Shear behavior of fiber-reinforced ultra-high performance concrete beams[J]. Engineering Structures, 2018, 168: 119-127.

MANFRED C, SPECK K. Ultra high performance concrete under biaxial compression, Proceedings of the second international symposium on ultra high performance concrete. Kassel, Germany[C]. Kassel university press GmbH, 2008.

MILLS L, ZIMMERMAN R. Compressive strength of plain concrete under multiaxial forcing conditions[J]. ACI Journal, 1970, 67: 802-807.

MURRAY Y D. Users manual for LS-DYNA concrete material model 159: No. FHWA-HRT-05–062[R]. Colorado Springs: Federal Highway Administration, 2007.

MURRAY Y D, ABU-ODEH A Y, BLIGH R P. Evaluation of LS-DYNA concrete material model 159: No. FHWA-HRT-05–063[R]. Colorado Springs: Federal Highway Administration, 2007.

OZBOLT J, SHARMA A. Numerical simulation of reinforced concrete beams with different shear reinforcements under dynamic impact loads[J]. International Journal of Impact Engineering, 2011, 38: 940-950.

PHAM T M, HAO Y, HAO H. Sensitivity of impact behaviour of RC beams to contact stiffness[J]. International Journal of Impact Engineering, 2018, 112: 155-164.

PRABHA S L, DATTATREYA J K, NEELAMEGAM M, RAO M S. Study on stress-strain properties of reactive powder concrete under uniaxial compression[J]. International Journal of Engineering Science Technologies, 2010, 2(11): 6408-6416.

REN G M, WU H, FANG Q, LIU J Z, GONG Z M. Triaxial compressive behavior of UHPCC and applications in the projectile impact analyses[J]. Construction and Building Materials, 2016, 113: 1-14.

REN G M, WU H, FANG Q, LIU J Z. Effects of steel fiber content and type on static mechanical properties of UHPCC[J]. Construction and Building Materials, 2018a, 163: 826-839.

REN G M, WU H, FANG Q, LIU J Z. Effects of steel fiber content and type on dynamic compressive mechanical properties of UHPCC[J]. Construction and Building Materials, 2018b, 164: 29-43.

SAATCI S, VECCHIO F J. Effects of Shear Mechanisms on Impact Behavior of Reinforced Concrete Beams[J]. ACI Structural Journal, 2009, 106(1): 78-86.

SAINI D, SHAFEI B. Concrete constitutive models for low velocity impact simulations[J]. International Journal of Impact Engineering, 2019, 132: 103329.

SCHWER L E. Demonstration of the continuous surface cap model with damage: concrete unconfined compression test calibration[R]. LS-DYNA Geomaterial Modeling Short Course Notes, 2001.

SIMO J C, JU J W. Strain and Stress Based Continuum Damage Model[J]. International Journal of Solids and Structures, 1987, 23(7): 821-840.

SOVJÁK R, VAVŘINIK T, ZATLOUKAL J, MÁCA P, MIČUNEK T, FRYDRÝN M. Resistance of slim UHPFRC targets projectile impact using in-service bullet[J]. International Journal of Impact Engineering, 2015, 76: 166-177.

STOUFFER D C, DAME L T. Inelastic deformation of metals: models, mechanical properties, and metallurgy[M] New York: John Wiley and Sons, 1996: 520.

WANG X, ZHANG Y, SU Y, FENG Y. Experimental investigation on the effect of reinforcement ratio to capacity of RC column to resist lateral impact loading[J]. Systems Engineering Procedia, 2011, 1:35-41.

WARDHANA K, HADIPRIONO F C. Analysis of recent bridge failures in the United States[J]. Journal of Performance of Constructed Facilities, 2003, 17: 144-150.

WANG W, WU C, LI J, LIU Z, LV Y. Behavior of ultra-high performance fiber-reinforced concrete (UHPFRC) filled steel tubular members under lateral impact loading[J]. International Journal of Impact Engineering, 2019a, 132: 103314.

WANG W, WU C, LI J, LIU Z, ZHI X. Lateral impact behavior of double-skin steel tubular (DST) members with ultra-high performance fiber-reinforced concrete (UHPFRC)[J]. Thin-Walled Structures, 2019b, 144: 106351.

WANG R, HAN L H, HOU C C. Behavior of concrete filled steel tubular (CFST) members under lateral impact: experiment and FEA model[J]. Journal of Constructional Steel Research, 2013, 80: 188-201.

WEI J, LI J, WU C. An experimental and numerical study of reinforced conventional concrete and ultra-high performance concrete columns under lateral impact loads[J]. Engineering Structures, 2019, 201: 109822.

WILLIAMS E M, GRAHAM S S, REED P A, RUSHING T S. Laboratory characterization of cor-tuf concrete with and without steel fibers[R]. Defense Technical Information Center, 2009.

WU H, REN G M, FANG Q, LIU J Z. Response of ultra-high performance cementitious composites filled steel tube (UHPCC-FST) subjected to low-velocity impact[J]. Thin-Walled Structures, 2019, 144: 106341.

WU C, OEHLERS D J, REBENTROST M, LEACH J, WHITTAKER A S. Blast testing of ultra-high performance fibre and FRP-retrofitted concrete slabs[J]. Engineering Structures, 2009, 31: 2060–2069.

WU H, REN G M, FANG Q, LIU J Z. Effects of steel fiber content and type on dynamic tensile mechanical properties of UHPCC[J]. Construction and Building Materials, 2018, 173: 251-261.

WU Y C, CRAWFORD E J, MAGALLANES J M. Performance of LS-DYNA concrete constitutive model. Proceeding of the 12th International LS-DYNA Users Conference, Dearborn, MI, USA[C]. 2012.

WU H, FANG Q, CHEN X W, GONG Z M, LIU J Z. Projectile penetration of ultra-high performance cement based composites at 510m/s to 1320m/s[J]. Construction and Building Materials, 2015, 74(15): 188-200.

XU J C, WU C Q, XIANG H B, SU Y, LI Z X, FANG Q, HAO H, LIU Z, ZHANG Y, LI J. Behaviour of ultra high performance fibre reinforced concrete columns subjected to blast loading[J]. Engineering Structures, 2016, 118: 97-107.

YOO D, BANTHIA N, KIM S, YOON Y S. Response of ultra-high-performance fiber-reinforced concrete beams with continuous steel reinforcement subjected to low-velocity impact loading[J]. Composite Structures, 2015, 126: 233-245.

YOO D Y, BANTHIA N, YOON Y S. Impact resistance of reinforced ultra-high-performance concrete beams with different steel fibers[J]. ACI Structural Journal, 2017, 114(1): 113-124.

ZHAO D, JIAN Y, KUNNATH S K. Shear Mechanisms in Reinforced Concrete Beams under Impact Loading[J]. ACI Structural Journal, 2017, 143(9): 4017089.

ZHAO D, YI W, KUNNATH S K. Numerical simulation and shear resistance of reinforced concrete beams under impact[J]. Engineering Structures, 2018, 166: 387-401.

ZHAN T B, WANG Z H, NING J G. Failure behaviors of reinforced concrete beams subjected to high impact loading[J]. Engineering Failure Analysis, 2015 56: 233-243.

Chapter 10
Residual Axial Capacity of UHPCC-FST Column Under Contact Explosion

10.1 Introduction

Aiming to produce the mass casualties, visually dramatic destruction, significant economic aftershocks and fear among the population, civil infrastructures have gradually been becoming the potential attacking targets in the modern terrorism activities (Fujikura and Bruneau 2008). For example, during the 80 years from 1920 to 2000, nearly 900 terrorist attacks were related to the above-ground transportation systems, especially in the later 30 years from 1970 to 2000 (Jenkins 2001). Comparably, the much easier accessibility for bridge makes it more prone to be the potential targets under close-range even contact explosion attacks, which has been recognized by both the engineering community and public officials, e.g., Federal Highway Administration (FHWA) and American Association of State Highway and Transportation Officials (AASHTO) (Fujikura et al. 2008; Williamson et al. 2010 FHWA 2003). For bridge structures, as the key load carrying members, the damage and even failure of bridge columns may induce the collapse of the whole bridge superstructures. Therefore, the studies on the post-blast residual axial capacity of bridge columns, particularly for some extreme scenarios (i.e., the contact explosion), are very important and essential for assessing the performance of the bridge structures against the potential terroristic or accidental explosions.

Most existing related works are mainly concentrated on the reinforced concrete (RC) columns, including the blast resistance (Xu et al. 2016; Kyei and Braimah 2017; Astarlioglu and Krauthammer 2014) and the post-blast behavior of the columns (Li et al. 2012, 2017; Wu et al. 2011a; Bao and Li 2010). It has been drawn that RC columns may suffer global damage such as flexure or localized damage such as spalling of concrete cover and diagonal shear cracks. Concrete-filled steel tube (CFST) columns has recently gained wide applications in bridge engineering due to their superior load-bearing capacity and seismic resistance than those of the conventional RC columns. The steel tube of CFST column, with different section profiles such as rectangular, hexagonal, circular and corrugated section which can help to enhance the ductility and local stability of the column (Farahi et al. 2016, 2017;

Q. Fang et al., *UHPCC Under Impact and Blast*,
https://doi.org/10.1007/978-981-33-6842-2_10

Nassirnia et al. 2016), can confine the core concrete and prevent the spalling damage of concrete under blast loading. Ultra-high performance cementitious composite (UHPCC) is a relatively new cement-based composite with prominent mechanical properties (both static and dynamic), low shrinkage and excellent durability. It is a promising construction material for protective structures (e.g., military fortifications, civil defense shelters, nuclear containments) to resist intensive impact and blast loadings. A CFST column infilled with UHPCC can take advantage of the high compressive strength and ductility of the core concrete which can delay or prevent the buckling of the steel tubes under close-in range and even contact detonations. Ultra-high performance cementitious composite filled steel tube (UHPCC-FST) has been widely applied as the load bearing members for long-span bridges, which are the potential bomb explosion attack targets in terrorism activities.

This chapter aims to perform the experimental and numerical studies on the residual axial capacity of UHPCC-FST columns under contact explosion (Wang et al. 2020a, b). Firstly, five UHPCC-FST cylinder columns are fabricated with the column height, outer diameter and core concrete compressive strength being 2000 mm, 203 mm and 131.5 MPa, respectively. Secondly, three of which are tested under contact explosion with the TNT charge weights of 1, 2 and 3 kg, in which the impact craters are formed and the integrity of the columns is maintained under 1 kg and 2 kg TNT explosions, while the outer steel tube is seriously ruptured and the core UHPCC is fully crushed for 3 kg TNT explosion. Then, the original axial capacity of intact columns and the residual axial capacity of blast-damaged columns are evaluated though the axial compression tests, and the relationships between the axial load and axial/lateral displacement are derived. It indicates that, (i) both the intact and blast-damaged columns under axial compression show diagonal shear failure, and the local bulging of steel tube is induced due to the low confinement of steel tube and high compressive strength of UHPCC; (ii) the quantitative dependence of column damage index on the charge weight is determined.

Furthermore, by utilizing the multi-material Arbitrary Lagrange-Euler (ALE) algorithm, Fluid–Structure Interaction (FSI) method, the restart input data method and element erosion algorithm implemented in the finite element (FE) code LS-DYNA, the numerical simulations corresponding to the above contact charge explosion and the following axial compression tests are carried out. The damage and failure modes of UHPCC-FST columns are reproduced and validated by comparing with the experimental data. Finally, the related parametric influences, e.g., thickness and strength of steel tube, compressive strength of core concrete and column diameter, on the local blast formed crater depth, the post-blast residual axial load bearing capacity, as well as the corresponding damage index of UHPCC-FST column under contact detonation are discussed. The influential degree of above parameters on the residual axial capacity of UHPCC-FST columns under contact explosion is clarified. The present work can provide helpful references for evaluating the post-blast performance as well as the design of UHPCC-FST column under contact explosion.

10.2 Review of the Existing Works

As for the CFST members subjected to the explosive loadings, Fujikura et al. (2008) experimentally examined the blast resistance of 1/4 scaled CFST bridge columns, in which the compressive strength of concrete was 42 MPa, the column height was 1.5 m, the column diameters were ranged within 102–152 mm, and the steel tube thickness was 3.2 mm. It was found that the CFST columns exhibit a ductile behavior under close-in range blast loadings without sectional breaching (punching through the full concrete thickness) or spalling of the core concrete. However, it should be noted that the detailed charge weight and corresponding standoff distance were not given due to the confidential issue. Fouché and Bruneau (2015) experimentally assessed the performance of concrete-filled double-skin steel tube (CFDST) columns subjected to the close-in range detonation, in which the standoff distance of explosion and detailed charge weight were also not disclosed. The compressive strength of core concrete was 37 MPa, and the outer-diameters of steel tubes were selected as 152.4, 168.3 and 203.2 mm with the corresponding inner-diameters ranging from 50.8 to 101.6 mm. Three typical failure modes were observed, i.e., local indentation, global deformation and outer steel tube failure, and it was found that the inner steel tube can effectively prevent the direct shear failure of the CFDST column. Li et al. (2018) further experimentally investigated the damage mechanism of CFDST columns filled with normal strength concrete of 40 MPa subjected to the contact detonation (0.6 and 1.0 kg TNT) with different outer steel tube thickness (6, 7, and 8.5 mm). It showed that the damage mode of CFDST column under contact explosion exhibited a localized pattern, including cratering around the detonation point, fracture or rupture failure of the outer steel tube, and breaching failure of the front sandwich wall. It was also derived that the blast resistance of CFDST columns was enhanced with the increase of the outer steel tube thickness.

Zhang et al. (2016) conducted the field blast tests on the ultra-high performance concrete (UHPC) infilled double-skin steel tube columns with circular and square cross-sections under close-in range detonation. The compressive and flexural tensile strength of core concrete were 170 MPa and 33.8 MPa, respectively. It was found that, subjected to 35 and 50 kg TNT detonated at 1.5 m distance (scaled distances $Z = R/W^{1/3}$ were 0.41 and 0.46, R is the standoff distance from the explosive center to the column surface, W is the TNT charge weight), there was no visible buckling on the CFDST columns, and only minor cracks occurred in the core concrete. The integrity of the column was maintained and both two typical CFDST columns exhibited excellent blast resistance. Furthermore, Zhang et al. (2015a) further conducted the residual axial compression tests on the post-blast CFDST columns. It was derived that, the outer steel tubes always sufferred local buckling failure at mid-span and/or column ends for the square CFDST columns, while for the circular CFDST columns, the steel tube ruptured and local buckling failure were more likely to occur under high and low damage level, respectively.

Due to the high cost and extensive resources are needed for the field blast test, it is more economic to perform the numerical simulation. Zhang et al. (2015b) established a finite element model of CFST column subjected to blast loadings with the scaled distance of 0.55 and 0.4 m/kg$^{1/3}$, and the CONWEP code in LS-DYNA (LSTC 2007) was employed to simulate the blast load. The numerical results showed that the infilled core concrete can significantly reduce the local deformation of steel tube, and the CFST column has better blast resistance compared to the RC column. Zhang et al. (2015c, 2017) further adopted the CONWEP code to perform numerical studies on the dynamic response of CFDST columns infilled with normal strength concrete (NSC) and UHPC under close-in range detonation ($0.35 \leq Z \leq 0.55$ m/kg$^{1/3}$), and the parametric influences indicate that increasing the compressive strength of core concrete significantly reduces the residual deflection of column, increasing the outer steel tube thickness reduces the mid-span deflection more significantly than increasing the inner steel tube thickness, and the blast resistance of CFDST column is mainly dominated by the shape of outer steel tube However, the shape of inner steel tube has a slight effect. Considering that the blast loadings generated by the CONWEP code deviate obviously for the close-in range explosions, Remennikov and Uy (2014) proposed a simplified engineering-level model for the very close-in range blast induced impulse to predict the dynamic response of CFST columns, and good agreement with the field blast test data is derived in terms of damage pattern and level of specimens. Furthermore, based on Arbitrary-Lagrangian–Eulerian (ALE) numerical algorithm to reproduce the interaction between the blast wave and the structures, Ngo et al. (2015) and Li et al. (2018) analyzed the dynamic response of CFST column specimens subjected to close-in range explosion and contact detonation, respectively.

Generally, the existing works still have the following limitations: (i) UHPCC has the prominent blast and impact resistance compared with NSC, but the studies on the blast resistance of CFST column, especially for the UHPCC-FST column, are relatively insufficient; (ii) considering the easy public accessibility of bridge structure, it is supposed that the terrorist explosive (suitcase or vehicle bombs) can be placed as near as possible to the bridge columns, i.e., the contact explosion. However, the existing studies mainly concentrates in the close-in ($0.05 \leq Z \leq 0.15$ m/kg$^{1/3}$) or near range ($0.2 \leq Z \leq 0.5$ m/kg$^{1/3}$) blast events, and the relevant studies on the contact explosion scenario is scarce; (iii) the residual axial load bearing capacity is vital to evaluate the stability and workability of the whole bridge superstructure, and the relationship between the damage level of column and the explosive charge weight is vital for the practical anti-blast design of bridge piers. However, the related studies are limited.

10.3 UHPCC-FST Columns

Aiming to investigate the failure mode and residual axial capacity of the UHPCC-FST bridge columns under contact detonation, considering the prototype of the UHPCC-FST columns with the diameter of 800 mm under the suitcase or vehicle bombs

Table 10.1 Similarity ratio for the scaled blast test

Quantity	Symbol	Similarity factor	Quantity	Symbol	Similarity factor
Scaling factor	λ_r	1/4	Column height	$\lambda_h = \lambda_r$	1/4
Sound velocity in air	λ_c	1	Column diameter	$\lambda_d = \lambda_r$	1/4
TNT mass density	λ_ρ	1	Cross section area	$\lambda_A = \lambda_r{}^2$	1/16
Time	$\lambda_t = \lambda_r/\lambda_c$	1/4	Steel tube thickness	$\lambda_{st} = \lambda_r$	1/4
Blast pressure	$\lambda_p = \lambda_\rho\, \lambda_c{}^2$	1	Strength	$\lambda_{fc} = \lambda_p$	1
TNT mass	$\lambda_M = \lambda_p\, \lambda_r\, \lambda_t{}^2$	1/64	Strain	λ_ε	1

attacks, the 1/4 scaled test was designed. For the blast test, the cube root scaling rule (Baker et al. 1973) was used to represent the blast wave. In order to reduce the size influence on the dynamic behavior of CFST columns, the height-to-diameter ratio and diameter-to-tube wall thickness ratio in the model should be kept the same as that in prototype (Mirmomeni et al. 2017). Hence, the material and structural properties of column can be further derived through the dimensional analysis. The similarity ratio for UHPCC-FST column in current study is given in Table 10.1.

The experimental program of the present test is given as follows: five 1/4 scaled UHPCC-FST columns were firstly fabricated, and three of the columns (C-1, C-2, C-3) were subjected to the field contact explosion with the TNT charge of 1 kg, 2 kg and 3 kg, respectively. Then, the above explosion damaged specimens and the other two intact columns (I-8 and I-9) were delivered into the laboratory to perform the axial compression test to check their post-blast residual and original axial load bearing capacity.

10.3.1 Fabrications

Five 1/4 scaled UHPCC-FST columns with the height of 2000 mm were fabricated, and the geometrical and sectional details are illustrated in Fig. 10.1. The steel tube was made of hot-rolled seamless steel tube according to the National Standard of China (GT/B 8162-2008) (GT/B 8162-2008 2008), which has an outer diameter of 203 mm and tube thickness of 6 mm. The bottom of the specimen was welded to a steel plate which has a dimension of 400 × 400 × 20 mm^3. Besides, four stiffening ribs were welded between the steel tube and the bottom plate to strengthen their connection. The steel tube was then filled with UHPCC and cured in the air at the room temperature for 28 days.

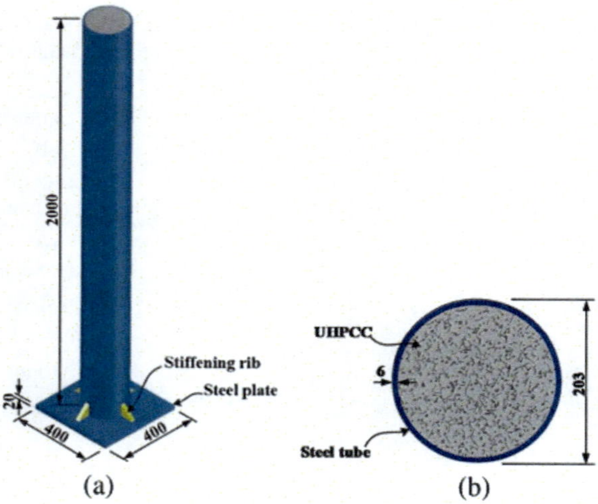

Fig. 10.1 UHPCC-FST column **a** geometric configuration **b** cross-section (*units* mm), reprinted from Wang et al. (2020a, b), copyright 2020, with permission from Elsevier

10.3.2 Steel Tube

As shown in Fig. 10.2a, five standard steel coupons of the steel tube were fabricated according to National Standard of China (GB/T228.1-2010 2010). By conducting the direct tension test with the MTS 810 material testing machine at the loading rate of 2 mm/min (Fig. 10.2b), the engineering tensile stress–strain curves of three specimens are obtained as shown in Fig. 10.2c. The yield strength f_y, ultimate tensile strength f_u and the elongation at the rupture of steel tube were determined as 318.5 MPa, 497 MPa and 26%, respectively.

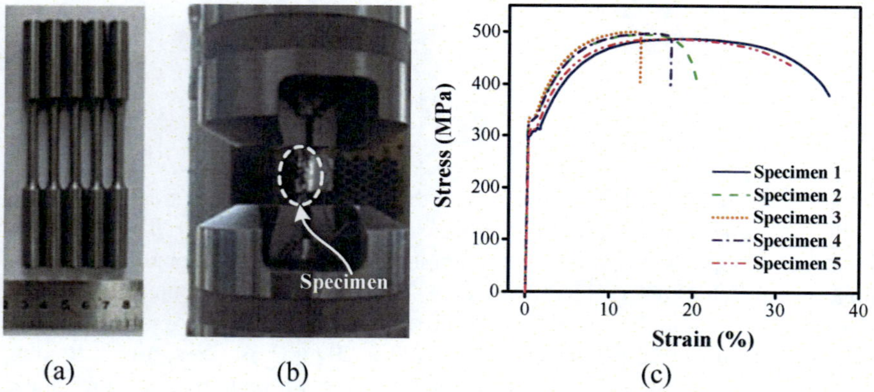

Fig. 10.2 Steel tube **a** specimens **b** direct tension test **c** tensile stress–strain curve, reprinted from Wang et al. (2020a, b), copyright 2020, with permission from Elsevier

Table 10.2 UHPCC ingredients (kg/m^3)

Cement	Admixture (fly ash and ultra-fine slag)	Sand	Steel fiber	HRWR	Water
700	250	1200	156	22.8	152

10.3.3 UHPCC

UHPCC material was prepared by State Key Laboratory of High Performance Civil Engineering Materials, Jiangsu Research Institute of Building Science in China. The mixing portion of UHPCC is given in Table 10.2, mainly including the Chinese standard Graded 52.5 P.II type Portland cement (20–30 μm in particle diameter), the natural river sand with maximal diameter of 2.5 mm, polycarboxylic type high-range-water-reducer (HRWR) with the water-reducing ratio larger than 35% and the admixture. The admixture is consisted of fly ash (density of 2.7 g/cm^3) and ultra-fine slag (particle size of 3–6 μm, density of 2.8 g/cm^3), and the specific proportions of the admixture is confidential for commercial security. The straight brass-coated steel fiber was incorporated with the volume fraction of 2.0%, the equivalent diameter, length and tensile strength of which were 0.175 mm, 13 mm and 3000 MPa, respectively. The slump and spread of UHPCC were about 230 mm and 610 mm respectively, which guaranteed the workability requirements of on-site constructions.

The present UHPCC was identical with that in our previous works (Ren et al. 2016, 2017, 2018; Wu et al. 2018), in which a series of uniaxial and triaxial static and dynamic compressive and spalling tensile tests were conducted to study the mechanical properties of UHPCC with various types and volumetric ratios of steel fibers. Also, the mixing and curing procedure are introduced in details.

The control specimens of UHPCC, including three cubes of 100 × 100 × 100 mm^3 and three prisms of 100 × 100 × 300 mm^3, were cast and tested according to the Chinese standard (GB/T 50081-2002 2003). Due to the lack of relevant common standards, the unnotched dog-bone specimens with an overall length of 500 mm referring to Kang and Kim 2011 were used in the direct tension test. The average cubic compressive strength, axial compressive strength, direct tensile strength, elastic modulus and Poisson's ratio of UHPCC are determined as 131.5 MPa, 121 MPa, 9 MPa, 42.9 GPa and 0.23, respectively.

10.4 Field Contact Explosion Test

10.4.1 Test Setup

The field blast test was conducted at the field laboratory of China Academy of Engineering Physics, and the test setup is shown in Fig. 10.3a and b. Considering that the axial compression ratio of most bridge columns is relatively low under

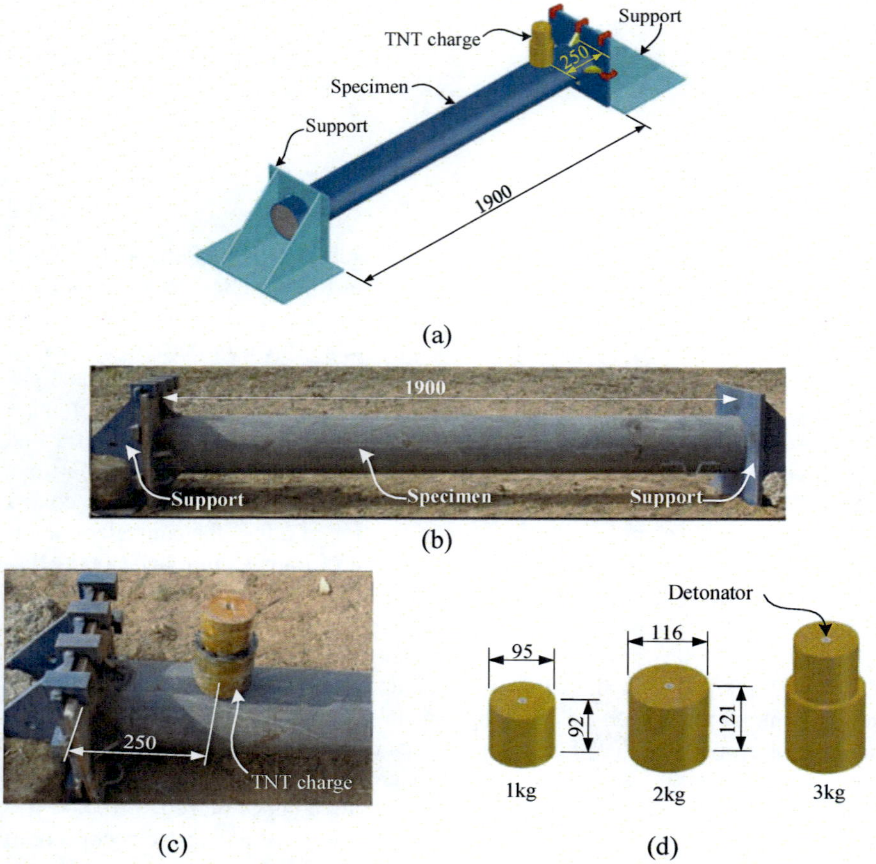

Fig. 10.3 Test setup **a** schematic view **b** photograph **c** explosive layout **d** explosive dimensions (*units* mm), reprinted from Wang et al. (2020a, b), copyright 2020, with permission from Elsevier

service load and only local response and limited global response are induced under contact detonation from previous works (Li et al. 2018; Yuan et al. 2017). The axial compressive force due to service load was neglected and the specimens were laid parallelly to the ground at present. The support devices were employed at the top and bottom ends of the column to simulate the pin and fixed constraints at the top and bottom ends, respectively. The clear span was 1900 mm, and the cylindrical TNT explosive was placed on the column surface with the center at 250 mm away from the bottom of column (corresponding to the prototypical burst height of 1 m for vehicle bomb), as shown in Fig. 10.3c. Two standard TNT cylinders, i.e., 1 kg, and 2 kg, were adopted, and the corresponding dimensions are shown in Fig. 10.3d. The explosive was ignited by the electric detonator, which was embedded at the top center of the TNT cylinder with the depth of 40 mm.

10.4.2 Test Results

10.4.2.1 C-1 Column

Figure 10.4 demonstrates the failure mode of C-1 specimen after 1 kg TNT contact explosion. It can be seen that C-1 specimen had no global deformations, and the steel tube was not fractured but an approximately elliptical crater was formed beneath the charge. The elliptical short axis (R_1) and long axis (R_2) were 105 and 110 mm, and the maximum depth of the crater (D_c) was 27 mm. In addition to the plastic deformation of the crater, the steel tube of C-1 specimen was dilated laterally with the dimension (R_3) of 209 mm, which was 6 mm greater than the original diameter. Due to the eccentricity of the initial dentonation point, the distance between the center of the crater and the bottom of the column was 255 mm, slightly larger than 250 mm. A crack was found in the connection part between the bottom plate and one stiffening

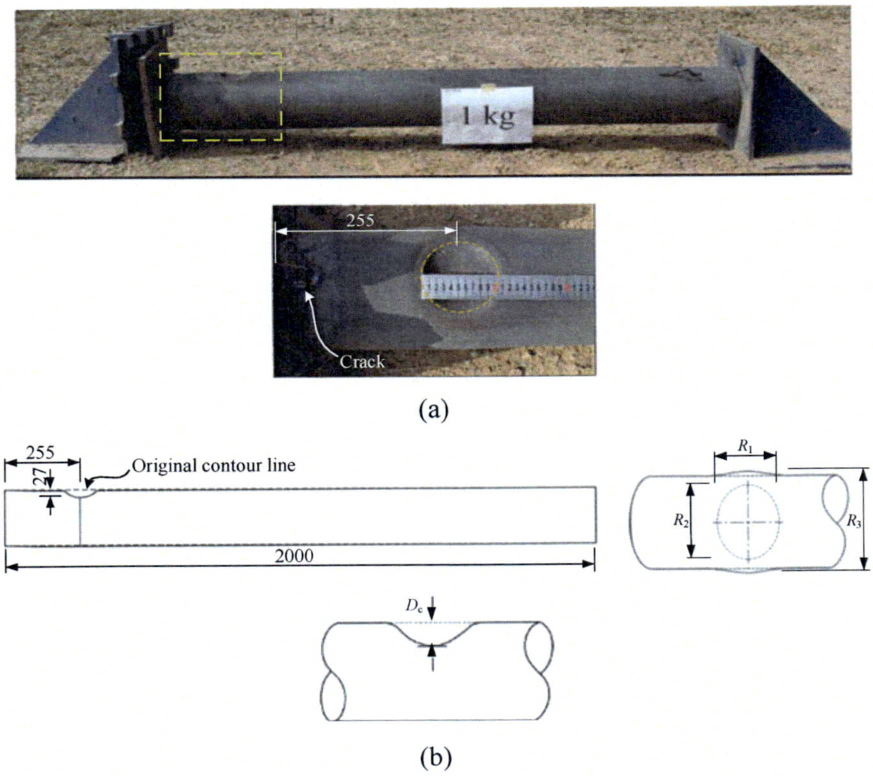

Fig. 10.4 Post-blast C-1 column **a** photograph **b** schematic view (*units* mm), reprinted from Wang et al. (2020a, b), copyright 2020, with permission from Elsevier

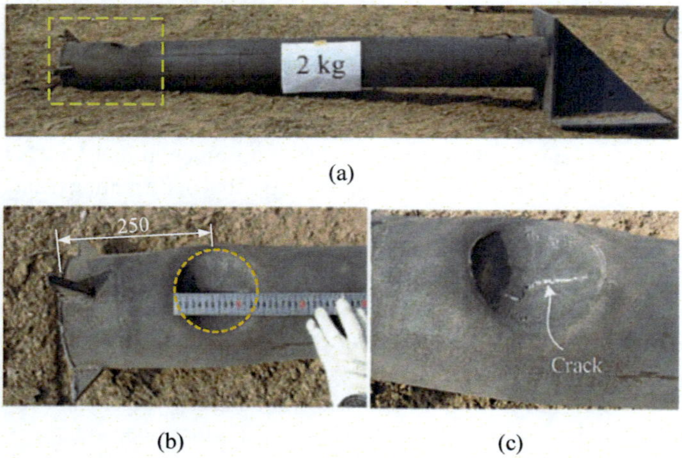

Fig. 10.5 Post-blast C-2 column **a** post-test column **b** local cratering **c** crater craking (*units* mm), reprinted from Wang et al. (2020a, b), copyright 2020, with permission from Elsevier

rib, probably attributed to the reflection of the blast wave on the steel plate at the end of the column as well as the blast induced vibrations of the specimen.

10.4.2.2 C-2 Column

Figure 10.5 shows the failure mode of C-2 specimen after 2 kg TNT contact explosion. Figure 10.5a indicates that the steel plate welded to the end of the column was separated from the column due to the more intensive blast loadings compared to that of 1 kg TNT explosion. There is no integral residual deformation of the column, and also only a localized crater was formed beneath the explosive. Figure 10.5b demonstrates that the crater also had an approximately elliptical shape with R_1 of 129 mm, R_2 of 141 mm, and D_c of 40 mm, respectively. The crater center was 250 mm away from the bottom of the column, which was consistent with the explosive location. The detailed local damage of C-2 specimen is shown in Fig. 10.5c, and it can be seen that the outer steel tube experienced rupture failure at the bottom of the crater with the crack length and width of 86 mm and 3 mm, respectively. The maximal lateral deformation of steel tube at the location of the crater was $R_3 =$ 220 mm, which was 17 mm greater than the original diameter.

10.4.2.3 C-3 Column

Figure 10.6 shows the failure mode of C-3 specimen after 3 kg TNT contact detonation, and it indicates that the bottom steel plate was separated from the bottom of column. In the vicinity of the detonation point, the steel tube was seriously fractured

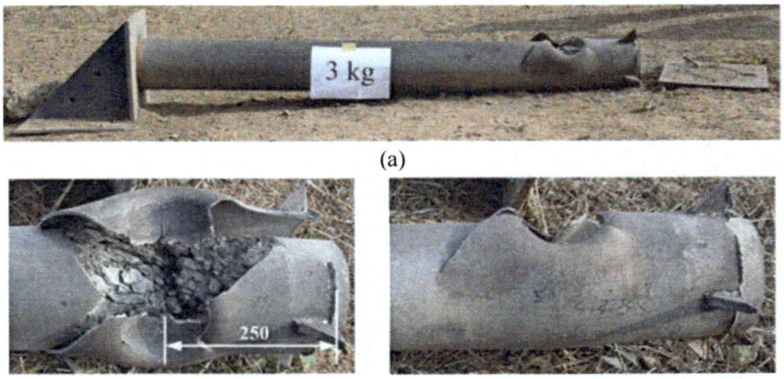

(a)

(b) (c)

Fig. 10.6 Post-blast C-3 column **a** post-test column **b** local damage **c** side view (*units* mm), reprinted from Wang et al. (2020a, b), copyright 2020, with permission from Elsevier

within the range of 450 mm above the bottom of column, and the core UHPCC was crushed. The specimen was almost breached. The depth of the crater in C-3 specimen was greater than 125 mm, the distance from the crater center to the bottom of the column was 250 mm, and the maximal lateral deformation of the fractured tube was increased by about 60 mm. Therefore, C-3 specimen can be regarded as completely losing its axial capacity, and the other C1 and C2 specimens were further tested under axial compression to qualitatively evaluate their damage level induced by contact detonation with different charge weights.

10.5 Axial Compression Test

10.5.1 Test Setup

Two UHPCC-FST columns (I-8 and I-9) in the control group were firstly tested to obtain their original axial load capacity, followed by the two post-blast columns C-1 and C-2. The 1000 t hydraulic machine was adopted to conduct the axial compression test as shown in Fig. 10.7a. The load was applied at the top of each column with the loading rate of 1 mm/min. To prevent the displacement transducers from falling and breaking due to excessive deformation or failure of the column, the loading would be terminated when the axial load dropped to about 60% the maximum axial load. Figure 10.7b shows the layout of instrumentation in the test, where two rope linear voltage differential transducers (LVDT) with the measurement range of 500 mm were installed at the middle height of the column, i.e., left and behind the column, to obtain the lateral deflections at the middle height of the column. Besides, four and two pointer LVDTs with the measuring capacity of 100 mm were arranged around the top and bottom of the column to obtain the vertical displacement of the column ends.

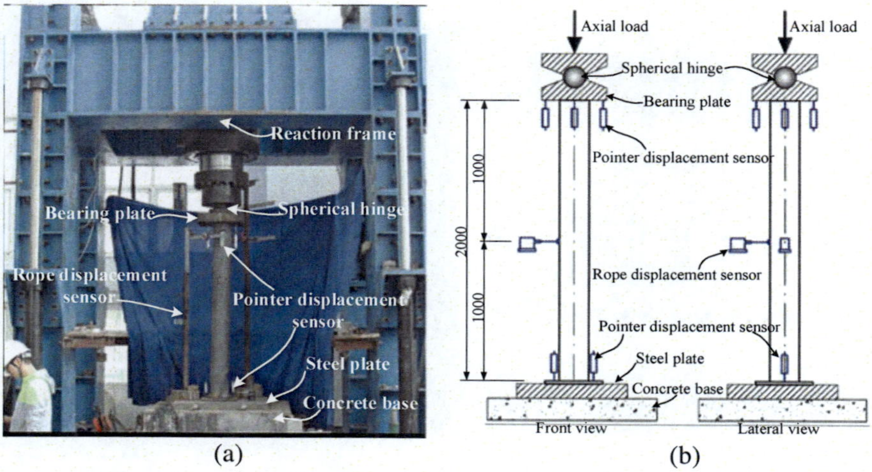

Fig. 10.7 Axial compression test **a** 1000 t hydraulic machine **b** layout of instrumentation (*units* mm), reprinted from Wang et al. (2020a, b), copyright 2020, with permission from Elsevier

Thus, the net vertical deformation of the specimen could be obtained by subtracting the bottom displacement from the top displacement.

10.5.2 Test Results

10.5.2.1 Intact Columns (I-8, I-9)

Under axial compression, the failure mode and the location of failure plane of intact I-8 and I-9 specimens are shown in Fig. 10.8. It can be seen that both two specimens were failed in a combined compression and shear mode. The reason lies in that, with the increase of the axial compressive load, the shear dilatation of core UHPCC would firstly occur to form the failure plane, then the failure plane was dislocated as the column being further compressed, and finally, the local bulging of steel tube was induced due to the low confinement of steel tube and high compressive strength of UHPCC. Figure 10.8a shows that the failure plane of I-8 column occurred at nearly one quarter of the column height from the bottom. The angle between the failure plane and the column axis was about 24°, and the distance from the bottom of failure plane to the column bottom was around 140 mm. Figure 10.8b illustrates that the location of the failure plane of I-9 column occurred at 3/4 of the column height from the bottom. The angle between the failure plane and column axis was about 25°, which was almost identical with that of I-8 specimen.

Furthermore, the curves of the axial load versus axial and lateral displacements for two intact specimens are illustrated in Fig. 10.9. In Fig. 10.9a, the two curves coincide

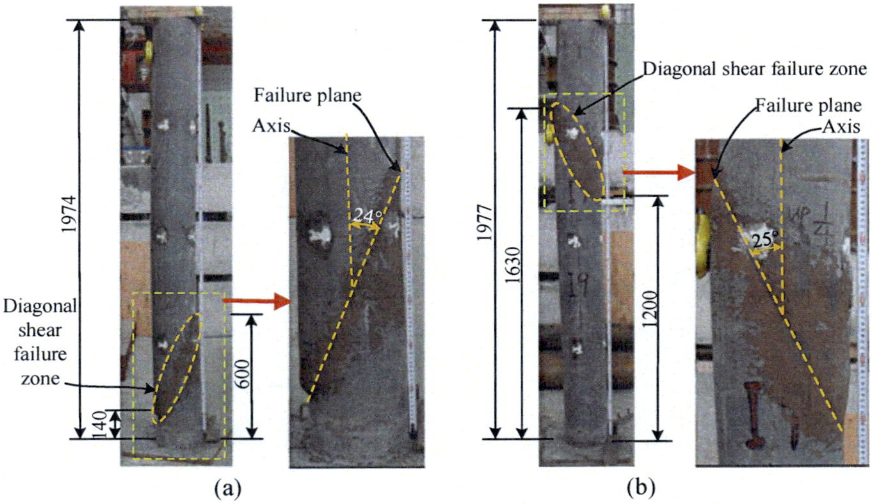

Fig. 10.8 Failure mode of intact columns under axial compressive load **a** I-8 **b** I-9 (*unit* mm), reprinted from Wang et al. (2020a, b), copyright 2020, with permission from Elsevier

with each other very well, including the linear ascending stage, the short platform stage and the following fast descending stage. The maximal bearing capacity of the two columns are 4713 kN (I-8) and 4588 kN (I-9) with the corresponding difference of 2.7%. In Fig. 10.9b, it can be found that there exists a certain deviation in the initial ascending stage for the two columns, probably due to the relatively larger initial eccentricity of I-9 column compared with I-8 column. However, in terms of their axial load capacity and failure modes, such discrepancy can be neglected, and the present axial compression test can be regarded repeatable.

The axial capacity of present UHPCC-FST columns is also evaluated by the existing design codes of CFST column, including the American AISC360 code (AISC) (AISC Committee 2010), European Eurocode-4 code (EUROCODE-4) (EN 1994-1-1 2004.) and Chinese GB50936 code (GB50936-2014 2014). Table 10.3 lists the calculated and averaged experimental axial capacity of present UHPCC-FST column. It is clearly shown that, the predictions of EUROCODE-4 agrees well with the experimental results. The axial load capacity obtained from AISC code is lower than the experimental result, probably because the confining effect of steel tube on the core UHPCC is not fully taken into account. Besides, the calculated axial capacity from the Chinese standard GB50936 is much larger than the test result. According to the Chinese Standard GB50936, the confining coefficient of steel tube for CFST column is suggested to be 0.5–2.0, while the actual confining coefficient for the present UHPCC-FST column is determined as 0.306 based on GB50936. Besides, the maximum compressive strength of core concrete recommended in GB50936 is 80 MPa, which is much lower than that of present UHPCC.

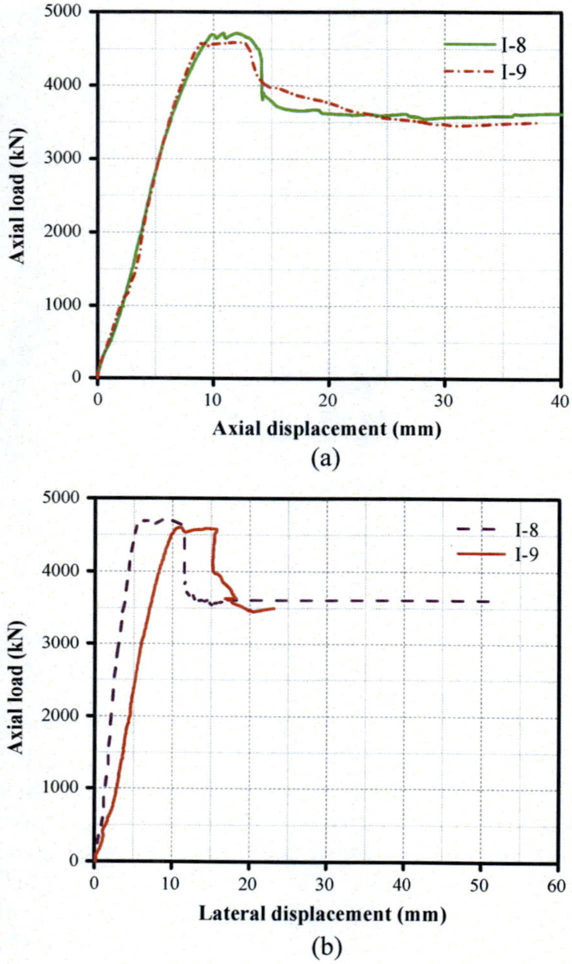

Fig. 10.9 Test results of intact columns **a** axial compressive load vs axial displacement **b** axial compressive load vs lateral displacement, reprinted from Wang et al. (2020a, b), copyright 2020, with permission from Elsevier

Table 10.3 Axial capacity of UHPCC-FST column (kN)

Test result	AISC	EUROCODE-4	GB50936
4650.5	4231.1	4660	5877.3

10.5.2.2 Post-Blast Columns (C-1, C-2)

In this section, the blast-damaged UHPCC-FST columns are examined for their residual axial capacity, and the corresponding test set-up and instrumentation are identical with those for the intact columns. It should be noted that, the lateral deflection of the column at the mid-height was not measured considering that the potential diagonal shear failure. Due to the local detonation damage of the column, the excessive compressive displacement was likely to cause brittle failure of the specimen, thus the axial compression test was ended when the axial load dropped to about 60% of the maximum axial load.

The failure mode and residual deformation of the blast damaged C-1 and C-2 columns after the axial compression test are shown in Figs. 10.10 and 10.11, respectively. It can be seen that the failure plane of C-1 column occurred at the damaged area induced by contact explosion. The angle between the failure plane and the longitudinal axis of column was about 29°, and the distance between the bottom of failure plane and the bottom of column was 140 mm. Under axial compressions, the shear dilatation of the core UHPCC firstly occurred at the location of the crater, and then the surrounding steel tube was suffered the external bulging. With the axial compressive load further increasing, the shear dilatation of core UHPCC led to the development of circumferential tensile stress at the surrounding steel tube. Such circumferential tensile stress then induced a long crack with 240 mm in length and 7 mm in width occurring at the steel tube. For the C-2 column, the failure plane is also found at the damaged area induced by detonation, and the corresponding deformation and failure mode were in a similar manner with those of C-1 specimen. The angle between the failure plane and the longitudinal axis of column was also about 29°, and the bottom of failure plane was 150 mm away from the bottom of column. The major diagonal crack with 250 mm in length and 50 mm in width was observed at the crater.

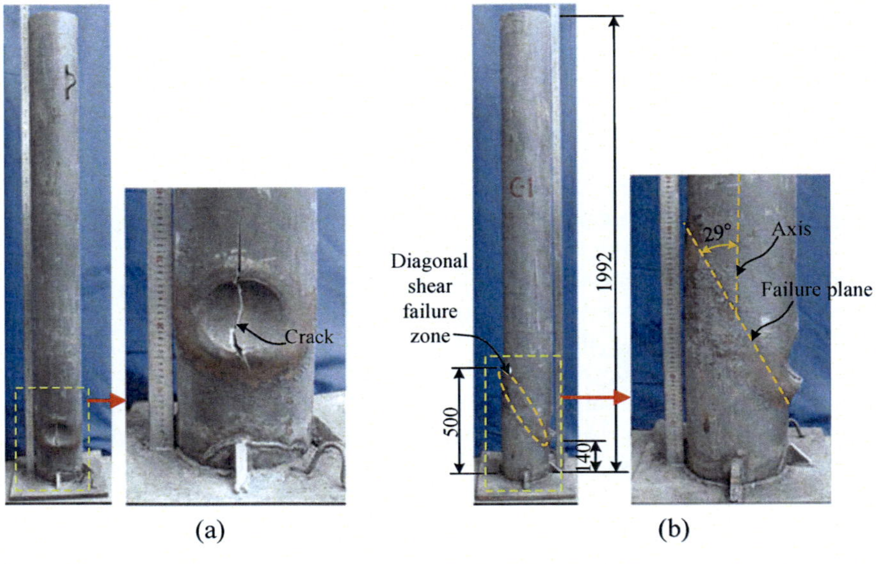

Fig. 10.10 Failure mode and deformation of C-1 column **a** front view **b** side view (*units* mm), reprinted from Wang et al. (2020a, b), copyright 2020, with permission from Elsevier

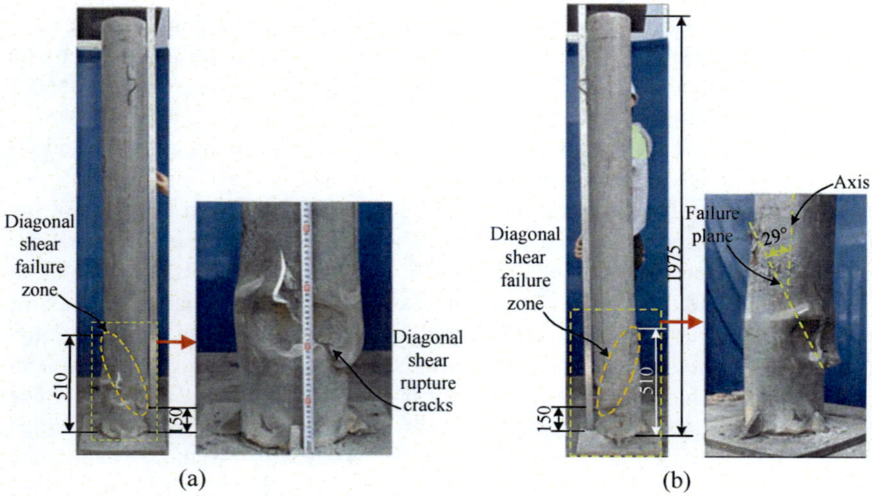

Fig. 10.11 Failure mode and deformation of C-2 column **a** front view **b** side view (*units* mm), reprinted from Wang et al. (2020a, b), copyright 2020, with permission from Elsevier

In general, the test results show that the damage degree of the column increases with the increase of TNT charge weight. In terms of failure mode, both the C-1 and C-2 columns exhibit the compression-shear failure, and the location of the failure plane is observed at the local crater formed due to the explosion. The blast-damaged specimens exhibit a much less ductile manner than the intact UHPCC-FST specimens under axial compressions. Both the intact and the blast-damaged columns are failed by the combination of compression and shear, and the corresponding angles between the failure plane and the longitudinal axis are almost the same. The main failure phenomenon for the UPHCC-FST column is the bulging of the outer steel tube, which is mainly due to the shear dilatation of core UHPCC.

Figure 10.12 further compares the axial load–displacement curves of the intact and blast-damaged columns. It can be seen that, (i) identical with the intact column, the curves of axial load versus axial displacement of the blast-damaged columns also exhibit the linear ascending stage, transient equilibrium stage and descending stage; (ii) the blast-damaged columns have lower axial stiffness and peak load than the intact columns due to the degradation of the steel tube and core UHPCC neighboring the explosion crater; (iii) the axial stiffness and load bearing capacity are deteriorated with the increase of charge weight. Compared with the average axial capacity of the intact columns (4650.5 kN), the residual axial capacities of C-1 and C-2 columns are dropped by 13.5% (4023 kN) and 27.3% (3380 kN), respectively; (iv) C-2 column shows a more ductile behavior after peak load compared to C-1 column, which can be explained by the fact that, after the contact explosion, C-1 column experienced no visual cracks at the crater while C-2 column had a crack with the length of 86 mm at the outer steel tube. Therefore, during the axial compression test, C-1 column

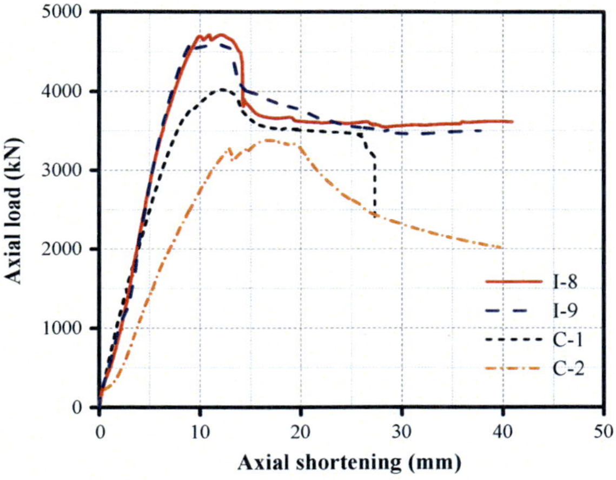

Fig. 10.12 Axial load versus axial displacement of the intact and blast-damaged columns, reprinted from Wang et al. (2020a, b), copyright 2020, with permission from Elsevier

suffered a rupture failure of the tube when the circumferential tensile stress caused by the shear dilatation of core UHPCC exceeded the tensile strength of steel tube, while C-2 column deteriorated gradually with the development of the crack.

10.5.2.3 Damage Index

At present, the damage index defined by Shi et al. (2008) is adopted to evaluate the damage degree of the present UHPCC-FST columns after contact explosions

$$D = 1 - \frac{P_{Residual}}{P_{Intact}} \tag{10.1}$$

where P_{Intact} and $P_{Residual}$ are the axial load capacity of the column before and after detonations, respectively.

The axial load capacity of present UHPCC-FST columns and corresponding damage indexes with various TNT charge weights are shown in Fig. 10.13. Considering that C-3 column is completed destroyed, the corresponding residual axial capacity and damage index are 0 kN and 1, respectively. It can be seen from Fig. 10.13 that the damage indexes of present UHPCC-FST columns after 1 kg and 2 kg contact detonations are 0.14 and 0.27, respectively. The damage index of the UHPCC-FST column is linearly enhanced when the charge weight increases from 0 to 2 kg and sharply aggravates when the charge weight further increasing from 2 to 3 kg.

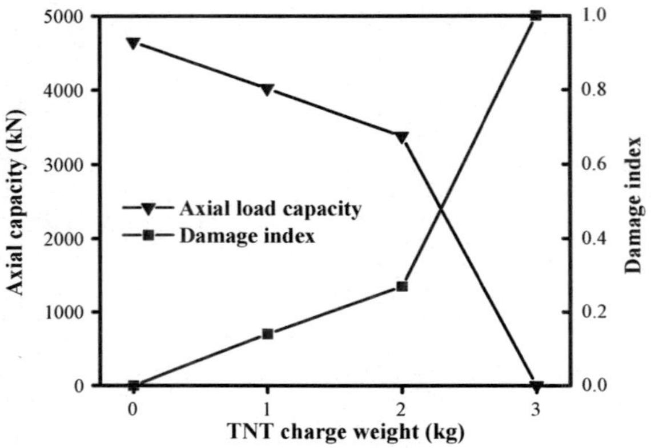

Fig. 10.13 Axial capacity and damage index of UHPCC-FST column under contact detonation with TNT charge weight, reprinted from Wang et al. (2020a, b), copyright 2020, with permission from Elsevier

10.6 Numerical Simulation

10.6.1 FE Model

The explicit finite element program LS-DYNA (LSTC 2007) is employed to establish the 3D numerical model of the UHPCC-FST column under contact detonation and the subsequent axial compression tests. The detailed view of the numerical model of UHPCC-FST column under contact detonation is shown in Fig. 10.14a. The numerical model consists of the UHPCC-FST column, steel support, air domain and TNT explosive charge, in which the TNT, air domain, concrete, steel plate and steel support are discretized with 8-node solid hexahedron elements while the steel tube is modelled using the shell elements. The concrete is meshed with 64 elements around the circumference and 250 elements along the column length, which results in an element size about 7.7 mm. The steel tube is divided with the mesh size identical with the concrete around the circumferential direction. The total number of the elements is 128,000 for the concrete and 16,000 for the steel tube. A perfect bond is assumed for the interface between the steel tube and core concrete by merging the contacting nodes of the concrete and steel tube elements, which has been proven reasonable by the previous studies for modeling CFST/(concrete-filled double-skin steel tube) CFDST columns under blast loading (Li et al. 2018; Zhang et al. 2015b, c, d, 2017). By performing the mesh convergence analyses, the predicted results of blast induced crater depth of columns are nearly unchanged when the average mesh size of the concrete is reduced smaller than 10 mm. Therefore, the mesh size of 7.7 mm is adopted to achieve a balance between accuracy and computational efficiency.

The ALE algorithm is adopted to realize the interaction between the blast wave and the column, the air and TNT charge are simulated with Euler element, and the Lagrange mesh is used to represent the structural component. The *CONSTRAINED_LAGRANGE_ IN_SOLID card is coded to provide the coupling mechanism for modeling the Fluid–Structure Interaction (FSI) between the ALE

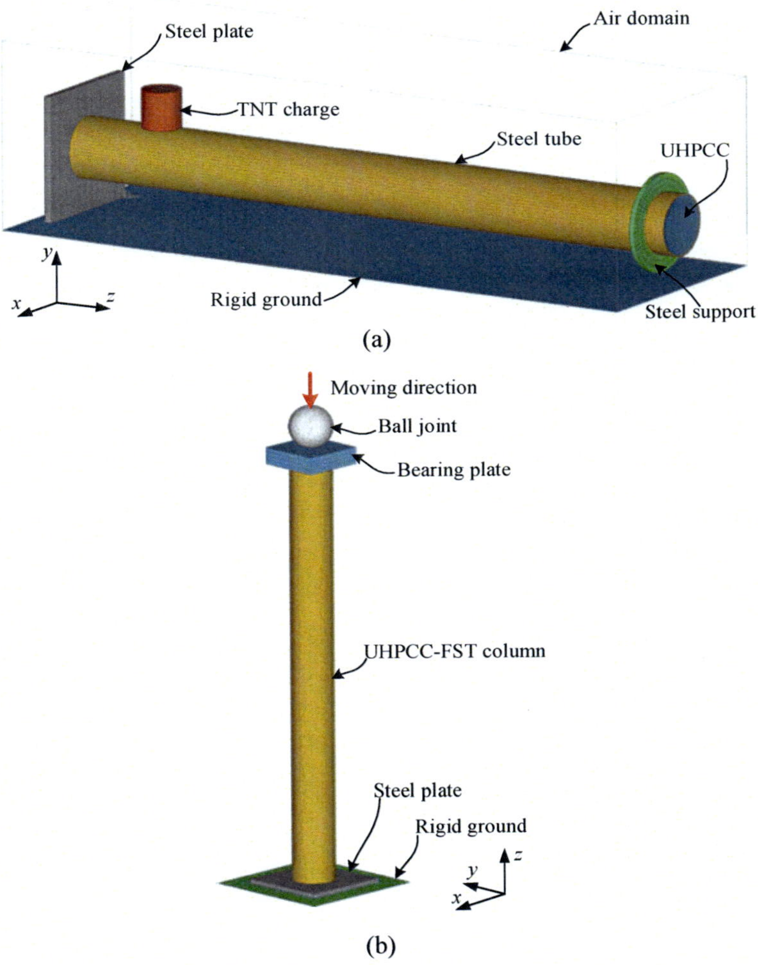

Fig. 10.14 FE models of **a** contact detonation test **b** axial compression test, reprinted from Wang et al. (2020a, b), copyright 2020, with permission from Elsevier

meshes and the Lagrange meshes. The dimension of the air domain is 2.1 m × 0.6 m × 0.6 m, and the non-reflection boundary is assigned to the air domain surfaces. The rigid ground surface is explicitly modelled by setting keywords (*RIGID-WALL_PLANAR) to consider the blast wave reflection off the ground. To be consistent with the boundary condition in the field test, the bottom steel plate of the column is fully fixed while the top of the column is constrained with the steel support against movement in x and y directions. To define the contact behavior between the steel support and column, the keyword *CONTACT_AUTOMATIC_SURFACE_

TO_SURFACE is adopted, in which the static and dynamic friction coefficients are set as 0.2 and 0.15, respectively.

Besides, the FE model of axial compression test of the intact column is shown in Fig. 10.14b. The numerical model consists of the UHPCC-FST column, bearing plate and ball joint, in which the UHPCC-FST column has the identical FE model and material model with the column illustrated in Fig. 10.14a. To be consistent with the setup in the field test, the UHPCC-FST column is placed on the rigid ground. The ball joint which is restrained in the horizontal direction (i.e., in x and y directions) pushes the bearing plate vertically and imposes an axial load on the top of the column with the displacement rate of 10 mm/s. The keyword *CONTACT_AUTOMATIC_SURFACE_TO_SURFACE is adopted to define the contact behavior between the surfaces of the steel plate and rigid ground, bearing plate and column, as well as the ball joint and bearing plate, in which the static and dynamic friction coefficients are set to be 0.3 and 0.2, respectively.

10.6.2 Material Model

10.6.2.1 Concrete

Several concrete-like material models subjected to dynamic loading have been developed and embedded in the commercial FE program LS-DYNA (LSTC 2007), such as Riedel-Hiermaier- Thoma (RHT) model (Riedel and Hiermaier 1999), Holmquist-Johnson–Cook (HJC) model (Holmquist et al. 1993), and Karagozian & Case (K&C) model (Malvar et al. 1997). Among these robust material models, K&C model is validated for capturing the complex concrete behaviors and predicting the structural dynamic responses under blast loadings with good accuracy (Zhang et al. 2015b; Jayasooriya et al. 2014; Wu et al. 2011b).

K&C model (Malvar et al. 1997) takes pressure hardening, strain hardening, strain rate hardening and strain softening into considerations. Three pressure-dependent surfaces, i.e., the maximum strength surface, yield surface and residual strength surface, are implemented to model material hardening and softening responses. K&C model can automatically generate the relevant materials parameters by simply inputting the unconfined compressive strength of concrete, thus it can be easily employed to analyze the static and dynamic behaviors of concrete structure. However, it should be noticed that the automatic parameter generation algorithm embedded in K&C model is based on test data from the normal strength concrete with the uniaxial compressive strength of 45 MPa (Magallanes et al. 2010; Mao et al. 2014). Considering that the strength and post-peak behaviors of present UHPCC are obviously different from those of normal strength concrete, the automatically generated parameters by K&C model in LS-DYNA (LSTC 2007) should be modified according to the available static and dynamic tests of UHPCC.

When the concrete-like material is subjected to the dynamic loadings, the post-peak behavior is determined by the fracture stain energy both in compression and tension, which further controls the damage pattern of material. In K&C model, the post-peak behavior of material is normally governed by the accumulated effective plastic strain, which can be illustrated as

$$\lambda = \int_0^{\overline{\varepsilon}^p} \frac{d\overline{\varepsilon}^p}{r_f(1 + p/(r_f f_t))^{b_1}} \text{ for } p \geq 0 \tag{10.2a}$$

$$\lambda = \int_0^{\overline{\varepsilon}^p} \frac{d\overline{\varepsilon}^p}{r_f(1 + p/(r_f f_t))^{b_2}} \text{ for } p < 0 \tag{10.2b}$$

where λ is the damage factor, f_t is the static tensile strength of material, $d\overline{\varepsilon}^p = \sqrt{(2/3)d\varepsilon_{ij}^p d\varepsilon_{ij}^p}$ is the effective plastic strain increment, and $d\varepsilon_{ij}^p$ is the plastic strain increment tensor, b_1 is softening factor for compression, b_2 is the softening factor for tension. r_f is the dynamic increase factor (DIF), which is defined as the ratio of dynamic-to-static strength under certain strain rate and commonly used to represent the strength enhancement under high strain rate.

It can be seen from Eq. (10.2) that the damage factors have different expressions under compression ($p \geq 0$) and tension ($p < 0$), respectively. The softening factor b_1 in Eq. (10.2a) is determined by the descending branch in compressive stress–strain curve, and the softening factor b_2 in Eq. (10.2b) controlled the fracture strain energy of material under tensile softening. Hence, in current study, the tensile strength f_t, the compressive and tensile softening factor b_1 and b_2 automatically generated in the original K&C model will be modified in order to simulate the post-peak compressive and tensile behaviors of UHPCC properly.

The FE model of the uniaxial compression specimen of UHPCC is established as shown in Fig. 10.15, the size of the prismatic specimen is 100 mm × 100 mm

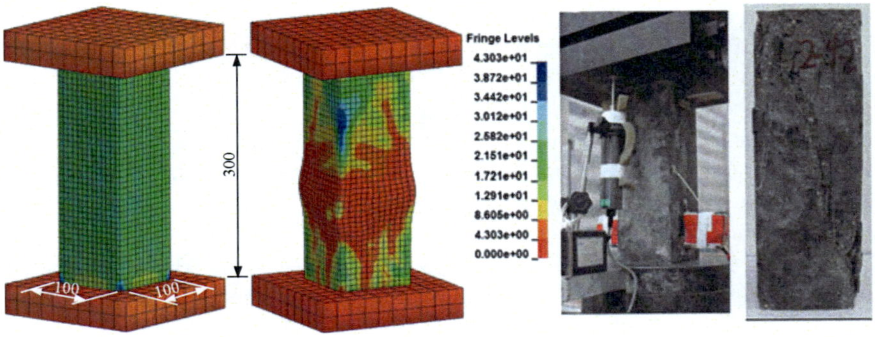

Fig. 10.15 **a** FE models and **b** test specimens before and after axial compression, reprinted from Wang et al. (2020a, b), copyright 2020, with permission from Elsevier

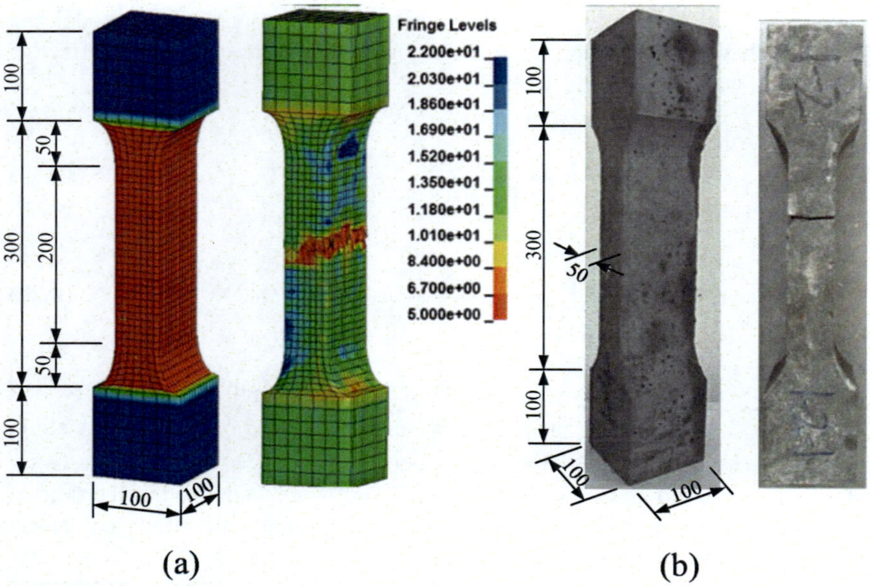

Fig. 10.16 **a** FE models and **b** test specimens before and after axial tension, reprinted from Wang et al. (2020a, b), copyright 2020, with permission from Elsevier

× 300 mm which is identical with that in the test, and the average mesh size of the specimen is 7.7 mm. There are two steel plates on the top and bottom of the specimen, where the bottom plate is fixed and the top plate is moving downwards with the displacement rate of 0.4 mm/s, the specimen is subjected to axial compressive load. The contact behavior between the surfaces of the steel plate and the specimen is described by keyword *CONTACT_AUTOMATIC_SURFACE_TO_SURFACE, in which the static and dynamic friction coefficients are set as 0.3 and 0.2, respectively. Besides, Fig. 10.16 further illustrates the FE model of the uniaxial tension specimen of UHPCC, i.e., a dog bone specimen. The geometric dimensions of the model are also kept identical with those of the test specimen. The initial lengths of the tensile region in the experiment and numerical model are both 300 mm (Fig. 10.16b), and the average mesh size in this region is 7.7 mm. The lower end of the specimen with a length of 100 mm is fixed, and the upper end of the specimen with a length of 100 mm is moved upwards by applying an axial tension load with the displacement rate of 0.4 mm/s.

The default and modified parameters of K&C model (*MAT_72R3) for the present UHPCC are given in Table 10.4. Figure 10.17 shows the corresponding experimental and numerical uniaxial compressive and tensile stress–strain curves, respectively. It indicates that: (i) the predicted compressive stress–strain curve by the default parameters is in good agreement with the experimental curve in the pre-peak stage, while the post-peak stage has large deviations. The corresponding numerically derived

Table 10.4 Default and modified values of K&C model parameters for present UHPCC

Parameters	f_c (MPa)	f_t (MPa)	b_1	b_2	Density ρ	Poisson's ratio υ
Default value	121	7.4	1.6	1.35	2480 kg/m^3	0.23
Modified values	121	9	1	−5.5		

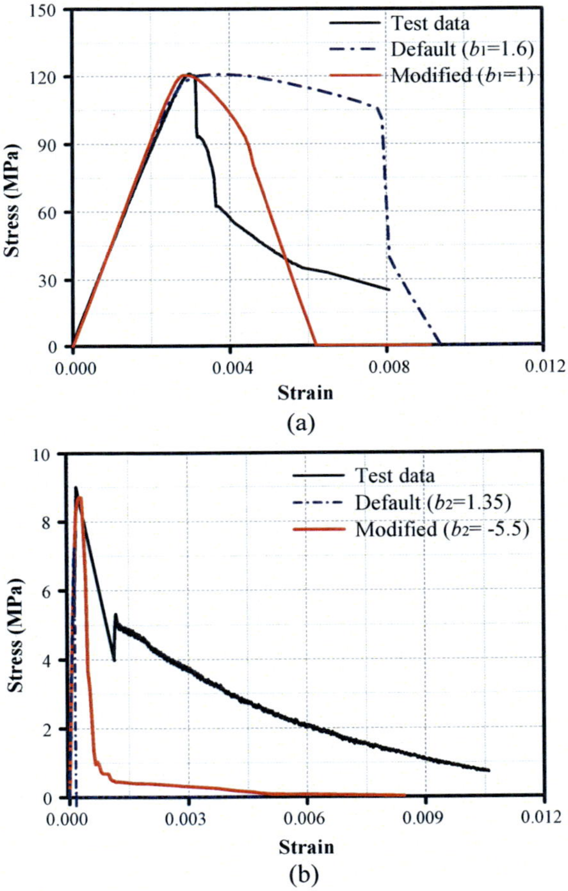

Fig. 10.17 Comparisons of experimental and numerical uniaxial stress–strain curves **a** compression **b** tension, reprinted from Wang et al. (2020a, b), copyright 2020, with permission from Elsevier

post-peak strain energy (the area under the post-peak stress-stain curve) is more than twice that of the test. In order to make the post-peak strain energy closer to the test, it is necessary to adjust the softening factor b_1 until the simulated post-peak strain energy is equal to the test data, the corresponding softening factor $b_1 = 1$; (ii) the simulated post-peak tensile strain energy of K&C model is almost zero, which is far different from the post-peak behavior of the experimental curve. By reducing the softening factor b_2, the post-peak strain energy of the curve can be approaching to the experimental curve and the post-peak behavior of the simulation curve can be improved. Further considering the prediction accuracy of the contact explosion and the following post-blast axial compression tests, by trial and error, and the softening factor b_2 of present UHPCC material is determined as −5.5.

In K&C model, DIF versus strain rate is given by a curve with the keyword *DEFINE_CURVE. Previously, we have conducted the dynamic compression and tension tests of UHPCC by using the split Hopkinson pressure bar (SHPB) (Ren et al. 2018; Wu et al. 2018), and the empirical DIF relations for dynamic compressive (DIF_c) and tensile (DIF_t) strengths of UHPCC were determined, which is given in Eq. (10.3) and illustrated in Fig. 10.18.

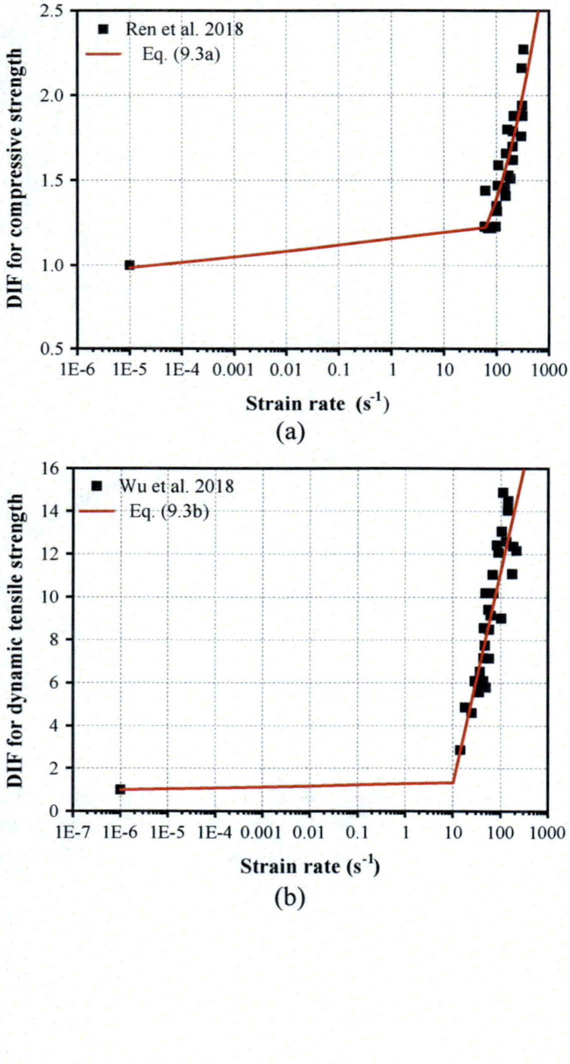

Fig. 10.18 Empirical DIF relations for UHPCC **a** compression **b** tension, reprinted from Wang et al. (2020a, b), copyright 2020, with permission from Elsevier

$$\mathrm{DIF}_c=\begin{cases}(\dot{\varepsilon}\times 10^5/3)^{0.014} \text{ for } \dot{\varepsilon}\le 30\mathrm{s}^{-1}\\ 0.5103(\log\dot{\varepsilon})^2-1.2301(\log\dot{\varepsilon})+1.6804 \text{ for } \dot{\varepsilon}>30\mathrm{s}^{-1}\end{cases} \tag{10.3a}$$

$$\mathrm{DIF}_t=\begin{cases}(\dot{\varepsilon}\times 10^6)^{0.018} \text{ for } \dot{\varepsilon}\le 30.0\mathrm{s}^{-1}\\ 11.561(\log\dot{\varepsilon})-10.446 \text{ for } \dot{\varepsilon}>30.0\mathrm{s}^{-1}\end{cases} \tag{10.3b}$$

It should be noticed that the concrete-like material tends to suffer localized failure such as cratering and spalling when subjected to contact or very close-in range detonation (Zhang et al. 2015b; Jayasooriya et al. 2014; Yuan et al. 2017; Beppu et al. 2010). Under such circumstance, the element representing concrete material in numerical model would perform severe distortion, which can cause numerical computing error. The erosion algorithm in LS-DYNA is a numerical technique to overcome the problems of mesh distortion, it is normally employed to delete failed elements from the calculation by pre-defined criteria. However, this algorithm could not truly reflect the physical phenomena, and thus it should be used with caution. Besides, a scaled damage indicator δ is used to describe the damage level of the simulated material, of which the value increasing from 0 to 1 to 2 indicates that the current strength surface moves from "yield surface" to "maximum strength surface" and to "residual strength surface" as the material being stressed. Therefore, the above damaged indicator implemented in K&C model is employed to illustrate the damaged level of core UHPCC in the steel tube under contact detonation, which has been successfully used in the previous work (Li et al. 2018; Jayasooriya et al. 2014).

10.6.2.2 Steel Tube

The Plastic-Kinematic model (*MAT_PLASTIC_KINEMATIC) in LS-DYNA is adopted to describe the dynamic behavior of the steel tube. It is an elastic–plastic model with kinematic hardening plasticity according to Von Mises yield criterion. The strain rate effect for steel tube is also taken into account by introducing the Cowper and Symonds parameters as follows (Ottosen and Ristinmaa 2005)

$$DIF_s = 1 + \left(\frac{\dot{\varepsilon}_s}{C}\right)^{1/p} \tag{10.4}$$

where $\dot{\varepsilon}_s$ is the strain rate under dynamic loading, C and p are the two constants for strain rate behavior, which are taken as 40.4 s^{-1} and 5, respectively (Deng and Tuan 2014). The parameters used for steel tube in current study are listed in Table 10.5. In order to model the crater and fracture of steel tube, the erosion algorithm is adopted to describe the physical fracture and failure process of the Lagrange steel tube material and handle the large distortion problem. The elements of steel tube will be deleted immediately when the material response reaches the failure strain of 0.26, i.e., the elongation at the rupture of steel tube.

Table 10.5 Parameters for steel tube, air and TNT

Steel tube (*MAT_PIECEWISE_LINEAR_PLASTICITY)					
Density	7850 kg/m^3	Young's modulus	206 GPa	Yield stress	318.5 MPa
Strain rate model C	40 s^{-1}	Strain rate model p	5	Poisson's ratio	0.3
Failure strain	0.26				
Air (*MAT_NULL, *EOS_LINEAR_POLYNOMIAL) (Cheng et al. 2013)					
Initial density ρ_0	1.29 kg m^{-3}	C_0-C_3	0	C_4	0.4
Internal energy E_0	2.5 × 10^5 J m^{-3}	C_5	0.4	C_6	0
Pressure cut-off P_c	0	Dynamic viscosity	0		
TNT (*MAT_HIGH_EXPLOSIVE_BURN, *EOS_JWL) (Cheng et al. 2013)					
Initial density ρ_0	1630 kg m^{-3}	CJ pressure P_{CJ}	21 GPa	Omega ω	0.3
Internal energy E_0	7 × 10^9 J m^{-3}	A	371.2 GPa	B	3.231 GPa
Detonation Velocity D	6930 m s^{-1}	R_1	4.15	R_2	0.95

10.6.2.3 Air and Explosive

The linear polynomial Equation of State (EOS) is adopted to describe the air as

$$p = C_0 + C_1\mu + C_2\mu^2 + C_3\mu^3 + (C_4 + C_5\mu + C_6\mu^2)E \tag{10.5}$$

where p is the pressure of the air, $C_0 - C_6$ are the user defined constants, E is the internal energy per unit volume, $\mu = \rho/\rho_0 - 1$, where ρ/ρ_0 is the ratio of current density (ρ) to initial density (ρ_0). Since air can be assumed as an ideal gas, Eq. (10.5) can be further simplified by setting $C_0 = C_1 = C_2 = C_3 = C_6 = 0$, and $C_4 = C_5 = \gamma - 1$, where $\gamma = 1.4$ is the ratio of the specific heats for air. Hence, the pressure for an idea gas can be expressed as

$$p = (\gamma - 1)\frac{\rho}{\rho_0}E \tag{10.6}$$

It should be noticed that the parameter of air density and E in Eq. (10.6) are related to the temperature, that is, the different values for air density and specific energy should be adopted according to specific local temperature.

The TNT explosive is described by Jones-Wikins-Lee (JWL) EOS as

$$p = A\left(1 - \frac{\omega}{R_1 V}\right)e^{-R_1 V} + B\left(1 - \frac{\omega}{R_2 V}\right)e^{-R_2 V} + \frac{\omega E}{V} \tag{10.7}$$

where A, B, R_1, R_2, and ω are empirically derived constants which are different in each explosive, V is the relative volume or the expansion of the explosive product, and E is the detonation energy per initial unit volume. The parameters of the idea gas and TNT explosive are given in Table 10.5.

10.6.3 Loading Scheme

Based on the present test, the UHPCC-FST column firstly endure the contact detonation and then the axial compressive load. Since the duration time of explosive load is very short, while the axial compression test is a quasi-static process, if this two processes are placed in a keyword file uninterruptedly, it will consume a lot of computing resources. The restart input data method can divide the complete analysis process into multiple stages. The analysis process can be interrupted at any time and a restart file is simultaneously produced, in which the calculated results (damage, volumetric strain and plastic deformation, etc.) at the terminal instant are stored. This method can enhance the flexibility and controllability of the numerical analysis, as well as reduce the elements number and improve the computation efficiency.

According to the modification degree of the original keyword file, the restart input data method can be classified into three categories, i.e., simple restart, small restart and full restart. Since the new parts of bearing plate and ball joint need to be added, the full restart method is adopted at present. As shown in Fig. 10.14a, the simulation of UHPCC-FST column under contact detonation is firstly performed and the corresponding restart files are generated at the terminal instant of UHPCC-FST column under contact detonation.

In the numerical simulation, it is difficult to stop the vibration of the UHPCC-FST column in a short time after the contact explosion. Considering that the present column is basically in the elastic vibration state after 0.015 s of explosion detonates, the keyword *DAMPING_GLOBAL is adopted to stabilize the column from vibrating quickly. The original keyword files are modified by adding the FE model, material constitutive model and corresponding parameters of the bearing plate and ball joint shown in Fig. 10.14b. Then, the analysis process restarts, the ALE elements (explosive and air) are deleted immediately and the bearing plate (Fig. 10.14b) in the modified keyword file pushes the top of the post-blast UHPCC-FST column (Fig. 10.14a) in the restart file with a preset displacement rate. It should be noted that, in the experiment, the axial compressive load was applied with the displacement control manner, i.e., the displacement rate was 1 mm/min, which would consume high computational costs in the numerical simulations. In the present numerical simulations, it is found that nearly identical results are obtained for the two displacement rates of 2 and 10 mm/s. Hence, in order to balance the computational accuracy and efficiency, the displacement rate of 10 mm/s is applied at the top end of the post-blast UHPCC-FST column.

10.7 Comparisons with Test Data

In this section, the above developed numerical model is compared and validated based on the experimental results in terms of the damage and failure modes, crater dimensions and the residual axial capacity. The verified numerical algorithm, constitutive models and the corresponding parameters are utilized to discuss the parametric influences on the blast resistance of UHPCC-FST column under contact detonation in Sect. 10.8.

10.7.1 Damage and Failure Modes of Columns

Figure 10.19 illustrates the numerically simulated damage pattern of C-1 column after 1 kg TNT contact detonation. It can be found that C-1 column sufferred local crater damage with no occurrence of overall deformation, and the steel tube in the vicinity of the explosive is deformed without rupture, which are in good agreement with the experiment (Fig. 10.4). Also, it can be seen that the center of the crater is 250 mm away from the bottom of the column (Fig. 10.19a), and the crater is approximately elliptical shape (Fig. 10.19b), which are consistent with the observations in the field

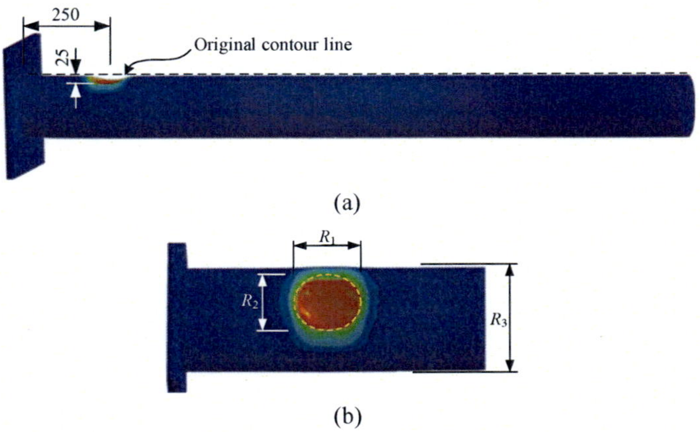

Fig. 10.19 Numerical simulation results for column C-1 **a** vertical view **b** horizontal view (*units* mm), reprinted from Wang et al. (2020a, b), copyright 2020, with permission from Elsevier

blast test. The detailed crater dimensions for C-1 column are listed in Table 10.6, in which "Exp", "Sim" and "Dev" denote the experimental data, numerically simulated results and the relative deviations. It is illustrated that the elliptical short axis R_1 and long axis R_2 are 109 mm and 106 mm, the maximal lateral dimension R_3 of steel tube at the location of crater is 206.9 mm, and the crater depth D_c is 25 mm, which are in good agreement with the experimental results.

Figure 10.20 shows the numerically predicted damaged pattern of C-2 column under 2 kg TNT contact detonation. It can be seen that, similar to C-1 column, C-2 column also exhibits no overall deformation but only a localized crater formed in the vicinity of detonation point, which is consistent with the experiment (Fig. 10.5). It is also shown that the crater center of C-2 column is 250 mm away from the bottom of the column, and the crater has an approximately elliptical shape, which agrees well with the test results. The detailed information of the simulated crater dimensions of C-2 column under detonation is given in Table 10.6, it shows that the simulated R_1, R_2, R_3 and D_c are nearly identical with those observed in the field blast test. It is noticed that two small pieces of steel tube elements are deleted in the crater area, while the field blast test shows a longitudinal crack (Fig. 10.5c). It is believed that if the element size is further reduced, a longitudinal crack would be developed due to the connection of the deleted elements, and hence a better steel tube fracture mode could be obtained, while the computational resource would be also greatly consumed.

The damaged pattern of C-3 column under 3 kg contact detonation obtained from the numerical simulation is given in Fig. 10.21. As shown in Fig. 10.21a, the overall deformation of C-3 column is slightly, and the elliptical crater is formed beneath the TNT charge. The outer steel tube suffers serious rupturing failure, in which three wider axial cracks and two narrower transverse cracks occur at the crater, as

Table 10.6 Comparisons of experimental and simulated crater dimensions

Column	R_1 (mm)			R_2 (mm)			R_3 (mm)			D_c (mm)		
	Exp	Sim	Dev (%)	Exp	Sim	Dev (%)	Exp	Sim	Dev (%)	Exp	Sim	Dev (%)
C1	105	109	3.8	110	106	−3.6	209	206.9	−1	27	25	−7.4
C2	129	136	5.4	121	123	1.7	220	213	−3.2	40	38	−5
C3	136	150	10.3	250	222	−11.2	263	221	−16	>125	53	>−57.6

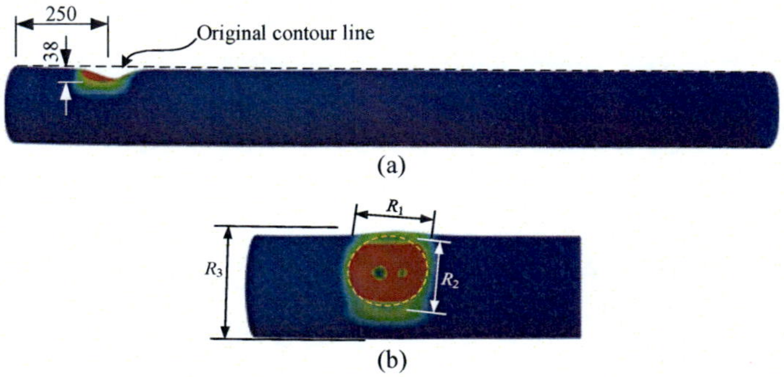

Fig. 10.20 Numerical simulation results for column C-2 **a** vertical view **b** horizontal view (*units* mm), reprinted from Wang et al. (2020a, b), copyright 2020, with permission from Elsevier

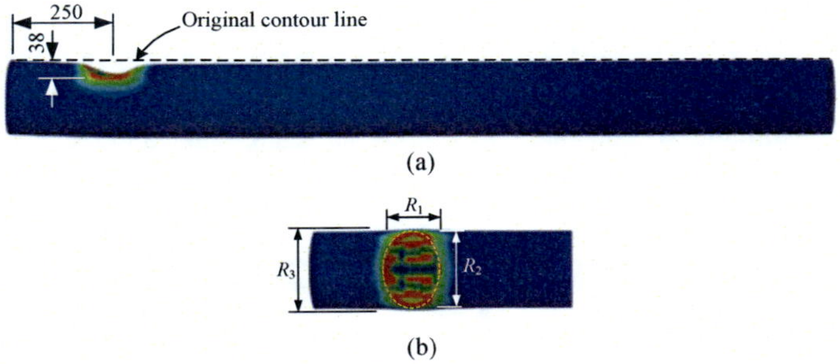

Fig. 10.21 Numerical simulation results for column C-3 **a** vertical view **b** horizontal view (*units* mm), reprinted from Wang et al. (2020a, b), copyright 2020, with permission from Elsevier

illustrated in Fig. 10.21b. The elliptical short axis R_1 and long axis R_2 of the blast-induced crater are consistent with those in the field blast test as comparison in Table 10.6. However, the crater depth D_c and the maximal lateral deformation of column R_3 are 57.6% and 16% lower than the experimental results, respectively. This might be due to the occurrence of the cracking at the steel tube in the field blast test. During the contact detonation, the high pressure from explosive is formed and tears away the steel tube, which may induce the stress concentration at the cracking. However, in the numerical model, the failure of the steel tube is realized by the erosion process, i.e., the deletion of the failed elements, and the cracking is ceased neighboring to the crater due to the sudden deletion of the nearby elements. Thus the stress tore away the crack cannot transmit to the nearby elements, which leads to less severe result in the numerical simulation than that in the field blast test.

It can be concluded from the above analyses that the present established numerical model can reproduce the damaged pattern and crater dimension of UHPCC-FST column under contact detonation quite well. However, due to the erosion technique adopted in the numerical simulation for the steel tube material, the fracturing behavior of the steel tube in the C-3 column is not fully reproduced.

10.7.2 *Residual Axial Capacity and Failure Mode of Columns*

10.7.2.1 Intact Columns (I-8, I-9)

Taking the intact UHPCC-FST column I-8 for example, Fig. 10.22a and b show the damage pattern of the intact column before and after axial compressive loading, respectively. Figure 10.22c further illustrates the damage pattern of the longitudinal section of the column after axial compressive loading. It can be seen obviously that the axial compressive load causes the diagonal shear fracture plane at the intact column, leading to the bulging failure of the external steel tube. The comparisons of the axial capacity of two intact columns from the test and numerical simulations are given in Table 10.7. Apparently, for the original axial capacity, the deviation of the numerical simulation results from the experiment is 8%, indicating the accuracy of the numerical simulation. The angle between the fracture plane and the longitudinal

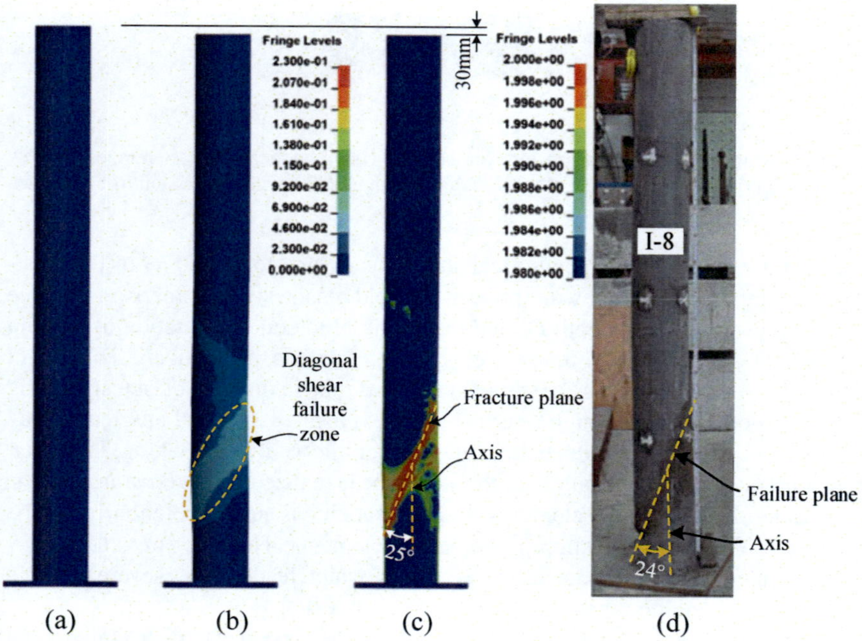

Fig. 10.22 Numerical simulation results for intact column I-8 **a** before compression test **b** after compression test **c** longitudinal section view **d** test observation, reprinted from Wang et al. (2020a, b), copyright 2020, with permission from Elsevier

Table 10.7 Comparison of the numerical and experimental axial capacity of intact columns

Column	Peak force (kN)			Angle of fracture plane			Height of the fracture plane center (mm)	
	Exp	Sim	Dev	Exp	Sim	Dev	Exp	Sim
I-8	4650.5	5022	8%	24.5°	25°	2%	370	403
I-9							1420	

axis is about 25° which is close to the angle observed in the test. It should be noticed that the distance between the center of the fracture plane and the bottom of the column predicted in the numerical simulation is slightly higher than that obtained in the experiment of I-8 column. The possible reason is that, the perfect alignment of axial load is assumed in the numerical simulation, while in the experiment the column may be subjected to the eccentric compressive load to some extent. In view of the identical failure mode of the intact column under axial compressive load between the numerical and experimental results, the accuracy of the established 3D model in current study is confirmed. Figure 10.23 further shows the comparisons of experimental and numerical axial load–displacement curves of two intact columns. It can be seen that, the numerically derived axial load–displacement curve includes the linear ascending stage and descending stage, the transient equilibrium stage occurred in the experimental curve is absent. Besides, in the ascending stage, the numerical curve has a larger slope than the experimental curve, the possible reason is explained in the following section.

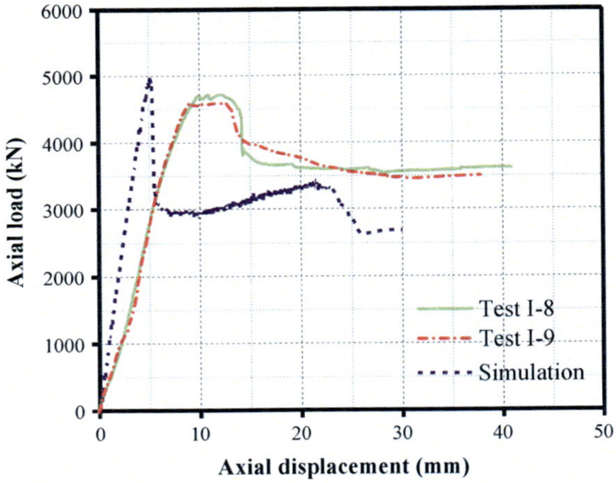

Fig. 10.23 Numerical and experimental axial capacity of intact columns, reprinted from Wang et al. (2020a, b), copyright 2020, with permission from Elsevier

10.7.2.2 Post-Blast Columns (C-1, C-2)

For the present test, the core UHPCC of C-3 column in the vicinity of TNT charge is completely crushed and suffered severe fracture after contact detonation, thus it is regarded as completed destroyed without any residual axial capacity. Only C-1 and C-2 columns are adopted to evaluate their residual axial capacities numerically in this section.

Figure 10.24a and b illustrate the damage patterns of the front view of column C-1 before and after axial compressive loadings, respectively. It can be found that under the axial compressive loading, the crack occurs at the location of the crater, and the surrounding steel tube is found to be extruded laterally, which is similar with that observed in experiment (Fig. 10.24g). Figure 10.24c and d show the corresponding damage pattern of the side view of C-1 column before and after the subsequent axial compression test, respectively. Figure 10.24e and f further report the damage pattern of the longitudinal section of the post-blast column C-1 before and after axial compressive loading, respectively. It is obvious that the axial compressive load causes the occurrence of fracture plane at the C-1 column, making the external bulging failure to the surrounding steel tube. The angle between the diagonal shear fracture plane and the longitudinal axis is about 28° which is close to the angle of 29° observed experimentally (Fig. 10.24h). As listed in Table 10.8, the numerically derived distance between the fracture plane center and the column bottom is 444 mm, which is larger than the experimental result. It may be also attributed to the existence of the slightly eccentric compressive loading in the experiment. However, both the

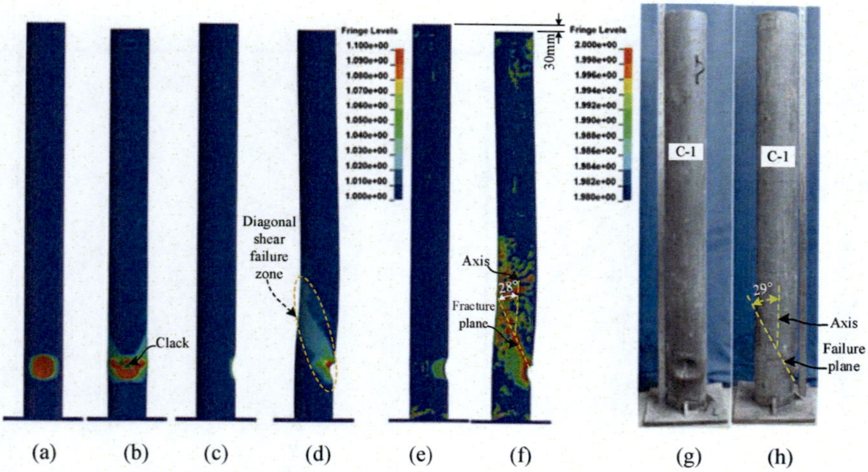

Fig. 10.24 Numerical simulation results for column C-1 **a** front view before and **b** after axial compression test **c** side view before and **d** after axial compression test **e** longitudinal section before and **f** after axial compression test; test observations **g** front view and **h** side view after axial compression test, reprinted from Wang et al. (2020a, b), copyright 2020, with permission from Elsevier

Table 10.8 Comparisons of the residual axial capacity of column after blast from numerical and experimental results

Column	Peak force (kN)			Angle of fracture plane			Height of the fracture plane center (mm)	
	Exp	Sim	Dev (%)	Exp	Sim	Dev (%)	Exp	Sim
C-1	4023	3883	−3.5	29°	28°	−3.4	320	444
C-2	3380	3193	−5.5	29°	33°	14	330	417

numerical and experimental results exhibit the identical failure mode of C-1 column under axial compressive loading.

Similarly, Fig. 10.25a–f further illustrate the damage pattern of the front view, side view as well as the longitudinal sectional view of the post-blast column C-2 before and after axial compressive loading. Figure 10.25g and h further show the corresponding experimental photograph. It can be seen that, (i) under the axial compressive loading, the severe damage with many vertical cracks occurs at the crater, and the surrounding steel tube is extruded laterally, which is similar with that observed in the experiment; (ii) the major shear fracture plane occurs at the same damaged area induced by detonation, which is consistent with the test; (iii) the angle between the diagonal shear fracture plane and the longitudinal axis from the numerical simulation is close to the experiment value. However, due to the same reason as C-1 column, the height

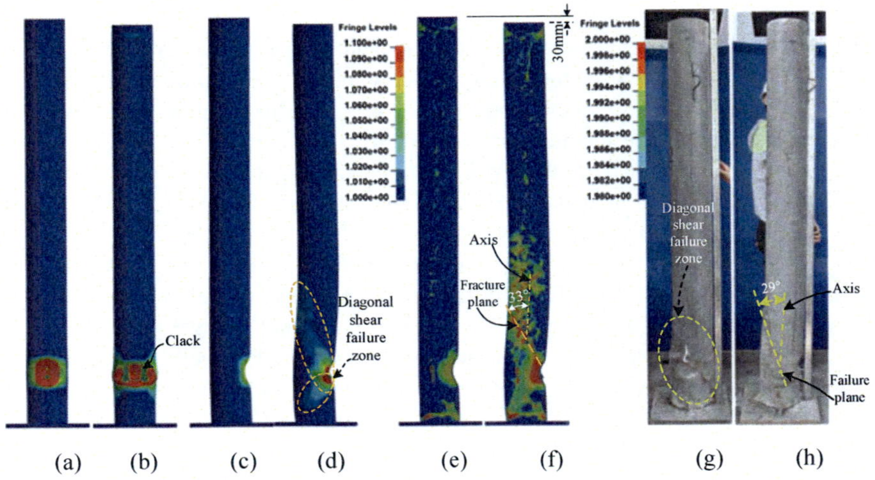

Fig. 10.25 Numerical simulation results for column C-2 **a** front view before and **b** after axial compression test **c** side view before and **d** after axial compression test **e** longitudinal section before and **f** after axial compression test; test observations **g** front view and **h** side view after axial compression test, reprinted from Wang et al. (2020a, b), copyright 2020, with permission from Elsevier

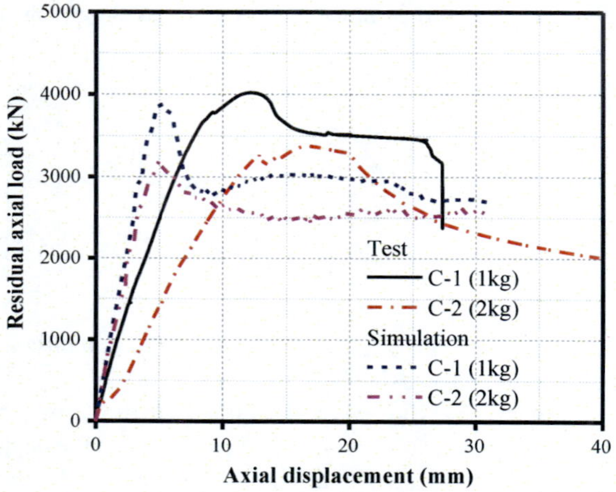

Fig. 10.26 Simulated and experimental residual axial capacity of C-1 and C-2 columns, reprinted from Wang et al. (2020a, b), copyright 2020, with permission from Elsevier

of the fracture plane center derived from the numerical simulations is slightly higher than the experimental result, as given in Table 10.8.

The experimental and numerical simulated axial load–displacement diagrams of two post-blast columns are further plotted in Fig. 10.26. The comparisons of the residual axial capacity of two columns are given in Table 10.8. Apparently, both for the simulated results and experimental data, the residual peak axial capacity of C-1 column (after 1 kg contact detonation) is larger than that of C-2 column (after 2 kg contact detonation), and the corresponding stiffness of C-1 column is greater than that of C-2 column. The deteriorated effect of contact explosion on the stiffness of column is reflected. As for the residual axial capacity, the numerical simulation results have −3.5 and −5.5 deviations from the test data for columns C-1 and C-2, indicating the accuracy of the numerical simulation.

It should be noted that for the axial load–displacement curves in Figs. 10.23 and 10.26, both numerical results exhibit higher stiffness in the ascending stage comparing with the experimental results. This might be due to the inherent drawback of K&C concrete model adopted in the numerical simulations, in which the Young's modulus cannot be artificially adjusted explicitly. At present, the Young's modulus of UHPCC (42.9 GPa) is larger than that of normal strength concrete, e.g., 33.5 GPa corresponding to the concrete strength of 45 MPa. It is well known that the pre-peak behavior of material is dominated by the initial bulk modulus depending on the young's modules. Hence, in order to describe the reasonable pre-peak behavior, the relationship between the yield strength surface and maximum strength surface in the K&C model should be further calibrated based on the systematic experimental results. However, this aspect is beyond the scope of current study and will be discussed in depth furthermore. The present chapter is mainly focused on the post-peak behavior and damaged pattern of UHPCC-FST column, by comparing with the contact TNT detonation and the subsequent axial compression tests, the current numerical simulation can provide reasonable predictions on the post-blast axial capacity of UHPCC-FST column. Therefore, the validations of adopted numerical simulation algorithms, constitutive models and the corresponding modified parameters are verified.

10.8 Parametric Study

In this section, based on the above verified numerical models, the parametric studies are carried out to further investigate the factors that may influence the blast resistance and residual axial capacity of UHPCC-FST column subjected to contact detonation, including the steel tube wall thickness and strength, core concrete strength and section diameter, respectively. The scenario of the present column C-2 under 2 kg TNT contact detonation is adopted as the baseline, and the detailed parameters used in the numerical model for C-2 column can be referred to Sect. 10.3.

10.8.1 Steel Tube Thickness and Strength

From the above analysis, it is found that the presence of steel tube can prevent the spalling of core UHPCC, thus the influences of the steel tube thickness and strength on explosion crater depth, residual axial capacity and damage index of UHPCC-FST column under contact detonation and the subsequent axial compression tests are discussed.

10.8.1.1 Thickness

Four different steel tube thicknesses, i.e., 4, 8, 10 and 16 mm, are further selected, while the other parameters, e.g., steel tube strength, compressive strength of concrete, and cross-sectional diameter of core concrete remain the same as those for column C-2.

Figure 10.27a illustrates the crater depth formed on the UHPCC-FST column with varied steel tube thicknesses under contact detonation. It can be seen that the crater depth decreases dramatically when the steel tube thickness increases from 4 to 6 mm, while the decreasing slope reduces when the steel tube thickness increases beyond 6 mm. Figure 10.27b further shows the dependence of residual axial capacity of the post-blast UHPCC-FST column on the steel tube thicknesses. For comparison purpose, the corresponding original axial capacity of intact column is also included. Figure 10.27(a and b) indicate that increasing the steel tube thickness could effectively improve the blast resistance of UHPCC-FST column, and thus enhance the residual axial load bearing capacity after the contact detonations. This mainly because that the thicker steel tube can provide prominent confinement to the core UHPCC.

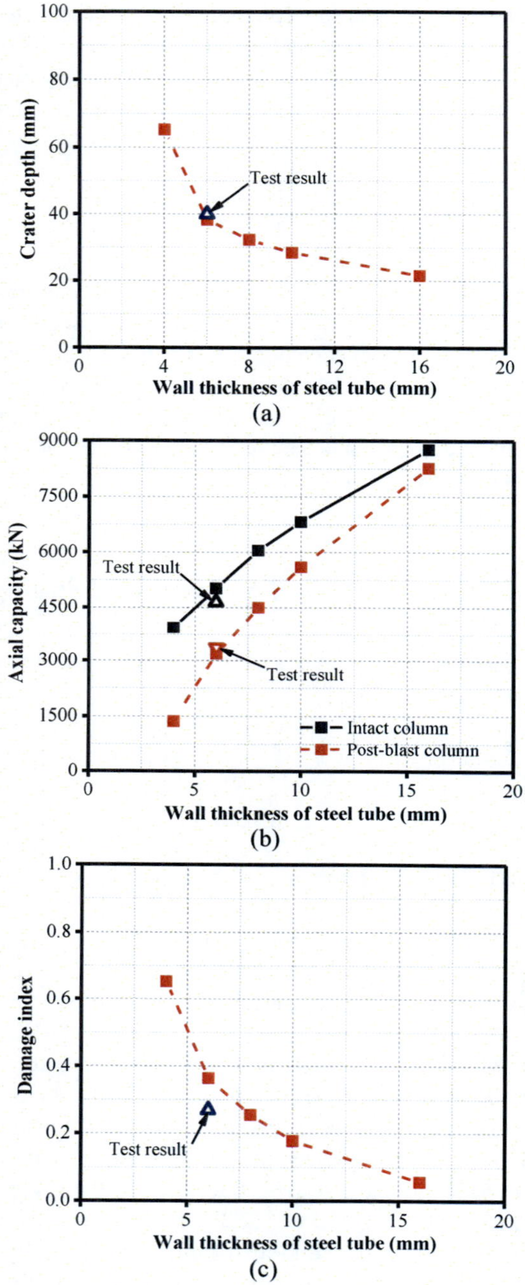

Fig. 10.27 Influence of steel tube wall thickness on **a** explosion crater depth **b** axial capacity **c** damage index, reprinted from Wang et al. (2020a, b), copyright 2020, with permission from Elsevier

Correspondingly, Fig. 10.27c plots the damage index for the present UHPCC-FST column with varied steel tube thicknesses. It is clearly that with the increase of the steel tube thickness, the damage degree of the column under contact detonation reduces significantly. This demonstrates that the thicker steel tube leads to higher residual axial capacity of UHPCC-FST column after enduring the contact explosion. However, it is also observed that, within the present discussed scenario, the strengthening effect of steel tube thickness is reduced when the steel tube thickness is greater than 6 mm.

10.8.1.2 Strength

Four various yield strength levels for steel tube are further considered, i.e., 235, 500, 800 and 1100 MPa, and the remaining parameters are kept identical with those of column C-2. Similarly, Fig. 10.28 further shows the blast formed crater depth, the residual axial load bearing capacity, as well as the corresponding damage index of present UHPCC-FST column with varied steel tube strength. It indicates that, (i) the crater depth decreases with the increase of steel tube strength, when the steel tube strength increases from 235 to 318.5 MPa, the corresponding crater depth is reduced by 17.4%, while such tendency becomes smoother beyond the threshold value of 318.5 MPa. This indicates that further increasing the steel tube strength cannot significantly reduce the crater depth; (ii) an approximately linear increasing curve in Fig. 10.28b indicates that higher strength of the steel tube enhances the axial load bearing capacity of UHPCC-FST column effectively; (iii) the damage index of the present column under contact detonation drops significantly within the strength range from 235 to 318.5 MPa. However, the effect of the steel tube strength on reducing the damage degree of UHPCC-FST column gradually weakens with further increasing the steel tube strength.

10.8.2 Core Concrete Strength and Cross-Sectional Diameter

10.8.2.1 Compressive Strength

Five strength levels of the core concrete are further selected, i.e., 30, 45, 60, 75 and 170 MPa, while the other parameters are kept identical with those of C-2 column. It should be noticed that the strength levels of 30–75 MPa represents the NSC, whereas the compressive strength of 170 MPa is considered as UHPCC. As mentioned above, UHPCC performed superior compressive and tensile fracture strain energy compared to the NSC. Therefore, the fracture strain energy of 0.12 MPa (i.e. the area under the modified numerical post-peak compressive stress-stain curve in Fig. 10.17a) is assigned to the UHPCC. Besides, according to CEB (CEB-FIP 2008), the fracture strain energy of the NSC is decreased from 0.041 MPa to 0.011 MPa when the concrete strength increases up to 110 MPa, respectively.

Fig. 10.28 Influence of steel tube strength on **a** crater depth **b** axial capacity **c** damage index, reprinted from Wang et al. (2020a, b), copyright 2020, with permission from Elsevier

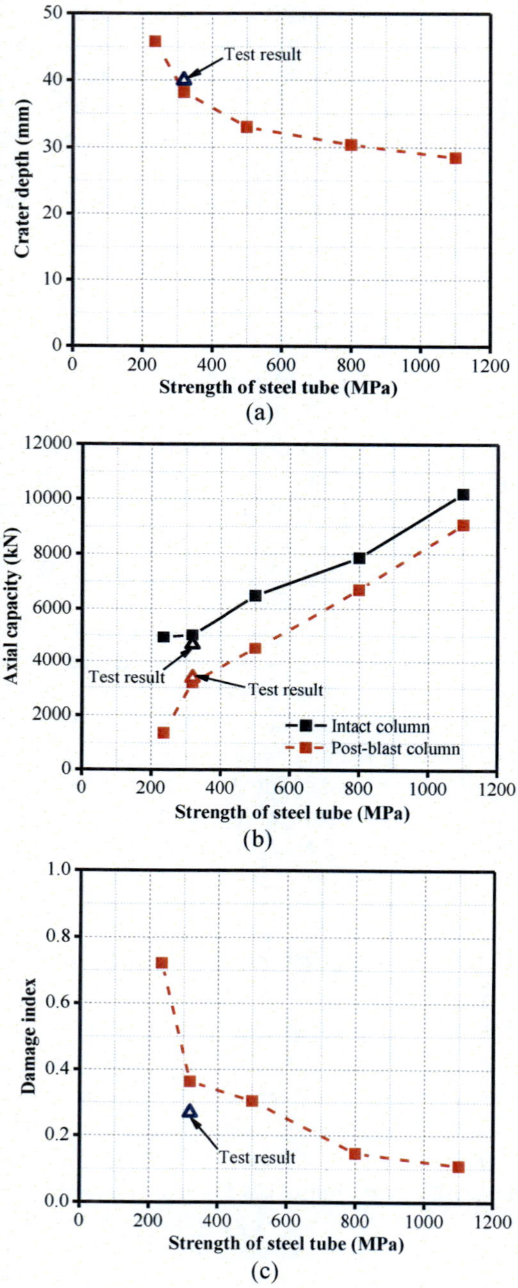

Figure 10.29 further shows the blast induced crater depth, the residual axial load bearing capacity, as well as the corresponding damage index of present UHPCC-FST column with varied core concrete strength. It indicates that, (i) with the increase of core concrete strength, the crater depth of the column reduces gradually, the corresponding reducing trend within the concrete strength range of 30–75 MPa is more prominent compared to that within the concrete strength range of 75–170 MPa; (ii) higher compressive strength of core concrete exhibits higher original and residual axial capacity of the column, the curve of residual axial capacity of the column exhibits linearly increasing trend, while the curve of the original axial capacity of the column shows firstly linearly increasing trend (below 75 MPa) and then enters into the plateau. It is probably attributed to the influence of the structural stability of the column under compression. When the core concrete strength exceeds a certain range, further increasing the core concrete strength do not have much contribution to the axial capacity of the column; (iii) the damage index of the post-blast column first increases for the concrete strength ranging from 30 to 75 MPa and then reduces within the concrete strength range of 75–170 MPa. The corresponding variation range of the damage index is from 0.3 to 0.4, which indicates that the damage index is not sensitive to the compressive strength of core concrete.

10.8.2.2 Cross-Sectional Diameter

The cross-sectional diameters of core concrete are selected as 88, 288, 388 and 488 mm. Considering the steel tube thickness of 6 mm, the corresponding cross-sectional diameter of column are determined as 100 mm, 300 mm, 400 mm and 500 mm, respectively. The other parameters, i.e., steel tube thickness, steel tube strength and compressive strength of core concrete, are kept identical with those of column C-2. Figure 10.30 further shows the blast induced crater depth, the residual axial load bearing capacity, as well as the corresponding damage index of present UHPCC-FST column under contact detonation with different cross-sectional diameters of core concrete. It should be noted that, the column with a diameter of 100 mm is breached in the vicinity of the detonation point and the crater depth was set as 100 mm. It indicates that, (i) the crater depth decreases with increasing the cross-sectional diameter of core concrete, e.g., the crater depth reduced by 61.8% when the core concrete diameter increases from 88 to 191 mm, while beyond 191 mm, further increasing the core concrete diameter has little effect on the crater depth; (ii) increasing the core concrete diameter can effectively improve the original and residual axial capacity of the column; (iii) the damage index reduces rapidly when the core concrete diameter increases from 88 to 191 mm, however, further increasing the core concrete diameter cannot significantly reduce the damage degree of UHPCC-FST column.

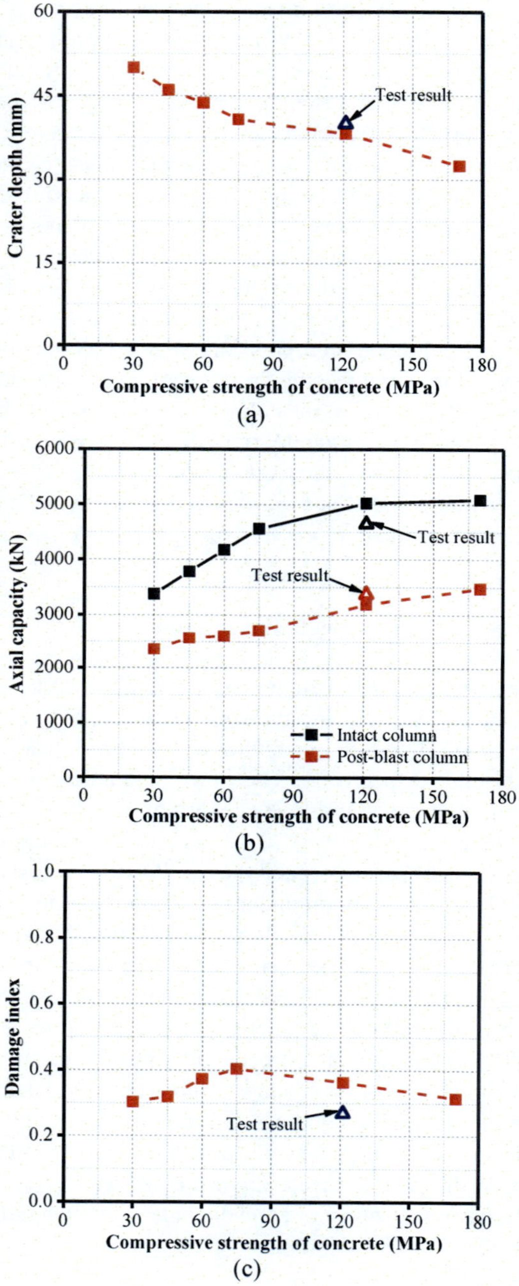

Fig. 10.29 Influence of compressive strength of core concrete on **a** crater depth **b** axial capacity **c** damage index, reprinted from Wang et al. (2020a, b), copyright 2020, with permission from Elsevier

Fig. 10.30 Influence of cross-sectional diameter of core concrete on **a** crater depth **b** axial capacity **c** damage index, reprinted from Wang et al. (2020a, b), copyright 2020, with permission from Elsevier

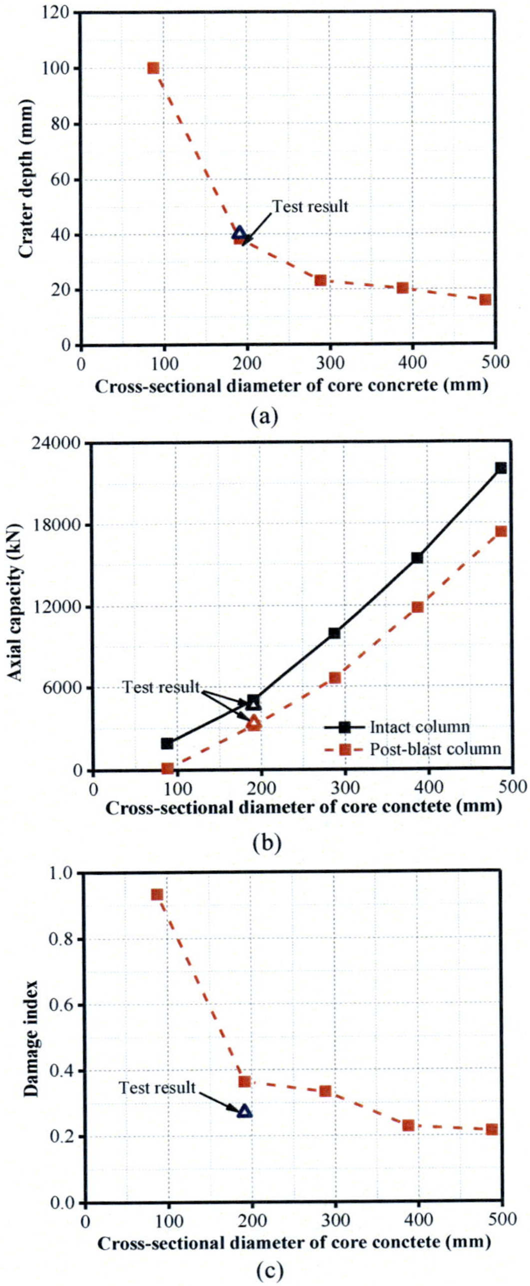

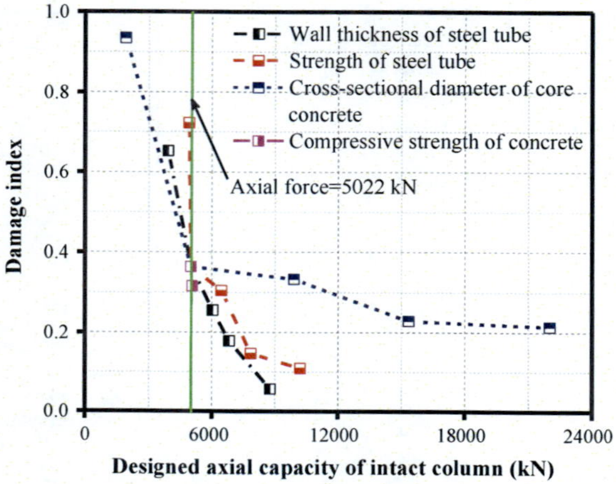

Fig. 10.31 Influences of varied parameters on the damage index of UHPCC-FST column under contact detonations, reprinted from Wang et al. (2020a, b), copyright 2020, with permission from Elsevier

10.8.3 Influence of Varied Parameters on Damage Index

Aiming to further examine the most influential factors on the enhancement of blast resistance of UHPCC-FST column under 2 kg TNT contact detonation, the damage indexes obtained from above parametric study are summarized and compared as shown in Fig. 10.31. It indicates that, the curves can be divided into two parts according to the original axial capacity of 5022 kN, which is the numerically derived axial capacity of intact columns as listed in Table 10.7.

When the axial capacity is less than 5022 kN, it can be observed from Fig. 10.31 that the slope of the curve that represents the effect of steel tube strength exhibits the steepest trend, followed by the steel tube thickness and the core concrete cross-sectional diameter. It is deduced that the steel tube strength has the highest effect on the damage degree of the post-blast UHPCC-FST column. The possible reason lies in that the steel tube with lower strength would suffer rupture failure under contact detonation, and hence the spalling of core concrete cannot be confined efficiently, causing a rapid decline of the axial load bearing capacity of column. On the other side, due to the confinement of steel tube, the damaged/crushed core concrete would be compacted under axial compressive loading and provide certain contribution to enhance the residual axial capacity of the column.

Considering the practical engineering constructions, further increasing the compressive strength of core concrete beyond 170 MPa is not concerned at present. As shown in Fig. 10.31, with the designed axial capacity of the column increasing beyond 5022 kN, the wall thickness of steel tube is the most influential factor on the damage degree of the post-blast UHPCC-FST column, followed by steel tube strength, while the cross-sectional diameter of core concrete has the least effect. The possible reason is that, with the increase of steel tube thickness, the bending stiffness of the column would be enhanced, resulting in the higher resistance to the localized deformation. Since the integrity of the steel tube is maintained under the contact detonation, the core concrete can be effectively confined although the slight spalling damage may occur, thus the higher residual axial capacity of the column will be achieved. Besides, for the explosion scenario, the generated blast wave would engulf the column instantaneously. Although increasing the core concrete diameter would improve both the original and residual axial capacity of the column, more blast energy would be absorbed by the core concrete due to the less blast wave diffraction of the lager sectional column, which may cause more extensive damage of the core concrete in the vicinity of detonation point.

10.9 Summary

In this Chapter, aiming to assess the performance of UHPCC-FST bridge column against the potential terroristic contact explosions, the residual axial capacity of UHPCC-FST columns under contact explosion is studied experimentally and numerically. The main works and conclusions are as follows:

(1) Considering the prototype of UHPCC-FST bridge columns with the diameter of 800 mm under the suitcase or vehicle bombs attacks, five 1/4 scaled UHPCC-FST cylinder columns were fabricated, with the average cubic compressive strength and direct tensile strength of UHPCC being 131.5 and 9 MPa. Three UHPCC-FST columns were tested under contact explosion with the TNT charge weights of 1 kg, 2 kg and 3 kg, respectively. The localized craters were formed and the integrity of the columns was maintained under 1 and 2 kg TNT explosions (corresponding to the prototype of 64 and 128 kg). The blast energy can be fully absorbed by the steel tube and the core UHPCC. The outer steel tube was seriously ruptured and the core UHPCC was fully crushed for 3 kg TNT explosion (corresponding to the prototype of 192 kg).

(2) The axial compression test was further carried out for both the intact and blast-damaged UHPCC-FST columns, and the relationship between the axial load and axial/lateral displacement was derived. Both the intact and blast-damaged columns exhibited the diagonal shear failure mode induced by the shear dilatation of core UHPCC and the bulging of steel tube. The shear failure plane occurred at the location of the blast induced crater, and the corresponding angles between the failure plane and the longitudinal axis were in the range of 24°–29°. The axial capacity of intact column predicted by EUROCODE-4 agreed well with the present test data, and the Chinese standard GB50936 was suggested to improve for UHPCC-FST. The quantitative dependence of column damage index on the charge weight was experimentally determined.

(3) The present established numerical model can reproduce the dynamic behavior of the UHPCC-FST column under low-yield TNT charge contact detonation very well in terms of damaged pattern and crater dimension. However, due to the adopted erosion technique for the steel tube material, the fracturing behavior of which in the C-3 column under 3 kg TNT contact detonation cannot be fully

reproduced. By comparing the numerical and experimental damage patterns, angle of diagonal fracture plane and the residual axial capacity of the post-blast UHPCC-FST column, the validations of the adopted numerical algorithms, FE model, and the modified constitutive model parameters for UHPCC are verified.

(4) The parametric influences are performed based on the validated models, and it is derived that increasing the thickness and strength of steel tube, compressive strength and cross-sectional diameter of core concrete improve the blast resistance of UHPCC-FST column under contact detonation, which helps to achieve less severe damage degree and higher residual axial capacity of the post-blast column.

The present work provides the comprehensive quantitative experimental data and verified numerical simulation approach for the blast resistance and residual capacity evaluations of UHPCC-FST column against the accidental and intentional contact explosions. Besides, the other aspects such as the seismic design regulation, clearance requirement and construction cost also should be considered comprehensively.

References

AISC Committee. Specification for Structural Steel Buildings (ANSI/AISC 360-10)[S]. American Institute of Steel Construction, Chicago-Illinois, 2010.

ASTARLIOGLU S, KRAUTHAMMER T. Response of normal-strength and ultra-high-performance fiber-reinforced concrete columns to idealized blast loads[J]. Engineering Structures, 2014, 61: 1-12.

BAKER WE, WESTINE PS, DODGE FT. Similarity Methods in Engineering Dynamics[M]. Hayden Book Company, 1973. ISBN 0-87671-564-1.

BAO X, LI B. Residual strength of blast damaged reinforced concrete columns[J]. International Journal of Impact Engineering, 2010, 37(3): 295-308.

BEPPU M, OHNO T, OHKUBO K, LI B, SATOH K. Contact explosion resistance of concrete plates externally strengthened with FRP laminates[J]. International Journal Protection Structures, 2010, 1(2): 257-270.

CEB-FIP. Constitutive Modelling of High Strength/High Performance Concrete[M]. International Federation for Structural Concrete (fib), 2008.

CHENG D S, HUNG C W, PI S J. Numerical simulation of near-field explosion[J]. Journal of Applied Science and Engineering, 2013, 16(1): 61-67.

DENG Y, TUAN C Y. Design of concrete-filled circular steel tubes under lateral impact[J]. ACI Structural Journal, 2014, 110(4), 691-701.

EN 1994-1-1. Eurocode 4: Design of composite steel and concrete structures-Part 1-1: General rules and rules for buildings[M] European Committee for Standardization, 2004.

FARAHI M, HEIDARPOUR A, ZHAO X L, AL-MAHAIDI R. Compressive behaviour of concrete-filled double-skin sections consisting of corrugated plates[J]. Engineering Structures, 2016, 111: 467-477.

FARAHI M, HEIDARPOUR A, ZHAO X L, AL-MAHAIDI R. Effect of ultra-high strength steel on mitigation of non-ductile yielding of concrete-filled double-skin columns[J]. Construction and Building Materials, 2017, 147: 736-749.

Federal Highway Administration (FHWA). Recommendations for bridge and tunnel security[S]. The Blue Ribbon Panel on Bridge and Tunnel Security, Washington DC, 2003.

Fouché PP, Bruneau M. Blast and seismic resistant concrete-filled double skin tubes and modified steel jacketed bridge columns[R]. 2015. Technical report MCEER-14-0004

FUJIKURA S, BRUNEAU M. Experimental and analytical investigation of blast performance of seismically resistant bridge piers[R]. Multidisciplinary Center for Earthquake Engineering Research (MCEER), 2008. Technical report MCEER-08-0028.

FUJIKURA S, BRUNEAU M, LOPEZ-GARCIA D. Experimental investigation of multihazard resistant bridge piers having concrete-filled steel tube under blast loading[J]. Journal of Bridge Engineering, 2008, 13(6): 586-594

GB/T 50081-2002. Standard for test method of mechanical properties on ordinary concrete[S]. Beijing: Standards Press of China, 2003.

GB/T228.1-2010. Metallic Materials-Tensile Testing-Part 1: Method of test at room temperature[S]. National Standard of China, 2010.

GB50936-2014. Design Code for Composite Structure[S]. National Standard of China, 2014.

GT/B 8162-2008. Seamless steel tubes for structural purposes[S]. Beijing: Standards Press of China, 2008.

HOLMQUIST TJ, JOHNSON GR, COOK WH. A computational constitutive model for concrete subjected to large strains, high strain rates, and high pressures. In proceedings of the 14th International Symposium on Ballistics, Quebec, QC, Canada, 26–29 Sepetember[C]. 1993: 591–600.

JAYASOORIYA R, THAMBIRATNAM D P, PERERA N J. Blast response and safety evaluation of a composite column for use as key element in structural systems[J]. Engineering Structures, 2014, 61:31-43.

JENKINS BM. Protecting public surface transportation against terrorism and serious crime: continuing research on best security practices[R]. San Jose, CA: Mineta Transportation Institute, 2001. MTI Report 01-07.

KANG S T, KIM J K. The relation between fiber orientation and tensile behavior in an Ultra High Performance Fiber Reinforced Cementitious Composites (UHPFRCC)[J]. Cement and Concrete Research, 2011, 41(10): 1001-1014.

KYEI C, BRAIMAH A. Effects of transverse reinforcement spacing on the response of reinforced concrete columns subjected to blast loading[J]. Engineering Structures, 2017, 142: 148-164.

LI B, NAIR A, QIAN K. Residual axial capacity of reinforced concrete columns with simulated blast damage[J]. Journal Performence of Constructed Facilities, 2012, 26(3): 287-299.

LI J, WU C Q, HAO H, LIU Z X. Post-blast capacity of ultra-high performance concrete columns[J]. Engineering Structures, 2017, 134: 289-302.

LI M H, ZONG Z H, LIU L, LOU F. Experimental and numerical study on damage mechanism of CFDST bridge columns subjected to contact explosion[J]. Engineering Structures, 2018, 159: 265-276.

LSTC. LS-DYNA Keyword User's Manual, Version 971[M]. 2007.

MALVAR L J, CRAWFORD J E, WESEVICH J W, SIMONS D. A plasticity concrete material model for DYNA3D[J]. International Journal of Impact Engineering, 1997, 19: 847-873.

MAGALLANES JM, WU Y, MALVAR J, CRAWFORD JE. Recent developments to release III of the K&C concrete model. In: Proceedings of the 11th international LSDYNA user's conference, Dearborn, MI[C]. 2010.

MAO L, BARNETT S, BEGG D, SCHLEYER G, WIGHT G. Numerical simulation of ultra high performance fibre reinforced concrete panel subjected to blast loading[J]. International Journal of Impact Engineering, 2014, 64: 91-100.

MIRMOMENI M, HEIDARPOUR A, ZHAO X L, AL-MAHAIDI R, PACKER J. Size-dependency of concrete-filled steel tubes subject to impact loading[J]. International Journal of Impact Engineering, 2017, 100: 90-101.

NASSIRNIA M, HEIDARPOUR A, ZHAO X L, MINKKINEN J. Innovative hollow corrugated columns comprising corrugated plates and ultra high-strength steel tubes[J]. Thin-Walled Structures, 2016, 101: 14-25.

NGO T, MOHOTTI D, REMENNIKOV A, UY B. Numerical simulations of response of tubular steel beams to close-range explosions[J]. Journal of Constructional Steel Research, 2015, 105: 151-163.
OTTOSEN NS, RISTINMAA M. The Mechanics of Constitutive Modeling[M]. Elsevier, 2005.
REMENNIKOV A M, UY B. Explosive testing and modelling of square tubular steel columns for near-field detonations[J]. Journal of Constructional Steel Research, 2014, 101: 290-303.
REN G M, WU H, FANG Q, LIU J Z. Triaxial compressive behavior of UHPCC and applications in the projectile impact analyses[J]. Construction and Building Materials, 2016, 113: 1-14.
REN G M, WU H, FANG Q, LIU J Z. Effects of steel fiber content and type on static mechanical properties of UHPCC[J]. Construction and Building Materials, 2017, 163: 826-839.
REN G M, WU H, FANG Q, LIU J Z. Effects of steel fiber content and type on dynamic compressive mechanical properties of UHPCC[J]. Construction and Building Materials, 2018, 164: 29-43.
RIEDEL W, HIERMAIER KTS. Penetration of reinforced concrete by BETA-B-500-numerical analysis using a new macroscopic concrete model for hydrocodes. In proceedings of the 9th International Symposium on Interaction of the Effect of Munitions with Structures, Berlin, Germany, 3–7 May[C]. 1999: 315–322.
SHI Y, HAO H, LI Z. Numerical derivation of pressure-impulse diagrams for prediction of RC column damage to blast loads[J]. International Journal of Impact Engineering, 2008, 35(11): 1213-1227.
WANG Z G, WU H, FANG Q, WU J. Experimental study on the residual axial capacity of Ultra high performance cementitious composite filled steel tube (UHPCC-FST) column under contact explosion[J]. Thin-Walled Structures, 2020a, 147: 106832.
WANG Z G, WU H, FANG Q, WU J. Numerical study on the residual axial capacity of ultra high performance cementitious composite filled steel tube (UHPCC-FST) column under contact explosion[J]. Thin-Walled Structures, 2020b, 153: 106832.
WILLIAMSON EB, BAYRAK O, WILLIAMS GD, DAVIS CE. Blast-resistant highway bridges: design and detailing guidelines.NCHRP report 645[M]. Washington DC (USA): Transportation Research Board, 2010.
WU H, REN G M, FANG Q, LIU J Z. Effects of steel fiber content and type on dynamic tensile mechanical properties of UHPCC[J]. Construction and Building Materials, 2018, 173: 251-261.
WU K C, LI B, TSAI K C. Residual axial compression capacity of localized blast-damaged RC columns[J]. International Journal of Impact Engineering, 2011a, 38(1): 29-40.
WU K C, LI B, TSAI K C. The effects of explosive mass ratio on residual compressive capacity of contact blast damaged composite columns[J]. Journal of Constructional Steel Research, 2011b, 67(4): 602-612.
XU J C, WU C Q, XIANG H B, SU Y, LI Z X, FANG Q, HAO H, LIU Z X, ZHANG Y D, LI J. Behaviour of ultra high performance fiber reinforced concrete columns subjected to blast loading[J]. Engineering Structures, 2016, 118: 97-107.
YUAN S, HAO H, ZONG Z, LI J. A study of RC bridge columns under contact explosion[J]. International Journal of Impact Engineering, 2017, 109: 378-390.
ZHANG F, WU C, LI Z, ZHAO X. Residual axial capacity of CFDST columns infilled with UHPFRC after close-range blast loading[J]. Thin-Walled Structures, 2015a, 96: 314-327.
ZHANG F, WU C, WANG H, ZHOU Y. Numerical simulation of concrete filled steel tube columns against blast loads[J]. Thin-Walled Structures, 2015b, 92: 82-92.
ZHANG F, WU C, ZHAO X, LI Z, HEIDARPOUR A, WANG H. Numerical modeling of concrete-filled double-skin steel square tubular columns under blast loading[J]. Journal of Performance of Constructed Facilities, 2015c, 29(5): B4015002.
ZHANG F, WU C, ZHAO X L, XIANG H, LI Z X, FANG Q, LIU Z, ZHANG Y, HEIDARPOUR A, PACKER J A. Experimental study of CFDST columns infilled with UHPC under close-range blast loading[J]. International Journal of Impact Engineering, 2016, 93: 184-195.
ZHANG H, LI Z, WU C. Investigation of blast effects on Double-Skinned composite steel tubular columns[J]. International Journal Protection Structures, 2015d, 6(3): 403-418.

ZHANG F, WU C, ZHAO X, HEIDARPOUR A, LI Z. Experimental and numerical study of blast resistance of square CFDST columns with steel-fibre reinforced concrete[J]. Engineering Structures, 2017, 149: 50-63.

Chapter 11
Experimental and Numerical Study of UHPCC-FST Columns Subjected to Close-Range Explosion

11.1 Introduction

From 1970 to 2017, there have been more than 180,000 terrorist attacks occurred around the world, including more than 88,000 bombings, 19,000 assassinations, and 11,000 kidnappings (Global Terrorism Database 2018). It is obvious that the bomb attack is the most common terrorist action. Besides, the public infrastructures have becoming the main attack targets by terrorists in recent years (Li et al. 2018). As one of the important public infrastructures, the bridges, especially the landmark and memorable ones are prone to be vulnerable to terrorist bomb attacks. Bridge columns are the main load-bearing members of bridges, of which the blast-resistance of which has great significance to study.

The exiting works are mainly concentrated on the dynamic responses of normal reinforced concrete (RC) building or bridge columns under explosions (Kyei and Braimah 2017; Shi et al. 2018; Wu et al. 2011; Li et al. 2017; Zong et al. 2017; Williamson et al. 2010, 2011a, b). Taking the bridge columns for instance, Zong et al. (Zong et al. 2017) performed the contact and non-contact explosion ($0.5 \leq Z \leq 2.56$ m/kg$^{1/3}$, $Z = a/W^{1/3}$ is the scaled standoff distance, a is the standoff distance from the center of the explosive to the surface of structure, W is the charge weight) test on eleven bridge column specimens. It was derived that the blast-resistance capacity of RC columns is enhanced by increasing the sectional area and employing the square section, spiral stirrup and steel fiber reinforced concrete. Supported by the National Cooperative Highway Research Program (NCHRP), Williamson et al. (2010, 2011a, b) conducted a systematic research to investigate the blast-resistance capacity of bridge columns, and it was observed that the concrete cover is easily cracked and spalled under blast loadings. Obviously, those failures have negative effect on the blast performance of the column.

In recent years, the concrete-filled steel tube (CFST) column has attracted lots of attentions due to its excellent strength, ductility as well as the fatigue, high temperature and impact resistance (Morino et al. 2001). As for the studies of CFST members under blast, Fujikura et al. (2008) conducted a blast test (scaled standoff distance is

Q. Fang et al., *UHPCC Under Impact and Blast*,
https://doi.org/10.1007/978-981-33-6842-2_11

not specified due to the secrecy considerations) on 1/4 scaled CFST columns (diameters of 102, 127 and 152 mm) with the capping beams to simulate the scenario of car bomb attack on bridge columns, and it was shown that the CFST bridge column has the prominent blast-resistance capacity. Li et al. (2013) conducted the field explosion test on twelve CFST column specimens, in which the effect of explosive charge weight (25–50 kg), standoff distance (3.5–5 m), axial compression ratio (7.6, 16 and 23%), concrete cubic compressive strength (37–68 MPa) and longitudinal reinforcement ratio (6.9, 8.3 and 9.9%) on the blast performance of CFST columns were systematically studied. It was derived that, the blast resistance of CFST column is enhanced with increasing the concrete strength and steel reinforcement ratio, as well as decreasing the axial compression ratio. By performing the contact explosion test on the concrete-filled double-skin steel tube (CFDST) columns (core concrete compressive strength is 40 MPa, charge weights are 0.6 and 1 kg) with different outer steel tube thickness (6, 7 and 8.5 mm), Li et al. (2018) derived that the blast-resistance capacity of CFDST column increases with increasing the outer steel tube thickness. Remennikov and Uy (2014) carried out the field contact and close-range explosion tests ($0.05 \leq Z \leq 0.15$ m/kg$^{1/3}$) on five square sectional CFST columns with the core concrete compressive strength of 46 MPa, the damage and failure of column specimens were recorded. Zhang et al. (2016) prepared twelve Ultra-high performance concrete (UHPC) filled (double-skin) steel tube columns with the compressive and flexural tensile strength of core UHPC averaging 170 MPa and 33.8 MPa, respectively. Then, an experimental study on above UHPC-FST columns subjected to close-range explosion (the scaled standoff distances were 0.41, 0.46 and 0.58 m/kg$^{1/3}$) was performed, in which the sectional shape of outer tube (circular and square), charge weight (17.5, 35 and 50 kg) and the axial load (0 and 1000 kN) were considered. Besides, under the close-range explosion with two scaled standoff distances of 0.19 and 0.14 m/kg$^{1/3}$, Sun (2013) and Cui et al. (2015) experimentally studied the dynamic response of CFDST columns (core concrete compressive strength of 45.8–55.2 MPa) as well as the overpressure distributions of blast wave acted on the columns, respectively.

As for the analytical and numerical studies of the column under close-range explosion, Zhang et al. (2015a, b, c) numerically studied the dynamic responses of CFDST columns subjected to close-range explosion based on the CONWEP codes embedded in the finite element program LS-DYNA. However, Remennikov and Uy (2014) pointed out that the CONWEP code is not suitable for most practical threats in near field, and then they proposed a simplified engineering-level model for the near-field blast impulse to predict the dynamic responses of structural members. Furthermore, based on Arbitrary-Lagrangian-Eulerian element method in LS-DYNA, Ngo et al. (2015) numerically reproduced the close-range explosion test on CFST specimens conducted in Remennikov and Uy (2014). Moreover, the equivalent single degree of freedom method has been widely used to predict the dynamic response of structural members under the blast loadings (Li and Meng 2002).

Generally, the existing studies have the following limitations: (i) Ultra-high performance cementitious composites (UHPCC) and UHPCC filled steel tube (UHPCC-FST) have been becoming the most prospective construction materials and bearing members for high-rise and long-span bridges. However, most of the blast tests mentioned above are concentrated on the normal strength concrete and RC members; (ii) Considering the accessibility of the bridge column and the threat of the terroristic car bombs, the close-range explosion (non-contact but $Z \leq 1$ m/kg$^{1/3}$) should be drawn much more attention. Besides, the detonation mechanism and blast loading of close-range explosion are more complex and different from that of the far-range explosion (Ngo et al. (2015)). However, the existing experimental studies in case of the close-range explosion are limited; (iii) There are no consensus for the efficient analytical approach to predict the dynamic response of structural columns subjected to the close-range explosion, although several methods have been proposed previously.

At present, the blast performance of UHPCC-FST columns subjected to close-range explosion is studied both experimentally and numerically. The close-range explosion test on four UHPCC-FST columns is firstly conducted with the scaled standoff distance of $0.12 \leq Z \leq 0.14$ m/kg$^{1/3}$. Then, three existing approaches are applied to analyze the dynamic responses of CFST columns in the present and existing close-range explosion tests, i.e., numerical simulations based on Arbitrary-Lagrangian-Eulerian elements method (ALE method) and velocity loading method (velocity method), as well as the equivalent single degree of freedom method (SDOF method). By comparisons of the experimental and predicted residual mid-span deflections, the applicability of above three approaches are discussed.

11.2 Explosion Test on UHPCC-FST Column

11.2.1 Specimens

Four circular UHPCC-FST column specimens were prepared to investigate the response of UHPCC-FST column under close-range explosion with different scaled standoff distances. The geometrical and sectional details of the specimens are illustrated in Fig. 11.1. The outer steel tubes of the specimens were made from Chinese 20# steel. By conducting the direct tensile test with the MTS 810 material testing machine at the strain rate of 2 mm/min, the true tensile stress-strain curves of five standard tensile coupons of 20# steel are derived, as shown in Fig. 11.2. The average yield strength, ultimate tensile strength and the elastic modulus of steel tubes can be determined as 318.5 MPa, 507.1 MPa and 206 GPa, respectively.

UHPCC was prepared in the State Key Laboratory of High Performance Civil Engineering Materials, Jiangsu Research Institute of Building Science in China. The mixture proportions of UHPCC are given in Table 11.1, in which the abbreviation HRWR denotes the high range water reducer. The detailed properties of the mixing

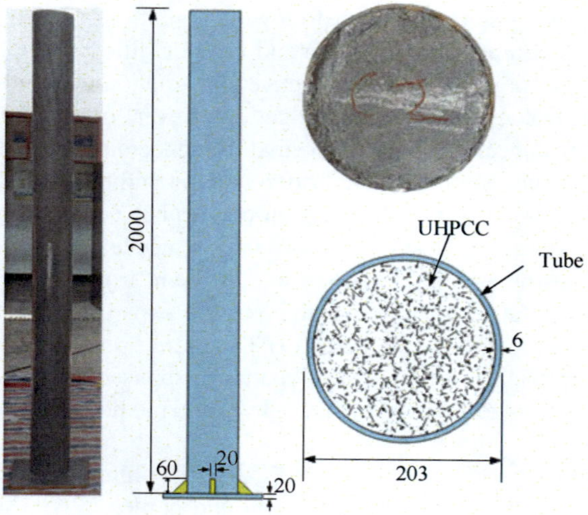

Fig. 11.1 Geometrical and sectional details of UHPCC-FST column (unit in mm), reprinted from Wu et al. (2020), copyright 2020, with permission from Elsevier

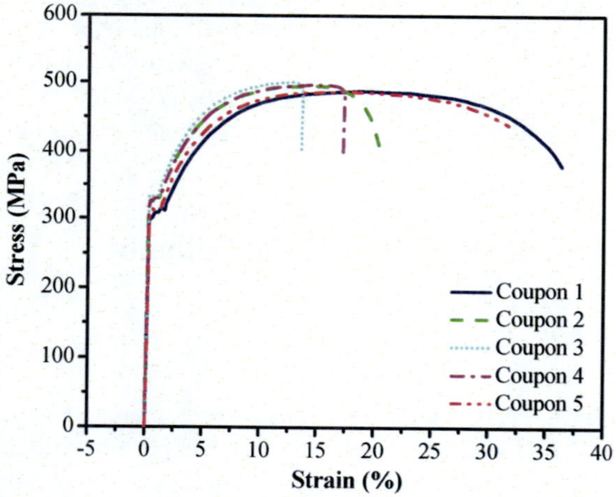

Fig. 11.2 Experimental stress-strain curves for 20# steel, reprinted from Wu et al. (2020), copyright 2020, with permission from Elsevier

Table 11.1 Mixture proportions of UHPCC (kg/m^3)

Cement	Silica fume	Ultra-fine mineral admixture	Sand	Water	HRWR	Fiber content
700	140	110	1200	152	22.8	156

gradients as well as the preparation procedure can be referred from our previous work (Ren et al. 2018a). The wet density, average quasi-static compressive strength, elastic modulus and Poisson's ratio of UHPCC are 2480 kg/m^3, 141.5 MPa, 43.8 GPa and 0.23, respectively.

11.2.2 Test Setup

Considering that the relative low axial compressive ratio (ratio of the axial force of column to the product of the cross-section area of column and cylinder compressive strength of core concrete) of the bridge columns, as well as the boundary constraints have negligible effect on the local deformations of the specimens subjected to the close-range explosions, in the present blast test, the UHPCC-FST columns were placed horizontally and the axial force is not considered. Two steel components were designed to support the column specimen away from the ground with a clear distance of 100 mm, as shown in Fig. 11.3a. The diagram and photograph of the experimental setup are shown in Fig. 11.3b and c, respectively. Four column specimens denoted as C-1–C-4 have a clear span of 1900 mm between the two steel supports. TNT explosive with the charge weight from 3 to 10 kg were placed in the mid-span and 100 mm above the top surface of the specimen. The cylindrical TNT explosive was used in the blast test because it is a standard high explosive deemed chemically safe and easy to cast, and it was initiated from the top by using the electric detonator. The experimental details for each specimen are given in Table 11.2.

11.2.3 Test Results and Analyses

The four specimens after close-range explosion are presented in Fig. 11.4. Generally, the four specimens are not heavily destroyed and the integrity of which are kept, i.e., the outer steel tubes were not cracked and the section of columns were not breached. Column C-4 suffers more severe damage than the other three columns due to smaller scaled standoff distance of explosive. Thus, the UHPCC-FST column can maintain the majority of its axial bearing capacity and exhibit the excellent blast resistance against the close-range explosions. It is observed that a local elliptical crater was formed beneath the explosive, as shown in Fig. 11.5, the dimensions and depth of the crater are denoted as D_x, D_y and D_p, respectively. By comparing the blast induced crater dimensions with the blast test data on normal CFST columns in Li et al. (2018), the superiority of UHPCC-FST column against the near field explosion is also validated qualitatively. It should be pointed out that, for the near range explosions, the pressure transducer and the displacement sensor could be easily damaged by the detonation wave and fire ball during the blast, thus only the residual deformations

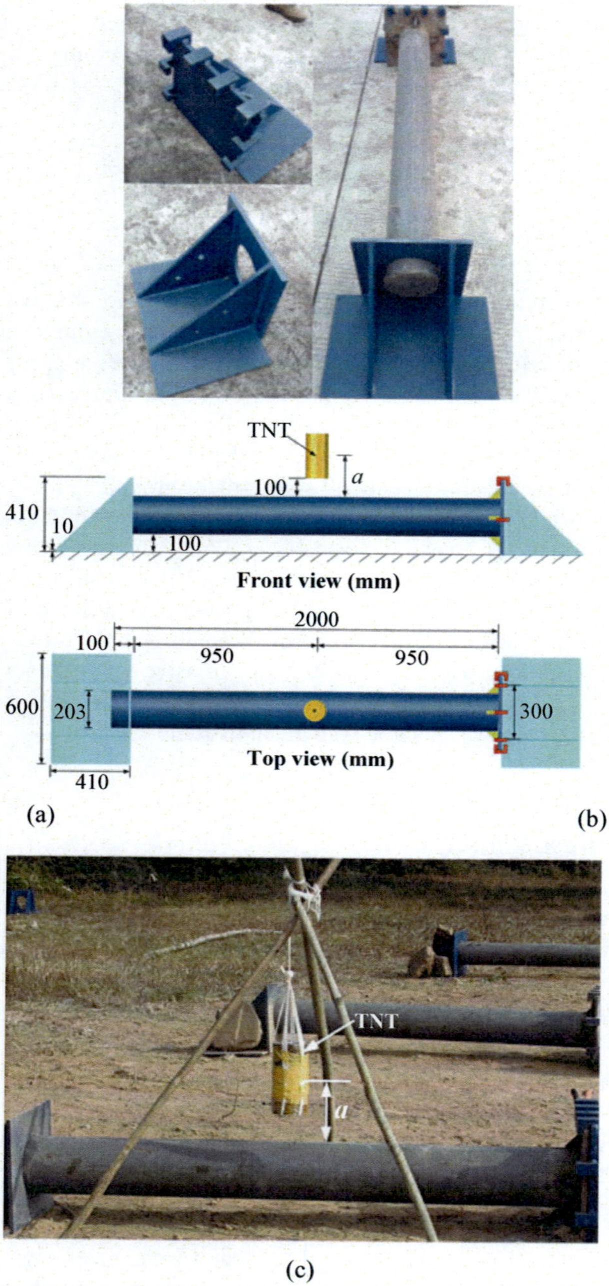

Fig. 11.3 Blast test setup **a** steel supports **b** diagram **c** photograph, reprinted from Wu et al. (2020), copyright 2020, with permission from Elsevier

Table 11.2 Experimental details of UHPCC-FST specimens

Specimen	TNT charge weight W (kg)	Standoff distance a (m)	Scaled standoff distance Z (m/kg$^{1/3}$)	Dx (mm)	Dy (mm)	Dp (mm)	Df (mm)
C-1	3	0.21	0.14	105	95	5	7
C-2	4	0.22	0.14	116	125	7	9
C-3	7	0.23	0.12	165	120	19	30
C-4	10	0.28	0.13	178	137	21	44

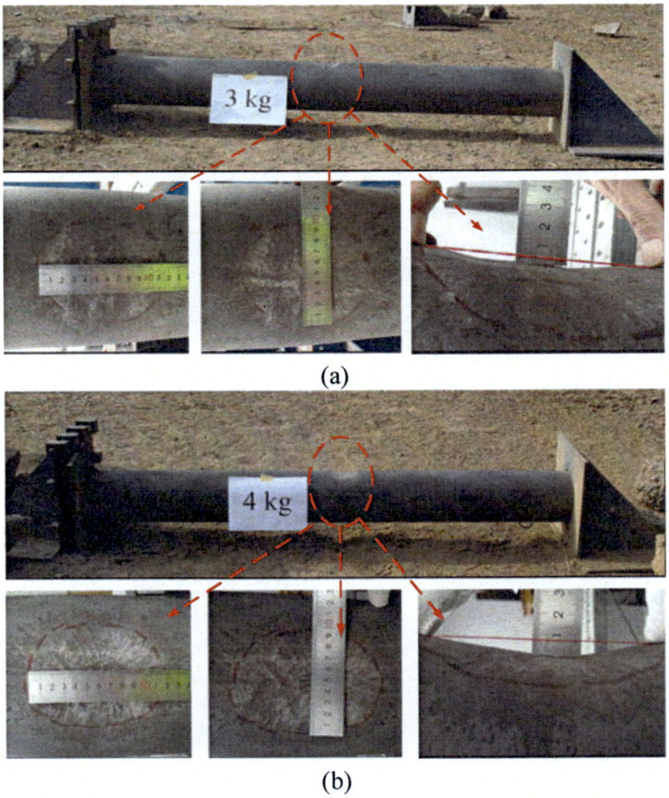

Fig. 11.4 Recovered UHPCC-FST specimens **a** C-1 **b** C-2 **c** C-3 **d** C-4

of the specimens (D_f) can be measured after the test. The detailed dimensions of the blast induced crater and residual deformations of the specimens are also given in Table 11.2.

(c)

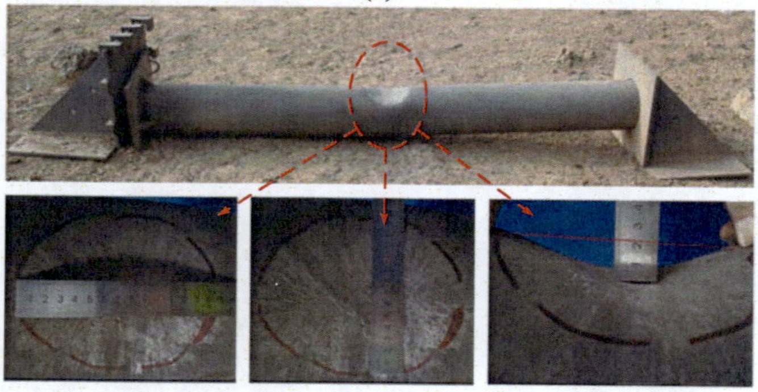

(d)

Fig. 11.4 (continued)

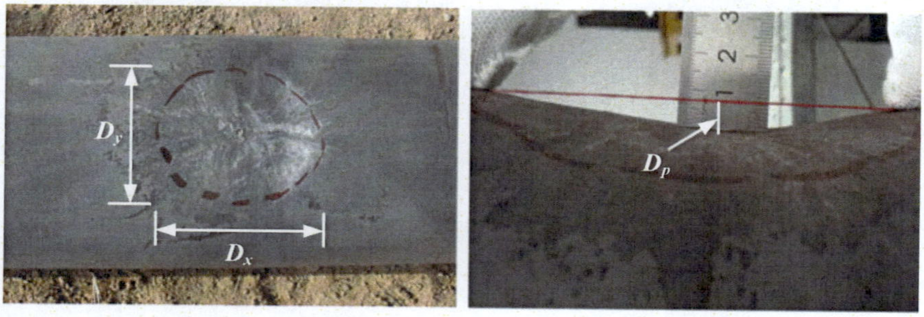

Fig. 11.5 Blast induced crater, reprinted from Wu et al. (2020), copyright 2020, with permission from Elsevier

11.3 Analytical Methods for Predicting the Dynamic Responses of UHPCC-FST Columns

There are commonly three main methods to analyze the dynamic responses of structural members subjected to the blast loadings. First is the numerical simulation method based on ALE elements (ALE method), the second method is a simplified numerical simulation method based on velocity loading proposed by Remennikov and Uy (2014) (velocity method), and the third one is the equivalent single degree of freedom method (SDOF method). In this section, the above three methods are adopted to predict the global deformations of UHPCC-FST columns subjected to close-range explosion presented in Sect. 11.2.

11.3.1 ALE Method

Based on the finite element program LS-DYNA (LSTC 2017), the 3D numerical simulations with the ALE method is performed and a detailed view of the finite element model is given in Fig. 11.6. The model consists of an air domain, steel supports, UHPCC-FST column specimen, ground and the blast source (TNT). The TNT, surrounding air domain, the infilled UHPCC and steel supports are modelled by 8-node solid elements while the steel tube is modelled by Belytschko-Tsay shell elements. TNT and air elements are described as fluid by using the Arbitrary-Lagrangian-Eulerian (ALE) mesh, while the other elements are described as structure by using the Lagrangian mesh. The Fluid-Structure Interaction (FSI) algorithm is used to simulate the interaction between fluid and structure. The *CONSTRAINED_LAGRANGE_ IN_SOLID card is utilized to provide the coupling mechanism between ALE meshes and the Lagrangian meshes.

For the contact between the components and the boundary constraints, it is assumed that the interfaces between the steel tube, the steel support and the UHPCC

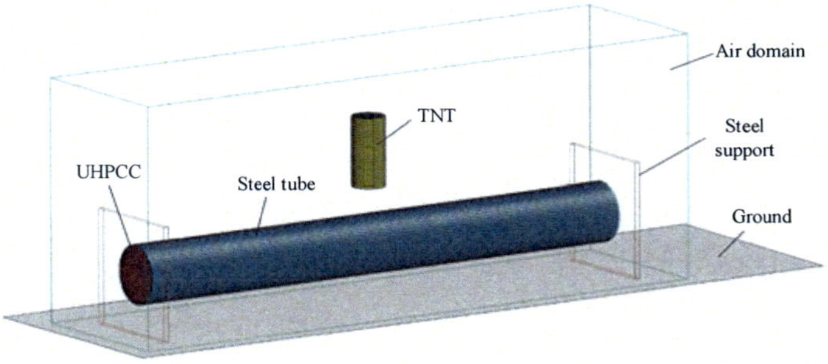

Fig. 11.6 Finite element model, reprinted from Wu et al. (2020), copyright 2020, with permission from Elsevier

have a perfect bond, which has been proved having adequate accuracy in modeling steel-concrete composite columns under blast loadings (Li et al. 2018; Remennikov and Uy 2014; Zhang et al. 2016). Therefore, the common joints between the steel tube, the steel supports and the UHPCC infill are defined. The ground is replaced by setting keywords (*RIGIDWALL_PLANAR). The steel supports are fixed to the ground by setting the *BOUNDARY_SPC card, which is constrained against vertical movement. Based on the mesh convergence study, a mesh size of 10 mm is found to be appropriate.

11.3.1.1 Steel Tube and Support

The steel tube is modelled by using the material card (*MAT_PIECEWISE _LINEAR_PLASTICITY) in LS-DYNA, which has been widely applied to predict the behavior of steel under large deformations [25]. By incorporating the Cowper Symonds strain rate parameters (C, P), the strain rate effect is accounted by using the dynamic increase factor (DIF_s) of steel as follows

$$DIF_s = 1 + \left(\frac{\dot{\varepsilon}_s}{C}\right)^{1/P} \tag{11.1}$$

where $\dot{\varepsilon}_s$ is the strain rate of steel. Two steel supports are used to support the column specimen against gravity and the negligible deformations are occurred in the test, thus the steel supports are treated as the rigid bodies and the material card (*MAT_RIGID) are utilized. The corresponding model parameters of steel tube and supports are given in Table 11.3.

11.3.1.2 UHPCC

UHPCC is modelled by using the K&C concrete model (*MAT_CONCRETE_ DAMAGE_REL3), which is particularly suitable for blast scenario since it is able to model the behaviors of concrete member subjected to active/passive confining stress and high strain rate effect. Two parameters, i.e., compressive strength and density listed in Table 11.3, are required to input, the other relevant material parameters are generated automatically by the embedded algorithm in LS-DYNA. It should be noted that, the above automatically generation algorithm is originally proposed for the normal strength concrete, the validation of which for the present UHPCC needs systematic static and dynamic tests, which is conducted underway and in the future.

The strain rate effect of UHPCC has been studied in our previous work (Ren et al. 2018b; Wu et al. 2018). The dynamic increase factor of UHPCC for dynamic compressive (DIF_c^c) and tensile (DIF_c^t) are given in Eqs. (11.2, 11.3) and illustrated in Fig. 11.7, respectively.

Table 11.3 Material parameters for steel tube, steel support and UHPCC

Material	Parameter	Value
Steel tube (*MAT_PIECEWISE_LINEAR_PLASTICITY)	Density	7750 kg/m^3
	Young's modulus	206 GPa
	Poisson's ratio	0.3
	Yield stress	318.5 MPa
	Strain rate parameter C	6844 s^{-1} (Li et al. 2018)
	Strain rate parameter P	3.91 (Li et al. 2018)
Steel support (*MAT_RIGID)	Density	7750 kg/m^3
	Young's modulus	206 GPa
	Poisson's ratio	0.3
UHPCC (*MAT_CONCRETE_DAMAGE_ REL3)	Density	2480 kg/m^3
	Poisson's ratio	0.23
	Compressive strength	141.5 MPa

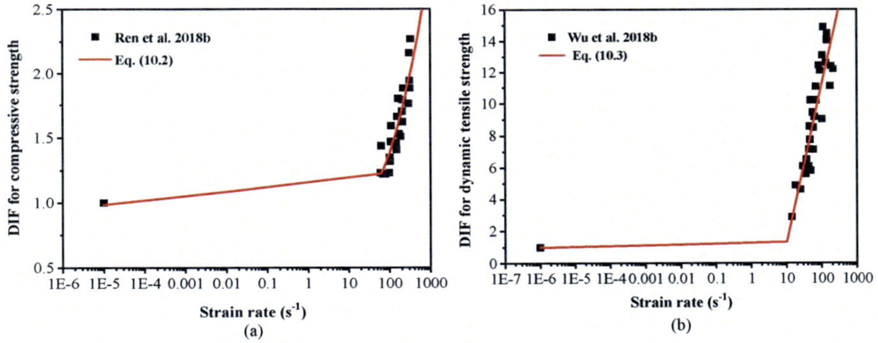

Fig. 11.7 Empirical *DIF* relations for UHPCC **a** compression **b** tension, reprinted from Wu et al. (2020), copyright 2020, with permission from Elsevier

$$\mathrm{DIF}_c^c = \begin{cases} (\dot{\varepsilon}/\dot{\varepsilon}_0)^{0.014} & \text{for } \dot{\varepsilon} \le 62\mathrm{s}^{-1} \\ 0.587(\log\dot{\varepsilon})^2 - 1.435\log\dot{\varepsilon} + 1.911 & \text{for } \dot{\varepsilon} > 62\mathrm{s}^{-1} \end{cases} \tag{11.2}$$

$$\mathrm{DIF}_c^t = \begin{cases} (\dot{\varepsilon}/\dot{\varepsilon}_0)^{0.018} & \text{for } \dot{\varepsilon} \le 10.3\mathrm{s}^{-1} \\ 9.959\log\dot{\varepsilon} - 8.718 & \text{for } \dot{\varepsilon} > 10.3\mathrm{s}^{-1} \end{cases} \tag{11.3}$$

Table 11.4 Material parameters and EOS for air and TNT (Cheng et al. 2013)

Air		TNT	
Initial density, ρ_0	1.29 kg/m^3	Initial density, ρ_0	1630 kg/m^3
Pressure cut-off	0	Detonation velocity	6930 m/s
Dynamic viscosity	0	CJ pressure P	21 GPa
Constant C_0-C_3	0	Constant A	371.2 GPa
Constant C_4	0.4	Constant B	3.23 GPa
Constant C_5	0.4	Constant R_1	4.15
Constant C_6	0	Constant R_2	0.95
Initial energy, E_0	2.5×10^9 J/m^3	Constant ω	0.3
		Initial energy, E_0	7×10^9 J/m^3

11.3.1.3 Air and Explosive

The air is modelled with the *MAT_NULL material model by combining the linear polynomial Equation of State (EOS) given in Eq. (11.4) (LSTC 2017).

$$P = C_0 + C_1\mu + C_2\mu^2 + C_3\mu^3 + \left(C_4 + C_5\mu + C_6\mu^2\right)E_0 \tag{11.4}$$

where P is pressure, C_0~C_6 are polynomial equation coefficients, E_0 is the internal energy per unit volume. $\mu = \rho/\rho_0 - 1$ represents the volumetric strain, where ρ and ρ_0 are the current and initial densities, respectively. The adopted parameters of material and EOS for air are listed in Table 11.4.

TNT explosive is modelled with the *MAT_HIGH_EXPLOSIVE_BURN material model card by combining the Jones-Wilkins-Lee (JWL) EOS given in Eq. (11.5) (LSTC 2017). The detonation of TNT is initialized by the *INITIAL_DET card, with the initial detonation point set at the top-center of the charge.

$$P = A\left(1 - \frac{\omega}{R_1 V}\right)e^{-R_1 V} + B\left(1 - \frac{\omega}{R_2 V}\right)e^{-R_2 V} + \frac{\omega E_0}{V} \tag{11.5}$$

where P is pressure, V is the relative volume, E_0 is the internal energy per unit volume, and A, B, R_1, R_2 and ω are coefficients related to the explosive type. The adopted parameters of the material and EOS for TNT are also listed in Table 11.4.

The present test is numerically simulated by ALE method and Fig. 11.8 shows the Von Mises stress contours of C-3 column specimen for example. The corresponding mid-span deflection time-histories for the present four specimens are shown in the following Fig. 11.16.

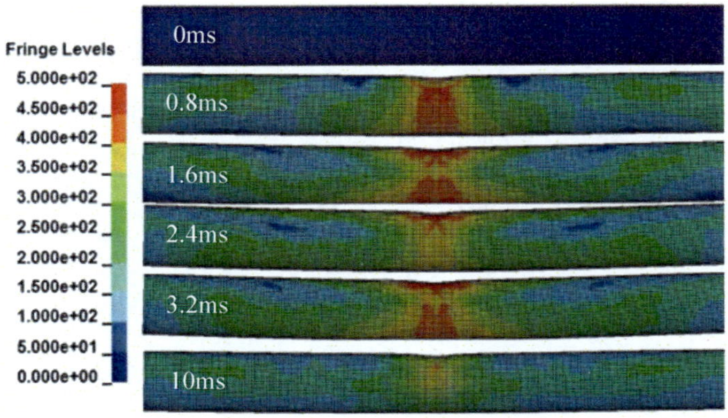

Fig. 11.8 Von Mises stress contours for C-3 specimen by ALE method, reprinted from Wu et al. (2020), copyright 2020, with permission from Elsevier

11.3.2 Velocity Method

In Remennikov and Uy (2014), a simplified numerical simulation method based on velocity loading is proposed to analyze the dynamic response of CFST column specimens subjected to close-range explosion, which is denoted as velocity method at present. Following this method, only the column specimens need to be modeled, the blast loading is equivalent to a velocity loads, and the specific velocities along the column are derived as follows.

As shown in Fig. 11.9, denoting x as the any sectional position along the column specimen with the clear span of l, the velocity of the corresponding position $v(x)$ can be obtained by

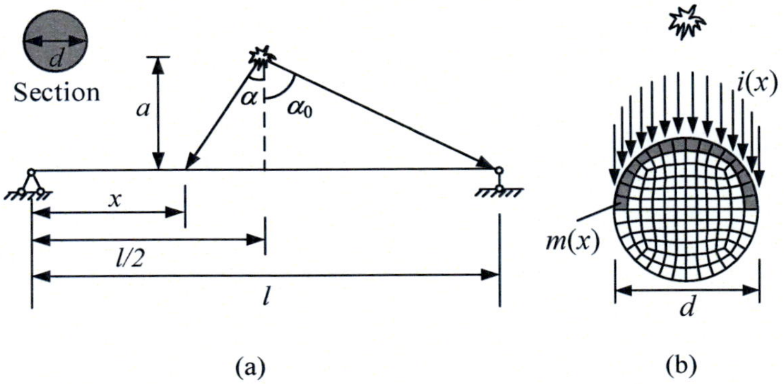

Fig. 11.9 Determination of elementary specific impulse on column **a** diagram of the column under blast impulse **b** cross section of the column under blast impulse, reprinted from Wu et al. (2020), copyright 2020, with permission from Elsevier

$$v(x) = \frac{i(x)}{m(x)} \tag{11.6}$$

where $i(x)$ and $m(x)$ are the blast impulse and the mass of top surface elements at the sectional position x shown in Fig. 11.9b.

The blast impulse at the corresponding sectional position $i(x)$ can be calculated by Henrych (1979) and Remennikov and Uy (2014) as

$$i(x) = \frac{S_c A_0 W d}{a^2} \cos^4 \alpha \tag{11.7}$$

where S_c is the reduction factor to account for the circle shape of the column and its value can be derived from Eq. (11.8). $A_0 = (u_0 + w_0)/4\pi$, u_0 is the particle velocity of the detonation products, w_0 is the displacement velocity of the outburst surface. u_0 and w_0 are dependent on the explosive type and the encountered material of the detonation wave. For TNT and the CFST column, $u_0 = 505$ m/s and $w_0 = 5150$ m/s are adopted at present. W is TNT charge weight, a is the standoff distance of the explosive, d is the side length of the rectangular sectional column or the diameter of the circular sectional column, and α is the incident angle of the blast wave, as shown in Fig. 11.9.

$$S_c = 0.019\left(\frac{a}{d}\right) + 0.39 \tag{11.8}$$

The incident angle α can be expressed as

$$\cos \alpha = \frac{a}{\sqrt{\left(\frac{l}{2} - x\right)^2 + a^2}} \tag{11.9}$$

Assuming that $\xi = x/l$, Eq. (11.7) can be written as

$$i(x) = \frac{S_c A_0 W d}{a^2} \frac{a^4}{l^4\left(\left(\frac{1}{2} - \xi\right)^2 + \frac{a}{l}\right)^2} \tag{11.10}$$

According to the $i(x)$ given by Eq. (11.10) and the relationship between the velocity and the impulse of any sectional position x given by Eq. (11.6), the sectional velocity distribution along the column specimen can be derived. Then, the above initial velocities are assigned to the corresponding nodes of the finite element model of column to reproduce the effect of blast loadings on the structural component. Figure 11.10 shows the distribution of the initial velocities for the present four columns specimens subjected close-range explosion.

For any section of the column, the nodes on the column surface suffered blast are grouped in the node sets and assigned the initial velocities with the keywords *INITIAL_VELOCITY. Adopting the mesh size of 10 mm, the initial velocities shown in Fig. 11.10 are assigned to the node sets on the top surface of the steel

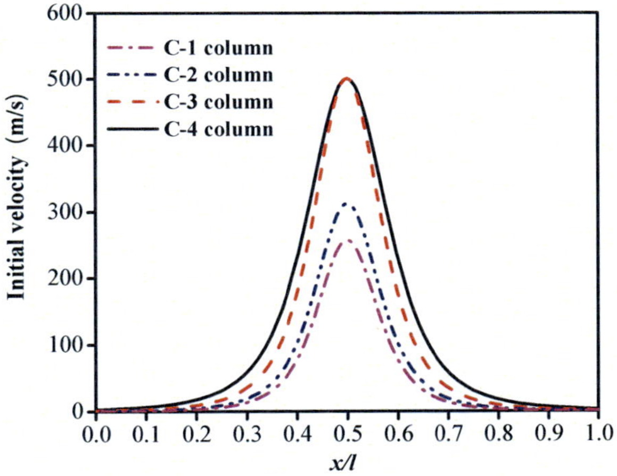

Fig. 11.10 Distributions of initial velocities along four experimental column specimens, reprinted from Wu et al. (2020), copyright 2020, with permission from Elsevier

tube. Figure 11.11 shows the finite element model of C-3 specimen with the initial velocities established in LS-DYNA. Compared with the finite element model by ALE method, the present model in velocity method does not need TNT and air domain elements. The material parameters of concrete, steel tube and steel support in this method are identical with those adopted in ALE method. The dynamic responses of four column specimens subjected to close-range explosion can be numerical analyzed. For instance, Fig. 11.12 shows the Von Mises stress contours of C-3 column specimen predicted by velocity method, and the mid-span deflection time-histories for the present four specimens are shown in Fig. 11.16.

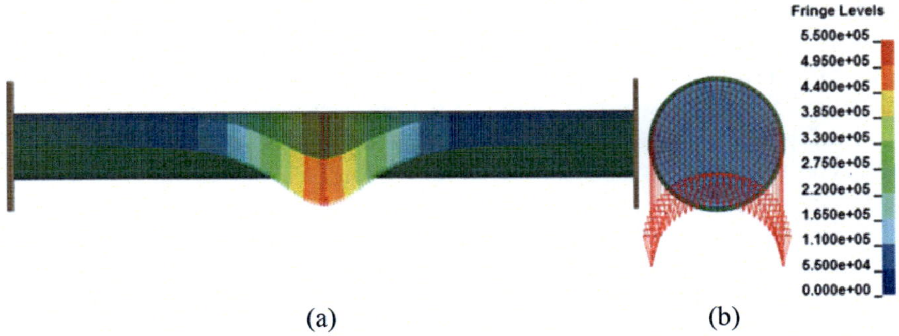

Fig. 11.11 Finite element model of C-3 specimen by the velocity method **a** front view **b** sectional view, reprinted from Wu et al. (2020), copyright 2020, with permission from Elsevier

Fig. 11.12 Von Mises stress contours for C-3 specimen by velocity method, reprinted from Wu et al. (2020), copyright 2020, with permission from Elsevier

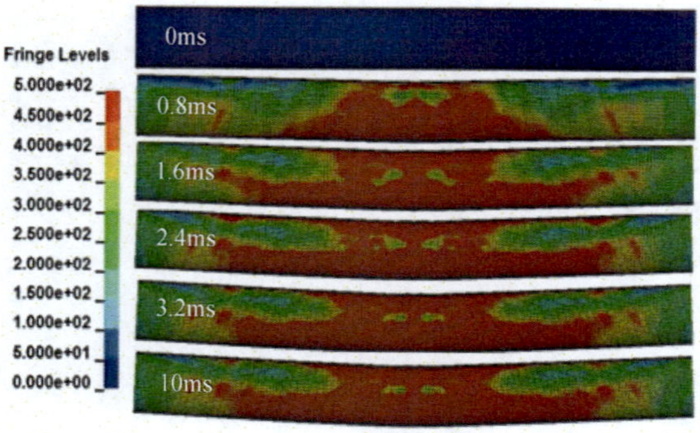

11.3.3 SDOF Method

Considering that the present UHPCC-FST column specimens are all kept their integrity and regarding them as the elastoplastic component, based on the SDOF method, the deformation of the column specimen can be divided into the following three stages assuming that plastic hinge forms at the mid-span of the column specimen, as shown in Fig. 11.13. When the blast impulse acted on the specimen, the elastic deformation is observed at first. Then, the plastic deformation appears in the specimen when the mid-span displacement reaches the elastic limit z_e, which indicates that the plastic hinge has formed at the mid-span. When the mid-span displacement reaches the ultimate z_m, the blast impulse dissipate and the elastic deformation of specimens will be resilient, which is denoted as the resilient stage. At the end of resilient stage, the mid-span displacement equals to the plastic deformation $z_p = z_m - z_e$.

UHPCC-FST column specimen subjected to close-range explosion can be described by an equivalent SDOF mass-spring system with a non-linear spring, as illustrated in Fig. 11.14. M_e and K_e are the total mass and spring stiffness of the equivalent SDOF system. $z(t)$ and $v(t)$ are the instantaneous deformation and velocity of the equivalent SDOF system of the mid-span displacement of specimen. For the column specimen subjected to close-range explosion, the duration of the applied blast load is far less than its natural period. Therefore, the blast load can be seen as an impulse applied on the column specimen. The principle of the equivalent SDOF system is that the deformation of the system equals to the mid-span deflection

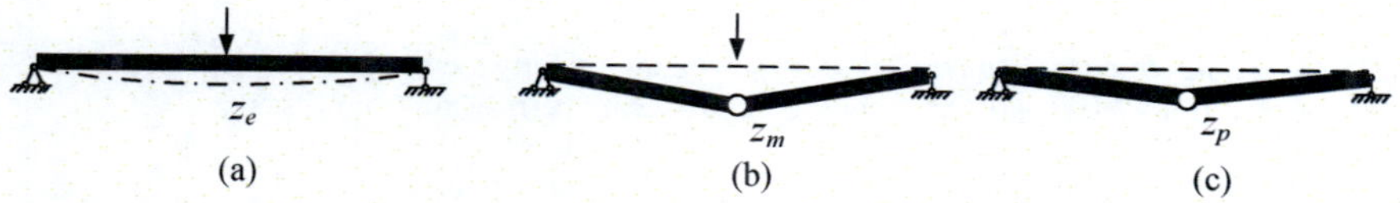

Fig. 11.13 Three deformation stages of the specimen **a** elastic stage **b** plastic stage **c** resilient stage, reprinted from Wu et al. (2020), copyright 2020, with permission from Elsevier

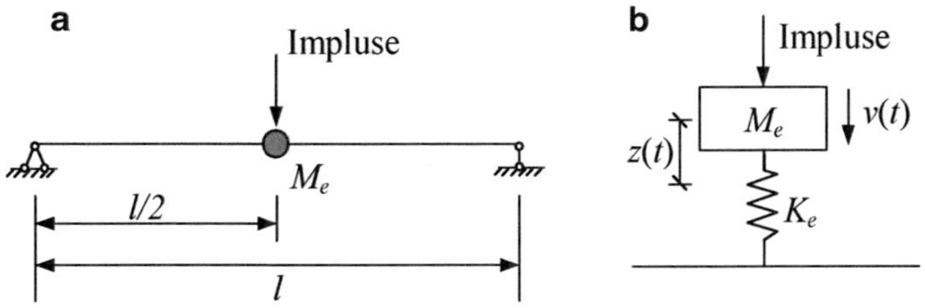

Fig. 11.14 Equivalent SDOF system **a** specimen loaded by blast impulse **b** equivalent SDOF system (Remennikov and Uy 2014)

of the specimen, thus the deformation of the equivalent SDOF system can be used to represent the mid-span deflection of the UHPCC-FST column specimen.

The SDOF system can be described by the following differential equation of motion

$$K_{LM}M\ddot{z}(t) + c\dot{z}(t) + R(z(t)) = 0 \tag{11.11}$$

where M is mass of the column specimen, c is the damping of the system and it can be ignored in elastic and plastic stages for the very short duration, R is the resistance of the system and K_{LM} is load-mass transformation factor of the equivalent SDOF system defined in (Department of Defence 2008). Table 11.5 presents the values of K_{LM} for different structural modes (Department of Defence 2008). The initial conditions (when $t = 0$) for Eq. (11.11) can be written as

$$z(0) = 0, v_0 = v(0) = \dot{z}(0) = \frac{\mathrm{d}z}{\mathrm{d}t} = \frac{I}{M} \tag{11.12}$$

Table 11.5 K_{LM}, k_M and R_u values for different structural modes (Department of Defence 2008)

Structural modes	Deformation stage	K_{LM}	k_M	R_u
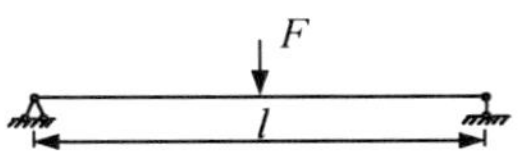	Elastic	0.49	$\frac{48EJ}{l^3}$	$\frac{4M_u}{l}$
	Plastic	0.33	0	$\frac{4M_u}{l}$
	Resilient	0.49	$\frac{48EJ}{l^3}$	–
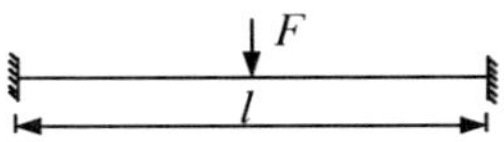	Elastic	0.37	$\frac{192EJ}{l^3}$	$\frac{8M_u}{l}$
	Plastic	0.33	0	$\frac{8M_u}{l}$
	Resilient	0.43	$\frac{192EJ}{l^3}$	–
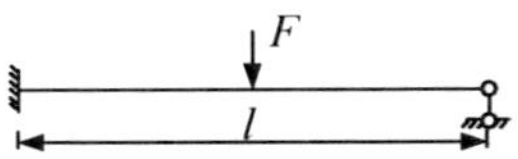	Elastic	0.43	$\frac{107EJ}{l^3}$	$\frac{16M_u}{3l}$
	Elastoplastic	0.49	$\frac{48EJ}{l^3}$	$\frac{6M_u}{l}$
	Plastic	0.33	0	$\frac{6M_u}{l}$
	Resilient	0.46	$\frac{107EJ}{l^3}$	–

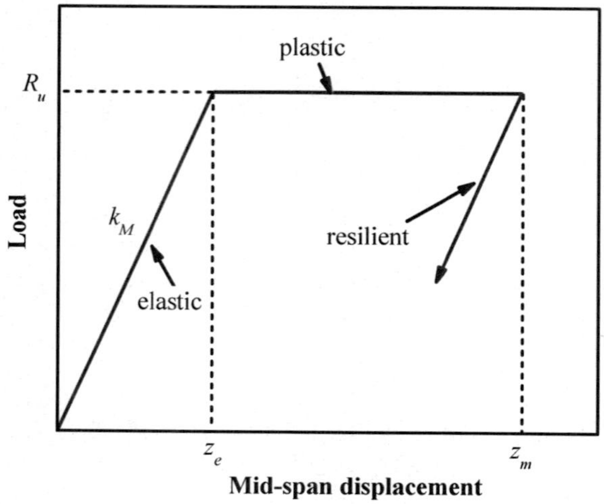

Fig. 11.15 Resistance-deflection relation for UHPCC-FST column specimen, reprinted from Wu et al. (2020), copyright 2020, with permission from Elsevier

where v_0 is the initial velocity of the system, I is the total impulse applied on the system which can be derived from Henrych (1979) and Remennikov and Uy (2014) as

$$I = 2\int_{0}^{\alpha_0} dI = \frac{S_c A_0 W d}{a}\left(\alpha_0 + \frac{1}{2}\sin 2\alpha_0\right) \quad (11.13)$$

where S_c, A_0, a, d have been introduced above and α_0 = arctan($l/2a$) is shown in Fig. 11.9.

The resistance R of SDOF system in Eq. (11.11) during different deformation stage is given in Eq. (11.14), and the resistance-deformation relationship for the system is shown in Fig. 11.15.

$$R = \begin{cases} k_M z(t) & \text{Elastic stage} \\ R_u & \text{Plastic stage} \\ k_M\left(z(t) - z_p\right) & \text{Resilient stage} \end{cases} \quad (11.14)$$

where k_M is the elastic bend stiffness, $z_p = z_m - z_e$ is the plastic deformation of the SDOF system and R_u is the ultimate resistance of the SDOF system. The values of k_M and R_u are given in Table 11.5, where EJ and M_u are the effective flexural stiffness and the ultimate flexural capacity of the member determined by the Chinese standard GB 50936-2014 (2014).

Therefore, the differential motion equation Eq. (11.11) of the equivalent SDOF system during different deformation stage can be written as follows.

When $0 \le t \le t_e$, t_e is the time instant when the system reaches the elastic limit, the system is in the elastic stage and its differential motion equation is given in

Eq. (11.15) and the corresponding initial conditions are given in Eq. (11.12).

$$K_{LM}M\ddot{z}(t) + k_M z(t) = 0 \tag{11.15}$$

When $t_e < t \leq t_m$, t_m is the time instant when the deformation of the system reaches the ultimate state, the system is in the plastic stage and its differential motion equation is given in Eq. (11.16), and the corresponding initial conditions are given in Eq. (11.17).

$$K_{LM}M\ddot{z}(t) + R_u = 0 \tag{11.16}$$

$$z(t_e) = z_e,\, v(t_e) = \dot{z}(t_e) = \left.\frac{\mathrm{d}z}{\mathrm{d}t}\right|_{t=t_e} \tag{11.17}$$

When $t > t_m$, the system is in the resilient stage and the corresponding differential motion equation is given in Eq. (11.18). In this stage, the damping should be considered to make the system stabilized and stopped. The corresponding initial conditions are given in Eq. (11.19).

$$K_{LM}M\ddot{z}(t) + c\dot{z}(t) + k_M\left(z(t) - z_p\right) = 0 \tag{11.18}$$

$$z(t_m) = z_m,\, v(t_m) = \dot{z}(t_m) = 0 \tag{11.19}$$

where z_m can be calculated by Eq. (11.16) of the system during plastic period.

The related parameters of the present test are given in Table 11.6, by substituting them into Eqs. (11.15), (11.16) and (11.18), the mid-span displacement time-history can be derived by solving the above differential equations. The corresponding mid-span deflection time-history curves for the present four UHPCC-FST specimens are shown in Fig. 11.16.

Table 11.6 Calculating parameters of SDOF method for the present test

Specimen	M (kg)	R_u (kN)	EJ (kN•m^2)	k_M (kN/m)	v_0 (m/s)
C-1	185.7	399.8	5.34 × 10^3	37.4 × 10^3	13.46
C-2					17.11
C-3					28.63
C-4					33.46

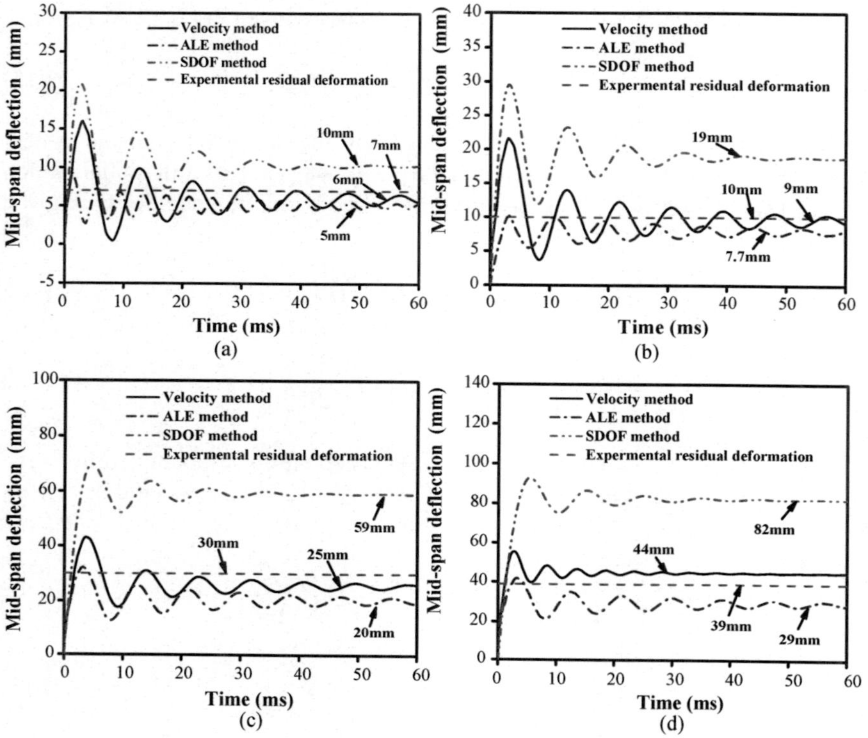

Fig. 11.16 Mid-span deflection time-history for the UHPCC-FST column specimen **a** C-1 **b** C-2 **c** C-3 **d** C-4, reprinted from Wu et al. (2020), copyright 2020, with permission from Elsevier

11.3.4 Comparisons of Predictions by Different Methods

Figure 11.16 shows the comparisons between the analytical mid-span deflection time-histories predicted by above three methods with the corresponding residual global deformation of four UHPCC-FST specimens in the present test, respectively. It can be found that, both ALE and velocity methods give the satisfied predictions of residual mid-span deflections of UHPCC-CFT column under close-range explosions. Although the SDOF method is more efficient than the other two methods since it only need to solve the differential equation of motion and the establishment of finite element model and the numerical simulations are not required, the SDOF method overestimates the results. Table 11.7 further gives the computation time costs of both the ALE and velocity methods. It indicates that the computation time cost by the ALE method is roughly more than 70 times of that by the velocity method. Therefore, considering both the accuracy and efficiency of computations, the velocity method is the optimal method to analyze the dynamic response of UHPCC-FST column subjected to close-range explosion.

Table 11.7 Computation time cost of ALE and velocity methods

Specimen	ALE method	Velocity method
C-1	22 h 54 min 36 s	20 min 19 s
C-2	27 h 24 min 46 s	21 min 28 s
C-3	31 h 23 min 28 s	18 min 59 s
C-4	28 h 30 min 07 s	23 min 46 s

11.4 Further Numerical Analyses and Discussion

Aiming to further evaluate the applicability of above three analytical methods in predicting the dynamic responses of structural members under close-range explosions, the existing close-range explosion tests on fifteen CFST columns conducted by Remennikov and Uy (2014), Zhang et al. (2016), Sun (2013) and Cui et al. (2015) are examined, in which the scaled standoff distance $0.14 \leq Z \leq 0.58$ m/kg$^{1/3}$. Table 11.8 lists the related parameters of each test, in which the notations of the specimens are kept identical with the original test. l is the clear span of the column specimen, D_o is the diameter or the side length of the outer steel tube for circle or rectangular sectional column, and t_o is the thickness of the outer steel tube. Especially for the CFDST, D_i is the diameter or the side length of the inner steel tube for circle or rectangular sectional column, and t_i is the thickness of the inner steel tube. f_y is the yield strength of the steel tube and f_c is the compressive strength of the core concrete.

Based on the above three analytical methods introduced above, the dynamic responses of those fifteen CFST column specimens subjected to close-range explosion are analyzed. The Von Mises stress contours derived by the ALE and the velocity methods for three typical column specimens (C5, S1A, HC-1) are shown in Figs. 11.17, 11.18 and 11.19, respectively. Figure 11.20 shows the comparisons between the analytical mid-span deflection time-histories for the above three columns with the test data. It should be pointed out that, the mid-span deflection time-history of column S1A is experimentally derived since the column is only front face loaded, the details can be found in Zhang et al. (2016).

Furthermore, Table 11.9 gives the comparisons of experimental and predicted maximum and residual mid-span deflections of thoes fifteen column specimens in Table 11.8. The values in bracket are the relative errors $(z_{cal} - z_{exp})/z_{exp}$, where z_{cal} and z_{exp} are the calculated results and experimental data, respectively. From Fig. 11.20 and Table 11.9, it can be also found that the results predicted by velocity method agree much better than the other two methods, thus the velocity method is proved again as a prior choice to analyze the dynamic response of CFST column specimen subjected to close-range explosion.

Table 11.8 Parameters of existing tests (Remennikov and Uy 2014; Zhang et al. 2016; Sun 2013; Cui et al. 2015)

Specimen	W (kg)	a (m)	l (m)	$D_o \times t_o$ (mm × mm)	$D_i \times t_i$ (mm × mm)	Sectional shape	f_y (MPa)	f_c (MPa)	Axial load (kN)
C5	2.61	0.15	2	100 × 5	–	Square	350	46	0
S1A	35	1.5	2.3	210 × 5	110 × 5	Square	360	170	0
S1B	50	1.5	2.3	210 × 5	110 × 5	Square	360	170	0
S2A	35	1.5	2.3	210 × 5	110 × 5	Square	360	170	1000
S2B	50	1.5	2.3	210 × 5	110 × 5	Square	360	170	1000
S3A	50	1.5	2.3	210 × 5	110 × 5	Square	360	170	1000
S3B	50	1.5	2.3	210 × 5	–	Square	360	170	1000
C1A	35	1.5	2.3	210 × 5	110 × 5	Circular	360	170	0
C1B	35	1.5	2.3	210 × 5	110 × 5	Circular	360	170	0
C2A	35	1.5	2.3	210 × 5	110 × 5	Circular	360	170	1000
C2B	17.2	1.5	2.3	210 × 5	110 × 5	Circular	360	170	0
C3A	50	1.5	2.3	210 × 5	110 × 5	Circular	360	170	1000
C3B	50	1.5	2.3	210 × 5	–	Circular	360	170	1000
SC-3	50	0.7	1.8	273 × 7	–	Circular	292.5	33.7	0
HC-1	50	0.5	1.8	273 × 7	50 × 4	Circular	292	33.7	0

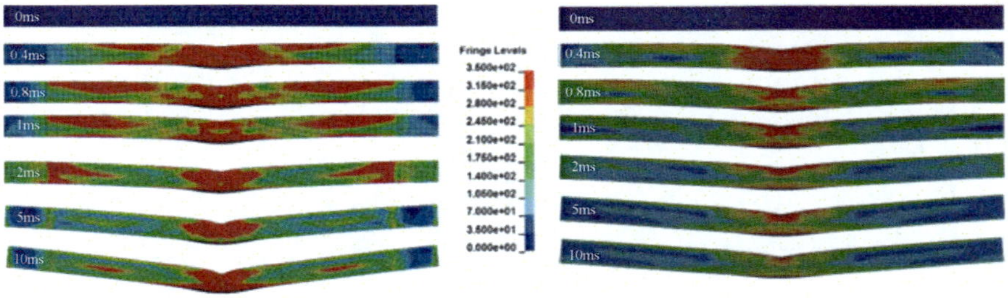

(a) (b)

Fig. 11.17 Von Mises stress contours for C5 specimen by **a** ALE method **b** velocity method, reprinted from Wu et al. (2020), copyright 2020, with permission from Elsevier

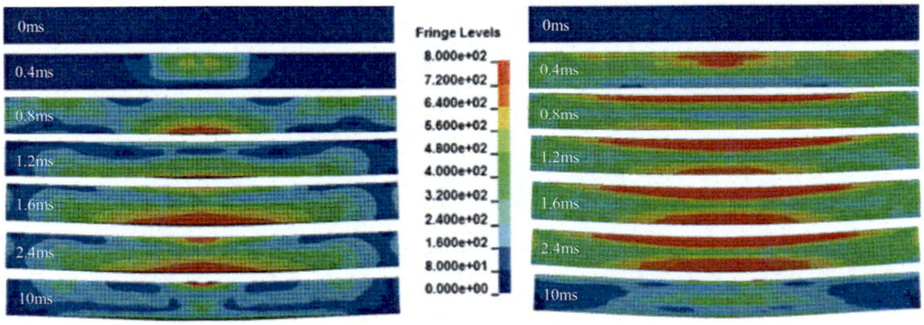

(a) (b)

Fig. 11.18 Von Mises stress contours for S1A specimen by **a** ALE method **b** velocity method, reprinted from Wu et al. (2020), copyright 2020, with permission from Elsevier

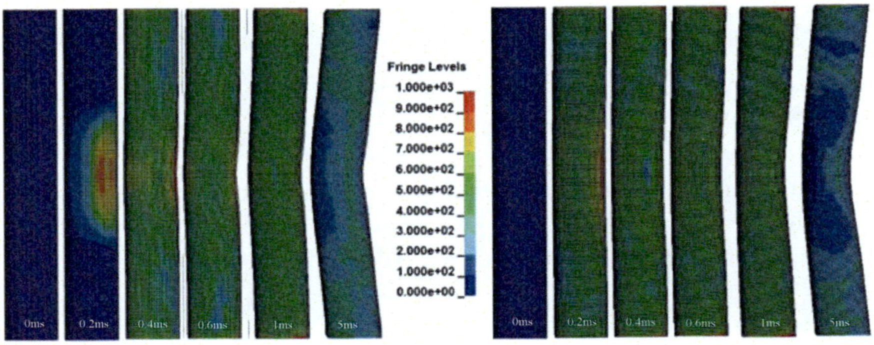

(a) (b)

Fig. 11.19 Von Mises stress contours for HC-1 specimen by **a** ALE method **b** velocity method, reprinted from Wu et al. (2020), copyright 2020, with permission from Elsevier

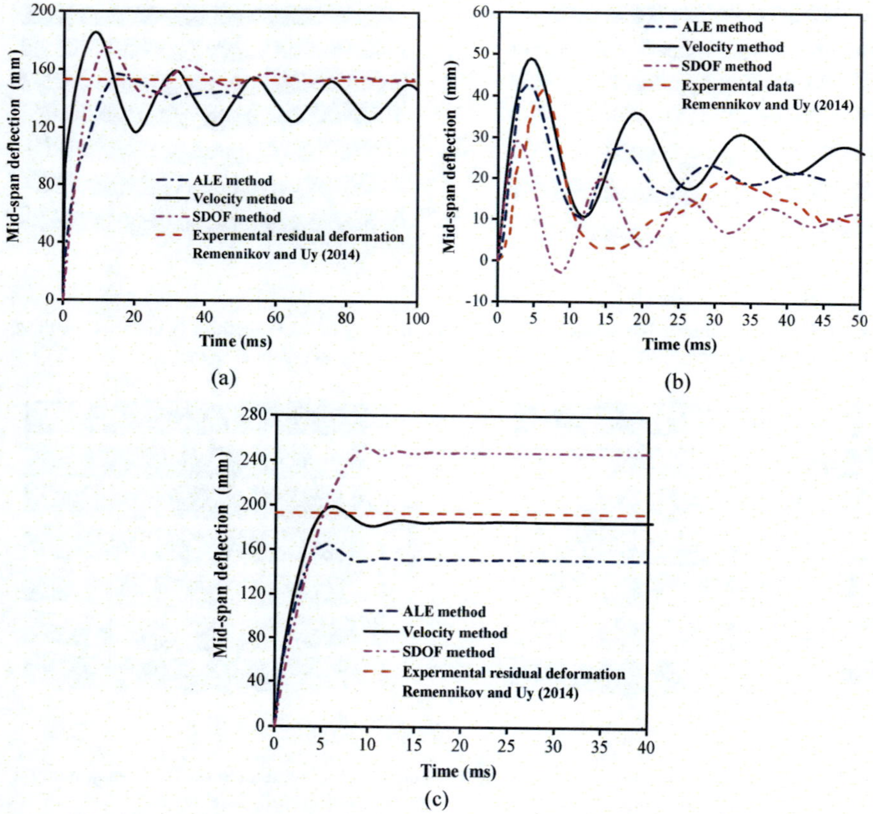

Fig. 11.20 Analytical mid-span deflection time histories for column specimens **a** C5 **b** S1A **c** HC-1, reprinted from Wu et al. (2020), copyright 2020, with permission from Elsevier

11.5 Summary

The blast performance of ultra-high performance cementitious composites filled steel tube (UHPCC-FST) columns subjected to close-range explosion were investigated experimentally and numerically in the present study. The main works and conclusions can be drawn as follows:

(1) The field test of four circular UHPCC-FST column specimens subjected to the close-in TNT charge explosion with the scaled standoff distance of $0.12 \leq Z \leq 0.14$ m/kg$^{1/3}$ were conducted, and the superiority of UHPCC-FST column against the near field explosion is validated. It shows that both the global and local deformations are induced by the close-range explosions, and the above deformations are becoming more serious with reducing the scaled standoff distance.

Table 11.9 Comparisons of experimental and predicted maximum and residual mid-span deflections

Specimen	Maximal mid-span deflection (mm)				Residual mid-span deflection (mm)			
	Test	ALE (%)	Velocity (%)	SDOF (%)	Test	ALE (%)	Velocity (%)	SDOF (%)
C5	–	–	–	–	153	141(−8)	135(−12)	154(1)
S1A	41	43(5)	49(20)	29(−29)	16	20(25)	25(56)	10(−38)
S1B	49	60(22)	81(65)	46(−6)	22	41(86)	53(141)	27(23)
S2A	36	41(14)	25(−31)	30(−17)	8	21(163)	18(125)	12(50)
S2B	–	–	–	–	14	28(100)	42(200)	32(129)
S3A	41	57(39)	41(0)	50(22)	8	29(263)	25(213)	32(300)
S3B	50	50(0)	54(8)	83(66)	10	23(130)	35(250)	74(640)
C1A	98	51(−48)	53(−46)	48(−51)	65	29(−55)	30(−54)	20(−69)
C1B	50	56(12)	39(−22)	50(0)	20	32(60)	31(55)	22(10)
C2A	–	–	–	–	39	30(−23)	16(−60)	25(−36)
C2B	35	39(11)	17(−51)	19(−46)	12	18(50)	6(−50)	0(−100)
C3A	91	76(−16)	60(−34)	88(−3)	52	53(2)	41(21)	67(29)
C3B	83	73(−12)	59(−29)	104(25)	43	51(19)	42(−2)	90(109)
SC-3	–	–	–	–	93	90(-3)	90(−3)	91(−2)
HC-1	–	–	–	–	193	151(-22)	185(−4)	247(28)

(2) Three common methods were applied to analyze the global dynamic responses of CFST columns under close-range explosion, i.e., ALE method, velocity method and SDOF method. By comparing with the present and existing explosion test on total nineteen CFST columns with $0.14 \leq Z \leq 0.58$ m/kg$^{1/3}$, the superiority of the velocity method is validated with considering both the accuracy and efficiency of computation.

References

CHENG D S, HUNG C W, PI S J. Numerical simulation of near-field explosion[J]. Journal of Applied Science and Engineering, 2013, 16(1): 61–67.

CUI Y, ZHAO J, ZHANG C, SUN S, CHEN B. Explosion-resistance behavior and damage assessment of a duplex hollow CFST column subjected to blast loading[J]. Journal of Vibration and Shock, 2015, 21: 188–193. (in Chinese).

Department of Defence. Unified facilities criteria (UFC) 3-340-02: Structures to Resist the Effects of Accidental Explosions[S]. Washington, D C, 2008.

FUJIKURA S, BRUNEAU M, LOPEZ-GARCIA D. Experimental investigation of multihazard resistant bridge piers having concrete-filled steel tube under blast loading[J]. Journal of Bridge Engineering, 2008, 13(6): 586–594.

GB 50936-2014. Specification for design and construction of concrete-filled steel tubular structures[S]. Beijing: Standards Press of China, 2014. (in Chinese).

Global Terrorism Database. Information on more than 180000 terrorist attacks[EB/OL].(2018). https://www.start.umd.edu/gtd.

HENRYCH J. The dynamics of explosion and its use[M]. Oxford: Elsevier Scientific Publishing Company, 1979.

KYEI C, BRAIMAH A. Effects of transverse reinforcement spacing on the response of reinforced concrete columns subjected to blast loading[J]. Engineering Structures, 2017, 142: 148–164.

LI Q M, MENG H. Pressure-impulse diagram for blast loads based on dimensional analysis and single-degree-of-freedom model[J]. Journal of Engineering Mechanics, 2002, 128(1): 87–92.

LI G, QU H, YANG T, LU Y, CHEN S. Experimental study of concrete-filled steel tubular columns under blast loading[J]. Journal of Building Structures, 2013, 34(12): 69–76. (in Chinese).

LI J, HAO H, WU C. Numerical study of precast segmental column under blast loads[J]. Engineering Structures, 2017, 134: 125–137.

LI M H, ZONG Z H, LIU L, LOU F. Experimental and numerical study on damage mechanism of CFDST bridge columns subjected to contact explosion[J]. Engineering Structures, 2018, 159: 265–276.

LSTC. LS-DYNA Keyword User's Manual, Version 971[M]. 2017.

MORINO S, UCHIKOSHI M, YAMAGUCHI I. Concrete-filled steel tube column system-its advantages[J]. Steel Structures, 2001, 1: 33–44.

NGO T, MOHOTTI D, REMENNIKOV A, UY B. Numerical simulations of response of tubular steel beams to close-range explosions[J]. Journal of Constructional Steel Research, 2015, 105: 151–163.

REMENNIKOV A M, UY B. Explosive testing and modelling of square tubular steel columns for near-field detonations[J]. Journal of Constructional Steel Research, 2014, 101: 290–303.

REN G M, WU H, FANG Q, LIU J Z. Effects of steel fiber content and type on static mechanical properties of UHPCC[J]. Construction and Building Materials, 2018, 163: 826–839.

REN G M, WU H, FANG Q, LIU J Z. Effects of steel fiber content and type on dynamic compressive mechanical properties of UHPCC[J]. Construction and Building Materials, 2018, 164: 29–43.

SHI Y, HAO H, LI Z. Numerical derivation of pressure–impulse diagrams for prediction of RC column damage to blast loads[J]. International Journal of Impact Engineering, 2008, 35(11): 1213–1227.

SUN S. Investigation on dynamic response of CFST columns subjected to blast loading[D]. Xi'an, China: Changan Univ., 2013. (in Chinese).

WU K C, LI B, TSAI K C. Residual axial compression capacity of localized blast-damaged RC columns[J]. International Journal of Impact Engineering, 2011, 38(1): 29–40.

WU H, REN G M, FANG Q, LIU J Z. Effects of steel fiber content and type on dynamic tensile mechanical properties of UHPCC[J]. Construction and Building Materials, 2018, 173: 251–261.

WU H, PENG Y L, FANG Q. Experimental and numerical study of ultra-high performance cementitious composites filled steel tube (UHPCC-FST) subjected to close-range explosion[J]. International Journal of Impact Engineering, 2020, 141: 103569.

WILLIAMSON E B, BAYRAK O, WILLIAMS G D, DAVIS C E. Blast-resistant highway bridges: design and detailing guidelines[R]. Washington DC, USA: Transportation Research Board, 2010. NCHRP report 645.

WILLIAMSON E B, BAYRAK O, DAVIS C, WILLIAMS G D. Performance of bridge columns subjected to blast loads. I: Experimental program[J]. Journal of Bridge Engineering, 2011, 16(6): 693–702.

WILLIAMSON E B, BAYRAK O, DAVIS C, WILLIAMS G D. Performance of bridge columns subjected to blast loads. II: Results and recommendations[J]. Journal of Bridge Engineering, 2011, 16(6): 703–710.

ZONG Z, TANG B, GAO C, LIU L, LI M, YUAN S. Experiment on blast-resistant performance of reinforced concrete piers[J]. China Journal of Highway and Transport, 2017, 09: 51–60. (in Chinese).

ZHANG F, WU C, WANG H, ZHOU Y. Numerical simulation of concrete filled steel tube columns against blast loads[J]. Thin-Walled Structures, 2015, 92: 82–92.

ZHANG F, WU C, ZHAO X, LI Z, HEIDARPOUR A, WANG H. Numerical modeling of concrete-filled double-skin steel square tubular columns under blast loading[J]. Journal of Performance of Constructed Facilities, 2015, 29(5): B4015002.
ZHANG H, LI Z, WU C. Investigation of blast effects on Double-Skinned composite steel tubular columns[J]. International Journal Protection Structures, 2015, 6(3): 403–418.
ZHANG F, WU C, ZHAO XL, XIANG H, LI Z X, FANG Q, LIU Z, ZHANG Y, HEIDARPOUR A, PACKER J A. Experimental study of CFDST columns infilled with UHPC under close-range blast loading[J]. International Journal of Impact Engineering, 2016, 93: 184–195.

Chapter 12
Experimental Study on the Residual Seismic Resistance of UHPCC Filled Steel Tube (UHPCC-FST) After Contact Explosion

12.1 Introduction

In the past two decades, the threat of terrorist around the world has been on the rise, in which explosive from car or luggage bomb has been the main means of terrorist attack (Global Terrorism Database 2016). Based on the statistics, from 1970 to 2017, there have been more than 180,000 terrorist attacks occurred around the world, including more than 88,000 bombings, 19,000 assassinations, and 11,000 kidnappings (Global Terrorism Database 2018). The vital buildings and infrastructures, e.g., bridges, are the main terrorist attacking targets. As the key load carrying members, the damage and even failure of building or bridge columns due to blast loading may induce the collapse of whole superstructures, which will further induce a large number of casualties and property losses. Therefore, the studies on the blast resistance and the post-blast behavior, e.g., the residual axial load bearing capacity and seismic resistance of above columns are very important and essential for assessing the performance of the building and bridge structure against the potential terroristic or accidental explosions.

Most of the existing studies were performed to examine the dynamic performance and failure mode of reinforced concrete (RC) column under various explosive scenarios, for both normal strength concrete (NSC) (Shi et al. 2008; Wu et al. 2011; Williamson et al. 2010, 2011a, 2011b; Kyei et al. 2017; Li et al. 2017a) and Ultra-high performance concrete (UHPC) (Xu et al. 2016; Li et al. 2015a, b, c, 2017b), respectively. It was derived that the RC columns may suffer global damage such as flexure or localized damage such as spalling of concrete cover and diagonal shear cracks. Furthermore, although the reinforced UHPC column exhibited better blast resistance and residual axial load capacity compared to normal RC columns, both NSC and UHPC materials are still prone to generate debris under blast loading, which might cause secondary damage to personnel and equipment. In recent years, concrete-filled steel tube (CFST) columns has been gained wide applications in building and bridge engineering due to their superior load-bearing capacity and seismic resistance and excellent ductility and fatigue properties than those of the conventional RC columns. The steel tube of CFST column, with different section profiles such as rectangular,

Q. Fang et al., *UHPCC Under Impact and Blast*,
https://doi.org/10.1007/978-981-33-6842-2_12

hexagonal, circular and corrugated section which can help to enhance the ductility and local stability of the column, can efficiently confine the core concrete and prevent the spalling damage of concrete under blast loading (Farahi et al. 2016; Nassirnia et al. 2016; Fujikura et al. 2008).

In the existing studies of CFST member under blast loadings, Fouche and Bruneau (2015) conducted a series of close-range explosion tests ($0.12 \leq Z \leq 0.27$ m/kg$^{1/3}$, $Z = a/W^{1/3}$ is the scaled standoff distance, a is the standoff distance from the center of the explosive to the surface of structure, W is the charge weight) on the concrete filled double steel tube (CFDST) column. Three types of failure modes, such as local buckling of outer steel tube, overall bending of the column, and failure of outer steel tube, were observed. Li et al. (2019) also performed the close-range explosion tests ($0.1 \leq Z \leq 0.18$ m/kg$^{1/3}$) on four 1/3 scaled CFDST columns, in which the influences of burst height and TNT charge weight on the blast resistance of the specimens were analyzed. The main failure mode of CFDST column under the near-field explosion was the local damage with slight overall deformation, and both the area and depth of the crater increased with increasing the charge weight. Li et al. (2018) carried out the close-in range explosion tests ($0.05 \leq Z \leq 0.06$ m/kg$^{1/3}$) on the CFDST column, and the effects of TNT charge weight and the outer steel tube thickness on the blast resistance of CFDST column were investigated. It demonstrated that the CFDST column exhibited localized damage under close-in range explosion, and there was no occurrence of overall response. Remennikov and Uy (2014) conducted the field contact and close-in range ($0.12 \leq Z \leq 0.15$ m/kg$^{1/3}$) detonation test on five CFST columns, and it was found that CFST column suffered the localized damage with the rupture of the steel tube and the crushing of the inner core concrete. Zhang et al. (2015a, b, 2016, 2017a, b) conducted a series of experimental and numerical studies on the dynamic performance of ultra-high performance concrete filled double steel tube (UHPC-FDST) column under blast loadings. Zhang et al. (2016) carried out the near-field explosion test ($0.4 \leq Z \leq 0.5$ m/kg$^{1/3}$) on six UHPC-FDST specimens with square and circular sections, and the influences of sectional shape, scaled distance of explosive, axial force on the structural failure mode of UHPC-FDST column were examined. It was derived that the circular UHPC-FDST column could take advantage of the high compressive strength and ductility of the core UHPC, which could delay or prevent the buckling of the steel tubes under the near-field explosion. In Wang et al. (2020a), the quantitative degradation of the axial load capacity of the Ultra high performance cementitious composite filled steel tube (UHPCC-FST) specimens as well as the dependence of column damage index on the charge weight is experimentally determined.

Generally, the existing studies have the following limitations: (i) Ultra high performance cementitious composite (UHPCC) is a relatively new cement-based composite with both prominent static and dynamic mechanical properties, and UHPCC filled steel tube (UHPCC-FST) have been becoming the most prospective construction bearing members for high-rise buildings and long-span bridges. However, the existing work for the blast resistance of UHPCC-FST is limited; (ii) Considering the accessibility of bridge pier and the threat of terroristic car bombs, the most extreme contact explosion scenario should be paid much more attention, while the existing studies were mainly focused on the near or far range ($Z > 0.1$ m/kg$^{1/3}$) detonation; (iii) The residual seismic resistance (RSR) of structural members under service in the seismic zone after blast loadings is also very important, while the related work is scarce. For instance, only the seismic capacities of CFST column and CFST column-to-beam joint were studied in (Elremaily and Azizinamini 2002; Liao et al. 2014; Ge and Usami 1996; Han et al. 2009), and the residual seismic performance of the structural members after environmental corrosion or exposure to fire were concerned in (Ma et al. 2018; Rajput and Sharma, 2018; Mosherf et al. 2017; Chung et al. 2008).

In this chapter, the experimental study on the RSR of UHPCC-FST column after contact explosion was conducted. Firstly, six UHPCC-FST columns were fabricated, and four of them were tested under the contact explosion with the TNT charge weight varying from 1 to 3 kg, from which the local deformations and failure modes of columns were examined. Secondly, the seismic resistance of the intact and blast-damaged columns was evaluated through the low-frequency cyclic loading test. The failure modes of the columns under combined action of contact detonation and cyclic loading were obtained. The seismic capacities of the specimens, such as load-carrying capacity, stiffness, and energy consumption, were discussed based on the hysteresis behavior of the UHPCC-FST columns. Finally, the composite damage index was proposed to evaluate the RSR of UHPCC-FST column after blast.

12.2 UHPCC-FST Specimens

12.2.1 Steel Tube

As shown in Fig. 12.1a, five steel coupons of steel tube according to the national standard (GB/T228.1-2010 2010) for direct tensile test were fabricated. Then, the

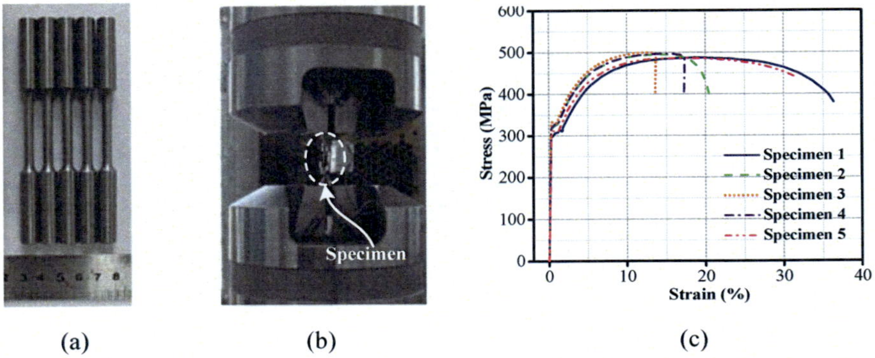

Fig. 12.1 Test setup and mechanic propriety of steel tube **a** steel coupon **b** direct tensile test **c** tensile stress-strain curve, reprinted from Wang et al. (2020b), copyright 2020, with permission from Elsevier

direct tensile test was conducted under the displacement control scheme with the speed of 2 mm/min via the MTS 810 hydro-servo machine (Fig. 12.1b). The stress-strain curve was then obtained as reported in Fig. 12.1c, and the average yield and ultimate tensile strength of the steel tube material are found to be 318.5 MPa and 497 MPa, respectively. From the tensile test results, it was found that the average elongation of the steel at break was 26%, and the elastic modulus and Poisson's ratio of the steel was 201 GPa and 0.3, respectively. It should be noted that, the mechanical treatment during the fabricating process caused the internal damage in specimens 2, 3 and 4, which led to the premature break of these specimens in the direct tension test. Hence, the average values of five samples were adopted.

12.2.2 UHPCC

The mix proportion of UHPCC is given in Table 12.1. The strength control specimens were cured at room temperature of 21.0 ± 2 °C with the relative humidity of 98.4% for 28 days and then tested for the axial compressive and tensile strength. The 100 mm^3 cubic compressive strength of UHPCC was obtained as 135.1 MPa. The corresponding axial stress strain curve of the 100 mm × 100 mm × 300 mm prism specimen is shown in Fig. 12.2a, the axial compressive strength, the elastic modulus and Poisson's ratio of UHPCC could be determined as 121 MPa, 42.9 GPa and 0.20, respectively. The stress strain response of the UHPCC derived from the direct tensile

Table 12.1 Mix proportion of UHPCC (kg/m^3)

Cement	Admixture (fly ash and ultra-fine slag)	Sand	Steel fiber	Water reducer	Water
700	250	1200	156	22.8	152

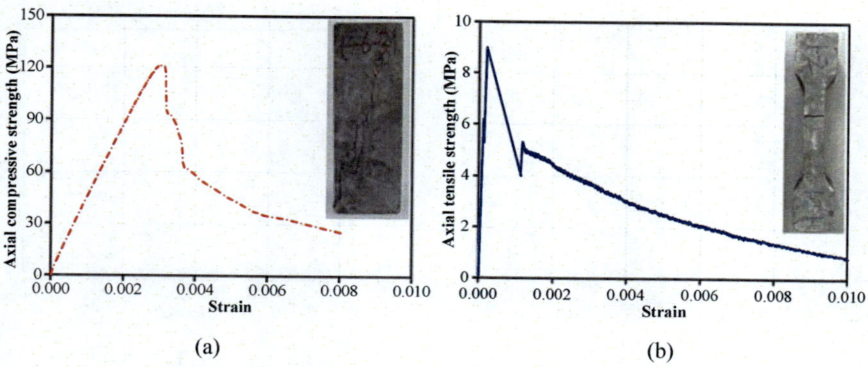

Fig. 12.2 Stress-strain curve of UHPCC **a** axial compressive test **b** direct tensile test, reprinted from Wang et al. (2020a), copyright 2020, with permission from Elsevier

test is reported in Fig. 12.2b, from which the tensile strength of the UHPCC was measured to be 9 MPa.

12.2.3 Fabrications

In this study, a total of six UHPCC-FST specimens were designed with the reduced scale of 1/4 according to the prototype of a height of 8 m. The specimens consisted of a reinforced concrete (RC) base block and a steel tube filled with UHPCC, and the specific dimensions are illustrated in Fig. 12.3. The total height of UHPCC-FST was 2420 mm, the bottom 420 mm length of which was buried in the base block. The bottom of the base block was welded with a steel plate which had a dimension of 400 mm × 400 mm × 20 mm. The outer diameter of the steel tube was 203 mm with the tube wall thickness of 6 mm. The base block was fabricated with the dimension of 800 mm × 800 mm × 420 mm, and cast by the ordinary Portland cement concrete with the grade of 46 MPa. The configuration of the steel rebar cage in base block is shown in Fig. 12.4, in which the reinforcing bars and stirrup were made of Q345 threaded steel with the diameter of 22 mm and 12 mm, respectively. For holes with the diameter of 50 mm were reserved in the base block for the following low-frequency cyclic test, and four anchor hooks were placed though the pre-fabricated holes for

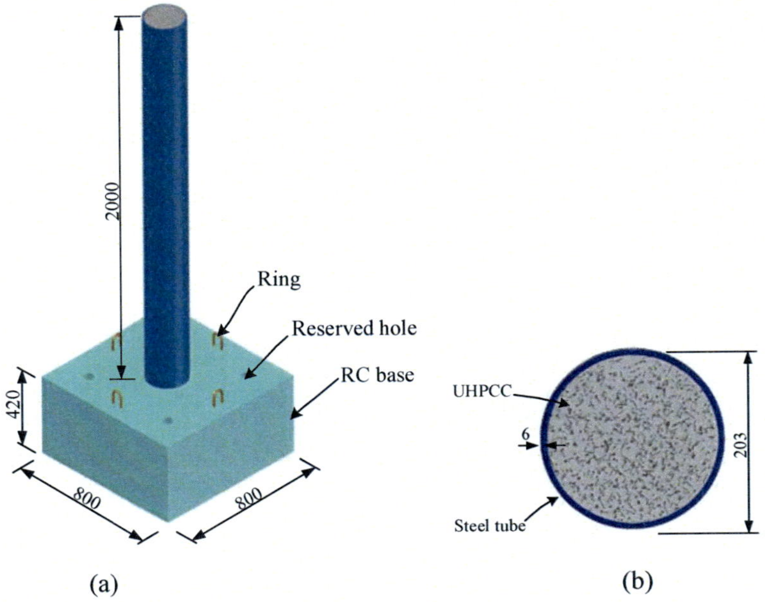

Fig. 12.3 UHPCC-FST column **a** geometric configuration **b** cross-section (*unit* mm), reprinted from Wang et al. (2020b), copyright 2020, with permission from Elsevier

(a) (b)

Fig. 12.4 UHPCC-FST **a** steel cage of base **b** specimen, reprinted from Wang et al. (2020b), copyright 2020, with permission from Elsevier

the lifting of the specimen during the test. The steel tube was filled with UHPCC and then cured at the room temperature for 28 days, with average relative humidity of 61%.

The test scheme is given in Table 12.2. Four specimens (C-11, C-12, C-13, C14) were first subjected to the contact explosion with the charge weight of 1–3 kg, and then conducted by the low-frequency cyclic loading test. Two another intact specimens (II-1 and II-2) were only tested under the low-frequency cyclic loading test, which was set as the control specimens. Besides, in order to investigate the influences of cyclic loading directions on the seismic performance of the post-blast specimen, the loading direction exerted on the C-11 specimen was parallel to the detonation face

Table 12.2 Test scheme

Specimen	Loading scheme	TNT charge weight W (kg)	Standoff distance R (mm)	Scaled standoff distance Z (m/kg$^{1/3}$)	Pushing direction
C-11	Contact explosion + Cyclic loading	1	46	0.046	Parallel
C-12		1	46	0.046	Perpendicular
C-13		2	60.5	0.048	Perpendicular
C-14		3	96	0.067	Perpendicular
II-1	Cyclic loading	–	–	–	–
II-2					

(denoted parallel in Table 12.2), while for other specimens, the loading direction was perpendicular to the detonation face (denoted perpendicular in Table 12.2).

12.3 Contact Explosion Test

12.3.1 Test Setup

Four UHPCC-FST specimens were first tested under the contact explosion. The setup for the field blast test is reported in Fig. 12.5. The reaction frame was designed and made of high strength steel, which was connected to the reinforcement concrete foundation via the steel bolts. The foundation had a cubic pit with the dimension of 1200 mm in length, 1000 mm in width and 420 mm in depth. After the specimen was placed into the designated position in the foundation pit, the gap between the specimen base and the foundation pit was filled by bricks and sands. The compaction was then performed manually by using a 10 kg drop hammer to achieve dense state in order to provide the fixed boundary for the specimen base block. The top of the specimen was clamped via the steel bolts, which was used to simulate simple supporting boundary. Hence, the net height of specimen was determined to be 1900 mm, which was the distance between the surface of the foundation and the center of the clamp. Considering that only local cratering damage and limited global response were induced under contact detonation from our previous works (Wang et al. 2020b), the axial loadings were not considered in the present test.

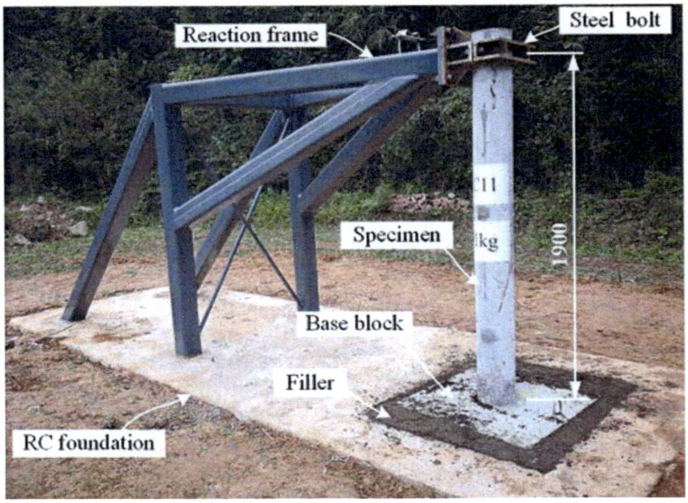

Fig. 12.5 Field contact detonation test setup, reprinted from Wang et al. (2020b), copyright 2020, with permission from Elsevier

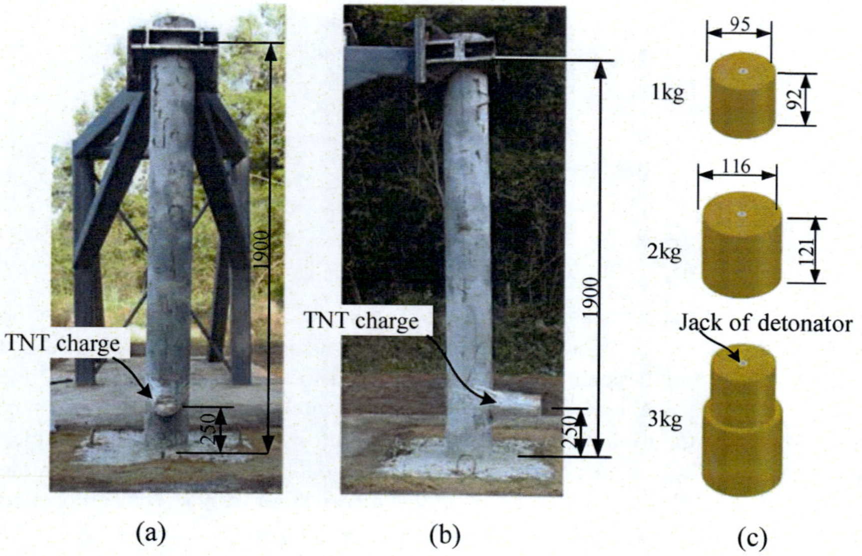

Fig. 12.6 Detailed setup **a** front view **b** side view **c** dimension of explosive, reprinted from Wang et al. (2020b), copyright 2020, with permission from Elsevier

As shown in Fig. 12.6a, b, during the field detonation test, the cylindrical TNT explosive was placed on the column surface with the center at 250 mm away from the bottom of column, which was designed corresponding to the prototypical height of 1 m for vehicle bomb explosion. Two standard TNT cylinders, i.e., 1 kg, and 2 kg, were employed and the corresponding dimensions are shown in Fig. 12.6c. The explosive was ignited by the electric detonator, which was embedded at the top center of the TNT cylinder with the depth of 40 mm. The experimental standoff distance R from the detonation center to the specimen surface, as well as the scaled standoff distance $Z = R/W^{1/3}$ are given in Table 12.2.

12.3.2 *Test Results and Discussions*

12.3.2.1 C-11 Specimen

Figure 12.7 summaries the test results of C-11 specimen after 1 kg TNT contact detonation. It is observed that C-11 specimen had no global deformations, and the steel tube was not fractured but an approximately elliptical crater was formed beneath the charge. From Fig. 12.7b and c, the axial and perpendicular dimension of the crater (R_x and R_y) were measured as 106 mm, and the maximum depth of the crater (D_c) was 27 mm. In addition to the plastic deformation of the crater, the steel tube of C-11 specimen was dilated laterally with the dimension (R_w) of 208 mm, which

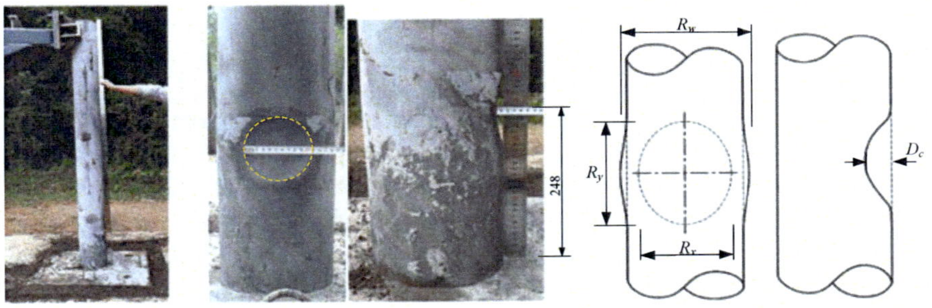

(a) (b) (c)

Fig. 12.7 Post-blast C-11 column **a** overall side view **b** blast induced crater **c** dimensions of crater (*unit* mm), reprinted from Wang et al. (2020b), copyright 2020, with permission from Elsevier

was 5 mm greater than the original diameter. Due to the eccentricity of the initial detonation point, the distance between the center of the crater and the top surface of the foundation was found to be 248 mm, which was slightly smaller than 250 mm.

12.3.2.2 C-12 Specimen

The test condition of C-12 specimen was identical to that of C-11 specimen, in which 1 kg TNT was detonated at the surface of the column with the burst height of 250 mm from the foundation ground to the charge center. Figure 12.8 shows the

(a) (b)

Fig. 12.8 Post-blast C-12 column **a** overall side view **b** frontal and side view of blast induced crater (*unit* mm), reprinted from Wang et al. (2020b), copyright 2020, with permission from Elsevier damage pattern of C-12 specimen after 1 kg TNT contact detonation. From Fig. 12.8b, the dimensions of the blast induced crater, i.e., R_x, R_y and D_c, were measured as 102 mm, 102 mm and 24 mm, respectively. The steel tube of C-12 specimen was dilated laterally with the dimension (R_w) of 209 mm, which was 6 mm greater than the original diameter. The distance between the crater center and the foundation ground was found to be 251 mm, which was very close to 250 mm. Compared with the C-11 specimen under the identical detonation scenario, the experimental damage dimensions of C-12 column were nearly the same, which indicates the reproducibility of the present field blast tests.

12.3.2.3 C-13 Specimen

Figure 12.9 illustrates the test results of C-13 specimen after 2 kg TNT contact explosive. From Fig. 12.9a, it can be observed that C-13 specimen had no global deformation, and only a localized crater was occurred beneath the explosive. Figure 12.9b demonstrates that the crater also had an approximately elliptical shape with short axis of 132 mm and long axis of 137 mm, respectively. The steel tube of C-13 specimen was dilated laterally with the dimension (R_w) of 218 mm, which was 15 mm wider than the original diameter. Besides, the maximum depth of the crater (D_c) was measured as 40 mm. Compared with the C-11 and C-12 specimens under 1 kg TNT contact explosions, it demonstrates that the larger weight TNT charge will induce more serious plastic deformation of the specimen. which was larger than that of specimens after 1 kg TNT contact explosive. Besides, a major vertical fracture crack occurred with the length and width of 105 mm and 6 mm, respectively. Several slight cracks were also formed at the periphery of the crater due to the intensive detonation

Fig. 12.9 Post-blast C-12 column **a** overall side view **b** frontal and side view of blast induced crater (*unit* mm), reprinted from Wang et al. (2020b), copyright 2020, with permission from Elsevier

energy. From Fig. 12.9b, it is found that the crater center was 248 mm away from the bottom of the column, which was slightly smaller than 250 mm. The concrete base located behind the column was partially crushed and peeled off due to the high magnitude of the compressive pressure from the explosive.

12.3.2.4 C-14 Specimen

Figure 12.10 shows the test results of C-14 specimen after 3 kg TNT contact explosive, and it is clearly shown that C-14 specimen suffered more severe damage compared to the aforementioned three columns. According to Fig. 12.10a, it is observed that the localized crater occurred beneath the explosive and slight deflection of 3 mm at the position of 250 mm from the foundation ground was induced. Figure 12.10b presents that the crater had an approximately elliptical shape with short axis (R_x) of 130 mm and long axis (R_y) of 148 mm, respectively. The steel tube of C-14 specimen was found to be dilated laterally with the dimension (R_w) of 226 mm, which was 23 mm wider than the original diameter. A major vertical fracture crack occurred at the bottom of the crater with the length and width of 120 mm and 16 mm, respectively. Meanwhile, the core UHPCC located at the crater was found to be crushed, which was mainly due to the high compressive pressure from the contact explosive. The crater center was 250 mm away from the bottom of the column. The maximum depth of the crater (D_c) was measured to be 47 mm, which was larger than that of the aforementioned three specimens. The concrete located at the back of the column base was partially crushed and peeled off mainly due to the high magnitude of the compressive pressure from the contact detonation. After lifting the specimen out of the foundation pit, many micro-cracks developed at the column base were found as shown in Fig. 12.10c, indicating that the partial detonation energy imparted

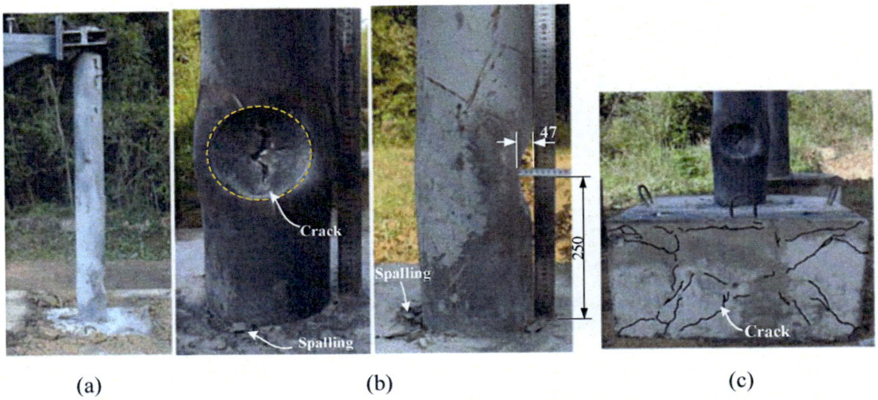

Fig. 12.10 Post-blast C-14 column **a** overall side view **b** frontal and side view of blast induced crater **c** column base (*units* mm), reprinted from Wang et al. (2020b), copyright 2020, with permission from Elsevier

from the explosive were transmitted into the foundation due to the larger charge weight of explosive.

From the above analysis, in the present contact detonation scenario, it can be concluded that the UHPCC-FST column suffered the localized damage with the formation of crater beneath the explosive, and there was almost no overall response in the columns. The repeatability of the present test is validated and the dimensions of the local crater are enlarged with increasing the charge weight. In the following section, the seismic resistance of these four post-blast UHPCC-FST and two intact column specimens was experimentally studied, and the influencing factors, such as damaged degree of specimens after blast and cyclic loading direction are considered.

12.4 Low-Frequency Cyclic Loading Test

12.4.1 Test Setup

In this section, the seismic resistance of the intact and post-blast UHPCC column under contact detonations was studied and compared experimentally under low-frequency cyclic loading test. The setup of the low-frequency cyclic loading test is illustrated in Fig. 12.11. The lateral loading was applied to the top of the column with the displacement control regime via the 500 kN hydraulic servo actuator, and the actual maximum stroke of the actuator was 200 mm. The push-out and pull-back forces in actuator are defined as the positive and negative phases, respectively. It

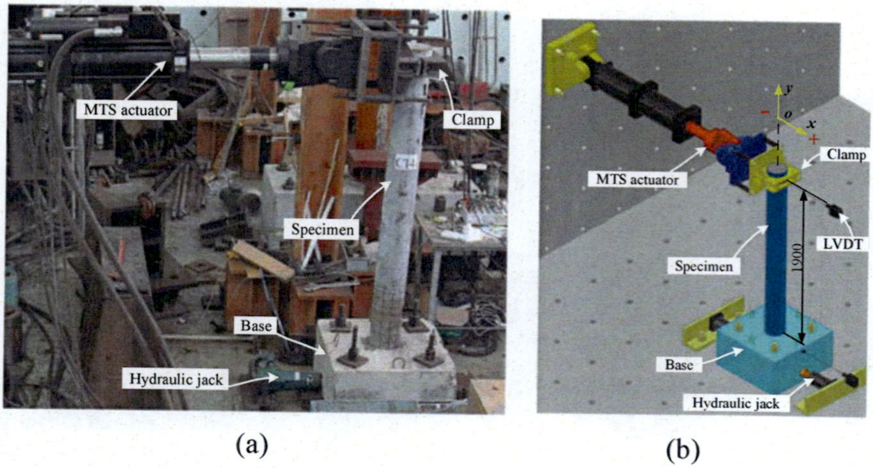

Fig. 12.11 Low-frequency cyclic test setup **a** photograph **b** schematic view, reprinted from Wang et al. (2020b), copyright 2020, with permission from Elsevier

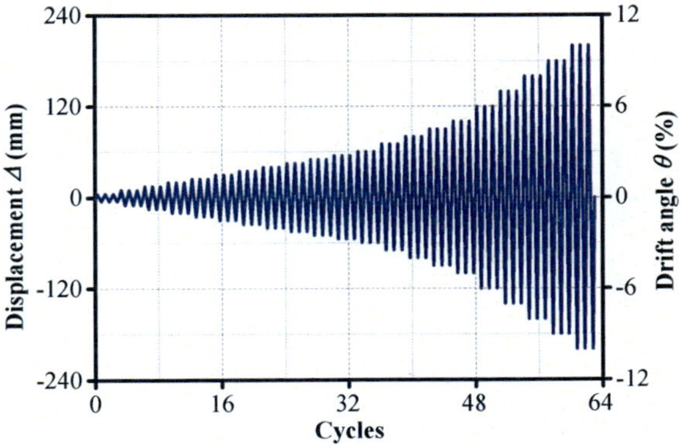

Fig. 12.12 Loading history in the cyclic test, reprinted from Wang et al. (2020b), copyright 2020, with permission from Elsevier

should be noticed that in the cyclic loading test, no axial force was applied at the top of the column.

A clamp adopted to waling the free head of column was connected to the actuator through the steel bolts. Hence, the net length between the center of the clamp and the top surface of column base was 1900 mm. A linear variable differential transducer (LVDT) with the range of 500 mm was attached at the head of column to record the horizontal displacement of the MTS actuator. To guarantee the stability of the test specimen, the reinforced concrete column base was attached firmly to the ground with four anchor bolts, and two hydraulic jacks were further installed at the two sides of column base to prevent the horizontal sliding of the specimen.

The loading history for the low-frequency cyclic loading test is shown in Fig. 12.12, in which the loading was applied in the form of lateral displacement (Δ). The applied displacement was initially started from 5 to 60 mm with the increment of 5 mm, and subsequently increased from 60 to 100 mm with the increment of 10 mm. Then the lateral displacement was continuously applied from 100 to 200 mm with the increment of 20 mm. Three cycles were imposed at each lateral displacement level. The drift angle (θ) was further defined as the ratio of input displacement to column height. Hence, the corresponding drift angles for the above varied lateral displacements are varied from 1/380 to 1/9.5. The test was terminated when one of the following two conditions was satisfied: (i) the lateral load P exhibited abrupt decrease; (ii) the specimen suffered obvious damage during the loading process.

12.4.2 *Test Results and Discussions*

12.4.2.1 Failure Pattern

The failure pattern of the present intact and post-blast UHPCC-FST column in the low-frequency cyclic loading test can be observed and shown in Fig. 12.13. It can be summarized as follows:

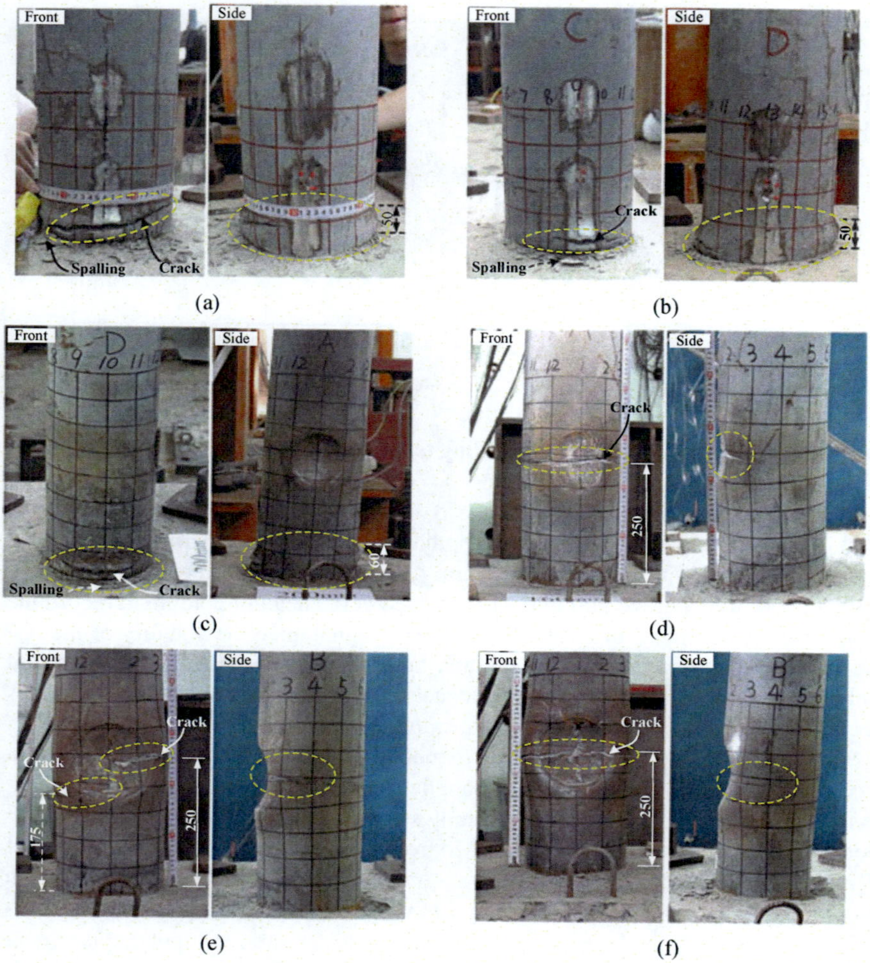

Fig. 12.13 Failure pattern of specimens **a** II-1 **b** II-2 **c** C-11 **d** C-12 **e** C-13 **f** C-14 (*unit* mm), reprinted from Wang et al. (2020b), copyright 2020, with permission from Elsevier

(i) II-1 specimen was an intact column, which was tested as the control column. At early stage of loading, the steel tube was in the elastic state, and nearly no obvious cracks appeared on the specimen. With the lateral displacement Δ increasing up to 140 mm, the steel tube exhibited local buckling at the bottom of the column. After three loading cycles at the same lateral displacement level, the plastic hinge further developed at the height of 50 mm from the top surface of the column base. As the lateral displacement increased, the degree of buckling enhanced intensively along the loading direction. It is also found that under the repeated tension and compression, an evident sign of cracking and crushing was observed in the vicinity of the column base. At the 2nd cycle of $\Delta = 180$ mm, the steel tube at the plastic hinge exhibited abrupt cracking, accompanied by the rapid drop in the lateral load. After one more loading cycle, the crack continuously developed, resulting in the lateral load dropping to 25% of the peak load. Hence, the test is terminated.

(ii) The test scenario of II-2 specimen was identical with that of II-1 specimen, while the corresponding damage pattern of II-2 specimen was slightly different. When the lateral displacement Δ reached 160 mm, the plastic hinge of the steel tube started to develop at the height of 50 mm from the top surface of the column base. With the lateral displacement further increasing up to 180 mm, the rupture of steel tube did not occur, while the buckling degree along the loading direction was more pronounced than that of II-1 specimen. Subsequently, at the loading stage of $\Delta = 200$ mm, the buckling part of the steel tube at the bottom of the column exhibited suddenly rupture with the lateral load dropping to 43% of the peak load. The ultimate displacements for II-2 and II-1 specimens were different, which might be attributed to the different tightness degrees of anchor bolts and the deviations of the material properties of specimens. However, the failure modes for both two intact UHPCC columns can be regarded as in the similar manner.

(iii) C-11 specimen is the post-blast column after 1 kg TNT contact detonation. Since the lateral displacement was applied parallel to the cratering face, the plastic hinge and buckling developed below the blast induced crater, i.e., at the height of 60 mm from the top surface of the column base, which was much similar with the intact specimens II-1 and II-2. After the 3rd cycle of $\Delta = 200$ mm level, the steel tube at the bottom of column did not exhibit breakage, while an evident sign of cracking and crushing was observed in the vicinity of the column base. Since the maximum stroke of the actuator was 200 mm, the lateral displacement cannot further increase, and hence the additional cycles were continually applied with the lateral displacement of 200 mm. At the end of the additional 2nd loading cycle, the buckling part at the steel tube exhibited suddenly breakage with the load decreasing to 54% of the peak load, and then the test is ended. Therefore, it can be drawn that when the cyclic loading direction was parallel to the cratering face, the seismic capacity of the post-blast column was not significantly affected by the damage induced by contact detonation. Although the core concrete might be damaged due to blast loadings, the confining effect provided by the outer steel tube was remained to guarantee the certain resistance to the lateral cyclic loading.

(iv) C-12 column was the post-blast specimen after 1 kg TNT contact explosive, which was identical with the C-11 column except that the loading direction of C-12 specimen was perpendicular to the cratering face. It is found that at the lateral displacement of 140 mm, the buckling of steel tube was emerged at the height of 250 mm from the top surface of the column base, and the plastic hinge tended to develop at the blast induced crater. At the 1st cycle of $\Delta = 160$ mm, a horizontal crack occurred at the crater with the lateral load abruptly dropping to 55% of the peak load, and then the test is ended.

It is found that the fracture of C-12 column occurred at the position of blast induced crater, while the rupture of steel tube occurred near the bottom of column base for C-11 specimen. It can be concluded that when the loading direction was perpendicular to the crater, the residual seismic resistance of the post-blast specimen was significantly reduced by the contact detonation.

(v) C-13 specimen was the post-blast column after 2 kg TNT contact detonation. In the cyclic loading test, the loading direction for C-13 specimen was perpendicular to the crater, which was identical with C-12 column. When the lateral displacement reached 140 mm, the plastic hinge occurred at the crater, not at the bottom of the column. At the negative loading phase of the 1st cycle of $\Delta = 160$ mm, one transverse crack occurred in the cratering position with the height of 250 mm from the top surface of column base. At the negative loading phase of the 2nd cycle of $\Delta = 160$ mm, the second transverse crack developed at the height of 175 mm with the lateral load dropping to 76% of the negative peak load. Up to the negative loading phase of the 3rd cycle of $\Delta = 160$ mm, the lateral load-carrying capacity of the column decreased to 38% of the peak load, and hence the test is stopped. During the test, two horizontal cracks subsequently developed at the crater to weaken the loading, and hence the lateral load-carrying capacity of C-13 column did not decrease rapidly. Since C-13 column sustained two more cycles at the lateral displacement of 160 mm compared to that of C-12 specimen, it exhibited more severe damage. However, it does not mean that the seismic performance of C-13 is superior to that of C-12, because under the same displacement condition, the resistance of C-12 is greater than that of C-13.

(vi) C-14 specimen was the post-blast column after 3 kg TNT contact detonation. The loading direction for C-13 specimen was perpendicular to the face of the crater, which was the same as that of C-12 and C-13 specimens. It is observed that when the lateral displacement reached 100 mm, the plastic hinge developed at the crater. At the negative loading phase of the 2nd cycle of $\Delta = 120$ mm, an obvious transverse crack occurred in the area of the crater at the height of 250 mm. Meanwhile, the lateral load-carrying capacity quickly dropped to zero, indicating the specimen was suffered total failure. Therefore, with increasing the charge weight, the residual seismic resistance of the post-blast column deteriorated continuously as expected.

Based on above experimental observations of specimens under lateral cyclic loading, it can be concluded that,

(i) II-1, II-2 and C-11 specimens exhibited the bending failure mode under the lateral cyclic loading. The critical section (or the weakest section) of the column was near the bottom of column where the plastic hinge continually developed. It is observed that before the failure of the column, the steel tube firstly exhibited dilated buckling at the plastic hinge under the cyclic loading. With elevating the loading level, the steel tube at the plastic hinge tended to suffer brittle fracture, where one major transverse crack developed near the bottom of the column. The heights of plastic hinge were about 50 mm for both II-1 and II-2 specimens, while it was about 60 mm for C-11 specimen. Therefore, when the loading direction is parallel to the cratering face, the position of the critical section or the plastic hinge is not affected by the previous blast induced damage.

(ii) The bending failure mode was also observed in C-12, C-13 and C-14 post-blast specimens. The location of the critical section (weak section) or plastic hinge in the column was shifted to the position of the blast induced crater, and the steel tube tended to fracture at the crater finally. By comparing C-11 with C-12 specimens, it is found that under the identical explosion scenario, the loading direction had a great influence on the occurrence of the critical section and the plastic hinge.

12.4.2.2 Hysteresis Curve

The experimental hysteresis curves of the lateral load and its corresponding displacement (*P*-Δ) of the present specimens are shown in Fig. 12.14. From Fig. 12.14a, it is found that the hysteresis curve of II-1 specimen showed the unexpected pinching phenomenon at early loading stage. This might be attributed to the excessive tightness of the anchor bolt during the test. Hence, in the following part, the hysteresis behavior of II-1 specimen was not discussed. From Fig. 12.14b to Fig. 12.14f, it is observed that the hysteresis curves of these specimens exhibited fusiform shape, implying that the column had good ductility and energy dissipation capacity. In general, the seismic behavior of each specimen can be divided into three stages, i.e., elasticity, elastoplasticity and plastic failure.

(i) When $\Delta \leq 30$ mm, all specimens were in the elastic state, in which the *P*-Δ curves showed linearly increasing trend.

(ii) With elevating the lateral displacement loading level, the specimen performed plastic deformation accordingly. The upper and lower ends of the three hysteresis loops at the same lateral displacement level were no longer coincided. It is postulated that the accumulated damage of the specimen under cyclic loading increased, resulting in the large area of yielding. During the three cycles at the same lateral displacement level, the peak load corresponding to the latter cycle was lower than that of the previous one, indicating the degradation of the stiffness of column. However, the unloading stiffness of the column exhibited slight variations, which was almost kept unchanged as its initial stiffness. The hysteresis curve of each specimen exhibits slight pinching phenomenon, indicating the occurrence of debonding and slippage between the steel tube and the core concrete.

(iii) As the lateral displacement continuously increased, the specimen stepped into the stage of plastic failure, in which an abrupt breakage failure of the steel tube was occurred.

Among three specimens of C-12, C-13 and C-14, it is found that II-2 specimen sustained more loading cycles than that of C-12, C-13 and C-14 column, which indicates that the pre-damage induced by the blast had significantly impact on the

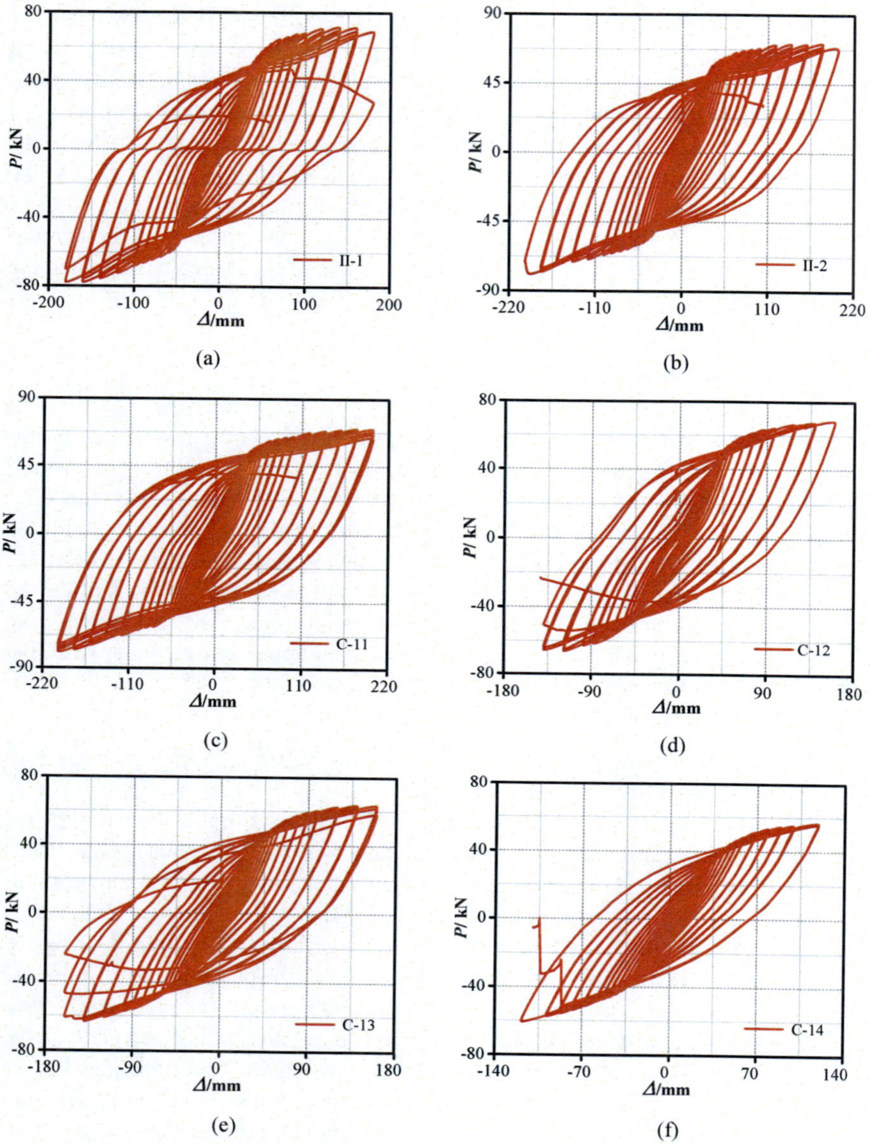

Fig. 12.14 Hysteresis curve of test specimens **a** II-1 **b** II-2 **c** C-11 **d** C-12 **e** C-13 **f** C-14, reprinted from Wang et al. (2020b), copyright 2020, with permission from Elsevier

residual seismic response of the column. It should be noticed that the number of cycles for C-13 specimen was more than that of C-12 column. This is because that at the plastic failure stage, two major horizontal cracks developed at the crater of C-13 column which delayed the occurrence of failure. Comparably, C-12 column exhibited the rupture of steel tube at the crater with the development of one major horizontal crack, resulting in rapid drop of lateral load-carrying capacity. Besides, when the cyclic loading direction was perpendicular to the cratering face, the hysteresis performance of the post-blast column was reduced with increasing the TNT charge weight.

In summary, when the cyclic loading direction was perpendicular to the blast induced crater, the plastic hinge tended to develop at the crater and further caused the buckling of the steel tube, resulting in the low seismic resistance of the post-blast column. When the loading direction was parallel to the cratering face, the plastic hinge tended to develop near the bottom of column, not at the location of crater, which indicates that the damage induced by the blast did not change the position of the plastic hinge of the post-blast column. For the post-blast specimens under the identical detonation scenario, the loading direction has significant influences on the residual seismic responses of column.

12.4.2.3 Skeleton Curve

Based on Figs. 12.14 and 12.15 further shows the corresponding skeleton curves of tested post-blast and control specimens, in which Fig. 12.15a illustrates the comparisons of specimens after 1 kg TNT contact detonation under different cyclic loading directions, and Fig. 12.15b illustrates the results of the specimens after varied TNT charge weights explosions for the loading direction being perpendicular to the cratering face. It can be drawn that,

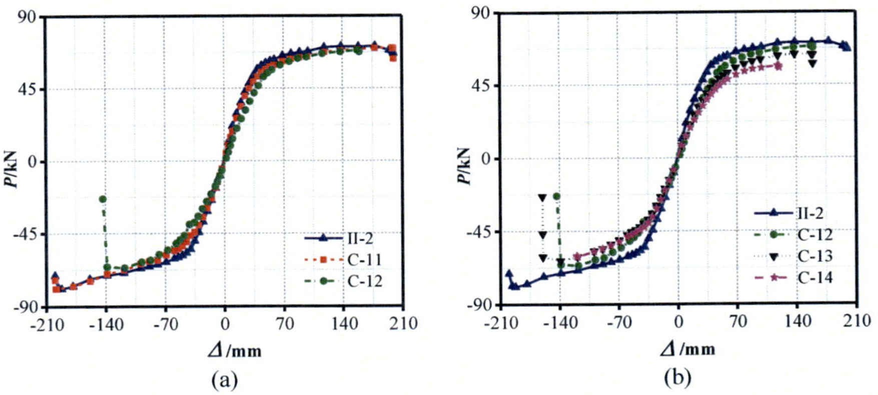

Fig. 12.15 Skeleton curves of (a) II-2, C-11 and C-12 specimens **b** II-2, C-12, C-13 and C-14 specimens, reprinted from Wang et al. (2020b), copyright 2020, with permission from Elsevier

(i) The shape of the skeleton curves is in a similar manner among all the specimens. However, it is found that for each specimen, the load-displacement relationship exhibits an asymmetric behavior in the push and pull phases. This may be caused by the imperfect fabrication of the specimen and the existence of the blast induced crater in the post-blast column.

(ii) Under the same lateral displacement level, the corresponding lateral load for the post-blast column is lower than that of the intact control column. For the cases of the same TNT charge weight and lateral displacement level, the lateral load of specimen with the loading direction being perpendicular to the cratering face is lower than that of loading direction being parallel to the cratering face. Meanwhile, it is clearly shown that with the increase of the TNT charge weight, the ultimate lateral load of the specimen decreases accordingly, implying that the seismic resistance of the post-blast column is significantly influenced by the pre-damage induced by blast.

(iii) The skeleton curve of C-13 specimen exhibites the inflection points at both the push and pull phases, it rapidly drops at the lateral displacement of ±160 mm, indicating the occurrence of failure of the specimen. For C-12 column, the inflection points are also observed at both the push and pull phases, and the rapidly dropping behavior are found at the pull phase when the lateral displacement increases up to 140 mm. II-2 and C-11 specimens have the similar tendency under the cyclic loading, in which the skeleton curves smoothly descends to the end of the test. C-14 specimen has the lowest lateral load capacity among all the specimens, where the sudden dropping point during the pull phase occurs at the lateral displacement of 110 mm. This is consistent with the above findings that the higher TNT charge weight causes much serious damage, which significantly deteriorates the residual seismic resistance of column when the loading direction was perpendicular to the cratering face.

Generally, although the maximum and minimum numbers of loading cycles on the present post-blast specimens are 64 and 52, and the loading period seemed to be shorter compared to that of the intact column. This is because that the existence of the crater in the post-blast column leads to the fast development of plastic hinge with increasing the loading cycles, and then causes the rupture failure of steel tube of the column. The inflection points of the skeleton curve are missing for some post-blast columns, since the brittle failure occurs at the pull phase of the loading cycle. This may be due to the absence of axial load on the specimen during the present cyclic loading test. It is known that the axial load was beneficial to enhance the bending stiffness at the tensile part of the column, and further delay the occurrence of the tensile failure. Therefore, the yielding at the compressive part of steel tube and core concrete occur earlier to exhibit inflection points in the skeleton curve.

12.4.2.4 Load-Carrying Capacity and Ductility

The load-carrying capacity and ductility of the present UHPCC-FST specimens can be discussed based on the energy equivalent method illustrated in Fig. 12.16, which is recommended by the Code for Seismic Tests for Buildings (JGJ/T101-2015 2015). The peak load $P_{\max}$ and its corresponding displacement Δ_m of each specimen can be determined from the skeleton curve, in which the peak load $P_{\max}$ is calculated as the average value of the maximum lateral loading both from the push and pull phases at the same lateral displacement level. The ultimate load P_u is determined as the 85% of the peak load $P_{\max}$. For the skeleton curve with the descending segment, the displacement corresponding to P_u is set as the limit displacement Δ_u. For the case of skeleton curve without descending segment, the limit displacement Δ_u is defined as the displacement at the fracture of steel tube. Additionally, the yield load P_y and its corresponding yield displacement Δ_y corresponding to the point F are determined by iteratively changing the slope of the secant line OC until the area of S_{OAB} equals to S_{ACD} ($S_1 = S_2$), then a vertical line draw from the point C is intersected with the load-displacement curve at point F. Therefore, the yield, peak and ultimate load of each specimen in the cyclic loading test can be determined and listed in Table 12.3.

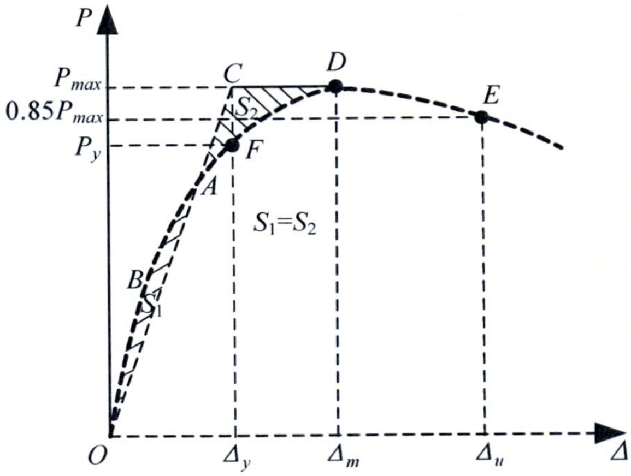

Fig. 12.16 Energy equivalent method, reprinted from Wang et al. (2020b), copyright 2020, with permission from Elsevier

Table 12.3 Parameters derived from the skeleton curves

Specimen	Nominal yield point		Peak load point		Limit point		μ
	Δ_y/mm	P_y/kN	Δ_m/mm	P_{max}/kN	Δ_u/mm	P_u/kN	
II-2	59.65	61.51	185.30	75.19	201.85	63.91	3.38
C-11	75.31	60.77	188.15	74.58	200.30	63.39	2.66
C-12	66.52	57.26	137.75	67.33	151.10	57.23	2.27
C-13	72.22	53.60	138.65	63.39	159.90	53.89	2.21
C-14	62.79	49.18	118.60	58.66	120.90	49.86	1.93

It can be seen that the peak and ultimate loads of C-11 column was almost the same as that of intact II-2 column, while the peak and ultimate loads of C-12 specimen was lower than that of II-2 specimen, indicating that the cyclic loading direction has significantly effect on the lateral load-carrying capacity of the post-blast column.

Figure 12.17 shows the variations of lateral load-carrying capacity of the post-blast column under various TNT charge weights when the cyclic loading direction is perpendicular to the cratering face, where the negative sign denotes the decrease of the lateral load-carrying capacity. Compared to the intact II-2 specimen, the lateral load-carrying capacity of three post-blast specimens (C-12, C-13 and C-14) decreased almost linearly with the reduction of 10.5%, 15.7% and 22%, respectively. It indicates that, under the same loading direction, the much serious damage induced by the blast will result in the higher degradation of the lateral load-carrying capacity of the post-blast specimen.

The ductility is an important index to evaluate the structural seismic performance, which is adopted to describe the deformation capacity of a structural member being stressed from yield to failure state. The displacement ductility coefficient μ is expressed as Δ_u/Δ_y, and Table 12.3 gives the corresponding displacement ductility coefficients of present specimens. Compared to the intact II-2 column, the displacement ductility coefficient of C-11 column decreases by 21.3%, while the displacement ductility coefficient of C-12, C-13 and C-14 specimens are reduced by 32.8%, 34.6% and 42.9%, respectively. Figure 11.17 depicts the reduction of displacement ductility coefficient of the column under varied TNT charge weights. It indicates that the displacement ductility coefficient of the post-blast column exhibits the descending trend with increasing the TNT charge weight. Additionally, the displacement ductility coefficient of the post-blast column for the loading direction being parallel to the cratering face is higher than that of being perpendicular to the cratering face. Hence, it can be concluded that when the loading direction was perpendicular to the cratering face, the higher degree of damage induced by the blast leads to the higher reduction in the displacement ductility capacity of the post-blast specimen in the cyclic loading.

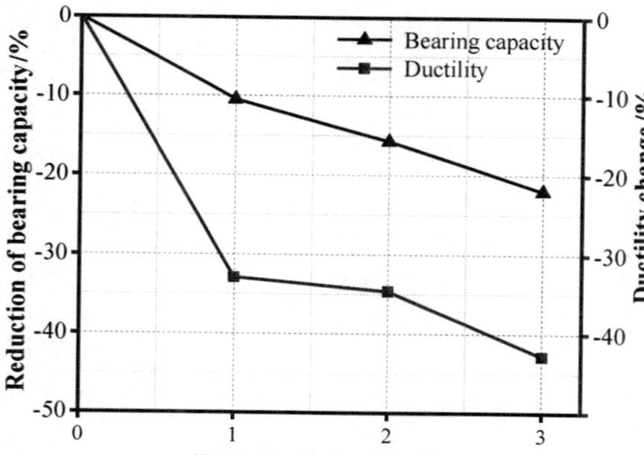

Fig. 12.17 Dependence of bearing capacity and ductility on TNT charge weight, reprinted from Wang et al. (2020b), copyright 2020, with permission from Elsevier

12.4.2.5 Stiffness Degradation

Based on experimental results, it is found that under cyclic loading test, the stiffness of the post-blast UHPCC-FST column was degraded. According to the Code for seismic test methods of buildings (JGJ/T101-2015 2015), the secant stiffness K_i of specimen under the ith displacement loading level is expressed as

$$K_i = \frac{\sum_{j=1}^{m} \left(\left|+P_{i,j}\right| + \left|-P_{i,j}\right|\right)}{\sum_{j=1}^{m} \left(\left|+\Delta_{i,j}\right| + \left|-\Delta_{i,j}\right|\right)} \tag{12.1}$$

where m is the mth cycle in the ith lateral displacement loading level, positive and negative sign represent the push and pull direction in the cycle, respectively. $P_{i,j}$ and $\Delta_{i,j}$ are the lateral load and its corresponding lateral displacement for the jth cycle at the ith displacement level, respectively.

Figure 12.18a shows the degradation of stiffness of the specimens at each lateral displacement level. It indicates that, (i) The stiffness of all the specimens degrade correspondingly with increasing the lateral displacement. The reason lies in that, with the increase of lateral displacement, the buckling of the steel tube tends to develop at the plastic hinge, and then the rupture of the steel tube occurs; (ii) Under the identical loading scenario, i.e., II-2, C-12, C-13 and C-14 specimens, the higher TNT charge weight causes the lower stiffness of the post-blast column. Under the same TNT charge weight, the stiffness of the post-blast column with loading direction being perpendicular to the cratering face (C-12) is lower than that of loading direction being parallel to the cratering face (C-11). This is because that when the crater located at the tension and compression sides along the loading direction, the stiffness of column is

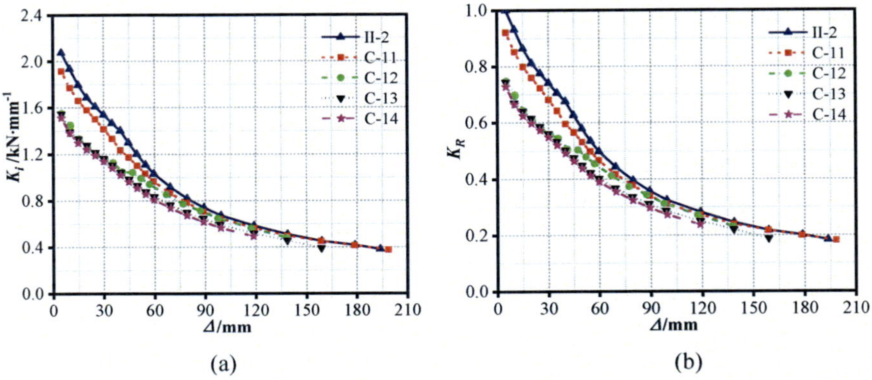

Fig. 12.18 Stiffness degradation of specimens **a** stiffness versus displacement **b** relative stiffness versus displacement, reprinted from Wang et al. (2020b), copyright 2020, with permission from Elsevier

more significantly influenced by the crater induced by blast. While for the horizontal cyclic loading applied is parallel to the cratering face, there was no damage along the loading direction at the beginning of the test, thus the seismic behavior of the post-blast column was mainly dominated by the initial stiffness of specimen; (iii) The stiffness of II-2 and C-11 specimens was higher than that of other three specimens at the early loading stage, while the stiffness of all the specimens became almost identical at the late stage.

Figure 12.18b further illustrates the degradation of the relative stiffness of each specimen at each lateral displacement level, in which the relative stiffness K_R is defined as the ratio of the stiffness of specimen to the initial stiffness of II-2. It is found that the initial relative stiffness of C-11, C-12, C-13, and C-14 specimens decreased by 7.7%, 25.2%, 25.5%, and 27.1% at the beginning of the test, respectively. With the increase of the lateral displacement level, the difference of relative stiffness K_R of each specimen decreases accordingly. C-12 specimen performed lower relative stiffness compared to that of C-11 at the same lateral displacement loading level, indicating that the cyclic loading direction significantly influences on the residual seismic capacity of the post-blast column. Furthermore, it is also found that all the specimens are failed at the relative stiffness of 0.2 due to the rupture of the steel tube.

12.4.2.6 Energy Consumption

For a concrete-filled steel tube (CFST) column, the earthquake input energy would be primarily dissipated by the damping capacity of the structural elastic behavior, and further mainly dissipated by the structural hysteresis developed within the inelastic stage. Hence, the hysteretic energy dissipation capacity can be used as an important index to evaluate the seismic performance of structural member. The energy dissipation capacity is characterized by the area enclosed by a complete load cycle in the hysteretic load-displacement curve. It is known that at certain condition the larger the area enclosed by the hysteresis curve is, the better energy dissipation capacity will be. Hence, in order to quantify the energy dissipation capacity of the specimen, the cumulative hysteresis energy E_{sum} of each specimen, which was the sum of the energy dissipation at each displacement level before the failure of the specimen, was adopted at follows.

Therefore, the cumulative hysteresis energy E_{sum} of specimens II-2, C-11, C-12, C-13 and C-14 can be derived as 372.798 kN m, 444.778 kN m, 176.395 kN m, 216.893 kN m and 103.579 kN m, respectively. It indicates that (i) the cumulative hysteresis energy $E_{\max}$ of C-11 specimen is 19% higher than that of II-2 column. The reason lies in that, when the loading direction is parallel to the cratering face, the existing blast induced crater is beneficial to the formation of plastic hinge to some extent. Furthermore, the formation of plastic hinge is helpful to dissipate energy at the bottom of column, and hence enhances the energy dissipation of the post-blast column. When the cyclic loading is applied perpendicular to the cratering face, the cumulative hysteresis energy of C-12, C-13 and C-14 specimens is lower than that of II-2 specimen; (ii) the trend of the cumulative hysteresis energy did not change monotonically with the increase of TNT charge weights. It can be explained that, at the early stage of cyclic loading, C-13 specimen first exhibits one major crack, and subsequently the second major crack is developed at the late stage of loading cycle to dissipate energy. While for the C-12 column, only one major crack occurs at the blast induced crater to sustain plastic deformation. Hence, the cumulative hysteresis energy of C-13 column is higher than that of C-12 column. However, with further increasing TNT charge weight, the cumulative hysteresis energy of specimen decreased rapidly.

Figure 12.19 further illustrates the cumulative hysteretic energy of specimens at each lateral displacement level. It can be seen that the cumulative hysteretic energy of each specimen is nearly equal to zero when the lateral displacement is below 30 mm, indicating that the specimens are in the elastic stage. As the displacement increases, the cumulative hysteretic energy increases as the J-shaped curve.

(i) The cumulative hysteretic energy of II-2 and C-11 specimens is larger than that of the other three specimens. When the lateral displacement is ranged from 30 to 180 mm, the cumulative hysteretic energy of II-2 specimen is slightly larger than that of C-11 specimen. However, at the lateral displacement of 200 mm, the cumulative hysteretic energy of C-11 specimen is larger than that of II-2 specimen. This is attributed to that, at the lateral displacement of 200 mm, II-2 specimen exhibits the rupture of the steel tube after one loading cycle, while C-11 specimen sustained three loading cycles before failure occurs.

(ii) For the case of the cyclic loading being perpendicular to the cratering face, it is found that the cumulative hysteretic energy of C-13 specimen showed the largest value, whereas the C-14 specimen exhibited the lowest the cumulative hysteretic energy attributed to the severe damage induced by the high charge weight of the explosive.

Fig. 12.19 Cumulative hysteretic energy dissipation versus displacement, reprinted from Wang et al. (2020b), copyright 2020, with permission from Elsevier

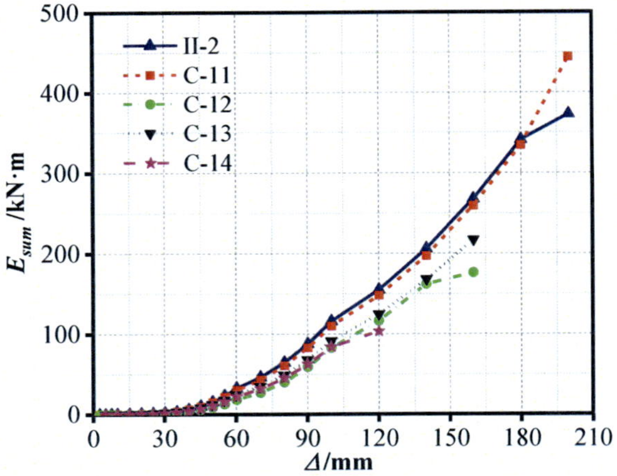

The equivalent viscous damping coefficient is another important index to describe the energy consumption of structural member under seismic loadings. At present, the energy dissipation capacity of post-blast column is further discussed by adopting the equivalent viscous damping coefficient. As shown in Fig. 12.20a, the equivalent viscous damping coefficient h_e can be defined as (JGJ/T101-2015 2015; Fan and Zhuo 2001)

$$h_e = \frac{1}{2\pi} \cdot \frac{S_{ABCDEFA}}{S_{OCG} + S_{OFH}} \tag{12.2}$$

in which, $S_{ABCDEFA}$ is the area enclosed by one hysteresis loop. S_{OCG} and S_{OFH} are the area of triangle of *OCG* and *OFH*, respectively. The points of *C* and *F* represent the peak load and its corresponding lateral displacement at the push and pull phases in one cycle, respectively. The sum of the areas of the two triangles represents the elastic strain energy stored in structural member. It is obvious that for the same lateral displacement level, the higher value of h_e is corresponding to the higher energy consuming efficiency.

Figure 12.20b shows the average equivalent viscous damping coefficient of the specimens under each lateral displacement level. It can be seen that with increasing the lateral displacement, the average equivalent viscous damping coefficient tended to increase as the S-shaped curve. When the lateral displacement is less than 30 mm, the average equivalent viscous damping coefficient of each specimen has slight change, indicating that all the specimens are in the elastic stage. The corresponding average equivalent viscous damping is mainly dissipated by the friction between the specimen and the loading equipment, e.g., actuator, anchorage, etc. As the lateral displacement increasing, the ascending tendency of the average equivalent viscous damping coefficient for both II-2 and C-11 specimens increase more quickly compared to the other three specimens, indicating that the energy dissipation of the plastic hinge at

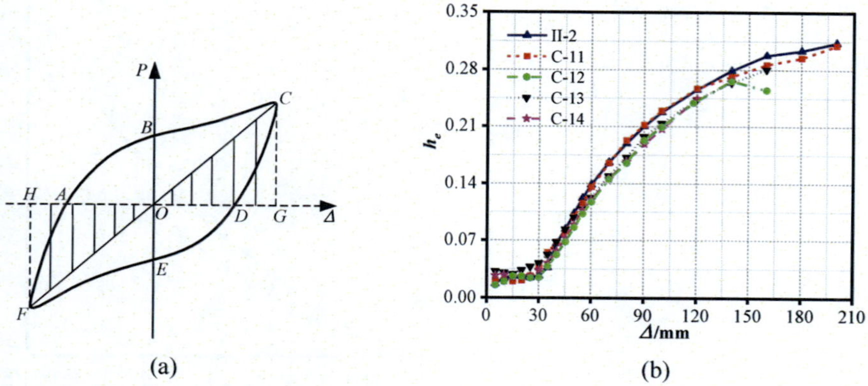

Fig. 12.20 Equivalent viscous damping coefficient **a** definition (Fan and Zhuo 2001) **b** $h_e \sim \Delta$ curve, reprinted from Wang et al. (2020b), copyright 2020, with permission from Elsevier

the column base is more prominent than that of the plastic hinge located at the blast induced crater. When the lateral displacement is in the range of 30 mm to 120 mm, the average equivalent viscous damping coefficient of II-2 column is almost the same as that of C-11 specimen. Then, the average equivalent viscous damping coefficient ratio of II-2 specimen is becoming larger, implying that the energy efficiency of II-2 specimen is kept much better than that of C-11 column. The average equivalent viscous damping coefficient of C-12, C-13 and C-14 specimens tend to increase with the increase of the lateral displacement, while the corresponding average equivalent viscous damping coefficients are close to each other. It can be concluded that the plastic hinge occurs at the location of blast induced crater is the main energy-consuming component, and hence the same level of the average equivalent viscous damping coefficient among the above three specimens is observed.

Therefore, it is further proved that the loading direction in the cyclic loading test plays an important role in the energy consuming of the specimen. It is also concluded that when the loading direction is parallel to the cratering face, the influence of loading direction is not obviously on the energy consuming capacity of the post-blast column.

12.5 Assessment of Residual Seismic Resistance of the Post-blast Column

Based on above analysis, it is found that after the extreme blast loading, the seismic resistance of the present UHPCC-FST specimens is obviously deteriorated. Hence, it is necessary to quantitatively evaluate the residual seismic performance of the post-blast column, which could provide helpful reference for the evaluation and strengthening of the post-blast column after accidental or malicious explosion attacks.

The Park-Ang model (Park and Ang 1985) is one of the most widely used damage assessment model for the structural members after earthquake. This model can consider both the hysteretic energy and the maximum deformation of the structural members by using the seismic damage index D described as follows:

$$D = \frac{\delta_m}{\delta_n} + \beta \frac{E_n}{P_y \delta_n} \tag{12.3}$$

$$\beta = (-0.447 + 0.073\lambda + 0.24n + 0.314\rho_l)0.7^{\rho_h} \tag{12.4}$$

where δ_n and δ_m are the ultimate displacements of the structural member under monotonic and cyclic loadings; P_y and E_n are the yielded shear force and cumulative hysteretic energy of the structural member; β is the energy dissipation factor of the structural member; n is the axial compression ratio; λ is shear span ratio of the structural member; ρ_h and ρ_l are the volume hoop ratio and the longitudinal steel reinforcement ratio of the structural member. Since the Park-Ang model applies essentially to RC structure, ρ_h and ρ_l need to be derived for the CFST column according to the longitudinal and circumferential equivalent effects (Zhou et al. 2019), which are as follows,

$$\rho_h = \frac{\sigma_h A_s}{\frac{\pi}{4} f_y (d - 2t)^2} \tag{12.5}$$

$$\rho_l = \frac{\sigma_l A_s}{\frac{\pi}{4} f_y (d - 2t)^2} \tag{12.6}$$

where t and d are the wall thickness and outside diameter of steel tube, f_y is the yield strength of steel tube; σ_h and σ_l are the circumferential and longitudinal stress of the steel tube, respectively. Considering the confining effects of the steel tube (Han 2004; Mander et al. 1988; Zhang et al. 2005), the σ_h and σ_l can be further calculated based on Eq. (12.7) to Eq. (12.11).

$$f_0 = \left[1 + (-0.054\xi^2 + 0.4\xi)\left(\frac{24}{f_c}\right)^{0.45}\right] f_c \tag{12.7}$$

$$f_0 = f_c\left(-1.254 + 2.254\sqrt{1 + 7.94\frac{f_r}{f_c}} - 2\frac{f_r}{f_c}\right) \tag{12.8}$$

$$f_r = \frac{2t}{d - 2t}\sigma_h \tag{12.9}$$

$$\sigma_l^2 + \sigma_h^2 - \sigma_l \sigma_h = f_y^2 \tag{12.10}$$

$$\xi = \frac{A_s f_y}{A_c f_c} \tag{12.11}$$

where f_0 is the ultimate compressive strength of the confined concrete; f_r is the lateral pressure acted on the core concrete; ξ is the confining coefficient; A_s, A_c and f_c are the area of steel tube, area of core concrete and axial compressive strength of concrete respectively.

It should be noticed that the ultimate displacement δ_n under the monotonic loading is the input parameter in the Park-Ang model. However, the ultimate displacement obtained in current study is derived from the cyclic loading. Therefore, as suggested by Fajfar (1992), at the failure state, the ultimate displacement of a structural member under cyclic loading is assumed to be equal to the ultimate displacement for the monotonic loading ($\delta_n = \delta_m$). The corresponding damage index of the present post-blast column can be derived from Park-Ang model as given in Fig. 12.21. For comparison, the seismic damage index for intact column is also given. It is clearly shown that the seismic damage index D of each specimen exceeded 1, which is higher than that of the original defined limit of the seismic damage index. This reason lies in

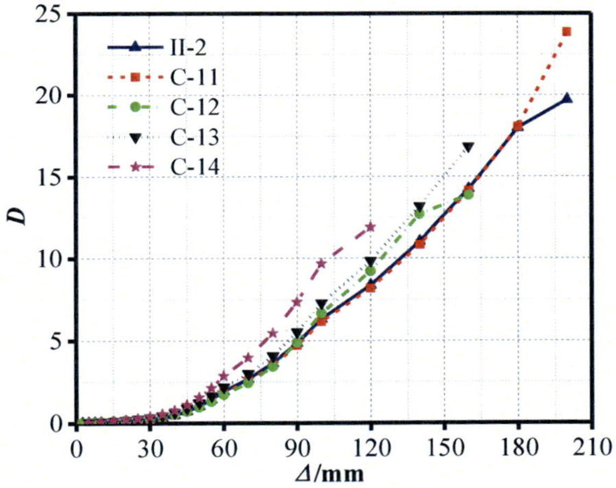

Fig. 12.21 Damage index derived from Park-Ang model, reprinted from Wang et al. (2020b), copyright 2020, with permission from Elsevier

that the Park-Ang model is proposed to evaluate the residual seismic behavior of RC members. Therefore, the Park-Ang model is not suitable for evaluating the residual seismic performance of the post-blast CFST column in current study.

Since that both the lateral load-carrying capacity, ductility and energy consumption are changing under the cyclic loading, it is more meaningful to employ the composite damage indicator which considering the above three seismic resistance indexes of the post-blast column. At present, the composite damage indicator is proposed as

$$D = 1-\sqrt[3]{d_1 d_2 d_3} \tag{12.12}$$

where $d_1 = P_{Residual}/P_{intact}$, $d_2 = \mu_{Residual}/\mu_{intact}$ and $d_3 = E_{Residual}/E_{intact}$ represent the damage index of the post-blast column corresponding to the load carrying capacity, ductility and energy consumption capacity of specimen, respectively. $P_{Residual}$ and P_{intact}, $\mu_{Residual}$ and μ_{intact}, as well as $E_{Residual}$ and E_{intact} are the maximum lateral load, ductility coefficient and cumulative energy consumption of the post-blast column and the intact column under cyclic loading, respectively. $D = 0$ and $D = 1$ are corresponds that the seismic performance of the structural member is not deteriorated and the structural member fully lost its residual seismic capacity.

Figure 12.22 shows the composite damage index of the present specimens under the seismic loading. It is illustrated that when the loading direction is perpendicular to the cratering face, the composite damage index generally increases with the increase of TNT charge weight, implying the reduction of the residual seismic capacity of the post-blast column. For example, for the C-14 specimen after 3 kg TNT contact detonation, the composite damage index was found to be 0.5, which meant that seismic capacity of the post-blast column decreases by 50%. Under such circumstance, the post-blast column is not safe if it is still kept using, and thus it is suggested to be strengthened. As for the influence of the cyclic loading direction, under 1 kg

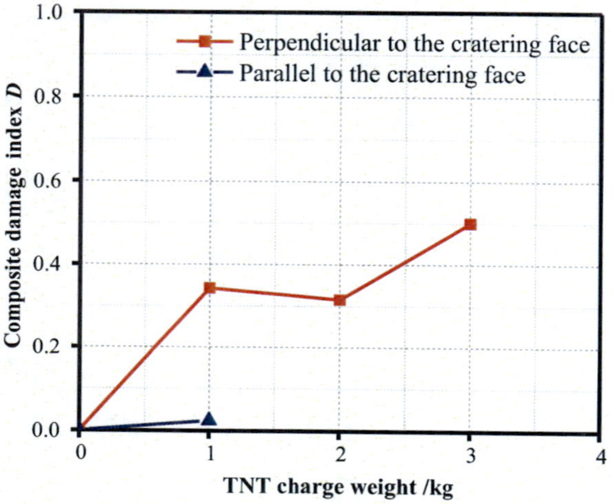

Fig. 12.22 Composite damage index versus TNT charge weight, reprinted from Wang et al. (2020b), copyright 2020, with permission from Elsevier

TNT contact detonation, the seismic performance of the C-12 column decreases by 34% compared with that of C-11 column only decreases by 2%, indicating that the seismic performance of the post-blast column is significantly influenced by the loading direction. However, it should be noted that during the earthquake, the seismic loading direction is random, and thus the worst scenario should be considered for the seismic design of structural members. Hence, it is suggested that the strengthening for the post-blast column should be made, e.g., welding steel reinforcement of the cratering region, fiber reinforced polymer retrofitting, to improve the residual seismic performance of the column after blast event.

12.6 Summary

In this chapter, the residual seismic resistance of UHPCC-FST specimen after contact explosion is studied experimentally. Six 1/4 scaled circular UHPCC-FST were fabricated, and four of which were tested under contact explosion with the TNT charge weights were 1–3 kg. Then, the low-frequency cyclic loading test was conducted to quantitatively examine their residual seismic resistance (RSR) in terms of the lateral load-carrying capacity, stiffness and energy consumption. The main conclusions can be drawn as follows:

(1) The field contact detonation test shows that the present UHPCC-FST specimen exhibits the localized damage with no overall response. With the increase of the charge weight, the dimension of the blast induced crater enlarges, resulting in more severe damage of specimen.
(2) Under the lateral cyclic loading, each specimen exhibits the bending failure. For the intact specimen, the bending failure is occurred near the bottom of column,

whereas the bending failure tends to develop at the position of the crater for the post-blast column. The hysteresis curve of each specimen exhibits the fusiform shape with no obvious pinching at early loading stage, and the specimens are failed due to a premature buckling of the steel tube at the final stage of the cyclic loading test.

(3) The loading direction in the lateral cyclic loading test plays an important role in the residual seismic performance of the post-blast specimen, such as load-carrying capacity, ductility, stiffness and energy consumption. When the loading direction is perpendicular to the cratering face, RSR of the post-blast column is continuously deteriorated with increasing the charge weight. While for the loading direction being parallel to the cratering face, the RSR of the post-blast specimen was almost the same as that of the intact specimen.

(4) The charge weight has significant effect on the RSR of the post-blast specimen. the corresponding load-carrying capacity, ductility, stiffness and energy consumption continuously decreases with the charge weight increasing.

(5) The composite damage index is proposed to quantitatively evaluate the residual seismic performance of the present UHPCC-FST column after the contact detonation, which could provide helpful reference for the evaluation and strengthening of the post-blast column after accidental or malicious explosion attacks.

(6) Numerical simulation has been becoming the reliable approach to study the dynamic responses of structural member under both blast and seismic loadings. However, for the present coupling loading effect as well as the dynamic constitutive model of UHPCC, the related work is very challenging.

References

CHUNG Y, PARK C, MEYER C. Residual seismic performance of reinforced concrete bridge piers after moderate earthquakes[J]. ACI Structural Journal, 2008, 105: 87–95.

ELREMAILY A, AZIZINAMINI A. Behavior and strength of circular concrete-filled tube columns[J]. Journal of Constructional Steel Research, 2002, 58(12): 1567–1591.

FARAHI M, HEIDARPOUR A, ZHAO X L, AL-MAHAIDI R. Compressive behaviour of concrete-filled double-skin sections consisting of corrugated plates[J]. Engineering Structures, 2016, 111: 467–477.

FAN L C, ZHUO W D. Seismic design of bridges[M]. Beijing: People's Communications Press, 2001. (in Chinese).

FAJFAR P. Equivalent ductility factors taking into account low-cycle fatigue[J]. Earthquake Engineering & Structural Dynamics, 1992, 21: 837–848.

FUJIKURA S, BRUNEAU M, LOPEZ-GARCIA D. Experimental investigation of multihazard resistant bridge piers having concrete-filled steel tube under blast loading[J]. Journal of Bridge Engineering, 2008, 13(6): 586–594.

FOUCHÉ P P, BRUNEAU M. Blast and seismic resistant concrete-filled double skin tubes and modified steel jacketed bridge columns[R]. 2015. Technical report MCEER-14–0004.

GB/T228.1–2010. Metallic materials-tensile testing-part 1: Method of test at room temperature[S]. National Standard of China, 2010.

GE H, USAMI T. Cyclic tests of concrete-filled steel box columns[J]. Journal of Structural Engineering, 1996, 122(10): 1169–1177.

Global terrorism database. National consortium for the study of terrorism and responses to terrorism (START)[EB/OL].(2016). https://www.start.umd.edu/gtd.

Global terrorism database. Information on more than 180000 terrorist attacks[EB/OL].(2018). https://www.start.umd.edu/gtd.

HAN L H. Concrete-filled steel tube structure-theory and practice[M]. Beijing: Science Press, 2004. (in Chinese).

HAN L H, HUI Q U, TAO Z, WANG Z F. Experimental behavior of thin-walled steel tube confined concrete column to RC beam joints under cyclic loading[J]. Steel Structures, 2009, 47(8): 847–857. (in Chinese).

JGJ/T101-2015. Code for seismic test of buildings[S]. Beijing: Ministry of Construction of the People's Republic of China, 2015. (in Chinese).

KYEI C, BRAIMAH A. Effects of transverse reinforcement spacing on the response of reinforced concrete columns subjected to blast loading[J]. Engineering Structures, 2017, 142: 148–164.

LI J, WU C Q, HAO H, SU Y. Investigation of ultra-high performance concrete under static and blast loads[J]. International Journal Protection Structures, 2015, 6(3): 403–418.

LI J, WU C Q, HAO H. Residual loading capacity of ultra-high performance concrete columns after blast loads[J]. International Journal Protection Structures, 2015, 6(4): 649–669.

LI J, WU C Q, HAO H. An experimental and numerical study of reinforced ultra-high performance concrete slabs under blast loads[J]. Materials & Design, 2015, 82: 64–76.

LI J, HAO H, WU C. Numerical study of precast segmental column under blast loads[J]. Engineering Structures, 2017, 134: 125–137.

LI J, WU C Q, HAO H, LIU Z X. Post-blast capacity of ultra-high performance concrete columns[J]. Engineering Structures, 2017, 134: 289–302.

LI M, ZONG Z, LIU L, LOU F. Experimental and numerical study on damage mechanism of CFDST bridge columns subjected to contact explosion[J]. Engineering Structures, 2018, 159: 265–276.

LI M, ZONG Z, HAO H, ZHANG X, LIN J, XIE G. Experimental and numerical study on the behavior of CFDST columns subjected to close-in blast loading[J]. Engineering Structures, 2019, 185: 203–220.

LIAO F Y, HAN L H, TAO Z. Behavior of composite joints with concrete encased CFST columns under cyclic loading: Experiments[J]. Engineering Structures, 2014, 59: 745–764.

MA G, LI H, HWANG H. Seismic behavior of low-corroded reinforced concrete short columns in an over 20-year building structure[J]. Soil Dynamics and Earthquake Engineering, 2018, 106: 90–100.

MANDER J B, PRIESTLEY M J N, PARK R. Theoretical stress-strain model for confined concrete[J]. Journal of Structural Engineering, 1988, 114(8): 1804–1826.

MOSHERF A, KHANMOHAMMADI M, TEHRANIZADEH. Assessment of the seismic capacity of main shock-damaged reinforced concrete columns[J]. Bulletin of Earthquake Engineering, 2017, 15: 291–311.

NASSIRNIA M, HEIDARPOUR A, ZHAO X L, MINKKINEN J. Innovative hollow corrugated columns comprising corrugated plates and ultra high-strength steel tubes[J]. Thin-Walled Structures, 2016, 101: 14–25.

PARK Y J, ANG H S. Mechanistric seismic damage model for reinforced concrete[J]. Journal of Structural Engineering, 1985, 111(4): 722–739.

RAJPUT A, SHARMA U. Corroded reinforced concrete columns under simulated seismic loading[J]. Engineering Structures, 2018, 171: 453–463.

REMENNIKOV A M, UY B. Explosive testing and modelling of square tubular steel columns for near-field detonations[J]. Journal of Constructional Steel Research, 2014, 101: 290–303.

SHI Y, HAO H, LI Z. Numerical derivation of pressure-impulse diagrams for prediction of RC column damage to blast loads[J]. International Journal of Impact Engineering, 2008, 35(11): 1213–1227.

WU K, LI B, TSAI K. Residual axial compression capacity of localized blast-damaged RC columns[J]. International Journal of Impact Engineering, 2011, 38(1): 29–40.

WANG Z G, WU H, FANG Q, WU J. Experimental study on the residual axial capacity of Ultra high performance cementitious composite filled steel tube (UHPCC-FST) column under contact explosion[J]. Thin-Walled Structures, 2020, 147: 106515.

WANG Z G, WU H, WU J, FANG Q. Experimental study on the residual seismic resistance of ultra high performance cementitious composite filled steel tube (UHPCC-FST) after contact explosion[J]. Thin-Walled Structures, 2020, 154: 106852.

WILLIAMSON E B, BAYRAK O, WILLIAMS G D, DAVIS C E. Blast-resistant highway bridges: design and detailing guidelines.NCHRP report 645[S]. Washington DC (USA): Transportation Research Board, 2010.

WILLIAMSON E B, BAYRAK O, DAVIS C, WILLIAMS G D. Performance of bridge columns subjected to blast loads. I: Experimental program[J]. Journal of Bridge Engineering, 2011, 16(6SI): 693–702.

WILLIAMSON E B, BAYRAK O, DAVIS C, WILLIAMS G D. Performance of bridge columns subjected to blast loads. II: Results and recommendations[J]. Journal of Bridge Engineering, 2011, 16(6SI): 703–710.

XU J, WU C, XIANG H, SU Y, LI Z, FANG Q, HAO H, LIU Z, ZHANG Y, LI J. Behavior of ultra-high performance fibre reinforced concrete columns subjected to blast loading[J]. Thin-Walled Structures, 2016, 118: 97–107.

ZHANG F, WU C, ZHAO X L, XIANG H, LI Z X, FANG Q, LIU Z, ZHANG Y, HEIDARPOUR A, PACKER J A. Experimental study of CFDST columns infilled with UHPC under close-range blast loading[J]. International Journal of Impact Engineering, 2016, 93: 184–195.

ZHANG F, WU C, LI Z, ZHAO X. Residual axial capacity of CFDST columns infilled with UHPFRC after close-range blast loading[J]. Thin-Walled Structures, 2015, 96: 314–327.

ZHANG F, WU C, ZHAO X, HEIDARPOUR A, LI Z. Experimental and numerical study of blast resistance of square CFDST columns with steel-fibre reinforced concrete[J]. Engineering Structures, 2017, 149: 50–63.

ZHANG F, WU C, ZHAO X, LI Z, HEIDARPOUR A, WANG H. Numerical modeling of concrete-filled double-skin steel square tubular columns under blast loading[J]. Journal of Performance of Constructed Facilities, 2015: B4015002.

ZHANG F, WU C, LI X, LI Z. Numerical derivation of pressure impulse diagrams for square UHPCFDST columns[J]. Thin-Walled Structures, 2017, 115: 188–195.

ZHOU L, SU L, JIANG L. Seismic performance and damage of concrete filled square steel tubular columns under inclined horizontal load[J]. Journal of Civil Engineering and Management, 2019, 36(01): 48–54. (in Chinese).

ZHANG S M, GUO L H, YE Z L, WANG Y Y. Behavior of steel tube and confined high strength concrete for concrete-filled RHS tubes[J]. Advances in Structural Engineering, 2005, 8(2): 101–116.

Chapter 13
Experimental and Numerical Studies on Dynamic Behavior of Reinforced UHPCC Panel Under Medium-Range Explosions

13.1 Introduction

Over the past several decades, the explosive effects that stem from terrorist attacks and industrial accidents have attracted increasingly more public concerns. The occurrence of explosions will threaten the safety of the surrounding building structures and lead to huge economic losses and extensive casualties. Therefore, it is of great significance to improve the blast-resistant performance of civil structures, especially for the strategically vital infrastructures, landmark buildings, and military or civil shelters, etc. Concrete is a widely applied construction material and the traditional normal strength concrete (NSC) structures have limited blast-resistance attributed to its low tensile capacity and energy absorption capacity. Comparably, ultra-high performance cementitious composites (UHPCC) is a relatively new type of cementitious material mixing with very low water-to-binder ratio, high amount of high-range water reducer (HRWR), fine aggregates and high-strength steel or organic fibers. Due to its prominent mechanical properties, i.e., high compressive and tensile strengths, high ductility as well as the high fracture energy, UHPCC is regarded as the most prominent construction materials for both civil and military structures to resist the intensive loadings, e.g. high-speed projectile penetration (Habel and Gauvreau 2008; Wu et al. 2015; Sovják et al. 2015), low-velocity drop hammer impact (Fan et al. 2019; Farnam et al. 2010; Wang et al. 2019a, b; Wei et al. 2019; Yoo et al. 2015) and blast (Barnett et al. 2010; Ellis et al. 2014; Foglar et al. 2017; Kuznetsov et al. 2006; Li et al. 2015a, b, 2016; Lin 2018; Mao et al. 2014, 2015; Ngo et al. 2007; Peng et al. 2019; Schleyer et al. 2011; Wu et al. 2009; Yi et al. 2012; Zhou et al. 2008). The blast-resistant performance of reinforced UHPCC panels is mainly concerned in this chapter.

From the detailed review of the existing related work in Sect. 13.2, generally, the existing related work still has the following limitations: (i) Based on the plenty of the test data, Orton et al. (2014) defined the scaled distance $Z \leq 0.4$ m/kg$^{1/3}$, $0.4 < Z \leq 1.0$ m/kg$^{1/3}$ and $Z > 1$ m/kg$^{1/3}$ ($Z = R/W^{1/3}$ is the scaled distance, R is the standoff distance from the explosive center to the surface of structure, W is the charge

Q. Fang et al., *UHPCC Under Impact and Blast*,
https://doi.org/10.1007/978-981-33-6842-2_13

weight of the equivalent TNT explosive) as the close-, medium- and far-range explosion scenarios, respectively. Most of the existing work is either concerning the local damage and failure of UHPCC panels under contact/close-in explosions or focusing on the dynamic performance of UHPCC panels under far-range explosions, the related studies for the medium-range explosions are insufficient; (ii) For the design of test setup, the existing test UHPCC panels were usually constrained by a ground frame which is completely exposed to air and could not eliminate the effect of the diffracted blast waves on the rear face of test panel. The boundary conditions of the test panel provided by the steel angles and wood stripes are not clear. Besides, attributed to the damage of the measuring sensors under intensive blast loadings, the systematic dynamic responses of panels, e.g. deflection- and acceleration-time histories, as well as the instantaneous incident and reflected blast overpressures are limited; (iii) K&C model (Malvar et al. 1997) (*MAT_CONCRETE_DAMAGE_REL3, *MAT_72R3) embedded in LS-DYNA (LSTC 2007) is validated for capturing the complex concrete behaviors and predicting the structural dynamic responses under blast loadings with good accuracy (Jayasooriya et al. 2014; Wu et al. 2011; Zhang and Jin 2012). The most significant user improvement provided by *MAT_72R3 is the model parameters auto-generated algorithm based solely on the unconfined compression strength of concrete. However, this parameter generation algorithm is proposed for NSC with the uniaxial compressive strength of 45 MPa. Therefore, the corresponding K&C constitutive model parameters for UHPCC should be comprehensively calibrated, while the related work is very lacked.

At present, in Sect. 13.3, three reinforced UHPCC panels and three NSC control panels were fabricated and tested on a specially designed box-like blast loading apparatus to avoid the effect of the diffracted blast waves and provide the simply-supported boundary constraints, and the scaled distances of 4 kg TNT charges were ranged from 0.5 to 1.0 m/kg$^{1/3}$. Various sensors were arranged to record the incident and reflected overpressure-time histories, deflection-and acceleration-time histories. Besides, by examining the damage of the post-blast panels, the blast resistance of NSC and UHPCC panels are evaluated quantitatively. Then, in Sect. 13.4, the refined FE models of the present test was established, and the parameters of K&C model for UHPCC are comprehensively calibrated. Finally, in Sect. 13.5, the applicability of the CONWEP method to predict the blast overpressures are firstly discussed, and then the simulated dynamic responses and damage of panels are compared systematically with the test results. The present conclusions can provide references in the design and analyses of UHPCC protective structures under blast loadings.

13.2 Review of the Existing Work

As for the existing experimental studies, Li et al. (2016) carried out the contact explosion test on seven reinforced NSC and UHPC slabs with the scaled distance Z of 0.043 m/kg$^{1/3}$ and 0.06 m/kg$^{1/3}$, respectively. The slabs with the dimensions of 2 m length, 0.8 m width, and 0.1–0.15 m thickness were tested on a steel frame provided one-way support boundary condition. By assessing the spalling and cratering damage quantitatively, it was concluded that UHPC panels displayed significantly improved blast-resistant performance than NSC panels, and increasing the slab thickness and reinforcement ratio is beneficial to enhancing the blast-resistant behavior of UHPC slabs. Foglar et al. (2017) conducted the field blast test on full-scale reinforced UHPC bridge decks with the dimension of 6 m × 1.5 m × 0.3 m, in which the test slabs were placed on fixed timber posts and subjected to blast loadings with the scaled distance of 0.154 m/kg$^{1/3}$. The dimensions of the front spalling and rear scabbing craters were quantified, and the permanent deflections of the post-blast slabs were derived. It was derived that increasing the fiber content and compressive strength could improve the blast-resistant performance of decks. Kuznetsov et al. (2006) presented an experimental study of one-way spanning reinforced NSC and UHPC panels tested under the detonations of 0.5 kg Composition B and Powergel charges at a distance of 0.125 m, in which the specimens with the dimension of 1.3 m × 1.0 m × 0.1 m were placed on a holding frame and clamped using steel angles and wood strips along the short edge. The scabbing damage on the rear face of the panels was examined, and it was derived that the reinforced panels suffer more severe damage when the pressure and impulse generated from the explosive are more intensive, and UHPC panels performed better than NSC panels. Mao et al. (2015) focused on the influence of fiber content and fiber type on the blast-resistant performance of small-scaled UHPC panels. Nineteen two-way fixed UHPC slabs with dimensions of 0.66 m × 0.66 m × 0.025 m were tested under blast loadings induced from 0.20 ~ 1.10 kg PE4 charge at the standoff distance of 0.5 m. By examining the peak 1/4 span deflections and slab cracks, it was indicated that increasing fiber content is effective in improving the blast-resistance of UHPC panels, while the effect of the adopted two typical steel fibers (straight and hooked) seemed to have slight differences. Wu et al. (2009) conducted the explosive test on eight slabs including plain and reinforced UHPC slabs, reinforced NSC slabs, and reinforced NSC slabs externally retrofitted with fiber reinforced polymer (FRP) plates ($0.37 \leq Z \leq 3$ m/kg$^{1/3}$), and the test boundary constraints of slabs were identical with that reported by Kuznetsov et al. (2006). By recording and examining the reflected overpressure- and mid-span deflection-time histories, it was proved that the blast-resistance of reinforced UHPC slab was superior to all the other tested slabs. Li et al. (2015a) conducted blast test on five 2 m × 1 m × 0.1 m one-way spanning panels including one control NSC slab and four UHPC panels reinforced with different reinforcement ratios (0.8% and 1.4%), yielding strength of reinforcing steel rebars (300 ~ 1750 MPa), and scaled distances (0.41 ~ 3.05 m/kg$^{1/3}$), respectively. The flexural failure in UHPC slab and brittle shear failure in NSC slab are observed, which further verifies the effectiveness of reinforced UHPC slabs against blast loadings. Ellis et al. (2014) performed the planar blast loading test on four UHPC panels by using the blast load simulator, in which the specimens with dimensions of 1.626 m × 0.864 m × 0.051 m were tested under approximate simply-supported boundary condition. By recording the blast impulse and examining the fracture patterns of slabs, the upper and lower bounds to the blast resistance of UHPC panels were assessed. Yi et al. (2012) carried out the explosion test on two control NSC slabs and four UHPC panels. The slab specimens with the dimensions of 1 m × 1 m × 0.15 m were clamped on all four sides and subjected to the detonations of 15.88 kg TNT or ANFO charges at the standoff distance of 1.5 m. The incident and reflected overpressure-time histories, the central deflection- and acceleration-time histories, as well as the strains of both concrete and rebars were recorded and examined. The outstanding blast-resistant capacities of UHPC was confirmed. Barnett et al. (2010) and Schleyer et al. (2011) studied the blast-resistance of one-way spanning plain and reinforced UHPC panels. The test specimens were fabricated with the dimensions of 3.5 m × 1.3 m × 0.1 m and suffered from the explosion loadings generated from 100 kg TNT equivalent explosive charge with the scaled distances ranging from 1.508 m/kg$^{1/3}$ to 2.585 m/kg$^{1/3}$, respectively. By comparing to the peak and residual deflections, it was manifested that the higher fiber content and the application of rebar improved the performance of UHPC slabs under explosions. Ngo et al. (2007) carried out the blast test on UHPC panels retrofitted with the prestressed tendons, in which the panels with the dimensions of 2 m × 1 m × (0.075 ~ 0.1) m were installed into a box-like concrete frame and tested against the blast loadings from 6 t TNT equivalent explosive charge at a scaled distance of 2.201 m/kg$^{1/3}$ and 1.651 m/kg$^{1/3}$, respectively. By collecting the blast pressures and deflections of panels, it was indicated that the reinforced UHPC panels with larger thickness performed better under blast loadings.

Besides, the numerical simulations have been becoming the efficient, cost-effective and safe approach to study the blast loading effects on concrete structures, which can be categorized into coupled (Li et al. 2015b; Peng et al. 2019; Schleyer et al. 2011; Zhou et al. 2008) and uncoupled methods (Foglar et al. 2017; Li et al. 2015a; Lin 2018; Mao et al. 2014, 2015). For the coupled method, the dynamic interactions between blast waves and structures are considered, while the blast loadings are determined by the empirical formulas and then applied on structures separately in the uncoupled method (Rasouli and Toopchi–Nezhad 2018). Comparably, the coupled method is generally more time-consuming. Li et al. (2015b) numerically studied the damage and failure of plain UHPC panels under contact explosion, in which the smoothed particles hydrodynamics (SPH) algorithm was adopted to reproduce the interactions between the explosion wave and slab. The Elastic-Plastic Hydrodynamics model (Hallquist 2006) was adopted for modeling UHPC material, and the feasibility and accuracy of the numerical model were validated. Similarly, Peng et al. (2019) adopted the SPH algorithm and Brittle Damage concrete model (Govindjee et al. 1995) embedded in LS-DYNA (LSTC 2007) to numerically study the influences of slab thickness and TNT charge weight on the performance of UHPC slab under contact explosions. Schleyer et al. (2011) numerically studied the dynamic responses and damage of UHPC panels under free air explosions with the program AUTODYN (2005), the interactions between the blast wave and panel was fully coupled by using Multi-Material Arbitrary Lagrangian-Eulerian (MMALE) methodology. The Riedel-Hiermaier-Thoma model (Riedel et al. 1999) for high strength concrete was calibrated and validated for UHPC. The dynamic responses of reinforced UHPC panels subjected to blast loadings was predicted by Zhou et al. (2008) through the program AUTODYN (2005), where the MMALE method for modeling blast loadings and modified Drucker-Prager model (Drucker and Prager 1952) for UHPC are verified by comparisons with the test reported by Kuznetsov et al. (2006). CONWEP (Hyde 1993) is an implementation of the empirical blast models of Kingery and Bulmash (Kingery and Bulmash, 1984; TM5-855-1 1986), which was originally incorporated into in DYNA by Randers-Pehrson and Bannister (Randers-Pehrson and Bannister 1997). In LS-DYNA (LSTC, 2007), CONWEP method is adopted to predict the blast induced overpressure through the keywords *Load_Blast or *Load_Blast_Enhanced. Foglar et al. (2017) numerically clarified the contribution of basalt mesh incorporated into UHPC panel under explosions, in which the uncoupled *Load_Blast_Enhanced function and Continuous Surface Cap model (Schwer and Murray 1994) were employed to model blast loadings and UHPC, respectively. Li et al. (2015a) conducted the numerical simulations of reinforced UHPC panels under free field air explosions, in which the blast loading was predicted by using the built-in *Load_Blast function, and the Elastic-Plastic Hydrodynamics model (Hallquist, 2006) without considering the strain rate effect of UHPC was validated. The *Load_Blast/*Load_Blast_Enhanced function and K&C model were employed by Mao et al. (2014, 2015) and Lin (2018) to reproduce the free air blast test on plain and reinforced UHPC panels, respectively. The strain rate effects of UHPC in tension and compression were incorporated, and the parameters controlling the softening behavior of UHPC was calibrated to fit a pre-designed stress-strain relationship (Mao et al. 2014, 2015) and the experimental uniaxial stress-strain relationships (Lin 2018), respectively. It was proved that K&C model can well describe the dynamic behavior of UHPC under blast loadings after careful calibrations of its constitutive model parameters.

13.3 Field Blast Test

13.3.1 Specimen

A total of six test specimens are fabricated in the current test, including three reinforced NSC panels and three reinforced UHPCC panels, which are referred as NSC panels and UHPCC panels hereinafter, respectively. The geometrical dimensions and the reinforcement of test panels are schematically shown in Fig. 13.1, in which two layers of 10 mm diameter mesh reinforcements with 100 mm spacing in major bending plane and 200 mm spacing in the minor bending plane were placed, and the yield strength of rebars is 480 MPa. UHPCC panels are prefabricated by mixing the Chinese standard Graded 52.5 P.II type Portland cement, silica fume, ultra-fine mineral admixture which consists of fly ash and ultra-fine slag, the natural river sand with a maximal diameter of 2.5 mm, polycarboxylic type high-range-water-reducer (HRWR) and 2% (volume fraction) straight brass-coated steel fiber. The mixture proportions are shown in Table 13.1. Based on Chinese standard (GB/T 50081-2002 2003), the compressive strength and splitting tensile strength of NSC specimens are 30.4 MPa and 3.36 MPa, respectively. Correspondingly, the compressive strength

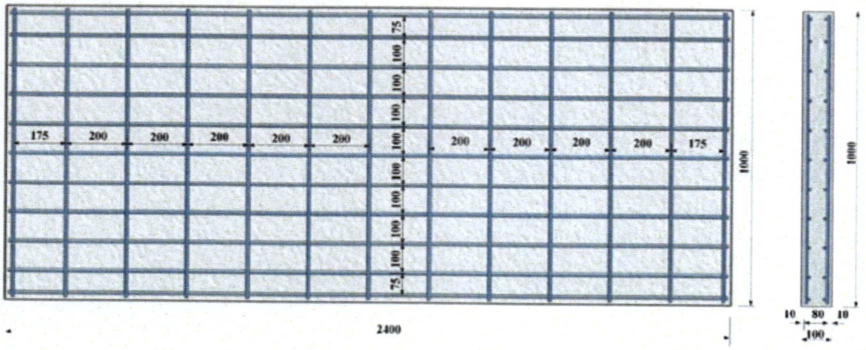

Fig. 13.1 Geometry and reinforcement of test panels (*unit* mm)

Table 13.1 Mixture proportions of UHPCC (kg/m^3)

Cement	Silica fume	Ultra-fine mineral admixture	Sand	Water	HRWR	Steel fiber
700	140	110	1200	152	22.8	145

and the direct tensile strength of UHPCC specimens are 125.27 MPa and 8.33 MPa, respectively.

13.3.2 Test Setup

Figure 13.2 shows the present field blast test setup. A box-like blast loading apparatus was designed, which consists of steel plates, angle steel, round rolls, I-steel with or without stiffeners, and channel steel, respectively. As shown in Figs. 13.2a, b, five steel plates were welded together to form the four walls and the bottom of the testing box. One of the four walls has an entrance and the bottom plate was anchored with the concrete foundation by bolts. The displacement and acceleration sensors were installed inside the testing box that was reinforced by brackets composed of I-steel with and without stiffeners. Angle steel and round roll were welded together to support the panels. Two reaction frames were set by bolting I beams and channel steel together to prevent the uplifting of panels during the test. In each shot, the panel specimen was arranged on the designed testing box buried underground with the top surface along to the ground level, so as to avoid the diffracted blast waves arriving at the bottom face of panel. Besides, to simplify the boundary conditions, as shown in Fig. 13.2b, the specimen was supported by four round rolls with a clear span of 2 m, i.e. the one-way simply-supported boundary conditions. TNT charges were suspended from a bamboo frame above the panels, the suspended rope was used to adjust the scaled distance of the explosive.

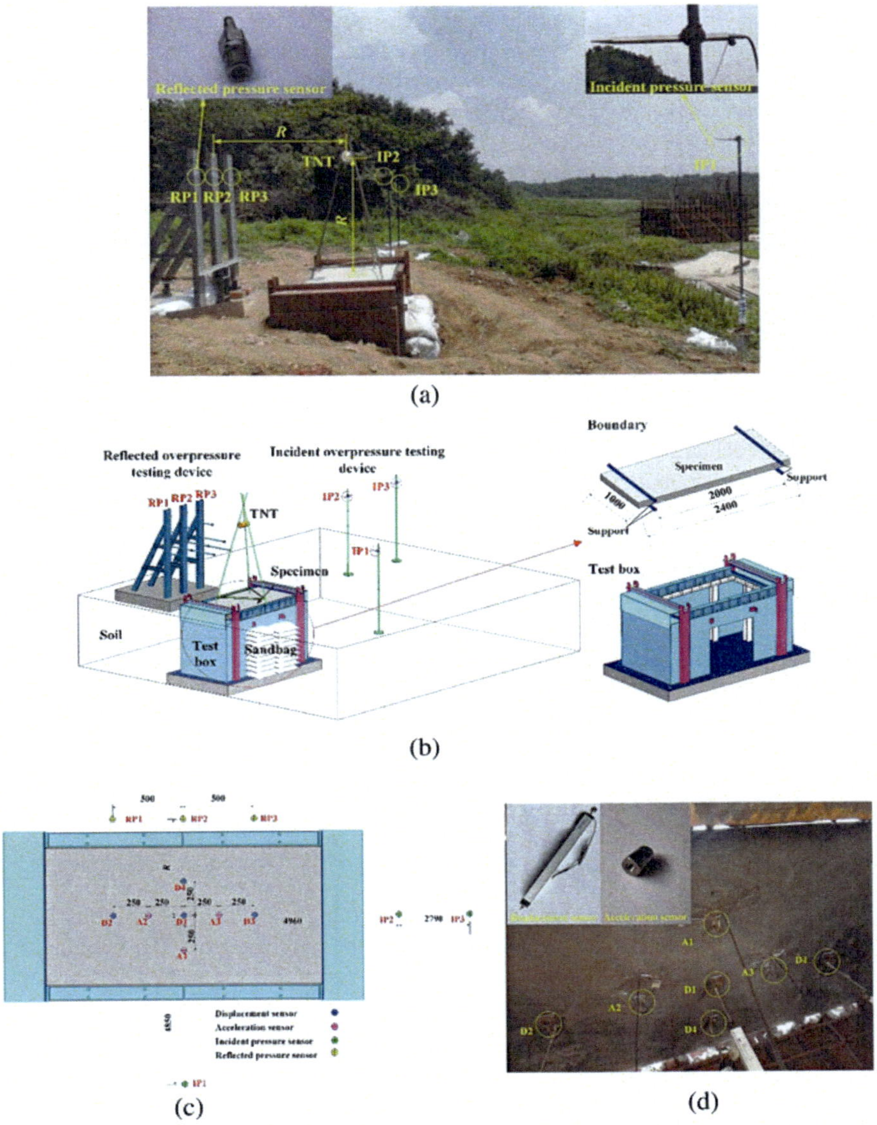

Fig. 13.2 Field blast test setup **a** photograph **b** test layout **c** plane-view of sensor arrangement **d** displacement and acceleration sensors (*unit* mm)

Systematic test data are designed to be measured, and the plane view of sensors arrangement is schematically shown in Fig. 13.2c. To obtain the blast overpressure-time histories, three incident pressure sensors (IP1, IP2, and IP3) and three reflected pressure sensors (RP1, RP2, and RP3) were arranged with the identical horizontal plane with the TNT charge center. The horizontal distances between the incident pressure sensors and panels were fixed in each shot, in which IP1, IP2, and IP3 were located at a horizontal distance of 4.85 m, 4.96 m, and 7.75 m from the panel center, respectively. Thus the identical incident pressure-time histories can be recorded in each shot ideally. As for the reflected pressure sensors, to avoid the breakdown of test devices during the blast, the reflected pressure sensors were installed on the frame aside the panel instead of on the surface of the panel. Three reflected overpressure sensors were placed at the same height and the relative position is unchanged (with a spacing of 500 mm) in each shot. However, the relative positions from reflected pressure sensors to the panel center varied with the different scaled distances. That is to say, in each shot, both the vertical and horizontal distances between reflected pressure sensor RP2 and panel center are equal to the standoff distances R, as shown in Fig. 13.2a, c, which guarantees that the distances from sensor PR2 to the charge center were consistent with the distance from the front face center of panels to the charge center. Therefore, the overpressure obtained from PR2 can be considered as the reflected overpressure of the front face center of panels. Accordingly, the reflected overpressures of RP1 and RP3 can be considered as the reflected overpressures of points D2 and D3 on the front face of panels, respectively. Besides, the displacement sensors (D1-D4) and acceleration sensors (A1-A3) were arranged at the bottom of the panel inside the testing box, as shown in Fig. 13.2d with the locations illustrated in Fig. 13.2c. During each shot, the locations of pressure sensors were firstly adjusted according to the scaled distances, then the inner displacement and acceleration sensors were installed. After the signals were checked, the sandbags were stacked at the entrance of the testing box to make sure that blast waves would not arrive at the bottom face of panels.

Total six shots were performed and the test program is summarized in Table 13.2, in which the panel specimens were tested under the explosion of 4 kg cylindrical TNT explosives with a diameter of 60 mm and a height of 224 mm. The standoff distances were ranging from 0.79 ~ 1.59 m, and the corresponding scaled distances are

Table 13.2 Test program

Shot No.	Specimen No.	TNT charge weight W (kg)	Standoff distance R (m)	Scaled distance Z (m/kg$^{1/3}$)
1	NSC-1	4	1.59	1.0
2	NSC-2	4	1.27	0.8
3	NSC-3	4	0.95	0.6
4	UHPCC-1	4	1.27	0.8
5	UHPCC-2	4	0.95	0.6
6	UHPCC-3	4	0.79	0.5

Fig. 13.3 Charge placement (*unit* mm)

0.5 ~ 1.0 m/kg$^{1/3}$. As shown in Fig. 13.3, all the cylindrical charges were positioned horizontally with their symmetry axis parallel to the concrete panel and perpendicular to its shorter axis of symmetry, which was initiated at one end by using an electrical detonator.

13.3.3 Test Results and Discussions

In this section, the explosion-induced instantaneous incident and reflected overpressures are assessed firstly, and the influences of scaled distance, charge shape and point of detonation are discussed. Then, the deflection- and acceleration-time histories as well as post-blast damage of two typical panels with different scaled distances are analyzed and compared.

13.3.3.1 Overpressure-Time History

Figure 13.4 shows the recorded incident overpressure curves, in which the data from IP1, IP2, and IP3 are denoted as P_{i1}, P_{i2}, and P_{i3}, respectively. The reliability of incident overpressure curves was proved by the good repeatability between the curves observed from different shots. The averaged peak values, time durations and impulses of the incident and reflected blast wave overpressures are listed in Table 13.3. It can be found that, the average peak overpressures of P_{i1}, $P_{i2,}$ and P_{i3} are 0.071 MPa, 0.058 MPa, and 0.034 MPa, and the corresponding average impulses are 0.090 MPa•ms, 0.088 MPa•ms, and 0.063 MPa•ms, respectively. The peak overpressure and overpressure-time histories resulting from the detonation of a cylindrical charge are dependent on the aspect ratio (length/diameter) of the cylinder, which is 3.73 in the present test from Fig. 13.3. The peak pressure of P_{i1} is significantly larger than that of P_{i2}, although they were measured with close scaled distances, i.e., 3.055

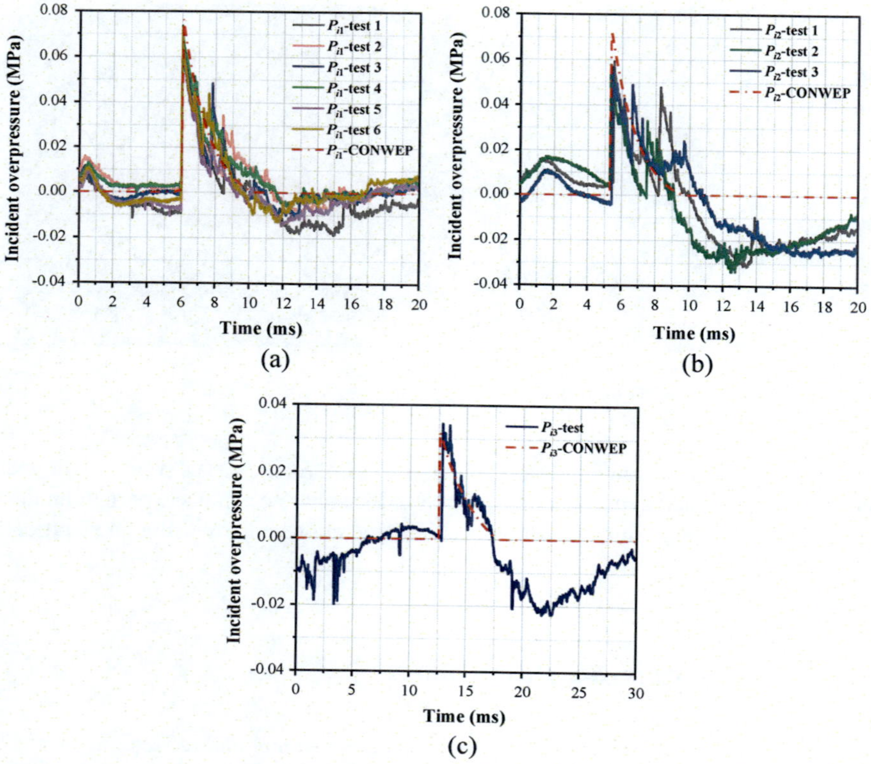

Fig. 13.4 Incident overpressure-time histories **a** P_{i1} **b** P_{i2} **c** P_{i3}

and 3.125 m/kg$^{1/3}$, which indicates that more energy was released in the radial direction of the cylindrical charge. It is consistent with the existing conclusions that more energy is directed in the radial direction of explosive with larger aspect ratios, while more energy is directed in the axial direction for the explosive with smaller aspect ratios (Wu et al. 2010; Sherkar et al. 2016).

Additionally, the representative reflected overpressure-time histories with different scaled distances recorded by pressure sensors are shown in Fig. 13.5, and the data from pressure sensors RP1, RP2, and RP3 are denoted as P_{r1}, P_{r2}, and P_{r3}, respectively. As listed in Table 13.3, the peak reflected overpressures, time durations and impulses are the average values of the test data from the reflected overpressure histories with the same scaled distance. It indicates that, the scaled distance has a significant influence on the reflected overpressure. In details, both the time durations and the arriving time of the blast wave to reach each sensor decrease with decreasing the scaled distances since the corresponding distances are reduced. Generally, the peak reflected overpressures and impulses increase with decreasing the scaled distances.

Table 13.3 Incident and reflected overpressures of blast wave

Z (m/kg$^{1/3}$)	Overpressure	Peak value (MPa)			Time duration (ms)			Impulse (MPa ms)		
		Test	CONWEP	Error (%)	Test	CONWEP	Error (%)	Test	CONWEP	Error (%)
0.5 ~ 1.0	P_{i1}	0.071	0.076	7.0	4.38	4.30	−1.80	0.090	0.099	10
	P_{i2}	0.058	0.073	25.8	4.44	4.36	−1.80	0.088	0.097	10.2
	P_{i3}	0.034	0.032	−5.8	4.58	5.25	14.6	0.063	0.065	3.2
1.0	P_{r1}	4.55	2.75	39.6	1.12	1.30	13.8	0.568	0.686	20.8
	P_{r2}	3.45	3.71	7.5	1.59	1.60	0.6	0.632	0.863	36.6
	P_{r3}	**0.88**	2.75	212.5	**0.68**	1.30	91.2	**0.210**	0.686	226.7
0.8	P_{r1}	5.90	4.50	−23.7	**0.48**	1.40	191.7	**0.502**	0.897	78.7
	P_{r2}	8.25	6.82	−17.3	1.11	1.30	17.1	1.226	1.255	2.4
	P_{r3}	4.12	4.50	9.2	1.12	1.40	25	**0.355**	0.897	152.7
0.6	P_{r1}	7.64	5.27	−31.0	0.57	1.00	75.4	0.526	0.848	61.2
	P_{r2}	17.85	9.59	−46.3	0.47	0.80	70.2	1.416	1.413	0.2
	P_{r3}	**2.44**	5.27	116.0	**0.55**	1.00	81.8	**0.236**	0.848	259.3
0.5	P_{r1}	6.74	6.79	0.7	**0.26**	0.70	169.2	**0.447**	0.986	120.6
	P_{r2}	20.49	14.70	−28.3	0.47	0.50	6.4	2.046	1.975	−3.5
	P_{r3}	**3.39**	6.79	100.3	**0.55**	0.70	27.3	**0.300**	0.986	228.7

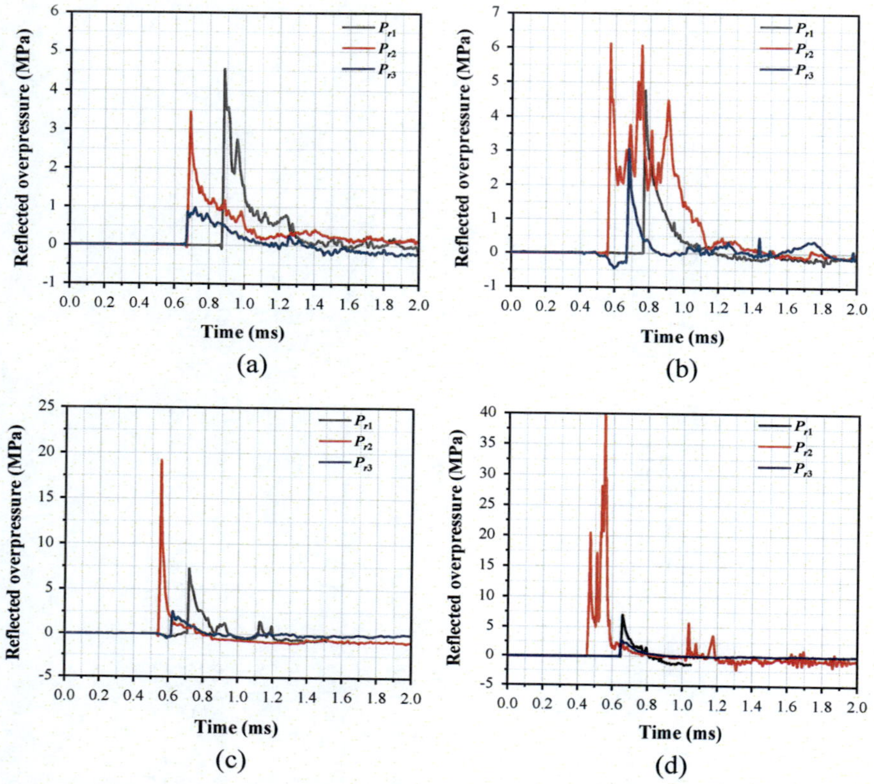

Fig. 13.5 Reflected overpressure-time histories **a** $Z = 1.0$ m/kg$^{1/3}$ **b** $Z = 0.8$ m/kg$^{1/3}$ **c** $Z = 0.6$ m/kg$^{1/3}$ **d** $Z = 0.5$ m/kg$^{1/3}$

Besides, the shape and point of detonation of charge affect the explosion-induced reflected overpressure obviously. In details, during each shot, the blast wave generally arrives at senor RP2 first, then RP3 and finally RP1, and the magnitude of the reflected overpressure recorded from different sensors is sorted as follows: $P_{r2} > P_{r1} > P_{r3}$. The distance between the senor RP2 and the detonation point of charge is the shortest, thus it takes the shortest time duration for the shock wave to reach the sensor RP2 and makes the peak value of P_{r2} the largest among the three reflected overpressures. The sensors RP1 and RP3 are symmetrically located on the left and right sides of the explosive center, and the detonation from the left end of explosive makes the shock wave travel faster along the right side than the left side. Therefore, the blast wave arrives at sensor RP3 earlier than at sensor RP1, and thus it results in $P_{r1} > P_{r3}$ since larger pressure caused by more detonation products as time went on. It demonstrates that the overpressures were significantly different at each end of a charge if the charge was detonated at one end only, which is consistent with the conclusion of Tham (2009).

13.3.3.2 Deflection-Time History

The deflection-time histories of each specimen are shown in Fig. 13.6, and the maximum and residual deflections of test specimens are listed in Table 13.4, in which the measuring data from sensors D1-D4 are denoted as D_1-D_4. Some displacement sensors located at the mid-span of panels were dropped attributed to the intensive vibrations of specimens under blast loadings, failing to obtain the maximum and/or residual mid-span deflections. However, the residual mid-span deflections can be measured manually after the explosions and denoted as 'Mid-span' in Table 13.4.

By comparisons of the deflection-time histories of present three NSC and three UHPCC panels, it can be drawn that both the maximum and residual deflections increase as the scaled distances of explosives decrease, indicating that the panels suffer more severe flexural deformations when the scaled distances are reduced. It is noted that UHPCC-2 was tested under a larger scaled distance than UHPCC-3 but suffered a larger deflection response, the possible reason may lie in the fabrication error of UHPCC-2 panel, e.g., the non-homogenous distribution of the fibers.

Compared with NSC specimens tested under the same scaled distances, UHPCC panels exhibit higher blast resistance under explosions. For instance, the maximum mid-span deflections of NSC-2 and UHPCC-1 are 43.48 mm and 27.86 mm, while the corresponding residual deflections (average value of D_1 and D_4) are 15.56 mm and 5.92 mm, respectively. The maximum and residual center deflection of UHPCC-1 are 35.9% and 62.0% smaller than those of NSC-2, respectively. Also, it should be noted that although UHPCC-2 exhibits relatively weak blast resistance among the three UHPCC specimens, it still performs better than the control specimen NSC-3. Furthermore, by comparisons of UHPCC specimens tested at smaller scaled distances and NSC specimens tested at larger scaled distances, it also can be found that the residual center deflection of UHPCC-1 and UHPCC-3 is 46.2% (5.08 mm) and 36.8% (16.32 mm) smaller than that of NSC-1 and NSC-3, respectively. It indicates that even subjected to more intensive blast loadings, UHPCC panels still show less damage than NSC panels. Based on the above quantitative comparisons of the present test data, we can draw the conclusion that UHPCC is a more outstanding concrete material than NSC to resist the blast loadings, and the superiority of UHPCC becomes more obvious as the scaled distance decreases.

13.3.3.3 Acceleration-Time History

For the NSC specimens, only one or two acceleration sensors were set to record data, while all three acceleration sensors were arranged for the UHPCC specimens. For UHPCC-3 specimen under the most intensive blast loadings among all tested specimens attributed to the smallest scaled distance, the measuring data is not reliable since that the recorded signals far exceed the measuring range of sensors. The experimentally recorded acceleration-time histories without filtering are shown in Fig. 13.7, and the recording data from sensors A1-A3 are denoted as A_1-A_3, respectively. It indicates that the test panels oscillate violently under the explosive induced

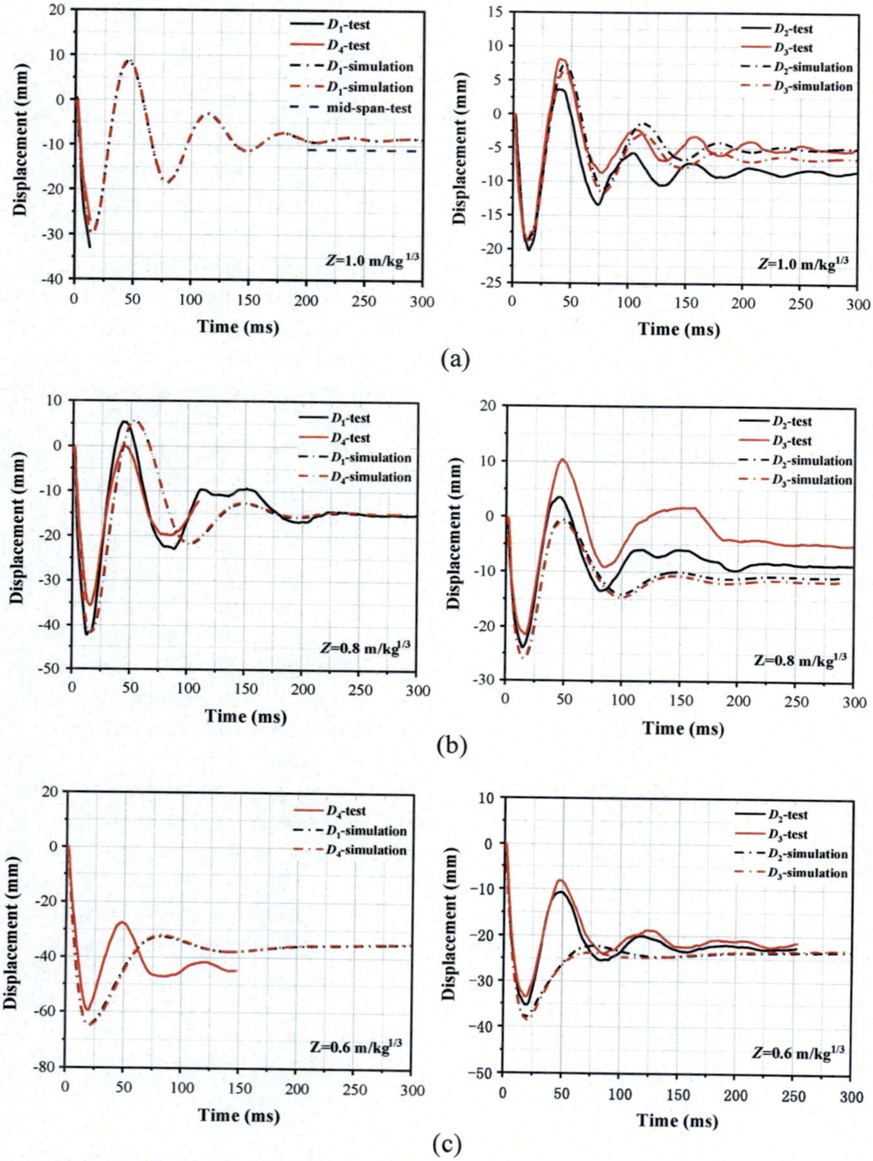

Fig. 13.6 Deflection-time histories of specimens **a** NSC-1 **b** NSC-2 **c** NSC-3 **d** UHPCC-1 **e** UHPCC-2 **f** UHPCC-3

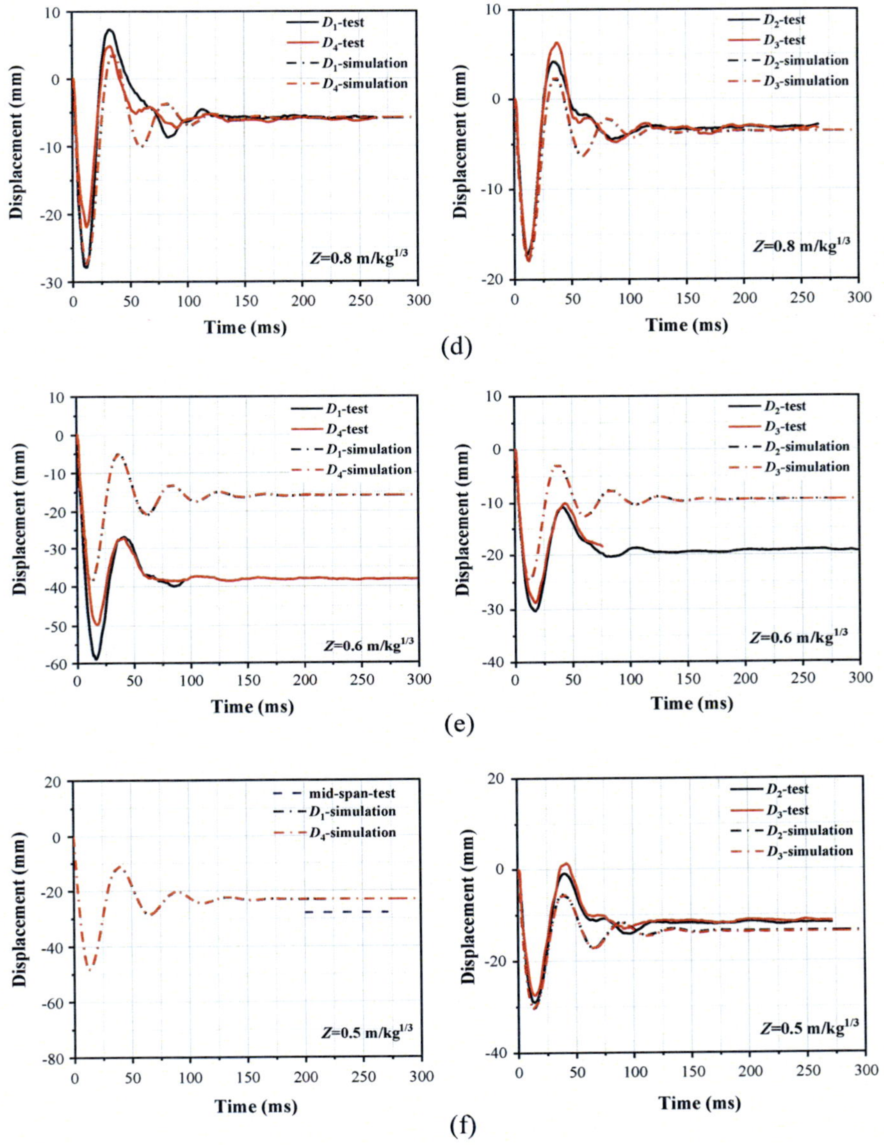

Fig. 13.6 (continued)

Table 13.4 Maximum and residual deflections of test specimens

Specimen	Z ($m/kg^{1/3}$)	Maximum deflection (mm)				Residual deflection (mm)				
		D_1	D_2	D_3	D_4	D_1	D_2	D_3	D_4	Mid-span
NSC-1	1.0	32.96	20.16	18.80	26.66	/	8.60	5.38	/	11
NSC-2	0.8	43.48	23.95	21.41	33.59	15.51	8.83	4.94	15.79	/
NSC-3	0.6	/	35.34	33.53	59.50	/	22.44	21.37	44.32	/
UHPCC-1	0.8	27.86	17.17	17.94	21.90	5.94	2.91	3.25	5.89	/
UHPCC-2	0.6	58.94	30.25	28.71	49.63	37.15	19.28	18.36	38.33	/
UHPCC-3	0.5	/	28.97	27.4	/	/	11.69	11.48	/	28

intensive loadings, the maximum acceleration generally becomes larger when the blast loadings are enhanced gradually, and the maximum accelerations even exceed 10,000 g in some scenarios. For example, as shown in Fig. 13.7b, c, the maximum acceleration of NSC-2 is approximately 6000 g while that of NSC-3, i.e., panel subjected to more intensive blast loading, is more than 10,000 g. When subjected to the explosions with the same scaled distances, the vibration of UHPCC panels is more obvious than that of NSC specimens due to the higher toughness and energy absorbing capacity of UHPCC.

Besides, the FFT (Fast Fourier Transform) spectrums of partial specimens are shown in Fig. 13.8. From Fig. 13.8a, c, it is concluded that the specimens with smaller scaled distances oscillate with larger amplitude and higher frequency attributed to more intensive blast loadings. Comparing the spectrums at different locations of identical specimen, i.e., Figure 13.8d–f, it also can be found that the locations suffered from more intensive blast loadings (the magnitude of blast loadings suffered by the three acceleration sensors is sorted as follows: A1 > A2 > A3) oscillate with larger amplitude and higher frequency. The comparison between Fig. 13.8b and d exhibits that the UHPCC panel oscillates with higher frequency than NSC panel under same scaled distance. This is due to the higher toughness of UHPCC panel compared to the NSC panel. Moreover, UHPCC panels oscillate with larger amplitude since the higher energy absorbing capacity of UHPCC.

13.3.3.4 Post-blast Damage

Figure 13.9 shows the post-blast damage of front, rear and side faces of test panels, and the test specimens mainly exhibit the bending failure mode. Under the present middle-range explosions, cracks firstly occur on the rear face of panels, and then develop along the height direction of panels. The cracks were marked to be more clear, and the solid lines represent the visible cracks while the dash lines represent the tiny cracks. Generally, with the decrease of scaled distance, both NSC and UHPCC panels suffered more serious damage, not only the total number of cracks but also the

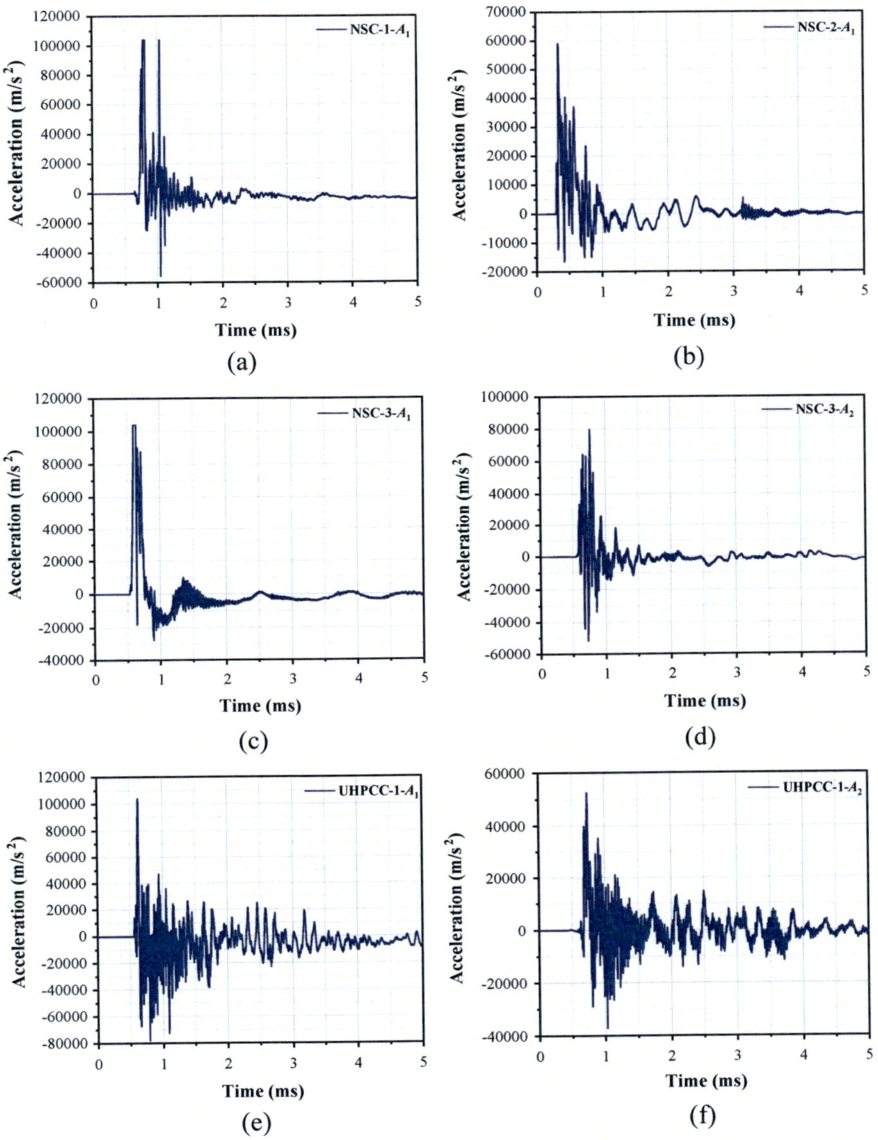

Fig. 13.7 Acceleration-time histories of specimens

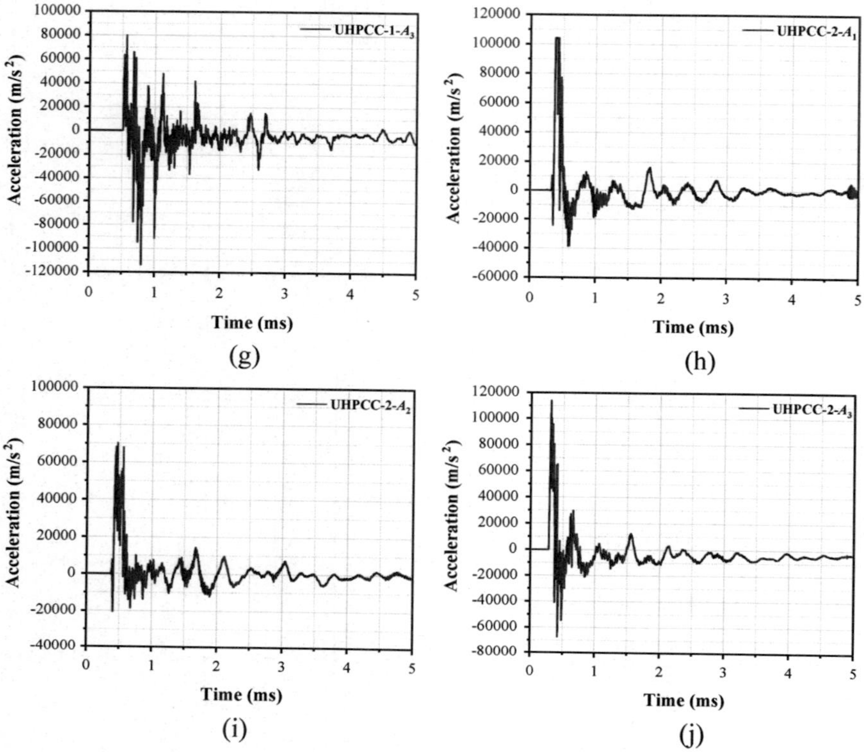

Fig. 13.7 (continued)

number and width of visible cracks increase, and the permanent deflections enlarge correspondingly. Besides, the concrete crushing even occurs on the front face of panels under more intensive blast loadings.

For the NSC panels, there is no obvious damage on the front faces of both NSC-1 and NSC-2 specimens, while the concrete crushing occurs on the front face of NSC-3 with a smaller scaled distance. Besides, two tiny cracks marked with dash line also exist on the front face of NSC-3. As the scaled distances varying from 1.0 m/kg$^{1/3}$ to 0.6 m/kg$^{1/3}$, the number of visible cracks on the rear face of NSC panels increases from one to four, the distribution range of cracks become wider, and the flexure of specimen becomes more obvious from the side view. Comparably, there is no obvious damage, i.e. concrete crushing and visible cracks, on the front face of UHPCC-1 and UHPCC-3 panels. The number of visible cracks on the rear face of UHPCC panels varies from one to three when the scaled distances are reduced from 0.8 m/kg$^{1/3}$ to 0.5 m/kg$^{1/3}$. Similarly, more cracks and larger deflections emerge on the UHPCC panels as the scaled distances decrease. Furthermore, the cracks of UHPCC panels tend to be shorter than that of NSC panels since the bridging effect of mixing fibers

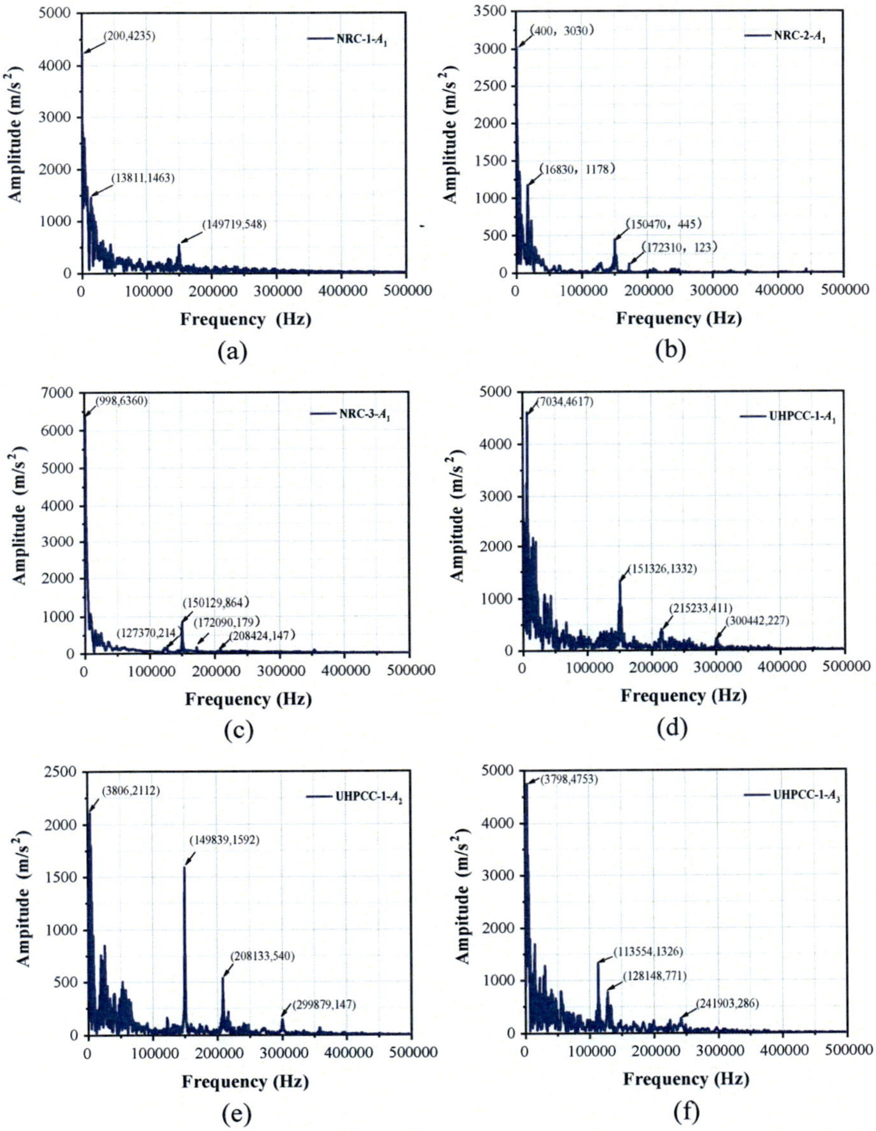

Fig. 13.8 FFT spectrum of specimens under explosions

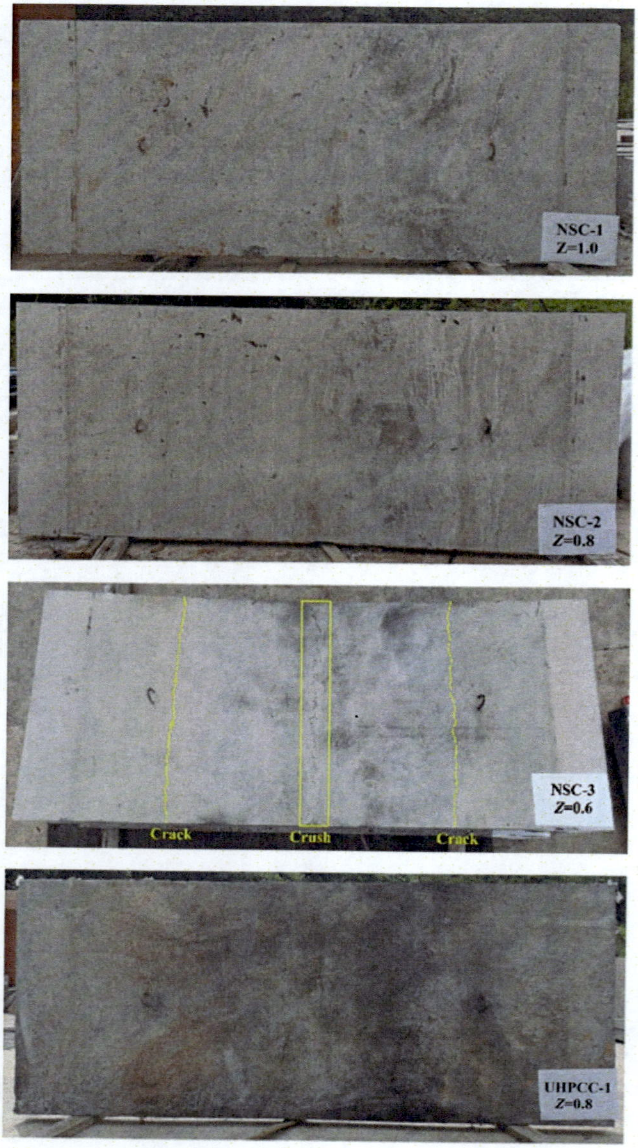

Fig. 13.9 Post-blast damage of panels **a** front face **b** rear face **c** side face

(a)

Fig. 13.9 (continued)

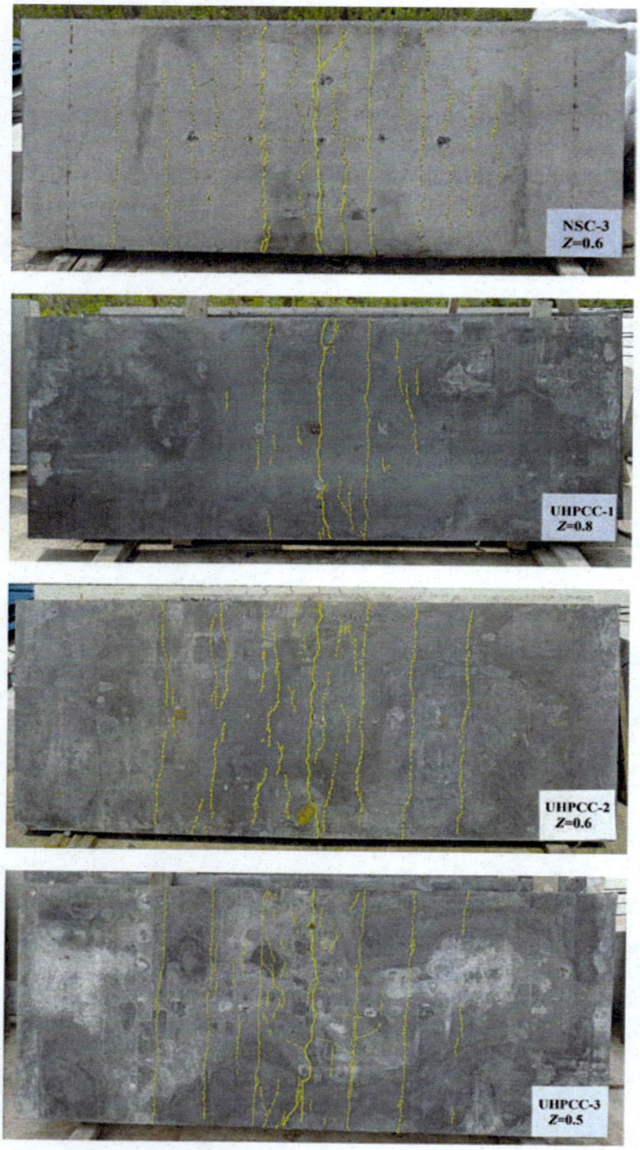

(b)

Fig. 13.9 (continued)

(c)

Fig. 13.9 (continued)

hinders the development of cracks efficiently. Therefore, UHPCC panels possess better blast-resistant performance than NSC panels.

13.4 Numerical Simulation

In this section, the 3D FE models of the present test are established by using the commercial dynamic explicit code LS-DYNA (Hallquist 2006). Firstly, the effectiveness of CONWEP method for predicting the blast loadings is validated by comparing with the experimental overpressure-time histories. Then the parameters of material models, especially for the K&C model of UHPCC are calibrated. Finally, the corresponding numerical simulations are performed, and the prediction results are compared with the test data.

13.4.1 FE Model

Figure 13.10 shows the finite model, in which single point integration 8-node hexahedral solid elements were used for both concrete panel and steel supports, and the Lagrangian formulation is applied in the analysis. Rebars were modeled using beam elements, i.e., Hughes-Liu with cross-section integration. The element size affects the strain-softening behavior of concrete, and it has been concluded

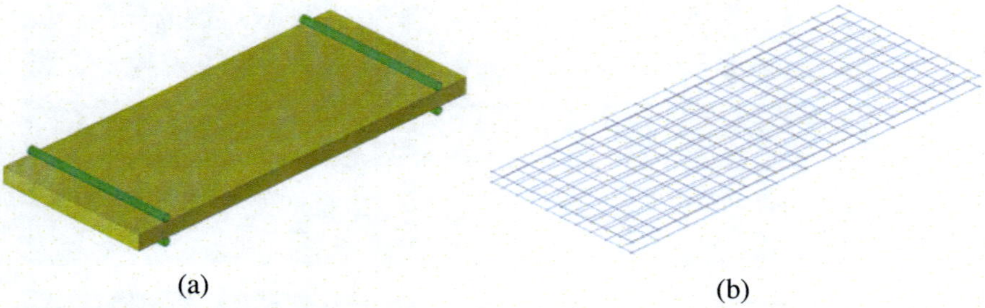

Fig. 13.10 FE model **a** concrete and steel supports **b** rebars

that the softening rates are relatively stable for the elements with size smaller than 25 mm (Lin and Gravina 2017). Therefore, elements with size of 10 mm are adopted for the concrete, steel rebars and supports. The contact algorithm *CONTACT_NODES_TO_SURFACE was applied to reproduce the contact interface between the concrete and the support. The static and dynamic coefficients of friction are 0.3 and 0.2, and the contact stiffness parameters SLSFAC, SFS and SFM are 0.01, 1 and 1, respectively. The keyword *CONSTRAINED_LAGRANGE_IN_SOLID was adopted to model the contact between the rebar and concrete, a perfect bond was assumed between steel reinforcements and concrete. The reason lies in that, for the instantaneous blast loadings with the time duration of several milliseconds, the bond and slip effect at the interface of rebars and concrete on the integral dynamic responses of plate are slight.

To apply the blast reflected overpressures on the panels, the uncoupled CONWEP method was adopted by using the keywords *LOAD_BLAST_ENHANCED and *LOAD_SEGMENT _SET. This option determines the blast pressure-time history based on the inputs of the equivalent TNT mass, location of detonation, and type of blast. The validity of CONWEP method is discussed in the following Sect. 13.5.1.

13.4.2 Material Model of UHPCC

For UHPCC, the K&C model (*MAT_CONCRETE_DAMAGE_REL3) (LSTC 2007) is used. Since the automatic parameter generation in the K&C model is based on test data from normal strength concrete, it could not reflect the different behavior of UHPCC that significantly different from NSC, e.g., strain hardening and softening, strain rate effect, etc. Therefore, the K&C model parameters for UHPCC must be calibrated systematically.

13.4.2.1 K&C Model

(1) Three failure surfaces
In K&C model (Malvar et al. 1997), three failure surfaces with consideration of damage and strain rate effects are defined:

$$\Delta\sigma_m = a_0 + \frac{p}{a_1 + a_2 p} \quad \text{(maximum failure surface)} \tag{13.1}$$

$$\Delta\sigma_r = \frac{p}{a_{1f} + a_{2f} p} \quad \text{(residual failure surface)} \tag{13.2}$$

$$\Delta\sigma_y = a_{0y} + \frac{p}{a_{1y} + a_{2y} p} \quad \text{(initial yield failure surface)} \tag{13.3}$$

where p is the pressure, a_0, a_1, a_2, a_{1f}, a_{2f}, a_{0y}, a_{1y}, and a_{2y} are eight strength surface parameters.

(2) EOS
The EOS describe the volumetric material behavior. In LS-DYNA (LSTC 2007), K&C model adopts the tabulated compaction EOS by using the keywords *EOS_TABULATED_COMPACTION, in which the current pressure p is expressed as a function of volumetric strain μ by

$$p = C(\mu) + \gamma_0 T(\mu) E_0 \tag{13.4}$$

(3) Damage evolution
In K&C model, after reaching the initial yield failure surface, the current strength surface is obtained as a linear interpolation between two of the three strength surfaces. In detail, for the hardening stage, the two surfaces are initial yield failure surface and maximum failure surface, while for the softening stage, the two surfaces are maximum failure surface and residual failure surface, which can be expressed as

$$\gamma = \begin{cases} r'\left[\eta(\Delta\sigma_m - \Delta\sigma_y) + \Delta\sigma_y\right] & \lambda \le \lambda_m \\ r'\left[\eta(\Delta\sigma_m - \Delta\sigma_y) + \Delta\sigma_y\right] & \lambda > \lambda_m \end{cases} \tag{13.5}$$

where the scale factor η varies from 0 to 1 depending on the accumulated effective plastic strain parameter λ, the transition from hardening to softening is denoted by $\lambda = \lambda_m$, where $\eta = 1$. In K&C model, the function $\eta(\lambda)$ is realized by inputting thirteen (λ, η) pairs, while for the values not provided by the data sets, the associated η value is obtained using interpolation. r' is the scale factor to make the deviatoric section typically transitions from triangular curves with smooth corners at low pressures to circular at high pressures.
The accumulated effective plastic strain parameter can be expressed as:

$$\lambda = \lambda\left(p/r_f\right)/r_f = \begin{cases} \int_0^{\bar{\varepsilon}_p} d\bar{\varepsilon}_p/r_f(1 + p/f_t)^{b_1} & p > 0 \\ \int_0^{\bar{\varepsilon}_p} d\bar{\varepsilon}_p/r_f(1 + p/f_t)^{b_2} & p \leq 0 \end{cases} \tag{13.6}$$

$$d\bar{\varepsilon}_p = \sqrt{1 + 2(\frac{\omega\gamma, p}{3})^2} d\bar{\mu} \tag{13.7}$$

where r_f is the strain rate enhancement factor, i.e., dynamic increase factor (DIF), $d\bar{\varepsilon}_p$ is the effective plastic strain increment, f_t is material tensile strength, b_1 and b_2 are parameters controlling the concrete softening behavior after peak stress under the uniaxial compression and uniaxial tension, respectively. $\bar{\mu}$ is the plasticity consistency parameter, and ω is the associativity parameter which controls shear dilatancy behavior.

(4) Strain rate effect

Under the dynamic loadings with a high stain rate, the strength of concrete will increase significantly and influence the performance of structures. Therefore, the strain rate effect needs to be considered for the reliable predictions of structural responses under the instantaneous blast loadings. In K&C model, the strength enhancement was implemented in general along the radial stress paths:

$$\Delta\sigma_e = r_f \Delta\sigma\left(p/r_f\right) \tag{13.8}$$

where $\Delta\sigma_e$ is the enhanced value of the failure surface at pressure p, $\Delta\sigma\left(p/r_f\right)$ is the unenhanced strength. The strain rate effects of material under compression and tension are considered separately in K&C model, and the strain rate enhancement factor r_f versus strain rate is given according to the keyword *DEFINE_CURVE in LS-DYNA (LSTC 2007).

13.4.2.2 Calibration of K&C Model Parameters for UHPCC

(1) Strength surface parameters

The strength surface parameters of UHPCC material were calibrated base on the test data of the uniaxial compression tests and conventional triaxial compression tests of UHPCC (Ren et al. 2016). By curve fittings of Eqs. (13.1) ~ (13.3), the revised failure surfaces of UHPCC with uniaxial compressive strength of 133 MPa are presented in Fig. 13.11 and the corresponding strength surface parameters are listed in Table 13.5. For UHPCC with other uniaxial compressive strength, the strength surface parameters could be calculated according to Eqs. (13.9) ~ (13.12) (Malvar et al. 1997). As for the present UHPCC with the uniaxial compressive strength of 125 MPa, the corresponding strength surface parameters are also given in Table 13.5.

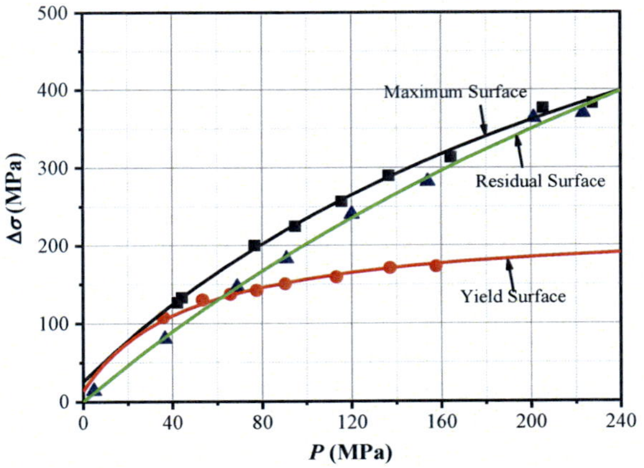

Fig. 13.11 Modified failure surfaces for UHPCC with uniaxial compressive strength of 133 MPa

Table 13.5 Strength surface parameters of K&C model for UHPCC

f_c (MPa)	a_0	a_1	a_2	a_{1f}	a_{2f}	a_{0y}	a_{1y}	a_{2y}
133	25.61	0.36	1.18e-3	0.42	7.70e-4	14.00	0.23	4.03e-3
125	24.07	0.36	1.26e-3	0.42	8.19e-4	13.16	0.23	4.29e-3

$$r = f_{\text{c,new}} / f_{\text{c,old}} \tag{13.9}$$

$$a_{0i} = a_{0i,\text{old}} \cdot r \tag{13.10}$$

$$a_{1i} = a_{1i,\text{old}} \tag{13.11}$$

$$a_{2i} = a_{2i,\text{old}} / r \tag{13.12}$$

(2) EOS

The EOS in HJC model (Holmquist et al. 1993) is applied in this study, in which the pressure-volume relationship is expressed in three stages. The first is the linear elastic stage ($0 \leq \mu \leq \mu_{\text{crush}}$), which is expressed as

$$P = K\mu, \quad P \leq P_{\text{crush}} \tag{13.13}$$

$$K = P_{\text{crush}} / \mu_{\text{crush}} = E / [3(1 - 2v)] \tag{13.14}$$

where K is the elastic bulk modulus, μ_{crush} is the ultimate volumetric strain, P_{crush} is the hydrostatic pressure corresponding to μ_{crush}, E is the elastic modulus and v is the Poisson's ratio.

The second is the transition stage ($\mu_{\text{crush}} \leq \mu \leq \mu_{\text{plock}}$), in which the air voids are gradually compressed out of the concrete, and the plastic volumetric strain produces the compaction damage until reaching the point (μ_{plock}, P_{lock}). The loading path in this phase is expressed as:

$$P = P_{\text{crush}} + K_{\text{lock}}(\mu - \mu_{\text{crush}}), \quad P_{\text{crush}} \leq P \leq P_{\text{lock}} \tag{13.15}$$

where $K_{\text{lock}} = (P_{\text{lock}} - P_{\text{crush}})/(\mu_{\text{plock}} - \mu_{\text{crush}})$ is the slope, and μ_{plock} is the volumetric strain at the compacting pressure P_{lock}. The unloading path is expressed as:

$$P = P_0 - [(1 - F)K + FK_1](\mu_0 - \mu) \tag{13.16}$$

where $F = (\mu_0 - \mu_{\text{crush}})/(\mu_{\text{plock}} - \mu_{\text{crush}})$ is the interpolation factor, and $(1 - F)K + FK_1$ represents the slope of the unloading path. P_0 is the pressure whesn the volumetric strain reaches μ_0 before unloading.

The third is the non-porous compact stage, in which all air voids are removed from concrete. During the loading state, the non-linear relationship is as follow:

$$P = K_1\bar{\mu} + K_2\bar{\mu}^2 + K_3\bar{\mu}^3, \ P \geq P_{\text{lock}} \tag{13.17}$$

where $\bar{\mu} = (\mu - \mu_{\text{lock}})/(1 + \mu_{\text{lock}})$ is the modified volumetric strain, and $\mu_{\text{lock}} = \rho_{\text{grain}}/\rho_0 - 1$ is the locking volumetric strain. ρ_{grain} is the grain density of the material without air voids. K_1, K_2, and K_3 are the material constants. During the unloading state, the relationship is as follow:

$$P = K_1\bar{\mu} \tag{13.18}$$

By fitting the hydrostatic pressure test data of high strength concrete, the relationship between the uniaxial compressive strength f_c and P_{crush} can be obtained. The test data of P_{crush} are the elastic limit pressure from the hydrostatic pressure-volumetric strain curves. In this study, the P_{crush} of UHPCC with uniaxial compressive strength of 125 MPa is 159 MPa, as shown in Fig. 13.12a. According to Eqs. (13.13) and (13.14), the first stage of EOS is determined. Besides, the second and third stages of EOS are determined by curve fitting of the existing flyer-plate-impact test data according to Eqs. (13.15) and (13.17), as shown in Fig. 13.12b. The related EOS parameters are listed in Table 13.6. Based on Table 13.6 and Eqs. (13.13) ~ (13.18), the tabulated EOS input data of UHPCC are obtained and given in Table 13.7.

Fig. 13.12 Equation of state

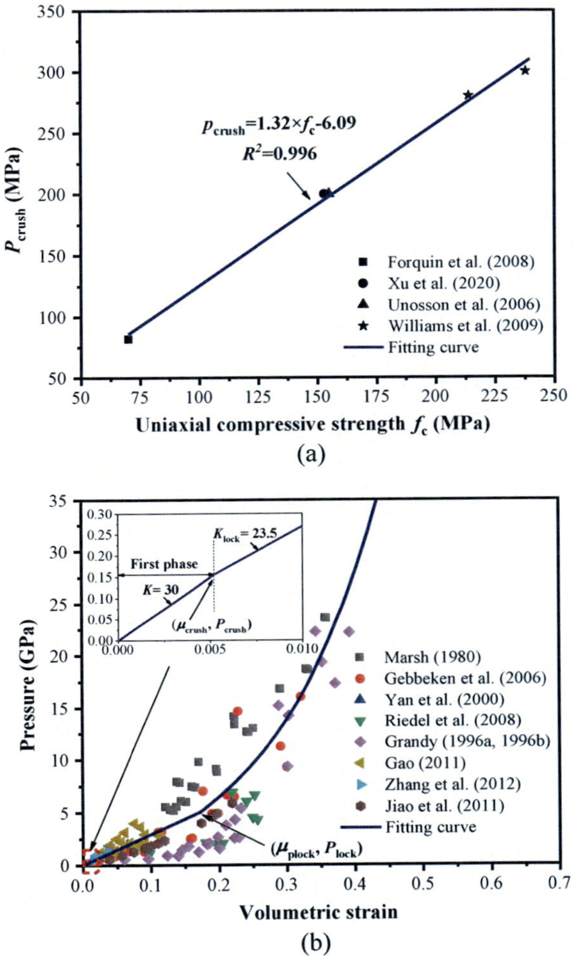

Table 13.6 Equation of state parameters

μ_{crush}	P_{crush} (GPa)	μ_{plock}	P_{lock} (GPa)	μ_{lock}	K_1 (GPa)	K_2 (GPa)	K_3 (GPa)
0.0053	0.159	0.17	5.01	0.078	69	−211	981

(3) Damage evolution parameters

To construct the relationship between the parameters η and λ, this study adopted the piecewise equation recommended by Yin et al. (2020), which has the form as

Table 13.7 EOS input data of UHPCC

i	1	2	3	4	5	6	7	8	9	10
m_i	0	−0.0053	−0.0101	−0.0305	−0.0513	-0.0726	-0.0943	-0.120	−0.174	−0.208
K_i (GPa)	30	30	30.13	35.97	40.89	45.94	51.07	57.16	69	69
P_i (GPa)	0	0.159	0.300	0.901	1.514	2.141	2.780	3.537	5.164	6.973

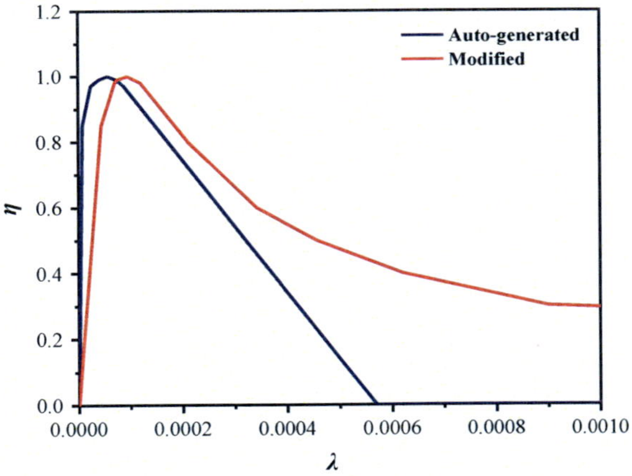

Fig. 13.13 Comparisons of damage evolution function $\eta(\lambda)$

$$\eta(\lambda) = \begin{cases} \alpha\frac{\lambda}{\lambda_m} + (3-2\alpha)\left(\frac{\lambda}{\lambda_m}\right)^2 + (\alpha-2)\left(\frac{\lambda}{\lambda_m}\right)^3, & \lambda < \lambda_m \\ \frac{\lambda/\lambda_m}{\alpha_c(\lambda/\lambda_m-1)^{\alpha_d}+\lambda/\lambda_m}, & \lambda \geq \lambda_m \end{cases} \tag{13.19}$$

where α is the parameter affecting the shape of the hardening stage of the stress-strain relationship, α_c and α_d are the parameters controlling the softening stage. The parameter λ_m corresponding to $\eta = 1$ is related to the peak strain. Based on the amounts of experiments, Yin et al. (2020) gave the recommended values of α, α_c, α_d, and λ_m, as 3, 0.381, 1.90 and 9.5×10^{-5}, respectively. Figure 13.13 shows the comparisons of the auto-generated and modified damage evolution function $\eta(\lambda)$.

In addition, the damage evolution parameters also include the associativity parameter ω, softening parameters b_1 and b_2. The default value of ω for NSC is 0.5, while the recommended value of ω is less than 0.5 for ultra-high-strength concrete with fine aggregate (Wu and Crawford 2015). In this study, $\omega = 0.1$ is adopted since UHPCC shows slight dilation. Based on the modified strength surface parameters, EOS parameters, damage evolution function, and associativity parameter, the numerical simulations of single element under compressions and tensions are carried out to determine the parameters b_1 and b_2. Finally, $b_1 = 0.55$ and $b_2 = -1.5$ are determined by comparing with the uniaxial compressive and tensile stress-strain curves, as shown in Fig. 13.14.

(4) Strain-rate effect

In this study, the empirical DIF relations proposed in our previous work (Ren et al. 2018; Wu et al. 2018, 2019) are adopted for UHPCC. The DIF relations for dynamic compressive (DIF_c) and tensile (DIF_t) strengths of UHPCC are expressed as

$$\text{DIF}_c = \begin{cases} (\dot{\varepsilon}/\dot{\varepsilon}_0)^{0.014} & \text{for } \dot{\varepsilon} \leq 62\text{s}^{-1} \\ 0.587(\log\dot{\varepsilon})^2 - 1.435\log\dot{\varepsilon} + 1.911 & \text{for } \dot{\varepsilon} > 62\text{s}^{-1} \end{cases} \tag{13.20}$$

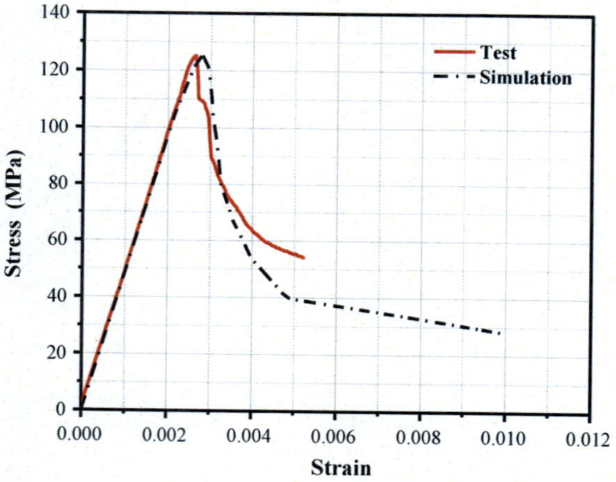

(a)

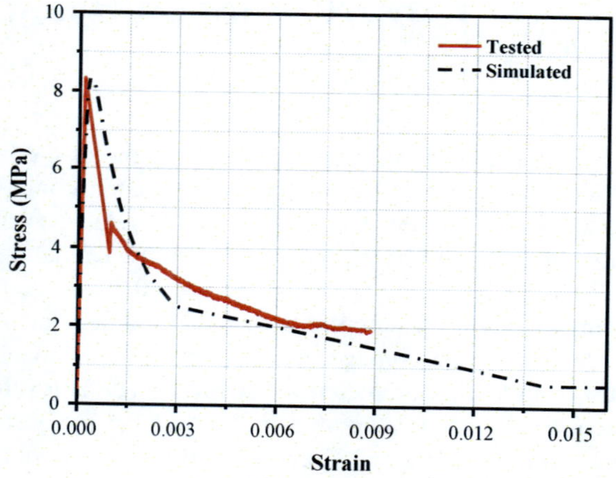

(b)

Fig. 13.14 Stress-strain relationships of UHPCC under uniaxial **a** compression and **b** tension

$$\text{DIF}_t = \begin{cases} (\dot{\varepsilon}/\dot{\varepsilon}_0)^{0.018} & \text{for } \dot{\varepsilon} \le 10.3\text{s}^{-1} \\ 9.959 \log \dot{\varepsilon} - 8.718 & \text{for } \dot{\varepsilon} > 10.3\text{s}^{-1} \end{cases} \tag{13.21}$$

where $\dot{\varepsilon}$ is the strain rate, and $\dot{\varepsilon}_0 = 10^{-6}\text{s}^{-1}$ is the static strain rate.

13.4.3 Material Model of NSC

At present, K&C model is also adopted for modeling NSC. The auto-generation function is applied to determine the strength surface parameters, EOS parameters, and 13 pairs of (λ, η) for NSC. Especially, based on the compressive stress-strain relationship and tensile fracture energy calculated according to CEB-FIP model code 1990 (CEB 1993), $b_1 = 3.042$ and $b_2 = 1.44$ are determined through the single element simulations. The DIF of NSC are referred from the CEB-FIP model code 1990 (CEB 1993).

For NSC in compression, DIF_c is calculated according to Eq. (13.22) when the strain rate is no more than 200 s^{-1}, and DIF_c remain unchanged when the strain rate is larger than 200 s^{-1}.

$$\mathrm{DIF}_c = \left(\frac{\dot{\varepsilon}}{\dot{\varepsilon}_s}\right)^{1.026\alpha_s} \quad \dot{\varepsilon} \le 200\mathrm{s}^{-1} \tag{13.22}$$

where $\alpha_s = 1/(5 + 9f_c/f_{cm0})$, $\dot{\varepsilon}$ is the strain rate, $\dot{\varepsilon}_s = 10^{-6}\mathrm{s}^{-1}$ is the static strain rate.

For NSC in tension, the DIF_t is calculated from Eq. (13.23).

$$\mathrm{DIF}_t = \begin{cases} \left(\frac{\dot{\varepsilon}}{\dot{\varepsilon}_s}\right)^{\delta} & \dot{\varepsilon} \le 1\mathrm{s}^{-1} \\ \beta(\dot{\varepsilon}/\dot{\varepsilon}_s)^{1/3} & \dot{\varepsilon} > 1\mathrm{s}^{-1} \end{cases} \tag{13.23}$$

where $\delta = 1/(1 + 8f_c/f_{cm0})$, $\log\beta = 6\delta - 2$, and $\dot{\varepsilon}$ is the strain rate ranges from 10^{-6} s^{-1} to 30 s^{-1}.

13.4.4 Material Models for Rebar and Support

The keyword *MAT_PLASTIC_KINEMATIC was used for describing the behavior of rebar in the panels, and the strain rate is accounted for by using the Cowper and Symonds model which scales the yield stress with the factor:

$$\mathrm{DIF} = 1 + \left(\frac{\dot{\varepsilon}}{C}\right)^{1/P} \tag{13.24}$$

where C and P are the strain rate parameters (LSTC 2007), and the values of 40 and 5 are adopted according to Stouffer and Dame (1996). Based on the true stress-strain data of rebar, the material parameters of rebar, i.e. density ρ_0, elastic modulus E, Poisson's ratio v, yield strength f_y, tangent modulus E_t, hardening parameter β, and strain rate parameters are listed in Table 13.8.

Table 13.8 Material parameters of rebar

*MAT_PLASTIC_KINEMATIC							
ρ_0 (kg m^{-3})	E (Pa)	υ	f_y (Pa)	E_t (Pa)	β	C	P
7850	1.84e11	0.3	4.80e8	1.42e9	0	40	5

Additionally, *MAT_RIGID was used for modeling the four round rolls supports, and the deflection and rotation of rolls in all directions was fixed.

13.5 Comparisons of Numerical Results with Test Data

13.5.1 Overpressures-Time History

As shown in Fig. 13.4 and Table 13.3, the CONWEP method can give a good prediction on the incident overpressure, for both the peak values and the time durations. Some deviations are induced by the shape of explosive, cylindrical charge is used in the test while CONWEP method is originally adopted for predicting overpressure induced by the centrally detonated spherical shaped explosive. The tested P_{i1} and P_{i2} were obtained at similar scaled distances in different directions. It is found that the prediction of the incident overpressure in the radial direction (P_{i1}) of explosive is more accurate than that in the axial direction (P_{i2}). The reason lies in that, the difference of explosive shape between the CONWEP scenario and field test is relatively smaller in the radial direction than in the axial direction. However, the influence of charge shape on the blast loadings decreases with the scaled distance increases, since that CONWEP method gives better predictions on P_{i3} than P_{i1} and P_{i2}.

Table 13.3 shows the comparisons of reflected overpressure data between the test results and the simulation results obtained through the keywords *DATABASE_BINARY_BLSTFOR. The bold test data in Table 13.3 might not be reliable considering the relatively large deviations in the explosion test. Generally, the simulation results of reflected overpressure show good agreement with the test data. However, the different charge shapes between the test and the ideal CONWEP scenario also lead to some deviations. Besides, the different point of detonation also causes deviations to some extent. The distribution of the numerical simulated overpressures are symmetry since the charge is centrally detonated in the CONWEP scenario. However, due to the asymmetric initiation, there exists obvious distinctions between the tested P_{r1} and P_{r3}. Overall, the prediction accuracy of peak reflected overpressure decreases with the decrease of scaled distances since the influence of charge shape and point of detonation become more obvious, while the prediction of impulses agrees well with the test results even under small scaled distances. Figure 13.15 further shows the comparison between tested and simulated reflected overpressure curves as an example.

Fig. 13.15 Experimental and simulated reflected overpressure histories ($Z = 0.8$ kg/m$^{1/3}$) **a** P_{r1} **b** P_{r2} **c** P_{r3}

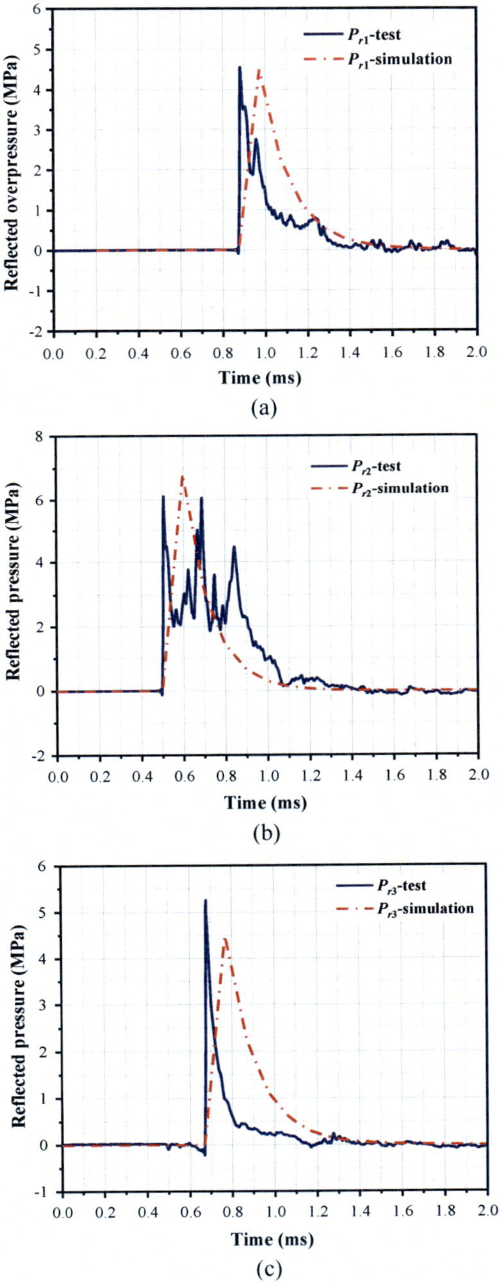

In summary, it is reasonable to use CONWEP method to predict the blast loadings from the present one-end-detonated cylindrical charge with the scaled distances of $0.5 \sim 1.0\ \mathrm{m/kg}^{1/3}$.

13.5.2 Deflection-Time History

Figure 13.6 illustrates the comparisons between the experimental and numerical simulated deflection curves, the rather good agreement can be derived, especially for the maximum and residual deformations of panels. Therefore, the reliability of the numerical model and the calibrated K&C parameters for UHPCC panels under blast loadings are validated.

However, there are some deviations between the tested and simulated results. For instance, the recorded maximum deflection of D_1 is larger than that of D_4 in the test while there is slight difference between the two simulated results. The possible reason may lie in that, compared with the FE model, the experimental panels were not ideally supported with the one-way constrained condition. Besides, there is a slight difference between the experimental deflection curves of D_2 and D_3, which are recorded from the two geometrically symmetric locations, while the simulated results of the two locations are almost identical, which may be attributed to the different overpressure distribution between the test and simulation. As mentioned above, attributed to the different detonated points, the overpressure distributions on panels are symmetric in the numerical model but asymmetric in the field test.

By comparing the deflection curves with various scaled distances, for both NSC and UHPCC specimens, it is found that the precision of simulation results drops as the scaled distance decreases, which is attributed to the reduction of precision of CONWEP method on predicting the reflected overpressure with decreasing the scaled distance. Comparably, for the small scaled distance, the numerical simulations give a better predictions for the UHPCC specimens, since that NSC has relatively low blast-resistance than UHPCC and the charge shape and point of detonation have more obvious influences on NSC panels.

13.5.3 Post-blast Damage

Figure 13.16 shows the numerically predicted post-blast damage contour of the slabs. By comparing with the test results shown in Fig. 13.9, it is concluded that the simulated post-blast damage of panels agrees well with the test results. However, the deviations of damage distributions between the test and simulation also exist to some extent, i.e. the simulation results show higher degree of damage. Three factors could account for the above deviations: (i) the charge shape and point of detonation in the numerical model differ from that in the field test; (ii) the heterogeneous concrete is

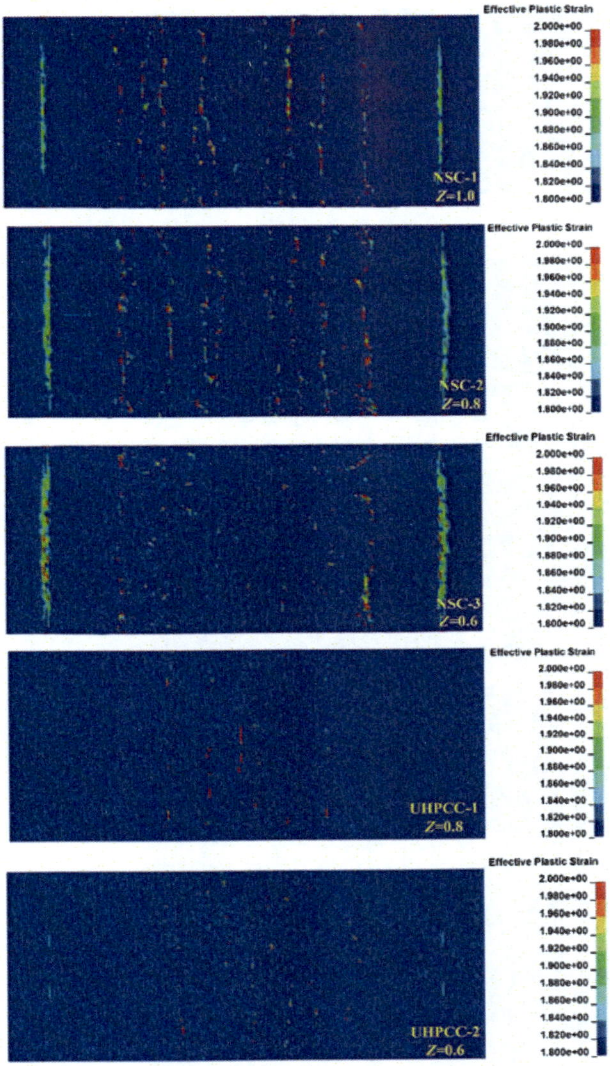

Fig. 13.16 Numerical simulated post-blast damage contour **a** front face **b** rear face **c** side face

modeled as homogeneous materials for simplification; (iii) some cracks on the real panels are invisible.

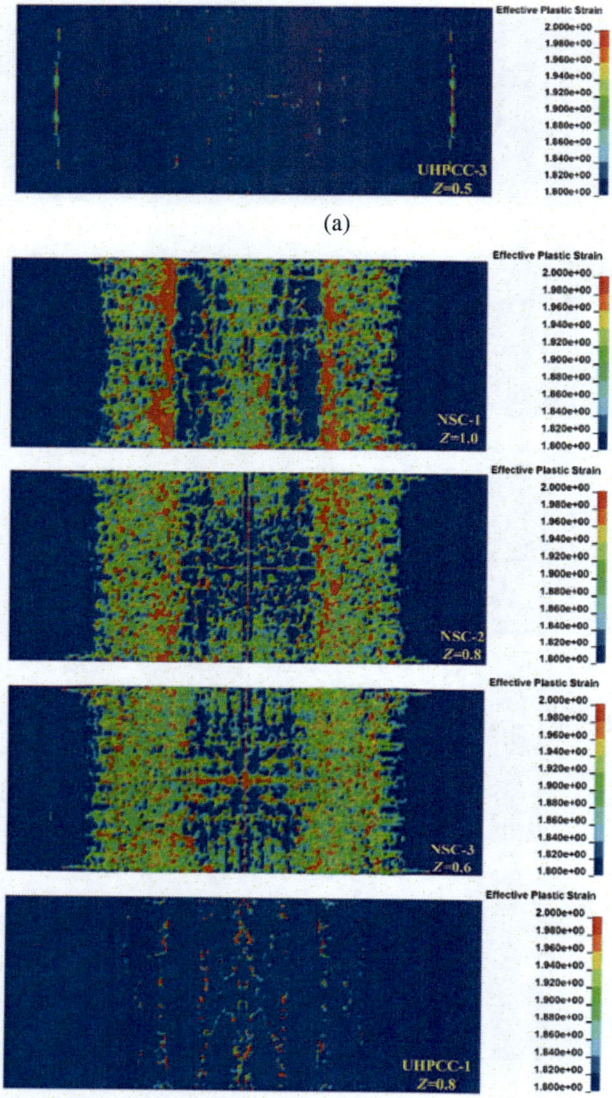

Fig. 13.16 (continued)

13.6 Summary

The blast-resistance of the reinforced UHPCC panels is studied experimentally and numerically in the present study, the main work and conclusions are drawn as follows:

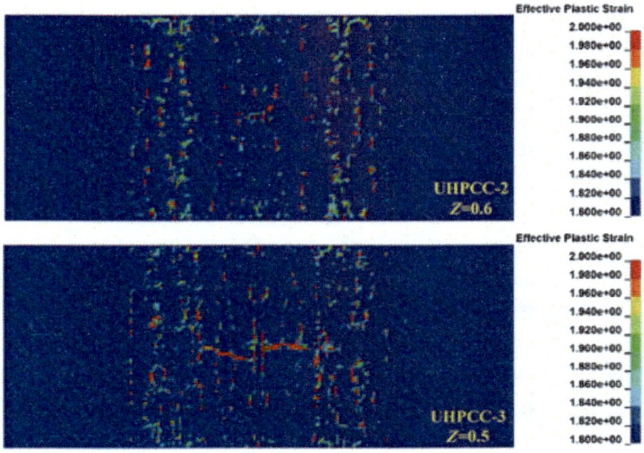

(b)

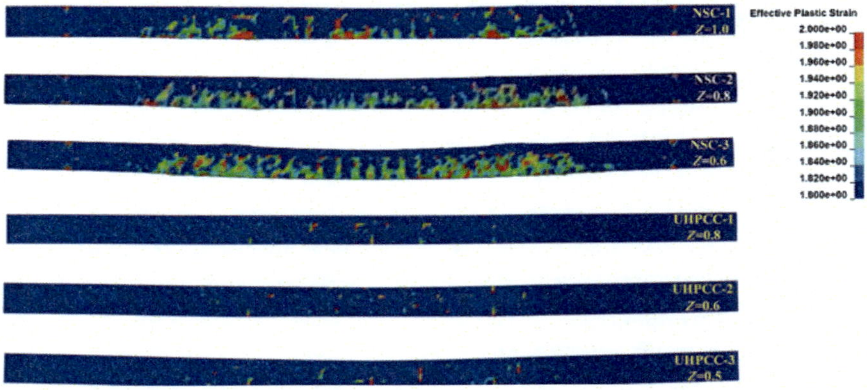

(c)

Fig. 13.16 (continued)

(1) A box-like blast loading apparatus is designed to provide the one-way simply-supported boundary constraints of the test panels and avoid the effect of the diffracted blast waves. Six reinforced UHPCC and NSC panels were fabricated and tested under medium-range explosions with the different scaled distances from 0.5 to 1.0 m/kg$^{1/3}$. The systematic data including the blast-induced incident and reflected overpressure-time histories, deflection- and acceleration-time histories, as well as the post-blast damage of panels were obtained.

(2) Under the medium-range explosion, the one-way test specimens mainly exhibit the bending failure mode, and the damage of panels become more serious when the scaled distance is reduced. Compared with the NSC panels, the UHPCC

panels show prominent blast-resistant performance. The outstanding mechanical properties of UHPCC, especially the existence of steel fibers hinders the development of cracks and improves the energy absorbing capacity of UHPCC.

(3) By comparing with the experimental incident and reflected instantaneous overpressures, the effectiveness of CONWEP method on predicting blast loadings, especially the applicability on predicting the reflected overpressure of blast wave induced by the one-end-detonated cylindrical charge with the scaled distance of 0.5 ~ 1.0 m/kg$^{1/3}$ is validated. It is also indicated that the charge shape and point of detonation affect the distributions of blast wave overpressure, and the corresponding influential degree increase with decreasing the scaled distance, which leads to a decrease in prediction accuracy of the post-blast behavior of panels.

(4) Karagozian & Case concrete (K&C) model parameters for describing the strength surface, EOS, damage evolution, and the strain rate effect of UHPCC were comprehensively calibrated by the existing basic mechanical properties test data, which are fully validated by comparing with the present test data, e.g. dynamic deflection-time histories and damage distributions.

References

AUTODYN. Century dynamics, theory manual[M]. 2005.

BARNETT S J, MILLARD S G, SCHLEYER G K, TYAS A. Briefing: Blast tests of fiber-reinforced concrete panels[J]. Construction Materials, 2010, 163(3): 127–129.

CEB. CEB-FIP model code 1990[M]. Thomas Telford Ltd., London, 1993.

DRUCKER D C, PRAGER W. Soil mechanics and plastic analysis or limit design[J]. Quarterly of Applied Mathematics, 1952, 10: 157–165.

ELLIS B D, DIPAOLO B P, MCDOWELL D L, ZHOU M. Experimental investigation and multi-scale modeling of ultra-high-performance concrete panels subject to blast loading[J]. International Journal of Impact Engineering, 2014, 69: 95–103.

FAN W, SHEN D, YANG T, SHAO X. Experimental and numerical study on low-velocity lateral impact behaviors of RC, UHPFRC and UHPFRC-strengthened columns[J]. Engineering Structures, 2019, 191: 509–525

FARNAM Y, MOHAMMADI S, SHEKARCHI M. Experimental and numerical investigations of low velocity impact behavior of high-performance fiber-reinforced cement based composite[J]. International Journal of Impact Engineering, 2010, 37(2): 220–229.

FOGLAR M, HAJEK R, FLADR J, PACHMAN J, STOLLER J. Full-scale experimental testing of the blast resistance of HPFRC and UHPFRC bridge decks[J]. Construction and Building Materials, 2017, 144: 588–601.

FORQUIN P, GARY G, GATUINGT F. A testing technique for concrete under confinement at high rates of strain[J]. International Journal of Impact Engineering, 2008, 35(6): 425–446.

GAO L. Research of high pressure equation of RPC[D]. Guangzhou: Guangzhou Univ., 2011. (in Chinese)

GB/T 50081–2002. Standard for test method of mechanical properties on ordinary concrete[S]. Beijing: Standards Press of China, 2003.

GOVINDJEE S, KAY G J, SIMO J C. Anisotropic modeling and numerical simulation of brittle damage in concrete[J]. International Journal for Numerical Methods in Engineering, 1995, 38(21): 3611–3633.

GRADY D E. Dynamic decompression properties of concrete from Hugoniot states 3 to 25GPa[R]. USA: Sandia National Laboratories, 1996.

GRADY D E. Shock equation of state properties of concrete. International Conference on Structures under Shock and Impact, Udine, Italy[C]. 1996.

HABEL K, GAUVREAU P. Response of ultra-high performance fiber reinforced concrete (UHPFRC) to impact and static loading[J]. Cement & Concrete Composites, 2008, 30(10): 938–946.

HALLQUIST J O. LS-DYNA Theory manual[M]. Livermore Software Technology Corporation, 2006.

HOLMQUIST T J, JOHNSON G R, COOK W H. A computational constitutive model for concrete subjected to large strains, high strain rates and high pressures. In proceedings of 14th international symposium on Ballistics, Quebec, Canada[C]. 1993: 591–600.

HYDE D W. User's Guide for microcomputer program CONWEP, applications of TM 5-855-1, fundamentals of protective design for conventional weapons: SL-88-1[R]. Vicksburg, MS: U.S. Army Corps of Engineers Waterways Experiment Station Instruction, 1993.

JAYASOORIYA R, THAMBIRATNAM D P, PERERA N J. Blast response and safety evaluation of a composite column for use as key element in structural systems[J]. Engineering Structures, 2014, 61: 31–43.

JIAO C J, JIANG G P, GAO L. Preliminary study on the high-pressure equation of state of concrete[J]. Advanced Materials Research, 2011, 261–263: 161-165.

KINGERY C N, BULMASH G. Air-blast parameters from TNT spherical air burst and hemispherical surface burst: ARBRL-TR-02555[R]. Aberdeen Proving Ground, MD U.S: Army Ballistic Research Laboratory, 1984.

KUZNETSOV V, REBENTROST M, WASCH J. Strength and toughness of steel fiber reinforced reactive powder concrete under blast loading. 1st International Conference on Analysis and Design of Structures Against Explosive and Impact Loads, Tianjin, China[C]. 2006: 70–74.

LI J, WU C, HAO H. An experimental and numerical study of reinforced ultra-high performance concrete slabs under blast loads[J]. Materials & Design, 2015, 82: 64–76.

LI J, WU C, HAO H. Investigation of ultra-high performance concrete slab and normal strength concrete slab under contact explosion[J]. Engineering Structures, 2015, 102: 395–408.

LI J, WU C, HAO H, WANG Z, SU Y. Experimental investigation of ultra-high performance concrete slabs under contact explosions[J]. International Journal of Impact Engineering, 2016, 93: 62–75.

LIN X, GRAVINA R J. An effective numerical model for reinforced concrete beams strengthened with high performance fibre reinforced cementitious composites[J]. Materials and Structures, 2017, 50(5): 212.

LIN X. Numerical simulation of blast responses of ultra-high performance fiber reinforced concrete panels with strain-rate effect[J]. Construction and Building Materials, 2018, 176: 371–382.

LSTC. LS-DYNA Keyword User's Manual, Version 971[M]. 2007.

MALVAR L J, CRAWFORD J E, WESEVICH J W, SIMONS D. A plasticity concrete material model for DYNA3D[J]. International Journal of Impact Engineering, 1997, 19(9–10): 847-873.

MAO L, BARNETT S J, TYAS A, WARREN J, SCHLEYER G K, ZAINI S S. Response of small scale ultra high performance fiber reinforced concrete slabs to blast loading[J]. Construction and Building Materials, 2015, 93: 822–830.

MAO L, BARNETT S, BEGG D, SCHLEYER G, WIGHT G. Numerical simulation of ultra high performance fiber reinforced concrete panel subjected to blast loading[J]. International Journal of Impact Engineering, 2014, 64: 91–100.

MARSH S P. LASL shock Hugoniot data[M]. California: University of California Press, 1980.

NGO T, MENDIS P, KRAUTHAMMER T. Behavior of ultrahigh-strength prestressed concrete panels subjected to blast loading[J]. Journal of Structural Engineering, 2007, 133(11): 1582–1590.

ORTON S L, CHIARITO V P, MINOR J K, COLEMAN T G. Experimental testing of CFRP-Strengthened reinforced concrete slab elements loaded by close-in blast[J]. Journal of Structural Engineering, 2014, 140(2): 04013060.

PENG Y, WU C, LI J, LIU J, LIANG X. Mesoscale analysis on ultra-high performance steel fibre reinforced concrete slabs under contact explosions[J]. Composite Structures, 2019, 228: 111322.

RANDERS-PEHRSON G, BANNISTER K A. Airblast loading model for DYNA2D and DYNA3D[R]. Aberdeen Proving Ground, MD: Army Research Laboratory, 1997. Report ARL-TR-1310.

RASOULI A, TOOPCHI–NEZHAD H. Coupled vs uncoupled analysis of one–way RC–slabs under nearby air explosions[J]. International Journal of Advanced Structural Engineering, 2018, 10: 421–437.

REN G M, WU H, FANG Q, LIU J Z, GONG Z M. Triaxial compressive behavior of UHPCC and applications in the projectile impact analyses[J]. Construction and Building Materials, 2016; 113: 1–14.

REN G M, WU H, FANG Q, LIU J Z. Effects of steel fiber content and type on dynamic compressive mechanical properties of UHPCC[J]. Construction and Building Materials, 2018, 164: 29–43.

RIEDEL W, THOMA K, HIERMAIER S, SCHMOLINSKE E. Penetrating of reinforced concrete by BETA-B-500 numerical analysis using a new macroscopic concrete model for hydrocodes. In proceedings of ninth international symposium of IEMS, Berlin[C]. 1999: 315–322.

RIEDEL W, WICKLEIN M, THOMA K. Shock properties of conventional and high strength concrete: experimental and mesomechanical analysis[J]. International Journal of Impact Engineering, 2008, 35(3): 155–171.

SCHLEYER G K, BARNETT S J, MILLARD S G, WIGHT G, REBENTROST M. UHPFRC panel testing[J]. Structural Engineer, 2011, 89(23–24): 34-40.

SCHWER L E, MURRAY Y D. A three invariant smooth cap model with mixed hardening[J]. International Journal for Numerical & Analytical Methods in Geomechanics, 1994, 18(10): 657–688.

SHERKAR P, SHIN J, WHITTAKER A, AREF A. Influence of charge shape and point of detonation on blast-resistant design[J]. Journal of Structural Engineering, 2016, 142(2): 04015109.

SOVJÁK R, VAVŘINIK T, ZATLOUKAL J, MÁCA P, MIČUNEK T, FRYDRÝN M. Resistance of slim UHPFRC targets projectile impact using in-service bullet[J]. International Journal of Impact Engineering, 2015, 76: 166–177.

STOUFFER D C, DAME L T. Inelastic deformation of metals: models, mechanical properties, and metallurgy[M]. New York: John Wiley and Sons, 1996: 520.

THAM C Y. Numerical simulation on the interaction of blast waves with a series of aluminum cylinders at near-field[J]. International Journal of Impact Engineering, 2009, 36(1): 122–131.

TM5–855-1. Department of the Army. Fundamentals of protective design for conventional weapons[S]. Washington DC, 1986.

UNOSSON M, NILSSON L. Projectile penetration and perforation of high performance concrete: experimental results and macroscopic modelling[J]. International Journal of Impact Engineering, 2006, 32(7): 1068–1085.

WANG W, WU C, LI J, LIU Z, LV Y. Behavior of ultra-high performance fiber-reinforced concrete (UHPFRC) filled steel tubular members under lateral impact loading[J]. International Journal of Impact Engineering, 2019, 132: 103314.

WANG W, WU C, LI J, LIU Z, ZHI X. Lateral impact behavior of double-skin steel tubular (DST) members with ultra-high performance fiber-reinforced concrete (UHPFRC)[J]. Thin-Walled Structures, 2019, 144: 106351.

WEI J, LI J, WU C. An experimental and numerical study of reinforced conventional concrete and ultra-high performance concrete columns under lateral impact loads[J]. Engineering Structures, 2019, 201: 109822.

WILLIAMS E M, GRAHAM S S, REED P A, RUSHING T S. Laboratory characterization of Cor-Tuf concrete with and without steel fibers: No. ERDC/GSL-TR-09-22[R]. Engineer Research and Development Center Vicksburg ms Geotechnical and Structures Lab, 2009.

WU C, OEHLERS D J, REBENTROST M, LEACH J, WHITTAKER A S. Blast testing of ultra-high performance fiber and FRP-retrofitted concrete slabs[J]. Engineering Structures, 2009, 31(9): 2060–2069.

WU C, FATTORI G, WHITTAKER A, OEHLERS D J. Investigation of air-blast effects from spherical-and cylindrical-shaped charges[J]. International Journal Protection Structures, 2010, 1(3): 345–362.
WU H, FANG Q, CHEN X W, GONG Z M, LIU J Z. Projectile penetration of ultra-high performance cement based composites at 510 m/s to 1320 m/s[J]. Construction and Building Materials, 2015, 74: 188–200.
WU H, REN G M, FANG Q, LIU J Z. Effects of steel fiber content and type on dynamic tensile mechanical properties of UHPCC[J]. Construction and Building Materials, 2018, 173: 251–261.
WU H, REN G M, FANG Q, LIU J Z. Response of ultra-high performance cementitious composites filled steel tube (UHPCC-FST) subjected to low-velocity impact[J]. Thin-Walled Structures, 2019, 144: 106341.
WU K C, LI B, TSAI K C. The effects of explosive mass ratio on residual compressive capacity of contact blast damaged composite columns[J]. Journal of Constructional Steel Research, 2011, 67(4): 602–612.
WU Y, CRAWFORD J E. Numerical modeling of concrete using a partially associative plasticity model[J]. Journal of Engineering Mechanics, 2015, 141(12): 04015051.
XU S, WU P, WU C. Calibration of KCC concrete model for UHPC against low-velocity impact[J]. International Journal of Impact Engineering, 2020, 144: 103648.
YAN S H, QIAN Q H, ZHOU Z S, LU Y S, YIN F L. Experimental study of equation of state for high-strength concrete and high-strength fiber concrete[J]. Journal of PLA University of Science and Technology, 2000, 1(6): 49–53. (in Chinese)
YI N, KIM J J, HAN T, CHO Y, LEE J H. Blast-resistant characteristics of ultra-high strength concrete and reactive powder concrete[J]. Construction and Building Materials, 2012, 28(1): 694–707.
YIN H, JIANG K, ZHANG L, HUANG L, WANG C. K&C Model of steel fiber reinforced concrete plate under impact and blast load[J]. Chinese Journal of High Pressure Physics, 2020, 34(3): 034205. (in Chinese)
YOO D, BANTHIA N, KIM S, YOON Y. Response of ultra-high-performance fiber-reinforced concrete beams with continuous steel reinforcement subjected to low-velocity impact loading[J]. Composite Structures, 2015, 126: 233–245.
ZHANG F, WU C, WANG H, ZHOU Y. Numerical simulation of concrete filled steel tube columns against blast loads[J]. Thin-Walled Structures, 2015, 92: 82–92.
ZHANG Z, JIN Y J. Experimental study of equation of state for concrete[J]. Applied Mechanics & Materials, 2012, 174–177: 566-569.
ZHOU X Q, KUZNETSOV V A, HAO H, WASCHL J. Numerical prediction of concrete slab response to blast loading[J]. International Journal of Impact Engineering, 2008, 35(10): 1186–1200.

Chapter 14
Constitutive Modelling of UHPCC Material Under Impact and Blast Loadings

14.1 Introduction

Ultra-high performance cement based composite (UHPCC) is a relatively new cement based composite with both prominent static and dynamic mechanical properties, such as the ultra-high compressive strength, high elastic modulus and good durability. Particularly, the incorporation of fibers can effectively hinder the propagation of microcracks and significantly improve its toughness, ductility and energy absorption capacity. These excellent performances make UHPCC a prospective material for protective structures subjected to intensive loading conditions such as impact and blast (Li et al. 2015b; Zhang et al. 2016; Verma et al. 2016; Wu et al. 2009).

Experiments on dynamic responses of UHPCC material and structures under impact and blast loadings have been experimentally investigated in recent years (Li et al. 2015b; Zhang et al. 2016; Verma et al. 2016; Wu et al. 2009, 2018; Ren et al. 2018a). However, these experiments are usually time-consuming, expensive, and sometimes difficult to perform, especially for parametric studies. With the rapid development of computational power, numerical simulation plays an increasingly important role in prediction of dynamic responses of UHPCC structures. Previous numerical studies (Guo et al. 2017; Xu and Wille 2014; Othman and Marzouk 2018; Xu et al. 2020) demonstrated that a sound material model and an accurate set of parameters for UHPCC are crucial. Under the intensively dynamic loadings, UHPCC material is subjected to a high strain-rate, high pressure and local failures including compression-dominated crater and tension-dominated scab are frequently observed at the frontal and rear surfaces of UHPCC structures, respectively (Fang and Wu 2017). Therefore, descriptions of the high strain-rate, equation of state (EoS) at high pressure and dynamic tensile behavior in a material model are vital.

In recent decades, several concrete material models have been proposed and are available in commercial hydrocodes, such as the Holmquist-Johnson-Cook model (HJC model) (Holmquist et al. 1993) and its modified versions (Kong et al. 2016; Liu et al. 2009), the Riedel-Hiermaier-Thoma model (RHT model) (AUTODYN 2013) and its modified versions (Leppänen 2006; Tu and Lu 2010) and the Karagozian and

Q. Fang et al., *UHPCC Under Impact and Blast*,
https://doi.org/10.1007/978-981-33-6842-2_14

Case model (K&C model) (Malvar et al. 1997) and its modified versions (Kong et al. 2017a; Hong et al. 2017). Reviews of these models mentioned above were made by Tu and Lu (2009) and Kong et al. (2018a), in which their advantages and limitations were clearly pointed out. However, these material models were originally developed for traditional concrete and cannot be applied to UHPCC material directly, which results from their different mechanical properties, especially the strain-rate effect, EoS and dynamic tensile behavior mentioned above. Frequently, the strain-rate effect is incorporated into the material model by defining the dynamic increase factor (DIF) at different strain-rates. Review of dynamic increase factor for tension (DIFt) of UHPCC material demonstrates that DIFt of UHPCC is higher than that of traditional concrete at a high strain-rate, and effects of fiber volume content, fiber type and fiber distribution on DIFt are negligible (Thaomas and Sorensen 2017). For the EoS that strongly affects the compressive behavior under high pressure, the porosity in materials has an obvious effect on the so-called transition stage where the air voids are gradually compressed out, beyond which all air voids are removed and materials are in a densification state. In the densification stage, flyer-plate-impact experiments indicate that UHPCC and traditional concrete have a similar EoS (beyond a pressure about 3GPa) mainly because they have a similar condense density (Neel 2018). However, since the porosity of UHPCC is always small, in the transition stage UHPCC should have a higher volumetric stiffness when compared with the traditional concrete. In particular, the tensile behavior especially the post-peak (softening) response of UHPCC is far different from the traditional concrete. The tensile-softening behavior of the traditional concrete has been extensively studied and a power function known as the "Hordijk-Reinhard expression" is used to define the softening curve (Kong et al. 2018a; Weerheijm and van Doormaal 2007). However, the tensile-softening of UHPCC exhibits multiple complex stages mainly due to the fiber bridging mechanism, which involves the first softening stage due to matrix cracking, the second platform where stress remains almost unchanged with the increase of crack width mainly due to the fiber bridging effect between the disconnected crack surfaces and the final failure softening stage associated with the pull-out or break of fibers (Kang et al. 2010). Modelling the tensile-softening behavior of UHPCC in a material model is still a challenging topic and such model has not been fully established (Othman and Marzouk 2018).

Considering the different material properties between traditional concrete and UHPCC discussed above, existing studies on the material model of UHPCC have been mostly conducted by calibrating parameters of concrete material models based on corresponding test data. Mao et al. (2014) modified the tensile-softening curve of the K&C model by adjusting the value of b_2, while holding other parameters unchanged. This modification leaded to a reasonable description of tensile behavior of UHPCC, but the other parameters, such as the strength surface parameters and EoS parameters still used the automatic generated ones originally suggested for the traditional concrete with an unconfined compressive strength (f_c) of 45.4 MPa. Recently, Xu et al. (2020) presented a systematic calibration of the K&C material model for UHPCC material, including the strength surface parameters based on triaxial compression data, the EoS parameters based on hydrostatic compression test data, the strain-rate effect parameters based on dynamic tensile and compressive test data and the damage parameters based on repeated trials. The calibrated set of parameters were then used to numerically simulate the low-velocity impact tests of UHPCC columns and beams and reasonable agreements with test data were observed. However, the damage parameters were obtained by adjusting the relation of the yield scale factor h and the damage function l, which affected both the compressive damage and tensile damage accumulation, leading to a dilemma situation that the compressive behavior cannot be well modelled if the h-l relation was adjusted according to the tensile behavior and vice versa. The other important issue is that in the K&C model the elastic modulus E is determined by a built-in formula ($E = 57000\sqrt{f_c/145}$, both E and f_c are in MPa units) and cannot be modified by users. This formula always underestimates the elastic modulus of UHPCC, which may result in underestimation of structural stiffness. To model the dynamic response of UHPCC subjected to a low-velocity impact, Guo et al. (2017) calibrated the parameters of continuous surface cap model (CSCM) (Murray 2007) in detail based on existing experimental data, such as the parameters for failure surface functions and the strain rate effects in tension and compression. However, for the intensively dynamic loading concerned in this study, the CSCM model has some limitations as pointed out in (Wu et al. 2012), e.g. the inaccurate description of compressive behavior under high pressure and inaccurate consideration of high strain-rate effect (Kong et al. 2017b).Othman and Marzouk (2018) presented a set of parameters of concrete damage plasticity (CDP) constitutive model available in ABAQUS for UHPCC material. The CDP model separately considers the compressive damage and tensile damage and supports user-defined damage input, consequently, it could predict both the compressive and tensile behavior well. The predictive capability of their calibrated parameters was validated by numerical simulation on two slabs subjected to repeated drop-weight impact. However, it is known that the CDP model was originally developed for structural responses at relatively low pressures. At high pressures, the EoS that governs the compaction behavior becomes important, and consequently, the CDP model may not be suitable for intensively dynamic loadings concerned in this study.

The present study aims to establish a material model for UHPCC material that can well capture its dynamic behavior under impact and blast loadings. This involves proposing a new tensile damage model for UHPCC which is then incorporated into a recently proposed concrete material model (Kong et al. 2018a) (named Kong-Fang model for simplicity), and calibrating parameters of this modified material model based on existing and our test data. After a brief description of the original Kong-Fang concrete material model, a new tensile damage model suitable for UHPCC material is presented in Sect. 14.2, followed by detailed parameter calibration of the modified material model. Then in Sect. 14.3, the improved performances of the proposed damage model and modified set of parameters are demonstrated by single element tests. Experimental validations and corresponding discussion will be presented in Sect. 14.4, which is followed by conclusive remark in Sect. 14.5.

14.2 UHPCC Material Model

The so-called Kong-Fang model recently developed (Kong et al. 2018a) is an elastoplastic-damage based material model, which was originally developed to simulate the responses and failures of the traditional concrete under impact and blast loadings and was recently extended to rock materials (Huang et al. 2020). In this section, a new tensile damage model for UHPCC with the consideration of fiber bridging effect is developed, which is then incorporated into the Kong-Fang model. Finally, based on existing and our test data, parameters for the modified material model are carefully calibrated.

14.2.1 *Brief Introduction of the Original Kong-Fang Concrete Model*

The original Kong-Fang model uses independent strength surfaces of compressive meridians, i.e. the maximum strength surface σ_m and the residual strength surface σ_r, which are defined as,

$$\sigma_m = \begin{cases} 3[P/(1-D)+T] & P \le 0 \\ 1.5/\psi(P+T) & 0 < P \le f_c/3 \\ f_c + \frac{P - f_c/3}{a_1 + a_2 P} & P > f_c/3 \end{cases} \tag{14.1}$$

$$\sigma_r = \begin{cases} 0 & P \le 0 \\ \frac{P}{a_1 + a_2 P} & P > 0 \end{cases} \tag{14.2}$$

where a_1 and a_2 are material constants, which are determined from a suitable set of triaxial compression (TXC) data. f_c and T represent the unconfined uniaxial compressive and tensile strengths, respectively. P denotes the pressure, and D is the damage index. ψ is the tensile-to-compressive meridian ratio, which is expressed as,

$$\psi(P) = \begin{cases} 0.5 & P \le 0 \\ 0.5 + 1.5T/f_c & P = \frac{f_c}{3} \\ 1.15/[1 + 1.3/(3a_1 + 2.3a_2 f_c)] & P = \frac{2.3 f_c}{3} \\ 0.753 & P = 3f_c \\ 1 & P \ge 8.45 f_c \end{cases} \tag{14.3}$$

The current failure surface is determined as,

$$Y(\sigma_{ij}, D) = \sqrt{3J_2} = r'[D(\sigma_r - \sigma_m) + \sigma_m] \tag{14.4}$$

where J_2 is the second deviatoric stress invariant. r' is the ratio of current meridian to compressive meridian and is expressed as,

$$r'(\theta, \psi) = \frac{2(1-\psi^2)\cos\theta + (2\psi - 1)\sqrt{4(1-\psi^2)\cos^2\theta + 5\psi^2 - 4\psi}}{4(1-\psi^2)\cos^2\theta + (1-2\psi)^2} \tag{14.5}$$

with θ being the Lode-angle.

To consider the strain-rate effect, the current failure surface is enlarged by the "radial enhancement" approach (Hartmann et al. 2010), which is not repeated here. In this approach, the instantaneous strain-rate is used which neglects the strain-rate history (Kong et al. 2020). To address this issue, a new corrected strain-rate was proposed by Kong et al. (2020), which is defined as,

$$\bar{\dot{\varepsilon}}_{n+1} = \begin{cases} \dot{\varepsilon}, & D \le 0 \\ \frac{\bar{\dot{\varepsilon}}_n + (\Delta t/\eta)\dot{\varepsilon}_{n+1}}{1+\Delta t/\eta}, & D > 0 \end{cases} \tag{14.6}$$

where $\bar{\dot{\varepsilon}}$ and c are the corrected strain-rate and instantaneous strain-rate, respectively. The subscript n and n+1 denotes the quantities at the beginning and end of current time step, respectively. η is a time scale governs the evolution of strain-rate, and is suggested as 0.1 ms (Kong et al. 2020). This corrected strain-rate concept is used in this study.

In the Kong-Fang model, the compressive damage and tensile damage are separately considered through a modified effective plastic strain λ defined as,

$$\lambda = \begin{cases} \lambda_t \\ \lambda_c \end{cases} = \sum \begin{cases} \frac{\Delta\bar{\varepsilon}_p}{r_f d_1 (T/f_c + P/r_f f_c)^{d_2}}, & P > 0 \\ \Delta\bar{\varepsilon}_1, & P \le 0 \end{cases} \tag{14.7}$$

where $\Delta\bar{\varepsilon}_p = \sqrt{(2/3)\Delta\varepsilon_{ij}^p \Delta\varepsilon_{ij}^p}$ is the effective plastic strain increment with $\Delta\varepsilon_{ij}^p$ being the tensor of the plastic strain increment in a time step. $\Delta\bar{\varepsilon}_1$ is the plastic strain increment in a stress state of unconfined uniaxial tension. d_1 and d_2 represent the damage constants. r_f is the DIF and is defined separately for compression and tension.

By introducing an empirical constant a, the compressible damage D_c is empirically defined as,

$$D_c = \frac{\alpha\lambda_c}{\lambda_c + 1} \tag{14.8}$$

The tensile damage D_t is defined as,

$$\mathrm{D_t} = 1 - \left(1 + \left(\mathrm{c_1}\frac{\lambda_t}{\varepsilon_{frac}}\right)^3\right)\exp\left(-\mathrm{c_2}\frac{\lambda_t}{\varepsilon_{frac}}\right) + \frac{\lambda_t}{\varepsilon_{frac}}\frac{\lambda_t}{\varepsilon_{frac}}(1 + \mathrm{c_1^3})\exp(-\mathrm{c_2}) \tag{14.9}$$

where ε_{frac} is the fracture strain, and c_1 and c_2 are constants. A volumetric tensile damage ΔD_t is used to resolve the situation that no damage is accumulated when the stress path is close to the tri-axial tensile path, which will not repeated here. The total damage index D is defined as

$$D = 1 - (1 - D_c)[1 - (D_t + \Delta D_t)] \tag{14.10}$$

For the complete description of a material model in hydrocode, a proper EoS is needed to relate pressure to volumetric strain. The tabulated compaction EoS (EOS #8) in LS-DYNA is used in the Kong-Fang model, which defines the current pressure P as a function of the volumetric strain μ,

$$P = C(\mu) + \gamma_0\theta(\mu)E_0 \tag{14.11}$$

where E_0 is the internal energy per initial volume, and γ_0 is the ratio of specific heat. $C(\mu)$ and $\theta(\mu)$ are the tabulated pressure and temperature as functions of volumetric strain, respectively. The term related to the internal energy is neglected in the present study. Therefore, ten pairs of pressure and volumetric strain, and unloading bulk modulus and volumetric strain should be defined. Further details of this EoS is referred to the keyword manual of LS-DYNA (LSTC 2007).

14.2.2 New Tensile Damage Model

Previous numerical studies on concrete material models demonstrated that the tensile damage model is crucial for prediction of tensile-dominated responses (Kong et al. 2016; Liu et al. 2009; Leppänen 2006; Tu and Lu 2010; Kong et al. 2017a; Tu and Lu 2009). Compared with traditional concrete, tensile crack and tensile softening behavior of UHPCC material are more complex due to the incorporation of fibers that can produce obvious bridging stress between the disconnected crack surfaces. Based on our experimental data (Ren et al. 2018b) and previous finite element inverse analysis (Kang et al. 2010), a four-stage tensile fracture model is proposed and shown in Fig. 14.1. Description and justification of each stage are discussed as follows.

The first stage is elastic described by the peak strength (T) and corresponding strain ($\varepsilon_{\max}$). In this stage, there are no visible major cracks, and the matrix and fibers can be regarded as an entire body. It should be pointed out that strain-hardening before peak strength is not considered. This assumption is acceptable since experimental data (Ren et al. 2018b) shows that strain hardening is characterized by a limited stress increase, and the fracture energy dissipated is small compared with strain-softening

Fig. 14.1 Four-stage tensile fracture model of UHPCC material

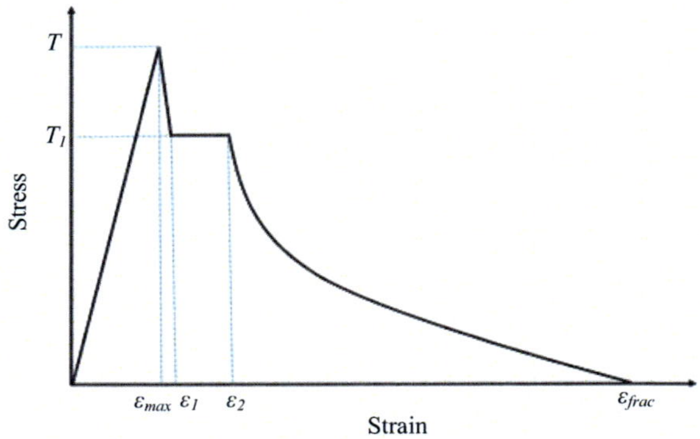

stage. Previous numerical study also demonstrated that the strain hardening can be ignored in the FE analysis of UHPCC materials with low strain-hardening behavior (Othman and Marzouk 2018). After the peak stress, strain-softening occurs, which is defined by three different stages as follows.

(i) In the first sharp softening stage, cracks are gradually produced and expanded in the matrix. When the cracks develop to a certain width, the bridging effect between the matrix and fibers is activated and the fibers with large inclination angle to the stress direction may be pulled out, beyond which the second stage occurs.

(ii) In the second stage, stresses are redistributed between matrix and randomly distributed fibers, and the fibers with little inclination angle to the stress direction start to bear loads. When the fibers reach their yield stress, cracks continue expanding while the resultant force remains almost unchanged, leading to a platform of stress-strain curve of UHPCC shown in Fig. 14.1. It should be pointed out that the existence of this platform stage may depend on the fiber type, fiber length and etc. If the bonding failure between matrix and fibers happens before the yield of fibers, the stress will gradually decrease with the increase of strain and consequently no platform branch exists. For the steel fibers concerned in this study, our experimental data (Sect. 14.3.1) shows that the platform stage indeed exists. Nevertheless, the present tensile damage model can also consider other UHPCC material without this stage, which can be easily done by adjusting corresponding parameters discussed as follow.

(iii) As the crack width increases, the fibers are gradually pulled out or broken depending on fiber type, fiber length, matrix and etc., and the final softening stage appears. This stage finishes when all fibers are pulled out or broken, and a nonlinear stress-strain curve is used to describe this stage. The proposed three tensile-softening stages are demonstrated by our previous direct tensile tests (Ren et al. 2018b), and finite element inverse analysis by Kang et al. (2010), which will be further validated in Sect. 14.3.1.

Based on previous discussion, the tensile stress-strain curve of UHPCC may be expressed as,

$$\sigma = \begin{cases} E * \varepsilon, & \varepsilon < \varepsilon_{\max} \\ \left(\frac{T}{T_1}\right) * (\varepsilon - \varepsilon_{\max})/(\varepsilon_{\max} - \varepsilon_1) + T, & \varepsilon_{\max} \le \varepsilon < \varepsilon_1 \\ T_1, & \varepsilon_1 \le \varepsilon < \varepsilon_2 \\ \frac{T_1 * (\varepsilon - \varepsilon_{frac})^2}{(\varepsilon_{frac} - \varepsilon_2)^2}, & \varepsilon_2 \le \varepsilon < \varepsilon_{frac} \\ 0 & \varepsilon_{frac} \le \varepsilon \end{cases} \tag{14.12}$$

where T is the tensile strength and T_1 is the platform strength. $\varepsilon_{\max}$, ε_1, ε_2 and ε_{frac} are cracking strain, first softening strain, second softening strain, and fracture strain, respectively. As observed in Eq. (14.12), the platform branch can be easily neglected by equaling ε_1 and ε_2.

To construct the dynamic tensile stress-strain curve, relations of the parameters in Eq. (14.12), such as ε_1, ε_2 and ε_{frac} with the strain-rate are needed. However, existing experiments on the dynamic tensile behavior of UHPCC mostly focus on relation of DIFt and strain-rate and such detailed information is always absent (Wu et al. 2018; Thaomas and Sorensen 2017; Kang et al. 2010). It is assumed that only the tensile strength T increases with the increase of strain-rate while other parameters remain unchanged. This intuitive assumption is supported by the following considerations. Firstly, the platform stage mainly results from the yield of fibers. Since fibers, e.g. steel fibers usually have small rate-sensitivity compared to the matrix, the platform parameters (ε_1, ε_2 and T_1) could be treated as rate-independent. Secondly, the fracture strain ε_{frac} is rate-independent inspired by the fact that fracture strain of traditional concrete is a constant under high strain-rate (Weerheijm and van Doormaal 2007; Schuler et al. 2006). The assumptions above make the fracture energy increase with the increase of strain-rate, which is also supported by recent spalling tests on traditional concrete (Weerheijm and van Doormaal 2007; Schuler et al. 2006).

Considering $\varepsilon_{\max}$ is relatively small compared with other softening strains (e.g. ε_{frac}), the tensile damage can be easily defined as,

$$D_t = \begin{cases} \left(1 - \frac{T_1}{T}\right) * \frac{\lambda_t}{\varepsilon_1}, & \lambda_t < \varepsilon_1 \\ 1 - \frac{T_1}{T}, & \varepsilon_1 \le \lambda_t < \varepsilon_2 \\ 1 - \frac{T_1 (\lambda_t - \varepsilon_{frac})^2}{T (\varepsilon_2 - \varepsilon_{frac})^2} & \varepsilon_2 \le \lambda_t < \varepsilon_{frac} \end{cases} \tag{14.13}$$

There are five parameters need to be carefully defined in the new tensile damage model, which are calibrated using our direct tensile experiments on dog-bone UHPCC specimen (Ren et al. 2018b). In this test, five volume fractions (0.5, 1.0, 1.5, 2.0 and 2.5) of micro-straight steel fibers (diameter of 0.2 mm, length of 13 mm and tensile strength of 2800 MPa) were considered and the tensile stress-strain curves were carefully measured. More detail of this test is referred to (Ren et al. 2018b).

Figure 14.2 presents the effect of fiber volume fraction on the five parameters, i.e., T, T_1, ε_1, ε_2 and ε_{frac}.

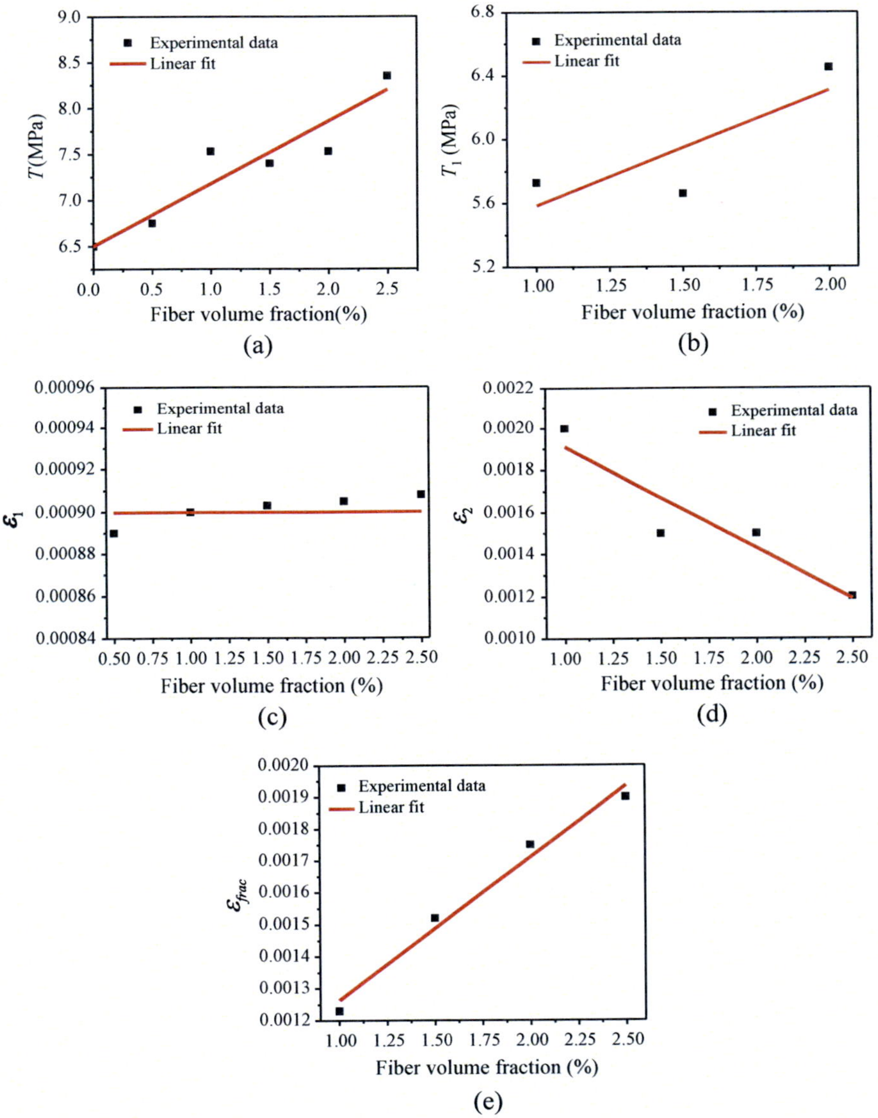

Fig. 14.2 Influence of fiber volume fraction on the tensile strength T (**a**), platform stress T_1 (**b**), first softening strain ε_1 (**c**), second softening strain ε_2 (**d**) and fracture strain ε_{frac} (**e**)

It can be observed that all of the five parameters almost increase linearly with the increase of fiber volume fraction V_f. Therefore, the following equations are proposed to determine these five parameters.

$$T = 80 \times V_f + 6.5 \quad (\text{unit in MPa}) \tag{14.14}$$

$$T_1 = 5.45 + 72 \times V_f \quad (\text{unit in MPa}) \tag{14.15}$$

$$\varepsilon_1 = 9\mathrm{e}^{-4} \tag{14.16}$$

$$\varepsilon_2 = 0.002 - \left(6\mathrm{e}^{-2}\right) \times V_f \tag{14.17}$$

$$\varepsilon_{frac} = 0.0125 + \left(V_f - 0.01\right) \times 0.5 \tag{14.18}$$

14.2.3 *Parameter Calibration*

Three set of parameters should be defined in the Kong-Fang model, i.e., parameters related to the failure surface, the damage and EoS. Based on existing and our test data, these parameters are carefully calibrated in this section.

14.2.3.1 Parameters Related to Failure Surface

As discussed in Sect. 14.2.1, parameters related to the failure surface include the elastic modulus E, the Poisson's ratio ν, unconfined uniaxial compressive strength f_c and tensile strength T, material constants a_1 and a_2, tensile-to-compressive meridian ratio ψ and dynamic increase factors for tension and compression (DIFt and DIFc). It should be pointed out that only UHPCC with steel fibers is considered.

Usually, the f_c, T and E are available in the impact and blast experiments. If E is absent, it can be approximately determined by cwith E_m and E_s being the Young's modules of matrix and steel fibers, respectively (Wille et al. 2014). E_m can be estimated as $E_m = k \times f_c^{1/3}$ (both E_m and f_c are in MPa units) with the value of k ranged from 8800~11,000 and E_s as 210GPa. Experimental data (Ren et al. 2018b) shows that the Poisson's ratio of UHPCC material is around 0.23, and it is almost unchanged with the fiber volume content, thus a fix value of 0.23 is suggested. Then the bulk modulus K and shear modulus G can be determined according to $K = E/3(1 - 2\nu)$ and $G = E/2(1 + \nu)$, respectively.

Material constants a_1 and a_2 should be determined by a suitable set of TXC data. In the original Kong-Fang model, $a_1 = 0.5876$ and $a_2 = 0.025/f_c$ are determined by a large amount of TXC data of traditional concrete. Figure 14.3 presents our

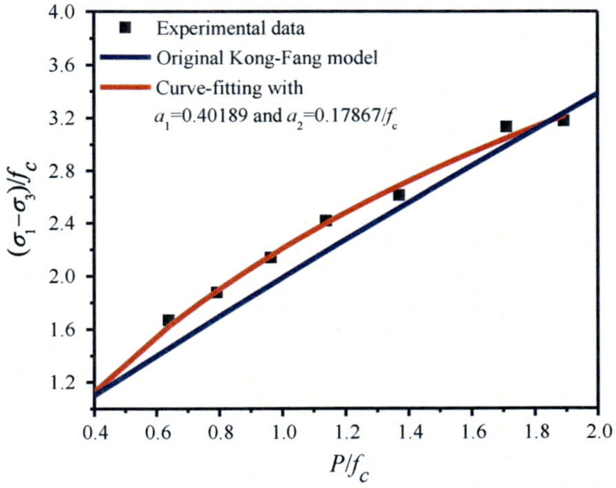

Fig. 14.3 Triaxial compression data and predictions by original and modified sets of a_1 and a_2

recent TXC data of UHPCC (Ren et al. 2016) along with the predictions by the original Kong-Fang model. It can be observed that the original Kong-Fang model underestimates the test data, and the deviation decreases with the increase of pressure. A new set of parameters is proposed by curve-fitting the data shown in Fig. 14.3 as a_1=0.40189 and a_2=0.17867/f_c.

For the UHPCC, the tensile-to-compressive meridian ratio ψ is modified as,

$$\psi(P) = \begin{cases} 0.5 & P \leq 0 \\ 0.5 + 1.5T/f_c & P = \frac{f_c}{3} \\ 1.07/[1 + 1.14/(3a_1 + 2.14a_2 f_c)] & P = \frac{2.14f_c}{3} \\ 0.753 & P = 3f_c \\ 1 & P \geq 8.45f_c \end{cases} \tag{14.19}$$

The first point ($\psi = 0.5$) is determined by the uniaxial tensile test corresponding to the stress state (P, Y)=(-T/3, T), the second by the uniaxial compressive test corresponding to the stress state (P, Y)=(f_c/3,f_c). The third point corresponds to the biaxial compression test, which suggests that the biaxial compression strength is 1.07f_c (Curbach and Speck 2010) so that the stress state (P, Y)=(2.14f_c/3, 1.07f_c) should be satisfied. For a high pressure, triaxial tension (TXE) data is needed, which is, however, limited in relevant literatures. Consequently, the original suggested values for traditional concrete are used here. It should be pointed out that in computational code, $\psi(P)$ is determined by a linear interpolation of the discrete points described in Eq. (14.19)

UHPCC is known as a strain-rate sensitive material, and the strain-rate enhancement is usually represented by the ratio of dynamic strength to static strength termed as the DIF. Review of the experimental data on DIFt of UHPCC material demonstrates that effects of fiber volume content, fiber type and fiber distribution on DIFt

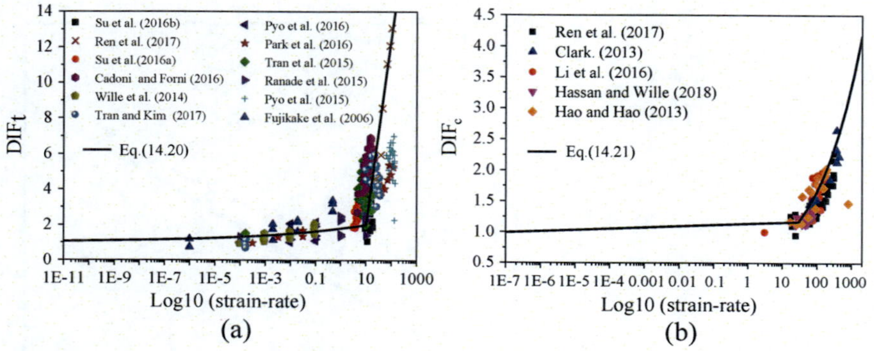

Fig. 14.4 Experimental DIFt (**a**) and DIFc (**b**) and fitted empirical formulas

are negligible (Thaomas and Sorensen 2017). Therefore, it is reasonable to collect existing data to construct empirical formulas for DIFt and DIFc. A large amount of experimental data on DIF-$\dot{\varepsilon}$ is shown in Fig. 14.4 (Ren et al. 2017; Su et al. 2016b; Cadoni and Forni 2016; Wille et al. 2016; Tran and Kim 2017; Pyo et al. 2016; Park et al. 2016; Tran et al. 2015; Ranade et al. 2015; Pyo et al. 2015; Fujikake et al. 2006; Clark 2013; Li et al. 2016; Hassan and Wille 2018; Hao and Hao 2013). By curve-fitting the test data, DIFt and DIFc for UHPCC material are separately defined as,

$$\text{DIFt} = \begin{cases} \left(\frac{\dot{\varepsilon}}{\dot{\varepsilon}_0}\right)^{0.01} + 1, & \dot{\varepsilon} < \dot{\varepsilon}_0 \\ 10.5 \log \dot{\varepsilon} - 8.5, & \dot{\varepsilon} \geq \dot{\varepsilon}_0 \end{cases} \tag{14.20}$$

where $\dot{\varepsilon}_0 = 10s^{-1}$ is the transition strain-rate for tension, and

$$\text{DIFc} = \begin{cases} \left(\frac{\dot{\varepsilon}}{\dot{\varepsilon}_1}\right)^{0.01} + 0.17, \dot{\varepsilon} < \dot{\varepsilon}_1 \\ \left(\frac{\dot{\varepsilon}}{\dot{\varepsilon}_1}\right)^{0.33} + 0.17, \dot{\varepsilon} \geq \dot{\varepsilon}_1 \end{cases} \tag{14.21}$$

with $\dot{\varepsilon}_1 = 30s^{-1}$ being the transition strain-rate for compression. To avoid overestimation of DIF, cutoff values of 10 and 4 are applied to the DIFt and DIFc, respectively. Note that during numerical implementation, the strain-rate uses the corrected strain-rate $\bar{\dot{\varepsilon}}$ defined in Eq. (14.6).

14.2.3.2 Parameters Related to Damage

There are six parameters, i.e., compressive-softening parameters a and constants d_1 and d_2, and tensile-softening parameters T_1, ε_1, ε_2 and ε_{frac}. The tensile-softening

Table 14.1 Parameters of the modified Kong-Fang model

Failure surface		Damage	
fc	–	T_1 (MPa)	$5.45 + 72 \times V_f$
T (MPa)	$80 \times V_f + 6.5$	ε_1	$9e^{-4}$
E (MPa)	$(1 - 100V_f)E_m + V_f E_s$	ε_2	$0.002 - (6e^{-2}) \times V_f$
K (MPa)	$E/3(1 - 2\nu)$	ε frac	$0.0125 + (V_f - 0.01) \times 0.5$
G (MPa)	$E/2(1 + \nu)$	α	1
υ	0.23	d1	0.03
a1	0.40189	d2	1.5
a2	$0.17867/ f_c$	d3	0.1

parameters are calibrated based on our previous test data in Sect. 14.2.4 and not repeated here.

The compressive-softening parameters are determined by trial-and-error method until the predicted uniaxial stress-strain curve agrees well with the experimentally obtained uniaxial stress-strain relations. $\alpha = 1$, d_1=0.03 and d_2=1.5 are suggested.

It should be pointed out that the introduce of strain-softening, especially the tensile strain-softening in a material model unavoidably results in the mesh size dependency (Kong et al. 2018b). To mitigate the mesh size dependency, the smeared crack approach is frequently used in previous studies (Guo et al. 2017; Othman and Marzouk 2018), in which the area under the tensile strain-stress curve is hold as G_f/h_c with G_f and h_c being the fracture energy and characteristic length of an element. This approach is appropriate for static loadings, but is inappropriate for dynamic loadings, which mainly results from the change of failure pattern from a major single crack under static loadings to multiple parallel cracks under dynamic loadings (Kong et al. 2018b). A new damage-based nonlocal model was recently proposed by Kong et al. (2019) to resolve this issue, which is, however, beyond the scope of the present study. The smeared crack approach is not used but mesh sensitive analysis is conducted before all simulations. Table 14.1 presents all the suggested parameters related to the failure surface and damage in the modified Kong-Fang model.

14.2.3.3 Parameters Related to EoS

As discussed in the Introduction, the EoS used to describe the compaction behavior of UHPCC under high pressure depends on the porosity in materials. Since the UHPCC material and traditional concrete have different porosity, the parameters related to EoS must be calibrated.

Figure 14.5 presents the pressure-volumetric strain test data for UHPCC materials obtained from hydrostatic pressure test (Erzar et al. 2016) and flyer-plate-impact tests

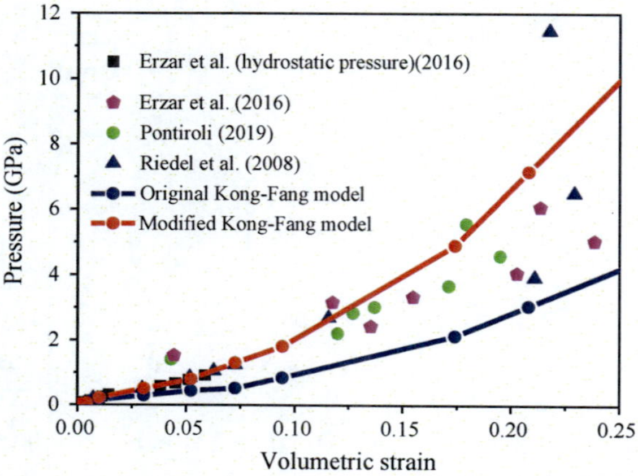

Fig. 14.5 Pressure-volumetric strain data and predicted EoS by original and modified sets of parameters

(Erzar et al. 2016; Pontiroli 2019) and high strength concrete obtained from flyer-plate-impact test (Riedel et al. 2008). The predicted EoS using the original set of parameters in the Kong-Fang model is also shown in Fig. 14.5. It is clearly observed that the original parameters greatly underestimate the pressure at a given volumetric strain, which mainly results from the low porosity of the UHPCC material. Using the test data, ten pairs of pressure and volumetric strain, and unloading bulk modulus and volumetric strain can be determined and listed in Table 14.2. Note that the unloading bulk modules are obtained by an empirical estimation since no test data is available. It should be pointed out that these EoS parameters can also be used for other material models, such as the HJC model, the RHT model and the K&C model.

Table 14.2 Ten pairs of EoS parameters for UHPCC material

Number	Volumetric strain	Pressure (GPa)	Unloading bulk modulus (GPa)
1	0	0	K
2	0.0015	0.04 (P2)	K
3	0.0043	1.96P2	1.014 K
4	0.0101	4.41P2	1.065 K
5	0.0305	11.8P2	1.267 K
6	0.0513	21.5P2	1.470 K
7	0.0726	32.0P2	1.672 K
8	0.0943	47.3P2	1.825 K
9	0.1740	128.0P2	4.106 K
10	0.2080	181.0P2	5.000 K

14.3 Single Element Tests

The proposed tensile damage model is incorporated into the Kong-Fang model and implemented into LS-DYNA through the user-defined material model (LSTC 2007). To demonstrate the improved performances of the tensile damage model and the modified set of parameters, numerical predictions by the original Kong-Fang model using the automatic generated parameters suggested in (Kong et al. 2018a) and the modified Kong-Fang model using the parameters in Table 14.1 and Table 14.2 at material point level are compared using a single element.

The single element model with size of 10 mm × 10 mm × 10 mm is employed, as shown in Fig. 14.6. For the unconfined boundary condition, three symmetry planes are used so that only 1/8 of a structure is modeled. When confinement is needed, the three free planes are applied with hydrostatic pressure. The top face is pushed down for unconfined uniaxial compression (UUC) and triaxial compression (TXC) tests, and pulled up for unconfined uniaxial tension (UUT) test.

14.3.1 Unconfined Uniaxial Tests

Figure 14.7 presents the comparison of the numerical predicted strain-stress curves of the UUC test with the test data by Ren et al. (2018b). It is observed that the plastic-softening behavior is well reproduced by the original and modified models. However, it is noted that the original Kong-Fang model underestimates the elastic modulus, which is determined by a built-in formula ($57000\sqrt{f_c/145}$, both E and f_c are in MPa units) and can be modified by users. In other material models, such as the K&C model, the elastic modulus is automatically calculated by the formula above and cannot be separately defined by users. Obviously, this will lead to underestimation of structural stiffness.

Recently, Ren et al. (2018b) conducted several sets of direct tensile experiments on UHPCC materials, in which five volume fractions, i.e. 0.5, 1.0, 1.5, 2.0 and 2.5%, of

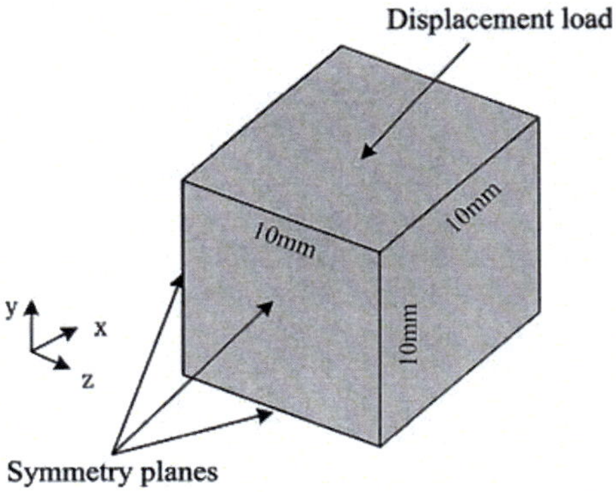

Fig. 14.6 FE model of single element test

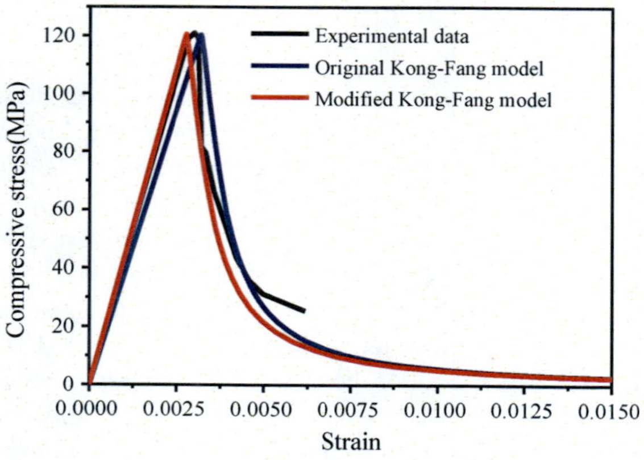

Fig. 14.7 Comparisons of the experimental strain-stress curve of UUC test with predictions by the original and modified Kong-Fang models

straight steel fibers were used. Figure 14.8 presents the selected experimental stress-strain curves along with the corresponding numerical predictions by the original and modified Kong-Fang models. It can be observed that the modified model shows a good agreement with the experimental data, and the first sharp softening branch, the second platform branch and the final softening branch are well reproduced. However, the original model has a larger softening gradient when compared with the test data. This will result in underestimations of energy absorption capability and structure resistance, which will be demonstrated in Sect. 14.4.

14.3.2 Triaxial Compression Test

The compressive behavior of high confinement is important for impact and blast loadings where materials near the projectile and explosive are subjected to a high pressure. The triaxial compressive behavior of UHPCC under high confining pressure (up to 100 MPa) was experimentally studied in our recent work (Ren et al. 2016) using the cylindrical specimens with a diameter of 50 mm and length of 100 mm. Figure 14.9 shows the experimentally obtained axial stress-strain curves under different confining pressures and corresponding numerical predictions by the original and modified models. It should be pointed out that the elastic modulus in the original Kong-Fang model is modified based on the test data to have comparable predictions. It can be observed that the modified model shows reasonable agreement with the test data in terms of peak stress and softening gradient. Although predictions by the original model have a similar softening gradient with test data, the peak stress is overestimated, which mainly results from the overestimation of the strength surface shown in Fig. 14.3.

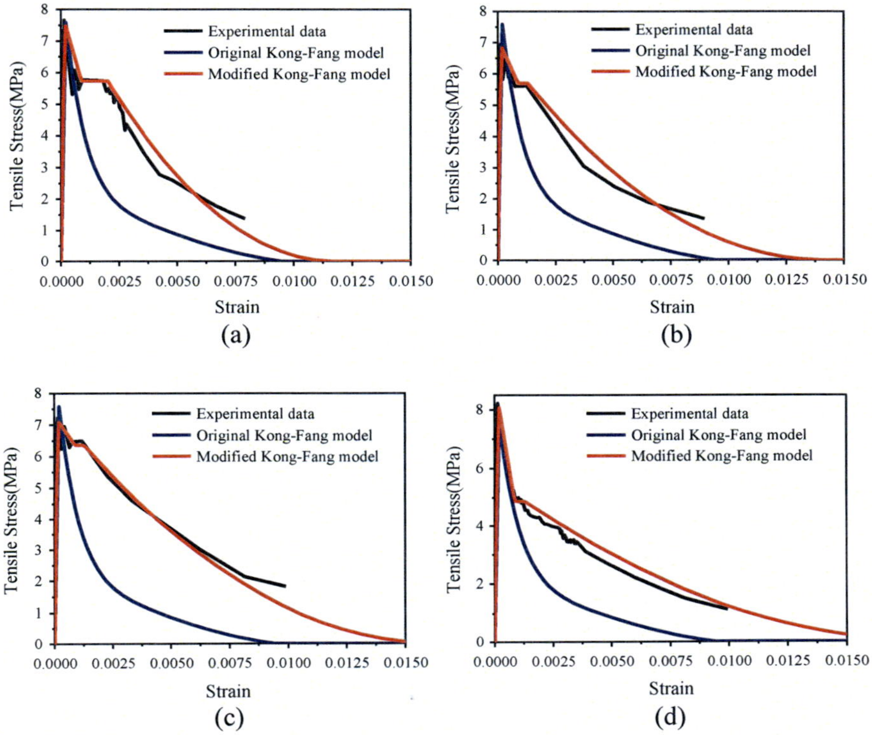

Fig. 14.8 Comparisons of tensile strain-stress curves predicted by the original and modified Kong-Fang models with the test data with steel volume fractions of **a** 1%, **b** 1.5%, **c** 2% and **d** 2.5%

14.4 Experimental Validation

To further validate and evaluate capability of the modified Kong-Fang model to predict dynamic responses of UHPCC structures subjected to impact and blast loadings, three selected experiments on UHPCC structures subjected to impact and blast loadings are numerically simulated and compared with corresponding experimental data. Mesh-size sensitivity analysis is conducted before all the simulations, and generally, numerical predictions are not very sensitive to the mesh size when the mesh size is smaller than a critical value. Unless stated otherwise, all of the parameters in the modified material model are determined by Tables 14.1 and 14.2. For a comparison purpose, numerical predictions by the original Kong-Fang model with the automatically generated parameters suggested in (Kong et al. 2018a) are also presented.

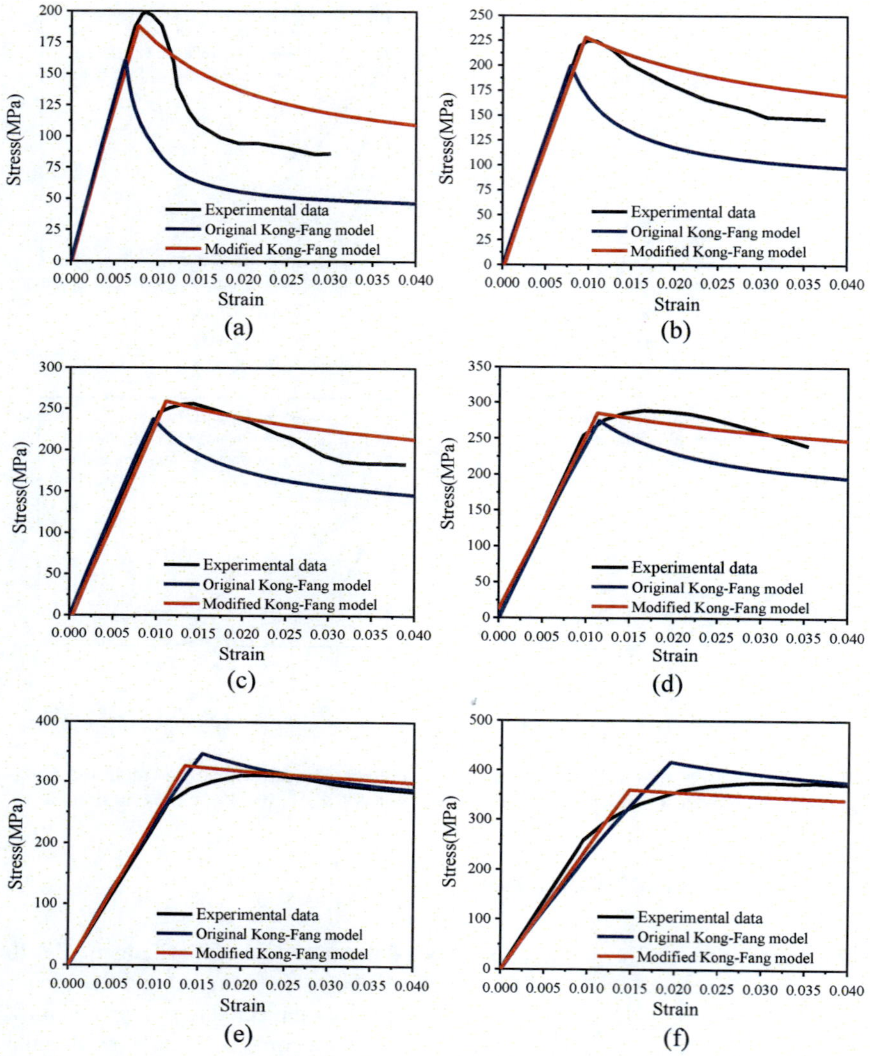

Fig. 14.9 Comparisons of compressive stress-strain curves between predictions by the original and modified Kong-Fang models and test data under a confining pressure of **a** 10 MPa, **b** 20 MPa, **c** 30 MPa, **d** 40 MPa, **e** 60 MPa, **f** 80 MPa and **g** 100 MPa

14.4.1 UHPCC-FST Column Subjected to Low Speed Impact

Wu et al. (2019) conducted experiments on dynamic behavior of UHPCC filled steel tube (UHPCC-FST) subjected to a drop-hammer impact. In this test, the UHPCC material had a cubic compressive strength of 141.5 MPa, elastic modulus of 43.8

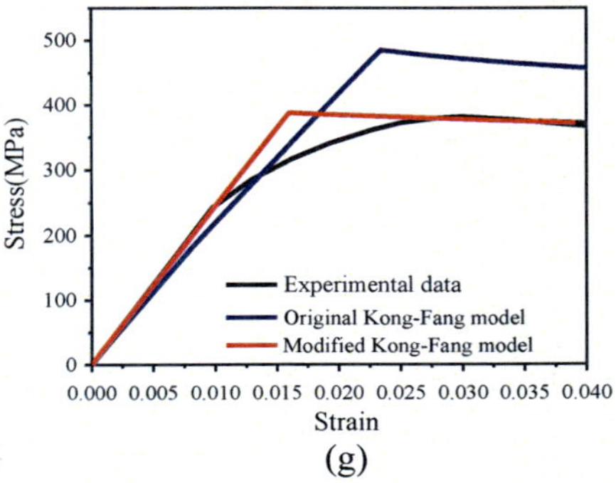

Fig. 14.9 (continued)

GPa and Poisson's ratio of 0.23, respectively. The micro-straight steel fibers with a diameter of 0.2 mm, length of 13 mm volume fraction of 2.0% were used. The UHPCC material was cast in a steel tube made from 20# steel, which had an outer diameter of 203 mm, length of 2000 mm and thickness of 6 mm.

A drop-hammer impact device was used to conduct the impact test. The total mass of hammer was 530 kg, including the hammer frame with a mass of 200 kg, the weight blocks of 6 × 50 kg and indenter of 30 kg. The specimen was simply supported by the frames with the clear span length of 1.6 m. The hammer was dropped freely along the guide rails at heights of 5.0 m, 8.0 m and 11.0 m once electromagnet release was triggered and the impact forces were recorded. Further details of the experiments can be referred to (Wu et al. 2019).

The finite element (FE) model is shown in Fig. 14.10. The steel tube, UHPCC, drop hammer and supporting system are modeled using the 3D solid elements, and the interface between the tube and UHPCC are completely bonded so that the possible slip between them is not considered. For all components, an approximate element size of 10 mm × 10 mm × 10 mm is used based on a mesh-size sensitivity analysis. The frictionless contacts between the UHPCC-FST column and hammer and supporting system are considered by the automatic contact algorithm available in LS-DYNA.

For constitutive models, the hammer and supporting system are modelled as rigid bodies since little deformations are observed after test. The steel tube is modelled by the bilinear plastic model with an elastic modulus of 206 GPa, tangent modulus of 0.8 GPa and yield stress of 318 MPa determined by the experimental stress-strain curve (Wu et al. 2019). The UHPCC material is modelled by the modified Kong-Fang model with parameters listed in Tables 14.1 and 14.2.

Only the test with the drop height of 8 m is simulated, and Fig. 14.11 presents experimental and numerically predicted impact force-time curves. It can be observed that the prediction by the modified Kong-Fang model agrees well with the test data in terms of peak force and response time. However, the original Kong-Fang model underestimates the peak force about 30% which mainly results from the inaccuracy of the original tensile damage model. As shown in Fig. 14.8, the original Kong-Fang

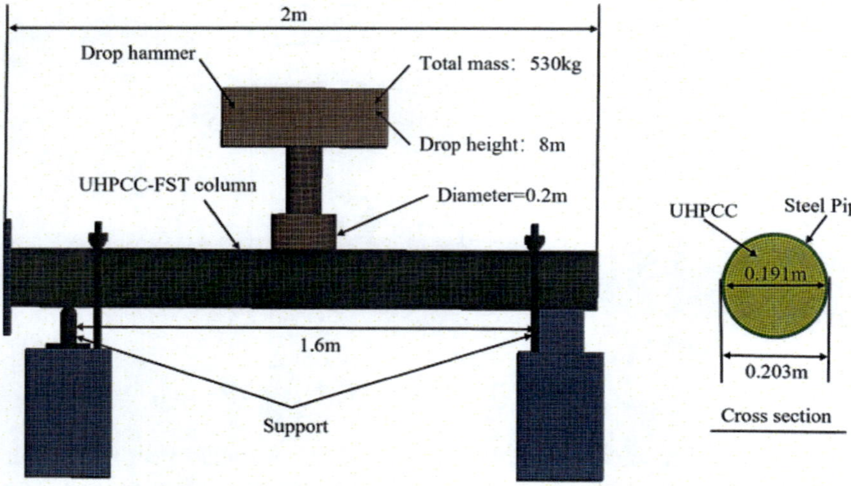

Fig. 14.10 FE model of drop-hammer test on UHPCC-FST

Fig. 14.11 Comparison of impact force between test data and numerical predictions

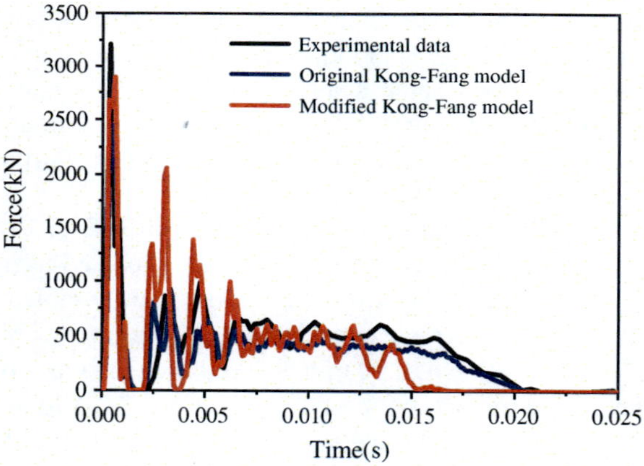

model underestimates the capability of energy absorption determined by the area below the tensile stress-strain curve, which leads to an underestimation of the peak reaction force shown in Fig. 14.11. Numerical predicted tensile, compressive and total damage distribution with the experimental deformation are shown in Fig. 14.12. It is observed that the compressive damage mainly occurs around the impact area while the tensile damage exists at the rear of impact area since it is mainly subjected to a tensile stress state. Numerical predicted deformation and damage distribution excellently agree with the experimental observation.

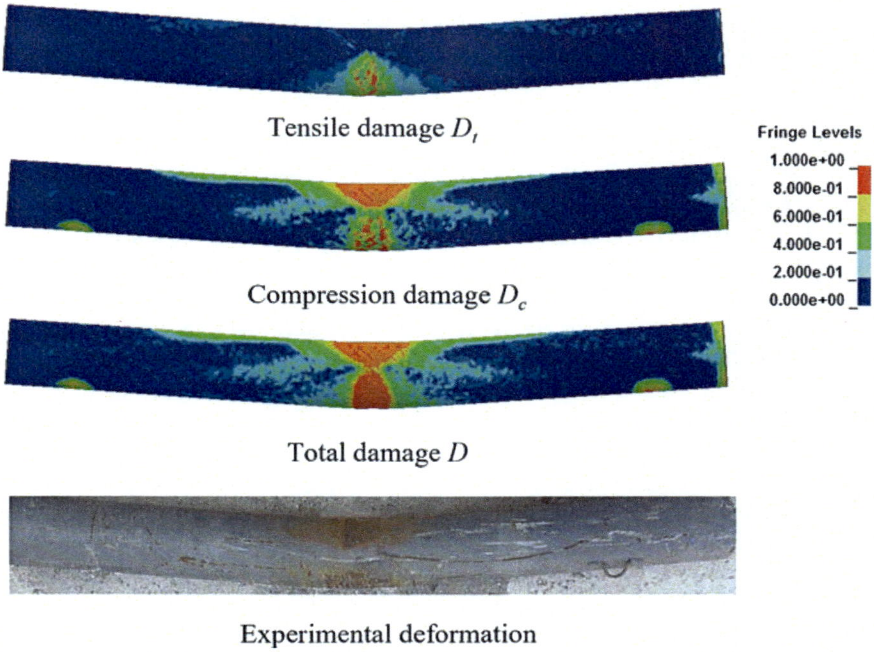

Fig. 14.12 Comparison of the numerically predicted tensile, compressive and total damage distribution with the experimental deformation

14.4.2 UHPCC-FST Column Subjected to Near Explosion

The dynamic response of UHPCC-FST column subjected to a close-in TNT explosion was conducted by Wang et al. (2020) and is numerically simulated in this section. The setup for this field blast test was reported in detail in (Wang et al. 2020) and is briefly shown in Fig. 14.13. The UHPC-FST column used in this test is exactly the same as that used in the drop-hammer impact introduced in Sect. 14.4.1 and not repeated here. The top of the column was clamped by a steel bolt to represent the

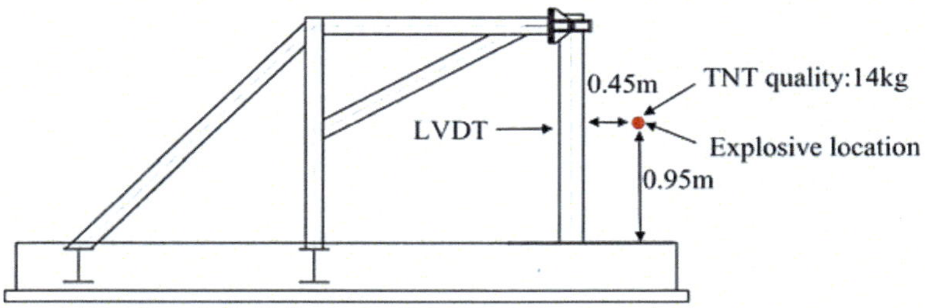

Fig. 14.13 Test setup for UHPCC-FST column subjected to a close-in TNT explosion

simple supporting boundary. The reaction frame was made of high strength steel and was fixed by the RC foundation. The cylindrical TNT charge had a diameter of 270 mm, height of 280 mm and mass of 25 kg, which was placed at the center of the column with a standoff distance of 0.45 m. The deflection-time history at the mid-span of the column was recorded using the LVDT transducer.

The so-called ALE method is always used to simulate the interaction of blast shock waves generated by explosive and structures, in which the explosive and air are modelled as Eulerian grid and structures as Lagrangian grid and their interaction are determined by an automatic coupling algorithm available in LS-DYNA. However, this approach usually involves a huge number of elements and time-expensive. An alternative approach is the so-called velocity method proposed by Remennikov and Uy (2014). In this method, the blast loading generated by explosive is simplified as a velocity loading applied on the structures as a boundary condition so that only the UHPCC-FST column is needed to be modelled, which greatly improves computational efficiency. Recent study by Wu et al. (2020) demonstrated that the velocity method is preferred for modelling the close-in explosion with the consideration of both computational accuracy and efficiency.

As shown in Fig. 14.14, the steel tube, UHPCC and steel bolt are modelled by the 3D solid elements with an approximate element size of 10 mm $\times$10 mm $\times$ 10 mm. The fix boundaries are applied at both the steel bolt and the bottom of the column. The steel bolt is treated as a rigid body as almost no deformation is observed after the test. The material models for the steel tube and the UHPCC material and corresponding parameters are identical to those introduced in Sect. 14.4.1. The

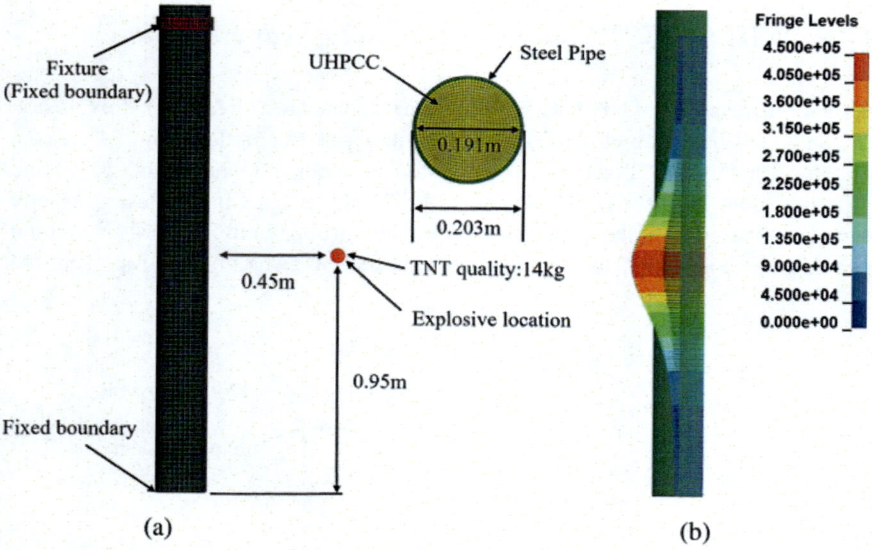

Fig. 14.14 FE model of UHPCC-FST column (**a**) and velocity boundary condition (**b**)

velocity field determined by the velocity method is applied on the column as a boundary condition, as shown in Fig. 14.14b.

Comparison of deflection-time history between experimental data and numerical predictions is shown in Fig. 14.15. The modified Kong-Fang model reproduces the test data well. However, the FE model of the original Kong-Fang model is destroyed at the time of 82 ms and the calculation is then stopped, which mainly results from the underestimation of energy absorption capability discussed above. Numerical predicted tensile, compressive and total damage distribution and the experimental deformation are shown in Fig. 14.16. It is observed that the compressive damage mainly occurs around the explosion side while the tensile damage mainly occurs

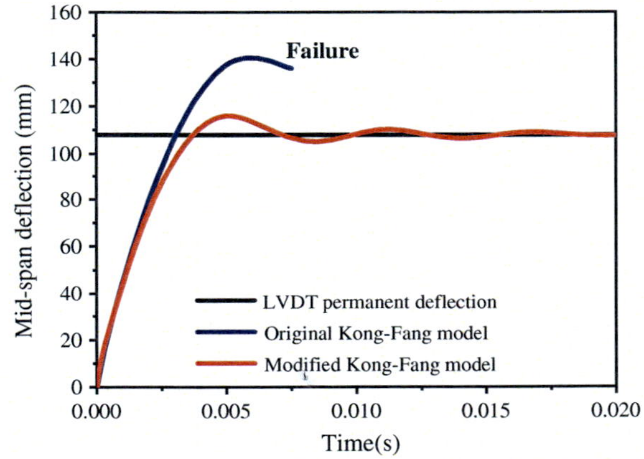

Fig. 14.15 Comparison of deflection-time curve between test data and numerical predictions

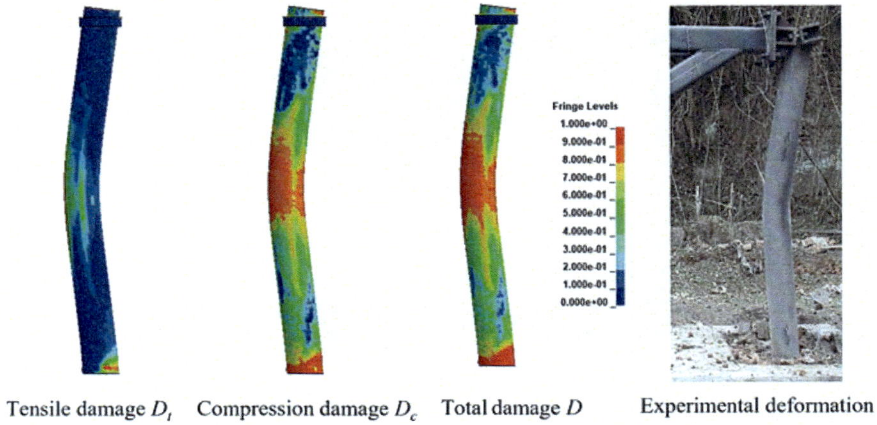

Fig. 14.16 Comparison of numerically predicted tensile, compressive and total damage distribution with experimental deformation

at the rear of explosion side since it is mainly subjected to a tensile stress state. Numerical predicted deformation and damage distribution excellently agree with the experimental observation.

14.4.3 Reinforced UHPCC Slab Subjected to Blast Loading

It is noted that the experimental validation in previous section are based on our own test data. To further validate the modified Kong-Fang model, the experiment on a reinforced UHPCC slab subjected to a blast loading conducted by Li et al. (2015) is numerically simulated. In this test, the UHPCC material had an unconfined compressive strength of 128.9 MPa, elastic modulus of 51.5 GPa, Poisson's ratio of 0.2 and steel fiber volume fraction of 2%. The slab had a size of 2000 mm (length) × 1000 mm (width) × 100 mm (height), and was longitudinally reinforced by rebar and its location is shown in Fig. 14.17. The short sides of the slab were supported by a designed frame while the long sides were free. A 14 kg-TNT explosive was suspended by a string over the slab at a height of 1 m (scaled standoff distance of 0.41 m/kg$^{1/3}$) and was detonated to generate blast loading. A pressure sensor was erected to record the free air reflected pressure and a LVDT transducer was used to record the deflection at the mid-span.

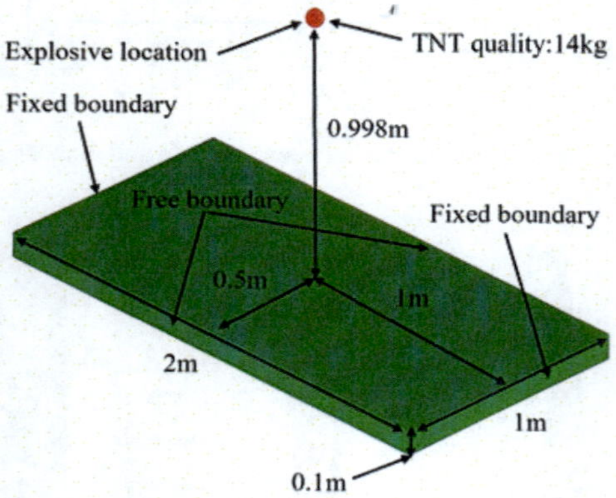

(a)

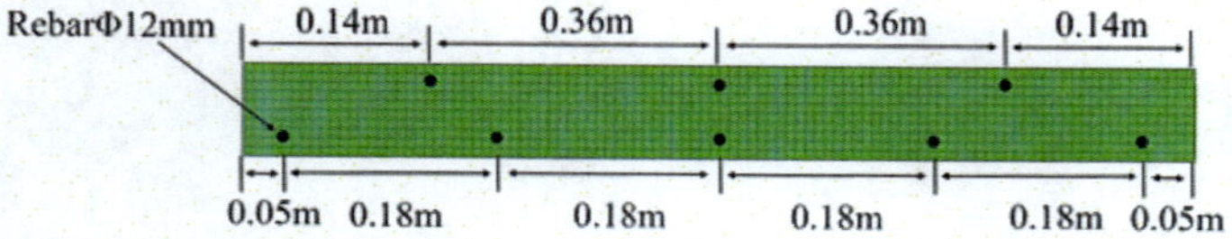

(b)

Fig. 14.17 FE model of reinforced UHPCC slab subjected to blast loading: (**a**) dimension of slab and location of TNT and (**b**) location of rebar

The FE model is shown in Fig. 14.17. The solid hexahedron element with a size of 10 mm × 10 mm × 10 mm is used to model the UHPCC and beam element is used for rebar. The rebar is modeled as the ideal elastic-plastic material for simplification with an elastic modulus of 206 GPa and yield stress of 600 MPa. The blast load is modelled by the Load_Blast function available in LS-DYNA, which is based on empirical models proposed in TM5-855-1 US army handbook (Baylot et al. 1984). Although this function was developed based on a spherical charge shape, the predicted reflected overpressure was compared with experimental recorded one and reasonable agreement was observed (Li et al. 2015).

Figure 14.18 presents the numerically predicted deflection-time curves along with the permanent deflection recorded by the LVDT. The modified Kong-Fang model well predicts the permanent deflection, while the slab is totally broken when using the original Kong-Fang model which is consistent with the same experiment (with the same slab dimension and TNT charge) on the normal strength concrete conducted in (Li et al. 2015). This demonstrates that the original Kong-Fang model with the original set of parameters is suitable for traditional concrete but not for UHPCC material. This also indicates that the material model originally developed for traditional concrete, e.g., the HJC model, the RHT model and the K&C model, without carefully parameter calibration or model modification, cannot be directly used to predict the dynamic responses of UHPCC structures subjected to intensive dynamic loading, since the use of these original material models may result in underestimation of structural stiffness even to a wrong prediction.

Figure 14.19 presents the final distribution of the total damage predicted by the original and modified Kong-Fang models along with failure pattern observed from experiments. It is clearly observed that the modified model can well predict the final deformation of the slab. The slab is totally broken when using the original Kong-Fang model, which mainly results from inaccuracy of the tensile damage model discussed in this chapter.

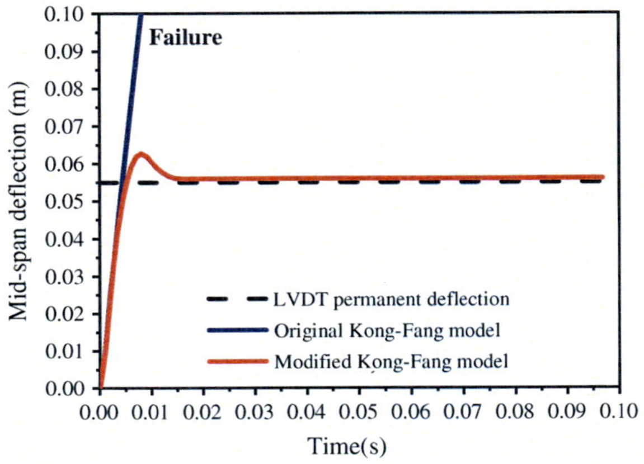

Fig. 14.18 Comparison of deflection-time curve between test data and numerical predictions

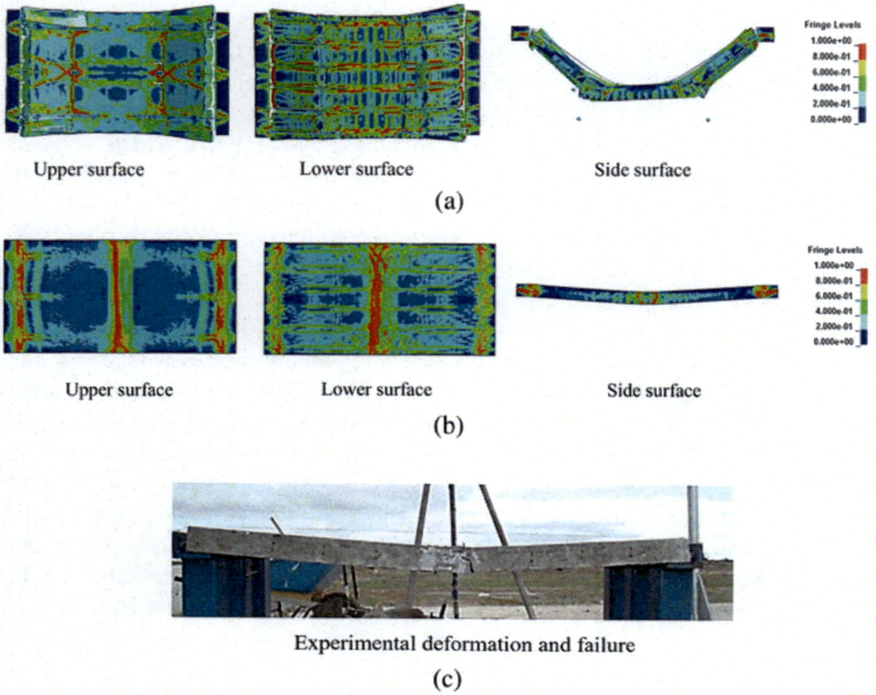

Fig. 14.19 Comparison of the numerically predicted total damage distribution by the original Kong-Fang model (**a**) and modified Kong-Fang model (**b**) with the experimentally observed deformation and failure (**c**)

14.5 Summary

A new constitutive model of UHPCC material under impact and blast loadings is developed based on the Kong-Fang material model recently developed, which involves two major steps. In the first step, a new tensile damage model is proposed for UHPCC considering the fiber bridging mechanism and is incorporated into the Kong-Fang material model. Then in the second step, all the parameters needed in the modified Kong-Fang material model are carefully calibrated based on our and existing test data. The modified material model is then implemented into the LS-DYNA through user defined material model, which is validated by single element tests and three selected experiments on UHPCC structures subjected to impact and blast loadings. Corresponding numerical predictions by the original Kong-Fang material model are also presented to demonstrate the improved performances due to the modification and parameter calibration. The main contributions and findings are as follows.

(1) A four-stage tensile fracture model for UHPCC material is proposed, including the linear elastic stage, the second sharp softening stage due to cracks in the

matrix, the third platform stage due to the fiber bridging mechanism and the final softening stage due to the pullout and/or broken of fibers. Then a corresponding dynamic tensile fracture model is presented based on the assumption that the strain-rate only affects the tensile strength similar to the traditional concrete. Finally, a new dynamic tensile damage model is established and incorporated into the Kong-Fang material model.

(2) All the parameters needed in the modified Kong-Fang material model, i.e., parameters related to the failure surface, damage and EoS are carefully calibrated using the static and dynamic uniaxial compressive and tensile tests, triaxial compression test, hydrostatic pressure test and flyer-plate-impact test, etc. It should be pointed out that the calibrated parameters, especially the DIF and EoS, can be directly used in other material models, e.g., the K&C material model.

(3) Comparison of the numerical predictions by the modified Kong-Fang material model with experimental data on UHPCC structures subjected to impact and blast loadings demonstrates that the modified Kong-Fang material model can well predict the dynamic response (impact force and deflection) and failure in UHPCC structures subjected to intensively dynamic loadings.

(4) Corresponding numerical predictions by the original Kong-Fang material model show that it cannot be used for UHPCC material. This indicates that material models originally developed for traditional concrete, without carefully parameter calibration or model modification, cannot be directly used for UHPCC material. The use of these original material models may result in underestimation of structural stiffness even to a wrong prediction.

References

AUTODYN. Theory Manual (Revision 4.3)[M]. Century Dynamics, Inc, 2013.

BAYLOT J T, BRITT J R, CARRE G L, COLTHARP D R, DAVIS L K, DRAKE J L, HUFF W L, KIGER S A, ROHANI B, ROOKE A D. Fundamentals of protective design for conventional weapons[M]. U.S. Army Engineer Waterways Experiment Station: 1984.

CADONI E, FORNI D. Experimental analysis of the UHPFRCs behavior under tension at high stress rate[J]. The European Physical Journal Special Topics, 2016, 225: 253–264.

CLARK J. Preliminary Investigation of ultra-high performance concrete behavior at high strain rates using the split-Hopkinson pressure bar[D]. Michigan: Michigan Technological Univ., 2013.

CURBACH M, SPECK K. Zweiaxiale Druckfestigkeit von ultrahochfestem Beton[J]. Beton-und Stahlbetonbau, 2010, 102(10): 664–673.

ERZAR B, PONTIROLI C, BUZAUD E. Shock characterization of an ultra-high strength concrete[J]. The European Physical Journal Special Topics, 2016, 225: 355–361.

FANG Q, WU H. UHPCC targets under projectile impact. In: concrete structures under projectile impact[M]. Singapore: Springer, 2017.

FUJIKAKE K, SENGA T, UEDA N, OHNO T, KATAGIRI M. Effects of strain rate on tensile behavior of reactive powder concrete[J]. Journal of Advanced Concrete Technology, 2006, 4(1): 79–84.

GUO W, FAN W, SHAO X D, SHEN D J, CHEN B S. Constitutive model of ultra-high-performance fiber-reinforced concrete for low-velocity impact simulations[J]. Composite Structures, 2017, 185: 307–326.

HAO Y, HAO H. Dynamic compressive behaviour of spiral steel fibre reinforced concrete in split Hopkinson pressure bar tests[J]. Construction and Building Materials, 2013, 48: 521–532.

HARTMANN T, PIETZSCH A, GEBBEKEN N. A hydrocode material model for concrete[J]. International Journal Protection Structures, 2010, 1(4): 443–468.

HASSAN M, WILLE K. Comparative experimental investigations on the compressive impact behavior of fiber-reinforced ultra high-performance concretes using split Hopkinson pressure bar[J]. Construction and Building Materials, 2018, 191: 398–410.

HOLMQUIST T J, JOHNSON G R, COOK W H. A computational constitutive model for concrete subjected to large strains, high strain rates, and high pressures. In proceedings of the 14th International Symposium on Ballistics[C]. 1993: 591–600.

HONG J, FANG Q, CHEN L, KONG X Z. Numerical predictions of concrete slabs under contact explosion by modified K&C material model[J]. Construction and Building Materials, 2017, 155: 1013–1024.

HUANG X P, KONG X Z, CHEN Z Y, FANG Q. A computational constitutive model for rock in hydrocode[J]. International Journal of Impact Engineering, 2020, 145: 103687.

KANG S T, LEE Y, PARK Y D, KIM J K. Tensile fracture properties of an ultra high performance fiber reinforced concrete (UHPFRC) with steel fiber[J]. Composite Structures, 2010, 92(1): 67–71.

KONG X Z, FANG Q, CHEN L, WU H. A new material model for concrete subjected to intense dynamic loadings[J]. International Journal of Impact Engineering, 2018 120: 60–78.

KONG X Z, FANG Q, CHEN L, WU H. Nonlocal formulation of the modified K&C model to resolve mesh-size dependency of concrete structures subjected to intense dynamic loadings[J]. International Journal of Impact Engineering, 2018, 122: 318–332.

KONG X Z, FANG Q, HONG J. A new damage-based nonlocal model for dynamic tensile failure of concrete material[J]. International Journal of Impact Engineering, 2019, 132: 103336.

KONG X Z, FANG Q, LI Q M, WU H, CRAWFORD J E. Modified K&C model for cratering and scabbing of concrete slabs under projectile impact[J]. International Journal of Impact Engineering, 2017, 108: 217–228.

KONG X Z, FANG Q, WU H, HONG J. A comparison of strain-rate enhancement approaches for concrete material subjected to high strain-rate[J]. International Journal Protection Structures, 2017, 8(2): 155–176.

KONG X Z, FANG Q, WU H, PENG Y. Numerical predictions of cratering and scabbing in concrete slabs subjected to projectile impact using a modified version of HJC material model[J]. International Journal of Impact Engineering, 2016, 95: 61–71.

KONG X Z, FANG Q, ZHANG J H, ZHANG Y D. Numerical prediction of dynamic tensile failure in concrete by a corrected strain-rate dependent nonlocal material model[J]. International Journal of Impact Engineering, 2020, 137: 103445.

LEPPÄNEN J. Concrete subjected to projectile and fragment impacts: modelling of crack softening and strain rate dependency in tension[J]. International Journal of Impact Engineering, 2006, 32(11): 1828–1841.

LI J, WU C Q, HAO H. An experimental and numerical study of reinforced ultra-high performance concrete slabs under blast loads[J]. Materials & Design, 2015, 82: 64–76.

LI J, WU C Q, HAO H. Residual loading capacity of ultra-high performance concrete columns after blast loads[J]. International Journal of Impact Engineering, 2015, 6(4): 649–670.

LI Q H, ZHAO X, XU S L, XIANG G. Influence of steel fiber on dynamic compressive behavior of hybrid fiber ultra high toughness cementitious composites at different strain rates[J]. Construction and Building Materials, 2016, 125: 490–500.

LIU Y, MA A A, HUANG F L. Numerical simulations of oblique-angle penetration by deformable projectiles into concrete targets[J]. International Journal of Impact Engineering, 2009, 36(3): 438–446.

LSTC. LS-DYNA Keyword User's Manual, Version 971[M]. 2007.

MALVAR L J, CRAWFORD J E, WESEVICH J W, SIMONS D. A plasticity concrete material model for DYNA3D[J]. International Journal of Impact Engineering, 1997, 19(9): 847–873.

MAO L, BARNETT S, BEGG D, SCHLEYER G, WIGHT G. Numerical simulation of ultra high performance fibre reinforced concrete panel subjected to blast loading[J]. International Journal of Impact Engineering, 2014, 64(64): 91–100.

MURRAY Y D. Users manual for LS-DYNA concrete material model 159[R]. Washington DC: Federal Highway Administration, 2007. Report No. FHWA-HRT-05–062.

NEEL C. Compaction and Spall of UHPC Concrete Under Shock Conditions[J]. Journal of Dynamic Behavior of Materials, 2018, 4: 1–24.

OTHMAN H, MARZOUK H. Applicability of damage plasticity constitutive model for ultra-high performance fibre-reinforced concrete under impact loads[J]. International Journal of Impact Engineering, 2018, 114: 20–31.

PARK S H, KIM D J, KIM S W. Investigating the impact resistance of ultra-high-performance fiber-reinforced concrete using an improved strain energy impact test machine[J]. Construction and Building Materials, 2016, 125: 145–159.

PONTIROLI C. Impact response of UHPC and UHPFRC: experimental study and numerical simulation. 10th International Conference on Fracture Mechanics of Concrete and Concrete Structures[C]. 2019.

PYO S, EL-TAWIL S, NAAMAN A E. Direct tensile behavior of ultra high performance fiber reinforced concrete (UHP-FRC) at high strain rates[J]. Cement and Concrete Research, 2016, 88: 144–156.

PYO S, WILLE K, EL-TAWIL S, NAAMAN A E. Strain rate dependent properties of ultra high performance fiber reinforced concrete (UHP-FRC) under tension[J]. Cement & Concrete Composites, 2015, 56: 15–24.

RANADE R, LI V C, HEARD W F. Tensile rate effects in high strength-high ductility concrete[J]. Cement and Concrete Research, 2015, 68: 94–104.

REMENNIKOV A M, UY B. Explosive testing and modelling of square tubular steel columns for near-field detonations[J]. Cement and Concrete Research, 2014, 101: 290–303.

REN G M, WU H, FANG Q, KONG X Z. Parameters of Holmquist-Johnson-Cook model for high-strength concrete-like materials under projectile impact[J]. International Journal Protection Structures, 2017, 8(3): 352–367

REN G M, WU H, FANG Q, LIU J Z, GONG Z M. Triaxial compressive behavior of UHPCC and applications in the projectile impact analyses[J]. Construction and Building Materials, 2016, 113: 1–14.

REN G M, WU H, FANG Q, LIU J Z. Effects of steel fiber content and type on dynamic compressive mechanical properties of UHPCC[J]. Construction and Building Materials, 2018, 164: 29–43.

REN G M, WU H, FANG Q, LIU J Z. Effects of steel fiber content and type on static mechanical properties of UHPCC[J]. Construction and Building Materials, 2018, 163: 826–839.

RIEDEL W, WICKLEIN M, THOMA K. Shock properties of conventional and high strength concrete: experimental and mesomechanical anlaysis[J]. International Journal of Impact Engineering, 2008, 35: 155–171.

SCHULER H, MAYRHOFER C, THOMA K. Spall experiments for the measurement of the tensile strength and fracture energy of concrete at high strain rates[J]. International Journal of Impact Engineering, 2006, 32: 1635–1650.

SU Y, LI J, WU C Q, WU P T, LI X Z. Influences of nano-particles on dynamic strength of ultra-high performance concrete[J]. Composites Part B-Engineering, 2016, 91: 595–609.

SU Y, LI J, WU C Q, WU P T, LI Z X. Effects of steel fibres on dynamic strength of UHPC[J]. Construction and Building Materials, 2016, 114: 708–718.

THAOMAS R J, SORENSEN A D. Review of strain rate effects for UHPC in tension[J]. Construction and Building Materials, 2017, 1153: 846–856.

TRAN N T, KIM D J. Synergistic response of blending fibers in ultra-high-performance concrete under high rate tensile loads[J]. Cement & Concrete Composites, 2017, 78: 132–145.

TRAN N T, TRAN Y K, KIM D J. High rate response of ultra-high-performance fiber-reinforced concretes under direct tension[J]. Cement and Concrete Research, 2015, 69: 72–87.
TU Z, LU Y. Evaluation of typical concrete material models used in hydrocodes for high dynamic response simulations[J]. International Journal of Impact Engineering, 2009, 36(1): 132–146.
TU Z, LU Y. Modifications of RHT material model for improved numerical simulation of dynamic response of concrete[J]. International Journal of Impact Engineering, 2010, 37(10): 1072–1082.
VERMA M, PREM P R, RAJASANKAR J, BHARATKUMA B H. On low-energy impact response of ultra-high performance concrete (UHPC) panels[J]. Materials & Design, 2016, 92: 853–865.
WANG Z G, WU H, WU J, FANG Q. Experimental study on the residual seismic resistance of ultra high performance cementitious composite filled steel tube (UHPCC-FST) after contact explosion[J]. Thin-Walled Structures, 2020, 154: 106852.
WEERHEIJM J, VAN DOORMAAL J C A M. Tensile failure of concrete at high loading rates: new test data on strength and fracture energy from instrumented spalling tests[J]. International Journal of Impact Engineering, 2007, 34(3): 609–626.
WILLE K, EL-TAWIL S, NAAMAN A E. Properties of strain hardening ultra high performance fiber reinforced concrete (UHP-FRC) under direct tensile loading[J]. Cement & Concrete Composites, 2014, 48: 53–66.
WILLE K, XU M, EL-TAWIL S, NAAMAN A E. Dynamic impact factors of strain hardening UHP-FRC under direct tensile loading at low strain rates[J]. Materials and Structures, 2016, 49(4): 1351–1365.
WU C, OEHLERS D J, REBENTROST M, LEACH J, WHITTAKER A S. Blast testing of ultra-high performance fibre and FRP-retrofitted concrete slabs[J]. Engineering Structures, 2009, 31(9): 2060–2069.
WU H, PENG Y L, FANG Q. Experimental and numerical study of ultra-high performance cementitious composites filled steel tube (UHPCC-FST) subjected to close-range explosion[J]. International Journal of Impact Engineering, 2020, 141: 103569.
WU H, REN G M, FANG Q, LIU J Z. Effects of steel fiber content and type on dynamic tensile mechanical properties of UHPCC[J]. Construction and Building Materials, 2018, 173: 251–261.
WU H, REN G M, FANG Q, LIU J Z. Response of ultra-high performance cementitious composites filled steel tube (UHPCC-FST) subjected to low-velocity impact[J]. Thin-Walled Structures, 2019, 144: 106341.
WU Y C, CRAWFORD J E, MAGALLANES J M. Performance of LS-DYNA concrete constitutive models. 12th International LS-DYNA Users Conference, Detroit[C]. 2012.
XU M, WILLE K. Calibration of K&C concrete model for UHPC in LS-DYNA[J]. Advanced Materials Research, 2014, 1081: 254–259.
XU S C, WU P T, WU C Q. Calibration of KCC concrete model for UHPC against low-velocity[J]. International Journal of Impact Engineering, 2020, 144: 103648.
ZHANG F R, WU C Q, ZHAO X L, XIANG H B, LI Z X, FANG Q, LIU Z X, ZHANG Y D, HEIDARPOUR A, PACKER J A. Experimental study of CFDST columns infilled with UHPC under close-range blast loading[J]. International Journal of Impact Engineering, 2016, 93: 184–195.